Advances in Intelligent and Soft Computing 163

Editor-in-Chief

Prof. Janusz Kacprzyk
Systems Research Institute
Polish Academy of Sciences
ul. Newelska 6
01-447 Warsaw
Poland
E-mail: kacprzyk@ibspan.waw.pl

For further volumes:
http://www.springer.com/series/4240

Wei Zhang (Ed.)

Advanced Technology in Teaching

Selected Papers from the 2012 International Conference on Teaching and Computational Science (ICTCS 2012)

Springer

Editor
Wei Zhang
Wuhan
China

ISSN 1867-5662 e-ISSN 1867-5670
ISBN 978-3-642-29457-0 e-ISBN 978-3-642-29458-7
DOI 10.1007/978-3-642-29458-7
Springer Heidelberg New York Dordrecht London

Library of Congress Control Number: 2012935246

Printed on acid-free paper

Springer is part of Springer Science+Business Media (www.springer.com)

Preface

2012 International Conference on Teaching and Computational Science (ICTCS 2012) is held on April 1–2, 2012, Macao.

Education in its broadest, general sense is the means through which the aims and habits of a group of people lives on from one generation to the next. Generally, it occurs through any experience that has a formative effect on the way one thinks, feels, or acts. In its narrow, technical sense, education is the formal process by which society deliberately transmits its accumulated knowledge, skills, customs and values from one generation to another, e.g., instruction in schools.

A right to education has been created and recognized by some jurisdictions: Since 1952, Article 2 of the first Protocol to the European Convention on Human Rights obliges all signatory parties to guarantee the right to education. At the global level, the United Nations' International Covenant on Economic, Social and Cultural Rights of 1966 guarantees this right under its Article 13.

Computational science (or scientific computing) is the field of study concerned with constructing mathematical models and quantitative analysis techniques and using computers to analyze and solve scientific problems. In practical use, it is typically the application of computer simulation and other forms of computation to problems in various scientific disciplines.

The field is distinct from computer science (the study of computation, computers and information processing). It is also different from theory and experiment which are the traditional forms of science and engineering. The scientific computing approach is to gain understanding, mainly through the analysis of mathematical models implemented on computers.

Scientists and engineers develop computer programs, application software, which model systems being studied and run these programs with various sets of input parameters. Typically, these models require massive amounts of calculations (usually floating-point) and are often executed on supercomputers or distributed computing platforms.

Numerical analysis is an important underpinning for techniques used in computational science.

The goal of ICTCS 2012 is to bring together the Researchers from academia and industry as well as practitioners to share ideas, problems and solutions relating to the multifaceted aspects of teaching and computational science.

I very sincerely thank the authors who have contributed to this book and referees who reviewed the quality of the submitted contributions. Our sponsors' support, either financial or moral, is gratefully acknowledged. The members of the Organizing Committee as well as other numerous individuals behind the scene are greatly appreciated for their tireless effort and dedication in the organization of the conference.

Wei Zhang

Organizing Committee

Honorary Chair

Chin-Chen Chang	Feng Chia University, Taiwan

Keynote Speakers

Chin-Chen Chang	Feng Chia University, Taiwan
Gerald Schaefer	Loughborough University, U.K

Scholarship Committee Chairs

Qiyuan Zhou	Jiangsu Teachers University of Technology, China
Wei Zhang	Huazhong Normal University, China

Program Committee Chair

Jian Lee	Hubei Second Normal University, China

Publication Chairs

Wei Zhang	Huazhong Normal University, China
Yanwen Wu	Huazhong Normal University, China

Program Committee

Hui Jiang	Wuhan Institute of Technology, China
Zhonghong Wang	Huazhong Normal University, China
Yan Huang	Huazhong University of Science and Technology, China
Jing Chen	Wuhan University of Science and Technology, China
Junqiao Xiong	Wuhan University, China
Huiyuan Li	Huazhong Normal University, China
Ying Ma	Huazhong Normal University, China
Yijin Wu	Huazhong Normal University, China
Zefang Feng	Yangzhou University, China

Contents

The Study of the Dynamic Model of Technology Innovation Decision Based on the Profit Forecast

Xia Li, Fanxing Kong, and Yingbo Qin

School of Management, Xi'an Polytechnic University, Xi'an, Shanxi, China
lixia1390@163.com, 362758539@163.com, 270402896@qq.com

Abstract. In this paper, technological innovation had been broken down into three stages, using price forecasting model market-oriented , the profit at different stages of technological innovation have been expressed with dynamic mathematical model. Subdivision of time variable on the third stage and calculation of equipment cost descript in detail changes of the options cost and profits in this important phase. Based on the principle of profit maximization, comparison of profits at three stages will supply basis for decision-making.

Keywords: technological innovation, dynamic mathematical model, options decision-making.

1 Introduction

With the United States sub-loan crisis and Dubai crisis, experts and scholars call for a return to the real economy, this is the only way to ensure that sustain and stable development of global economy, the real economy is manufacturing industry, manufacturing industry is facing the biggest challenge which is the problem of technical innovation.

Technical innovation has become a important strategic choice of enterprises to participate in the competition and promote enterprise's sustainable development. Decision of technological innovation has not only the relationship with corporate investment income, but also affects the sustained and stable development of enterprises. The problem of enterprise technology innovation decision, in recent years, The decision-making direction technical innovation can be divided into three types. The first is the choice of technological innovation strategies, such as Tether Bruce (2002), the main method is the Mathematics Analysis to study the choice of three kinds of enterprise technological innovation strategy; the second method is to study the decision problems of technology innovation investment, such as the Han T. J. Smit & Lenos Trigeorgis (2003), Guo Qiong Yang Deli and Chi Guotai (2005), mainly using the real option method of technology innovation investment to make decision analysis, focus on technological innovation in the value chain value analysis, emphasizes the static analysis of system; third method is the study of the technological innovation of internal evaluation and decision system, such as Zhiyong Yang and Feng Jia (2009) ,Who's main reaserch is technical evaluation method of technical innovation, technology assessment and decision process etc..

W. Zhang (Ed.): Advanced Technology in Teaching, AISC 163, pp. 1–7.
springerlink.com

The content of this article belongs to the second kind, but which is different from previous research, which from the perspective of market value and only for technical innovation each stage profit of the single enterprise mathematical model to predict, which will be combination of certain variables and economic indicators - profits to adequately show the technological innovation economic characteristic, play a direct role in the process of carrying out the value. Wang and Gerchak (2001) establish a supply chain coordination model of demand decide shelf placed volume, which relates to the supplier's R & D investment caused cost variable model, in this paper, the concept of the cost of changing combining with time variable has been assigned dynamically to technological innovate three stages, each stage of the profit has been forecasted effectively.

The specific scientific research scholars has made many contributions in technology forecasting, American biologist and demographer Raymond Perle (Raymond pear) proposed pear curve method, the British mathematicians and statisticians proposed Gompertz curve. some scientists go a step further to except some mathematical models based on technological substitution, Fisher and Pu Lai established Fisher - Pu Lai model (1971), and put forward and the envelope curve method based on system theory and so on, this article will use these methods and models.

The entrepreneur as the leading factor to promote technological innovation has been the core of innovation theory research and focus, and entrepreneurial technological innovation decision making on technology innovation activities is important, not all entrepreneurs understand the technical index, but they are very sensitive to economic indicators, so in this paper, the calculation of profit will show intuitively decision-making basis before the policymaker.

2 Model

2.1 Modeling Methods

Management of process of technological innovation, technological innovation implementation of the technical demand side is divided into four stages: opportunity recognition→decision making of introducing→technology implementation→operation and improvement. In this paper, with reference to the four stage, combined with the information process , technological innovation process is divided into three stages, from information stimuli to produce the actual technology creative to R & D investment , this period of time is defined as the period of study, mainly study on the technology market value, from R&D investment into research and development production , this period is defined as research and development, for the period, mainly make technical test, There is mainly R&D investment ,in this stage, study the possibility of the realization of the technology. After put into production, to achieve benefit, to training employees to master technology and skilled operation, technology began to spread, and while technology application will be improved continuously, this stage is called technology development. In the competitive market, technology is inevitable, regardless of competitor or the original technologists are all seeking technical progress, technology to replace have happen in the long-term. Since the using customary of original technology products and the scale of the industry, and new technology is not mature, even though the original technology to achieve the growth bottleneck and will have a period of time. By the above describe the conceptual model as:

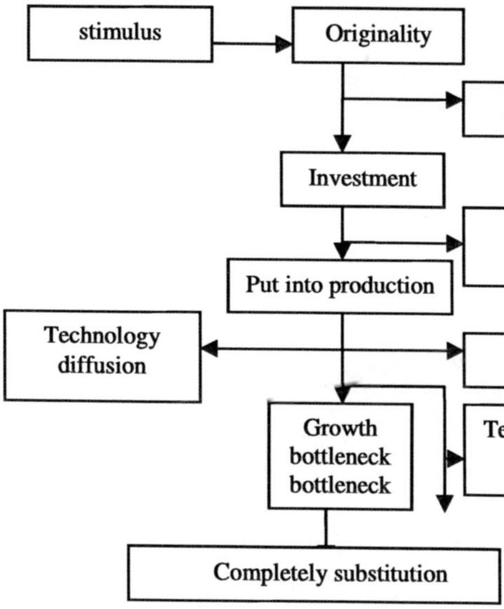

Fig. 1. Model of periods of technological innovation

2.2 Model Hypothesis

(1) technological innovation to change production capacity (yield), cost and price;
(2) the technology diffusion will also change the production capacity, cost and price, but do not need to continue invest;
(3) the market share by production decision;
(4) technical substitution does not influence the original technology product market share.

2.3 Mathematical Model

Profit forecast model in the various stages of organization
(1) organization profit calculation in the period of study:
Product has not through technical innovation in this period, production and quality has not changed, the unit cost of the product (C) did not change, the unit product price (P (T)) only changes with the market supply demand relations, the product output per unit time (Q) has unchanged (behind relates to Q with the same definition) of the organization.

The profit per unit time of study period:

$$\omega_1 = T_1[P(t)Q - CQ] \tag{1}$$

According to the demand of data we can for the future of social total demand ($Q^d(t)$) is predicted, based on microeconomic theory $Q^d(t) = f(p(t))$, demand will reflect and impact the change of a price, so put

$P(t) = f^{-}(Q^{d}(t)$ into (1) type:

$$\omega_1 = T_1[f^{-}(Q^{d}(t))Q - CQ] \tag{2}$$

(2) Profits of technological innovation development:
From the R & D investment to research and development production, because this time has been invested, so the cost of the product must be considered in this part, this part of the investment in the development process must get compensation, with the technology innovation of the theory of real option.

$$\omega_2 = T_2[P(t)Q(t) - CQ(t)] - I_t \tag{3}$$

I_t is the investment of R & D stage research; T_2 is time length of the R & D stage;

To accept the price of market guiding so $P(t) = f^{-}(Q^{d}(t))$, Because technical innovation is not complete so output unchanged $Q(t) = Q$.

(3) Profit after successfully put into production:
The general profit formula

$$\omega_3 = PQ - CQ \tag{4}$$

The cost of this phase to consider two factors, first is technical innovation investment brought about the unit cost , according to the findings of Wang and Gerchak

$$C(I_t) = C_0 - a[I_t / Q(t)]^b .$$

C_0 —Investment in previous unit cost; a —The unit product R & D sensitive coefficient, $a > 0$; b —The incremental cost of unit product investment elasticity, $1 > b > 0$. Also consider the equipment investment will lead to increased cost, must allocate the costs to new products, and this needs to be considered technology development time, in the growth period yield and price will also change, the type can be expressed as:

$$\bar{C} = \frac{U}{\sum_{t_1}^{t_m} t_i(Q + \bar{Q}_{t_i}^{+}) + (T - t_m)(Q + Q_m^{+})}$$

U — the total investment in equipment;—Calculation formula should consider the effect of technology diffusion, by Ridenour curves.

$$t = \frac{\ln \dfrac{L}{N\left[1 + \left(\dfrac{L}{N_0} - 1\right)\right]}}{b};$$

Among them, N—。 the number of familiar with the techniques; N_0—the initial technical knowledge workers; L—the technical performance index limit value; B—constants. $\sum_{t_1}^{t_m} t_i(Q + \bar{Q}_{t_i}^{+})$ — t_i from the production start (t_1) to the technology growth(t_m) (P_t^+ and Q_t^+ limit value is the one moment of t_m) in the middle period, originally technology growth curve should be the curve, considering the difficulty of computational is discrete as the time, $\bar{Q}_{t_i}^+$ is the average yield per unit time increment during period of t_i relative to the Q, the formula to express the meaning of that the reason of technology diffusion and growth is the product output is increased per unit time, but also has been changed always until technical limit of growth and diffusion termination;

T—time of technology instead, namely time from production to the technology is completely replaced, with Fisher flutter model to predict: $\frac{\ln \frac{y}{1-y}}{2a} + t_0 = T$

Among them, y —already replacement ratio; a —constant; t_0 —time of replacement ratio y = 50%.

$$C = C(I_t) + \bar{C} \tag{5}$$

Put(5) into (4) type:

$$\omega_3 = \sum_{t_1}^{T} (P_{t_1} + P_{t_i}^{+})(Q + Q_{t_i}^{+}) - [C(I_t) + \bar{C}]\sum_{t_1}^{T} t_i(Q + \bar{Q}_{t_i}^{+}) \tag{6}$$

2.4 Model Summary

Table 1. Model Summary

Technical innovation phase	The calculation of phase profit ω
Period of study	$\omega_1 = T_1[f^-(Q^d(t))Q - CQ]$
Technology innovation research	$\omega_2 = T_2[f^-(Q^d(t)) - C]Q - I_t$
The successful development of production	$\omega_3 = \sum_{t_1}^{T}(P_{t_1} + P_{t_i}^{+})(Q + Q_{t_i}^{+}) - [C(I_t) + \bar{C}]\sum_{t_1}^{T} t_i(Q + \bar{Q}_{t_i}^{+})$

2.5 Using the Model to Make Decision

The first step of decision-making:

The profit of technology innovation in second phases $T_2\omega_2$ <0, But beyond the business capacity, When it is necessary to choose the financing, or give up. As the conditions to convince investors, only through the profit $T_3\omega_3$ of the third stages.

The second stage of decision:

if $\omega_1 + \omega_2 + T_3\omega_1 < (\omega_1 + \omega_2 + \omega_3)$, make technological innovation, or give up.

3 Model Simulation

A textile school, needle Market Demand Research on ramie, ramie technology improved, now seeking venture capital, a company now this technology is deciding whether to invest in profit prediction.

The company now uses coarse linen costs 5000 RMB / T, the processing cost is 500 RMB / T, ramie yarn and the selling price is 40000 RMB / T, 4 tons of ramie can produce 1 tons of ramie yarn, the year sales volume is estimated to be 190T, new technical modification made selling price of ramie yarn estimate increase to 50000 RMB / T, and 0.5% annual rate of increase. The company study for half a year, the school 's research and development work for half a year, R & D investment about 2000000 RMB, unit product sensitivity coefficient of the technology is $a = 0.1$, the investment flexibility of incremental unit cost of product R&D is $b = 0.3$, improve the new technology after the machine investment of the 6380000 RMB, technology diffusion of technology under the long-term for 3 years, time of technical substitution for 5 years, and after the new technology production, within five years ,expected sales will be 210t, 389t, 551t, 936t, 936t.

The profits is $\omega_1 = 144$ million RMB After calculating the survey during half, the profit of technical research phase is $\omega_2 = -56$ million RMB, during the technology to be replaced the profit $C(I_t) = 0.4184$ million RMB, $\bar{C} = 0.211$ million RMB, then $\omega_3 = 9590$ million RMB. In second types of decision making $1728 < 9648$ which shows the profit after this technological innovation after six years is far greater than making no technical innovation, if can take losses 560000 RMB and 6830000 RMB investment of half a year, technology investment can be made.

4 Conclusion

In this paper, according to practical of technical innovation problems the single company faced, the technical innovation is subdivided into three periods, Then the problem organization of technical innovation concerned profit of technology innovation at all stages, and the third stages of technical innovation, computational grasp, make technical innovation value prediction more intuitive and strong persuasion.

Prediction formula of the time and yield relates to this paper are all applied by economists through long time, Through the years the simulation has demonstrated its effectiveness. The model can not only predict the formula, and the profit model can also be used to do the evaluation, as long as replaced the predictor variables with the actual variable.

The deficiencies of this paper lies in that just only relation to a single investment model of technological innovation, the reinvestment of innovation in the process has not discuss, this will continue to work hard in future studies.

References

1. Bruce, T.: Who Cooperates for Innovation, and Why an Empirical Analysis. Research Policy (31) (2002)
2. Smit, H.T.J., Trigeorgis, L.: R&D Option Strategies. Working Papers Eramus University. University of Chicago Graduate School of Business (2003)
3. Guo, Q., Yang, D., Chi, G.: Contract Coordination Model in Supply Chain Based on The Option. Systems Engineering 23(10) (2005)
4. Yang, Z., Jia, F.: Enterprise Technological Innovation Ability Evaluation Research Based on The Structural equation model. Progress of Science and Technology and Countermeasures 12, 119–121 (2009)
5. Wang, Y., Gerchak, Y.: Supply Chain Coordination when Demandis Shelf-Space-Depe-ndt. Manufacturing & Service Operations Management 3(1), 82–87 (2001)

The Empirical Study of Firm's Social Capital and Technology Innovation Ability

Xia Li, Tingting Wang, and Yingchun Liu

School of Management, Xi'an Polytechnic University, Xi'an, Shanxi, China
lixia1390@163.com, ttingwang@126.com, gxj608@163.com

Abstract. The new economic times requires firms to improve the enterprise's technological innovation capability and competitive advantage by integrating a variety of resources. as an important component of firms capital, Social capital is a new perspective to improve technological innovation. In this paper, I combined theoretical and empirical study methods. Through the study of Social capital theory, combining theoretical development of technology innovation and knowledge from three dimensions of Social capital, using quantitative methods for statistical analysis, obtain the social capital of the relationship between technological innovation ability.

Keywords: Social capital, technological innovation, factor analysis, regression analysis.

1 Introduction

With the increasing international competition, high-tech enterprises become more aware of the importance of good technological innovation to enterprises. To gain competitive advantage on technical innovation is the key to get sustainable competitive advantage variables. Competitive advantage depends not only on its own internal resources, but also on all kinds of social relations embedded various resources and capabilities in the network that is difficult to imitate by competitors.

With the deepening of social capital theory and the role in the social and economic development continues to improve, social capital is conducive to the development of technological innovation, as a important variables in China social and economic development in 21st century, technological innovation in businesses should be highly valued. Jin Chen and Fanghua Zhang claim that competitiveness of enterprises depends on how companies deal with internal resources and whether external resources have be integrate effectively, aimed to improve corporate knowledge creation and technological innovation capacity. The cooperation with external companies and alliances are built on good foundation of social capital. With increasing competition among enterprises, Fujita argues, the increasing internationalization of business and product life cycle has been shortened, technological innovation for the survival and development of a company become increasingly important.

W. Zhang (Ed.): Advanced Technology in Teaching, AISC 163, pp. 9–15.
springerlink.com

2 Literature Review

Social capital is knowledge-based, assets owned by the enterprise, and can bring value to the enterprise. John•Keneth• Galbraith believes that social capital is not just a static intangible assets of an enterprise, but also the process of using knowledge effectively and achieving long-term goal. China scholar Qinghong Yuan thought social capital is a potential application of knowledge, skills and ability to create value. For the composition of social capital, domestic and foreign scholars have different views, Stewart claims the company's social capital come from relational dimension, structural dimension and customer capital, called "H-S-C" structure. Edvinsson and Malone thought social capital is the difference between market value and the book value in the enterprise, including relational dimension and structural dimension. Consolidated existing knowledge, combining with the research focus on this paper, we defined social capital of the access and remain competitive for enterprises that link relational dimension, structural dimension and cognitive dimension. Relational dimension is owned by employees refer to their knowledge, skills, experience, loyalty and desire for cooperation such as collection; structural dimension is has been institutionalized and coded knowledge, ability of organize architecture and organizational culture; cognitive dimension are mainly refers to the relationship between capital, enterprise, including the formation cooperation process of internal and external social relations, social relations.

Technological innovation summarize in two ways, a intensive technological innovation is the enterprise for potential market, through research and development activities, create new products, processes, production, operation and management methods, operating conditions of production, Combined elements and organizations to further optimize, established the stronger performance, more efficient, lower cost of production and management systems. Broad technological innovation can be seen as a entire process of "research and development—a narrow technological innovation—innovation diffusion,", which includes not only new technology research, development, acquisition and mastery, but also the proliferation of new technologies, transfer, penetration and opening the market. Technological innovation capability of enterprises aimed alter or create market demand to use resources, configuration, integration, and create new products , changes in production technology, open up new markets, new forms of organization, and to gain greater economic benefits to increase the possibilities competitive advantage.

3 The Establishment of Model and Assumptions

Based on social capital research, this paper argues that social capital from three dimensions made corresponding impact to improve technological innovation capabilities. Currently, as the competition is fierce in science and technology, the accumulation of social capital is a key way to maintain a competitive edge and preserve the greatest resource. Based on analysis of social capital theory, this paper established the model of relationship between technological innovation capabilities and social capital, showing in Figure 1.

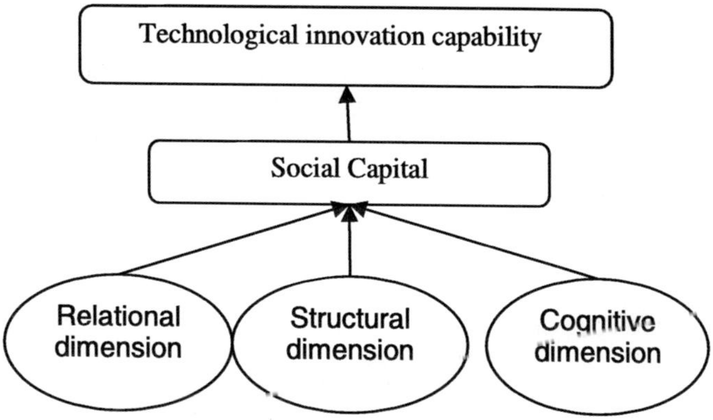

Fig. 1. Model of relationship between technological innovation capabilities and social capital

4 Analysis

4.1 The Proposed Hypothesis

Relational dimension is based on individual, including human qualities, intelligence, knowledge, skills, adaptability, vision and other indicators, for the technology industry, the era of relational dimension in the economy, especially means to the important strategic resource, it has a greater impact to the enterprise innovation capacity.

Hypothesis 1: Relational dimension on technology innovation has a positive influence. Corporate structure capital is reply of relational dimension's power and specific, is supporting infrastructure of relational dimension, is an ability organized, Including the the physical system which dissemination and storage social capital. It includes four elements: system, structure, strategy and culture.

Hypothesis 2: Structure Capital on technology innovation has a positive influence.

Enterprise relationship capital is the value of the relationship which is formed during the communication between corporate and other organization. Include the supply chain's up and down and the organizations such as customer, government, research institute and bank. Hypothesis 3: Cognitive dimension on technology innovation has a positive influence.

This study collected data by provide questionnaire to the technology enterprises in the Xi'an High-tech Zone, using Principal Component Analysis and Regression Analysis through the SPSS to analyze the collected data .For the various sub-indicators of the social capital and technological innovation which we should pay more attention to, analyzed by the Principal Component Analysis and classified index. Analyze the Correlation coefficient between the indexes of social capita and variables of technological innovation by Pearson correlation analysis, investigated whether the variables were significantly related and test the hypothesis.

4.2 Factor Analysis

First of all, the social capital , the explanatory variables is constitute of relational dimension, structural dimension, cognitive dimension .The indicators related to Relational dimension includes the level of personnel quality, education, age, technical ability, business adaptability; Structural dimension, including corporate culture, organizational control systems, strategic business planning, business contact and other indicators; three indexes of cognitive dimension is the relationship between customer, supplier, and other corporate, the relationship between research institutes, universities and technology intermediary organizations, the relationship between government departments, banks, trade associations. Analyze the data of the indicator by Reliability analysis and KMO test, Reliability values were greater than 0.7 certificate that the collected data has better reliability, the value of KMO greater than 0.8, certificate that the data is suitable for factor analysis. (Show in table 1 and 2.)

Table 1. Reliability test

Indicators	Cronbach's Alpha	Variables numbers
Relational dimension	**0.889**	**12**
Structural dimension	**0.802**	**20**
Cognitive dimension	**0.832**	**12**

Table 2. KMO Test

Main variables	Kaiser-Meyer-Olkin
Relational dimension	0.818
Structural dimension	0.801
Cognitive dimension	0.811

Using the Principal Component Analysis in the factor analysis, analyze the indicator which represents the external social capital by the factor analysis, extracted three common factors: x1、x2 and x3. x1 explain the relational dimension dimension of social capital,x2 explain the structural dimension, x3 explain the cognitive dimension. Show on the table 3.

Table 3. Factor analysis of social capital variables explain

Component	Initial Eigenvalues			extraction of Squared Loadings			Rotation Sums of Squared Loadings		
	Total	%of Variance	Cumulative%	Total	%of Variance	Cumulative%	Total	%of Variance	Cumulative%
X1	7.287	34.699	34.699	7.287	34.699	34.699	5.826	27.743	27.743
X2	3.954	18.829	53.528	3.954	18.829	53.528	3.745	17.835	45.578
X3	1.926	9.173	62.700	1.926	9.173	62.700	3.596	17.122	62.700

Similarly, analysis by SPSS, the data shown in table 4, each sub-index technological innovation capability of enterprises was standardized to produce a single factor variety Y, then the variety as the sample of interpreted variables related to access to the latter analysis.

Table 4. Factor analysis of technological innovation to explain the total variable

Component	Initial Eigenvalues			extraction of Squared Loadings			Rotation Sums of Squared Loadings		
	Total	%of Variance	Cumulative%	Total	%of Variance	Cumulative%	Total	%of Variance	Cumulative%
y1	4.772	68.166	68.166	4.772	68.166	68.166	3.018	43.120	43.120
y2	0.560	7.996	76.163	0.560	7.996	76.163	2.313	33.042	76.163

4.3 Regression Analysis

Established a multiple regression model:

$$Y=b_0+b_1\bullet x_{1i}+b_2\bullet x_{2i}+b_3\bullet x_{3i}++c,$$

$$(i=1,2,3\cdots n) \tag{1}$$

Then calculated influences about the various samples of social capital and the three dimensions that human, structure, and relations to technology innovation. As explanatory variables the three dimensions of social capital, technology innovation as the dependent variable analysis of variance (shown in table 5).

Table 5. Variance Analysis

Model		Sum of Squares	df	Mean Square	F	Sig.
	SSR	18.794	6	3.132	27.952	0.000
1	SSE	10.422	93	0.112		
	SST	29.216	99			

Table 5 shows that, the significant is as small (0.000) that the 1% level in the multiple regressions has a significant effect and the influences of various dimensions of social capital on technology innovation has a significant impact. And regression coefficients could be drawn down in Table 6.

According to Table 6, the explanatory variables and the coefficient variable could be substituted into regression equation (1) and so get the regression equation (2):

$$Y=0.286\bullet x_{1i}+0.204\bullet x_{2i}+0.202\bullet x_{3i} \tag{2}$$

From the above results, we can see the regression coefficients of every three dimensions of social capital are positive, that is, all the various dimensions of social capital have a positive influence on technology innovation, in which effective of relational dimension is more significant than structural dimension and cognitive dimension.

Table 6. Regression coefficients

	Regression coefficient	Standard regression coefficient	T	P
constant	0.000		0.000	1.000
X1	0.286	0.286	3.334	0.001
X2	0.204	0.204	1.878	0.064
X3	0.202	0.202	2.581	0.011

4.4 Correlation Analysis

To a further explain the relationship between three dimensions of social capital and the technological innovation, we carried out correlation analysis of the four, as shown in Table 7, the correlation coefficient are: 0.674,0.563,0.531. The results showed that the three dimensions of social capital get a significant correlation for technology innovation. Relational dimension related to the technological innovation stronger, certainly, relational dimension have more impact on technological innovation than any other.

Table 7. Correlation Analysis

		Relational dimension	Structural dimension	Cognitive dimension	technology innovation
Technology innovation	Pearson	0.674	0.563	0.531	1
	Sig.	0.000	0.000	0.000	
	Num.	110		110	110

5 Conclusion

The empirical results show that the social capital of companies can promote the capability of technology innovation. They are positively correlated between social capital and technological innovation, and the three dimensions of social capital have a positive influence on all different levels of technological innovation. While, relational dimension has more significant and positive influence on the technology innovation than structural dimension and cognitive dimension.

The conceptual model and empirical results show that in high-tech industry companies should emphasize the knowledge, innovation, and update rate, especially the importance of relational dimension in social capital, while enterprises can cooperate with external organizations to effectively integrate the internal and external resources for further improving the capability of technology innovation. Therefore, China's enterprises, in order to improve the capability of technological innovation, must have a good social capital, that is, mutual trust and cooperation. It should gain competitive advantage through enhancing the capability of technology innovation, and it also must pay attention to the cooperation with external organizations. Enterprises should continuously strengthen the cooperation with customers and suppliers, and understand and master varieties of information about the market demand and technical development in time. High-tech enterprise should put social capital management on an important strategic position. Through the use of effective human resources management and modern technology, it could optimize the organizational structure and share the corporate culture. In addition, effective management of social capital can bring companies the promotion on technological innovation and maintain the competitiveness of enterprises.

Acknowledgment. My deepest gratitude goes to my first promoter, Professor Li, who guided my research steps from the very beginning. With the valuable advice given by her, and great support from others, e.g. Miss. Liu as my classmate, who takes some time for the careful reading of my thesis and puts forward his valuable suggestions. Thus, this paper can be completed in time. Without your continuous commitment this thesis would have never come to this point! The author hereby would like to take this opportunity to thank all who have been providing supports during and after the preparation of this paper.

References

1. Chaminade, C., Roberts, H.: Social Capital as a Mechanism: connecting knowledge within and across forms. Leaning and Capabilities, Greece (2002)
2. Freeman, C., Soete, L.: The economics of industrial innovation, 3rd edn. printer, London (1997)
3. Fujita, A.: Strategy for Corporate Innovation. Asian Productivity Organization, Tokyo (1997)
4. Coleman, J.: Social Capital in the Creation of Human Capital. American Journal of Sociology 9, 95–120 (1998)
5. Putnam, R.: The Prosperous Community: Social Capital and Public Life. American Prospect 13 (1993)
6. Powell, W.W., Koput, K.W., Smith-doerr, L.: Inter organizational Collaboration and the locus of innovation: Network of leaning in biotechnology. Administrative Science quarterly (1996)
7. Nahapiet, J.V., Ghoshal, S.: Social capital, intellectual capital, and the organizational advantage. Academy of Management Review 23, 242–266 (1998)

Discussion of Improving Student Ability through Electro-hydraulic Experiment*

Guo Jinjin*,**, Zhang Jiageng, and Yuan Xu

School of Mechanical Engineering , Tianjin University of Technology, Tianjin, China
Guojinjin97@163.com

Abstract. The focus of this paper is how to improve the students' comprehensive ability. The electro-hydraulic experiments were designed, which include the professional knowledge application, such as the hydraulic transmission and electrical control. It can train the students' abilities of comprehensive system design. During these experiments, simulation experiments were done at the same time. The design, practice and computer applying ability of students were strengthened when they take advantage of simulation tools to analyses the electro-hydraulic control system and perfect the design scheme and complete the experiment design and assemblage of these circuits system. All of these can make students to have good skill for working in the future. It also can meet the demands of mechanic-electronic-hydraulic talents for the community.

Keywords: hydraulic, experiment, simulation, talent education.

1 Introduction

In the university during four years, students have learned the mechanical principles, hydraulic, PLC control technology and other relevant courses. The knowledge in relevant fields is interpreted profoundly in each course, but graduated students often face with system works sometimes. A complex system includes multiple professional spheres. How to provide an integrated system platform and cultivate the comprehensive capacity of the students from the perspective of system is one of the training goals. In this paper, comprehensive experiment about mechantronics and hydraulics integration was designed, combining the existing experimental laboratory equipment. Computer simulation was introduced to the experiment to improved practicl skills and ability of students who study the mechanical and electrical control. It is not only to develop comprehensive ability of the student, but also to improve the ability of computer application. This article introduces the component of the experiment equipment, the design and the practices of content of experiment.

* Project Support: The paper is supported by the education reform project of the Office of Academic Affairs of Tianjin University of Technology.

** Author introduction: GUO Jinjin (1968-), female, doctor, professor, mainly engaged in the fluid power transmission and control, CAD and other aspects of teaching and research work.

W. Zhang (Ed.): Advanced Technology in Teaching, AISC 163, pp. 17–23.
springerlink.com

2 The Component of the Experiment Equipment

Laboratory bench is shown in Figure 1. Including mechanical, electrical, hydraulic types of components, it can achieve the control of systems about mechatronics and hydraulics integration. It is composed of modules, including the control modules, hydraulic modules and mechanical modules. Modules can be combined depending the need of experimental. Among control modules, different control methods are included such as relay control panels, PLC programmable controller board, multi-function power control box, stepper motor, CNC programmable stepper motor control board. For example, the control module and hydraulic module are combined to achieve hydraulic drive and control; the control module and the mechanical module (two-degrees cross slider, as shown in Figure 2) can be combined to achieve the control of two-degrees Cross Slider.

Hydraulic technology is one of the key factors to realize modern drive and control, and the development of hydraulic industry has caught great attention from many countries in the world. With the development of electrical and electronic technology, hydraulics is closely combined with the mechanical, electrical, electronic, computer and network technology to form a mechatronics and hydraulics integration technology, in order to overcome the shortcomings of the hydraulic technology, exerting the superiority of hydraulic technology in further. In this paper, we take this electro-hydraulic experiment for an example to illustrate the process of carrying out comprehensive experiment.

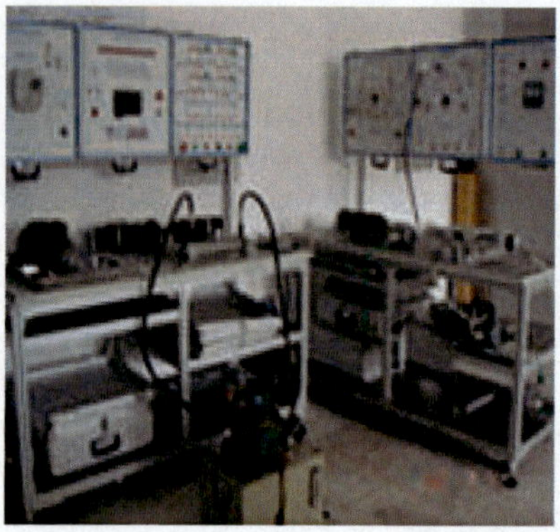

Fig. 1. The experiment equipment

Fig. 2. Mechanical module

3 Design and Implementation of Electro-hydraulic Experiment

Hydraulic module is consisted of a pump station and hydraulic actuators. Hydraulic power station includes oil tanks, hydraulic pumps, motors, 3-position 4-port electromagnetic Directional valve, relief valve, hydraulic restrictor, hydraulic gauge and other components. Hydraulic actuators include hydraulic cylinder and hydraulic motor. The output pressure of hydraulic pump station can be adjusted through the relief valve. In order to ensure the safety of the experiment, the pressure of hydraulic oil is generally controlled between 2MPa and 3MPa. The principle and appearance of Hydraulic pump station is shown in Figure 3, Figure 4.

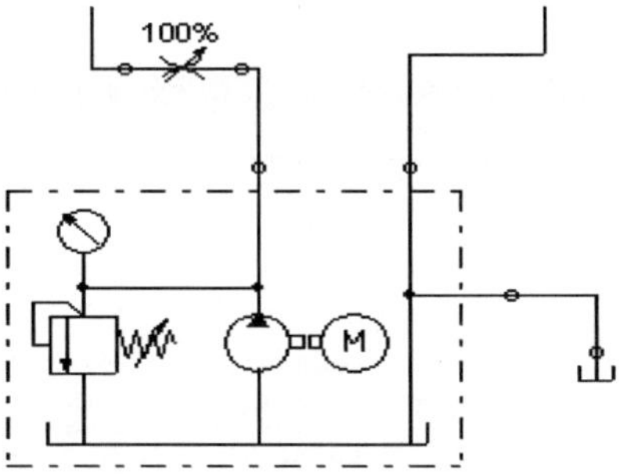

Fig. 3. The principle of Hydraulic pump station

Fig. 4. Appearance of Hydraulic pump station

Based on the components of the laboratory bench, the hydraulic control module and the module will be organically connected, in this way , more than eight integrated electro-hydraulic experiments can be provided, for example: hydraulic cylinders can be controlled by the relay , and the reciprocating movement of hydraulic cylinder can be achieved by the relay limit switch. The hydraulic cylinder running with carwl can be controlled by programmable controller, and the reciprocating movement of hydraulic cylinder can be achieved by the programmable controller. The hydraulic cylinder movement experiment can be controlled by the time counting function of the programmable controller. In the experiment, the start of pump and the action of hydraulic valve needs the control of the electrical control loop, so the experiment contents include the applied training of hydraulic transmission and control. The application trainings of the relays, PLC and other control technology are also included. Electro-hydraulic combination will enable students to understand how the complex systems is connected by the specialty knowledge, and it provide guidance for professional learning.

In the experimental implementation, the list of facilities should be provided to students in the first. Second, based on the component of the bench, students should draw the hydraulic principle and control circuit which is designed by themselves. Figure 5 shows that the hydraulic cylinder is carwling which is controlled by relay. Figure 6 is the schematic of electrical control system and Figure 7 is the experimental system designed by students.

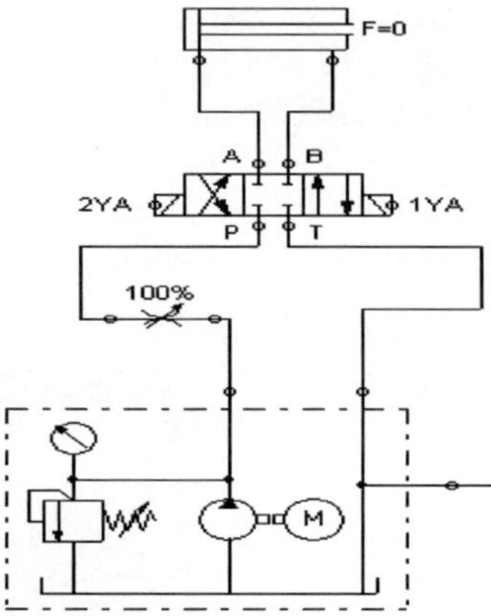

Fig. 5. Hydraulic schematic

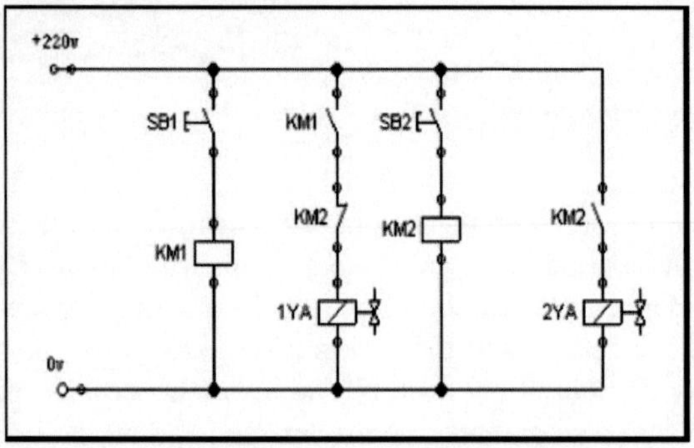

Fig. 6. Electrical control system schematic

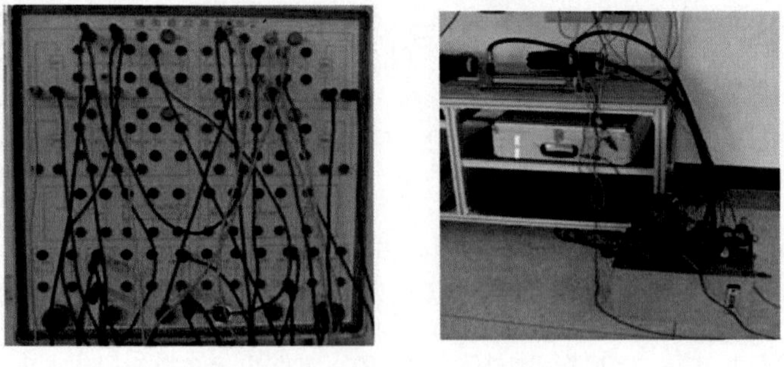

Fig. 7. Experimental system

The direction of the hydraulic cylinder can be changed by position switches and relays to make the cylinder realize automatic reciprocating motion. The cylinder can also be controlled by the PLC. Students compile the PLC control program to complete the control functions. In practice the students can be divided into several groups in which include 2 or 3 students. Each group can design 2 different experimental programs and compare the features of programs. In the experiment, the students, according to their schematic design of the hydraulic and control circuit, connect hydraulic and electrical control system loop circuit by themselves under the guidance of teachers. When encountering problems, they can find the circuit failure or failure of the hydraulic circuit to solve it. By the practice session, the system design capabilities of the students can be trained. The understanding of electro-hydraulic working system of students can be enhanced .Connecting the physical circuit by themselves, the students can master the practical skills of hydraulic, electrical circuit, such as hydraulic hose connections, quick-change connector structure and installation method, PLC programming and control circuit connections and so on.

4 The Combination of Computer Simulation and Experiment

With the development of computer technology and the popularity of the application, the demand of computer applications ability of students have become increasingly .To combine the professional application software with the professional learning, this article introduces the computer simulation. It requires students to do the simulation for the results of the first design , using software tools to analysis the reasonableness of the design before the physical experiment. FluidSIM is the used computer simulation software. Its biggest feature of the software is the integration of hydraulic and pneumatic and its matched electrical control. It make up hydraulic and pneumatic teaching defects, which students saw only hydraulic (pneumatic) but not the electrical circuit loop, and thus do not understand a variety of switches and valves action process. At the same time the software has a circuit simulation function. It can improve the capacity of understanding of the electro-pneumatic and electric hydraulic and the capacity of practical application.

FluidSIM software is developed by the German company Festo, which runs on Microsoft Windows operating system. Software includes three main modules: graphics module, the system simulation modules and integrated presentation modules. FuidSIM software can help designers draw the hydraulic system and electrical control systems by providing the basic functions of circuit diagrams and component symbols, thus allow them to focus more on system design. Graphics module contains 100 kinds of standard hydraulic, pneumatic, electrical components, including necessary pneumatic and electrical components when we creating a new circuit. We can use it to design and draw the module hydraulic, pneumatic and electrical circuit. Users can drag the required components or circuit diagram to drawing area on the desired position directly from the library to layout component and build. The circuit should be build before the system simulation.

Figure 8 shows the relay to control the hydraulic cylinders for the simulation experiments. Figure 8 (a) shows the hydraulic circuit diagram which is connected by the students, and the direction of the hydraulic cylinder is controlled by a 3-position

4-port electromagnetic directional Valve. Figure 8 (b) shows the electrical control schematics, and relays 1YA, 2YA control magnetic action, so it can control the working status of 3-position 4-port electromagnetic directional valve. In figure the bold line displays that when the relay 2YA is with power on, the hydraulic and circuit is connected. 2YA is charged, and 3-position 4-port directional valve left position access. The oil goes through the 3-position 4-port directional valve from P port to B port, and arrives in the right chamber of the cylinder. It push the piston left , so it can achieve the control of the cylinder. Figure 8 (c) is that FluidSIM figured out the components of system automatically. By simulation, students intuitively understand the work process of hydraulic circuits and electrical circuits without the physical experiments, appreciating how the interaction of electro-hydraulic system to achieve system control. FluidSIM has other functions. For example, it is available as option for students who is interested in learning applications in depth.

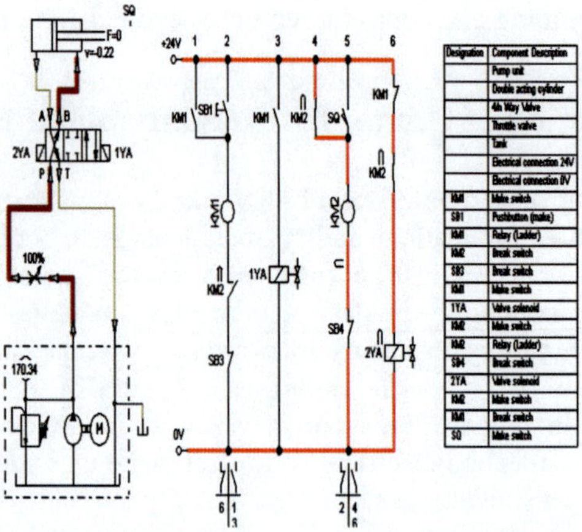

Fig. 8. Simulation experiments based by FluidSIM

5 Conclusion

This paper introduces an integrated experimental which includes mechanic, electronic, hydraulic in the practice teaching. In the experiment, the students personally to complete the hydraulic, electrical control system design, building, developing and improving their design capabilities, practical ability and comprehensive ability to apply. It plays an important role in strengthening the design of experiments and the integrated use of the practical aspects. It lay a good foundation for students, meeting the social pressing needs of electro-hydraulic integration of personnel.

While carrying out physical experiments, the introduction of the computer simulation to teaching has the following characteristics:

A. In the process of explaining the basic hydraulic circuits, FluidSIM can be used to simulated the working course of the circuit to help students understand circuit working principle and the working process. While doing the simulation, you need to build hydraulic transmission and electrical control circuit, so this link can enable students recognize the hydraulic and electrical control relationships, improving their comprehensive ability.

B. Before carrying out the hydraulic test, the students should do some related simulation tests to prepare for the physical experiments.

C. It combined computer application with professional training to improve their computer application skills and master CAD applications.

References

1. Guan, Z.: Hydraulic transmission system. Machinery Industry Press, Beijing (1989)
2. Germany FESTO company, FluidSIMple Pneumatics Simulation design software instructions (2002)
3. Zuo, J.: Hydraulic and Pneumatic Transmission. Machinery Industry Press, Beijing (2000)
4. An, L., Lian, Z.: Optimization based on dynamic simulation of valve AMESim. Fluid Power Transmission and Control (2009)
5. Xie, W., Zhou, H.: Pilot Study on Water Hydraulic Relief Valve. Hydraulic Pneumatic and Sealing, 21–26 (2004)
6. Fu, Y.: AMESim system modeling and simulation, pp. 36–42. Beijing University of Aeronautics and Astronautics Press, Beijing (2006)
7. Guo, J.: Performance Analysis of A Screw Two-way Cartridge Valve. Machine and sHydraulic (2010)

Method Study on Demarcation Line of Professional Qualification Attainments

Zhao Hui[1] and Wang Xue-qing[2]

[1] School of Management, Qingdao Technological University, Qingdao, Shandong, China
zjj45@126.com
[2] School of Management, Tianjin University, Tianjin, China
liuhx626@126.com

Abstract. Test scale of professional qualification is unceasingly expanded. As to the current method of demarcation line of professional qualification attainments, many people are dissent. In order to solve this question, relying on theories and methods of educational survey and statistics, considering intelligent request of professional qualification and observing objectivity, the paper introduces a method on demarcation line of professional qualification attainments and demonstrates the method combining an example for reference.

Keywords: Professional qualification, demarcation line of professional qualification attainments, method.

1 Introduction

In order to adapt to the development of basic construction in China, the state implement various qualification certificate system in the field of construction, and organize the unified national examination by the Ministry of Personnel and the Ministry of Construction every year. With the implementation of the system, the test scale expands unceasingly, for example, the national level 1 build division in the field of construction qualification examination in 2009,the number of applications have more than 350,000. Can we really select the talent conformed to our country's construction and development in many of candidates for exam, will have an important impact on the development of China's construction, it's self-evident. At present, the selected mode is still using the "passing line" approach, that is delimit out a passing line before the text, meet or exceed the line of qualification examination in the engineering construction field will have the qualification, below will not have the basic level, for this passing line, has been the focus of public attention. With 60 points or 60% total score for multiplied by passing line is the current practice, for this kind of delimiting method, all have objections. In order to safeguard the reputation of the national qualification examination in the field of engineering construction, and ensure the quality of the talented person selected from the national text, this paper aimed at this problem presents a demarcated method of qualification examination passing line in the field of engineering construction, and demonstrates the method combining with cases, I hope that this method can provide useful reference to the demarcation line of professional qualification attainment in the field of engineering construction.

W. Zhang (Ed.): Advanced Technology in Teaching, AISC 163, pp. 25–31.
springerlink.com

2 Assumptions

The demarcation of qualification examination passing line in engineering construction field should be basic on the results of examination candidates, and it will be more reasonable with the results of all candidates. This is because the demarcation of passing line before examination always bases the case of "sample candidate", as we all know, the accuracy of "sample candidate" is high but it is biased with actual situation. Therefore, the scientific of the demarcation line after examination is higher. In view of this, the passing line is designated after examination in this paper.

In addition, "demarcation after exam" is based on the results of candidates, whether the text scores reflect the true level of candidates, which has a significant influence to the rationality of the demarcation passing line, it is self-evident. According to the modern psychological and educational measurement theory-item response theory (IRT), proposed by the U.S measurement expert Lord in 1952, its basic idea can be simply described as: when the difficulty of exam topics are more comfortable with the level of candidates, text scores can provide greater amount of information, and therefore more able to estimate the true level of candidates, the error of examination is smaller. Conversely, when the difference the difficulty of the exam topics and the level of examinee is father, text scores can provide smaller amount of information, and the error is greater. Based on this theory, in order to make the results of examinee reflect their true level as much as possible, and make the passing line defined more rationality, the difficulty of exam topics should be strictly controlled, so how much should the question difficulty index be? According to educational measurement theory, when difficulty value close to 0 or close to 1, that all answer correct or all answer wrong in the same question, it will not provide any information on individual differences. And only when the difficulty value close to 0.5, the subject can make the greatest degree of distinction to the examinee, so most of questions' difficulty value should be close 0.5 in exam. At present, the vision of qualification examination in our country engineering construction field is to let the papers provide text the examinee's true level as much as possible, the results of papers and the true level of examinee similar or consistent, namely text difficulty is also basically "examination difficulty about 0.5" for target. Although it has deviation in specific practice, from the analysis of qualification examination papers, on the whole, the will of setting subjects will be effectively achieved in engineering construction filed. Here, we assume that "the vision of setting subjects" can be achieved.

3 Demarcation Passing Line

Demarcation passing line of professional qualification examination in engineering construction is both based on the examinees' scores and the qualification requirement of professional qualification for itself. In addition, to make the demarcation passing line more scientific and reasonable, we should try to eliminate human factors of interference, and make it more objective.

3.1 Statistics of Candidates' Achievement

According to educational and statistic theory, in a appropriate difficulty and reliable examination, candidates' achievement should be close to normal distribution, that is, when the results of candidates close to the normal distribution, then the examination has basically reached requirement. Therefore, in "the will of setting subjects" to achievement better, we can get candidates' achievement closed to the normal distribution. As shown in Figure 1.

According to the nature of the normal distribution, combined with the candidate's achievements, we can compute the data needed in delimiting passing line.

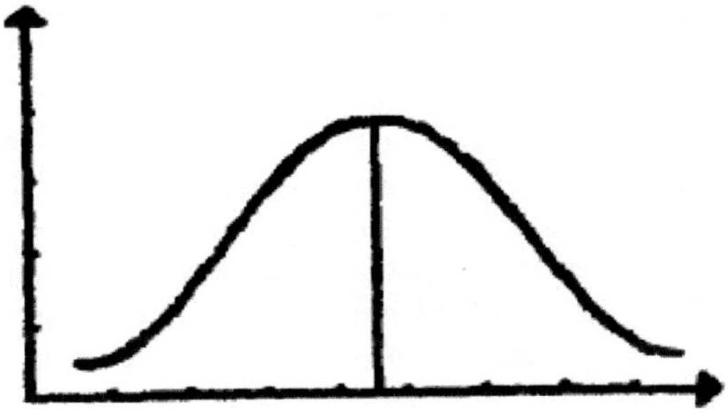

Fig. 1. Normal Distribution Curve

3.2 Preliminary Demarcation of the Passing Line in Engineering Construction Field

According to the introduction of "Educational Measurement Program" authored by U.S. Ebel and "Modern Educational Measurement" authored by Song Zhaohong etc., when a group of scores subject to the normal distribution of the average ($\bar{X}$) and standard deviation (S) ,according to the five-level score, you can set above $\bar{X}+1.5S$ excellent, accounting for 7% of overall; Scores in ($\bar{X}+0.5S,\bar{X}+1.5S$) as well, approximately about 23~24% ; Scores in ($\bar{X}-0.5S,\bar{X}+0.5S$) for medium, approximately about 38~40% ; Scores in ($\bar{X}-1.5S,\bar{X}-0.5S$) to pass the examination, approximately about 23~24% ; Scores below $\bar{X}-1.5S$ for failed the exam, approximately about 7%.

Professional qualification is different from general qualification which is starting point standard engaged in a certain professional (type of work) knowledge, skills, and ability, is the general requirements to employee, and isn't related to civic right and licensing problems about the distribution of interests. But professional qualification is an access control implemented to some profession (type of work) which is heavy responsibility, high universality, related to public interests, and open a private practice in accordance with law or is necessary standard engaged in professional (type of work) knowledge, skills, and ability. It's mandatory industry access control means based on clear law, in a sense, it discharges the prohibition to a few people had special conditions which is on the basis of universal banned to employees, is restriction of civil rights, and affects the distribution of benefits to citizens. Therefore, relative to working qualification, professional qualification have higher requirement to aptitude. Based on this, we will be here with the score $\left[\bar{X}+0.5S, \bar{X}+1.5S\right)$ for pass. It's not difficult to find that the results of eligible are not less than good grades in this way, and guarantee effectively the quality of eligible.

3.3 Final Demarcation Passing Line of Professional Qualification Examination in Engineering Construction Field

Proposition persons of examination always hope that the text results are the best measure of candidate ability, and they tend to make interpretation of text results absolute, that the text results reflect absolutely the candidates' level of ability. It is understandable, because such explanations may make the text results more "compelling", and the demarcation passing line seems to be more fair and justice. However, text results are not absolutely perfect index of the examinees' level from the perspective of text measurement, because the text measurement itself has limitation. Text measurement limitations are mainly in two aspects: one is the definition of measurement target, and the other is the proposition and scoring limitations. It can be found through "assumption" that "the limitations of the definition of measurement target" and "the limitations of the proposition" have been very good to avoid, however, the limitation of the marking is not resoled, it will inevitably have negative impact on the objectivity of examination results and the rationality of passing line. Therefore, the limitations of the marking should be taken into consideration in demarcation passing line.

At present, professional qualification examination questions are generally composed of objective questions and subjective questions in engineering construction field, and some are only composed of objective questions or subjective questions; the marking forms of objective questions and subjective questions have big difference, when marking, objective questions are commonly with "machine readable card", but the subjective questions are still marked by teachers. Practice proves that the marking form of "machine readable card" which is using modern science and technology has high accuracy and the scoring error is almost minimal micro, by contrast, because of human factors to the on-site marking teacher, the accuracy will wore. In order to make our demarcated passing line more reasonable, it is necessary to put the "marking error"-"marking standard error" produced by human factors into consideration.

The so-called "marking standard error" means that it will appear error even the teachers are in strict accordance with scoring criteria. Difference quantity relates to the reliability of the examination, the lower the reliability, the larger the difference quantity (the scope of score distribution); smaller difference, and higher reliability. High reliability proves that we have more confidence, that is, the results of examination have high accuracy.

Formula is usually used to estimate "marking standard error" value of σ.

$$\sigma = s\sqrt{1-\gamma}$$

Where s is standard deviation of the text scores, r is the coefficient of reliability.

$$s=\sqrt{\frac{1}{n}\sum_{i=1}^{n}(x_i-\bar{x})^2}$$

Where x_i represents i examinee's score, $\bar{X}$ is the average score of all candidates.

$$\gamma=\left(\sum_{i=1}^{n}x_i y_i-n\overline{xy}\right)\Big/\left[(n-1)s_x s_y\right]$$

Where x and y are two measurement samples, n is the number of candidates; $\bar{x},\bar{y}$ are respectively the mean value of x, y; s_x, s_y are the standard deviation of x, y.

From the above equation, we only know the reliability and standard error of the professional qualification exam results in engineering construction field, then can calculate the "marking standard error" value, which is easy to get in the text statistics.

So, combined with the preliminary demarcated passing line and "marking standard error", the final standard can be designated as: $\left[\bar{X}+0.5S+\sigma,\bar{X}+1.5S-\sigma\right)$。

4 Presentation of Methods

Source of material: the text results of the subject 《XXXX》 in professional qualification exam of a project construction field 2009. For good knowledge in the delineated methods of professional qualification exam in engineering construction field, for simple, we assume that the exam questions difficulty has been effectively controlled, and the general distribution of scores is the normal distribution(here we just plain simple presentations using data).

First, overall $\bar{X}$ and overall S can be calculated combined with the candidates' overall scores.

Its computation formula is:

$$\bar{X}=\frac{1}{N}\sum_{i=1}^{N}X_i$$

Where N is the number of all candidates, X_i is the score of every candidate (the sum of objective questions and subjective questions).

$$s=\sqrt{\frac{1}{n}\sum_{i=1}^{n}(x_i-\bar{x})^2}$$

Where X_i is the score of i examinee, $\bar{X}$ is the average score of all candidates. Here the score is the sum of objective questions and subjective questions.

Calculated that the overall average score and variance of subject《XXXX》.As shown in Table 1.

Table 1. The Overall Average Score and Variance of Subject 《XXXX》

The score of subject XXXX	$\bar{x}$	s
	69	36

Second, to the results of subjective questions of subject 《XXXX》, calculate "marking standard error" value of σ. Note, here the formula is

$$\sigma = s\sqrt{1-\gamma}\cdot$$

Where the value of s and γ is based on the results of subjective questions, and not based on the total scores (the sum of the objective questions and the subjective questions), which should keep in mind.

Into the subjective questions, "Marking standard error" related to formula:

$$\sigma = s\sqrt{1-\gamma}$$

$$s=\sqrt{\frac{1}{n}\sum_{i=1}^{n}(x_i-\bar{x})^2}$$

$$\gamma = \left(\sum_{i=1}^{n} x_i y_i - n\bar{x}\bar{y}\right) \Big/ \left[(n-1)s_x s_y\right]$$

Calculated that marking standard error σ of the subjective questions, standard deviation s, and the reliability γ of the subjective questions of subject 《XXXX》.As shown in Table 2.

Table 2 Marking Standard Error σ of the Subjective Questions, Standard Deviation s, and the Reliability γ of the Subjective Questions of Subject 《XXXX》

Subjective problems results of subject 《XXXX》	σ	s	γ
	13..38	26	0.7351

Final, combine with Table 1 and Table 2, into the formula $\left[\bar{X}+0.5S+\sigma,\bar{X}+1.5S-\sigma\right)$,we can get passing domain $[100.38,109.62)$ of subject 《XXXX》.

5 Summary

This method is based on the theory and practice of education measurement and statistics, effectively meet the qualification requirements, and avoid the interference of human factors in the process of demarcation passing line. These make the method with stronger scientific rationality, and the demarcation passing line with the method is not in the traditional sense of "line" but "regional".

References

1. Wang, X.Q., Yu, G.: Study of construction project bidding based on the BP neural network improved by GA. China Civil Engineering Journal 40(7), 93–98 (2007)
2. Hu, S.S., He, Y.Q.: Study on rough sets theory and its application. Beihang University press, Beijing (2005)
3. Montazer, G.A.: Intelligent parameter reduction using rough sets theory and sensitivity analysis. WSEAS Transaction on Systems 6(3), 623–630 (2007)

Teaching Reforms and Practice on Applied Undergraduate Course of Computer Organization and Architecture

Ren Xiangmin and Jia Zongfu

School of Software, Harbin University, Harbin, Heilongjiang Province, 150080, China
min0070@sina.com, sjjzf@hrbu.edu.cn

Abstract. Combined with the school-running characteristics of Harbin University which is a local comprehensive university, and the orientation of applied talent cultivation, this paper researches teaching reforms of Organization and Architecture of Computer (OAC) for the applied undergraduate, generalizes briefly the teaching reforms from the aspect of teaching contents, teaching methods, teaching means, examination method, practice teaching, examination database and so on. The experience is beneficial to the teachers of embarking on the same course.

Keywords: Organization and Architecture of Computer, teaching content, teaching methods and means, examination method, practice teaching.

1 Introduction

Obviously cultivation of talents in applied undergraduate colleges should stress "application" [1], which is the soul and university-running stand point, as well as scientific orientation of the applied undergraduate education. Cultivating students in applied undergraduate education should abide by its objective, follow its own way to train students, simultaneously highlight practice and application. We can't follow regular undergraduate teaching pattern and make it a research collage just because what we cultivate is undergraduate students; we can't reduce basic theory teaching because of our focus on application. Instead we should bridge the gap between application and theory, focus on training students' ability of application, strengthen the parts of application in basic curriculum, make the whole course of teaching application-oriented, and enhance students' operative ability and applied ability, meanwhile strengthen basic theory teaching, widen students' knowledge, and laid firm theory basis for them. We should consider continual development of students, as well as the cultivation of their applied ability, and build gradually scientific and reasonable curriculum system in applied undergraduate colleges.

Organization and Architecture of Computer (OAC) is a professional required course for the students of computer major, and an important link in computer hardware curriculum. For the applied undergraduate, we should attach importance to the introduction of hardware structure and principle, thus students can have a more complete knowledge of computer internal structure and its principles, at the same

W. Zhang (Ed.): Advanced Technology in Teaching, AISC 163, pp. 33–38.
springerlink.com

time, students' practical ability should be emphasized. This is essential for students' further studies and very important for fostering students' theoretical interest, scientific spirit, approaches, spirit of innovation and ability of practice [2].

OAC is difficult to learn for many students with its abstract theories and concepts, most students studies it with lower efficiency and poor perception. This brings more difficulties to students who want to further their studies. In order to achieve the teaching objective of solid theoretical foundation and strong practical ability, we should systematically consider each teaching link, and carry out teaching reform [3][4][5].

2 Teaching Reforms and Practice

2.1 The Reform of Teaching Contents

Teaching contents should be carefully chosen, reorganized and updated by dynamically modifying syllabus. Based on talent cultivation target and development of computer science, the syllabus need to be made clear. Teaching purpose, teaching requirements and the depth and width of teaching content, scientific class-hour distribution and time phasing (including proportion of theory class hour and practice one) should be carried out. The requirements on students' ability, training measures, and relationship of OAC with other courses should be clarified according to the nature and characteristics of the curriculum. Great importance has been attached to training students' ability of logic, theory and practice, especially their ability of application, strengthening the parts of application in basic curriculum teaching in OAC syllabus. The organization of teaching contents should follow the principles of: (1) proceeding from the simple to the complex, (2) from easiness to difficulty, and (3) from the concrete to the abstract.

With the rapid development of computer technology, new concept, new technology, new theory, and new discovery appear continuously, textbook of new edition and more contents should be applied according to OAC syllabus. The key factor in the reform of OAC is how to transmit clearly the basic knowledge, put new knowledge in teaching within a limited time, and make students solve new problems by using new knowledge during the course of study, so as to enhance their self confidence and interests in the study, improve their comprehensive abilities of application, and make them feel they can make full use of what they learned in the future.

2.2 The Reform of Teaching Method and Means

Modern education has been changed from one form to diversity, school education is the core of traditional education, and class education is main method in the organization of teaching. The wide use of modern information technology provides us with flexible organizations of teaching; class education is not only limited in the classroom, space and time of teaching activities have been both expanded.

2.2.1 Reform of Teaching Methods

Improvement of teaching method is the key to improving the teaching quality. The traditional forced-feeding model must be changed, which will be the focus of our reform. Teaching method of heuristic teaching, "discussion" teaching, "research" teaching, self-regulated learning etc should be used, especially "3•5" teaching method and "essential teaching and practice in the best quality" method (ETPBQ) are suitable for applied undergraduate.

a) "3 • 5" teaching method

"3•5" teaching method should be used in the theoretical and practical contents of OAC. "3" indicates three links in theoretical teaching: preparation before class, heuristic instruction in class, reading and practice after class. "5" represents five links in practical teaching: preparation, instruction and analysis, experiment operation, discussion, comment and summary.

b) Essential teaching and practice in the best quality method(ETPBQ)

ETPBQ can make students deepen their understanding of the basic concepts and basic theory. For example, students can draw inferences about other cases from one instance by analysing the case of several solutions to one problem.

c) Flexible application of teaching methods

In accordance with the characteristics of OAC course, the application of different teaching methods will greatly stimulate the motivation among students, enhance a variety of their abilities, and therefore bring about a more desirable result and reach the teaching goal.

For instance, you can use heuristic teaching method when explaining DMAC. Teachers can raise a question in class for discussion according to the correlative relations of DMAC, CPU and Interrupt Controller that had been lectured, arouse the students to think, and make them to summarize the functions of DMAC by themselves.

Various external memory devices appear on the market now, such as memory stick, DF, CF, TF, USB flash disk, optical disk, magnetic disk and so on. How should we introduce these to students?

- Student must consult literature materials by Internet and study what they need to know by group discussion,
- They must introduce external memory in terms of outward appearance, storage capacity, characteristic, principle etc,
- They should summarize a complete memory system,
- Finally, teachers expounded external memory theory to the class.

Our students can not only learn external memory theory form textbook, but also know the latest of memory technology by self-regulated learning.

2.2.2 Reform of Teaching Means

The traditional forced-feeding model must be changed into "Learning and Leading" Teaching. Teachers should respect the leading status of students and create the suitable environment for the students. Teachers should also advocate cooperation study and creative study for them, and pay attention to developing the ability of students. So it's urgently needed to reform teaching means by utilizing the modern education technology.

a) Teaching with the help of multimedia technology

Multimedia technology manifests abstract theory in visual and audio visual form, makes up the insufficiency of class teaching, promotes students' learning interest, and makes students grasp the content of the curriculum objectively. Simultaneously, it can promote the students' abstract thinking ability and enhance the teaching effect of OAC.

For example, when we introduce the process of carrying out an instruction by a computer, students can study it by using Demo System (computer animation simulation), which makes students understand the process better.

b) Teaching with CAI Software

c) Teaching with the platform of network

Network teaching breaks the limits of space and time of class education. Teachers can prepare lessons, lecture, tutor, correct students' papers on the platform of network teaching; students can also study, review, question, submit assignment, exercise, take test etc on it at anytime and anywhere.

2.2.3 Reform of Examination Method

Examinations are a necessary way of checking on the performance of students and teaching effect. In order to highlight the positive role of test in teaching, examination materials should be selected according to OAC syllabus. The test should check students' grasp of elementary theory and basic knowledge, moreover examines the students ability of discovering problems, analyzing and solving problems. The contents of final examination should cover the whole teaching contents, with moderate difficulty, and more than 7 testing questions such as multiple-choice, blank-filling and true/false questions etc.

We can use various testing methods such as class testing, chapter testing, final examination etc in order to avoid the use of final examination scores as the last result, emphasizing the whole process of studying. For example, students' studying attitude; their assignment, question-asking, experiment and the result of class testing should be combined with final examination; the percentage is 10%, 10%, 20%, 60% respectively. Such testing method can make students stimulated during the study of OAC course, ensures the continuity of knowledge students acquired, and promote the teaching quality.

2.2.4 Strengthening Practical Teaching

Practical teaching is an important part of OAC course teaching, helpful to train students' ability of discovering and thinking, to cultivate their working style of preciseness and perfection, to mold their good quality of being practical and faithful, to create their group spirit of dividing the work and coordination as well as the concept of hard work. [7].

OAC is also a hardware course which has a strong sense of technicality, practicality, engineering. The reforms are as follows:

a) Theory and experiment period arranged in the proportion of 2:1. Increasing experiment period is the key for improving students' practical ability, and it can promote students more direct-viewing and more thorough understanding of ALU, controller, memory, and computer system.

b) Setting up a open laboratory, which is open to the students full-time. The students can do experiments in their spare time, and students will be provided with propositions which be discussed in groups, then they can get creative credit and award credit according to it.

c) It can improve the using rate of experiment devices, and increase the proportion of comprehensive experiment and designing experiment (50%). It is an effective means to improve the student's practical ability.

d) Strengthening experiment instruction. The students are limited to 3 in each group, and a teacher can coach no more than 3 groups.

e) Training the operating ability of students, improving the method of experiment examination through every teaching step.

2.2.5 No Paper Test and Examination Separated from Teaching through Improving Examination Database System

With the further developments of the teaching reform and gradual accomplishment of curriculum construction, OAC examination must be standardized, systematized, scientific and intelligent. Examination database has two special advantages in the teaching practice. The first shows in teaching management, the characteristics of efficiency, economy, flexibility and secrecy of the examination database. The second shows in educational measure the database can produce controllable test paper with high quantity, and equivalent comparability. According to two characteristics metioned above, examination database system should be introduced in examination link, consequently, achieving no paper test and examination separated from teaching [6].

2.2.6 Other Links

- High quality teaching plan has an great effect on class teaching.
- To make a standard teaching schedule.
- Strengthen teaching staffs construction.
- Develop teaching research frequently.

3 Conclusion

Teaching reforms of OAC course will be a long process, which need to explore and practice continuously. Through the 8 years' teaching practice, we have gained some helpful experience for your reference.

Acknowledgment. The authors want to thank the helpful comments and suggestions from the anonymous reviewers. This work was supported by excelsior curriculum program of Harbin University "Organization and Architecture of Computer".

References

1. Hong, L., Wang, A.-J.: Reformation and Innovation of Practical Teaching in Application Oriented Undergraduate Courses. Reseach and Exploration in Labaoratory 23, 5–9 (2004) (in Chinese)

2. Xu, G.-P., et al.: Exploring On the Computer Architecture Teaching Reform. Computer Knowledge and Technology (3), 862–863 (2007) (in Chinese)
3. Xu, A.P., et al.: The educational reform and research of the course of computer organization's principle. Journal of Architectural Education in Institutions of Higher Learning 13(1), 53–55 (2004) (in Chinese)
4. Wang, H., et al.: On Teaching Computer Organization's Principle. Journal of Beijing Polytechnic College 17(12), 101–103 (2008) (in Chinese)
5. Wang, Z., et al.: The Reform and Practice of Computer Composition Principle Curriculum Teaching. Computer Knowledge and Technology (13), 251–252 (2007) (in Chinese)
6. Gu, H.: Research and Implementation of intellectualized WEB Examination Database System. Northeast Petroleum University, Paper (2004)
7. `http://dj.iciba.com/coordination`

The Aided Teaching System Based on Web for Engineering Blasting

HongJian Li[1] and Hong Hu[2]

[1] School of Civil Engineering, Shijiazhuang Tie Dao University, Shijiazhuang, China
lhjhh73@163.com
[2] Department of Foreign Language, Shijiazhuang Institute of Railway Technology Shijiazhuang, China
zhimahu009@163.com

Abstract. In order to improve teachers' perfonnance of engineering blasting, the Aided Teaching System Based on Web for Engineering Blasting was built. There a lot of materials for helping students' study in this system. In this system students can talk each other or with their teachers, ask questions to teacher about this course. And teachers can set and correct students' homework. The system also provides some examinations to students. As a aided teaching system, it has given lots help to students. simultaneously it resolved the relationship between web teaching and teacher teaching.

Keywords: engineering blasting, web-based, teaching system.

1 Introduction

Civil engineering is the production activities and engineering technologies which people used it to build house, railway, road, bridge, tunnel, canal, dam and port. In those civil activities, a lot of soil or rock need excavate, such as the excavating of house base, tunnels and cutting of railway and road. The blasting technology is one of the engineering technologies which use energy of explosive embedded in rock to shape, damage, move and throw the rock to achieve the engineering aim, while the energy of explosive was translated to mechanical work. So the blasting engineering is closely with the civil engineering from its begin to its development, and was used in railway, highway, mine and metallurgy widely. Blasting engineering is still be using in many engineering for its adaptable ability, simple construction equipment, lower cost and high construction efficiency. Blasting engineering construction has the important place in mining, cutting mountain and excavation of railway tunnel and cutting, special in some project which has a lot of stones and soil to be excavated. Now through the Mechanical excavation method has a large development, in a long time the blasting construction method will be a main construction technology of civil engineering. Visioning the 21st Century, with the implementation of the strategy of sustainable development in China's economic construction, the economic development main battlefield is now extended from east to west slowly. So lots of development and infrastructure construction project in communications, energy and mine resource will be constructed in the hilly and mountains region, and the engineering scale of

W. Zhang (Ed.): Advanced Technology in Teaching, AISC 163, pp. 39–43.
springerlink.com

earthwork will be more than its before. The blasting engineering task will be more heavy and harder. At the same time, in the national emphasis on the development of resources and environment protective policies, some old technology will be eliminated, and some new low consumption, high efficiency and harmless control blasting technology will appear, used in railway, highway, large factories, power station and long distance water diversion project.

Engineering blasting is a professional basic course for underground direction of civil engineering, is also a best-course of Hebei province. To students of underground direction, the blasting engineering course is very important. This course has a number of knowledge to be learned, involving rock mechanics, fracture mechanics and fluid detonation dynamics. This course is the required course for underground and communications students. The main contents of blasting engineering are explosive basic theory, commonly used blasting materials and basic initiating method, basic principle of rock blasting broken, tunnel blasting technology, deep hole blasting, chamber blasting, controlled blasting technology and blasting safety and control technology and so on. With the development of blasting practice, test technology and new technology, there will be created lots of new knowledge. From above, it can be got that students must master lots of knowledge in 32 class hour, so there is a contradiction that must be solved. In order to solve this contradiction, the aided teaching system based on web for engineering blasting must be set up to help students learning. Thus student can learn this course outside class on internet. So teachers can only teach the main and hard knowledge in class, students should learn knowledge through internet web outside class.

2 The Design of Teaching System

On internet, teachers can using browser software on teaching system to issue the teaching courseware and teaching materials, to set and correct homework, to release information and to answer questions that were from students. In the teaching system, students can also learn freely, test themself, ask questions and discuss the questions among teachers and students. The aim of this system is to help students autonomic learning and to expand students' scope of knowledge outclass. and then students can do some design on blasting engineering under the guide of this teaching system. In the internet web, students and teachers can shear their information, which they fell helpful to learning blasting knowledge, and learn from each other.

The teaching system takes advantage of internet to manage the teaching contents of blasting engineering, including electronic notes, electronic books, CAI courseware, teaching video, experimental teaching video and some blasting engineering projects videos, thus students can teach themself freely on web system, they can surf web page or download the resource materials. there is a web site for students to ask questions anytime when they are online. Teachers can also answer the questions of students', when they are convenient. Through web system, students can evaluate teachers' teaching effect and feedback some advices to teachers about how to improve the teaching method or teaching what contents. Teachers can absorb in students' advices to adjust their teaching method, so the teaching effect can be improved. The system offered some test questions in every chapter to test students' learning effect, after they finishing the task of every chapter.

3 The Functions and Structure of Teaching System

After several discussing with teachers and students, the structure of teaching system is shown in figure 1.

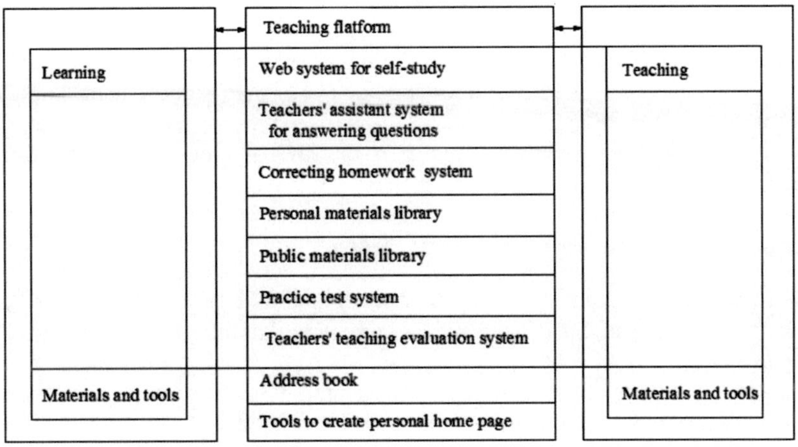

Fig. 1. The structure of aided teaching system based on web

3.1 Internet Self-study System

Internet self-learning system and teachers' class teaching is a complementary system. Students can freely select knowledge to learn, according to their level, difficult knowledge and important section. Of course, they can also learn the knowledge, which their teacher ask them to study. The internet system supplied a lot of resource for student to learn according to their knowledge structure, which included some supplementary basic knowledge, some extensional knowledge of this course, some design example of actual blasting engineering.

3.2 The Auxiliary Question-Answering System for Teachers

The system can provide quick and convenient function which students can question to teachers about learning and surf questions and their answers, even talk each other on internet. The system also provided the search function for students to look up a kind of questions by words. The biggest advantage of the system is that it can relieve labor intensity of teachers, and avoid to answer same question repeatedly. To some difficult question, students can talk with teacher online to solve the questions by MSN messenger. What teachers just must be do is to receive students' e-mail, to sum up students' questions, to give some hints or answers to students' questions and finally put them in questions library on server.

3.3 Teaching Resources Library

The resources library is designed for providing learning materials to students. The materials can be research results of teachers or public resources that teachers selected for students. The personal resources library can include teaching program, teaching aim, teaching schedule of this course, and researching projects and teachers' papers. In the public resources library there are teaching video, experimental teaching video, video and pictures of engineering blasting, the latest development about blasting in domestic and overseas, and some lessons from one's experience got from some losing blasting projects. For a good resources library, it can help student learn knowledge easily, provide a large of teaching materials, and extend the students' knowledge. Thus Teaching resources library can give full play to the advantage of the internet and extend the teaching space.

3.4 Internet Examination (Simulation) System

It is necessary that internet teaching system uses internet exam method. If the system can not test learning effect of student, the automatic learning is not realized. When setting up teaching system, the questions bank can be built according to some previous exams and key contents, which is listed by chapter. When testing, test questions are selected randomly by computer. In order to offer reference to the students, the test scores are kept in personal information.

3.5 Teachers' Teaching Evaluation System

The aim of teachers' teaching evaluation system is build the feedback mechanism of teaching effect. And then through the feedback mechanism, students give evaluation of class teaching to help teacher improve their teaching methods, then teachers can improve their teaching effect according to students' advices. The system is a path for exchanging their thought between teachers and students, and teachers can knew what are students want to learn and how to teach. Thus teachers can get the best teaching effect, and students can get most knowledge in class teaching.

4 The Notice Items When Using Teaching System

The aided teaching system based on web for blasting engineering is only a assistant teaching method for class teaching, so when using, teachers and students can not depend on it excessively. It must be known that class teaching is the main teaching measures, and function of the system is help students to practice, go over outclass, and extend knowledge of the course. When using the system, the following items must be payed attention to:

1) When teaching in class, teachers must give some room for web teaching. Assign some homeworks to student, and let students learn some knowledge, practice, even do some designs according to what teacher have taught in class teaching through web system.

2) Students must know the importance of class teaching, can not depend on the web teaching system unduly. Teachers should let students know that the class teaching and web teaching system have the various function, which can not be replaced each other.

3) When teaching, teachers should give play to the advantages of web teaching. Through internet, teachers can talk with students, and answer students' questions, and correct students' homeworks, even guide students to research or design some teachers' project. When students completed the research or design task, teachers can comment students' results, even to discuss with students online. Thus can improve the practical ability of students, and arouse students' learning enthusiasm.

4) Teachers should usually maintain and update the web teaching system, including adding some blasting news or new research progresses, and some teachers' researching results or teaching materials. So students can timely know the latest blasting progresses or new achievements. Then students' learning interest through internet will be added.

5 Conclusions

The web teaching system for blasting engineering is an assistant base for students' learning which used the advanced internet technologies. The system absorbed in large of materials, such as blasting videos, pictures and web courseware, which can help student to teach themself, replenish class teaching, and bring into play the internet advantages. Through internet, teachers and students can talk and help each other. Teachers can answer students' questions online, and can assign and correct homeworks, even guide students to research or design some teachers' project. Certainly students can timely know the latest blasting progress or new achievements. In a words, the system is convenient to students' learning, at same time it is the windows for students to know he latest blasting progresses or new achievements. But when using this system, the relationship between class teaching and web teaching must clear. Finally the most important is that the system must be usually maintain and update, so that it can attract students' interests and give play to its functions.

References

1. Liu, X.: Design and implementation of course network teaching system based on Web. Journal of Science of Teacher's College and University 28 (March 2008)
2. Chen, Y.: Design and implementation of network teaching system based on Web. Journal of University of Electronic Science and Technology (5), 52 (2007)
3. Yang, B.: Design and implementation of network Teaching and examination system. Journal of Jilin University (1), 39 (2006)
4. Chiu, D., Wang, P.: An approach for interoperable and customizable web-based mathematics education, Puerto Vallarta, Mexico, pp. 80–87

Web-Based Inquiry of Autonomy in Foreign Language Learning as a English Major in Higher Vocational College

Hong Hu and Zhan Du

Department of Foreign Language, Shijiazhuang Institute of Railway Technology, Shijiazhuang, China
zhimahu009@163.com, daoyi8100@163.com

Abstract. Web-based autonomy in foreign language learning is a new model and trend in English teaching. The author in this paper designed the model of autonomic learning, pointed out several aspects needed to be paid attention to and made a summary of first fruits in experimental class.

Keywords: network, autonomic learning, constructivist strategies.

1 The Concept of Self-studying and Web-Based Autonomous Learning

In today's knowledge-explosion society, the concept of lifelong learning is deep in the people's hearts. Self-directed studying ability has become one of the necessary qualities to adapt the "learning society" . Self-directed study refers to a way that learners in learning activities have acquired subject consciousness and self-awareness, and constantly stimulate study passion or enthusiasm, giving full play to the initiative in the learning process. Self-directed study includes several parts: the study habits, interest, method and ability, which is different from what we called self-study. Self-study is just the a component in it. It has the following features: Language learners participate in designing learning goals, laying down the learning schedule under the guidance o teachers, actively looking for learning strategies when coming up against problems, having internal power support during the process, gaining positive emotion experience from the study, self monitoring and make corresponding adjustment according to the cognitive activities.

In recent years, the rapid development of the modern information technology based on computer multimedia and network technology has had a huge impact on traditional teaching mode. CAI(Computer Assisted Instruction) and WBI(Web-Based Instruction) not only bring education technology on revolution, but also a teaching idea and education thoughts on revolution. The brand new autonomous learning mode is also breeded from this kind of education idea.

The web-based autonomous learning refers to the study that learners use and regulate metacognition motives and behavior for network course of study. "To make the students become independent, independent and effective learners is the ultimate aim of education".(LiShuping, 2009) Multimedia network teaching will no doubt has

W. Zhang (Ed.): Advanced Technology in Teaching, AISC 163, pp. 45–48.
springerlink.com

positive influence on widening the field of vision, enriching students' language learning material, assimilating the outstanding culture linguistic data and improving students' comprehensive quality. However, it is very easy to create the "information overload" and "information lost" due to the network resources' complexity. Therefore, it requires learners and teachers to make the overall evaluation of the cyber resources and carry on the reasonable selection and distribution.

2 Based on Constructivism, Autonomous Learning Design for Higher Vocational English-Major

The key of the teaching-reform in domestic and foreign school is: whether you can break the traditional teaching mode--"with the teacher as the center, speaking little practice", which neither can stimulate students' interest in study, nor restrain students' creative thinking. In order to change this situation, many of the educators come from domestic and overseas did a lot of research and exploration, and constructivism theory is the efforts of the main research.

Constructivism theory's content is very rich, student-centered is the core. The four elements or four attributes in Constructivism learning theory is that "environment", "cooperation", "conversation" and "construction of sense", which cleverly unite to the network improving students' autonomous learning by twice the result with half the effort.

2.1 Circumstances

Multimedia network create lifelike and colorful information world for students. "we should make the students dip in the actual language oceans thus free the teachers and the students from the boring grammar and detailed analysis. The teaching process must be conscious, purposeful". But we need teachers to guard the pass for searching for the essence of the material, which puts forward the new requirements for teaching design. That is, teaching design not only to consider the teaching goal, but also consider the construction meaning, and create the situation as the most important content of the teaching design.

2.2 Cooperation

Students' autonomous learning is not the so-called "self-study", it needs teachers' effective guidance to propagate the close cooperation between students and teachers. Collaboration occurs in the whole process. It matters a lot on students' material collection and analysis, the proposal and verification of the assumption, and the learning achievement evaluation.

2.3 Conversation

Conversation is the indispensable link in the process of cooperation. Study group members must through the conversation to discuss how to finish the set task. Moreover, collaborative learning process is also a process of conversation, in which each student's wisdom is shared for the study group.

2.4 Sense-Making

Construction significance is to help students to achieve a deep understanding to nature and rule from the current study content. However, the current situation is as following: because the multimedia web teaching is new to teachers and students, its characteristic and cognitive process can not be known in depth to teachers and students, and there were few English Learning Strategies in multimedia web teaching. Students had not english web Learning means and had no plan for learning. So students must got effective learning methods from english multimedia learning, and make up a learning group to help each other under the teachers' guide.

Based above characters of english multimedia learning, the self-studying process for web multimedia teaching was designed, and was shown in figure 1.

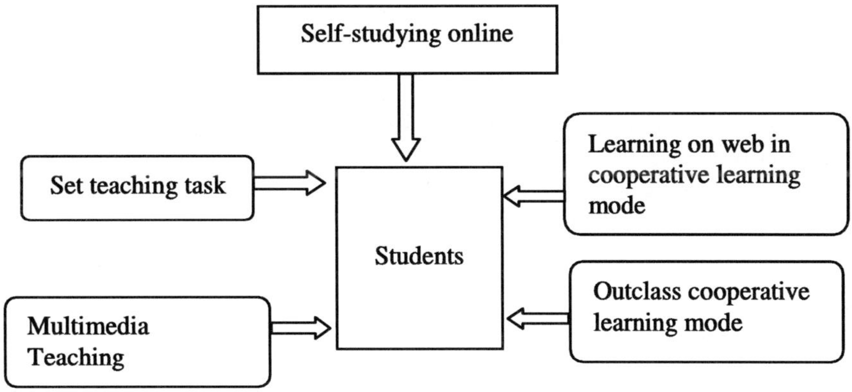

Fig. 1. Self-studying process for web multimedia teaching

3 The Notice Items for Web Based Self-studying

1) In general, the learners can be divided into four types, auditory Learning, visual Learners, kinesthetic type, haptic type, some students may be belong to synthesizing type. So teachers should select the learning contents for students according to various types.

In the beginning of web teaching, teachers should guide students how to learn, according to below three stages. First stage is to develop the thoughts of self-studying and cooperative learning, second stage is to finish the homework of web learning set by teachers and to solve questions that met in web learning. Third stage is students' entire self-studying guided by teachers.

The web learning must has a learning aim or task. When finishing the learning, there should has a exam to students to check their learning effect. in a words, in web sel-learning, teachers should manage students efficiently.

Teachers can set homeworks to students, give key contents, and upload some learning resources for students through class blog. In order to give play to internet function, the web should be a exchange platform between teachers and students. On this platform, students can freely help each others, and talk with teachers and others student.

4 The First Step Self-studying Achievement Web Based Learning of Higher Vocational English Major

From the self-studying researching in class 50911, department of foreign language, shijiazhuang institute of railway technology, from June 2010 to September 2010, it had got that students' learning initiative have be improved immensely, and the self-studying ability has be enhanced. From the testing, students' learning effect is very nice, special in students' abilities of listening and speaking. All of students of testing classes have past the 'A grade examination'. The passing rate of CET-4 is 14.7% more than previous class. Even some students got good achievement in state english match. Through the web teaching drillings, most of students have got the skills to operate the teaching web system, including researching learning resources, sending and receiving e-mail for exchanging with other person, discussing in internet web with students and teachers, and finishing their homeworks. The students' learning attribute has become to be active learning form previous passive learning. The web teaching system has aroused students' enthusiasm to learn tremendously.

5 Conclusions

In web teaching system, students got not only a lot of linguistic data but also teachers' guide. So the learning efficiency had a large improvement. Self-studying supported the diversified teaching mode, which teaching based on web, students is the center guided by teachers, and there are teachers' teaching and students' sel-studying outclass when learning a course. The web teaching extend class teaching to sel-studying outclass. The new teaching mode can switch teachers' role from teaching knowledges for students to teaching students how to learn.

References

1. Qi, J., Li, X., Wang, G.: Activity Construction—The Teaching Reform of innovate education. Shandong education press, Jinan (2004)
2. Pang, W.: Independent Study—Principles and Strategies of Study and Teaching. East China Normal University Press, Shanghai (2003)
3. Shi, X.: Network Aided Independent Study and English Learning Strategy Training. Foreign Languages Research (3) (2006)
4. Holech: Autonomy in Foreign Language Learning, vol. 3. Pergamon, Oxford (1981)
5. Benson, P.: Teaching and Researching Autonomy in Language Learning, vol. 47. Longman, London (2001)
6. Little, D.: Learner Autonomy 1: Definitions, Issues and Problems, vol. 49. Authentik, Dublin (1991)

Design and Implementation of Electronic Experimental Platform in General Technology Courses

Ying Li[1] and Lianmin Hu[2,*]

[1] College of Maths and Information Science, Leshan Teachers College, Leshan Sichuan
[2] College of Physics and Electronic Engineering, Leshan Teachers College, Leshan Sichuan
1238905470qq.com

Abstract. Described design of general technical courses experiment platform based on single-chip in the senior middle school. Seven pilot projects in electronic design to make up the experiment platform which is designed to fit for the high school students in the general technical courses based on the General technical requirements of the course and focus on the electronic control technology module in the General technical courses. During the design process, we fully use of the single-chip technology, sensor technology, as the main elements of the design. In addition, we give full consideration to the actual characteristics of the high school students, as far as possible the simple design, the performance of the effect, easy to operate and so on. The ultimate hope is that this General technical courses experiment platform can actually be used in the actual teaching practice, so that teachers can teach the General technical courses knowledge easily and also the students can learn better knowledge from General technical courses.

Keywords: General Technology, Single-chip technology, Electronic control technology, Experimental platform.

1 Technology of Electronic Controlling

Information technology courses and general technical courses are made to be a main way to bring up students' elements technology and improve technical ability in senior middle school new courses innovation. General technical courses is a main part of new courses, which is combined with theory and practice including information communication and processing, production design and application, general technical course aims to improve students technical ability, individuality, innovation consciousness and personal programming ability[1][2].

With the development of electronic technology and computer, the single-chip which is combined with CPU, RAM, ROM, and I/O, is developed quickly. As the controller is made of the single-chip, electroonic equipments which is applied simply and changeably with the help of software, are more and more intelligent and precision, smaller volume, lower cost[3][4]. It is important for students to study single-chip in general technical courses.

* Corresponding Author.

W. Zhang (Ed.): Advanced Technology in Teaching, AISC 163, pp. 49–53.
springerlink.com

In the new courses standards in senior middle school, students study information and energy transmiting in electonic controlling courses, after studying electronic controlling chapter, students will try their best to solve questions which are more intersting and valuable in daily life based on the electronic system design. After studying this chapter, students will know structure of electronic system, learn to design and fix electronic controlling circuits. Students ' ability will be improved by analysing principle of electronic controlling and resolving bugs. In this chapter, teacters should focus on application and amelioration of electronic system, teaching students how to solve questions wihich are in daily life and typical facts by the way of systems integrating and controlling. The chapter is made up of four parts, whiich are showed as follow in Fig.1.

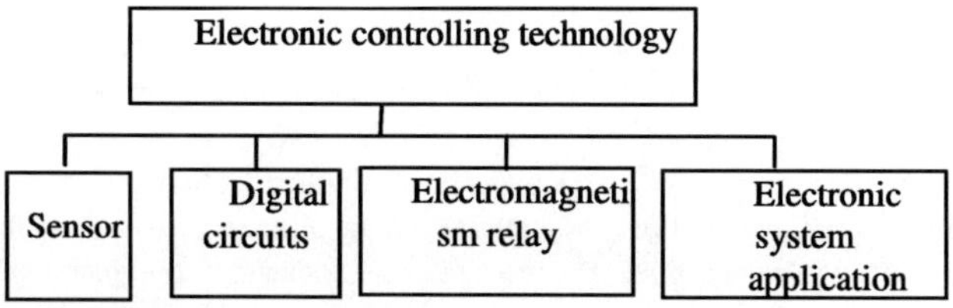

Fig. 1. Electronic controlling technology

In this paper, we described ways of design and implementation of electronic experimental platform which is composed of actual products and guide according to new courses standards in senior middle school in general technical courses, products are fit for students and will be finished easily. Students' intersting and ability will be increased in this course.

2 Design of Electronic Experimental Platform

Seven pilot projects which are composed of actual products and guide are finished in experimental platform, such as multifunction clock, facility calculator, digital thermometer, controlling system of traffic lights,and etc, which are showed as follow in table 1.

Table 1. Seven projects in electronic platform

ID	module name	uide	component	Core of Design
1	Multifunction clock	Yes	Yes	Application of system
2	Facility calculator	Yes	Yes	Application of system
3	Digital thermometer	Yes	Yes	Application of system
4	Controlling system of traffic lights	Yes	Yes	Application of system
5	Multiple responder	Yes	Yes	Application of system
6	Acoustooptic controlled switch	Yes	Yes	Sensor
7	Alcohol concentration testing instrumentr	Yes	Yes	Sensor

Students can learn how to design and control electronic system by one of the seven pilot projects. In general technical courses, teachers teach students single-chip technology, sensor technology and ways which used in data collecting, processing and feeding back. Students select one project in electronic experimental platform where they learn principle of electronic module, jointing and debugging, finally, they finish one product which displays some information clearly. Students learn general technical purpose in practice. Seven pilot projects are showed as follow in Fig.2.

Fig. 2. Figure of Experimental Platform

3 Design and Implementation of Multifunction Clock

3.1 Function

In this chapter, design and implementation of multifunction clock which displays date,time and temperiture, etc is described in detail. After initialization instructions are sent to chip DS1302 and DS18B20 from, Single-chip AT89C52, AT89C52 collects information of temperature form DS18B20 and year, month and date from chip DS1302[5].Data is displayed based on 128×64 LCD. Design of multifunction clock is composed of three modules, such as data collecting, processing and displaying. Stucture diagram of multifunction clock is showed as follow Fig.3.

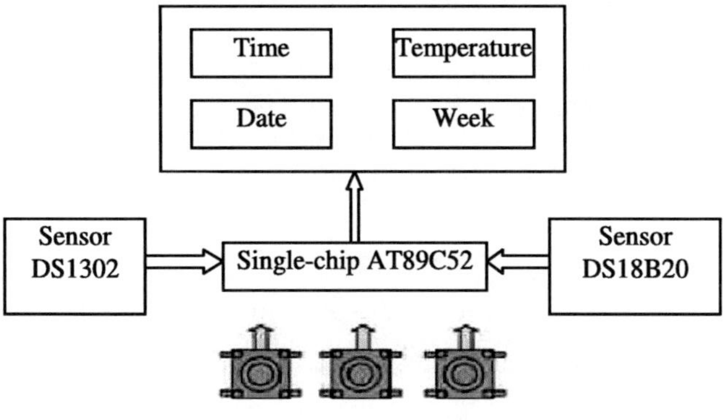

Fig. 3. Stucture of multifunction clock

3.2 Programming

Program is debugged in Wave 6000, in which C language is used,. Students understand programm easily because notes to programm are presented. Program struture diagram are showed as follow in Fig.4.

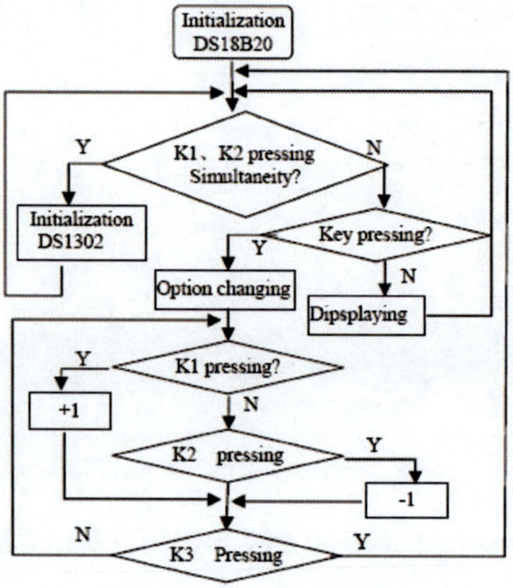

Fig. 4. Program struture diagram

After system is electrified, DS1302 and DS18B20 are initialized firstly, and then main interface which is compsed of four parts such as time, date, week and temperature is displayed on LCD. If key K1 and K2 are both pressed for two second, system will be initialized on 1,1,2008. Date, time, week can be changed by key, k3 is used for options and reset, K1 is used for increasing, K2 is used for decreasing, in addition, multifunction clock will alarm in days and temperature whose value are set up in progarmming. Product of multifunction clock is showed as follow in Fig.5.

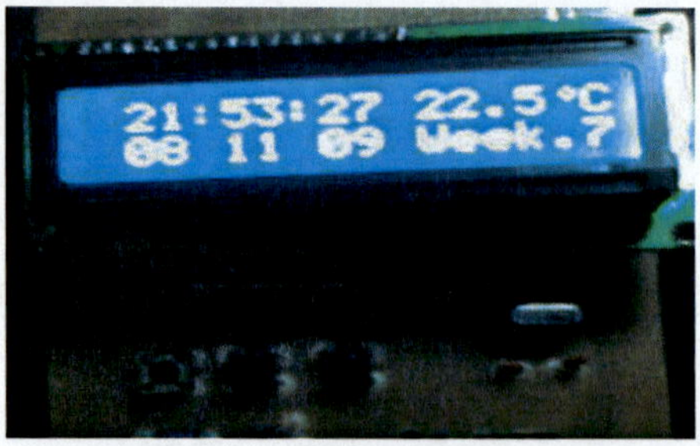

Fig. 5. Product of multifunction clock

4 Conclusion

Electronic experimental platform is composed of seven pilot projects, which can be used in general technical courses according to new course standard in senior middle school. As products, guide and components are presnted, products which are usually in daily life in electronic experimental platform will be finished easily. It is proved that interest and operation ablity are improved by products being finished.

References

1. Minstry of Education. Standards of technology courses(experiment) in senior middle school (2003)
2. Fu, H.-P.: Analysising teaching characteristics of general technology course in senior middle school. Primary and Middle School Educational Technology 32(1), 9 (2009)
3. Zhao, J.-L.: Development guidance based on AT89051 SCM. Pubnishing house of electronics industry, Beijing (2008)
4. Wang, F.-R.: Designing of testing and controling system based on SCM. Press of beihang univetsity, Beijing (2000)
5. Fong, W.: Displaying system of time and temperature based on AT89051 SCM. Advanced Display 15(12), 52–56 (2008)

Web-Based ESP Teaching in Higher Vocational College

Hong Hu and J.-W. Song

Department of Foreign Language, Shijiazhuang Institute of Railway Technology, Shijiazhuang, China
zhimahu009@163.com, daoyi8100@163.com

Abstract. This paper descried the teaching background using ESP in higher vocational college and indicated three disadvantages existed in current teaching process. It discussed how to improve ESP teaching quality through using web-based ESP method on strengthening listening and oral skills, accelerating materials development, building up a ESP corpus, ensuring personalized studies and enhancing ESP professional quality.

Keywords: network, ESP teaching, higher vocational education.

1 The Higher Vocational ESP Teaching Background

The main teaching direction in higher vocational En glish teaching is to foster students' language ability, strong English practical ability, especially the professional English ability, which decide whether the students' can truly adapt to the relevant professional technical field or professional post office of the request. Therefore teaching content should be based on a certain professional technology field and relevant professional post need, which completely have the same view with the core of ESP. The professional English (ESP), "English for Specific Purpose", is referred to a specific profession, course or purpose- related English, which in our country are also known as the Special Use English. Its purpose is to foster the students' English communication ability in a certain working environment, such as business English, tourism English, law English biological English, engineering English, and medical English. Because of its clear purpose and high practical value, it has got the love of the learners from the countries all over the world. At the same time the special demand gradually increase in professional field as English has further improved its international status. In this background, the specific profession or occupation of the ESP came into being and gradually formed the unique English teaching area.

At present, domestic college English course teaching has the trend of using ESP method. ESP teaching is the 21 st century English teaching development trend.(LiuRun qing,1996:1) ESP teaching should be gradually become the mainstream of the college English teaching.(CaiJigang, 2004:22) European countries and America ESP teaching research has developed fully systemization. There are a lot of the experts, book, and special research for the ESP journal. In Japan, ESP has in fact become the main model of college English teaching.

W. Zhang (Ed.): Advanced Technology in Teaching, AISC 163, pp. 55–58.
springerlink.com

2 The Existing Problems in ESP Teaching

2.1 Teachers' Professional Level Needs to Be Improved

We know, ESP teachers should have double teacher quality---that is higher level of English and relevant and professional knowledge and experience. Teachers should be competent in professional training, the employment organization and guidance. They teach students the knowledge and skills in related industries, such as marketing strategy, negotiation skills, social etiquette, etc, and guide students to practice activities. At present ESP courses are taken either by the foreign language teachers or professional teachers. Teaching's width and depth are limited because most of the foreign language teachers only understand language but don't understand profession, what's more, some have no any working experience. And for most professional teachers, they have shallow language foundation, poor communication ability and do not quite understand advanced language teaching theory. Therefore the teaching method is obsolete and boring and can't effectively develope the students' language learning skills.

2.2 Students in Different Level Bring Difficulty in Teaching

Students' English level in Higher vocational technology cannot compare with ordinary undergraduate. They come from great disparity and their application ability is uneven. So teachers may design, adjust good teaching level according to different students in order to highlight professional post and language skills in listening, speaking, reading, writing and translating.

2.3 Teaching Material Lacking of Practical Guide

As ESP teaching material, the books must have clear teaching goals, complete teaching system and related professional authority. But China ESP teaching books at present is not standard and quality intermingled. Most of the teaching books focuses on knowledge theory, don't pay attention to the training of the students' practical ability. Exercises of the text are mostly strengthening grammar and lack of practicability and interesting, which can't link corresponding professional scene of the work. And the key is teaching ESP is to emphasize the combination of language and professional communication ability, but most of the teaching materials can not meet this.

3 ESP Teaching Based on the Network

So in the present condition, how to overcome the above shortage and set up effective teaching mode? ESP network teaching is undoubtedly a beneficial attempt, whose course construction, faculty ratio, learning resources, learning style or the evaluation methods have flexible and open characteristics. The so-called ESP network teaching mode, is to handle text, image, audio and video media information relying on the network. In this way, more information establishes a logical connection and integrates as a interactive system. In English teaching, it has further broken English teaching space and time limit, fully realize the resources sharing. It also can use E-mail, BBS, blogs, SNS, RSS, Wiki software to organize network course.

3.1 Use Network Resources to Create Real Context, Strengthening the ESP Teaching in Listening

With the development of world economy, all works of life and international exchanges increase the need of outstanding industry representatives with excellent oral English. So more and more students hope to improve English ability in order to achieve the future international exchange level. Because of class hour or faculty reasons, the listening and speaking parts in ESP teaching in our country are very weak. This weak link can be made up through the network. Teachers can provide students with some websites with rich learning material such as BBC, VOA, China Daily, CRI English, where The materials are beneficial to stimulate students' learning enthusiasm and save students' time for finding learning resources, improving the efficiency of the autonomous learning. In addition, the teacher can create real language environment and bring the real social scene into classroom by using the network multimedia.

3.2 To Promote Teaching Books' Efficiency and Sustainable Development

ESP teaching material should organically integrate language and the content engaging in professional activities based on reality. The feasible books would be good to highlighted its positions characteristics. Most colleges use the press materials, also a few schools use the original text. According to statistics the books written by teacher takes up only 6.8%. ESP teaching materials should be update and constantly supplement latest language on each industry. As the network language is mainly in English, many professional web site will become ESP teachers advantageous material databank. And online various electronic press can also provide many free study materials for teachers and students. In the process of editing, it can not only improve teachers' professional level, and improve professional theory in domestic and international, making the teacher walking in front of the professional technology.

3.3 Build Professional Foreign Language Corpora with Students

Corpus as a product of the network times, subvert traditional professional English from the learning form and content. Due to the using of the computer technology, Corpus can hold more rich content and can be added, supplied and the maintained. The purpose of using the context of the corpus is to guide the students in contact with a lot of real, meaningful language and find the striking features and rules of the language. Corpus is of modular management and provide learners targeted teaching content, so each learner can make one's choice according to oneself and the ability. we advocate the student actively participate to establish the single-target corpus, and during the process they will understand the professional English language structure, composition and word-formation theory, which help them break the bottleneck and greatly enhance the vocabulary.

3.4 Teaching Students in Accordance of Their Aptitude, Using Network to Meet Their Individual Learning Needs

Interactive technology of network provides active participation for students, which give full play to the student's enthusiasm. And students can choose study content, arrange the learning process, independently solve all kinds of problems and get knowledge according to their different levels. ESP teachers also can design the real task, then let students use real network material to complete the task through the cooperation. In this way it cultivate students' autonomous learning ability and the ability to solve problems, and also to improve the students' co-operative spirit.

3.5 Enhancing the Professional Teachers' Quality and Allocating Teacher Resources

As is known to all, a team of qualitied ESP teachers is the core to improve the ESP teaching quality. ESP teachers should not only have a good foreign language level, but also has authoritative professional knowledge. Therefore, the teachers' training is rather crucial. But due to the insufficient teacher resources on normal teaching, not to mention such consuming training on ESP teachers. In the present conditions, the network teaching can provide a very effective transition plan, that is to promote the professional cooperation based colleges and intercollegiate communication, especially the cooperation between foreign language teachers and professional teachers. In this way, teachers could help each other on professional knowledge and language knowledge. Its purpose is to realize professional English teaching through the English teachers' right guidance.

4 Conclusions

The ESP make the"target scene" analysis or "need analysis" as the starting point and the center of teaching. The essence is to meet the different needs of different learners. It provides a feasible method for higher vocational English learners efficiently get language exchange form required profession. So it is suitable for higher vocational students' the objective reality. But its implementation has to face such series of problems and challenges as the formulation of syllabus, compiling the teaching material, teachers' training, and training mode. Therefore, teachers should actively explore and work together to improve students' professional capability and make students be versatile talent of "Professional plus foreign language".

References

1. Hutchinson, T., Waters, A.: English for Specific Purpose: A Learning-centered Approach. CUP, Cambridge (1987)
2. Steven, P.: ESP after twenty years: a re-appraisal. In: Tickoo, M. (ed.) ESP: State of the Art. SEAMEO Regional Language Center, Singapore (1988)
3. Liang, X., Chen, L., Lu, Y.: Problems and Strategies in Building the English Courses for Specific Purposes. Foreign Language World (4) (2006)
4. Zhang, J., Gaoqin, Wang, B.: Study on Corpus and English Courses for Specific Purposes(ESP) vocabulary Teaching. Foreign Language World (3) (2009)

Network Video Teaching Based on Learning Autonomy Taking Computer Teaching in the University as Example

Tang Lifang

Department of Computer Engineering, Cangzhou Normal College,
Cangzhou, Hebei Province, China
Tiandoudou2008@126.com

Abstract. With the development of information technology, the Video Teaching will become an important means of video network platform, which enable students study by themselves in mode of the video study. By this way, students can gain more knowledge while promote and improve teaching methods. FLV video teaching technology provides a simple, convenient and easy way to publish, browse solutions the video.

Keywords: Learning Autonomy, video network platform, network video, FLV.

1 Introduction

In 2006, the Ministry of Education Computer-based Education Steering Committee issued a "some opinions for the computer-based teaching " (White Paper), computer education for university should be based on "Learning", "competency-based" principle, the teachers should break the traditional "teaching" and, establishment of an independent study, collaborative learning, classroom lectures, laboratory training, etc. It relies on a combination of teaching and network teaching platform to achieve. Self-learning is a necessary requirement for modern college students, students can give full play to the initiative, to mobilize the students interest in learning. Based on the internet, the network teaching platform can provide a flexible time for the student to finish his study. The standards of the network platform are: to support the use of shared resources, to provide a quality assurance and an open environment. [1].

2 The Problems in Teaching Network Platform

National University network platform construction began from 1998 to 2001.Although it began late, level is usually high. There are 87 colleges in Hebei Province, including 3 central governmental universities, 25 provincial common colleges and 59 vocational colleges. 84 colleges of them can be successfully visited, only 3of them cannot be visited. From this we can see that our hardware construction in Hebei college network can meet the basic requirements. The problems found in the research will be listed here:

W. Zhang (Ed.): Advanced Technology in Teaching, AISC 163, pp. 59–63.
springerlink.com

2.1 The Level of Teaching Network Is Generally Low. 80 Percent of the Colleges Cannot Provide Plenty and Useful Teaching Information. and Most of the Resources are Documents or Pictures, they are Quite Dull. Some of them Still need to be Further Analyzed by the Students after Class, which Costs them too Much Time and Blunders their Learning Motivations

2.2 The Information Shared on the Teaching Network is Seldom Shared and Refreshed. in the Research We Found that 60% Information of the Net Work was Closed to the Outside. For the other Net Work which can be Visited, they are Seldom Refreshed. Students cannot find New Information there so it Cannot Show the Advantages of the High Technique

2.3 Lacking the Communicational Function. 90% of the Colleges just Simply Employ Chatting Room or BBS. they Lack the Function of Communications and Teachers cannot Immediately Answer Students' Questions

2.4 Models are Too Limited and Videos are not Fluent. 80% of the Colleges have their Own Competitive Lessons and Videos but they cannot be Fluently seen on the Net. And the Videos only Contain Parts of the Knowledges, they are not Complete

3 Advantages of Video Instruction

Video resources has an unparalleled advantage over other teaching resources. While bringing vigor and vitality for online teaching, the resources contributes considerably to students' self learning initiative. Main features:

3.1 Rich Video Resources, Strongly Visible

Video resources sets images, sounds texts in one, giving a strong visual impact. Video resources stimulates students multi-sensory, making boring teaching process lively and active. The resources can deepen students' understanding and memory of knowledge they have learned, which will help to mobilize students to learn initiatively and enthusiastically. Electronic texts can only put a lot of classroom information optionally "simplified", thus status in class can not be presented, which makes authenticity a question. Its difficult for electronic texts to reprocess the cases from different angles, so the teaching and research lack their own vividness and situation. In addition, the text capacity is limited for communication. [2].

Comparison between text and video resources:

class	information	Authenticcity	theme	Ingredients	space
Electronic text	general	Easy to suspect	Prominent	clear	small
Video resources	large	reliable	general	general	large

For example, in explaining computer hardware components in University Computer-Based, we frequently use the method of combining text and picture. Computer hardware components include input device, output device, controller, memory and calculator. When it comes to memory, we can show a few pictures of memory. Perhaps this will give students an impression, but do not meet teaching scenarios. If we record the actual objects and show them to students, the effects are better. Another example, when explaining computer assembly process, we can record the process with the voice explanation and then we place it online, providing students to consult any time. Therefore students may feel to assemble a computer is not a mystery but a relaxation, and they even assemble computers themselves by watching the video.

3.2 Scene Representation; Breakthrough Teaching Difficulty and Consolidation of Students' Learning Interest

University Computer Basic Knowledge has too many contents so it is difficult to learn. This course only has 64 hours in a semester, Students seldom have chance to practice. Students can't adapt to the teaching content, their concentration can not last long and lack enough time to think and understand. The students whose basic knowledge is not good will complaint that there are too many concepts.

Video teaching, by its scene representation, crossing space and time, flexible interception, etc, gives full play to the irreplaceable advantages. Every link in class including "leading, teaching, learning and practicing" can connect well each other. According to their own actual need, Students choose the content they are interested in, such as the teaching video together with the appropriate web page in the text, which fits students of different level so consolidates the effect of classroom teaching.

Teachers and students have a discussion for what they learn through the system Settings of communication platform. They can find and solve problems in time. This is not only beneficial to understanding between students and teachers, also helpful for students to learn and interactive learning autonomy. For example the teaching of the senior screening in office software EXCEL and the use of the function in triple if, which are common problems in computer level exam. The results are not satisfied even if teachers have explained many times. If we record the explanations of the difficult points and release them in network platform, students can learn after class as long as the computer can connect the internet which stimulates the students' potential memory, reduces the burden of teacher guidance. So video teaching becomes an effective method to teachers in teaching process. At the same time it fully embodies the learner-centered thoughts.

3.3 Video Resources Provide a Royal Road to Students to Get Information

It is an information era today, video resources provide a royal road to students to get much information. Video resources almost include all teaching content of the computer hardware and software. Computer makes easier to accept and understand what the students want to learn. It not only shortens the teaching time but also improves teaching efficiency.

3.4 Video Resources, with High Speed for Innovation, Has a Good Transmissibility

Video resource is the extension of classroom teaching. Recording the process of the lecture, editing it with Video Software, adding the explanation, then putting it online, it is easy to learn. Teachers, through simple training, could make good video resources. Colleges have the good network facilities, which supply the necessary hardware support to sharing video resources. Video resources could be achieved Short-range sharing through LAN or remote sharing through internet.

4 The Feasibility of Implementation of Video Teaching

4.1 FLV Streaming Media Technology

For a long time, affected by network bandwidth, transmission rate, and it's characteristics of huge dada and complex processing, watching the video on line is seldom used on network platform. Recently, with the development of broadband and Streaming Media Technology, the problem of transferring multimedia information on the Internet is solved, which enriches the information resources on internet, improved the application efficiency and value. FLV is short for Flash video, which is the fastest-growing and the most widely used Video Transferring System that is used by most Video Websites. With new image encoding, FLV get the high compression rate and make small files. One-minute clear FLV is about 1MB, even a film only 100MB,1/3 of the size of normal file. Now more than 80% videos on internet are FLV. [3].The main feature of FLV is achieving transmission and playing at the same time. FLV make it easy to check out video with Flash Player or to Merger video to web or flash document. The extensive application of Flash Player will ensure that most visitors could check out the video without downloading and installing plugins, from which video transmission has benefited a lot. [4].

4.2 The Make-Up of FLV

In the versions of Flash 8, there is a independent FLV conversion tool: Flash 8Video Encoder. It can convert the common type video into the type of FLV. Flash 8 Video Encoder supports various forms of videos, such as AVI、WMV、MPEG、ASF、DV、MOVetc, the interface of this software is rather easy. It just needs a few steps to complete it.

After getting FLV files, we can not use them in the web page directly. We need to engraft them into flash animation. As we saw in the various video sites, it needs playing control. How to make it? Click on the menu [File] → [import] → [import video], enter the dialog box, and Select "on your computer." Click the "Browse" button, open the dialog box through the system to load FLV files we just generated, click on "next" button to continue. Next select the "deployment" as: "progressive download from a Web server", enter "Appearance" option. In order to play FLV videos, Flash 8 has about ten playing controller which have different appearances and controlling options. You can select one proper playing controller according to your needs. At last, click "Done" button to finish. Now the FLV which can be played on the web page is done.

4.3 Increase Video Teaching Demand Modules in the Network Platform

For the students' convenience of searching and checking material resources, plenty of excellent video resources are combined on the network platform. Additionally, the video teaching demand modules should be established independently with the classified categories and core steaming media. The modules can be requested and played in any occasion and played in the models of forward, rewind, paused, and locate.

5 Conclusion

Some teachers' Unfamiliarity with the developed teaching technology, the equipment incompleteness of some colleges, insufficient bandwidth, and unavailable video material on teaching platform, all the elements influence the popularizing rate of applying video resources. But in the future, with the development of technology, the internet video can provide more video with high definition. With the specific feature, this module can provide service for teaching and promote students' automatic learning.

References

1. Wu, W.-M.: Network teaching platform and the localizationstrategy. Distance Education Journal (4), 18–19, 59 (2003)
2. Liu, Z.-G., Zhang, Y.-M.: On FLV Video Format. Computer Knowledge and Technology (August 25, 2008)
3. Liu, Z.-H., Zhou, Z.-J.: Teaching application analysis based on the campus network VOD VOD system. Journal of Teaching and Management (3), 25 (2009)
4. Hou, Z.-Q.: Progress of Internet Video Technology. ZTE Communications, 04 (2008)

Exploration and Reform on Programming Course for Non-computer Majors

Yunwei Jia[1], Dan Wang[2], and Qingsong Zhu[3]

[1] School of Mechanical Engineering, Tianjin University of Technology, Tianjin, China
yunweijia@163.com
[2] School of Computer and Communication Engineering, Tianjin University of Technology, Tianjin, China
wangdan@tjut.edu.cn
[3] Research Department, Beijing Vocational College of Electronic Science and Technology, Beijing, China
qingsongzhu@126.com

Abstract. The major of the teaching was turned to program design, because the purpose of studying programming course for non-computer majors was only to use it to solve practical problems of other fields. Different teaching methods and estimate methods were used in order to let the students more capable and more creative. Experiments showed that the reform was effective.

Keywords: Exploration, reform, programming course, Non-computer major.

1 Introduction

As the development of science and technology, applied computer technology has become a very important technology in research and industry manufacture, its superior performance benefit many aspects include living, manufacture, etc. Even for non-computer major students that master a programming language is the key to be a good graduate. But effected by traditional teaching ideas, teaching results are not satisfied, because students often weak in learn and programming to solve practical problems[1-3]. So based on the teaching purpose of programming course for non-major students, we have made many research and exploration to improve their ability to learn and program.

2 Teaching Content Reform

2.1 Teaching Content Focus on Programming

The purpose of programming course for non-computer major students is to practice and solve problem, but not research. So programming language is only a tool for students to finish task and solve problem, student only need to know how to use it, but not all the syntax of program. Traditional teaching content focus on teaching student about the syntax, teacher will teach students how to program until students have learned the

W. Zhang (Ed.): Advanced Technology in Teaching, AISC 163, pp. 65–68.
springerlink.com

syntax, but this teaching method more fit to computer major students, because they have more teaching hours, more detailed syntax training, and more in-depth study, and they can design and realize a more complex program, or even design a new algorithm. But for the non-computer major students, they have less teaching hours, so both study purpose and interesting are program to solve problem and finish task, so the teaching point is to teach students how to plan and design, how to program, and how to solve practical problem.

2.2 Refresh Teaching Content in Time

Computer technology has a fast develop speed, so teaching content must be refreshed in time. For example, the pointer of C language is an important but difficult knowledge point. It's very difficult for students to understand, but C++ language has a better command to instead of pointer, this command is easier for student to understand and use. And for 8031 SCM, it has no program memory, so it will take a long time to teach students how to expand program memory, but now 89 series SCM has a 4KB to 32KB program memory, it need not to be expand any more, so we only need a short teaching hours to teach students about this.

3 Teaching Method Reform

3.1 Teaching with Examples

Teaching with examples means that choose special examples to teach students, through explain examples to teach students knowledge points, and teach them how so solve analogous problems. Traditional teaching method is that teach students about syntax first, and then explain student a example to validate the syntax and teach student how it was be used. This kind of teaching method is more fit to computer major students, but not non-computer major students. For non-computer major students, we can choose special examples with special knowledge point and through explain examples to let students master knowledge points. By this way can let students know how the syntax works in the example and how to use several syntaxes to solve analogous problems.

3.2 Teaching with Metaphor

For teaching, the highest level is to make students like and desire to learn. For the programming design course, it's always bored and difficult to the students, but use metaphor can let the bored knowledge point became colorful and vivid, student can learn knowledge point in an relaxed and lively atmosphere. For example, we find that it's difficult for student to learn the relationship between class member and access right, so we compared this as a family of feudal society, public member compared as father, private member compared as children, and protected member compared as mother, by laughable animation and detailed explain let student master the relationship between class member and access right.

3.3 Teaching with Inspiration

Teaching with inspiration is to let student study more positively and actively, to improve the ability to analyze and solve problem. For example, when teaching an example, teacher often analyse and explain it step by step, the students listen and think passively. If teach with inspiration, encourage students to analyse the example themselves or explain it to other students, what the teacher need do is answer the question or give students some cues. By this way can mobilize enthusiasm of students and get better teaching effect.

Furthermore, when prepare for the course, teacher should design each example and question carefully to make sure that each student can gather themselves together to think and discuss.

3.4 Teach Individually

Different student has different ability, teacher should teach different knowledge to different student based on their interest and ability. If a student like program and has the ability, we can encourage and lead the student to do some research work or solve some complicated problem. If a student like program but only has a limited ability, we can encourage and lead the student to solve some simple problem. But to the students who are not interested in program, we only need to teach them basic knowledge.

4 Practice Teaching Reform

Traditional practice teaching is to consolidate knowledge, most of the practice is to validate the program, students copy the program from book and check the running result of it, this kind of practice has less help to train the ability of innovation, and many students even don't know how to program of their own[4.5]. So practice training should pay more attention to train the ability to analyze and program, train the students how to debugging and how to compile. Our opinions meet to requirement of ministry of education that attention to practice, train practical ability and program ability.

4.1 Design and Program Ability Training

In order to train the ability to design and program, increase confidence and interesting, students should do more design and program training. Practice training is to train and test the ability to design and program, too many validate training is adversely to the program ability and interesting. Education psychology show that if student don't know how to use knowledge to solve practical problem, or they have no sense of accomplishment, that will kill student's interesting to study[6]. Student can acquire and interested in program even when they can design and program to solve practical problem themselves. So for practice training, teacher should design a question bank, which is from easy to difficult and from basic to application. If student design and program themselves which has passed the compile and got running result, it will encourage the student and increasing interesting to study the course.

4.2 Debugging Ability Train

Our investigation shows that many students don't know how to use error information and debug tools if program failed to pass compile, most student choose check program step by step, it's difficult and has a low efficiency especially to the complex program. So teach student how to debug is very important. To master programming software, one must know how to use compile information and how to use error report, how to use debug tools, for example, the break point, MSDN etc.

5 Exam with Diversified Way

By diversified way to exam has become more and more popular, tradition method always use the result of final exam as the result of one course, but this is unfair to many students. So some educators advocate that exam with diversified way to make sure that education always service for growth of students[7]. With diversified way, the result of one course was divided into two parts, one part is the final exam result, the other is composite result, include lab result, homework result, attendance result, etc. All those compose of the result of one course. This way is more reasonable to the students.

Above is the summary of our research on C++ programming course reform to non-computer major, and we have acquire some achievements, for example, some students can use C++ language to program a scholarship management system all by themselves, and some student model to machine hand with C++ language to identify damage on workpiece. But teaching reform is a long way which filled with exploration and improvement, we will keep on research to improve teaching reform for non-computer majors.

References

1. Gao, M., Yang, Z.H.Q.: Exploration and practice on teaching reform of C/C++. Computer Age 11, 8–10 (2005)
2. Liu, X.P.: Exploration on computer courses for non-computer majors. Education and Vocation 11, 212 (2010)
3. Wang, X.L.: Exploration on teaching reform of visual C++ course design. Computer Learning 3, 99–100 (2009)
4. Hu, Z.H.J., Shu, Z.H.D.: Exchange roles for experimental teacher on course reform—Exploration on perspective on ontology. China Education Research 6, 89–91 (2007)
5. Zhang, H.,T.: Exploration and practice on training system for applied talents. China Education Research 2, 86–88 (2008)
6. Liu, R.D.: Psychology base. Publishing Company of Education Science, pp. 55–69 (2007)
7. Zhong, Z.H.X.: Ten directions of reform for college education. China Education Research 1, 88–91 (2007)

On the Cultivation of Job-Hunting Competitiveness of Students in Logistics Department—Chongqing Jiaotong University in Case

Zhang Yun and Li Li

Department of Management, Chongqing Jiaotong University, Chongqing, China
npzyhy@yahoo.com.cn, haotian860119@163.com

Abstract. It is the competitiveness on the employment that needed for college graduates to stand out in the fierce competition of employment. Consistent with the trend of the age, undergraduates should improve their abilities of competition on the employment. Based on characteristics of the market demand about the Logistics Professional, this paper analyzed formation factors of the professional competence. With the instance of Chongqing Jiaotong University, paths and countermeasures of improving their abilities of competition on employment are brought forward in this paper.

Keywords: Logistics Professional, Ability Formation, Competitiveness.

1 The Three Decades Course of Logistics of Reform and Opening

1.1 China's Logistics Industry Has Experienced to Explore the Theory and Start to Practice, Develop Comprehensively

From the eve of the Third Plenum in 1978, the Administration of State Materials organized a National Planning Commission, Ministry of Finance, Shandong Province and other relevant government departments and some colleges to study materials management of Japan, so the concept of "logistics" was first introduced into China. In January 2008, the restructuring and rejuvenation program of ten major industry, which are the automobile, steel, textile, equipment manufacturing, shipbuilding, electronics, petrochemical, light industry, non-ferrous metals and logistics, has been approved in principle in the State Council executive meeting. China's logistics industry has experienced theoretical exploration, practice started, a comprehensive development process. China's logistics industry has experienced to explore the theory and start to practice, develop comprehensively.

1.2 In the Following Several Aspects, the Logistics Industry in China Made Outstanding Achievements

First, the logistics industry has realized the rapid growth; Second, logistics enterprise has realized the rapid growth; Third, manufacturing and trade communities has implemented modern logistics management; Fourth, logistics infrastructure

W. Zhang (Ed.): Advanced Technology in Teaching, AISC 163, pp. 69–74.
springerlink.com

construction has developed rapidly; Fifth, logistics information and technology innovation has stepped on to new step; Sixth, logistics industry basic work system has been established; Seventh, logistics research, science and technology, and education have made important achievements; Eighth, the policy environment of the development of logistics industry has been improved

2 The Market Demand of Logistics Management Profession

Logistics is a comprehensive discipline, and is the edge discipline of combination of technology and economy. The logistics industry is both cross-trade and cross department compound industry, and at the same time, is the labor-intensive and technology-intensive combination of industry. So from the perspective of market demand, China's future demand characteristics of the logistics professional talent mainly displayed in the following three aspects.[1]

1) Logistics industry job groups were widely distributed and different types of enterprise set up different posts. Logistics post groups mainly distributed in enterprises, planning and consulting departments, scientific research and teaching organizations and so on.

2) Logistics profession was administrative levels. In the market economy, the graduates who were a type or a level educated were no longer a single position level, but were corresponding to the position with a level.. The every link of logistics operation needed management talents with certain knowledge and professional skill.

3) The features of Logistics professional became compound and specialized. That is the logistics industry's requirements for knowledge and ability have the features of chaining, integration and information.

3 Important Abilities of Logistics Management Professionals

The characteristics of the market demand about the logistics professional determine the logistics professionals must have some aspects of the important capabilities. According to the result of the survey in logistics business in Germany and Europe by German INWENT organization and the combination of domestic situation, summed up the capabilities of logistics management professionals consisting of the following:

The characteristics of the market demand about the logistics professional determine the logistics professionals must have some aspects of the important capabilities. According to the result of the survey in logistics business in Germany by German INWENT organization and the combination of domestic situation, summed up the capabilities of logistics management professionals consisting of the following fig.1: [2]

Abilities according to their order of importance are as follows(1=not important, 5=important):

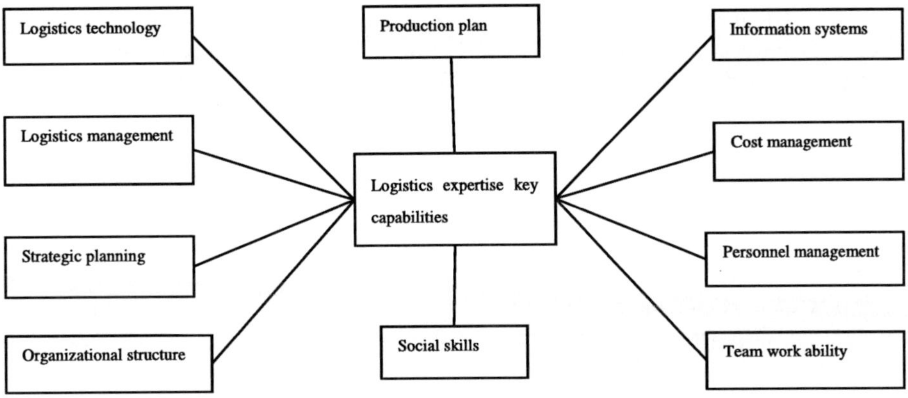

Fig. 1. Logistics expertise key capabilities

4.5	Team work ability
4.3	Logistics management
4.2	Information systems
4.1	Social skills
4.1	Personnel management
4.0	Organizational structure
3.8	Logistics technology
3.8	Cost management
3.7	Strategic planning
3.0	Production plan

1=not important,5=important

Fig. 2. Logistics professional's important abilities to sort

4 Logistics Students' Competitive Capacity-Building on Employment

4.1 Strengthen the Teaching Reform Study

1) Actively explore the establishment of scientific and rational logistics curriculum. Improving the traditional disciplines of the three-stage[3]: course content of basic courses, professional foundation courses, specialized courses, gradually establish a network system which focuses on training students' comprehensive qualities, to adapt to the need of integrated management talent in China's industry market and with the development of logistics.

2) Strengthen the teaching sense of reform. Timely track the dynamic forefront of education reform at home and abroad. From time to time invite Well-known university experts to introduce advanced teaching theory and methods. Combined with characteristics of professional courses, reasonably apply the new teaching methods to teaching practice.

3) Active in education reform research. Strive for all types and sources of education reform research projects. Mobilize professional teachers working closely with teaching practice, lessoning learned and achievements, actively write and publish papers on reform.

4) Promote education reform discussions. Faculty meetings held regularly to discuss developments in education reform and education reform program and measures of the discipline teaching. Analysis teaching problems and propose solutions.

4.2 Revised and Improved Personnel Training Programs

In order to meet the country's logistics industry development, combined with the characteristics of high integrated capacity requirements in logistics management, revise and further improve the training program.

1) Actively establish a scientific and rational logistics curriculum. Improving the traditional disciplines of the three-stage: course content of basic courses, professional foundation courses, specialized courses, gradually establish an integrated network system by the basic courses, professional basic courses, professional courses, elective courses, including practical aspects of training courses, cognitive internships, graduate internships[4].

2) Promote the construction of the core professional courses , and to cultivate compound talents who are proficient at modern information means, logistics business and the laws of logistics operation.

3) Enhance the practical ability of students through strengthening the construction of practical courses and spending more time on practice link, such as the comprehensive and innovative experiments.

4) Actively introduce foreign advanced teaching concepts and resources, explore the establishment of new courses and cultivate logistics talents who are close to market demands.

4.3 Strengthen the Construction of Practical Teaching System, and Improve the Innovative Ability and Comprehensive Quality of Students in Practice

1) Increase investment in laboratory, enhance school-enterprise cooperation and improve the professional skills of students.

Build a number of new design type and comprehensive type experiment items which have professional characteristics of logistics.

Further extend the number of practical teaching bases in this special and constantly increase the type of them on the basis of maintaining and consolidating primary ones. Make the practical bases include different types of enterprises, such as production enterprises, storage and transportation enterprises, professional distribution enterprises, shipping enterprises and so on. Improve the mode of practice: Arranged to practice in enterprises' various position, students could work in a real working environment. They can contact with real professional work early, widen their knowledge range, strengthen their perceptual knowledge, exercise their comprehensive application in professional knowledge and basic skills, and eventually improve their practical ability.

2) Enhance the construction of innovation platform, and improve creative ability and comprehensive quality of students

To promote the gradual undergraduate thesis reform, transform the current academic graduation thesis to the thesis design which focuses on training the ability of student how to use theories to analyze and solve problems. Improve the competition mechanism of logistics and expand its influence in the universities in Chongqing, and make this mechanism become institutionalized, standardization and long acting to cultivate students' innovative thinking and ability of logistics solution design. We also active in various social investigation activities, strengthen students' communication ability, interpersonal skills and team spirit, to improve students' comprehensive qualities.

4.4 To Strengthen the Management of Students, and to Enhance the Students' Comprehensive Quality

1) Implementation of undergraduate tutorial system, so that the students have the guidance of specialized teachers throughout from the entrance to graduate.

2) Enhance learning guidance, and make students to correct their attitude towards learning, clear learning objectives, mobilize the enthusiasm of learning.

3) To strengthen the education of students' success, to motivate students by positive model, according to the different characteristics of each grade, to design and organize different series and types of reports, lectures;

4) Through the holiday social practice, community activities, promoting college students' business plan competition, participate in the challenge cup contest and such way to training extracurricular practice ability, so as to improve the students' comprehensive quality.

4.5 Improve Rules and Regulations, Strengthen Education Management

Under the current teaching management system, to further improve the school, institution, department three-level teaching management structure and system, to clear responsibilities and strengthen supervision; To further improve the school, institution, department three-level teaching quality supervision and inspection system, to establish the whole process of teaching quality control and assurance system;To further improve the teaching quality network evaluation system, to strengthen the information technology system of the education management.

1) To strengthen the management of teaching files.Meticulous management of Teaching plans, course syllabi, teacher lesson plans, lesson plans, record lectures, teaching and research activities records, papers, teaching tasks, experimental guide books, thesis mission statement, student time sheets and Examination paper analysis table.

2) Improve the teaching manager's service consciousness, recognize the inherent law of teaching management, and break the old management model of bondage, using new scientific methods and modern scientific methods, and improve the effectiveness of teaching management. To establish a comprehensive services consciousness which regard the "student" and "teaching" as the center;

3) To strict teacher's approval and clear the responsibility such as lesson preparation, teachers in class, counseling and correct students' papers;

4) Fully mobilize the students to evaluate teaching quality network and publishing;

5) Strengthen quality control of the whole process which include the selection of graduation design topic, selection of guiding teacher, the guidance of graduation design and thesis defense, ensure the quality of practice teaching;

6) Continue to focus on the cultivation of the student's comprehensive ability, to actively contact the logistics companies and related institutions, to create a good practice and training opportunities for students.

References

1. Dong, Y.: Strengthen practical education of logistics management major. Educating and Training of Logistics 6, 11–13 (2007)
2. Hao, G.: Applications of 3D Simulation training system in practical education of logistics. Markets of China 11, 124–125 (2007)
3. Charness: Attribution and reciprocity in an experimental labor market. J. Lab. Econ. 22, 665–688 (2004)
4. Chen, S.: Thinks about Logistics Talent Education in Universities. University Education of China 5, 8–23 (2006)

The Scientific Research of Independent Colleges' Actuality and Foreground

Lou Su[1], Du Jiangyi[2], and Feng Jiajia[1]

[1] WenZhou University, WenZhou, ZheJiang Province, China
[2] Department of Computer Science, Hubei University of Technology, Wuhan, Hubei Province, China

Abstract. As regional college, the Independent Institute faces lots of challenges and difficulties in the development of academic research. That's because of the relying on the higher-up university and late starter. This article analyzes the difficulties of the development of scientific research and proposes the rational act to benefit the development of the Independent College in the future.

Keywords: Independent College, the development of academic research, challenges.

1 Introduction

Independent Colleges is a new initiative of higher education model, which evolved as the result of the private college enrollment "Secondary School" in the late 20th century. It has a unique position in the structure of higher education and also the result of the development of higher education. Independent College as a special school of the current model has made outstanding contributions to carrying out the strategy of invigorating the country through science and advancing higher education in a popular style. After nearly a decade of development, the independent college gradually formed a set of academic teaching management is really different with the traditional universities. However, some issues haven't been resolved such as how to deal with the relationship between teaching and research, how to carry out the academic research in the independent college. Those will be a key wether the independent college develop sucessfully or not.

This article analyses the challenges and difficulties in the scientific research of the present situation and put forward some act which benefit to developing the college based on the reality.

2 Actuality of Scientific Research

As a new model university, independent college starts late in research and faces many difficulties, the most prominent exhibitions are:

W. Zhang (Ed.): Advanced Technology in Teaching, AISC 163, pp. 75–81.
springerlink.com

2.1 The Young Teachers' "Working" Notion Restricts and the Senior Teachers' "Provide for the Aged" Notion Restricts

Independent Colleges generally implement the employment, the employees are the masters or doctors who are just graduated from school or the senior teachers. As a result of the employment, the young stuff of employees in college didn't put all the energy in teaching and research. Although they sill stay in the college. If they get a good chance which benefits their future, they will go. In addition, the young teachers have higher education and have ardor in working but they are also impetuous, inexperienced and profit-oriented thinking. Farther, the have just graduated from universities and the low position and the low probability of hitting the target of his scientific reasearch project, which affect more or less on the young teachers' confidence and enthusiasm of scientific research. Then, the result will directly restrict the research and development of independent college. On the other hand, independent college usually retain some experienced professors and specialist from the retired senior teachers. Although they have enough enthusiam, strong sense of responsibility in their work, they can't entirely devote themselves to the scientific research because of physical aging physical exhaustion and the dufficulties in mastering the modern technology. Sometimes, most of the retired senior teachers have no intention of scientific research, they just regard the work as a translation. Therefore, the two notions impact the development of the scientific research of independent college.

2.2 The Irrational Stucture of Procession, Lack of Projects

Teachers' age structure and possition structure are not very reasonable because of the short history of independent college. The teachers are older high-grade or younger low-grade. Independent college currently doesn't construct a consummate and rational team in the aspects of age and possition. Many teachers apply for the projects by themselves and lack of the communication with others. It is diffcult to form a strong research team because of the randomness of the team members. The teachers have fewer opportunities to Participate in international academic exchanges and have fewer instructions of heavy weight divisions. It is also very diffcult to get a significant vertical project because of the teachers' liberty and incontinuity.

2.3 Limited Outlay for Running a School, Shortage of School Funding

Condition takes a look on that from at present most independent academies running a school, a lot of independent academy has been ignored study of science this once important field. Limited school funding was put into the construction of school hardware and software, the rest of money for research into rare. There are no basic conditions for teachers to develop the research. There is no specialized laboratories, there is no funding support. The teachers enthusiasm in research reduced, which restrict the development in a deep degree.

2.4 Colleges Pay More Attention to Teaching Than Research, Teachers Spend Fewer Time on Research

At the present, the vast majority of independent colleges are putting teaching in the main position and regarding the research as far form even the auxiliary position. That quality of teaching height, the administrative system teaching are complete or not being a prerequisite for ensuring that the independent academy swear to complete for advantage, is to ensure that the independent academy absorbs the basis growing a source, is the key ensuring that independent academy student strengthens competition in complicated social environment. The amount of work as a result, teaching has become the bearing picket judging the independent academy teacher imperceptibly. But, more than the correct or required number teaching amount of work makes a teacher dynamic having no way to spare more time and energy going to be engaged in study of science, to absorb new knowledge, this vicious circle, to lead to scientific research inevitably being weakened.

2.5 Backing Mother's Body School, the Project Are Diffcult to Declare

Independent academy teacher scientific research item declaration is restricted by the mother's body is not had right to speak, more needless to say power to make decision. Because mother's body academy base number is big, the various project every academy mixing passes, the quota of people distributing to independent academy already has been as trifling as it is. Therefore, on one's own the academy teacher displays the rank declaring a project obviously low on scientific research project application, information is deficient kind is unitary, funds is weak, the face being short of a guarantee, applying is narrow wait for a characteristic. Besides, the independent academy inter poses the go into waiting for a pertinency's, carrying out also seldom according to self's discipine characteristic property, special field. This status leads to the independent academy inside scientific research item very difficult to interpose a few. As for some big longitudinal problems, declare the successful chance being furthermore pretty uncertain.

That to sum up, on one's own, at present , the academy asks development scientific research to return certain difficulty back to existence, overcomes difficulty, breaks a bottleneck, gets the breakthough on scientific research, is problem the independent academy must think as developing.

3 Independent Academy of Research and Development Prospects

Judging from independent academy overall current situation teacher scientific research consciousness is light, research and development management chaos is harmful for independent academy scientific research to develop on very big degree as and stimulating mechanism faultily. Need to change the independent academy at present scientific research situation, the author brings forward the following actions:

3.1 The Academy Needs to Reinforce Policy Guiding, Supporting Scientific Research on Policy

One job branch carrying out, being completed as well as process control all needs policy, sate giving first place to teaching at present to independent academy, necessary moment returns back to the incline ought to give policy. The academy can interpose self scientific research fund for instance, the support office holder is cower or the scientific research ability is weaker teacher, give their certain scientific research starting funds; At the same time, the academy can come into being some encouragement policy, the teacher to the problem applying to arrive at high rank gives funds forming a complete set; Also can give certain encouraging and rewarding on scientific research amount of work allowance; Give certain amount of work allowance to aged teacher drive the young teacher to carry out scientific research. The academy can attempt to break tradition's engaging way , engage connection to get up with the personal scientific research ability and post, in engaging post is appropriate consider result of scientific research, be created certain scientific research pressure by teacher, this pressure must can change into open-minded scientific research of teacher driving force. Be created by that the academy asks to reinforce the scientific research environment, builds a fine scientific research atmosphere for a teacher. Can often invite some stronger scientific research ability teacher or scientific research director to hold a lecture , bring some scientific research information , the artifice teaching some scientific research to some scientific research information , artifice teaching some scientific research teacher , provide artifice and method that the scientific research item declares. At the same time, the academy returns the place room back to the scientific research responding to mother's body school similar to setting up a speciality's , the active and local government cooperates with enterprise , the problem recommending independent academy teacher participation studies the difficult problem unites with the technology tackling key problems.

3.2 Perfect Scientific Research Management System , Build Scientific Research Guarantee Organization

Perfect management system is the basis ensuring that scientific research proceeds as planned and without a hitch. Also ought to extend the same treatment to all to scientific research while taking the job teaching seriously the academy leader,the building-up ought to take scientific research management system seriously especially, uses system to ensure the scientific research proceed as planned and without a hitch being in progress. Independent academy scientific research is to be in progress on the basis relying on the mother's body basically , scientific research of the mother's body policy has certain applicability , the academy responds to the particularity being aimed at oneself scientific research , works out the scientific research management system according with independent academy characteristic . The system that fundamental for example academy scientific research regulation on administration , scientific research encourage and reward , teacher scientific research and the system fostering a student to carry out ability union system , scientific research ability and teacher post union judging engaging . Use system a train of to do standard scientific research, the activating teacher does the scientific research interest , ensure the scientific research proceed as planned and without a hitch being in progress.

3.3 Lighten Teacher Teaching Burden , Provide Sufficient Time to the Teacher Carrying Out Study of Science

That the teacher teaching bears overweight is one of the open-minded direct cause affecting scientific research. The colleges and universities , faculty-student ratio that the independent academy expects because of being development are bigger , the academy should rearrange a teacher rationally to carry out study of science for plentiful of teacher time proportion. The simultaneous academy can distribute teaching body reasonableness , the teacher by open-minded proper teaching bears the corresponding teaching mission, but the teacher by open-minded proper study of science bears the corresponding research mission specifically for the speciality being unlike a teacher.

3.4 Take Always Bringing the New along , Using the Scientific Research Teaching as Breach , Encourage the Young Teacher to Carry Out Study of Science , Prompt Improve the Young Teacher Scientific Research Ability , Foster One Sustainable Development's , Rational Structure Research Team

The teacher who depends on the forefront teaching, the easiest to produce impressions to problem such as content of courses , teaching way , can let a teacher grasp intravenous drip thinking in depending on the process teaching therefore, the research teaching from the basis grasps up, scientific research is independent academy scientific research breach with teaching. The academy is not bad according to discipline interpose, appropriate introduce one who takes the lead or plays a leading role who has achievement on academic research , drive a batch of young teacher to carry out scientific research. Can found the group who studies at the same time , also, the difficulty that the collective comes across in the process consulting in connection with research , the direction some artifice and method carrying out a job's by a young teacher, discuss middle in the collective gaining the ability studying practice. Forming one to become old as well as the team who studies having arrangement of ideas's , sustainable development newly, change shape scientific research structure , expand scientific research relating to face , bring the collective strength into play. Many data counting indicates particular research results , the scientific research ability and result of scientific research output that scientific research organizes overtop all over. Therefore, the building-up scientific research group is also that the independent academy improves scientific research current situation's having one of approaches.

3.5 The Application Keeping a Foothold , Sharpening School Anxiously Expect the Channel Cooperating to Broaden Research Actively

The academy needs to reinforce the connection with the external world, except the secondary mother's body is middle besides gaining scientific research information, the academy asks Tuo to stretch a channel , can not stay in tier of longitudinal problem face , ask the application keeping a foothold , sharpening school to anxiously expect connection, active Tuo shows the lateral problem, the lateral problem of development, sharpening "study of product study " a job. The problem encouraging a teacher to participate in institution actively studies, correct teacher, but convenient conditions providing whose with research participating in the lateral problem. The advantage that the academy asks to make use of special field platform, helps the teacher having be bound to study capability to find appropriate institution, forms such as building a research institute together , declaring a problem combinedly by both sides. One aspect helps enterprise to resolve the pertinent business difficult problem , another aspect also helps an academy to found one team who studies , improves the teacher research ability.

3.6 Platform Builds , Takes Sharpening Scientific Research Information Seriously Scientific Research Encourages and Rewards , Improves Step by Step the Scientific Research Mass

Need to make use of network information platform , to develop the network resource actively, effective implement making use of a network to come to enearth , spread information , let a network become scientific research information propagation's. Announcing building independent academy scientific research information platform, carries out communicating with on information in time with the teacher. Work out scientific research encouraging and rewarding regulation , reward high level other scientific research item and the thesis, from matter teacher having outstanding contribution with encouraging and rewarding pair of scientific research spiritually, encourage a teacher to carry out high level other study of science, improves scientific research mass thereby.

3.7 "Person Is This for Setting Up" Perfect Scientific Research Administration , Improve Research and Development Management Team Quality

The research and development management organization managing idea is organization unit providing scientific research information , administration scientific research process for teacher, if organization perfect the development having direct impact to teacher scientific research. Another , director's quality also have direct impact to administration mass. The director needs to have "people oriented; people foremost " "to be others compose marry the thought dressing ", has the noble quality serving in the people , provides one kind of the personalization administration pattern. Such , the teacher are therefore likely to carry out the job studying in one kind of harmonious atmosphere , can make teacher state of mind delighted to carry out a job , to carry out to scientific research forehead have promote an effect.

4 Conclusion

With independent academy strength expanding and the rise affecting, we ought to face difficulty in scientific research development squarely, difficult position on analytical objective Qian Jin road, resolve it's the problem developing process appearing in time. By the teacher and director's unremitting efforts, adopt the leader takes measure seriously , system ensures , policy supports , team building , perfecting administration and so on certainly needs, that independent academy scientific research develops definitely is able to get notable achievement , organic reaching teaching and scientific research thereby unites.

References

1. Feng, X.-D.: The system support of the new development in the independent colleges. Journal of Higher Education 27(10) (October 2006)
2. Li, J.-M.: On the Science Research Strategy in the Intensive Development of Independent Colleges. Technology and Innovation Management 30(13) (May 2009)
3. Duan, W.: Study on Incentive Systems for Research Teachers in Private Colleges. Based on Contract Theory 25(2) (April 2007)
4. Ma, Z.-Y.: Analysis on the Scientific Research of Teachers in Independent Colleges. Economic Research Guide, Serial No. 36(17) (2008)

The Research and Design of Personalized Teaching Platform Based on Knowledge Relations Network

Du Jiangyi[1], Wen Zhaohai[2], and Chen Yanjun[1]

[1] Department of Computer Science, Hubei University of Technology, Wuhan, Hubei Province, China
jydu1980@sina.com, jydu1980@sina.com

[2] Faculty of Mechanical & Electronic Information, China University of Geosciences, Wuhan, Hubei Province, China
jydu1980@sina.com

Abstract. With the widespread application of Web technology, all kinds of teaching platform which used network to realize many kinds of teaching way are able to expand rapidly. However, behind the network education prosperous, we also discovered that the remote teaching platform at present has not achieved the expectation effect in the actual used quality and the depth and still hides various crises and the contradiction. we proposed the construction that the personalized teaching and learning platform based on the knowledge relations network to construct the more reasonable learning environment and the appraisal environment.

Keywords: Knowledge point, Relation between knowledge node, platform for teaching and learning.

1 Introduction

From 1990s to now, with the widespread application of Web technology, all kinds of teaching platform which used network to realize many kinds of teaching way are able to expand rapidly. Through discussion of theories and practice in these many years, everybody achieve the mutual recognition to the importance of using Web technology to realize the network teaching: fully developing many kinds of interactive modes on-line teaching can make students and teachers, students and students as well as teachers and teachers learn more from network.

However, behind the network education prosperous, we also discovered that the remote teaching platform at present has not achieved the expectation effect in the actual used quality and the depth and still hides various crises and the contradiction. One of the most prominent question is lacking the personalized teaching environment of "teaches students in accordance with their aptitude".

In addition, there are also the following questions:

1. The teaching resources form provided by network is so unitary that unable to attract the student. In addition, students can obtain many teaching resources from other ways so that many students feel that they can solve some problem not by net

W. Zhang (Ed.): Advanced Technology in Teaching, AISC 163, pp. 83–90.
springerlink.com

and why can we study only by net? So in many situations, the network platform teaching becomes a mere formality.

2. The students can not appraisal the independent study situation in time and lack the on-line independent incentive mechanism. Studying on-line is lack of pointed and the method of selecting the study reference quickly according to own study situation.

3. The teachers are lack of the monitoring method of students' on-line independent learning and the tools of appraise students' on-line learning process. Usually the network remote teaching platform used hits and on-line time to statistics the students' on-line study situation, but from the practice effect, is not ideal, and even sometimes can not reflect the real condition of student on-line study.

4. The teaching management department is lack of monitoring method to on-line teaching and unable to grasp the teaching activity situation on-line between the teacher and the student is also unable to appraise accurately the on-line teaching quality and the on-line study effect.

In view of the above questions, we proposed the construction that the personalized teaching platform based on the knowledge relations network. The essential research target of this system is that taking the knowledge topic which is the most basic unit of composed the system knowledge as the foundation, formulate the knowledge and knowledge system according national computer rank test (second-level C) and study how to fully, completely, dynamic express the discipline curriculum knowledge system and how to fully manifest "the personalization" and "teaches students in accordance with their aptitude" and construct the more reasonable learning environment and the appraisal environment.

2 Knowledge Topic and Knowledge Relations Introduction

2.1 What Is Knowledge Topic

The so-called knowledge topic is the fundamental unit of transmitting teaching information in the teaching active procedure which includes theory, principle, concept, definition, model, conclusion and so on.[1].It may be regarded as a whole teaching unit that elaborates something, some teaching unit, simultaneously contains some related practices and presents way as well as related knowledge chain. The knowledge topic which is usually taken as whole to consider and to deal with in the teaching system in is the fundamental unit of teaching information organization in the teaching active procedure. If it is necessity, some chapter, even a definition, theorem of one course can be as a knowledge topic. The smaller the knowledge divided, the higher the reusability is. Connecting knowledge according to certain order becomes a course.

The division principle of knowledge topic is based on the organization of textbook. It uses the thoughts of refinement to divide the domain knowledge. The division way and the granularity play a domain role to this domain knowledge system in the knowledge application decisive function. Generally the knowledge topic division principle is: (1) follows the general teaching rule; (2) meets the corresponding teaching needs, in order to realize the heuristic teaching and the individualizing teaching; (3) maintains the integrity and uniform of knowledge topic [1].

The knowledge topic structure may divide into such several parts generally:

Knowledge topic = {ID, subject, content, exercise, discipline classification, retrieval word, guidance chain}.ID is the only symbol to distinguish a knowledge topic. Subject is the main title of this knowledge topic. Content is the narration content of this knowledge topic. Exercise is some related exercises with this knowledge topic. Discipline classification which is taken as one of search channels which as well as retrieval word is the classification attribute of this knowledge topic. Guidance chain is a super link which linked with some other related knowledge topic. The guidance relations determination plays an important role in the whole study system, because the study process of the whole study system is based on the study of knowledge topic, and which knowledge topic is taken by students as the beginning of study, and which knowledge topic is to be learned next step according the degree of grasping the knowledge topic, which are depended on the judgment and recommendation of guidance system. Moreover, it may also add some knowledge explanation which is regarded as some certain creativity by teachers.

2.2 The Relations of Knowledge Topic

In the actual learning process, we know: The relation among knowledge is not isolated, and the different knowledge relation is dissimilar, therefore the structure and the order of study are also dissimilar. It must consider around the order of studying knowledge, as well as consider around the integrity of studying knowledge. Considering the needs of guidance as well as the knowledge library established, there are four relations for the division of knowledge relation [2].

1.Seniority in relationships relations. Whether the son's knowledge can study is depended on whether the father's knowledge can study. For the knowledge guidance strategy of relations, it must study all the knowledge topic of organizational connections so as to continue to study and guarantee the study integrity of knowledge.

2.Dependence. Whether knowledge topic can study is usually depended on whether other knowledge topic have studied, or the latter is the former's preparation knowledge; Regarding dependence's knowledge guidance strategy, it must be kept the order of study that it is must study all premise knowledge topic to continue study.

3.Reference relations. The domain knowledge of a knowledge topic is related with other knowledge. The knowledge usually has a part or completely the same keyword, but their dependence is not clear; regarding the reference relations' knowledge guidance strategy, providing link functions for someone interested.

4.Brothers relation. It refers to the relations among sub-knowledge of all brothers relations which belong to some compound knowledge in the above knowledge division process. These sub-knowledge topic elaborate some partial domain knowledge from the different side, the different angle, the different category, and their descriptions are close. For the knowledge guidance strategy regarding brothers, learners must be taken a whole to carry on the study. Once it starts some brother knowledge study, you must study all brothers knowledge of this knowledge before studying other knowledge.

3 The Character of Humanity Study Decides the Necessity of Using Knowledge Teaching

3.1 The Human Carries on the Study as the Unit

Teachers carry on the teaching according to the knowledge topic which is one by one teaching. The human also study based on the knowledge topic. All these are completely because if human wants to learn, it must take the knowledge topic as the unit and one by one knowledge study.

Knowledge is the reflection of attribute and relation about objective, and is the subjective reflection in human brain about objective world. Human's study is take knowledge topic as the unit. If you want to study effectively, you must carry on study one by one [3]. Regarding the way of knowledge organization, reference [4, 5] have studied the expression, the character, the classification and the organization of knowledge topic, and thought that knowledge spot includes atomic knowledge topic and compound knowledge topic which is composed of atomic knowledge topic and compound knowledge topic. The knowledge compound creates a knowledge granularity thickly repeatedly thin and the organization is quite perfect theoretically but it does not benefit for the organization of network teaching and education resource, as well as it is not convenience for student's personalized independent study courseware's production.

Reference [6] is thinner to the knowledge organization granularity and thought that the knowledge should divide into discipline, knowledge area, knowledge unit, knowledge topic and so on, and has carried on the analysis and the organization to the computer discipline's knowledge. It is convince for educator to organize the course content that dividing the knowledge according to knowledge's level clearly, but it is from teacher's angle not favor the resources the organization and learner's independent study, because simultaneously is excessively thick regarding a knowledge definition, also causes the teacher not easy assurance knowledge central content.

In fact, the person who is engaged in the teaching profession knows that any independent character, the word, the concept, the theorem, the law, the formula, the rule, the viewpoint, and so on is knowledge topic. Moreover, teachers in teaches also teaching one by one according to knowledge topic. Therefore, carrying on the teaching based on the knowledge topic, conforms to the anthropology custom choice inevitably, and also raises the teaching efficiency, and is the inevitable choice to obtain the better teaching effect.

In intelligence teaching system, the knowledge organization must be not only benefit for student study but also convince for the teaching organization and the system management, as well as conform to the scientific knowledge development present situation so as to benefit for the knowledge expansion and resources issue [7]. In view of this, according to the curriculum, we has redefined the knowledge topic and proposed a structure based on knowledge topic relations, simultaneously produced personalized curriculum suit for students according to the contemporary university student's knowledge structure and the study character.

3.2 The Inevitability of Which Human Learn Knowledge as a Unit

Learning new knowledge as a unit is decided by the character of knowledge composition and human cognition.

The integrity leads to that human learn knowledge as a unit [8]. The so-called knowledge topic is integrity is that each knowledge topic is consisted of a number of information blocks which linked with each other. If a complete knowledge topic is divided into several separate blocks artificially, then each of these blocks do not have the significance of they possess the overall composition. In the objective, this feature of knowledge topic needs us to understand that learning knowledge must point to knowledge as a whole, that is, knowledge topic must be a unit of knowledge. Every point is a whole of knowledge, so taking any knowledge topic out of their knowledge system will not affect the correct understanding to their expression.

Leading to the cognitive character that human learn knowledge as a unit is that human short-term memory is very limited. Modern constructivist research shows that people acquire knowledge by using information obtained to construct new knowledge [9]. According to the integrity of knowledge topic, in the process of construct new knowledge, some relevant information must be recalled in short-term. In this way, whether the activities constructing new knowledge can be completed depends on whether the amount of information knowledge can over the human capacity, which about seven blocks. Many studies have shown that if you need a few short blocks of information over this limit, the person will not remember the information, and accordingly, using the information to construct activities can not be completed. In fact, the vast majority of knowledge topic which we learned is composed of a number of blocks of information.

Therefore, the above describe that the human must learn as a unit of knowledge topic, what's more, it is difficult for someone to learn more than one knowledge topic simultaneously.

4 The Design of System Model

According to the relationship based on knowledge topic, the system consists of server and client. The system in server is composed of system management module, knowledge design module and teaching module. Knowledge design module includes content module, exercise module and guidance module, and teaching module is divided into student module and teacher module. Specific system model design is shown in Figure 1.

Among them, the content module includes curriculum design module and knowledge topic editing module. According to the teaching experience, there are two classifications about the logical relationship among knowledge topic: one is classified according to the course materials section, the other is the inherent cross-sections logical relationship of knowledge topic. Guidance module should make use of retrieval word of knowledge topic to index so as to make students find knowledge topic quickly.

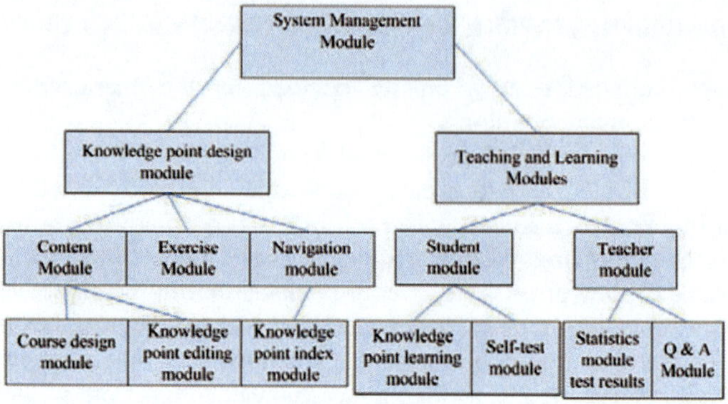

Fig. 1. System model design

Student module is divided into learning module and Self-test module. Teachers adjust teaching content through student self-test. This will realize in the test scores module of teacher module. The question module is also an important part.

According to the system design, the knowledge system design is shown in Figure 2. There are knowledge, knowledge unit and knowledge topic. If any adjacent layer in the Figure was connected, there will be a generational relation between them. And in the 3-layer structure, any underlying concepts have only one point to the upper edge, in others words, one upper concept has more than one lower concept.

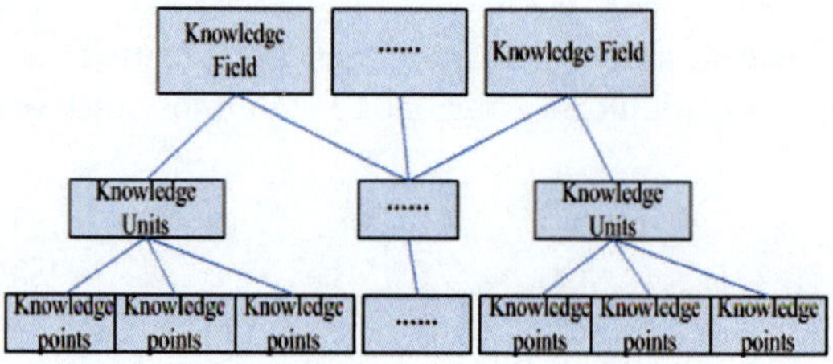

Fig. 2. Knowledge system design

5 System Implementation Tools

I use the Internet techonology and multimedia technology to design the system. Users can go to anywhere by the net, so can use the system very easily by using the Web browsers ; The system knowledge is extremely easy to update and upgrade by updating the content, so any student can learn directly to the latest content.

The system selecte CLIENT／SERVER as the model, the windows XP system as the paltform, and the ADO component to access data and use the C++Builder 6.0 to develop.

The system use the ADO component to achieve the data access, the theory about data,which apply the program to vist as figure 3. The data of the system are reserved in the SQL SEVER data table, which use the ASP script. And the system visit the database through the ADO components, the efficiency of system to database access can be improved rapidly

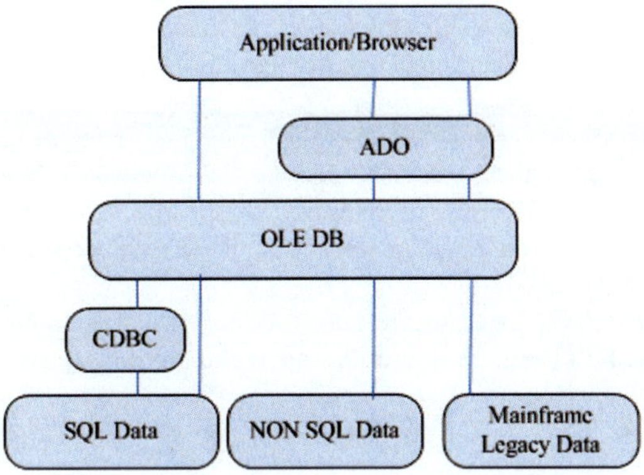

Fig. 3. ADO data acces

2.C++ Builder which developed by Borland is a visual C++ programming environment under Windows, usually referred to as BCB. C++ Builder 6.0 has strong graphics, image and multimedia processing functions; it provides a good support for the OLE, COM, ActiveX technology on Windows; in database programming, BCB offers a variety of connectivity options and multi- layer distributed data processing and other advanced technologies; BCB has strong compatibility. It can compile and link some source code or object module which developed by Turbo C, Borland C + +, VC + +, Delphi and other tools.

6 Conclusion

Modern education is developing from teaching as the center to learning as the center. Students transfer themselves from passive recipients and inculcation of knowledge objects into acceptance and processing of information and knowledge of the subject active constructors of meaning, while teachers transfer themselves from imparting knowledge, instilling into the guidance, helper and improver in the process that student construct knowledge initiative. How to express the curriculum knowledge topic fully, completely and dynamic and how to use the existing network system to evaluate student independent study and study results will be an important information technology education research in a very long time. The system based on the relations among knowledge topic strive to realize a better teaching platform for teaching, and constantly improve according to the teaching of the actual situation.

References

1. Jiang, Z.: The Application Study of Knowledge Point Relation and its Structure Diagram and Knowledge Network. Journal of Anshan Normal University 7(5), 99–101 (2005)
2. Hu, N., Xie, S., Jiang, H.: A Construction of Knowledge Model in IHMCAI System. Computer Engineering and Applications 38(12), 97–99 (2002)
3. Zhu, Z., Wu, K., Wu, Z., Chen, Y., Gao, M.: Personalized Curriculum Origanization Method Based on Knowledge Topic Ontology. Computer Science 36(12), 124–128 (2009)
4. Xie, S.: Analysis of the Properties of Knowledge Points and Their Nerworks. Journal of Software 9(10), 785–789 (1998)
5. Shi, Y., Zhang, S., Xiang, C.: Study on expression and connection technology of knowledge point in engineering network course. Journal of Zhejiang University (Engineering Science) 37(5), 508–511 (2003)
6. Engel, G., Roberts, E.: Computing Curricula 2001. IEEE Computer Society, Association for Computing Machinery (2001)
7. Jean-Marc, R., Gracia, G.: Experimental study on the reuse of learning objects and teaching practices. In: Proc. Int'l Conf. on Education and Technology, Calgary, Canada, pp. 107–112 (2005)
8. Chen, Z., Sui, G., Pi, X.: Knowledge-Point is a Cognitive Unit. Psychological Science 25(3), 369–370 (2002)
9. Li, Z., Wang, J.: Research of Computer Application Teaching Methods Based on Constructivism. Computer Education 3, 124–126 (2010)

Reform and Exploration of Sample of Mechanical Engineering Teaching Team Based on the Integration Concept

Feng Xianzhang[1], Wang Dan[2], Wang Lihong[1], and Jiang Zhiqiang[1]

[1] Department of Mechanical Engineering, Zhengzhou Institute of Aeronautical Industry Management, Zhengzhou, Hennan, China
Phdfxz@163.com
[2] Shenma Tyre Fabrics Co., Ltd, Zhongping Energy Chemical Group, Pingdingshan, Hennan, China

Abstract. Aiming at many problems of independent college curriculum settings, confusion of arrangement schedule, according to the specialized characteristics of mechanical design manufacture and automation specialty in Zhengzhou Institute of Aeronautical Industry Management, it can supply the ways to cultivate innovative talents in new era for the excellent engineering college students, by the methods of reorganizing the teaching team, carrying out the integration within the groups for mechanical basis curriculum system.

Keywords: course setting, course system, integration, teaching team, innovation ability instruction.

1 Introduction

It is the key problem how to creative machinery talents and enhance innovation ability, effectively improve the quality of teaching, around the course and depth practice of teaching reform. it order to adapt to rapid and sustainable development of China's higher education goals, it become the common facing problems for the professional teachers, strengthen and enhance the innovative capability of mechanical talent, training highly qualified personnel to meet the needs of the community.

These the problems were being discussed that teaching basic courses lead to mechanical repetitive using the traditional teaching mode in the paper. To further improve teaching effectiveness, educators have done a lot of domestic exploration. However, the basis for the mechanical integration in terms of the overall curriculum, there are few reports [1]

About features of more courses and the categories of the mechanical discipline-based curriculum. And intends to subject the entire mechanical system for integration of basic courses, to break the status of belong to different teaching and research programs, emphasizing the principles of practical courses, in reducing the burden on students under the premise of the school to further improve my overall level of engineering majors.

W. Zhang (Ed.): Advanced Technology in Teaching, AISC 163, pp. 91–96.
springerlink.com

2 The Features of Teaching Team

The so-called team refers to have certain skills, willing for the common goal of mutual cooperation by the formal group composed of individuals. For the definition of team, different people have different interpretations, the U.S.management expert Stephen• Robbins think that team is the organization by composed of two or more, interaction, interdependence of individual, specific goals to combine according to certain rules. American scholar Jon R Katzenbach is said: "There is a team, with complementary skills, willing to work together for a common goal of a formal group composed of individuals", Teaching team is currently no authoritative definition as a form of the team. [2, 3]

The Mechanical basis of professional and technical courses teaching team is this a teaching group, under the realistic background of the change for teaching function of profound, and continuously improve the quality of teaching, the courses as the platform for the mechanical design and manufacturing and automation technology, backbone teacher as members in teaching and research section has the courses, relying on the features, taking the principle of combined with teaching and research, and help each other, the mutual transformation, the mainly purpose to breaking the single curriculum reform, the mechanical basis of courses by module teaching reform, to achieve mechanical configuration based science courses, the overall optimization goal for the construction.

The Team persevere the teaching idea of scientific research into the classroom, teaching immaterialist, to teaching and research with a long goal. At the same time, continue to develop team's research strengths, services, local economic development efforts to explore the channels. [4]

3 Integration of Mechanical Courses System

The mechanical courses system has a very important role in the field for the engineering concepts to students, innovation, specialized courses of study, etc. and also it can plays a very important role for the mechanical professionals training program. The course is also very close links among the curriculums, the content has a certain convergence. There are many questions of the unreasonable for convergence of the part of course content, content overlap, insufficient for the interaction among teachers, inadequately to play the overall advantages. It is necessary to teaching reform for teaching team during reforming of a single course, improvement of personnel training requirements and the continuous reduction of teaching hours with the deepening of teaching reform.

It can realize the further improvement of the overall level for the foundation courses. Play to the full advantages of team teaching by use the method of multi-course approach and to break a single course of construction and integration. The guiding ideology of discipline the main line, based on teaching, the fundamental educating people, and can make greater contributions in the field of cultivate innovative talents, disciplines construct, curriculum development, teaching materials, teaching and research and teaching reform and research, etc. [5-8]

The building of teaching team is a long-term work, and the curriculum integration long way to go, the teaching team is expected in the following aspects for the breakthrough point to promote curriculum integration process:

3.1 Research of Integration Courses System

Mechanical discipline-based curriculum group includes mechanical design, mechanical principles, engineering drawing, engineering mechanics, measurement tolerances, materials, mechanical engineering and other related courses.

The relationship between curriculum groups as shown in Figure 1

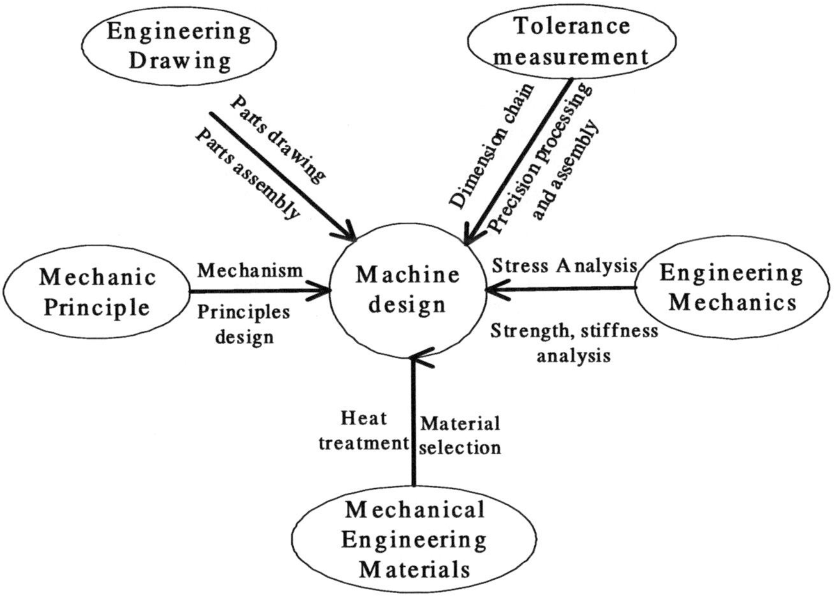

Fig. 1. Relationship of courses system

In Figure 1, the curriculum group to "mechanical design" course is the core position, other courses to provide the appropriate support.

Among them, "Engineering Drawing" can provide the knowledge points for drawing and assembly, "Tolerance measure" can provide the knowledge points of dimension, size, chain, etc."Engineering Mechanics" can provide the knowledge points of stress analysis, strength analysis, stiffness analysis and other parts of the design criteria, etc."Mechanical engineering materials" can provide the knowledge points of s selection, heat treatment, etc."Mechanical principle" can provide the knowledge points of design and exercise principles, etc. "Machine Design" is core place in foundation courses of the system, a mechanical device in the design process, involving the "Engineering Drawing"," Tolerance measurement"," Engineering Mechanics", "Mechanical Engineering Materials" and other related course content, the reducer gear design as example to illustrate.

It is a teaching task to make the reduce gear as the design objects for engineering university students in China. It is mainly main line for the mechanical system design, and purpose to improve the student's knowledge and proficiency and basic skill. To familiarize students with the basic mechanical design methods and general procedures, and gradually develop its integrated design and structural design capabilities. And mechanical design of the basic ability to work training, including stress analysis, strength and stiffness, drawing ability and use of relevant manuals. It is the integrated use of training engineering students, "Machine Design", "Mechanical principle", "Engineering Drawing", "Tolerance measurement", "Engineering Mechanics", "Mechanical Engineering Materials" and other related course knowledge, skills and problem-solving ability is an important part of the project for engineering students to enhance other aspects of the concept is particularly important.

The assembly drawings of a typical one gear shown in Figure 2.

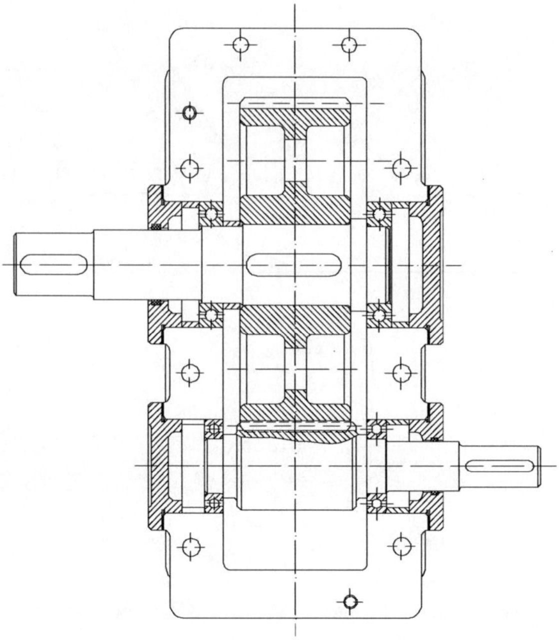

Fig. 2. The assembly drawings of a typical one gear

The Figure 2 shows that the components involved reducer including: shafts, gears, keys, pins, sleeves, bearings, seals, thread and so on, in which axis is a non-standard parts, the design has its own characteristics, that is, while drawing, while the calculation, while the design. And taking into account various factors before finalizing its size of any section of a shaft (such as radius, length, etc.).With the process of mechanical analysis the axis, the axis is generally reduced to simple beam model, which in turn related to materials, mechanics and other aspects.

The drive shaft parts drawing of the low-level reducer gear shown in Figure 3.

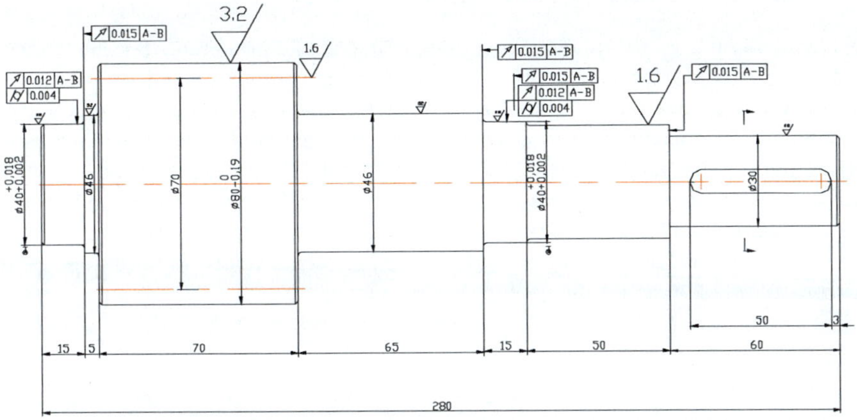

Fig. 3. The drive shaft parts drawing of the low-level reducer gear

In order to further study the integration of course group system, requiring team members take the same instance of the model in the teaching process. To the same model, students can master the different knowledge, with a view to establish an overall sense for the mechanical design. In order for team members to experience the intrinsic relationship among the various programs, it can select a specific chapter as a strong attempt to change teacher to teach during the reforming of the course content. And also can more effectively optimize the teaching content among the courses by teacher communication and exchange after class.

3.2 Research of Curriculum Design of Mechanical Design

Curriculum design is familiar with the general design process of mechanical devices, it can learn integrated application of knowledge, training design, computing power, etc."Mechanical Principle" and "Machine Design" are backbone technology-based courses of the mechanical design and manufacturing of automation. The course design of "Mechanical Theory" and "Machine Design" is two practical courses for the most suitable mechanical program to develop student's ability to innovate, and it is essential practical teaching process for the Engineering and technical training. The engineering colleges have separate class, the overall time is 3-4 weeks.

For the long time, all schools try a different teaching reform in the curriculum design of "Mechanical theory" and "Mechanical design".

The teaching reform of mechanical design curriculum is entering a new period in the 21st century. They were studied in a deep-going way to the purpose of curriculum design, content, methods, topics, forms of organization for ours college. Explore how to solve problems of practical engineering and innovative design and ability for the students, and proposed goals to provide the implementation of the program for innovation and personnel training curriculum design. The teaching practice supply theoretical preparation for the curriculum design teaching reform.

3.3 Improve the Quality of Teaching Staff

The team draws up the team development plan and appropriate teacher training program, based on the status of young teachers of teaching team of mechanical engineering, implementation guidance by respected teachers. Continue to do the work of training young teachers, striving to research and research-class team of teachers is a central pillar. The establishment the teaching team with sustainable development, encourage young teachers to participate actively in the reform of teaching and research work to continuously improve teaching and research capacity; create the conditions for sending young teachers to visit or study in well-known schools, to encourage young teachers to further enhance the doctorate degree level.

4 Conclusion

The integration of mechanical engineering courses of system is a long-term, we do some exploring reforms for the teaching methods in our college. It is impartment issue how to improve the enthusiasm for learning and culture innovative ability for higher education reform.

Acknowledgment. The research is supported by Program for Science&Technology Innovation Talents in Universities of Henan Province"HASTIT"), Assistance Scheme of Young Backbone Teachers of Henan Province Colleges and Universities (2010GGJS -147), Science and Technique Foundation of Henan Province (102102210152,112102210359); Science and Technique Foundation of Office of Education of Henan Province (2010A450001); The National Natural and Science Foundation of China (50905168). The 2010 Teaching Team of the Zhengzhou Institute of Aeronautical Industry Management.

References

1. Hu, M., He, Y.: On the Practical Teaching System Construction of Mechanic Specialty in Secondary Vocational Schools. Vocational and Technical Education 5, 72–74 (2006)
2. Ye, Y., Li, W.: Research to explore of mechanical theory and design of the integration course design. The Basis of Theoretical Discussions 28, 437–437 (2009)
3. Robbins, S.: Organizational Behavior. China Renmin University Press, Beijing (1997)
4. Hayes, N.: Successful team management, Yang, P. (trans.). Tsinghua University Press, Beijing (2002)
5. Ji, A.: On the Establishment of Teaching and Research Section in University. Journal of Anhui University of Technology (Social Sciences) (3), 145–146 (2008)
6. Feng, X.: Construction and Practice of Teaching Team for General Colleges. In: The 2rd International Conference on Education Technology and Training, Sanya, China, pp. 3–6 (2009)
7. Yan, Q., Peng, X.: Design and Implementation of Animation Simulation Courseware Based on VBA. Modern Educational Technology 1, 124–126 (2010)
8. Zhang, Z., Zhang, Y.: The Automatic Generation of the Complex Word Format Examination Paper Based on ASP and VBA Technical. Modern Educational Technology 6, 117–119 (2009)

Service Type Recognizing Arithmetic in WCDMA Iur Interface Implementation

Fang Cheng

Department of Computers, Wuhan Polytechnic University, Wuhan 430023, China

Abstract. This paper introduces the basic principle and protocol structure of WCDMA Iur interface. When Iur interface is implemented, how to recognize its service type is a challenge to developers. The matching-arithmetic is introduced in this paper to solve service type recognition puzzle. The max-rate-matching-arithmetic is proposed for TFS type recognition and eigenvalue-matching-arithmetic for TFCS type recognition. The core principle of these two matching arithmetics is that first constructing static mapping tables of TFS or TFCS, second computing max rate of TFS or eigenvalue of TFCS, finally service type recognition is achieved through matching max rate or eigenvalue with the static mapping tables. The matching arithmetic achieves service type recognition of DRNCs developed by different manufacturers and is significant to implement an opening WCDMA Iur interface.

Keywords: Iur, service type, recognition, WCDMA, TFS, TFCS, DRNC.

1 Introduction

Iur interface is the interface between RNCs in WCDMA systems. It's an open interface designed by 3GPP organization for smooth connections among different devices produced by different manufacturers in WCDMA[1].

The protocol structure of Iur interface is shown in Fig.1[2].

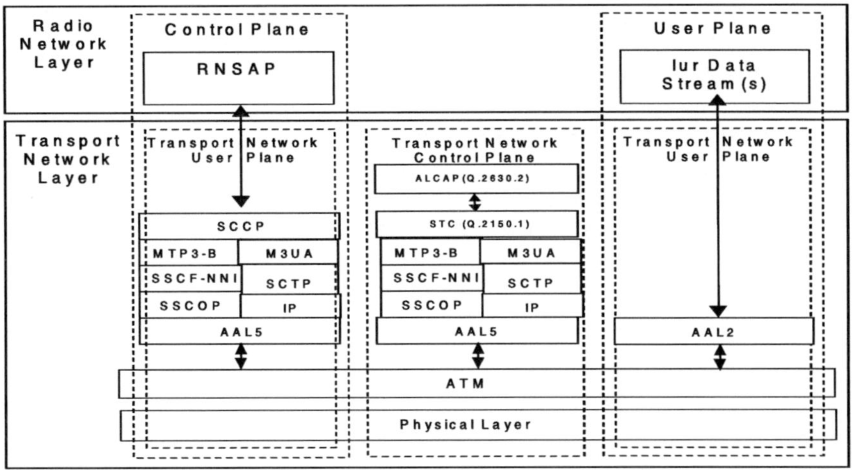

Fig. 1. Protocol Structure of Iur Interface

W. Zhang (Ed.): Advanced Technology in Teaching, AISC 163, pp. 97–102.
springerlink.com

Radio Network Layer defines the interaction procedure between one PLMN and two RNCs and it consists of Control Plane and User Plane. Transport Network Layer defines the establishing procedure of physical connection between one PLMN and two RNCs[3].

The following basic principles of Iur interface are proposed by 3GPP organization[4]:

(1) Iur interface should be open;

(2) Iur interface should support signaling communication between two RNCs and Iur data stream;

(3) Iur interface is a logic end-to-end interface between two RNCs. The logic interface should be usable even if there is no phycical connection between the two RNCs.

Therefore, the Iur interface between different manufacturers can provide more choices of different technologies and devices, which provides more flexibilities in rapid deployment and implementation of 3G networks.

2 Service Type Recognition in Iur Interface Implementation

In RNC implementation of UT company, service type is an important parameter. It exists all through the RRM_RAC, RRM_PS, RCP modules. The Inter_RNC flow diagram of RCP is shown in Fig.2 and Fig.3, the functions of these entities are described as following[5]:

(1) UE CENTER: UE Center is a dispatching and controlling center of UE signal flow in S-RES board. It has direct signal or API interactions with such entities as RANAP, RRC, PHY, GRMDATA, IU BEAR, RNSAP, RAB, RLC\MAC.

(2) PHY ENTITY: PHY Entity is used to collaborate with entities as RNSAP, NBAP, TBS to complete configurations of phycical layer data in base stations. Functions of PHY entity in SRNC and in DRNC is some different. In SRNC, PHY entity is used to complete creation, deletion and modification of Radio Link and DCH and to complete creation of TBS and TB (TBS is used to free them). But in DRNC, RNSAP is used to complete creation, deletion and modification of Radio Link and DCH. PHY entity is used to collaborate with NBAP and TBS entity and to create TBS and TB (TBS is used to free them).

(3) RNSAP ENTITY: RNSAP Entity is used to manage protocol procedure of RNSAP and to maintain signal connections of Iur interface.

(4) RLC/MAC ENTITY: RLC/MAC Entity is used to implement interface functions between S-RES and RLC/MACD, S-RES and RLC/MACC.

(5) RRC ENTITY: RRC Entity is used to manage messages of RRC layer of protocol stack.

(6) GRMDATA ENTITY: GRMDATA Entity is an entity connected to GRM, RRM and is used to manage resources distributed by RRM.

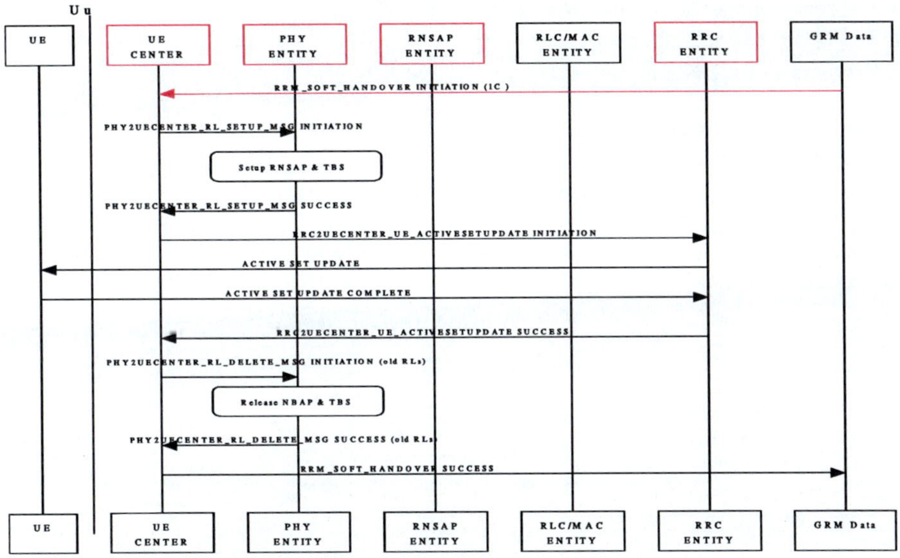

Fig. 2. Flow Diagram of switch procedure from Inter-RNC to SRNC

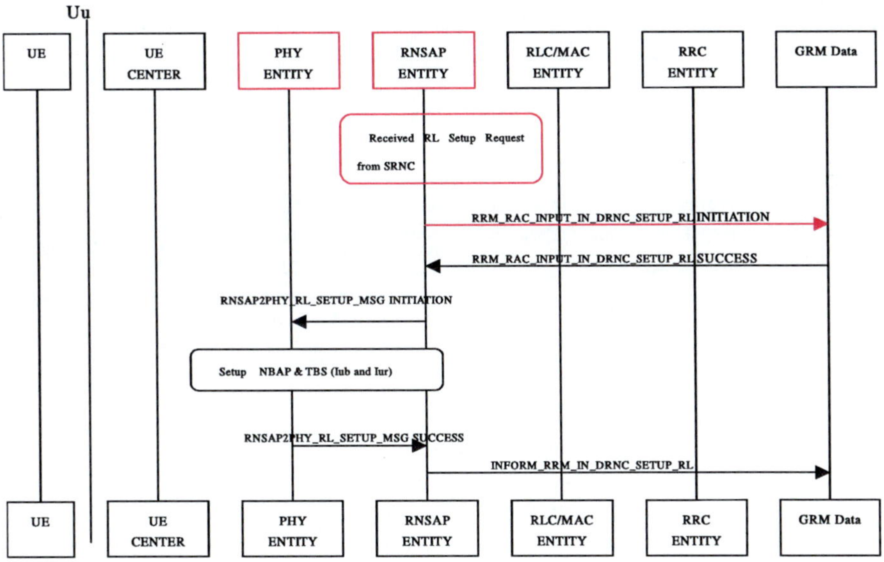

Fig. 3. Flow Diagram of switch procedure from Inter-RNC to DRNC

3 Recognition Technology of Service Type

3.1 Service Type Recognition Technology of SRNC

When a RNC is a SRNC, service type of UE is determined by following procedures[6]: First, RAB parameters and IU-UP user interface messages are determined by CN based on characters of RAB to be built and capacity which CN itself can support. Second, CN sends these messages to RNC by IU interface. Third, RAC module of RNC is used to determine service type according to these parameters and current network traffic. The recognition technology is shown in Fig.4:

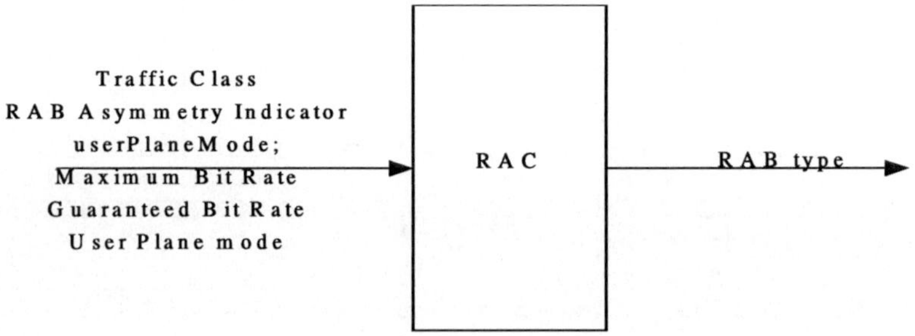

Fig. 4. Service type recognition in SRNC

3.2 Service Type Recognition Technology of DRNC

When a RNC is a DRNC, service type recognition is totally different. DRNC only exists in FP-Iub and FP-Iur layer. when DRNC receives a RNSAP_RL_SETUP_REQUEST message or a RNSAP_RL_RECONFIG_ PREPARE message, it only knows the format of TFS and TFCS, so an arithmetic is need to recognize their type from the format of TFS and TFCS. Furthermore, the flexibility of the arithmetic is critical to the opening degree of Iur interface.

Recognition of service type completes by following two steps: (1) type recognition of TFS; (2) type recognition of TFCS.

3.2.1 Type Recognition of TFS

TFS is a transmission format assembly of transmission channel. Protocol 34.108 gives typical value of TFS in various transmission channels and various manufacturers can take these typical value of TFS or its subclass. However, the tranmission format with maximum rate must be used because max user rate of service is determined by it.

The rate of one tranmission format is calculated as following: supposing that the transmission block of the transmission format is trchSize bites, number of transmission blocks is trchNum, TTI of the transmission channel is tti millisecond, then if DCH is chosen as transmission format, its transmission rate is: trchSize * trchNum / tti (kbps). Data is transferred from higher layer through RLC, MAC to FP and need some protocol overhead, so rate of tranmission format will be greater than (such as PS service) or be equal to (such as signalling) user rate. However, the protocol overhead is constant to a constant tranmission rate, so the max rate of a tranmission format is also constant. The Max-rate-matching-arithmetic is proposed in this paper based on the principle to recognize TFS type.

The Max-rate-matching-arithmetic follows three steps:

(1) constructing TFS static mapping table: constructing mapping between various TFS types to their max rate. Meanwhile, service factor is designed to distinguish different services of the same rate, defined as following:

```
typedef enum
{SRB_FACTOR = 1;
CS_FACTOR = 2;
PS_FACTOR = 6
}WRNSAP_SERVICE_FACTOR;
```

Different service type is determined by its typical transmission blocks. To make computing more convenient, max rate in static mapping table is defined as following:

tfsMapTable[tfsType] = actualRate*100*SERVICE_FACTOR

tfsType is defined as following:

```
typedef enum
{TFS_SRB3_4K = 2,
TFS_RAB_64K_CS = 7,
TFS_RAB_12_2K_CS_SUBFLOW1 = 12,
TFS_RAB_64K_PS = 55
}RNC_DCH_TFS_TYPE;
```

(2) computing max rate according to TFS to be recognized:

Different rates of different TF in TFS to be recognized are computed using formulas above, then max rate is chosen through comparison.

(3) the type of TFS is recognized through matching

Matching the max rate computed through step2 with static mapping table through step1, then TFS type is achieved.

3.2.2 Type Recognition of TFCS

Type recognition of TFCS is achieved through eigenvalue-matching-arithmetic. Eigenvalue of TFCS is computed as following:

```
typedef enum{
UL_FACTOR    = 5,
DL_FACTOR    = 7
}WRNSAP_DIRECTION_FACTOR;
tfcsMapTable[tfcsType]=
(tfsMapTable[ulTfsType1]+tfsMapTable[ulTfsType2]+...)*UL_FACTOR+
(tfsMapTable[dlTfsType1]+tfsMapTable[dlTfsType2]+...)*DL_FACTOR
```

First, max rate of up-link service and max rate of down-link service are computed respectively. Second, max rates multiply up-link factor and down-link factor respectively, then TFCS eigenvalues of up-link and down-link are achieved. Third, plus these two eigenvalue, then TFCS eigenvalue is achieved. A static mapping table of TFCS is also need to record different eigenvalues of different types of TFCS.

Giving a Conversational / speech / UL:12.2 DL:12.2 kbps / CS RAB + Interactive or background / UL:64 DL:64 kbps / PS RAB+ UL:3.4 DL: 3.4 kbps SRBs for DCCH service as an example, TFCS type is defined as following:

```
typedef enum
{… …
RL_12_2K_CS_AND_64K_PS_AND_SRB = 91,        /*Supported in SR2.5*/
… …
} RNC_RL_TYPE, RNC_DCH_TFCS_TYPE;
```

Then the service type is RL_12_2K_CS_AND_64K_PS_AND_SRB, 91, its eigenvalue is:
tfcsMapTable[91]=
(tfsMapTable[TFS_SRB3_4K]+tfsMapTable[TFS_RAB_12_2K_CS_SUBFLOW1]+ tfsMapTable[TFS_RAB_64K_PS])*UL_FACTOR+ (tfsMapTable[TFS_SRB3_4K]+tfsMapTable[TFS_RAB_12_2K_CS_SUBFLOW1]+ tfsMapTable[TFS_RAB_64K_PS])*DL_FACTOR = (tfsMapTable[2]+tfsMapTable[12]+ tfsMapTable[55])*5+ (tfsMapTable[2]+tfsMapTable[12]+ tfsMapTable[55])*7 = (370 + 2440 + 40320)*5 + (370 + 2440 + 40320)*7 = 517560

4 Conclusion

This paper introduces the basic principle and protocol structure of WCDMA Iur interface. The matching-arithmetic is introduced in this paper to achieve service type recognition of DRNC. The service type recognition technology is significant to implement an opening WCDMA Iur interface for different manufacturers.

References

1. 3GPP TS 25.423: UTRAN Iur interface RNSAP signalling
2. 3GPP TS 25.426: UTRAN Iur and Iub Interface Data Transport & Transport Layer Signalling for DCH Data Streams
3. 3GPP TS 25.427: UTRAN Iur and Iub Interface User Plane Protocols for DCH Data Streams
4. 3GPP TS 25.212: Multiplexing and Channel Coding (FDD)
5. 3GPP TS 25.221: Physical Channels and Mapping of Transport Channels onto Physical Channels (TDD)
6. 3GPP TS 25.213: Spreading and modulation (FDD)

A Method for Developing Software Based on Role Modeling and RSI

Fang Cheng

Department of Computers, Wuhan Polytechnic University, Wuhan 430023, China

Abstract. UML modeling has been widely applied in software engineering. Use Case method has become a common method to define and analyze application requirements. However, there are several disadvantages with Use Case method. To improve UML modeling, the technology of role modeling (RM) and requirement-service-interface (RSI) are introduced in this paper. In the end, an example of the development procedure of Internet Bank is given to illustrate how to use these two technologies synthetically in the analysis and design of application systems. This method is very practical in financial systems based on Internet.

Keywords: UML, Use Case, Role modeling, RSI, Internet Bank.

UML modeling has been applied widely in software development recently years. Use Case, as a method to define and analyze system requirements, is a description of the interaction between user and software system to achieve a specific goal. But it is actually very difficult to choose and define use case properly in system analysis. Traditional UML Use Case methods have some shortcomings as following,

(1) The granularity of Use Case is very difficult to define.

(2) It is very difficult to develop system analysis and development for software developers especially when Use Cases are very complicated.

(3)There is a wide gap between software analysis and design when applying Use Case methods.

Many extended methods have been proposed to overcome these shortcomings, for example, Role Modeling and R(Requirement)S(Service)I(Interface) structured Use Case. These two methods have improved Use Case methods to some extent, but they are not yet perfect. This paper introduces how to integrate Role Modeling methods and RSI structured Use Case methods through an example of an Internet bank system profile. In the early stages of system analysis, we use Role Modeling methods to identify and define Use Cases and interactions between them. In the late stages of system analysis and early stages of system design, we use RSI methods to refine Use Cases and control the granularity of Use Cases. So we can overcome the shortcomings of traditional UML Use Case methods and achieve a natural transition from system analysis to system design.

W. Zhang (Ed.): Advanced Technology in Teaching, AISC 163, pp. 103–109.
springerlink.com

1 Role Modeling Method (RM)

The role of an object is to describe the reason and function of its existence. The first job of identifying an object is to identify the role of the object. Role is also a kind of object and is named Role Object. One role object may have many roles. In a harmonized object model, roles of an object should match the roles of the harmonized object. All of the role objects make up the role model, which is used to describe role objects and the interaction between them.

2 RSI Method

RSI method provides a structural model for Use Case requirement analysis. The method also provides a framework to analyze potential Use Cases and the interior relations between them. It also provides a framework for considering the granularity, structure and content of Use Cases.

2.1 Structure of RSI Method

The Meta model of RSI method is described in Figure 1.

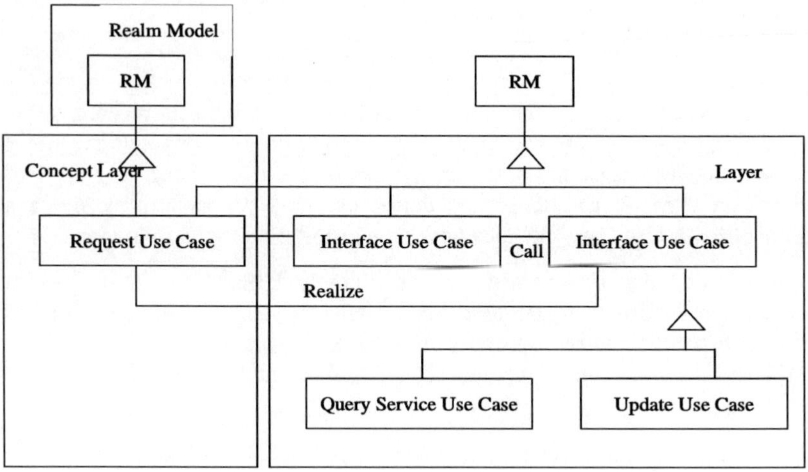

Fig. 1. The Meta Mode of RSI Use Case Method

RSI method partitions Use Case into three types, which are named 《Requirement》, 《Interface》 and 《Service》 Use Cases (they will be called R Use Cases, I Use Cases and S Use Cases for short later) respectively.

(1)R Use Cases document those business processes which need automatic support. R Use case documents system requirements in detail but doesn't define the system function.

(2)I Use Cases document those interfaces which reveal to system users and the functions of these interfaces.

(3)S Use Cases document a detailed description of the basic functions of the system, and these functions are independent of specific interfaces.

2.2 Use Case Analysis Process of RSI Method

The analysis process of RSI method is showed in Figure 2.

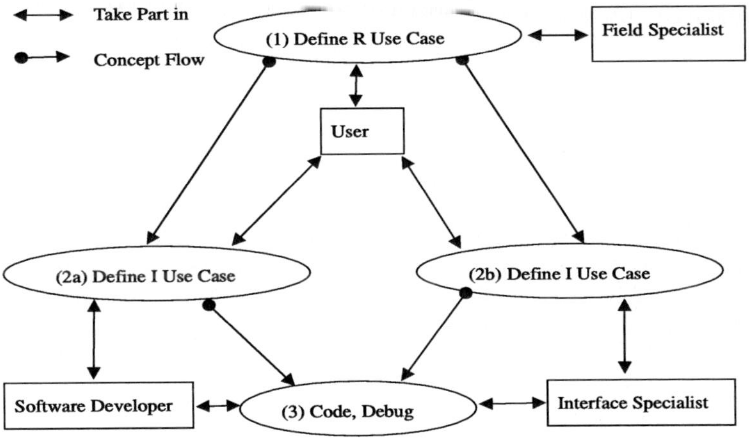

Fig. 2. The Analysis Process of RSI Use Case

The process can be partitioned into three stages. In Stage 1, system users and domain specialists cooperate to define and refine high-level R Use Cases. In Stage 2, S Use Cases and I Use Case are defined by software developers and interface designers, and these two kinds of Use Cases can be defined simultaneously. In Stage 3, coding and debugging are carried out by software developers.

3 Mutual Expansion between Role Modeling and RSI Methods

RSI methods lack of means to capture conceptual R Use Cases, which can be captured by Role Modeling. Meanwhile, Role Modeling lacks of the mechanism to properly identify the granularity of Use Cases, and RSI methods provide a framework for defining the granularity and contents of Use Cases. So we use Role Modeling to discriminate and define R Use cases, and use RSI methods to control the granularity of Use Cases in the following refining process. The relations between Role Modeling and RSI Methods are showed in Figure 3.

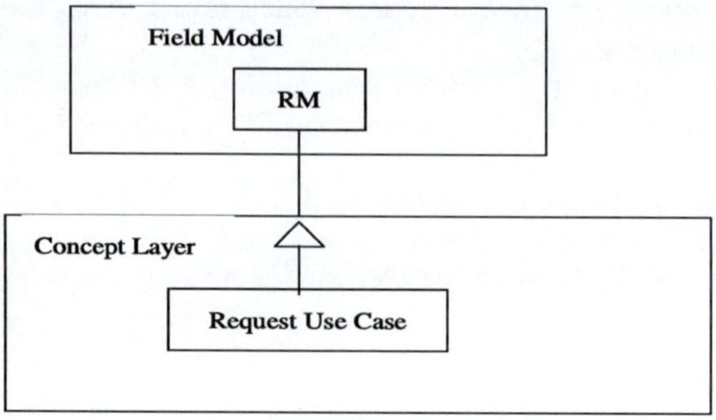

Fig. 3. The Relation between RM and RSI

4 Developing Internet Bank Systems Based on Role Modeling and RSI Methods

We will use the Internet Bank system development process as an example to show how to use Role Modeling and RSI methods to overcome the shortcomings of traditional UML Use Case methods and to achieve the natural transition from system analysis phase to system development phase.

4.1 Introduction to Internet Bank (Profile) System

The Internet Bank system is composed of four basic functions, which are purchasing function, returning goods function, query function and checking function. The component structure of the system is showed in Figure 4.

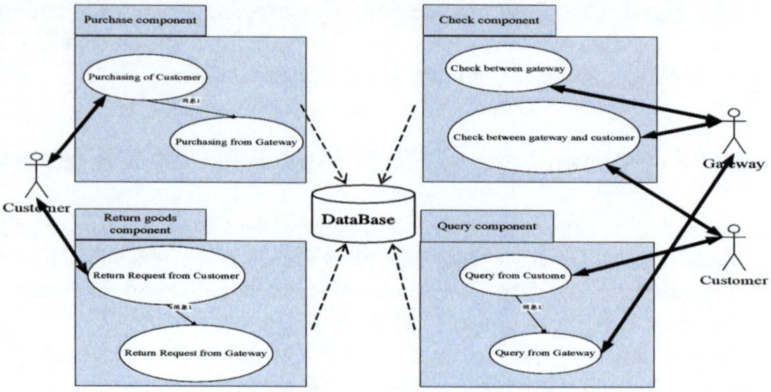

Fig. 4. System Component Structure

The Internet Bank profile system is composed of four components, which include 58 R Use cases. The system is refined into 16 I Use Cases and 54 S Use case in the development phase.

In the system analysis phase, we focus on refining R Use cases, using Role Modeling as the conceptual model, defining objects according to their roles and constructing object models through Role Modeling. In the system development phase, we focus on refining S Use cases and I Use cases.

4.2 Defining Use Cases in the Analysis Phase

Roles describe the existence reason and function of objects in the system. Users often perceive those use cases which are useful to them at first glance. So it's very natural to define Use Cases by means of objects. The definition of Use Cases of Internet Bank system in the initial analysis stage is showed in Figure 5.

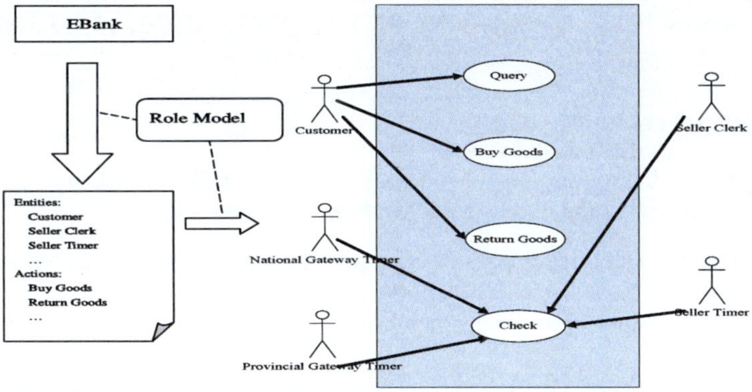

Fig. 5. Define Use Cases in the first stage of System analysis

According to the characters of Internet Bank and its functions on differently business processes, we define its role model and basic Use Cases which are showed at the right of Figure 5. They are composed of 4 Role Use Cases, which are named Query Use Case, Purchasing Use Case, Returning Goods Use Case and Checking Use Case.

The four role use cases are the R Use Cases in the RSI Methods. The granularity of those Use Cases is too large and should be refined in the later stage of system analysis phase.

4.3 Refinement of Use Cases in the Analysis Phase

We will introduce the refinement process of Use Case taking the Purchasing Use Case as an example. The process is showed in Figure 6, the dotted line starts from the use case which will be refined and ends at the refined result.

In the refinement process, we use Role Modeling and RSI methods. We should pay attention to controlling the granularity of Use Cases, which provides convenience to define I Use Cases and S Use Cases in the development phase and realize natural transition from system analysis to design. A principle of the refinement process is that there should be one and only one I Use Case corresponding to every final R Use Case.

4.4 Definition of I Use Cases and S Use Cases in the Development Phase

We will introduce how to define and refine I Use Cases and S Use Cases, taking the R.1.1.2.8 Process BuyGoods Order at National Gateway (they will be called R.1.1.2.8 for short later) as an example. The process is showed in Figure 7.

This Use Case will be composed of one I Use Case, which is I.1.1.2.1 Payment Bill Use Case, (they will be called I.1.1.2.1 for short later) and three S Use Cases, which are S.1.1.2.3 Save BuyGoods Order Use Case (they will be called S.1.1.2.3 for short later), S.1.1.2.4 Update Customers Net Account Use Case (they will be called S.1.1.2.4 for short later) and S.1.1.2.5 Form Payment Bill Use Case (they will be called S.1.1.2.5 for short later).

The I1.1.2.1 is the sole I Use Case corresponding to the R.1.1.2.8, whose requirements will be accomplished by S.1.1.2.3, S1.1.2.4 and S.1.1.2.5 through I.1.1.2.1.

4.5 Advantages of Our Method

Our method that based on Role Modeling and RSI methods has the following advantaged compared to the traditional UML Use Case methods.

(1) It can help system developers discriminating and defining Use Cases via Role Modeling.

(2) It provides a framework for deciding appropriate granularity and content of Use Cases.

(3) It supports separation between roles. Nontechnical analysts will use this method to define R Use Cases, and designers will use this method to define S Use Cases, and Interface specialists will use this method to define I Use Cases.

(4) It achieves natural transition from system analysis to system design. Each I Use Case and S Use Case corresponds to a specific class in the system. Through RSI methods we can achieve I Use Cases and S Use Cases, which can be put into program codes directly and leading to natural transition from system analysis to system design.

5 Summary and Future Work

This paper analyzes the disadvantages of the traditional UML Use Case methods and introduces the technology of Role Modeling (RM) and RSI methods. At last it gives an example of the development of Internet Bank to introduce how the use these two technologies synthetically in the analysis and design of the system, making a natural transition from analysis to design. The method has a good application value in finance system base on Internet. We will concentrate on developing software supporting RM and RSI technologies in the future research.

References

1. Unified Modeling Language Version 1.3 Specification. Object Management Group (OMG), MA01701, USA (1999)
2. Jacobson, I.: Object-oriented Software Engineering: A Use Case Appoach. Addison-wesley Press, Reading (1992)
3. He, K., He, F., Yin, S.: Role use case: a more complete analysis method for uml. Journal of Computer Research and Development 38(9) (2001)
4. He, K., Jiang, H., He, F.: Extended UML with Role Modeling. Wuhan University Journal of Natural Science 6(1-2), 175–182 (2001)
5. Collins-Cope, M.: The RSI Approach to Use Case Analysis. C++ Report (1999), http://www.creport.com/html/from-pages/feature.shtml

A Study of the Effects of "Management" Bilingual Teaching on English Proficiency and Subject Knowledge*

Zhang Hong

School of Economics and Management, Zhejiang Sci-Tech University
Hangzhou, Zhejiang Province, China
zhh96941@zju.edu.cn

Abstract. "Management" is the basic professional course in business management major; its bilingual education is of great significance. This paper research on the influence "Management" bilingual teaching to student's English proficiency and subject knowledge, it aims at increasing the teaching effectiveness of bilingual education, and better achieving the goal of bilingual education. We adopted questionnaire, classroom observation and interviews to investigate the status quo of "Management" bilingual teaching, and using the Independent Sample T-test analyzed the data of empirical study. Finally, based on the findings, some recommendations for bilingual teaching are made in the end.

Keywords: Management, bilingual teaching, English proficiency, subject knowledge.

1 Introduction

Bilingual education primarily means using a foreign language in a classroom for teaching, currently English is used frequently. Bilingual education is not a language teaching, but to content-centric, and subjects knowledge in the teaching are the main line. Bilingual education's aim is to use subjects as a carrier to help students understand and master the related terminology and expressions, use English as a tool to obtain information and achieve interpersonal communication, training students to think in a foreign language, and ultimately achieve the dual purpose that better grasp the language and subject knowledge.

This article uses the empirical research method, takes the student of "Management" bilingual teaching as investigation object, analyzes the impact of bilingual education on students' English proficiency and professional knowledge, and discusses the ways and means how to develop bilingual education.

* This work is partially supported by Zhejiang Sci-Tech University bilingual teaching model construction program "Management" (112133B3B07003); Zhejiang Sci-Tech University teaching research fund "Case teaching research on business administration oriented by student's employability" (yb1030); 2010 Teaching quality and reform project of economics and management school, Zhejiang Sci-Tech University.

W. Zhang (Ed.): Advanced Technology in Teaching, AISC 163, pp. 111–118.
springerlink.com

2 Bilingual Teaching Research Design for Management

The author does a field research lasted 16 weeks on the "Management" classes in Zhejiang Sci-Tech University, through classroom observations, interviews and experimental study, investigates the status of "Management" bilingual education, researches on the influence of bilingual teaching to students English proficiency and academic knowledge.

A. Study Object

In order to guarantee the research's reliability and validity, we firstly identified the appropriate study object only influenced by bilingual teaching, 09 Business management specialty and 09 financial specialty, both of these classes as parallel classes. Two classes of students have learning English more than 6 years, the age lie between 19-21. In same semester, we teaches "Management" separately, adopt bilingual teaching in business management class (experiments group), and Chinese teaching in the financial class (control group).

The reasons for selecting these two classes are, first of all, the two classmates were the first time to accept "Management" course learning, they haven't received other professional or bilingual courses; Secondly, the two classes taught by the same teacher, which can reduce the potential impact from different teachers; Again, it's also a same English teacher for the experimental class and control class, then the bilingual education in the experimental class will be the only reason for the difference of two classes; Finally, the independent samples T-test revealed that there is no significant difference in English and management knowledge level between them (see Table 1, Table 2).

Table 1. T-test of professional English's pre-test score

Score	Class	N	Mean	Std. Deviation	t	df	Sig. (two-tailed)	mean difference
	Experimental class	59	68.90	7.104	0.021	118	0.983	.029
	Control class	61	68.87	8.316				

Table 2. T-test of management knowledge's pre-test score

Score	Class	N	Mean	Std. Deviation	t	df	Sig. (two-tailed)	mean difference
	Experimental class	59	71.85	7.155	0.123	118	0.903	.175
	Control class	61	71.67	8.422				

Table 1 shows that t statistic significant (two-tailed) probability $p=0.983>0.05$, hence there are no significant differences of English tests level between the two classes. According to Table 2, there are no significant differences of management knowledge between the two classes.

B. Research Methods and Research Procedures

The implementation phase of this teaching experiment is from September to December in the year 2009. During this period, "Management" course used bilingual teaching in the experimental class, the textbooks is English original edition published by Tsinghua University Press, and in the control class used the Chinese teaching materials and teaching in Chinese. During the same period, the students attend normal college English course of study, two classes' English teacher is the same one. At the end of the semester, two classes of students also simultaneously participate in the examination of management and English courses. Both qualitative and quantitative research methods are used in our research, in the experimental period, we conducted interviews, classroom observation and examinations of bilingual teaching situations, etc.

1) Classroom observation on the bilingual teaching: observation is the basis to assess people's skills and behavior, teachers can assess what students have learned and what have not yet learned based on the observation, infer certain learning method may facilitate or hinder the learning outcomes, evaluate the effectiveness of a teaching method, and make certain what teaching activities and materials do the students like, and so on. Classroom observation provides the foundation for assessment of existing teaching activities and determination of future teaching methods. In order to fully understand the practice of bilingual education, have a clear concept on the bilingual education classes, this study implemented a classroom observation.

2) Interviews on the bilingual education: qualitative data collection is also very important; the interview results will support the experimental findings of the study. We interviewed two bilingual teachers based on the interview outline prepared in advance, acquired some details on the actual situation in "Management" course bilingual education. For example, the principles of selecting textbooks, requirements of teachers to students, the language and the ratios used in the classroom, the difficulties and obstacles faced by teachers and students, and so on. This information is very valuable for the analysis of experimental results and recommendations to improve teaching effectiveness.

3) Quasi-experimental study of bilingual education: In the study, in accordance with the "pretest-posttest non-reciprocal control group design", we used single-blind method, carried out two English tests, one management knowledge test and one "Management" course test. The two English tests separately arranged before and after the study, the first English test arranged in the previous semester final exam to test the student's English proficiency before participate in bilingual education. The second English test scheduled after the "Management" course to assess the influence of bilingual teaching. The examination for "Management" courses in the final semester are used to measure student's professional knowledge and ability, the examination paper is professional papers from my school annually revised question, has a high reliability and validity. Experimental classes adopted English papers, and answered in English, the control class answered the Chinese version of the paper in Chinese. Test results were statistically analyzed using SPSS.

3 The Influence of "Management" Bilingual Teaching to English and Subject Knowledge

A. Result of "Management" course examination

In accordance with the above research design, we carried out the "Management" courses inspection on the experimental class and the control class, and compared the test results (Table 3).

Table 3. Independent SAMPLE T-test of SCORE OF management knowledge

score	Class	N	Mean	Std. Deviation	t	df	Sig. (two-tailed)	mean difference
	Experimental class	59	78.83	5.160	-2.906	110	0.004	-2.465
	Control class	61	81.30	4.043				

Table 3 shows that the mean scores of the experimental class (78.83) lower than the mean scores of the control class (81.30), the mean difference is 2.465, and it seems that the control class's scores is better than the results of the experimental class. Results of two-tailed significance test 0.004 <0.05, indicate there is significant differences between the results of the two classes, that is to say, the "Management" bilingual education affect student's mastery of professional knowledge in certain aspects.

B. English proficiency test results

Bilingual education at least can stimulate the student's enthusiasm to learn English, through providing language environment and practice to affect the student's English proficiency. Table 4 is the test results.

Table 4. Independent SAMPLE T-test of professional English's post-test score

score	Class	N	Mean	Std. Deviation	t	df	Sig. (two-tailed)	mean difference
	Experimental class	59	74.02	6.999	2.163	118	0.033	3.017
	Control class	61	71.00	8.212				

Table 4 shows that the average scores of the experimental class (74.02) is higher than the average scores of the control class (71.00), and the significance of two-tailed test 0.033 <0.05, indicate the results of the two classes are significantly different, the experimental class have a greater progress at the English capacity.

4 The Main Findings of the Study

A. The classroom observation, interview and experimental studies have shown that the vast majority of bilingual teachers and students have a positive attitude toward bilingual education; they consider the bilingual education is necessary and meaningful. On the one hand, bilingual education has a certain level of contribution in promoting student's English proficiency, especially on English listening, reading comprehension and vocabulary, etc. On the other hand, the frequent contacts on English supplementary information and the aspirations of learning more advanced professional knowledge, force students and teachers to accept and be used to bilingual teaching.

B. The results of independent sample T-test showed that "Management" bilingual teaching can improve student's English ability, which may result from the following reasons: firstly, the bilingual education offers more opportunities for the practical application of English and thereby increasing the learner's English proficiency; Secondly, bilingual education model is a good incentive to promote student learning English; Furthermore, in the bilingual classroom, teachers use both English and Chinese as the medium of instruction, teaching students to read English version of the textbooks, read English slides, do English exercises, and discuss in English, so that students receive a lot of English inputs, their English skills will gradually progress.

C. According to the data analysis, there is a significant difference of the subject knowledge mastery level between the experimental class and the control class, that is, bilingual education hinders the student's mastery of subject knowledge to some extent. "Management" is the basic course in itself, opening in the lower grades, some students reflect that they are used to their way of Chinese language teaching, have a psychological fear on the bilingual education and English textbooks, someone even can not correctly comprehend the exam questions in English.

Bilingual teaching has two objectives, obtain subject knowledge and promote English skills, we should consistently give top priority to acquire specialized knowledge to ensure that students understand and master professional knowledge.

D. The findings also indicate that there are still some problems in the current bilingual education.

1) Student's English standard is not balanced. We find from the classroom observations that only a very small number of students have good English language ability, be able to fully integrate into the bilingual teaching. Most of the students are not active in the classroom teaching, because they can not fully understand the contents of the teacher teach, and fear to communicate with the teacher or students in English, to some extent, all this limited the acceptance and absorption for the teaching content.

2) Teacher's English level needs to improve. Bilingual teachers are the keyman who organize, guide and implement the successful bilingual education, whose language fluency and accuracy seriously affect the attitude of students in bilingual classes. Therefore, bilingual teacher's excellent language skills and profound professional knowledge is very important for students to overcome mental barriers in bilingual teaching.

*3) Some difficulties existing in textbooks.*Course materials should be prepared in accordance with China's education system, the basic conditions of Chinese students and the national schools curriculum design. English textbooks often need more length when describing same knowledge contents, on the one hand, a lot of teaching contents is not introduced deeply, on the other hand, the long rehearsal increase the preview difficulties and mentality burden, then bring difficulties to class teaching. Another aspect, because of the language obstacles, Chinese teachers are also more difficult to compile or rightly translating suitable English textbooks.

4) The classroom teaching is lack of interaction. Through field observations, we found that most of the bilingual education classes are lack of interaction, in addition, students related to business administration are generally more active in thinking; they have a strong sense of being the teaching subjects and involving teaching activities, they are not welcome "cramming education" or passive learning. More importantly, they have higher requirements to understand and explore practical issues under the certain theoretical guidance.

5 Recommendations to Improve Effectiveness of Bilingual Education

Based on the above findings, and from the perspective of academic knowledge learning and English skills improvement, we submit some recommendations aimed at improving the quality of bilingual education.

A. Stress interaction in classroom teaching

Interaction is a necessary criterion for measuring any good course, particularly important to the bilingual education classes. Specialized course bilingual education is specially difficult, and only sufficiently mobilize the student's learning enthusiasm, develop their subjective awareness, mutually interact between teachers and students, students actively participate in the whole teaching process, then will achieve good teaching effect. Therefore, in order to achieve the success of bilingual education, teachers and students need to work together.

Teachers should emphasize the display of student's subjectivity in the teaching process, mobilize their learning enthusiasm, and know every student's level, then based on student's current knowledge and understanding to carry out the teaching of new knowledge; provide sufficient situational support for the use of non-native language, for example, through body language, rich gestures, facial expressions and movements, etc.; use frequent and diversified methods to continuously get how do students master the content of teaching, and estimate the student's language level; use a large number of repetition, summary repeat to ensure that students keep up with the rhythm of teaching and understanding of the teacher's guidance; timely motivate students, encourage them to use non-native language to express and guide the students think by themselves, learn to think, clever at thinking, actively understand and master knowledge.

Students should fully preview and regularly review course contents, and actively engage in classroom teaching.

B. Professional course teaching and English languages teaching complement each other

Student's English proficiency to a large extent influences the effectiveness of bilingual education. Field observations and interviews indicated that student's current English proficiency is not ideal, there is a serious lack of English communication skills and listening ability, which directly affects the bilingual teaching of professional courses. Therefore, to strengthen the cooperation between English languages teaching and bilingual teaching of professional courses is very important.

First of all, the curriculum structure should be reformed. Traditional English teaching is too focused on the study of linguistic knowledge, such as grammar, vocabulary, syntax items, while ignoring the comprehensive language proficiency of students, thus limiting the development of bilingual education. In order to achieve better teaching effectiveness of bilingual professional courses, the English teachers should give consideration to both English teaching and specialized vocabulary; improve student's language skills, particularly communication skills with non-native language, increase student's familiarity with the professional language, and then improve their absorption effect on the knowledge from the bilingual classroom.

Secondly, teachers can use "immersion teaching" method in the bilingual education, thus mix the learning of subject knowledge and language ability together. They also can execute content-based instruction, create a good environment for language learning, strengthen the interaction of bilingual classroom, give students the opportunity to achieve the use of the language, then naturally learning the language while studying academic content.Finally, the students can obtain progress in the academic level, language level and language application, and achieve the improvement both in subject knowledge and language skills together.

C. Enhance the level of Bilingual Teachers

Teachers are the key to the successful implementation of bilingual education. Compared to traditional teaching mode, bilingual teacher should specialize in a particular area of expertise, has a wealth of professional knowledge, academic update capabilities, education research capacity and modern education technology skills. At the same time, also requires bilingual teachers having pretty good foreign language skills, proficiency in English teaching organization, have keen insight and good perception on the characteristics and needs of students, and have good communication skills with students and teaching enthusiasm and so on.

Bilingual teachers from their own terms, on the one hand, they should always observe and note the differences situation and extent between students, adopt appropriate teaching methods according to the student's differences in language ability and expertise knowledge, develop targeted teaching programs and adjust teaching schedule. On the other hand, bilingual teachers also should continuously enrich professional knowledge, accumulate and summarize teaching experience, train clear and accurate pronunciation, and deliberately adjust the speech rate, language habit and diction difficulties according to the student's reaction.

From the school point of view, normal and the other institutions can consider to set up bilingual teaching speciality, systematically train high level of teachers by assessment of bilingual teaching qualification. The schools that are opening the bilingual curriculum, should fully utilize and exploit the available teachers, strengthen the training of bilingual teachers, and strive to cultivate relatively high quality bilingual teachers in a relatively short period of time. Through a variety of ways with foreign institutions to conduct regular or ad hoc teaching and academic exchanges; also regularly send bilingual teachers to study abroad to promote their professional knowledge updates, and to build strong language skills, enabling them to become the backbone of bilingual education; at the same time, can give full play to the foreign teachers, arrange them to direct the bilingual teachers, this will definitely improve the communication skills of bilingual teachers, and also can increase access to the true western culture and proper English expression.

D. select and use right teaching materials

Textbook is not only an important teaching reference, and is the most important part in the course construction, its influence to bilingual education effect is enormous. The normal functioning of bilingual teaching must rely on foreign original edition textbooks; bilingual teachers should choose the textbook having certain authoritative and applicability, choose study materials with moderate difficulty and language specification based on students' language ability, and carry out certain degree adjustment on the curriculum and content again according to student's recipient degree and feedback. In order to expand students' horizons, help their independent study and research, teachers also need to rearrange and integrate some necessary references combing to different teaching contents, to help students understand, and is conducive to teaching and learning activities in an orderly manner. At the same time, it is recommended for students in a couple of famous Chinese reference materials to ensure that students better understanding subject without possible impacts of language.

Besides, having bigger degree of difficulty because of bilingual education, teacher needs more energy to prepare English teaching handouts than Chinese teaches, spend considerable time preparing and teaching, and their overall progress has a certain influence, but teaching effect is often lower than expected, allowing teachers to reduce the feeling of accomplishment. Therefore, in order not to affect the enthusiasm and commitment, bilingual teachers need to constantly self-encourage and motivate from the service students and lifting self's angle, school also needs to carry out certain encouragement and guidance on bilingual teachers. Ultimately, under the guidance of bilingual education, firmly put the specialized courses on the first study, supplemented improve students' English standards, and achieve real success of bilingual teaching.

References

1. Chen, H., et al.: The adaptability and countermeasures of undergraduate students on bilingual teaching: four years of longitudinal study. Modern Education Management (1), 80–82 (2010)
2. Deng, K.: On the application of the bilingual teaching method in the instruction for economic management major undergraduates. Journal of Hainan Normal University (Social Sciences) (3), 142–145 (2007)
3. Hu, D., Liu, L.: Study on the problems and solutions of the language in bilingual teaching. Education Research Monthly (6), 104–106 (2009)
4. Ning, H., Li, Y.: Research on the interactive teaching mode in college bilingual teaching. Chinese Adult Education (9), 127–128 (2009)
5. Weng, W.: Research on the university bilingual teaching practice in economics and management discipline. Management Observer (20), 107–108 (2008)

Practical Teaching Design of Basis of College Computer Engineering Based on Blended Learning Model

Huang Lin-na

Department of Computer Engineering,
Cangzhou Normal College, Cangzhou, Hebei Province, China
Hln0322@163.com

Abstract. It is essential for college students to improve their IT attainment by learning basis of computer engineering, whose practical teaching has a direct effect on college students' abilities of computer application. Therefore the teachers should pay more attention to both practical teaching of college computer engineering and students' practical ability. While Blended Learning Model, developed on the basis of online education, not only combines advantages of traditional teaching and web-based instruction , but also offers new method of teaching reform and is very suitable for practical teaching of basis of computer engineering. This paper discusses how to improve teaching effect by applying Blended Learning Model and how to design teaching activity by combining features of Blended Learning Model during the process of teaching of basis of computer engineering in higher colleges.

Keywords: Blended Learning, basis of computer engineering, design of teaching activity.

1 Concept of Blended Learning

Professor He Kekang points out, "Blended Learning refers to combination of advantages of traditional learning methods and that of E-Learning, developing not only the teachers' leading roles such as guidance, inspiration and monitoring, but also the main learning body—students' initiative, enthusiasm and creativity. Only through this combination can the best learning efficiency be achieved [1]."

Driscoll (2002) once has a comprehensive statement about Blended Learning saying "Blended Learning means the following learning processes, such as integration based on Web technology (like virtual class, collaborative learning, streaming media and texts) to achieve some teaching goals, integration of various teaching methods (like constructivism, behaviorism and cognitivism) and teaching technologies (or non-teaching technologies) to achieve perfect teaching efficiency, integration of any teaching technologies (like video, CD-ROM, training and movies based on Web) and face-to-face teaching or training, integration of teaching technology and specific work task [2]. "

In conclusion, Blended Learning is a kind of thought and strategy, a learning method or learning idea, an integration of on-line learning and face-to-face learning, a multi-integration under the conditions of the learners, designers, teachers and

W. Zhang (Ed.): Advanced Technology in Teaching, AISC 163, pp. 119–123.
springerlink.com

materialization technology. This paper shows that Blended Learning is, firstly, an integration of learning theories, and then an integration of learning surroundings and learning methods. During designing some specific teaching activity, such integrations as teaching medias, learning styles should be taken into account, play to their strengths and complement with each other with their advantages. An integration of learning methods can be interpreted as a combination of audiovisual medias (slide projectors, audio and video recordings) and traditional classroom teaching, a combination of CAI (CAI, internet courses) and traditional unitary teacher-centered teaching, a combination of students' self-directed learning and collaborative learning, etc.

2 Problems Existing in Current Practical Teaching of College Computer Basis

Strong practice is the key feature of College Computer Basis. But the present problems are conflicts of more knowledge and less machine hours. The main task of this course is how to improve teaching method in order to give consideration to different levels of students for them to grasp basic skills in limited time and how to develop their self-directed learning ability, creativity, team work spirit.

A. Problems about Students

Take our school for instance, a survey of students' entry level shows, on one hand, there are great differences of students' computer skills because of gap between town and city, gap between schools, gap between families; on the other hand, although the students have computer applied-courses in middle schools and most of them are at "non-zero level", they have low practical level under the situation of exams with less attention to computer. In addition, some students, especially new comers, don't think computer skills would help their future specialty, which leading to no good use of multi-medias to self-teach and pay less importance to computer courses. So facing difficulties, it is easy for them to be afraid and give up learning.

B. Problems about Teachers

Current problems of teachers' teaching exist like this: (1) Don't understand students' needs and learning interests, which leads to poor practical teaching with old cases and can't advance with the times. (2)Don't take effective measures to teach poor-level students, allow no flexibility in teaching without "level classes". (3) Do not guide students' self-directed learning, not arouse students' learning motivation, not inspire students' emotion, not instruct learning with effective modern teaching methods. (4) Do not combine teaching with students' practice such as arts, sciences or different specialties. All teaching is finished with only one model. (5)Pay attention only to passing rates of Level I Exams without attention to students' practical application ability.

C. Problems about Learning Resources

The students can't make good use of internet to do self-teaching because schools' internet is forbidden for fear of less broadband or viruses. While in internet bars outside school, the students have no good circumstances to search something or read something. Internet bars have no software available for students to solve problems existing in their learning. Moreover, some learning resources are not easy to get for its high repetition, low utilization factors, slow speed and often interrupted internet connection.

3 Practical Teaching Design of Basis of College Computer

Teaching design means a program that, in order to achieve some teaching aim, the teachers analyze all factors of teaching process, create some situations, make up appropriate teaching plans by combining course resources, analyzing learners' features and using modern teaching theory, which can help carry out teaching smoothly.

A. Principles of Practical Teaching Design

1) Principle of Integration: means that teachers, from points of microscopic views and whole structure of courses, should consider all teaching tasks, aims, contents, teaching organization and teaching methods to make all factors coordinate with each other, improve each other and go forwards together.

2) Principle of Subjectivity: In teaching process, the teachers--leading parts, should become favorers, orienters and organizers of the students—center parts. Subjectivity can be seen from three aspects: independence, initiative and creativity. Thus during designing teaching, the teachers should consider students as central parts of learning. All teaching contents and design should be finished for students' integrated development and personal development. Teaching strategy should be changed from only educating knowledge to emphasizing process experience and to emphasizing students' development, from emphasizing teaching to emphasizing learning, from emphasizing conclusions to emphasizing process.

3) Principle of Hierarchy: At first, by taking good advantages of internet, the teachers should, according to teaching contents and teaching aims, provide convenience for students' independent learning and individual learning in order to change teaching from traditional standard training to individual learning. For instance, as for the same learning subjects, the teachers should provide not only rich relevant resources, but also different materials to show the same ideas. Thus students can choose different resources with different levels to learn according to their own abilities, needs and interests[3].

4) Principle of Openness: Openness of classes refers to openness of course contents, course aims and course implementation. Such openness must show features like immediacy, order and selectivity. During designing teaching activities, the teachers should consider interactive and flexible learning methods, harmonious and happy atmosphere to build perfect learning system where students can have access to course contents and learning resources. Thus relevant course information with high quality can be achieved from various learning platforms, can be integrated into courses without limitation of time, space and transfer modes in order to realize free information-transfer, information-share, information-organization and information-store.

5) Principle of Interactivity: It, as an asynchronous learning, is connected with fine internet atmosphere which is not limited by time and space. While interactivity, which can improve efficiency of asynchronous learning, is the core and soul of asynchronous learning. In domestic internet education, importance of internet interaction has been popular widely. But the study blank space includes design of its microcosmic activities, like how to group students, how to communicate on internet, how to control the time in actual communication. how to have post comments ,how to design a perfect evaluation system to combine interaction itself and interactive learning [4]. From this point of view, how to carry out specific interactivity needs to be studied further.

B. Design Schedule of Practical Teaching Activities

Blended Learning determines two necessary components of design schedule of practical teaching: face-to-face teaching and on-line teaching.

1) Design of Face-to-face Teaching

①Design of process of face-to-face teaching: In this mode, face-to-face teaching is classified into two types, that is, face-to-face-lecture and after school program. The lecturers and students are in the same space to have face-to-face education. Their tasks are as followings: first to select appropriate teaching methods according to teaching aims, and second to guide learning, coach, discuss after class and solve problems after class between teachers and students.

②Design of synchronized and real-time communication: Communication between teachers and students is an important link in face-to-face education. The teachers and students are in the synchronized space and time. This communication includes face-to-face questions and answers, real-time discussion. In practical teaching computer basis based on Blended Learning, teach only the essential and ensure plenty of practice. Each time, the teachers should leave more time to guide students, to do face-to-face questions and answers, organize students to discuss, improve collaborative learning, or organize students to show their works, evaluate students reasonably and accurately. In a word, make students grasp the content of the classroom and adjust teaching speed and teaching methods.

2) Design of On-line Teaching

Constructivism regards students as meaningful constructors, while teachers are students' helpers, promoters instead of imputers of knowledge. In on-line teaching, teachers should adopt flexible teaching methods to help students finish meaning making. This mode gives first place to self-control and research study, and second place to group collaborative learning, learning websites, on-line video, network courses.

① Self-control and research study: It is a learning method where students can arrange their own learning time, learning place, control learning schedule and make self-evaluation, according to their own learning features, learning desires and learning styles. Such learning advocates self-plan, self-organization, self-evaluation, which breaks the limitation of time and space without group learning in a set time and place, and develops students' creativity with their full desire and initiative. But this mode is also a test of students' self-control. Because the freshmen couldn't be used to this mode, the teachers should guide them what to learn and what degree to learn. The new students would gradually adapt to this mode to achieve self-control and research study with teachers' letting go.

② Design of asynchronous interaction: Internet provides new learning space for students—resource libraries, electric libraries, virtual classrooms. Under internet circumstances, students can learn selectively and obtain on-line help and resources according to their own situations. On-line asynchronous interaction, a way to solve students' problems, seems essential because of no face-to-face education. In this mode, asynchronous interaction mainly includes messages on our website, E-Mail, QQ, Blogs, etc. We set up a teaching website depending on campus website to not only supply learning resources, exhibit students' works, submit homework, etc., but also play an important role in asynchronous interaction between teachers and students. During the process, please pay more attention to students' messages and response without delay to improve efficiency and timeliness of interaction.

4 Conclusion

Computer application is a necessary part of students' qualification education. Basis of College Computer Engineering, as a general course for freshmen, plays an important role in promoting students' IT ability. The practical teaching of Basis of College Computer Engineering has a direct influence on students' computer application ability. So the teachers should emphasize more practical teaching and foster students' practical ability. Blended Learning is a new teaching method guided by Blended Learning theory with the help of IT. It is good for computer practical teaching and is spread widely in teaching process.

References

1. Wu, M.-N.: Design of teaching. Higher Education Press, Beijing (1994)
2. Li, K.-D., Zhao, J.-H.: General Theory and Application Model of Blended Learning. E-education Research, 3–8 (July 2004)
3. Liu, C.-K.: Study of Self-learning Activity Methods of Students Based on Internet Environment. Teaching and Management, 52–53 (June 2007)
4. Pelz, B.: My Three Principles of Effective Online Pedagogy. Open Education Research, 30–37 (December 2007)

Principal Problems & Counter-Measures Existing in Interactive Teaching Mode for Basis of College Computer under Internet Environment*

Huang Lin-na

Department of Computer Engineering,
Cangzhou Normal College, Cangzhou, Hebei Province, China
Hln0322@163.com

Abstract. This paper, on the basis of practical teaching, discusses the current principal problems existing in Interactive Teaching Mode for Basis of College Computer and presents the counter-measures of constructing favorable teaching interaction under internet environment.

Keywords: computer basis education, interactive mode, problems, counter-measures.

1 Significance of Interactive Teaching

Interaction means communication activities between two parties. Wagner defines it as "mutual effect between individuals and groups which needs behaviors of at least two sides or two parties". While interaction in teaching refers to inter-communication between the learners and other classmates or teachers through media [1].

Interaction is the basic feature of teaching while interaction runs through all teaching [2]. It reflects the nature of modern teaching as a precondition of harmonious teacher-student relationship, reflection of students' subjectivity, a necessary method of positive class-culture. But in reality, especially in teaching of high schools, importance of interactive teaching is ignored to different degrees. It is common that teachers talk fluently while students act as they please. So it is necessary to rebuild interaction relationships between teachers and students, to set up interactive class-culture and to establish new interactive ideas between teachers and students.

2 Teaching Interaction under Internet Environment

Teaching interaction is divided into two types: direct interaction and indirect interaction. The former means direct interaction through language, movements and expression. The latter means interaction through internet [3]. Direct interaction is also called synchronous interaction because of the same time and space of the interaction subjects. While indirect interaction is named non-synchronous interaction because of

* 2009 Scientific Research Project of Ministry of Education of Hebei Province (No. Z2009201).

W. Zhang (Ed.): Advanced Technology in Teaching, AISC 163, pp. 125–129.
springerlink.com

un-limitation of time and space. Teaching interaction under internet environment means all inter-communication and interaction between students and teachers, between students and classmates, between students and learning recourses through internet [4]. Interaction mode of internet teaching aims to seek the best connection point between internet and courses in order to realize interaction between students and teachers, between students and machines, between internet and courses, which can make more space for students to act, to think, to interact, to explore the world, to explore themselves and further to strengthen subjectivity and develop creativity.

3 Problems Existing in Synchronous Interaction

A. Lack of Independent Activity of the Students

This lack shows: classroom becomes a special room for teachers' talk by controlling the initiative of the class. The teachers decide the teaching contents, teaching speed. For instance, in teaching computer basis in our college, the teachers give too much talk with few questions while students just listen and imitate mechanically. Whether it is task-driving or case-teaching, the students only operate the procedures without any personal views or different ideas, which shows that students lack independent learning in a state of passive receiving. Students' autonomous development and autonomous thinking activity are ignored, leading to the whole lack of students' autonomous learning in class.

B. Lack of Emotional Interaction

Class teaching is a special place where teachers and students have a communication of soul and spirit with the features of "democracy, equality, respect, trust and love [5]". In teaching, emotional interaction can help educate through lively activity, teach with emotions, mingle teachers' and students' feelings. So-called "loving teachers and believing their thoughts" means only teachers attract students in emotion, can students accept teachers' thoughts, teachers' knowledge and teachers' dispelling doubts. But in modern teaching design, especially in computer teaching, too much attention is paid to cognition and skill goals based on system cybernetics and cognition psychology. And also this teaching lays stress on dealing with problems and teaching process with views of "technology" while emphasizes "material" factors such as teaching methods and media, ignoring ones' spiritual value and emotional interaction .According to a survey of last term, the grade results of teachers evaluated by the students show that the top teachers are those who can have a good emotional interaction with students in daily teaching ,not only by their knowledge, operation skills and language expression.

C. Lack of Creativity of the Students

Poor creativity of the students is common in our teaching as following: students pay too much attention to books or teachers'" authority" while lack of criticism and doubts. The teachers regard knowledge as highest and ones' spiritual world is left away. Class teaching is boring without energy and creativity. Lack of creativity in computer teaching can be seen in students' imitating teachers mechanically and have no flexible solutions to different questions. The students can't have a deep practice on base of teachers' lecture and can't have a "beautiful" creativity and further application.

4 Problems Existing in Non-synchronous Interaction

Computer basis teaching can't go on without internet whether it is LAN or Internet. Interaction is the major method to finish internet teaching. The design of interaction decides the final effect of teaching. Non-synchronous is the major teaching method under internet also is the most popular teaching interaction in internet courses and excellent courses. This is mainly because the tool technology of non-synchronous is too easy. The mainly tools of non-synchronous include E-mail, message board, BBS and wiki, etc. Problems existing in non-synchronous interaction are as follows:

A. Imperfect Platform of Non-synchronous Interaction and Unsmooth Interaction Channel

Internet in very popular with modern college students and more teachers like to have a "zero-distance" communication with students through internet. But in our college, we haven't taken good advantages of campus internet because of the insufficient network bandwidth and weak internet environment. So we are short of digital resources and smooth internet interaction channel, which leads to failure of three-dimension interaction system that influences non-synchronous interaction between teachers and students. Because campus internet can't be used fully, we have no places like forums, messages to communicate. Instead, we have to use phone messages, QQ, E-mail. But some teachers are not willing to inform their personal information like phone number, QQ to students or some teachers have no QQ, students couldn't find their teachers to ask for help or discuss some questions. Let alone internet learning .

B. Insufficient Timeliness of Interaction

A great deal of teaching internets have rich courses but simple functional design of interaction. Even some excellent courses have no interactive parts. And teachers' less participation and delayed responses to the messages have a great influence on timeliness and efficiency of interaction.

C. Insufficient Guidance from Teachers to Students about How to Use Internet

Internet application is one of the core courses of computer basis. While some teachers only emphasize explaining technology and ways of using tools. How to obtain useful information from internet, how to learn more from internet have not been taught enough, making students unknown about how to search information, how to download and how to select better websites.

D. Lack of Humanistic Concern ,only Contents of Interaction

Interaction based on contents is one important part of internet learning [6]. In internet teaching, learning contents are key parts of students' visiting internet. So internet teaching is the core of internet courses, including teaching goals, teaching programs, teaching plans, self-study plans, video courses, mock examination questions, learning recourses. The conveying of contents is one-way. In internet courses, interaction process should be strengthened namely strengthening interaction function, like allowing learners to change, add comments or ideas. Forum, message boards, humanistic emotion should be added to allow learners to give questions , to learn from each other ,to share experiences or to discuss some topics etc. In addition, teaching knowledge and educating students are connected closely which can't be seen from internet based on contents. Lack of humanistic concern or emotion would have few influence or guidance on modern college students' thinking and morality.

5 Measures of Good Interaction

A. Strengthen Consciousness of Teaching Interaction, Build up Value of Teaching Interaction

"Today's teaching reform is not only to change traditional teaching theory, but also thousands of teaching ideas, and change teaching behaviors that they are doing in daily life [7]." Only changes of ideas and consciousness can change behaviors. For teachers of college computer, they should set up and strengthen consciousness of teaching interaction to give play to IT or internet platform. All new students from all departments learn computer basis because of its basic feature and application from the first term when students are adjusting themselves from senior schools to colleges, from senior school teachers' "teaching by holding his hand" to college teachers' "hands-off". Many students could not face these changes thus don't know what to do, how to do, lose learning goal and let matters drift. So the teachers should guide them appropriately, including how to plan college life, how to build life goal, how to deal with human relationships, how to develop self-learning ability or creativity. The author has been teaching computer for many years, knowing importance of interaction and interacts with students with the help of media or internet platform. For instance, at beginning, make a beautiful PPT to have a self-introduction , inform more information or contacts to students in order to keep contact with each other ; give clear requirements to students and sometimes assign homework; give timely comments on their homework or immediate response to their messages or e-mail; understand them as much as possible including computer basis, term learning plan; encourage students to learn hard or remind them of their learning problems; make good use of their blogs to search good essays to enjoy "chicken soup for the soul"; ask students to choose topic freely when teaching "word" or "ppt", to collect materials after class and show their electronic works in class and give them comments . By the end of the courses, ask students to summarize achievements, disadvantages or self-examination, or to give some advice to teachers. Teachers would write a letter "final-term words" to every student about their learning interaction and give some wishes and good advice to them which are popular with students. All student give high praise to teachers with achievements of not only computer but also methods of communication. Practice shows that good interaction is the precondition of good teaching while teachers' consciousness of teaching interaction and value of teaching interaction are the key.

B. Combine Synchronous Interaction and Non-synchronous Interaction

Synchronous interaction and non-synchronous interaction can complement each other's advantages [8]. Synchronous interaction can perfect class language which can make it easy to express emotion, to communicate, to strengthen sense of reality of interaction situation and timeliness of interactive information, to response immediately to specific learning problems and to make interaction more interesting. While non-synchronous interaction can break limitation of time and space to extend class teaching. Thus students can arrange their learning well, have a deep discussion about some problem in free time, share experiences and do more independent learning, to remove differences caused by different levels to enjoy the same chance of learning, to avoid some face-to-face pressure, to speak freely, to relax tense human relationships. Moreover, from point of meaning construction, learners must achieve information from all channels and different channels have different time. So in order to finish perfect teaching, we should combine synchronous interaction and non-synchronous interaction to set up a three-dimensional interactive system where class interaction is first and internet platform is second.

C. Combine Contents Interaction and Interpersonal Interaction

Contents interaction comes down to students' knowledge or skills while communication between teachers and students is full of richness or humanity. On one hand, teachers should regard teaching as life activate, as personality revealing, as wisdom impact. Teachers should make classroom into a spiritual place where students can search leaning, make widely known, perfect personality and develop oneself. On the other hand, non-synchronous interaction shouldn't be boring and dull. Teaching activity is, in fact, a cultural activity, is an "acculturation" under the teachers' "teaching culture" [9]. The teachers should set up equal, democratic teaching interaction culture, develop learners' ability to communicate with others, combine contents interaction, human interaction and emotional interaction by breaking single, old and closed teaching mode and relying on new teaching ideas, teachers should create new features in integration, interaction, cooperation and exploration from many points of view to study teaching reform.

References

1. Wang, C.-B., Wang, Y.-Y.: Interaction in Web Teaching in Perspective of Initial Teaching. Modern Educational Technology, 108–110 (November 2009)
2. Liu, Z.: On Development of Interaction Teaching in Class from View of Instructional Media. E-education Research, 103–105 (July 2009)
3. Niu, H.-L., Ma, Z., Qiao, Y.-M., Lu, S.-J.: Investigation and Practice on Online Interactive Teaching Mode. Computer Education and Educational Informationization, 5979–5980 (July 2009)
4. Ding, X.-F.: Distance Education. The Publishing House of Beijing Normal University, Beijing (2001)
5. Li, C.-P.: Interaction Characteristics of Teaching & Its Realization. Theory and Practice of Education, 40–42 (July 2008)
6. Wang, C.-B., Wang, Y.-Y.: Interaction in Web Teaching in Perspective of Initial Teaching. Modern Educational Technology, 108–110 (November 2009)
7. Ye, L.: Arousing Vitality for Life in Classroom. Education Studies, 3–8 (September 1997)
8. Qian, X.-S.: How to Implement Interactive System for Middle-school Students in Construction of Information Technology Course. E-education Research, 87–90 (October 2009)
9. Li, C.-P.: Interaction Characteristics of Teaching & Its Realization. Theory and Practice of Education, 40–42 (July 2008)

400m Running Training Methods and Means of Track and Field Athlete

Guangzhi Huang

College of Physical Education, Hainan Normal University, 571158, China

Abstract. This article discussing 400m running training methods and measures through literature material method, logic analysis, descriptive research, in an attempt to train optimization, scientific, and providing theoretical support for athletes and athletics lovers.

Keywords: Track and field, 400m running, training methods, training means.

1 Introduction

400m running is a sprint project with speed as characteristic and speed endurance as foundation, it is the longest distance and hardest project of short sprint running. The key of 400m running success lies in how to manage their speed and strength the most reasonable and effective. Its difficulty lies not only in athletes to have high level requirements of speed, speed endurance ability, still need to master reasonable running techniques and indomitable character. In order to get good grades and normal training level of play more important is to master the reasonable 400m running rhythm.

2 The Functions and Features of 400m Running

From the point of biological chemical process, 400m running power supply is decided by adenosine triphosphate (ATP), phosphoric acid creatine (CP) and sugar content. 400m running muscle work time in 43 seconds to one minute, it belong to anaerobic metabolism offer. Muscles necessary energy to finish the work from three types of energy generation: one is high-energy phosphate compound decomposition (20% ~ 25%); two is ferment sugar (55% ~ 60%); three is ATP aerobic generation (15% ~ 20%).

In 400m running project, most of body energy athletes need is provided by ATP-CP system. In process of running, muscle movement process 50% or even 60% in view of anaerobic lactic acid metabolization conditions, required energy first provided by adenosine triphosphate and phosphoric acid creatine (ATP-CP). Its features: be synthesized ATP energy from CP anaerobic decomposition, namely lactic acid in metabolites, not produce lactic acid substances. ATP-CP system capacity is limited, and the biggest release energy effect time in 30 seconds, role within time limit, equivalent to more than 10 seconds from athletes run up to 200 meters-270 metres, 400m running end or 270 meters-400 meters distance, human body must rely on anaerobic metabolism system powering from lactic acid. That is glycolysis, Its

W. Zhang (Ed.): Advanced Technology in Teaching, AISC 163, pp. 131–135.
springerlink.com

energy-supply feature: use candy muscle without oxygen, the final ferment products as lactic acid, and can put more quickly, can maintain about 35 seconds. Therefore, if there is enough candy, and can keep muscle contraction ability quickly under produce ferment lactic acid, improve body tolerance lactic acid, plays a decisive role of 400 meters result quality.

Therefore, 400m running ATP-CP system is mainly, glycolytic metabolic system supplemented competitive sports items. 400m running to athletes, the key of improve performance to grasp the following four points: (1) strengthen ATP-CP system powering ability; (2) shorten lactic acid anaerobic metabolism system generation time; (3) to slow down lactic acid anaerobic metabolism system offer speed; (4) improve body lactic acid ability and eliminate lactic acid ability.

3 400m Running Training Methods and Means

3.1 Improve Phosphoric Acid System Ability Training

That improve body lactic acid metabolize ability. Improve phosphate system function ability training aims to increase ability of rapid outbreak athletes, enhance human body early motion start and speed not yet the largest sports ability, improve human movement absolute speed and speed reserve and improve body ATP-CP content of biochemical.

400m running speed endurance training should be adopted close to special distance or super special distance and high strength, density large batch run and repeat run method, etc. Use sports time less than 10 seconds ultimate tensile strength repeat training and intermittent training method, physical conditions well relatively, not great exercise, intermittent arrangement must be reasonable, highlight exercise intensity.

Specific training means mainly:

1) (Interval run up (2°-3°) 60m) x 6 x (2-3) groups: 1 minutes rest between group rest 5 minutes;
2) Downhill run 80m x 6 times x 2 groups: speed by 80%-90% 1 minutes rest;
3) 20m x 6 times x 4 group (1-1.5 minutes practice intermittent, group intermittent 8-6 minutes);
4) 50m x 4 times x 3 groups (practice intermittent 3 to 5 minutes, group intermittent 8 to 10 minutes);
5) 80m x 3 times x 2 group (practice intermittent 5-6 minutes, group intermittent 8-10 minutes);
6) (60m blunt run + 60 meters go) x 8 times x 2-3 group (group intermittent 8 minutes);
7) 100m intermittent running. Arrange a group of four 100 meters three to four groups. Every 100 meters intermittent 30 seconds, decide strength according to athletes level and different stage of training;
8) Run steep hillside 10 x 30 meters, rapid jump rope 10 x 10 seconds, enhance explosive force development and muscle contraction speed.

3.2 Improve Glycolysis System Ability Training

Namely improve body lactic acid anaerobic metabolism ability. Glycolysis system function rate after phosphoric acid original system, but offer time is more than the latter, so it is main source of energy. Adopt intermission training method or repeat training method that time limit strength and the ultimate strength controlled in 11 seconds-80 seconds. Practice distance should choose mainly 200m- 300m, training intensity should not be very big, intermittent time controlled strictly.

Improve ability of body produce lactic acid is the key to improve ability of lactic acid energy supply, increase lactate tolerance, meantime strengthen ability of body eliminate lactic acid.

1) Enhancing biggest lactic acid content

Great strength of sports training can improve ability of body produce lactic acid, to achieve high levels of blood lactic acid level, repeat practice limit load. The strength of training should be master in the greatest strong 70% above, usual training methods:

- coming-and-going run 100m x 4 times x (2-4) group; Practice intermittent 20-60 seconds, group intermittent 8-6 minutes;
- (150m run + 50m go) x 4 times x (3-4) group; Group intermittent 8-12 minutes, Speed 70-80%;
- Ultimate strength run of 150-300m x 4 times x 3 groups, rest intermittent for 3-5 minutes, group intermittent 8-6 minutes;
- Change rhythm 150-180m x 4 times x 3 groups run quickly + slow run + run quickly, rest intermittent for 3-5 minutes, group intermittent 8-6 minutes;
- (200m run + 200m walk) x 3 times x 5 group (practice intermittent 1 minute, group intermittent 6-10 minutes);
- 200m x 8 times x 1 group (practice intermittent 1-3 minutes);
- 300m x 6 times x 1 group (practice intermittent 1-3 minutes);
- 60-200m high run (4-10) times (practice intermittent 4-10 minutes).

2) Enhance ability of lactic acid tolerance

A long time glycolysis powered result is substantial accumulation of blood lactic acid, and blood lactic acid concentration increases will lead to acidosis, affects nervous muscle feel ability of acid material incentive, reduce working ability of brain cells, or even accelerate emergence and development of fatigue. Increase lactate tolerance training is to human body can bear pain of lactic acid poisoning, improve athlete's pain threshold, enhance human body tolerance to injury, pain and acidosis.

The best method of improve body resistance lactic acid ability is speed endurance training, strength should be more than 90% of biggest oxygen, practice time about 1 minute. The main means:

- 300-600m run, repeat 4-10 times;
- Variable speed running. Such as 200m fast + 200m slow, 200m fast + 100m slow + 300m fast + 100m slow;
- 200-300m x (10-15) times intermittent run can be used training before, middle period, strength greater than 90% of fastest speed, intermittent 1-2 minutes;
- (100 + 200 + 300 + 400 + 300 + 200 + 100)m x 3 combination run can be used training before, middle period, strength close to top speed.

3.3 Improve Body Aerobic System Powering Ability Training

Although aerobic metabolic output power minimum, take scale also 5% in the 400-meter run, but it is foundation under load and ensure that athletes sports health. Commonly training method for used training method. Commonly used method:

1) Intermittent run 800-1000m x (4-6) times, practice intermittent 3-4 minutes;
2) 6-10 km cross-country race (heart rate for 150 times/points-160 times/points);
3) Circulation practice 6-8 stand x (15-30) x 6 group (including general physical training, auxiliary physical training or special strength training, practice intermittent 30-60 seconds, group intermittent 3-5 minutes);
4) 8-10 km cross-country race, accelerate run distance for 100-1000 meters, accelerate heart rate for 170 times/points-200 times/points;
5) Sand, hilly land cross-country race which increase difficulty.

3.4 400 Meters Special Quality Training

400m four 100m speed distribution. The most reasonable should not be slower one by one, but the second fastest, followed the third 100 metres and the first 100 meters, the last 100 meters slowest.

3.4.1 400 Meters Special Speed Endurance Development

1) Development glycolysis without oxygen ability with running speed. Adopt 100 ~ 150m run repeatedly, its strength training should be close to fastest speed, intermittent time 1 minute 10 seconds ~ 1 minute 30 seconds is advisable. Don't put too much number at the beginning, with 10 ~ 15 times;
2) Development glycolysis anaerobic ability adopt "short range" or "long distance" run. Through 90% ~ 95% of strength 200m repeatedly run training, repeat 8-12 times, intermittent time 2 minute 30 seconds is advisable. Also can adopt 85% ~ 90% of strength 600m repeatedly run training, repeat 3-5 times, intermittent 4-5 minutes;
3) Combination with trapezoidal run development glycolysis anaerobic ability. (50m + 100m + 150m + 150m + 100m + 50m) x 4-8 groups, or (100m + 200m + 300m + 200m + 100m) x 3-5 groups, walk the same distance as a break when each paragraph run out.

3.4.2 Development 400m Special Strength Endurance

1) Barbell training method.Special technical training after barbell strength training; jump practice after barbell strength training and then do some quick rhythm run practices; running practice fast rhythm after barbell strength training;
2) Artificial resistance training. Coaches or companion provide appropriate resistance against to practice of athletes according to need , to improve muscle strength of purpose;
3) Intermittent run (200m + 200m) x (3-5) groups (practice intermittent 1 minute, group intermittent 6-10 minutes);

4) Intermittent run (200m + 200m) x (3-5) groups (practice intermittent 1 minute, group intermittent 6-10 minutes); Intermittent run (300m + 300m after driving leg running) x (3-4) groups (practice intermittent 1 minute, group intermittent 10-15 minutes);
5) Intermittent run (100m x 4) x (3-5) groups (practice intermittent 1 minute, group intermittent 8-10 minutes);
6) Uphill gearshift to run 150m x 6 x (2-3) groups (practice intermittent 1.5 minutes, group of intermittent five minutes).

4 Conclusion

Phosphate system powering 20%-25% when 400m running, glycolysis ferment 55%-65%, aerobic powering accounted for 15%-25%, so training should be combined with its characteristics and related powering important way in sports level, reasonable arrangement of training means proportion. In addition, according to the 400 meters reasonable rhythm to design training program targeted. The most important is close to combination of special technical training, or it may backfire.

References

1. Huang, Q., Hu, X.: 400m running famous coach Clyde Hart from American training thought probing. Sports Culture Guide (March 2010)
2. Wang, W., Hao, J., Zhao, C.: 400m running rhythm and training. Training encyclopedia
3. Hui, Z.: 400m training were discussed. Education Tittle-Tattle (144) (2009)
4. Ou, J.: Thinking of 400m running. Teaching Innovation (2009)
5. Hou, F.: Man 400m running airframe energy supply characteristics and training methods. Shiyan Technical College Journals 17(1) (2004)
6. Bu, R.: 400m running training. Sports Research 25(4) (2004)
7. Wang, W.: High school sports 400m training methods are discussed. Heilongjiang Technical Information (22) (2009)
8. Yan, S.: Discussing 400m running special strength training methods and principles. Shandong Talents Academy Journals 6(2) (2010)

Research and Practice on the Blog-Assisted Audit Teaching

Luqiang Yang

Department of Accounting
Zhengzhou Institute of Aeronautical Industry Management
Zhengzhou, Henan Province, China
yangluqiang525@126.com

Abstract. The characteristics of the audit subject require that students read, exercise, communicate, and practice more. However, traditional audit teaching is low in capacity and lack of interactivity, which reduces the efficiency and effectiveness of teaching. Information technology, especially the blog, is a good solution to the problems in the initiative and interest of student learning. This article analyzes the characteristics of the audit subjects and its teaching requirements. Compared to the current situation of teaching audit, it puts forward the idea of the blog-assisted audit teaching, and then it discusses in detail the audit-teaching function of the blog. Finally, it proves by practice that it is feasible and effective to apply blog to the audit teaching.

Keywords: the blog, audit teaching, informative teaching, platform.

1 The Requirements and Current Situation of Audit Teaching

A. The characteristics and teaching requirements of audit

Audit is a kind of managerial behavior of political activities and economic life, and science of auditing belongs to Management Science. Audit should address the problem of accountability relationship formed from social, political and economic life, so the science of audit belongs to applied science and social science as well. The basic properties of audit make it possess coexistent properties such as theory and practice, marginality and comprehensiveness, rationality and sensibility, dominance and recessiveness.

The embodiment of the above mentioned properties must be emphasized in the audit teaching. Audit science needs supports from a multidisciplinary system. In the training process of compound audit talents, the school should provide conditions for the growth of compound talents, develop scientific, high-quality training programs and curricula, and offer related courses and detailed teaching. Audit has put much emphasis on rationality and audit science has a complete logical structure, therefore students should read widely around audit, focus on training their own comprehensive quality. The audit teaching should focus on practice, and promote to learn about audit and to improve one's ability through practice. Therefore, there should be a large number of cases in teaching, have laboratory practice, and also have internship experiences of actual audit programs outside school.

W. Zhang (Ed.): Advanced Technology in Teaching, AISC 163, pp. 137–142.
springerlink.com

In addition, students should also do a large number of exercises, especially the practical operational exercises, to improve perceptions on audit and the professional judgment ability of audit. If you want to really master audit, you have to go through long-term practice and summary of experience. The auditing learning process need to read more, see more, communicate more and practice more.

B. Current situation of audit teaching

Since the audit curriculum itself has a wide coverage, involving multi-knowledge areas such as management, psychology, accounting and so on, there exists widespread difficulties in the organization of contents, organization and implementation of experimental teaching, and evaluation of teaching effectiveness. The current audit teaching often focuses on classroom teaching, basically confined to explaining of the audit theory, and the teaching methods are confined to infusion. Limited by the learning hours, case teaching is not deep enough, and the examples we give usually center on minor details of audit. The chance for students to practice is also few. Therefore, most of students think the concepts of audit are too abstract to understand, they don't know how to apply them even if small parts of students can understand them. The initiative and interest of students haven't been developed, and the ability hasn't been improved.

We have been pondering on such a problem: how to enhance the guide to students in teaching so as to improve the learning initiative and interest of students, to fully mobilize their potential to obtain new knowledge and technology, and to cultivate their sense of innovation. It has been proved by practice that it is far from enough to only rely on theory teaching in class to acquire a subject. Today, with the increasing development of information technology, to fully mobilize various resources to serve for audit teaching is the road we must follow. The development of network information technology has offered technical supports for solving this problem better, especially the blog which has risen in recent years. The current social environment marked by digitization makes the "non-textbook" teaching resources more and more abundant, and it's easier to access these resources. The borderless nature of the learning scene and the emergence of hypertext and superlinear information media also promote us to look on the audit teaching with new eyes. With regard to how to make students master audit courses, the author thinks that we should establish a "practice-oriented, student-centered, resource-cored" teaching idea, and create a audit teaching model which is based on web environment. Paul • O • Chak Koski, who is the former president of Information Industry Association of America, pointed out that people who has information literacy are those who can use a lot of information tools and main information sources to solve practical problems. The application of information technology has provided broad space to carry out audit teaching activities.

2 Theoretical Discussion on Blog's Application in Audit Teaching

A. Introduction of blog

With the development of network information technology, blog is the forth way of network communication following the birth of E-mail., BBS, ICQ / QQ. "Blog" appeared in the United States in 1998, and became popular in 2000. It originally means visit log files of visitors on web server, and means webpage which is used to express personal ideas. People apply different network space online, and make full use of various network technologies to freely publish and manage their own resources, including articles, essays, collections, albums, etc. And arrange the content according to the reverse order on release time, and keep regular updates. This form of published articles on the network is the blog.

B. The function of audit teaching in the blog

Its application and popularity have provided a new tool and platform for further promoting the reform of education and teaching. Blog can filter and reconstruct information, and quickly transform the fragmented invisible information into explicit knowledge that has use value when the Internet became the ocean of information and knowledge. Compared to the traditional personal website, the advantages of Blog lie in zero technology, zero time lag and zero cost. More and more educators introduce it into network teaching, and it becomes a new network application model which is after information teaching models such as courseware, resource library and education theme website.

As a resource and tool of audit learning, Blog can achieve the objective of audit teaching well, improve the content and methods of audit teaching, and realize teaching informationization. The functions of blog in Audit teaching are as follow:

1. Blog is the precious resource library of Audit course

Blog is an excellent platform for knowledge management, which people can use to carry out knowledge management such as knowledge collection, knowledge extraction, and knowledge sharing and knowledge application. The functions of being easy to use, classification of a variety of documents, retrieval and searching which is provided by Blogs, can facilitate the construction of teaching resources. Teachers can provide some reference links through blog log or in the blog, transmit information that has been filtered to the students, which can help students absorb valuable information, expand the learning resources, and greatly improve the efficiency for students to access information.

The current situation of audit teaching is that the capacity of classroom teaching is relatively small; the capacity of teaching materials is also limited; the content of textbooks is relatively simple and concentrated; there are few introductions on some of the related causes and consequences and the background information. Audit educators can make use of the blog as their teaching tool, establish a complete teaching system for their own courses, collect all materials that are related to course teaching, such as reference materials, courseware handouts, experimental materials, learning experiences, development history, related anecdotes, industry news, emphases and difficulties discussions and exercises. In this case, courses will turn out to be very rich and full. And teachers can also display their own research direction, hobbies, research results, papers for the students and stimulate their interests, so that students who are interested in will participate in. It can greatly increase the interest and efficiency of student learning. Blog fully support RSS (Really Simple Syndication) technology. Teachers can subscribe student blogs, the expert blogs, and all kinds of news via RSS. The link function of Blog can save a variety of information, webpage and websites that are related to the course build a virtual library, achieve the expansion, improvement and digitization of curriculum resources.

Blog contributes to creating a digital learning environment, creating active learning scenarios, and ultimately achieving the purpose of improving the learning.

2. Blog is the air classroom of audit

Interactivity is one of the characteristics of Interpersonal communication exchange. The effective communication between teachers and students is the strong momentum to promote students to improve learning and form independent thinking. However, in our current situation of audit teaching, classroom instruction lacks interaction seriously, due to the classroom is restricted by time and space. The teaching activities should be “advanced in teaching and learning" almost become "indoctrination". And "information" flows from teachers to many students in one-way has reduced the efficiency and effectiveness of teaching.

Blog is an open platform with strong interactive function which provides exchange platform of audit teaching activities for teachers and students. Audit teachers can use this platform to put the learning task and learning requirement on their own teaching blogs, and organize learning activities. Students can focus on the theme which teachers have released to do some work and publish the theme learning log. While teachers can instruct students' theme exchange activities, publish theme teaching logs, and answer the problems which are rise in students' log, instruct them to collaborative learning, and evaluate their performance. When learners want to express their ideas, they must experience logical thinking and imaginable thinking, so blog use cultivate learner's intellect to a certain extent. Blog can be used as a homework submission system. Students can use blog as a management tool of learning materials and learning outcomes, collecting materials, recording the learning process, completing the exercises and homework, and communicating with many teachers and students. With the help of blog platform, they gain much more knowledge and resources. It records the learner's ideas, inspiration and experience in details, and makes the exchange reach to a higher level.

In addition, the blog can also serve as a tool for teachers' reflection; teachers can record their teaching experience and successful instructional syllabus design and so on, and reflect their own ideas and teaching methods continuously to improve themselves. Students can also put forward their views on teachers' teaching via blog, and come up with suggestions of teaching which should be improved. What's more, students can interact with teachers and classmates on emotion or life problems anonymously. In short, the blog communication platform is an effective way to promote exchange and communication between teachers and students.

3. Blog is an audit conference room

The cooperation and exchange between teachers and students on the blog is completely equal, and they can communicate extensively and deeply. Here, someone can put foreword his or her views, and the others can also carry out discussions around this theme and promote discussion-based learning. Teachers can release some research problems that need learners to answer through thinking, and students can find interested topics, find inspiration from other people's thinking and research, and focus on the research questions. The learners is determine by thinking on their own when they answer questions, the depth of this kind of communication is obviously much different from the effect of passive learning.

Collaborate learning can also be conducted through the blog, namely, teachers pose topics via blog, the students finish it in coordination way. In the process of collaborative learning, learners organize learning by a group or team, group members work collaborate is an integral part of the achievement learning objectives, the organization of blog which is order by time has reflect the process of collaborative learning and record the process of member's learning and thinking. So as to facilitate members to reflect, review and refine the results on their whole learning process.

The teachers of same discipline or same major communicate with each other usually and extensively can make the young teachers grow and mature quickly, and make excellent teaching ideas and methods be promoted and popularized timely, facilitate the knowledge share and enhance teachers' professional competence.

4. Blog is an archive of audit course

Blog can conduct a longitudinal management by time as a vertical axis and a horizontal management by classification as an abscissa axis and a depth management by group as a vertical axis. It is very easy for teachers and students to sift, manage, search and share the knowledge and resources. Information of course can be sifted and retrieved by using blog to increase the use value and efficiency of information. The basic principle of blog management is the process of accumulation, sharing and exchange. And the content in the blog is the essence that formed from deep consideration and impact of ideas and condensation by teachers and students' knowledge and experience. Through the blog, teachers can record the research results, teaching cases which have significant effect, successful teaching methods and teaching strategies at any time including the experience and notes into the theme knowledge base at any time too.

3 The Practice of Blog Asssting Audit Teaching

The "audit "is a specialized course of our accounting, finance management department. This course with the basic concept, the basic method and the basic procedures of modern audit for the frame structure elaborates the basic ideas and basic methods of modern audit. It introduces audit procedures and methods around accounting statements audit. This course is the necessary premise to begin further studying professional audit course, such as "the internal auditing ", "performance auditing". The junior began to learn it. Through the study of this course, the students should be able to master the basic theories, basic audit learning principles and basic methods, the practice operation methods of financial statements, and understand the present development situation of the modern audit and its trend. Perfect accounting, financial management and professional knowledge structure of students can lay the foundation for students engaging in related financial work. But due to the broad contents of this curriculum, numerous students, miscellaneous major, students' differences result in the schedule inadequacy. During the class, the teachers can only teach basic theoretical knowledge. Therefore, the practice guidance is difficult to carry out deeply. The contents involved with the development of present situation, development foreground of this course update quickly, so the students are difficult to get the latest news and begin deeply thinking combined with reality. Therefore, it is necessary to organize the students targeted to do cooperation and discussion work according to the task. Many teaching function of the blog provide new ideas and methods for solving these problems.

This study selected parts of our senior undergraduates as the research object. The blog carried out the audit course application research in practical teaching of the first semester of 2009-2010 year. The teaching textbook is finance public audit which is published by Northeast University and the editor is Professor Liu Minghui. In the teaching activities by use blog assisted course, it has expanded Audit learning resources and channels of learning resources, generated Audit learning resources, and formed the model that teachers and students build together, organized theme learning and carry out group collaboration by use blog. In addition, teachers offered online guidance and students helped each other online by applying blog for learning evaluation. The subject-based learning and collaborative learning methods in specific blog mainly applied to promote collaborative inquiry learning of students. It has been demonstrated that the network as an assistant tool, has enhanced learning effectiveness of Audit course and improved the efficiency of the teaching in practice.

We may predict optimistically that blog has the extremely broad possibilities in the audit education teaching information management and it also has important application value in the future of the audit education teaching.

References

1. Shi, A.: The Subject Attributes of Audit and Teaching Requirements. Audit & Economy Research (7) (2005)
2. Wang, A.: Development Teaching—The New Idea of Teaching University Audit. Consume Guide (12) (2009)
3. Gao, L.: The Design of Teaching Activities of Audit Case Discussion Teaching Under the Network Environment. Occupation (3) (2010)
4. Deng, G.: The Research of Blog Science Applied to Education and Teaching. Hunan Normal University (2009)
5. Schwartzman, R.: Refining the Question: How Can online Instruction Maximize Opportunities for All Students. Communication Education (1) (2007)

The Relation between Magnetic Field Strength and Ultraweak Luminescence*

Li-xin Li

Department of Physics, Dezhou University
lilixin116@163.com

Abstract. The paper studies the impacts on the ultraweak bioluminescence and POD activity of the mung beans at the different intensities of the magnetic field for certain time through experimental. The results show that the magnetic field strength will boost the luminous intensity and POD activity of the mung beans. When the magnetic field strength is 0.6 T, the performance of the activity of mung beans POD and the luminous intensity is the most obvious. But when the magnetic field strength is more than a certain range the luminous intensity of the mung beans weakens. This shows the increasing of the magnetic intensity of the beans has played an inhibitory effect. At the same time as the magnetic intensity of mung beans over-increases, activity of the POD does no longer increase, but declines. External stimulation can affect plant leaves on the use of solar energy capacity; and the increase of surplus- energy of the leaves can inhibit the light-emitting and affect the photosynthetic performance of leaves.

Keywords: ultraweak biophoton emission, luminous intensity, POD activity.

1 Introduction

Delayed luminescence emission, a unique ultra-weak luminescence phenomenon of living organisms, is a light-emitting phenomenon which is maintained for certain time by organisms exposed to the outside light (electromagnetic field, etc.) for a period of time. It takes place when the organism continuously change materials, energies and information with the outside world and it is the important way for the light to absorb and emit. It is closely related with differentiation of organism cells, signal transmission, value-added control, and the internal sequence of organism. Delayed luminescence is a window to react the function of organisms[1-3].

Biomagnetism is a Frontier Science on study the relationship between magnetic fields and biology. We can get the information about biological macromolecules, cells, tissue and organ structure through the bio-magnetic research. we can understand the performance and the role of bio-magnetic in material transport, energy conversion and transmission of information during life activities through it. There are a lot of research about the effects of magnetic field on organismsmagnetic at home and abroad. They obtained some valuable conclusion[4-7]. The magnetic field stress is one of the important factors to affect plant growth. Research has shown that plant seeds after

* This work was supported by The Foundation of De Zhou university.

W. Zhang (Ed.): Advanced Technology in Teaching, AISC 163, pp. 143–148.
springerlink.com

treatment by the magnetic field, can affect their the growth and metabolism such as Photomorphogenesis and photosynthesis etc[8]. Experiment has confirmed that magnetic treatment can stimulate the growth of plant roots and hypocotyls ,It also can Increase in mitotic indexand root vigor. Magnetic field effects can provide more energy for the life of plants and it also can provide a number of intermediate production for the important biosynthesis of plants which is benefit for the formation of seedlings and growth of new organs[9-11].The experiment takes mung beans as material to measure the impact of different intensity of magnetic field stress on the leaves' delayed luminescence. It can tell the impacts of the internal physiological of leaves with different magnetic field without damaging the original organizational structure of biological samples, and it can provide certain reference to the study of the impact of magnetic field stress on plant leaves by bio-physical method same as in the MS Word font size points. Times New Roman is the preferred font.

2 Experimental Equipment and Methods

A. Experimental Equipment

Ultra-weak luminescence was determined by the BPCL luminescence-based measuring instrument, which was developed by Chinese Academy of Sciences biophysics institute. Its structure diagram is showed in Figure 1.

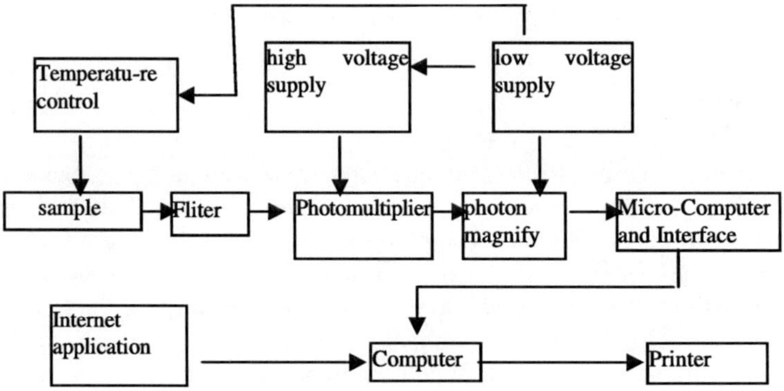

Fig. 1. Diagram of BPCL

B. Experimental Material

Select the fresh-harvest, plump-eared and uniform size mung bean as experimental material.

C. Experimental Method

First, select six groups of mung bean (Each group includes 200 particles) in the Petri dish with water immersion 5 ~ 6 hours, and then process the mung bean on different field strength of the magnetic field for 20 minutes. After each measurement six groups of petri dish will be added pure water and then be put into the calorstat settled to 28 °C. Measure the bud length, luminous intensity and POD in turn of six groups mung bean in petri dish at 8:00,14:00,19:00 everyday. Each time select five from each group to measure the bud length with a ruler and the luminous intensity with BPCL-4. When measuring the POD, randomly select 2 from the 5 to weigh their fresh weight, and to obtain enzyme by grinding, centrifugation.. Add the quantitative drugs and enzyme into the test-tube in turn to get the solution. There are seven test-tubes (one of them is taken as a comparison). Keep them under the fluorescent lamp in a half-hour while avoiding the interference of extraneous light. When the colors become blue, measure their absorbance . The total test need to measure eight times, and is done in three days

3 Experimental Results and Discussion

A. The Change of the Bud Length Stimulated by Different Magnetic Field Strength with Time

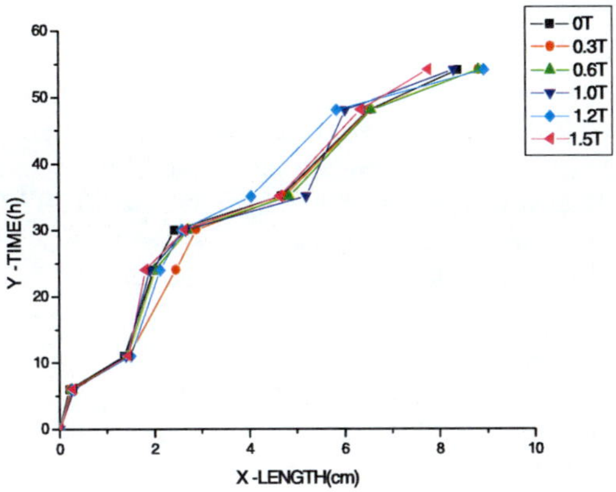

Fig. 2. The change of the length of bud of the seeds under different magenetic treatment with different time

It can be seen from Figure 2: The Growth Status of mung beans is not show significant difference at the beginning previous measurements with the different magnetic field. The bud length with H=0T is in a middle level on t=48h and t=54h. The Growth Status of mung beans treated by magnetic field show significant difference. When H=0T, 0.6T,1.0T, the bud length is longer. The bud length is becoming smaller as magnetic field strength continues to increase.

These results indicate that a certain magnetic field strength of stimulation can promote the growth of mung bean, but when the stimulus intensity exceeds a certain range will inhibit their growth.

B. The Change of Luminous Intensity Stimulated by Different Magnetic Field Strength with Time

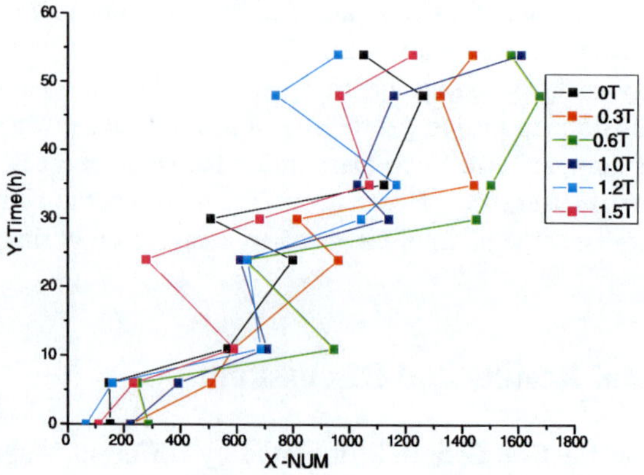

Fig. 3. The change curve of seed luminous intensity stimulated by different magnetic field strength with time

It can be seen from figure3: luminous intensity of mung bean during germination treated by different magnetic field strength shows a certain regularity. The luminous intensity of mung bean with no treatment and treated by H=1.2T, 1.5T is smaller; while the luminous intensity of mung bean treated by H=0.3T, 0.6T, 1.0T is greater. It is just like the curve shows the intensity is lower at the both sides and higher in the middle. The highest luminous intensity appears when H=0.6T.

These results indicate : Luminous intensity of mung bean during germination treated by different magnetic field strength is remarkably affected. The luminous intensity of mung beans increases as magnetic field strength increases at beginning. However, when the magnetic field continues to increase, the luminous intensity of the mung beans are no longer increasing, but are decreasing. This shows a certain magnetic field strength of stimulation can promote the luminous intensity of mung bean, but when the stimulus intensity exceeds a certain range will inhibit their luminous intensity.

C. The Change Curve of Absorbency Stimulated by Different Magnetic Field Strength with Time

It can be seen from figue4: The light absorbency of Mung bean has an obvious difference at different time of growth. In the measurement the light absorbency of

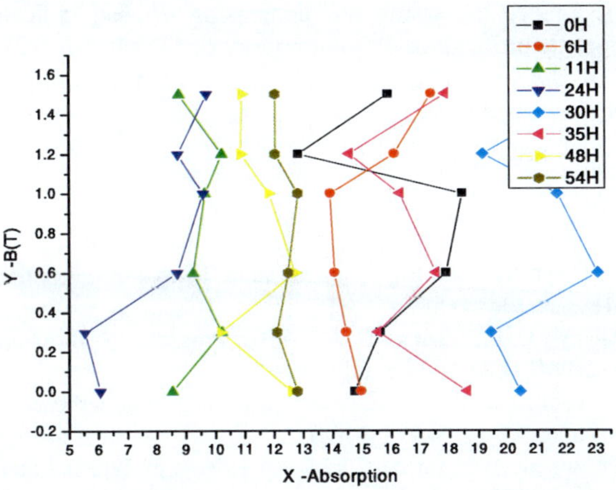

Fig. 4. The change curve of absorbency of light stimulated by different magnetic field strength with time

each group after treatment by different magnetic field also shows a significant difference. When t = 0h, the highest light absorbency of the two groups are the group with H= 0.6T and the group with H=1T. When t = 6h, the light absorbency of each group is in the same level, showing no obvious regularity. When t=11h, the light absorbency of the middle groups is larger; while that of the group with no treatment and the group treated with H=1.5T by magnetic field is smaller. When t=24h, the light absorbency treated with H=1.5T by magnetic field is the biggest. When t=30h , the light absorbency treated with H=0.6T by magnetic field is the biggest. When t=30h , t=48h and t=54h , the light absorbency treated with H=0.6T by magnetic field is the biggest. The POD activity varies directly with the absorbance of biological tissue, that is to say, the absorbance of sample is larger, the POD activity is stronger. The light absorbency in test have been obvious affected by the magnetic field, which shows the POD activity also has been affected by the magnetic field. A certain magnetic field strength of stimulation can promote the POD activity, but when the stimulus intensity exceeds a certain range will no inhibit their POD activity.

4 Conclusion

From the analysis about luminous intensity and POD activity during germination of mung bean treated by by different magnetic field strength, it can be proved that the response of mung beans to stimulate from outside magnetic field is obvious.The results of experiment show that : a certain magnetic field strength of stimulation can promote the growth, the luminous intensity and the POD activity of mung bean .When H=0T, 0.6T, 1.0T, the bud length is longer. The luminous intensity and the POD activity of mung bean treated with H=0.6T is most noticeable , but when the magnetic field strength exceeds a certain range , the luminous intensity of mung bean is weakened. It shows that the uninterrupted increasing magnetic field will inhibit their luminous intensity. With the increase of the magnetic field, the POD activity of Mung bean is no longer increasing but decreasing. It also indicates that external stimulate can affect the utilization of solar energy of the plant leaves, and can let the surplus- energy of the leaves increase so that inhibit the emitting and affect the photosynthetic performance of leaves.

References

1. Van Wijk, R.: Bio-photons and Bio-communication. Journal of Scientific Exploration 15(2), 183–197 (2001)
2. Lv, K.-C., Zhang, C.-P., Zhang, G.-Y.: THE Progress of Biophoton study. ACTA Photonica Sinica 26(12), 1123–1129 (1997)
3. Li, S.-S., Zhu, Y.-B., Liu, S.-H.: Study on Ultra-weak Luminescence of biology. ACTA Laser Biology Sinica 7(3), 223–225 (1998)
4. Mao, W.-H.: Magnetic field on the Biological Effects of Wheat Research Progress and Application. Journal of Triticeae Crops 17(1), 51–53 (1997)
5. Xu, A.-Q., Ding, X.-H., Jing, Y.: Experiment For The Trenting Seeds With Magnetic Field. Journal of ShanDong University 33(1), 58–62 (1998)
6. Yi, Y.-L., Zhang, D.-G., Xie, X.-H.: Effect of Seeds with Magnetic Field on the Biological Characteristics of Wheat. Journal of Shen Yang University 32(5), 333–338 (2001)
7. Li, A.L.: Effect of Gradient Magnetic Field on Growth of Stem Pearls of Dioscorea opposita During Seedling Stage. China Journal of Chinese Ma Teria MEDICA 25(6), 341–342 (2001)
8. Cheng, Y., Liao, Y., Chen, Z., He, X., Chen, Y.: Summary on Responses of Crop with Magnetic Field Treated. World Sci-tech R & D 26(6), 57–59 (2004)
9. Xi, G., Fu, Z.: Study on the Effect of External Magnetic Field on Seed Germination and Growth and the Mechanism Underlying It. Physics 22(10), 610–614 (1993)
10. Xia, L.H., Zhang, C.Y.: Effect of magnetic treatment seed of tomato on seed viability and seedling growth. Journal of Northeast Normal University (3), 69–74 (1999)
11. Xi, G., Yang, Y., Xue, S.: Research on the Effect of Difference Upon Germination and Growth of Wheat with Different Resistance in the Treatment of Magnetic Fields. Agricultural Resesch in the Arid Areas 13(2), 86–87 (1995)

The Application of Virtual Reality Technology in Teaching Reform

Lin Xiaoying and Su Xianbo

School of Energy Science and Engineering, Henan Polytechnic University
Jiaozuo, Henan Province, China, 454000
Lxy2002199@126.com

Abstract. Virtual reality (VR) technology, as a new teaching media, has been widely applied to teaching in colleges and universities. This paper introduces the concept of VR technology and features, and analyses the application of VR technology in teaching reform in higher education from several aspects, such as teaching methods, teaching contents, teaching measures, teaching management, and so on. And the advantages of VR technology applied to teaching are discussed. The application of the technology is favorable to improving the independent innovation capacity, self-learning ability of students, at the same time, be able to make up for shortage of teaching conditions.

Keywords: Virtual reality technology (VR), teaching concepts, teaching contents, teaching measures, teaching management.

1 Introduction

Since the 80's of 20th century, virtual reality (VR) technology has rapidly developed with the support of multi-media technology, man-machine interface technology, sensor technology and artificial intelligence and other disciplines [1]. It has been applied in many aspects such as economic, cultural, military, and people's lives, and so on. VR technology has become a hot topic the international scientific community concerned in recent years. At the same time, VR technology also has infiltrated the field of teaching at the rapid speed, and provided an excellent opportunity for the development of education. VR technology has been successfully employed in educational applications [2-6]. For education, VR technology can clearly express ideas of three-dimensional space that will enable learners directly and naturally interact with the various objects of the virtual environment, and participate in the development of changes through various forms, thereby get the most freedom controlling and operating throughout environment [7]. This paper will discuss the application of VR technology in teaching reform.

2 Virtual Reality

Virtual reality can be broadly defined as the use of computational methods to propel users into a multimedia environment that simulates reality. Through the combination of human--computer interfaces, graphics, artificial intelligence, haptic (touch and

W. Zhang (Ed.): Advanced Technology in Teaching, AISC 163, pp. 149–156.
springerlink.com

pressure feedback) technology, high-end computing, and networking, current virtual-reality systems allow the user to become immersed in and interact with an artificial environment. There are four essential components in any virtual-reality simulator: a virtual world, immersion, sensory feedback, and interactivity.

A. Characters of VR

Virtual reality has been defined as "Immersion-Interaction-Imagination" [8]. Virtual reality (VR) is understood as the use of 3D graphic systems in combination with various interface devices to provide the effect of immersion in an interactive virtual environment [5]. In order to allow learners to interact with VR environments, it is necessary to use special interfaces designed to input a learner's commands into the computer and to offer feedback from the simulation to the learner. Examples and functionalities of various commonly used VR interfaces are listed in Table 1[8-9].

Table 1. Examples and functionality of various interfaces

Interfaces	Examples	Functionality
Input devices		
User monitoring	position tracking, videometric (optical) tracking, body tracking (posture and gestures)	Monitoring a users interaction with virtual world
VR navigation	Tracker-based navigation/manipulation(3D mouse), trackballs	Offering more functionality than measuring position/orientation
Output devices		
Visual displays	Stationary displays (projection VR), Head-based displays, Hand-based displays (Palm VR)	Providing visual feedback from the simulation in response to users input
Auditory displays	Stationary auditory displays (speakers-based 3D sound), Head-based auditory displays (headphones)	Providing synthetic sound feedback to users interacting with the VR world
Input/output devices		
Haptic displays	Tactile displays, end-effector displays, robotically operated shape displays	Achieving physical touch and force an virtual object to user operators

(Sources: Burdea & Coiffet, 2003; Sherman & Craig, 2003).

VR technology can simulate a variety of natural environment people longing for, where the real feelings and experiences obtained by vision, hearing and touch. VR has the characteristics of immersion, interaction, construction [8], etc.

1) Immersion

Users can feel it when they are in stimulant environment. The unearthly stimulant environment can make the users not tell the differences between this and reality and for this, the users indulge themselves in the 3-D virtual environment which is created by computer.

2) Interaction

Users can get to know the maneuverability of the materials in the environment and can get the feedback from the environment.

3) Construction

The VR can not only review the environment which exists in the real world, but also construct the environment which does not exist or never happen.

B. Types of VR System

Ivan Sutherland has introduced in a seminal paper the key concepts of immersion in a simulated world, and of complete sensory input and output, which are the basis of current virtual reality research. His challenge was to set the screen is a window through which one sees a virtual world to make it looks real, acts real, sounds real, and feels real [9]. Although it is difficult to categorize all VR systems, most configurations fall into three main categories and each category can be ranked based on the different forms of participation of users in VR as well as various degrees of immersion. These categories include non-immersive (Desktop) systems, semi-immersive projection systems and fully immersive systems [10], which are shown in Table 2[11].

Table 2. Types of VR systems

VR system	Non-immersive VR	Semi-immersive VR	Fully-immersive VR
Input devices	Mice, keyboards, joysticks and trackballs	Joystick, space balls and data gloves.	Gloves and voice commands
Output devices	Standard high-resolution monitor	Large screen monitor, large screen projector system, and multiple television projection systems	Head mounted display , CAVE
Resolution	High	High	Low–medium
Sense of immersion	Non-low	Medium–high	High
Interaction	Low	Medium	High
Price	Lowest cost VR system	Expensive	Very expensive

(Sources: Burdea & Coiffet, 2003; Sherman & Craig, 2003).

1) Non-immersive systems

Users take part in the activity in the virtual environment through the computer screen and they can interact with the virtual reality environment by various input devices, such as a mouse. Users can see the virtual environment in all the aspects by the screen and control the things in the environment. In this system, users can not put themselves into the environment because of lackness of facticity. But this is very cheap, so this system can be used widely.

2) Fully immersive systems

Users can use an input device to interact with the virtual environment, such as a wired glove, the Polhemus boom arm, or omnidirectional treadmill. Because of the advanced equipment, users can indulge themselves into it. The feeling is perfect. However, it is very expensive, so it is not used widely.

3) Semi-immersive projection systems

Many users can interact with each other in a same virtual environment through computer which is created by a service. They can work together to achieve a common goal. The client can be the desktop system or immersion system. [6]

3 The Application of VR Technology to Teaching Reform

VR technology has injected new vigor and vitality to the field of education. Because of its many characteristics, VR technology provides a useful help to education innovation for the education sector [10]. VR plays an active role in promoting the reforms of teaching concept, teaching methods, teaching contents, etc.

A. Changes in Teaching Concepts

With the development of VR technology, virtual classroom, virtual labs, virtual library, virtual campus and so on emerged in the field of education, which injected fresh blood for the development of education, thus promoted changes in teaching concepts.

VR technology can simulate teaching, that is, teachers can build a variety of teaching content that their students need through VR technology, so the students can learn self-study. On the other side, VR technology can also simulate the vivid, life-like learning environment for students, which will help students to gain knowledge from a broad field of subject. The realization of simulation functions have changed the previous teaching methods that "to teach" give priority to the "student-centered" and "teacher-led, student-centered" teaching mode. Through the establishment of the virtual teaching platform or application of VR technology, it mobilizes the initiative of students and is conducive to student learning and reflection. With the use of the interactive feature of VR technology, students can be targeted to teaching guide based on the characteristics of each student, and will make great progress in academic performance.

B. Changes in Teaching Methods

Teaching methods are generalized as the behavior adopted by teachers and students during teaching and learning activities in order to achieve teaching objectives and teaching requirements. Workable teaching methods include lectures, conversation, demonstration, visit, experiment, practice, discussion, reading guidance, exercise and other methods. The above-mentioned methods can not be a good role to play because of limitations of weak teachers, lack of education funding, inadequate space in colleges and universities.

VR technology can be a virtual teacher and counseling, that is, intelligent Agent can be used as virtual teachers in the learning environment, and it can take on the responsibility of "navigation" and "answer" to guide and help students to gain the learning resources needed. VR technology can prevent "information" from filtering and "resources" from losing. It can answer students' issues according to the network teaching resources. The realization of the virtual counseling make learners receive counseling in the absence of teacher and enhance self-learning ability. The emergence of virtual teachers and counseling enable students and teachers to learn, communicate and discuss from the constraints of learning time and place. VR technology will increase the fun and user-friendly color, so as to improve the effectiveness of teaching.

C. Changes in Teaching Contents

According to various teaching programs in different colleges, instruction on the content is different. The teaching conditions are an important factor affecting the organization of teaching content. In many colleges and universities, some courses can not be offered; parts of the course content already offered can not be instructed as a result of limited teaching conditions. Boring, abstract and difficult to understand teaching contents are the widespread phenomenon of college teaching. That will result in that students are not interested in learning and do not understand the contents.

The introduction of VR technology in teaching makes the teaching content have a lot of changes in both the external form and internal structure. The external form of teaching can use three-dimensional virtual scene to simulate reality. The changes not be saw, the objects or dangerous places not be touched, or even the events could not exist in natural world or real life, can be shown through VR technology. In addition, VR technology enable a variety of media information combined, organize and display the content, build a knowledge structure by the way of the characteristics of human cognitive approach during the process the organization of teaching content. This network information organization is a non-linear structure, which can effectively combine the organization of information formation and the diversity, complexity of information content together, and provide students with dynamic, open, structural cognition forms.

D. Changes in Teaching Measures

The traditional teaching measures can not meet the age requirements for education owing to the increasing amount of teaching task and teaching effect demands. VR technology can make the teaching measures to be scientific and cost-effective. VR technology, as a means of direct transmission of information, provides visual, multi-sensory audio-visual materials. It is introduced into teaching as a new type of instructional media, and provides virtual reality scenarios for teaching, so that students feel personally on the scenes, and the enthusiasm of students are mobilized. VR technology provide an interactive platform for teachers and students and create a kind of "self-learning" environment, in which teachers play roles as the main guide, and students obtain knowledge and skills through interaction of the information environment and themselves. In addition, VR technology can vividly show the abstract concepts, principles of teaching, create a "virtual" learning environment to help students grasp the essence of the concepts. VR technology allow students to demonstrate around the assumption, close to or grasp the truth through analysis, synthesis, comparison, induction, reasoning and other high-level thinking skills, which will be favorable for formation of discovery learning styles and cultivation of high-level thinking skills of students. That is one of the important contents of education for all-around development.

E. Changes in Teaching Management

Monitoring and evaluation of teaching quality is the main contents of teaching management. At present, the higher education quality monitoring and evaluation system is mainly from the following parts: evaluating teaching of students, student feedback, inspection of teaching manager, testing period of teaching, attendance classes of leadership, evaluating teaching of teachers, teaching feedback. The implementation of these measures have brought the main problem is spending a large number of human, material and time.

VR technology can be applied to the teaching quality monitoring and evaluation system. Through the creation of a virtual evaluation system, teachers and students can evaluate teaching and open the virtual forum in virtual space. With virtual characters, one can carry out the evaluation of teaching with impunity, which has greatly enhanced the authenticity, objectivity of evaluation and reduce man-made factors. VR technology can create virtual classrooms. The image, sound system can be directly transferred to examination system through the installation of monitors in the classroom. The examination personnel can roam in virtual inter-classroom, and keep abreast of the situation of teaching in each classroom. These random samplings will not only be not aware by teachers and students, but will not affect the normal teaching order. So teachers will aboratively prepare each lesson, and the quality of teaching will be improved.

4 The Advantage of VR Technology Applied to Teaching

A. To Stimulate the Creative Thinking of Students; to Cultivate the Creative Ability of Students

Students learn knowledge by the use of virtual reality systems. Its application has two aspects. First, VR technology can reproduce natural phenomena or the changing course of things that can not be observed in real life, provide vivid, life-like l learning materials for students, and help students overcome their learning difficulties. For example, complex physical phenomena such as nuclear fission, semiconductor conductive mechanism can be demonstrated to the students by the use of VR technology during the process of learning physics knowledge. On the other hand, VR technology can simulate various assumptions put forward by students in the learning process. The assumption result or effect can be directly observed through the virtual system. For example, in a virtual the chemical system, students can combine the different molecules together in term of their own assumptions, so computer will dummy the combinative materials. Through these exploring learning styles, students may come up with new material. The use of VR technology to explore the study will help stimulate the creative thinking of students and cultivate the creative ability of students.

B. To Break the Space, Time Constraints; Cultivate Self-study Ability of Students

The limitations of space can be broken by use of VR technology. The students can observe the internal objects as large as the universe celestial bodies or as small as atoms and particles. For example, students can enter the virtual power plant, inspect running of each generator and the linkages among components, and understand the entire power generation process. That is not match for video media and practicality media. VR technology can also break through time constraints. Some changes that will take decades or even centuries to observation can be shown to the students to observe in a through VR technology. For example, bio-genetic law of Mendel is testified by experiments with fruit flies often will take a few months, but can be achieved in a virtual class.

C. To Make Up for the Lack of Teaching Conditions, to Enable Students to Fully Learn

A number of teaching experiments often can not be set up because of laboratory equipment, laboratory space, teaching financial, and so on in practice teaching. The use of virtual reality systems can make up for deficiencies in these areas, students will be able to do a variety of experiments staying at home and gain the same experience as the real experiments, which will enrich perceptual knowledge and deepen the understanding of the teaching content. In addition, simulation experiments also avoid the harmful effects of the real operation. The experiments that are dangerous or hazardous to human health are anciently replaced by the general way of video, so students can not directly participate in the experiment to obtain perceptual knowledge. The use of VR technology for virtual experiments can remove this concern. Students in a virtual lab environment can safely do all kinds of danger or harm experiments to human. For example: a virtual chemistry experiments can avoid combustion chemical reactions generated, and the risk of explosion.

5 Conclusions

VR technology, as a new instructional media, is applied in teaching, and adds a new direction for teaching display and simulation. With virtual reality, teaching will be more individualized, and teachers and students can operate and control learning environment by new and effective ways. VR plays an important role in the exploration and development of modern educational thinking, improvement of educational technology level, transformation of the experiment environment, optimization of teaching process, and promotion of higher education reform. However, it should be said that VR technology in China is still in its infancy [6], there are still many unresolved theoretical issues and not overcame technical obstacles. In addition, its expensive price, cumbersome equipment, poor graphics and other factors also affected popular application in the teaching.

References

1. Burea, G.C., Coifet, P.: Virtual Real Technology. John Wiley & SonS. Inc., USA (2003)
2. Chittaro, L., Ranon, R.: Web3D technologies in learning, education and training: motivations, issues, opportunities. Computers & Education 49, 3–18 (2007)

3. John, N.W.: The impact of Web3D technologies on medical education and training. Computers & Education 49, 19–31 (2007)
4. Monahan, T., McArdle, G., Bertolotto, M.: Virtual reality for collaborative e-learning. Computers & Education 50, 1339–1353 (2008)
5. Pan, Z., Cheok, A.D., Yang, H., Zhu, J., Shi, J.: Virtual reality and mixed reality for virtual learning environments. Computers & Graphics 30, 20–28 (2006)
6. Rauch, U.: Who owns this space anyway? The Arts 3D VL Metaverse as a network of imagination. In: Proceedings of ED-MEDIA, pp. 4249–4253 (2007)
7. Burdea, G.C.: Haptic feedback for virtual reality. Keynote Address of Proceedings of International Workshop on Virtual Prototyping, France, Laval (May 1999)
8. Burdea, G.C., Coiffet, P.: Virtual reality technology, 2nd edn. John Wiley & Sons, New York (2003)
9. Sherman, W.R., Craig, A.B.: Understanding virtual reality. Morgan Kaufmann Publishers, New York (2003)
10. Gu, B., Wan, H., Yang, L.: Thoughts of Virtual Reality Technology in Higher Education. China Modern Education Equipment (7), 42–44 (2008) (in Chinese)
11. Mujber, T.S., Szecsi, T., Hashmi, M.S.J.: Virtual reality applications in manufacturing process simulation. Journal of Materials Processing Technology 156, 1834–1838 (2004)

The Analysis and Research of the Recultured of the Computer Application Ability of Sports Master in China

Fang Qi[1] and Shiwei Yang[2]

[1] Department of Computer Technology, Tianjin University of Sport, Tianjin 300381, China
qitang0818@163.com
[2] Department of Graduate, Tianjin University of Sport, Tianjin 300381, China
shiwei429@yahoo.com.cn

Abstract. Sports master extremely lack of computer application ability, and it is difficult to adapt to the demand of the sports development for talents in information society. Education can shoulder the responsibilities of recultured of the computer application ability of Sports master. This article is based on characteristics of sports and the "White Paper" on the talents in university requirements of computer application ability, propose that sports master should have computer knowledge and application ability. And based on the characteristics of sports master education, it has constructed the system of little hours and modular elective courses for the cultivation of the computer application ability. Therefore practical program about the cultivation of ability by courses teaching is proposed.

Keywords: Sports master, sports master's education, computer application ability, cultivation, courses system.

As information technology continues to develop in depth, the process of information in all areas of society continues to accelerate. In the field of sports, especially in sports science, sports management, physical education and other aspects, information technology has become the most important work. This requires that relevant personnel engaged in such work should have a strong computer application skill. Currently, these workers are mostly young people, the majority of which are Sports master. Therefore, the computer application ability of sports master does influence the promotion and implementation of the concept of sports science information, management information, education information, which affects the development of entire sports in the information society as a whole.

1 Analyze the Necessity of the Recultured of Computer Application Ability

If education is seen as a system, which is composed of input, output and intermediate dispose process, in that way, input is the basic knowledge of students, and output is the national, social and personal needs or the knowledge and ability to identify, the disposal

W. Zhang (Ed.): Advanced Technology in Teaching, AISC 163, pp. 157–162.
springerlink.com

processes is specific implement process of education. Nowadays what society needs is education-oriented. Essence of education lies in bridging the gap between social needs and knowledge and ability of the training targets. Then whether it has the gap and how much does the gap between computer application ability of sports master and society needs?

First let's look at the output. After searching for 50 sports master graduated in 2004-2008, we have found that those students, who possess the necessary computer skills, most have received more attention at present. After analyzing the demand to be made by the desired employer in 2004-2008, it shows that in 2004, priority requirement in proficiency on the office software; in 2006, in addition to the skilled use of office software needs, it concerns more about web production capacity. At the same time, some scientific research units of sports have made a request of predominating professional research aids. For example, whether master of sports training can edit the game video and extract key frame images or not. In 2008, in sports research, sports information management, physical education information in the computer applications involved, more specific requirements on application ability are needed. For instance, motion simulation, psychological test software development, small-scale information management system development and management, and so on. This shows that as the increasing extent of the information society, the requirements of employer for sports master is limited not only to the requirements of professional competence, and their computer skills, but also to the ability of obtaining information and supporting research capacity have become increasingly demanding.

Next let's look at the input. Sports master are mostly from undergraduate related to majors of physical education. Therefore, information technological education, which is at undergraduate of sports level, has a direct impact on computer knowledge and ability of sports master. In the year of 2008, the first "Computer Basic Course about University Forum", research on computer education for undergraduates is only one paper in the 166 special topic reports and papers for written communication, less than 1%. Because of undergraduate education in the physical education level extremely lack researching on information technology, information technology in undergraduate education level have not its own independent and unique programs and educational materials. The frustration has led to follow the trend of the teaching objectives, whether it has been implemented is not clear, just by not more than 56 hours of teaching content of basic computer course [1], and has always been not to change for a few years. At the same time, students are not enthusiastic because of the irrationality of evaluation methods.Due to the basis of undergraduate education, the sports master lack basic knowledge of computer and use of information technology to carry out targeted secondary research, graduate students are even more difficult.

Look at the disposal process, at present, education stage of graduate students, few schools will set the courses of computer applications to minor course in the graduate education level, in addition to the corresponding computer disciplines. There is almost no chance to make up the defects of knowledge through the courses.

To sum up, there are significant differences among the lower basis points of computer application of the sports master with the increasing demand of the employers. To bridge the differences, it is only completed in the stage of graduated education. Otherwise, it is not only negative for the future employment of sports master, but also will bring obstacles to sports development of China. This process of education, which will cover and improve the computer application ability of sports master, is called recultured of the computer application ability.

2 Analyze the Approach of Recultured of the Computer Application Ability

Recultured of information technology for the ability of sports master has the following two ways: firstly, by course teaching to achieve the goal of recultured. Secondly, in order to achieve the goal of recultured through research and practice in the project, there is an absolute advantage in course teaching for the general improvement of the computer application ability of sports master, However, the direction of their research for students lack specificity, and it is impossible to set undergraduate course teaching for this. Through independent study in the research process of project, although targeted, but for most sports master, their basic computer knowledge and skills are weak, inadequate and difficult. Therefore, to construct the system of little hours and modular elective courses has become the most appropriate way to effective realization of the re-cultivation for the computer application ability of sports master.

3 The System of Little Hours and Modular Elective Courses

Little hours and modular elective courses program is base on 《 Basic Requirements of Computer Basic Courses Teaching for Non-computer Profession in the Colleges and Universities》 , which is from the " White Paper", in line with the principles of little hours, more choices, strong function, large information and high practical to plan. The embodiment of courses is modular form, to maintain the relative independence of modules in the horizontal and maintain the knowledge gradient of modules in the vertical. The demand of the development of sports is a guideline of setting up the courses. The nature of the elective, allowing students to study according to their different researching directions, selected to study. The system of little hours and modular elective courses consist of three modules of 18 hours of courses. Module 1 is a basic skills training modules and its training objective is basic ability to handle multimedia material; Module 2 as a training module of knowledge framework, a creation of static web page and facture multimedia courseware, its training objective is to establish a basic knowledge framework, which is required by the production of static web pages, based on the multimedia material and the concept of basic knowledge framework,which is required by the production of multimedia courseware. Module 3 is a module of basic applications for the database, its training objective is the application of database.The knowledge of three modules laid the basis for scientific research, management and information in education of sports. Although there can be no more time to teaching the knowledge-depth, the course allows students have a comprehensive and summary understanding of the demand of computer applications. When the students need it in the future, they know what to learn and where to start learning.

3.1 The Module of Culture Basic Skills

In this module, students can carry on basic process of the various types of information, related in the study and research in the future and be better understanding of the selection and application of material by the effective use of 18 hours of teaching. Specific teaching objectives and content as follows:

① Grasp the installation methods of many typical Office, graphics, file compression, conversion of file format, audio and video editing softwares such as Offices、Photoshop、 Fireworks 、WinRAR、WinAVI、Video Converter、Ultra Video Converter、Ulead Video Studiod.

② Learn to complete the creation and editing of the article directory by use of Word software; Complete the conversion and insertion of the indicate symbol of sports; Complete the basic mathematical statistics by use of Excel.

③Learn to set the sizes of pictures in Photoshop or Fireworks, learn to the local modification, editing and image synthesis of pictures.

④ Learn to use Flash and other two-dimensional animation softwares, producing animated graphics to indicate—stick picture.

⑤ Learn and acquaint the basic use of some softwares, such as WinRAR、WinAVI Video Converter、Ultra Video Converter、Ulead Video Studio and so on, for example, the compression and uncompressing of files; the conversion of different files format; the basic editing of audio and video.

⑥Acquaint the features of Baidu, Google and Sohu and other search engines and learn quickly search by keyword and the method of quickly download and preservation the necessary materials.

To consolidate and improve the ability of processing word and form by learning the various types of software and learn the editing methods of multimedia material, a more profound understanding of the effect of multimedia material in scientific research and education of sports. To judge the reasonableness of the application based on the principles of selection and application. To achieve the culture of the demand of basic computer applications ability in sports science, sports management and physical education.

3.2 Cultivate Module to Form the Knowledge Framework of the Creation of Static Web Page and Multimedia Courseware

Web production is the basis of the network management of information. Although it was added as an obligatory course of computer basic course that the course of the production of static web in the "White Paper" [2], it is not be implemented in some colleges of sport in China. Because of this, it is still blank that the knowledge and skills of the production of web pages for most of the sports master after graduating. At the same time, the capacity of the production of multimedia courseware becomes more and more important, in terms of the pendent's of the direction of physical education, most of them will become teachers of physical education in the future. In despite of some colleges or universities of sport set up elective courses for students to the production of multimedia courseware in the stage of undergraduate education [1], but for various reasons, the effect of teaching is not satisfactory.

To make up the knowledge, two basic contents in this module are the production of the static web page and multimedia courseware. It is impossible to achieve the target that grasp the two kinds of knowledge and the production capacity through 18 teaching hours. How can students acquire more knowledge and make knowledge more malleable only in 18 hours? Which becomes the most important issue in designing this modular? The best solution is to locate this module to the establishment of the knowledge framework, which allows students to fully understand the production of web page and multimedia courseware. Those enable students to understand the role of various types of software and the relationship between various materials in the process of creating web pages and courseware production. Establish the main function and give attention to the effect.

In the course teaching of creating web pages part, it makes the Dreamweaver web pages as the main structure and the application of the shaped function code as complimentary parts, and gives students the understanding of the production of web page. It can tell the students how to use the color, composition of a picture, and the rational use of multimedia elements by comparing demonstration to make the students form a design concepts of clearset framework, full-featured, aesthetic web interface design, and to understand its design method. At the same time, graduate students for physical education, courseware creation also follows the teaching methods of creating web pages, through the interpretation of Authorware software, the most basic use of six icons are given. Through reading a lot of cases, so that students can develop a concept of courseware creation.

3.3 The Module of Basic Application Ability of Database Technology

Database is the basis of information management technology, and it is the core and the foundation of information technology. Currently, any kinds of systematic information management are inseparable from the support of database technology. Therefore, the database was regarded as a basic teaching of computer applications course in one of the five technology applications course in the "White Paper", by the Education Ministry teaching steering committee in of computer science and technology [2]. However, many of our institutes and universities of sport has not implemented the spirit of "White Paper" spirit at present. As a result, it is very deficit that the database technology of sports management master and it is difficult to implement the research of informational management of sport. It is imperative to make up the knowledge of database technology. Database course is a strong theoretical course rather than purely technical course, but as for sports master it is a very difficult task to simply accept the theory, therefore, this module will explain and teach, based on the development and use of a small information management system, built by Access database. By conceptual model of the database and common data model to enable students to master the principles, procedures and methods of relational database systems. At the same time, for the basic database operations, such as insert, delete, modify and so on carry on intensive training, and introduce the functions and operations of index and sorting,so that students master the basic operations proficiently after grasping basic concepts, and knowing the complex operation based on basic operations.

4 Summaries

In today's knowledge society, learning is life long for all walks of life. In the culture of computer application ability, "To give the fish no better than to give the method "; "it is better to teach the way of thinking to learn than to teach the technology". At present, it has become the objective that culture specialized talents of research and teaching and a large number of high-level applied talents in the new era of master education after the analysis and research of the culture objectives of master education. The cultivated targeting of most graduate students must be applied and compound. One expert pointed out that the social demand for high-level talents is diverse, so the graduate cultivation objectives should be diversified. The embodiment of diversity not only at the vigorously development of all types of professional degrees, but also, in the each degree, flexibility to develop goals and set the course according to social needs [3]. It will help to the cultivation of information literacy of sports master that the recultured of the computer application ability of sports master in China, making the students be applied and compound talents in information society [4]. The most reasonable and feasible options and approaches to achieve culture objectives are the construction and implementation of the system of little hours and modular elective courses. It will also be a booster for the reform of graduate education.

References

1. Undergraduate teaching program in Tianjin university of Sport. Office of Academic Affairs in Tianjin university of Sport, pp. 76–204 (2008) (put up)
2. Comments about futher Strengthening the computer basic teaching in Colleges and universities. Education Ministry Teaching Steering Committee of Computer Science and Technology (16) (June 2006)
3. Concerned about graduate education: Training objectives to be clear Degree system should be separate (EB) China Youth Daily (August 2001), http://edu.sina.com.cn
4. Graduate education programs of Tianjin university of Sport. Graduate Departments in Tianjin University of Sport (102) (2008)

Research of PCF for Real-Time Multimedia Services in WLAN

Shuai Xiaoying[1,2], Lian Dong-ben[3], and Huang HaiSheng[2]

[1] School of Computer Science and Technology, Nanjing University of Science and Technology, Nanjing 210094, China
[2] Dept.of Computor Science, Chizhou College, Chizhou Anhui, 24700, China
xyshuai@163.com
[3] Shenyang Institute of Computing Technology, Chinese Academy of Sciences, Shenyang Liaoning 110004, China

Abstract. WLAN MAC layer implements Point Coordination Function (PCF) for real-time traffic. The polling overhead and null packets wasted bandwidth when the polled station has no data to send. To increase the channel utilization and reduce the delay of real-time packets, presents a modified polling scheme. Adaptive polling scheme adopts a dynamic polling list to reduce unnecessary Polls and Null packets, and uses moredata list with extended PCF scheme to utilize the remaining CFP. Unpolled station utilizes real-time uplink protocol to transmit real-time multimedia packets in CP. Simulations show that the proposed scheme reduces the poling overhead and the delay of real-time traffic, and improve the throughput of real-time traffic.

Keywords: WLAN, PCF, Real-time, Multimedia.

1 Introduction

IEEE 802.11 Wireless LAN (WLAN) has become a widely used Internet access technology. The 802.11 MAC incorporates two access methods: the mandatory Distributed Coordination Function (DCF) and the optional Point Coordinator Function (PCF) [1].

DCF is a contention-based protocol, which uses the Carrier Sense Multiple Access with Collision Avoidance (CSMA/CA) protocol for supporting asynchronous traffic. Since wireless stations must content to access the wireless medium in the DCF, the medium access delay for each station cannot be bounded due to high load. Thus, DCF was especially used to support best effort data transfer.

In order to support real-time traffic such as audio and video, the PCF has been included in 802.11 to provide sufficient delay guarantees. PCF is based on a centralized polling procedure controlled by a Point Coordinator (PC) at the Access Point (AP). PC periodically polls stations and gives them an opportunity to transmit packets. PCF provides contention-free service for real-time traffic. The biggest disadvantage of PCF is that a lot of bandwidth is wasted by sending polls and NULL data when stations have no data to send. In order to reduce the overhead and increase the channel utilization, some works have investigated the performance of the PCF when used to support real-time traffic, and modified the standard PCF [2-11].

W. Zhang (Ed.): Advanced Technology in Teaching, AISC 163, pp. 163–169.
springerlink.com

AURA GANZ[4] etc design the SuperPoll, which includes list of station that will be polled during a current CFP. And utilize the chaining machine in which each packet resends in its turn the SuperPoll message appended to its packet to improve the performance of multimedia applications. SuperPoll message append in the header of every sent packet, increase overhead. The work in [5] proposed a modified version of PCF called M-PCF for improving QoS, but it can't resolve the hidden node and Null packet problem.

Another modified PCF was introduced in [6, 7], modified PCF reduces the channel under-utilization due to polling overheads and null packets that occurs in the standard PCF. A. Kanjanavapastit [9] proposes a modified PCF with priority scheme, which reduces the channel under-utilization due to polling overheads and null packets that occurs in the standard PCF, and increases the utilization of channel. Meanwhile, the dynamically polling strategy is adopted to decrease the overhead due to NULL package and CF-Poll when stations have no data to send [10, 11]. These algorithms don't avoid unsuccessful polling. When a voice packet arrives, the station may transmit it in the next PCF. Waiting for the next polling cycle increase the delay, if the voice sent in the PC, which might result in a NULL packet in the next CFP cycle.

To improve the efficiency of the PCF, we propose a modified polling scheme, which consists of the adaptive polling list and real-time uplink protocol. Polling list is adaptively adjusted by the PC, which resumes the remaining PCF by extending polling list. Unpolled station that is not in the polling list and pollable tag is FALSE, it uses real-time uplink protocol to transmit real-time multimedia packets in CP, then adds into polling list. Simulation results show that proposed scheme achieves better performance than the standard scheme in 802.11 WLAN for real-time multimedia applications.

This paper is organized as fellow: section 2 provides a brief description of IEEE 802.11 PCF protocol. Section 3 presents the algorithm of modified polling scheme for real-time multimedia services in wireless local area networks. The performance simulation is demonstrated in section 4. Finally, the paper is concluded in section 5.

2 Overview of PCF

In order to support real-time traffic, the IEEE 802.11 MAC adopts an optional access method called a PCF. The PCF is based on a centralized polling protocol where a point coordinator located in an access point provided contention-free services to the wireless station associated with a polling list. The AP defines a periodic superframe composed of a Contention Free Period (CFP) and Contention Period (CP), as show on Figure 1.

PC controls the medium by broadcasting a Beacon. At the beginning of every CFP, the PC sends a Beacon frame to all stations in the basic service area (BSA) after the AP confirms that the medium is idle for PCF-inter frame space (PIFS). PIFS is smaller than a DIFS period, but longer than the SIFS period. Beacon frame contains the information on the maximum duration of the CFP, beacon interval, and BSS identifier. All stations in BSS set their NAV and not to send any packet in the CFP after receiving a Beacon. During the CFP, the AP polls the stations present in the polling list according to the order and periodicity defined by the scheduling mechanism. The PC sends a DATA+CF-poll frame or CF-poll frame to each station in its polling list. The station responds by sending a DATA+CF-ACK frame if it has data to send or a Null packet (CF-ACK) frame if it has no data to send at that time. If the PC receives a DATA+CF-ACK frame, it can send a DATA+CF-ACK+CF-Poll frame, or CF-ACK+ CF-Poll frame. If a station receives a CF-Poll from the PC, it can respond to the PC with DATA frame or a NULL frame. The PC continues to poll each station until it reaches the maximum duration of the CFP and the PC can terminate the CFP by sending a CF-End frame. The transmissions during the CFP are space by SIFS periods. But PC waits for a PIFS period and moves on to the next station in the polling list if it fails to receive an ACK for a transmitted frame.

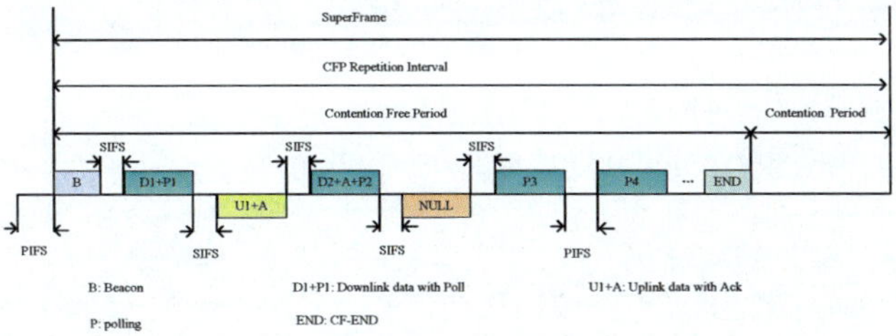

Fig. 1. SuperFrame

3 Modified Polling Scheme

It's well known that sending CF-Polls and Null packets waste a lot of bandwidth when stations have no packets to send. In order to overcome the under-utilization problem we propose a modified polling scheme.

3.1 Adaptive Polling List

1) Remove a station from polling list. In order to count the number of Null packet sent by each station, AP sets Null counters (Ncounter) for each station in polling list, each Ncounter is initialized by 0. If a station transmits a Null packet in the current PCF round, PC adds the correlative Ncounter by one and shifts it into the tail of the polling list in next CFP, otherwise set Ncounter to 0. When Ncounter is achieved or superior to the threshold, PC will remove the station from polling list.

2) Record a station in polling list. When real-time packets such as voice arrived at a station that isn't is polling list, the station transmits the first voice packet to AP in CP in current SuperFrame. When the AP receives the packet, PC records the station in header of the polling list and initializes correlative Ncounter by 0. The station will be polled starting from the next CFP. To avoid unbound-time delay of real-time packet in CP when CP is very congested, we tell the different priority level by different inter-frame space between real-time packets and non-real-time packets, see section 3.2.

3) Extended polling list. The throughput of PCF degrades because of waiting for CFPMaxDuration to expire when there are only few stations in the polling list. To

alleviate the idle time on waiting for CFPMaxDuration to expire, two solutions can be used: to terminate the CFP immediately or to resume the remaining CFP until the CFPMaxDuration elapsed [8]. Terminate the CFP immediately is not an efficient means, we resume the remaining time of the CFP and extend the polling list. We extend the CFP by adding the station into the tail of polling list which has more packets to transmit in current SuperFrame. The algorithm as follows:

```
If (polled_station->moredata==TRUE)
{insert(polled_station, moredata_list);}
If (polled_station==polling_list_last)
{  If (CFP_Duration<CFPMaxDuration)
   { copy (polling_list);
     Append (polling_list, moredata_list);
      }
   }
```

The algorithm of adaptive polling list as follow figure 2:

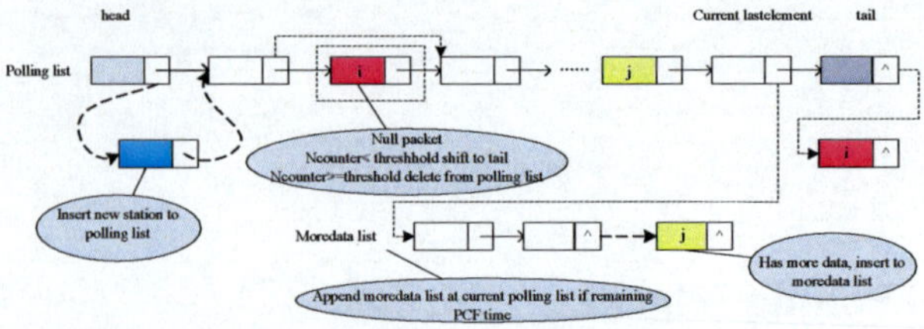

Fig. 2. The algorithm of adaptive polling scheme

3.2 Real-Time Uplink Protocol

1) Every station in the BSS maintains a counter (Null-counter), which is used to count the number of Null packets. When a station is polled in CFP, it transmits Null packet to ACK if there is no real-time packet in the station. The initial Null-counter value is 0 and it is increased by one with consecutive unsuccessful polling. When the counter equals its threshold, the station will be removed from polling list by AP, so the pollable flag in the station is tagged by FALSE.

2) Real-time packets transmitted in CP. When one real-time packet arrives at a station, the station contents the channel to transmit the real-time packet in CP if the pollable is FALSE. The real-time packet is superior to non-real-time packet. To possess minimum time delay of real-time packets, we adopt the different inter-fame space to tell different priority. If the channel is idle for PIFS, the station that gets real-time packets transmits RTS to content channel. The AP receives the RTS then responses CTS to ACK. The real-time station receives the CTS, after waiting for SIFS, it sends real-time packets. The process of real-time transmitted in CP as shown in figure 3. The AP receives the real-time packets, records the station in the polling list. The station sets the pollable to TRUE.

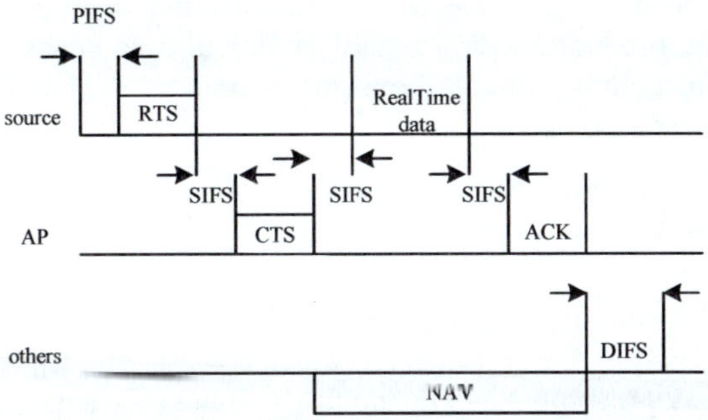

Fig. 3. Process of real-time transmitted in CP

3) Set More Data field. If any polled station that has real-time packets, the station needs to transmit more than one packet. It sets the "More Data" field in the header of the transmitted packet to alert the AP. When the AP detects the "More Data" field in a data packet, the PC inserts the station into the moredata list. The PC knows that there is more real-time data to transmitt, it will not end the CFP if the remaining CFP is not zero. The first station in the moredata list continues sending after the last station in the polling list has finished and then the second etc.

4 Simulation and Result

We evaluated the performance of the modified polling scheme by comparing it to the standard PCF. This paper uses all the parameters for Direct Spread Sequence Spectrum (DSSS) physical layer used in IEEE802.11b.

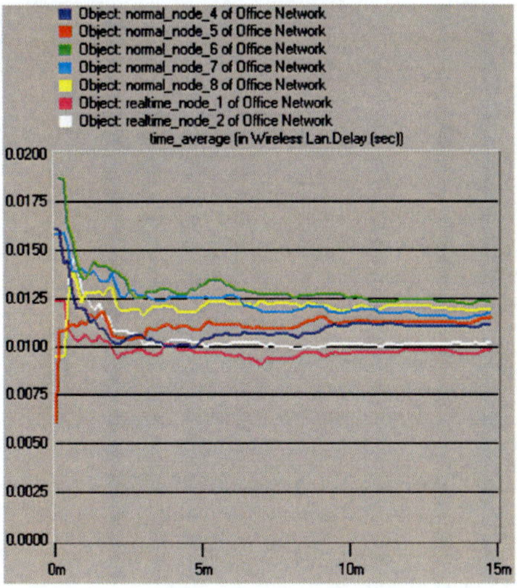

Fig. 4. Average delay

Fig.4 depicts the delay of some normal stations and real-time stations. Normal nodes transmit real-time data with standard PCF, real-time nodes transmit real-time data with modified polling scheme. From the figure, we see the delay of real-time nodes less than the normal nodes.

5 Conclusion

To increase PCF performance of IEEE802.11 WLAN, an efficient polling scheme based on adaptive polling list is presented, which consists of adaptive polling scheme and real-time uplink protocol. Adaptive polling scheme uses a dynamic polling list to reduce unnecessary Polls and Null packets, and uses moredata list with extended PCF scheme to utilize the remaining CFP, alleviates the wasted time in PCF on waiting for CFPMaxDurartion to expire. Station maintains Ncounter to count the consecutive Null packets. Any station which is not in polling list uses real-time uplink protocol to transmit multimedia data in CP. If the station successfully transmits the real-time data, the AP records it in the polling list. Station sets More Data field when has more than one real-time packet to notify the AP. Simulation results show that proposed scheme achieves better performance than the standard scheme in 802.11 WLAN for real-time multimedia applications. The proposed schemes reduce the poling overhead and the delay of real-time traffic, and improve the throughput of real-time traffic.

References

1. IEEE Std 802.11, Wireless LAN Medium Access Control (MAC) and Physical Layer (PHY) Specifications. IEEE (1999)
2. Veeraraghavan, M., Cocker, N., Moors, T.: Support of voice services in IEEE 802.11 wireless LANs. In: Proceeding of IEEE Computer and Communications Societies, pp. 488–497 (2001)
3. Coutras, C., Gupta, S., Shroff, N.B.: Scheduling of real-time in IEEE 802.11 LANs. Wireless Networks, 457–466 (June 2000)
4. Ganz, A., Phonphoem, A., Ganz, Z.: Robust SuperPoll with Chaining Protocol for IEEE802.11 Wireless LANs in Support of Multimedia Applications. Wireless Networks, 65–73 (July 2001)
5. Zhao, L., Fan, C.: M-PCF Modified IEEE 802.11 PCF Protocol Implementing QoS. Electronic Letters, 1611–1613 (November 2002)
6. Kanjanavapastit, A., Landfeldt, B.: A modified point coordination function in IEEE 802.11 wireless LAN. In: IEEE ICON 2003, pp. 561–566 (2003)
7. Kanjanavapastit, A., Landfeldt, B.: An Analysis of a Modified Point Coordination in IEEE 802.11. In: 14th IEEE Proceedings, vol. 2, pp. 1732–1736 (2003)
8. Hsu, M.-C., Chen, Y.-C.: Enhanced PCF Protocols for Real-time Multimedia Services 802.11 Wireless Networks. In: Proceedings of 26th IEEE International Conference on Distributed Computing Systems Workshops (ICDCSW 2006) (2006)
9. Kanjanavapastit, A., Landfeldt, B.: A Performance Investigation of the Modified PCF with Priority Scheme. ECTI Transactions on Electrical ENG., Electronics, and Communications 3(1), 2–9 (2005)

10. Ziouva, E., Antonakopoulos, T.: A dynamically adaptable polling scheme for voice support in IEEE802.11 networks. Computer Communications 26(2), 129–142 (2003)
11. Kawata, T.: Using Dynamic PCF to Improve the Capacity for VoIP Traffic in IEEE 802.11 Networks. In: Wireless Communications and Networking Conferernce, vol. 3, pp. 1589–1595. IEEE (2005)

The Reform and Research on the Teaching of Microwave Technology

Xing Xu, Yue Xuejun, and Lin Tao

College of Engineering, South China Agriculture University, GuangZhou, GuangDong Province
xuxing@scau.edu.cn

Abstract. Microwave technology is a compulsory course for the students of electronic engineering, and it is also one of the most difficult and abstract acknowledged courses for the students. Therefore, it is very important to inspire students' interest, promote their studying efficient, and change their passive ways of studying. In this paper, in order to improve the understanding and innovative capability of students, the reform and research on the teaching of microwave course is proposed.

Keywords: microwave technology, teaching reform, teaching research.

1 Introduction

In modern society, microwave and radio-frequency technology have been spreaded to the every side of our lives. There are a lot of business applications, including cellular phone, Wireless Local Area Network(WLAN), millimeter wave collision avoidance radar in car, communication satellite, global positioning system(GPS), radio-frequency identification technology, ultra-wide band wireless communications, which are closely and inseparably related to the development and research of microwave and radio-frequency new technology.

As the rapid update and popularization of wireless communications technology, microwave and radio-frequency technology will produce a marked effect in the electronic information fields in the foreseeable future.

Microwave is a basic course of electronics and information engineering and is also an important part of our major trains talented people with knowledge system. In this course, students will first meet some new knowledge in the microwave and radio-frequency engineering, such as the theory of distributed parameter circuit、field and microwave network, all kinds of microwave device and so on. These content are too theoretical and too much new and abstract concepts make our teaching difficult.

The teaching result shows that what the students have learned after the microwave course are just a bunch of microwave formulations. Their understanding for microwave still remains on the level of abstract and can't apply the theory to practice which incarnates in thier specific work is lacking of professional thinking and the capability of practice and innovation.

W. Zhang (Ed.): Advanced Technology in Teaching, AISC 163, pp. 171–175.
springerlink.com

In this paper, the author will introduce how to improve the students' interest to study and understand this subject, and innovative capability in teaching of microwave course.

2 Building Up the Foundation

Microwave technology is a course that involves a series of professional course knowledge, such as functions of complex variables, linear algebra and circuits, especially the electromagnetic fields and magnetic waves which is the basic of microwave technology studying. Therefore, if we spend two classes reviewing the theory about circuits、 electromagnetic fields and magnetic waves before the course begin that the knowledge system will more consistency, and the students can lay a firm foundation for understanding microwave technology.

As there are too many theories and formulations, it's not easy for the students to establishes the physical concept and it makes them feel that this course is not practical so that they are not interested in it. Therefore, we need to introduce some background information consciously, like introducing the radio frequency technology and its importance in mobile communication, radio-frequency identification(RFID) and its broad prospect to be applied. In this way, students' understanding and interest can be improved and it also makes the course content advance with times and introduce the latest developments and results to students effectively.

3 Choosing the Key Points of Teaching

In the teaching process of the microwave technology course, we should avoid inculcating students with a lot of new contents and formulations which make them confuse and arise the feeling of being feared. Thus, to help students have a good grasp of the basic of learning and standing, we should focus on the important formulations and contents. For example, in the course of metal waveguide, as the shapes of metal waveguides are different, we can get different formulations from electromagnetic field, but what the important points we should introduce to the students is: the electromagnetic analysis of metal waveguides are based on the basic of Helmholtz equation by variable separation. Different shapes of metal waveguides lead to different boundary conditions, then the electromagnetic expression of metal waveguides is different in different boundary conditions. As a result, the students will have a solid foundation of learning and standing which helps them memorize and analyze different kinds of metal waveguides.

At the same time, the emphasis also should be placed on developing students' thinking capacity in the teaching of the microwave technology. For example, in fact, we can analyze the microwave transmission line accurately by the way of analyzing electromagnetic field, but the analysis for electromagnetic fields so complex that it can't be used in practice, the equivalent circuit model is usually be used which makes the microwave transmission line equivalent to a circuit, and then using the method of circuit analysis to study the transmission line. This is the idea of changing the electromagnetic field to circuit and it's a key point that should be noted in the teaching of microwave transmission line. We can further lead students to extend the idea to their future learning and research. For the study of complex problems, the strategy that we adopt to analyze and study is usually reducing the problem gradually, then excluding the secondary contradiction and grasping the principal contradiction, what we can is the approximate solution of the complex problem at last.

4 Using all Kinds of Teaching Methods Helpful

Microwave technology is built and developed on the basic of the theories electromagnetic fields and electromagnetic waves, as they are invisible and intangible in practice that it's so difficult for students to understand. Therefore, to better understand the microwave technology, a variety of teaching methods need to be used, like using the multimedia technologies and the RF tools of Matlab(a version of 7.0 or higher) to show the contents of the course vividly by the way of animation or image. These contents, such as the travelling wave of transmission line、the field distribution of waveguide、the mixer、the design and simulation of filter, can be vividly expressed by dynamically graphic, which greater improve the understanding and interest of the students.

The teachers should give full play to the initiative of the students and develop interactive teaching, get the students to discuss in class for some key point and then to guide them think and speak. In the meanwhile, encouraging the students to ask the questions that they don't understand or be interested in. Besides, discussing with the teachers by network should also be encouraged, like building a QQ group、providing the e-mails of the teachers, which make the students can communicate with the teachers in advance. These measures can promote the learning activeness and confidence of the students and increase their interest and confidence for taking up a job of microwave in the future.

5 Helpful Hints

The learning and use of microwave CAD software is the direction of the development of microwave course in the future. Students can verify the result of problems assigned in the text readily by microwave CAD software, the feedback will make them confident and know that their efforts in the learning of microwave are rewarded, their interest of learning and design will be aroused. Besides, using the microwave CAD software in class develop their skills which help a lot when they after graduation. So far, a host of software manufacturers come up with special microwave CAD software, but the business edition of these software are very expensive. Fortunately, some software like university teaching use can have a discount, even offer free edition for students. The software we used at present is the ADS designed by AGILENT company, by applying for the use of free trial edition, students can get a month free trial right. Fig. 1 and Fig.2 are the schematic diagram and territory designed by students by using ADS.

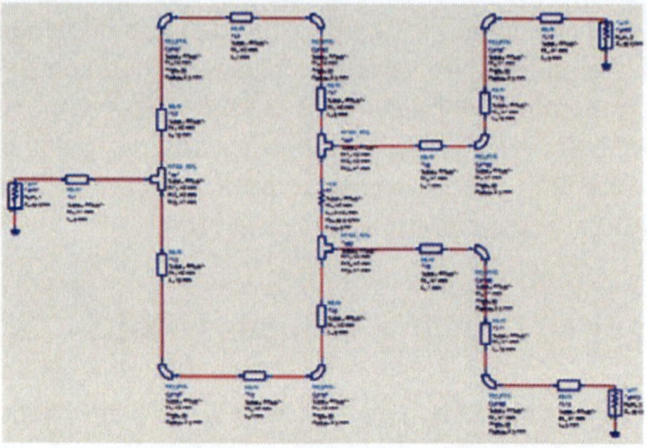

Fig. 1. The schematic diagram designed by students

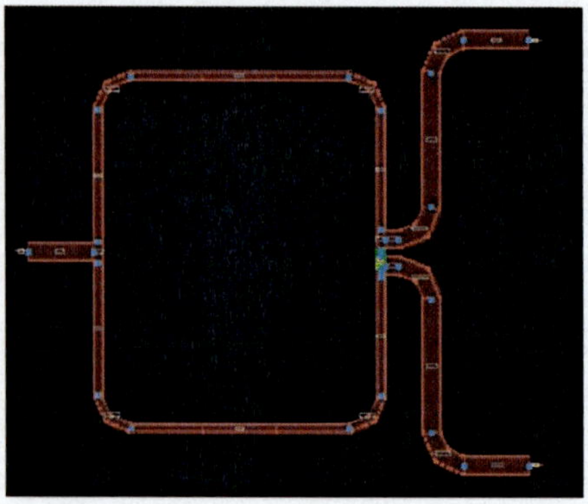

Fig. 2. The territory diagram designed by students

6 Attaching CAD Software

The content of microwave course is difficult and abstract, it's put the cart before the horse if the examination concentrates on the memorization of the contents and formulations, even the students can get 100 points. The examination is better to focus on the understanding and applying of the fundamental concepts of microwave and the design capacity. The teachers can assigned a few designing homework, then the students will study and write some short treatises after they look up some information. In this way, the thinking ability and innovative ability of students can be trained effectively. Not only the exam results but also the usually results should be calculated when judging the results of students, and the final exam should try to choose open－book or Semi-open-book examination. As a result, the situation that students only focuses on remembering formulations and ignoring the fundamental concepts and theories can be gradually redeemed.

7 Summary

In this paper, we introduce some ways and measures to improve teaching quality and deepen the reform. The microwave technology is still in the rapid development, and then the teaching reform will be a long-term hard work.

Our teaching reform and research are still at the beginning stage, more work should be done in the future.

As a microwave teachers, in addition to teaching, we should actively participate in the research of relevant fields and tracking cutting-edge issues. In this way, with the practical experience, our teaching will more interesting, and then the students feel real and kind, the course will have good effectiveness.

Acknowledgment. The paper thanks for the supported of Guangdong College of outstanding young creative talents cultivation project of LYM10035, Guangdong Provincial Agricultural Mechanization Scientific Research Projects, Public sector (Agriculture) scientific research special funds project of 20090323-06.

References

1. Pozar, D.M.: Microwave Engineering, 3rd edn. Electronic Industries Press (2006)
2. Wang, X., Li, P.: Microwave Technology and Antenna. Electronic Industries Press (2003)
3. Wang, Z., Xie, Y.: Discussion on Electromagnetic and Microwave Technology Experiment. Journal of China Education Innovation Herald 456, 128–128 (2007)

The Design of 2.4GHz Microstrip Band Pass Filter

Xing Xu, Yue Xuejun, and Lin Tao

College of Engineering, South China Agriculture University,
GuangZhou, GuangDong Province
xuxing@scau.edu.cn

Abstract. According to the problem existing in the teaching of microwave technology, such as the course was abstract, difficult and complex. The microstrip BPF (Band Pass Filter) design is adopted in the practical teaching of microwave technology, including calculating, soft simulation, optimization, layout design and simulation. The students can easily understand and enhance their professional ability.

Keywords: microwave technology, band pass filter, software simulation, practical teaching.

1 Introduction

Microwave technology is a compulsory course for the students of electronic engineering, and it is also one of the most difficult and abstract acknowledged courses for the students. The experiment is an important bridge that connects the theory and practice, how to change the abstract knowledge to clearly understanding is an important work in the practical teaching of microwave technology.

In this paper, the author will introduce some experience that adopting the microwave circuits design in the practical teaching of microwave technology, including calculating, soft simulation, optimization, layout design and simulation, which makes the students can easily understand and enhances their ability of understanding and creativity.

2 The Principle of Microstrip BPF

BPF or "band pass filter" is a device that allows the passage of frequencies of a certain desired range and rejects or blocks those that go below or above the particular range, which is in contradiction to the concept of band stop filter.

The microstrip BPF is a type of microwave filter which is widely be researched, different kinds of microstrip BPF have different properties. It's one of the important components in modern electronic system.

According to the theory of transmission, each microtrip can equivalent to a short segment of series inductance and parallel capacity. Coupling capacitors and

W. Zhang (Ed.): Advanced Technology in Teaching, AISC 163, pp. 177–183.
springerlink.com

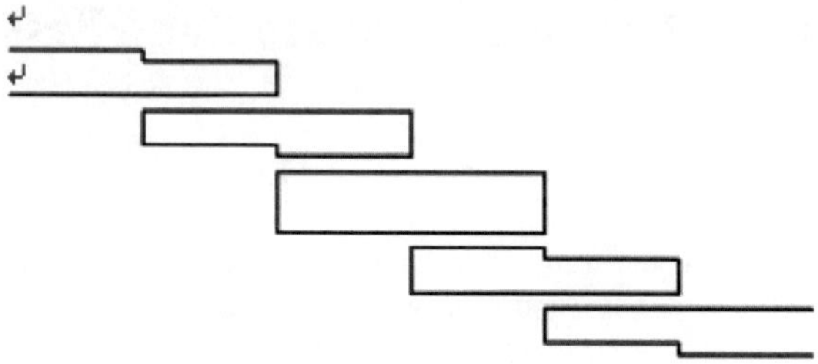

Fig. 1. The typical structure of the microstrip filter

inductances also need to be considered in parallel coupled line. Suppose each microtrip's characteristic impedance is Z0, the length of mutual coupling part is L, the width of microstrip is W, the distance between the every two microstrips is S, even mode characteristic impedance is ZEa and odd mode characteristic impedance is ZO. As single cell circuit can not get good frequency, we can get that by the way of symmetrical cascade as shown in Fig. 1. It consists of connected parallel coupled-lines which are also symmetrical. Each length of coupled-line is about a quarter of wavelength (for the center frequency) and constitutes the resonance circuit.

3 The Design of Microstrip BPF

To match the needs of modern electronic information industry more closely, and according to the future trends, we assign the design specifications for students, as follows:

1. Pass-band is 2.4－2.5GHz;
2. The insert loss of pass-band should less than 2dB and the fluctuations less than 1dB;
3. The loss of below 2.2GHz and over 2.7GHz should more than 40dB;
4. The port reflection coefficient should less than -20dB;

According to the design specifications, we can choose parallel coupled microstrip to constitute the filter which is chebyshev. The detailed parameter design includes the following steps:

1. As the pass-band is between 2.4 to 2.5GHz and the insert loss of pass-band should less than 2dB, choosing the order of the filter N as 7 by looking up the diagram that the chebyshev filter attenuations with frequency.

2. As N=7 and pass-band fluctuations less than 1dB, identifying the value of ripple low-pass filter prototype components are g0, g1, g2, g3, g4, g5, g6, g7, g8.

3. According to the parameter of prototype, writing programs by Matlab and calculating the odd and even mode characteristic impedance of each section. As follows:

ze1 = 61.4509, zo1 = 42.2398,
ze2 =52.2620044800, zo2 =47.9260044800 ;
ze3 =51.8214168050, zo3 =48.3024168050 ;
ze4 =51.7610041800, zo4 =48.3550041800 ;
ze5 =51.7610041800, zo5 =48.3550041800 ;
ze6 =51.8214168050, zo6 =48.3024168050 ;
ze7 =52.2620044800, zo7 =47.9260044800 ;
ze8 = 61.4509, zo8 = 42.2398.

4. According to the odd and even mode characteristic impedance of each section, identifying the microstrip parameter of each section by using the calculating tools of microstrip LineCalc in ADS.

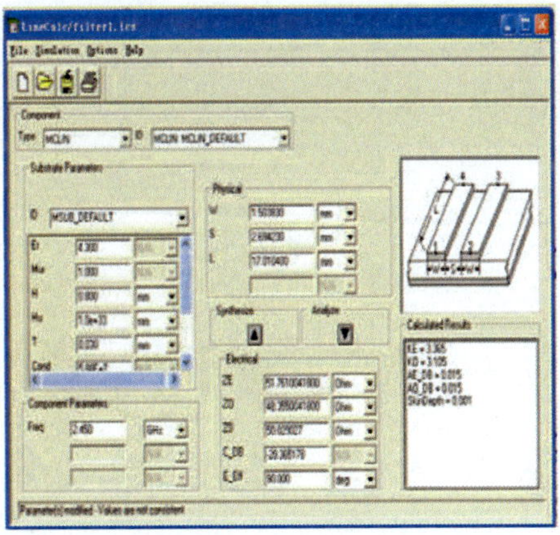

Fig. 2. Calculating the parameters of the parallel coupled

By using the calculating tools of microstrip LineCalc as shown in Fig. 1,we can get the each section of microstrip's parameters W,L and S(As symmetrical, just identifying the previous 4 sections only.). As follows:

W1=1.375, S1=0.514544, L1=17.2257 ;
W2=1.500750, S2=2.21198, L2=17.0049 ;
W3=1.50348, S2=2.6216, L3=17.0094 ;
W4=1.50383, S4=2.69423, L4=17.0104 ;

5. Setting board parameters. We choose the common FR4 as circuit board, its details as follows: H: the length of the substrate(0.8 mm) ; Er: the relative dielectric constant of the substrate (4.3) ; Mur: permeability(1); Cond: metal conductivity(5.88E+7) ; Hu: package height (1.0e+33 mm) ; T: metal thickness (0.03 mm) ; TanD: loss tangent (1e-4) ; Rough: surface roughness (0 mm). From the circuit board parameter and the center frequency, we can know that: W =1.52 mm, L=1mm.

4 The Simulation of Microstrip BPF

We can design and simulate the schematic after calculating all the parameters of the microstrip BPF. Adding every component and its parameters, and then we can get the microstrip BPF schematic in ADS software. As the following figure shows:

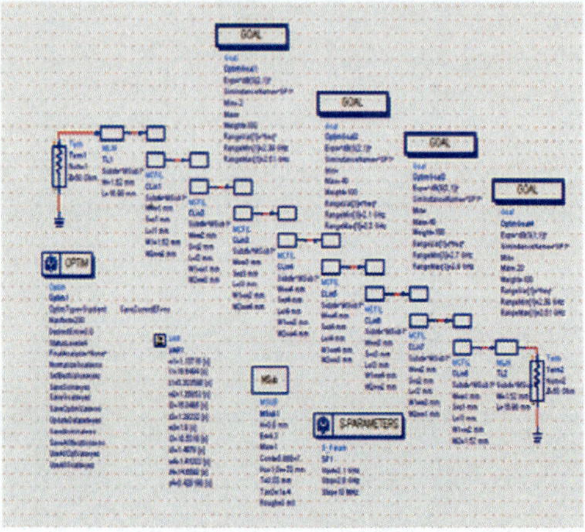

Fig. 3. Microstrip BPF schematic

At the simulation of the schematic, to match the design target, a simulation optimization is needed, which using the W, L, and S we calculate as initial and choosing their ranges as the optimization parameters. Finding the parameter values which meet the requirements by constantly simulating and optimizing, and then we can finish the design. Fig. 4 is the S parameter results of band pass filter after simulation and optimization.

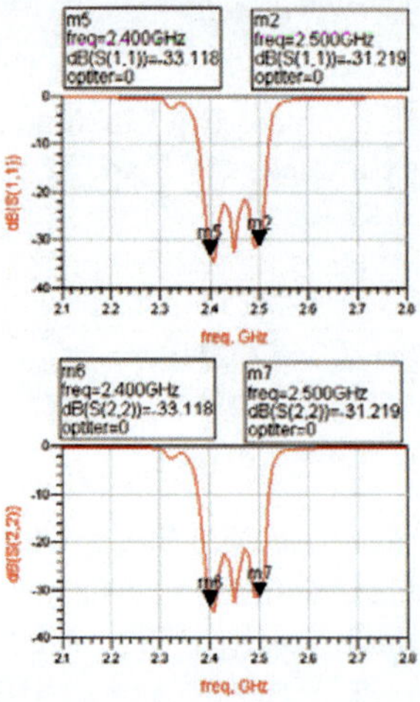

Fig. 4. The simulation and optimization of microstrip BPF

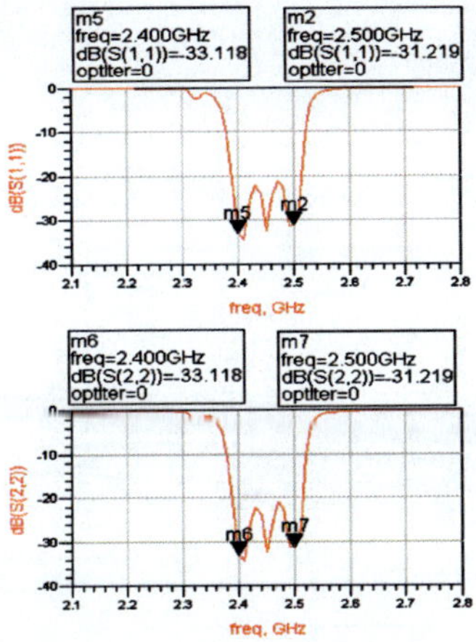

Fig. 4. *(continued)*

From the Fig. 4, we can know that band attenuation is 1.372dB and less than 2dB, which meets the design requirements. At 2.2GHz the attenuation is 71.066dB which is more than 40dB, and below this frequency the attenuation is larger, meeting the design requirements. At 2.4GHz, the reflection coefficient is -33.118dB which is less than 20dB, meeting the design requirements. At 2.5GHz, the reflection coefficient is -31.219dB which is less than 20dB, meeting the design requirements. The reflection coefficient of band pass is also less than -20dB. Therefore, the simulation results of the schematic seem to be very ideal.

5 The Layout and Simulation of BPF

Even thought the schematic design gets a ideal result after many simulations, the result of the schematic design may be different from the result of circuit test in practice. The reason is that we can the design and simulations of the schematic under ideal conditions, but many other factors need to be considered actually, like the influence of the edge capacitance of SMD for electromagnetic, dielectric loss. Thus, layout and simulation need to be done after the design and simulation of schematic so that the design of band pass filter will come quite close to the reality. Fig. 5 is the layout of band pass filter based on the schematic design.

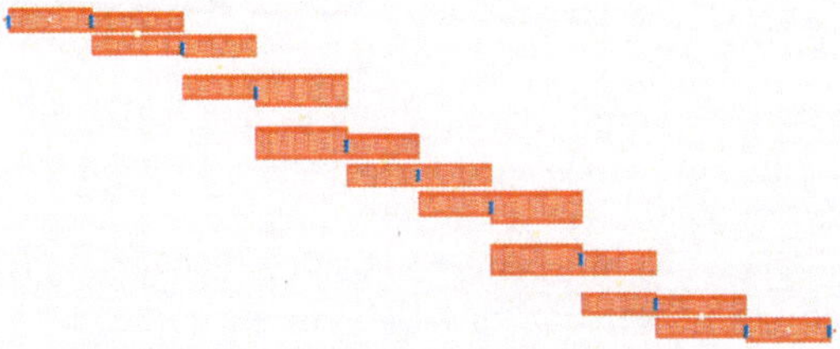

Fig. 5. The layout of microstrip BPF

Simulating the layout by the way of moments, we can get the S parameter results. Its simulation results as follows:

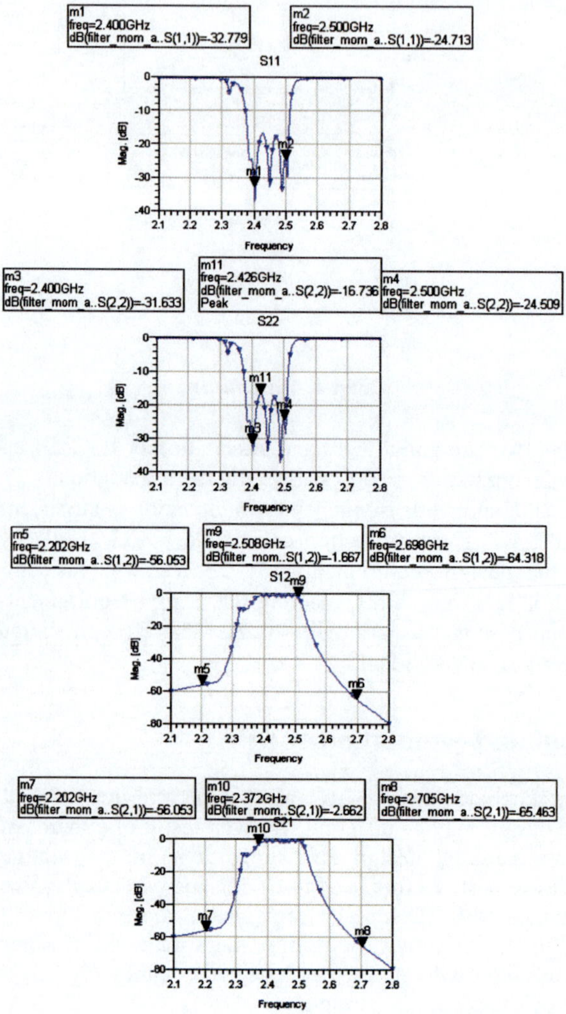

Fig. 6. The layout simulation results of the microstrip BPF

From the Fig. 6, we can know that band attenuation is 0.916dB and less than 2dB, which meets the design requirements. At 2.2GHz, the attenuation is 56.053dB which is more than 40dB, and below this frequency the attenuation is larger, meeting the design requirements. At 2.7GHz, the attenuation is 65.463dB which is more than 40dB, and over this frequency the attenuation is larger, meeting the design requirements. At 2.4GHz, the S11 is -32.779dB and S22 is -31.633dB, both of them are less than 20dB, meeting the design requirements. At 2.5GHz, S11 is -24.713dB and S22 is -24.509dB, both of them are less than 20dB, meeting the design requirements. The reflection coefficient of band pass is also less than -15dB.

From the contrast, we can know that the simulation results of the layout is worse than the schematic, however, the simulation results of the layout still match the design specifications.

What we need to point out is that if the simulation results of the layout can't match the design requirements, we should go back to the design stage of the schematic and modify the design and corresponding parameters.

6 Summary

Microwave technology is a course that is theoretical and practical, what we need to learn is not only the theories but also the practical skills, and the theoretical design, electronic circuit design and the design experience need to be improved. Therefore, the teaching of microwave technology is not only the theory teaching but also the practice teaching, which makes the students easier to understand. In this course, we arouse the students' interest and improve their professional ability and practical ability by such three theories foregoing and design practice.

Acknowledgment. The paper thanks for the supported of Guangdong College of outstanding young creative talents cultivation project of LYM10035, Guangdong Provincial Agricultural Mechanization Scientific Research Projects, Public sector (Agriculture) scientific research special funds project of 20090323-06.

References

1. Xu, X.: ADS2008 Radio Circuit Design and Simulation. Electronic Industries Press (2009)
2. Zhang, X., Luo, F.: Research on Teaching of Radio Circuit Course. Journal of EEE 28(3), 21–23 (2006)
3. Wang, Q., Lin, D.: New Experiment System of Electromagnetic and Microwave Technology. Journal of Experiment System and Management 22(2), 110–113 (2005)

A Report on Cognitive Processing of Chinese Characters between Hearing-Impaired and Normal Children

Zhao Zhang[1] and Zongfu Yu[2]

[1] School of Law, Shandong University of Technology, Zibo, Shandong, China
zzjoy5608@sina.com
[2] Department of Student, Shandong University of Technology, Zibo, Shandong, China
yuzongfu@126.com

Abstract. Based upon the showing three traits of words of Chinese as stimulus, the RT functions for the judgment involved in the verbal recording were compared between the hearing-impaired and hearing children. The results showed that the function of graphic code in visual processing was strongly correlated with recognition of words of Chinese characteristics than the phoneme code and semantic code in both the hearing –disabled and hearing children. For hearing–disabled children, the alternative was the direct access that a semantic code was produced directly from a translation of the graphic code. However, the hearing children used the strategies of grapheme-phoneme conversion rules to interpret the meaning of words. The analysis of various results showed that no significant differences of mental processing were found between the hearing-impaired and hearing children. Perhaps both the direct access and the grapheme-phoneme conversion would be a comprehensive information processing activated each other.

Keywords: Cognitive processing, Chinese characteristics, verbal receding.

1 Background

During the last the years, many cognitive psychologists have turned their attention to phenomenon of reading processes of alphabetic language system and logograph language system. From the view of information processing approach, the reading process can be divided into three states such as the decoding literal comprehension and comprehension monitoring. According to Ehrib(1982) , there are two main decoding processes: One is matching the printed words to a known pattern for the words, which activate the words meaning in long-term memory. Matching is used to recognize words in one's sight vocabulary. The other decoding process is recording which is involved in "sounding out" a word. So psychologists developed some hypothesis of the pattern recognition of speech signals, grapheme-phoneme conversion rules, direct access position and model of mutual activation although these findings have been criticize [1].

The purpose of the present study was to explore the relationships among graphic code, phonemic code and semantic code involved the verbal recording of Chinese

W. Zhang (Ed.): Advanced Technology in Teaching, AISC 163, pp. 185–188.
springerlink.com

characteristics basic assumption of this was that the different strategies would be used by the heating-disabled children and heating children.

2 Method

Subjects:
20 heating-disabled students were selected from the Sixth-Eighth Grade classes of the Xi'an Second Deaf School (10 boys and 10 girls, mean age - 17). They have received the formal audio-oral training fight after entering school study. However, the manual language in their main skill developed and practiced in their daily life. 28hearing students were selected from the Third Grade of Ninth Middle School (14boys and 14 girls, mean age = 15).

Materials:
Stimulus were generated by a computer and all the stimulus materials included 20 standard words and 60 matching. target words and disturbing words were chosen from "the Text Book of Deaf Students" and "Table of Frequency for Chinese Characteristics".

Procedure:
Both heating-disabled and hearing groups followed the some procedure. They were instructed after a standard word showed on the screen at 2 second, the 3target words and 3 disturbing words then would been presented. The task of subjects was to determine whether the target words and disturbing words conformed to the standard word according with three !rails of similar grapheme ,similar phoneme or similar semanteme by pressing the left, middle or fight hand key. For example, the subjects had to make a judgment that whether the target words"头"、"者""手"would be conformed to the standard word "首" .

Results:
Combined data of heating-impaired and hcaring groups used to compute the RT function for information processing were illustrated in Tables 1 and 2. The analysis indicated that there were differences under three conditions of verbal recoding for the heating-disabled groups (F [2.417] =5.66p = 0.044), but the fast reaction was the semanteme recoding the middle reaction was grapheme and the slow one was phonemic. A similar analysis for the verbal recoding of semantic judgment for both hearing-impaired groups and hearing groups did approach the significance level of .05. As Figure 1 and 2 summarized, there were significant differences which could be seen between hearing-impaired groups and hearing groups on the cognitive processing.

Table 1. Comparison of RT on the Verbal Recoding Between the hearing-impaired and hearing Subjects

	Hearing-impaired		Hearing		
Conditions	M	S	M	S	***P***
Similar Semanteme (义近)	734.6	353.3	803.1	360.0	0.05
Similar Grapheme (形似)	747.7	346.0	712.4	293.0	0.084
Similar Phoneme (音同)	846.6	390.2	782.5	305.0	0.091

Table 2. Results of Analysis of Variance for Three Conditions

	Source	SS	MS	F	*P*
	Factor	1499108	749554	5.66	0.004
Hearing-Disabled	Error	79000880	132330		
	Total	8049984			
	Factor	538435	269217	2.62	0.044
Hearing	Error	42819428	102684		
	Total	4335758			

3 Discussion

Comparing the function of verbal recoding of hearing, disabled and hearing children may have important implications for the cognitive psychology. Although it is widely believed that the hearing-impaired children visual processing may be enchanted and the speech of semantic recoding bay be slow because of auditory deprivation. But, the results from the experiment could not support traditional idea and our hypothesis. Why was the TR. of the semantic recoding of hearing-impaired subject shorter than that of the hearing subjects? Why was the RT of the graphemic recoding longer than the hearing subjects?."

First, the findings from the experiment demonstrated that stimuli of words can be coded in a variety of ways of 15honeme, grapheme and semanteme, and the capacity of information operation and depended on the material and strategies choose by the person. As the hearing-impaired person's memory system may be combined more closely than hearing person. Therefore, the hearing-impaired subjects can use the direct access do search the semanteme of word from the graphemic code, while the hearing persons use the strategies of grapheme-phoneme conversion rules to charge the semantic meaning of the verbal recoding. If the RT of the direct access position is shorter than the grapheme-phoneme conversion, the speed of verbal recoding of semanteme in hearing-impaired person would be faster than that of hearing person[2].

If the information stored in memory system has been divided as the semantic network and lexical network, there would be more semantic network than lexical network in hearing-impaired person[3]. Comparing the mental dictionaries, the level of hearing-impaired person may be more simplified than hearing person, so that the hearing-impaired person are easier to search for the semanteme from the mental dictionaries than that in hearing person. In present experiment, the high frequency Chinese characteristics were been used as stimulus, so that heating-disabled children could make the judgment. Next, if we use the low frequency Chinese as stimulus, maybe we would get more data to support it.

As many psychologists noted earlier, the cognitive processing involved in the verbal recoding is a very complex processing, the direct access and the grapheme-phoneme conversion are not taking the place in paralleled, but it is comprehensive information processing activated each other.

4 Conclusions

1. Although the logograph system such as Chinese and alphabetic system such as English and French are different language systems, all the three trails of phonemic code graphitic code and semantic code are involved in the cognitive processing of he verbal recoding. There is no essential distinction for the internal information processing between the logograph language system and alphabetic system, and between hearing-disabled and hearing children.

2. Hearing-disabled children may be strongly dependent upon the graphemic recoding and may use the strategies of the directed access position involved in the verbal Chinese characteristics. But the hearing children are more dependent upon the phonemic recoding and they use the grapheme-phoneme conversion (GPC) rules to interpret the meaning of the words of Chinese such as the Western students.

3. There are different levels of the mental dictionaries between the hearing-disabled children and hearing students. The hearing-impaired students have more simplified mental dictionary than the hearing students.

4. As the characteristics of Chinese united with the orthography-phoneme and semanteme the memory construction would be organized well to enhance the memo ability of materials and some phonemic recoding strategies could been learned by the hearing-impaired children.

References

1. Chan, C.K., Siegel, L.S.: Phonological processing in reading Chinese among normally achieving and poor readers. Exp. Child Psychol. 80(1), 23–43 (2001)
2. Stone, G., van Orden, G.C.: Strategic control of processing in word recognition. Exp. Psychol. Hum. Percept. Perform. 19(4), 744–774 (1993)
3. Carreiras, M., Perea, M., Stranger, J.: Effects of orthographic neighborhood in visual word recognition: cross-task comparisons. Exp. Psychol. Learn. Mem. Cogn. 23(4), 857–871 (1997)

On the Development and Prospect of Multimedia Assisted FL Learning in China*

Dexi Chen

Foreign Language School, Anhui Polytechnic University, Wuhu, Anhui, China
howard19720808@163.com

Abstract. The development of Multimedia Assisted FL Learning has brought about many changes to Chinese FL classrooms since the 1980s. Both the pedagogical approach and the organization of the FL classes have undergone extensive modifications and changes. Based on the facts observed and data collected, this paper reviews the development of Multi-media Assisted FL Learning in China from an analytical viewpoint, in the hope that a more substantial and adequate discussion on teacher-learner relations, technology implementation and so on under the multimedia circumstances would shed light on future inquiry. Finally the paper concludes with suggestive strategies for coping with constantly changing telematic technologies.

Keywords: Multimedia Assisted FL Learning, Present Development, Prospect.

1 Call in Mainland China from the 1950s to the 1990s

From the 1950s to 1980s,CALL in Mainland China arguably has no history. Computers deployed for educational purposes are only a very recent event. Although AV centres were emerging in the late 1980s, they were set up not as self-access centres, but as means of delivering audio-and video-based teachings. So they were not genuinely self-access centres of an analogue generation. In 1993 there was a CALL conference being organized in Beijing Foreign StudiesUniversity, only a dozen participants to be attended. In 1996 the then School of English Language Communication, BFSU, established a LAN-supported self-access centre consisting of 25 networked microcomputers. The resources available at that time were extremely limited, and the access was not genuinely free, but rather carefully allocated.Computer work stations began to appear in major universities, but CALL programs, if installed at all, have remained anecdotal and experimental in nature.

2 Web-Based CALL at the Turn of the Millennium

Mainland China seems to have bypassed the pre-network CALLs and stepped straight way into Web-based CALLs. In this section the writer critically reviews this cyber rush.The upsurge of the latter, as shown later, is closely associated with the social, political and educational situations of the 1990s.

* This work is partially supported by The Education Department of Anhui Province Humanities and Social Sciences Grant #2010sk541.

W. Zhang (Ed.): Advanced Technology in Teaching, AISC 163, pp. 189–196.
springerlink.com

A. The National Initiative: Establishment of Online Education Faculties

Web-based CALLs at the nationwide scale started in 1994 with 4 universities(Tsinghua University, HunanUniversity, Zhejiang University and Beijing Post and Communications University) authorized by the Ministry of Education(the MoE), each of which set up a piloting faculty that would provide e-learning programmes. The 4 universities were handpicked by the MoE to run the piloting faculties in order to gain experiences for the later wide-scale launching, which did take place in 1999. During the subsequent 4 years, 64 more piloting faculties(known as online education colleges, or e-learning institutes, or online education institutes) had been established by the host universities. Business sectors were allowed to be stakeholders providing funding for some piloting faculties[5].This cyber rush was motivated by a desire for solutions to the four major problems that could be identified to characterize the macro social-educational environment towards the end of the last century in China.

(1)There are too many people with too little education---the mass illiteracy problem;

(2) There are too many people with inadequate education---the inadequacy problem;

(3) Schools,colleges and universities are extremely under-funded---"Too many monks with too little porridge"problem, or the insufficiency problem;

(4) University graduates are no longer assigned by the government with life-long jobs---the iron rice bowl-broken problem[4].

B. Technological Infrastructure

Following Targowski [11], I shall use the term telematic technologies as an umbrella term that covers information technology(computers), telecommunication networks(satellite, fiber optics, wire or wireless telephone) and television(video). The term infohighways will be used to refer to the physical infrastructures that transport the multimedia(voice, data, graphics, video) converged signals of telematic technologies through paths of amatrix of networks(e. g. LAN, WAN).Cyberspace, incontrast, refers to"information-based space, that is the dispersed, infinitc constellation of electronic(digital)files, data bases, home pages, bulletin boards, directories,menus, where humans, with a password, interactively navigate in order to create, update, exchange, and retrieve information traces."The infrastructures(.i e. including telematic technologies and infohighways) for the 68 online education faculties are in large part provided by the central or local governments. China Education and Research Network(CERNET), for instance, is a government-sponsored national network that is open to high education sectors all over China. It is to the cyberspace that 68 faculties make their own contributions. It is also the cyberspace where learning and instructions take place. It is impossible here to give a comprehensive review of the cyberspace managed by 68 faculties. All I can realistically do is focus on the courseware design.

C. Six Types of Web-Based CALL Design

There are six types of Web-based CALL design emerging from the cyberspace.

(1)Print-textbook-transfer Design:This refers to the Web-based publication of an existing print textbook verbatim. The design, if any, will consist of changing the print pages to Web pages plus extra navigation buttons. The content of the former print textbook remains intact during the transfer process.

(2)Audio-supplement Design:This refers to the practice of adding audio clips to the print-textbook-transfer design.

(3)Video-supplement Design:This refers to the practice of adding video clips to the print-textbook transfer design. Note that video clips naturally include audio input. As it may have become apparent to the reader, the three designs so far all keep the content of the print-textbook intact. Yet the print textbook was originally written for classroom use, mediated by a human teacher, and designs like these have overlooked the fact that study with a computer is substantially different from face-to-face study with a teacher in the classroom.

(4)Classroom-teacher-model Design:This refers to the practice of video-taping a teacher delivering a lecture and publishing online the compressed video images accompanied by Powerpoint presentations. This design entails an asynchronous broadcast of a teacher talking to an imagined class———in fact:he is talking to a video camera in a studio!

(5)Multimedia-rich Design:This refers to the practice of integrating hyper texts with Web-based multimedia.In this design, the Web-based version is in some places considerably different from the print textbook version. The print textbook version is significantly enriched with still images, audio and video clips, hyper text links, and so on.

(6)Learning-process-model Design:This refers to the practice of selecting and constructing the content according to a projected learner, .i e., a default learner, who is engaged in a goal-directed learning process. Unlike the previous five designs, this design isnotbased on a traditional print textbook. The units and their contents reflecta learning process through which a default learner, together with his or her fellow study mates, undertakes planned study.Activities and specific learning tasks are constructed to reflect real life situations.

3 CALL in Retrospect and Prospect

CALL has been with us for about half a century(.i e.in the global context). Its path has never been smooth. It never has been short of enthusiasts as well as Luddites.Nowadays there are teachers who cannot teach without using telematic technologies in one way or another. There are also teachers who keep clear of them by all means possible. In this final section I shall look at CALL in retrospect and prospect by asking a very basic question: What differences does CALL make to learning and teaching in general?

A. Laws and Principles of Learning

It is a historical fact that learning can take place without CALL of whatever kind. It is also a fact that learners can learn without a teacher. For ease of reference I call it pure learning so that it can be contrasted with mediated learning, .i e. the learner being assisted in learning. Regarding pure learning, there are some fundamental "facts", which might very well be couched in terms of law.

The Law of Process:Learning is a process, which means that to learn something has a start point, an end point, and some time interval in between. This law, an obvious fact, has some theoretical as well as practical implications. Theoretically speaking, it rules out inborn abilities as non-learning, e.g. human babies crying, and sucking as non-learned behavior. Moreover, temporality is considered as an intrinsic property of learning. Practically speaking, learning always means time consumption, which can be very short or can be very long, depending on what is being learned.The Law of Internalization:Since learning is not something inborn(from the Law of Process), it has to be internalized by the learner, e. g. memorization, motor-automatization,etc.The Law of Irreplaceability:Just as no one can eat for someone else, so no one can learn for the other. Put it in another way, no one can internalize learning for someone else.The Law of Erasibility:Learning is not something that, once achieved, can be owned by the learner forever.It can be erased.With the four laws in place, there are three general principles that can be derived from them.The Principle of Temporality:(1) Learning increments with time———the more time one spends on learning,the more the learner learns, hence life-long learning. In view of human mental capacity, there seems to be no point at which the learner s brain is too full to learn. (2) Learning decays over time———human memory and physiological built-up change over time, which affects what has been learned. Non-stop practice or constant use resists natural decay over time.The Principle of Cognitive Resources:(1) The brain/mind is potentially unlimited in intake/output; (2) The brain/mind is limited in intake/output in a given timespace.The Principle of Multimodality:It is a norm rather than an exception that the learner learn multimodally. That is, in a natural learning environment, there are two or more senses of organ being activated and mobilized in learning.

B. CALL as Medium of Learning

Let us come back to the question raised earlier: What differences does CALL make to learning? This question now can be answered by asking another question: In what way does CALL affect the four laws and three principles of learning discussed in 5.1 above? The law of irreplaceability dictates that CALL of whatever kind cannot learn for the learner, which means that CALL is destined to act as an assistant, a learning-enhancing artifact to learning. CALL is therefore destined to be operative in the learner s learning process(the law of process). Does CALL enhance or diminish the learner s internalization? This is one of the areas where CALL is considered to be a two-blaze sword: It can either enhance or diminish internationalization, depending on the way it is designed and adapted to. Take the pint textbook-transfer-design for example. It is found that the majority of learners prefer the print textbook version to the Web-based version. The main reason for this preference is reported to be"notused to it","easy to forget".The reason for those preferring the Web-based version is reported to be"Don't have to carry the book" [6].

Mayer[7]explores the relation between multimedia presentation and learning. Although the study is not directly concerned with CALL, it is highly relevant, since multimedia presentation has become an important component in CALL design. His"cognitive theory of multimedia learning"is akin to the principle of multimodality.

A cognitive theory of multimedia learning assumes that the human information processing system includes dual channels for visual/pictorial and auditory/verbal processing, that each channel has limited capacity for processing,and that active learning entails carrying out a coordinated set of cognitive processes during learning[7].

Elsewhere[2][3]four hypotheses have been laid down concerning the Principle of Multimodality.

(1) It is easier to learn when the input and output are in the same modality;

(2) Given the circumstances, appropriate alterations of modality result in positive reinforcement———it is called modality congruence;

(3) Given the circumstances, multimodal learning fosters better memorization than otherwise, thus enhancing retrieval;

(4) Natural multimodal learning———learning in anatural social situation——— fosters better memorization and enhance retrieval.

It is worth reminding that 4 hypotheses of Mr.Gu's need to be verified through empirical investigation.

The greatest attraction CALL has to both policymakers and educationalists is its capacity of transcending physical space and time. That is, it offers opportunities for lifelong learning and continued learning. So if CALL is said to have any long-term effects on learning, it is its effect on the principle of temporality. The former sharp division between learning first and work afterwards for the rest of life time is seriously undermined by CALL.

So far we have dealt with CALL s relations with all the laws and principles of learning except for two: the law of erasibility and the principle of cognitive resources. Does CALL have any affect on them? Can CALL help the learner recover what has been erased? Can CALL act as an extra cognitive device that enlarges the learner s cognitive resources? I postpone exploring these two questions until I turn to CALL in prospective later in 3.5 below.

C. CALL as Medium of Teaching

What differences does CALL make to teaching? Historically, interest in CALL seems to be originated in CALL s potential in giving instructions rather than in assisting learning. In 1958, Skinner, for instance, published apaper in Science with the title"Teaching Machines" [9]. Ten years later in 1968, he published a book with the title The Technology of Teaching. He observes:Even in a small classroom the teacher usually knows that he is going too slowly for some students and too fast for others. Machine instruction would permit each student to proceed at his own rate[10].A computer capable of delivering machine instructions was enthusiastically envisaged as being able eventually to'teach'a lot of things and to replace human instructors.With hindsight, we now see clearly that he was over-ambitious. Even today, half a century later, with such advancement of telematic technologies as no one had ever dreamed of ,a machine that can replace a human instructor still does not exist!Is it the case that humans still just have to wait for it with more patience? Later I will explore this question again. Perhaps it is not a question whether it could ever exist, but whether it is desirable or not to have such a machine in the first place.

Against the theory of pure learning as depicted above, the role a human teacher plays in the learner's learning is similar to that of CALL, that is, to help the learner learn. As just no teacher can eat for the learner, so the teacher cannot learn for the learner. As simple as that! Over the years, a pattern of print-textbook-driven learning has emerged as the dominating mode. A(series of) textbook(s) is chosen as representing the learning object/content. What the teacher does is prepare him/herself and"teach"it to a classroom of learners, lesson by lesson, week by week, and so on. The fourth design———classroom teacher model design———discussed above, is actually a Web-based delivery of such teaching. This is not a genuine case of computer-assisted language teaching(CALT). It is more an extension of a traditional classroom teaching from a physical setting to cyberspace. A genuine case of CALT is what is known as integrated mode of teaching. Roblyer[8] has this to say about integration:Perhaps the most important———and the most difficult———challenge is for teachers to identify specific teaching and learning problems that technology can help address or how it can create important educational opportunities that did not exist without it. As part of this process, teachers decide what they need to make these changes occur. This process of determining where and how technology fits is known among users of educational technology as integration(italics original).Roblyer is mainly concerned with the use of telematic technology in campus class at primary and secondary education. He provides teachers with a whole range of integration strategies. Tomei's[12] edited collection of papers is also on integrating telematic technology into the classroom, but with focus on tertiary education.

Integration touches upon a more deep-seated issue of the relation between learning process and teaching process.Nowadays learners may find themselves having access to two learning environments: the traditional campus environment(TCE) and the virtual learning environment(VLE).In the TCE, they have two general modes of learning:face-to-face classroom learning with a teacher, and self study with the resources materialized in the traditional media(.i e. print books, audio and video cassettes). In the VLE, on the other hand, the learning resources are first of all materialized in digitalized media, and made accessible via the computer and the Internet. Learners access to the resources can be made inside the classroom with a teacher, or outside the classroom with or without a facilitator, .i e. facilitated mode of learning, or free self-access mode of learning.

How is this CALT related to other resources in VLE? Let me illustrate this issue with a specific example. There is a CALL program on teaching English phonetics to false beginners. Both the teacher and learners have access to it. The classroom integration may result in repeating what learners have already learned with the program. In my field interviews with some teachers Iwas told that the teacher felt that they had nothing else to do, as if their job of teaching had been pre-emptied by the course ware. The crux of the problem lies in the relation between teaching process and learning process: Should the teacher embed his or her teaching process as part of the learner s learning process, rather than the other way round, the problem can easily be solved. But this means that the teacher has to have a clear picture of at which stage his or her students learning is. Learning process monitoring becomes crucial. Tome the teaching process embedded in the learner's learning process is a genuine step towards a genuine learner-centred education, and this is where telematic technologies have a lot to offer[1].

D. Future CALL

It is very difficult to predict what future CALL will look like, because the technology has been developing so fast that even technology makers themselves become bewildered. David Denby observes:"... Are we unwittingly putting ourselves in thrall to a system that will dazzle us with choices yet dislocate us, pull us apart, even consumeus?" [8]. What I shall do here, instead of prophesying what changes will be, is explore what cannot be changed, and suggest some strategies to cope with the unpredictable changes.

(1)"Change is the only constant,"says the Roman philosopher Lucretius. Telematic technologies are here and stay, and will keep changing. So the best thing we can do is take changing as representing a normal state, rather than something annoying or worrying. I may call it an attitudinal strategy.

(2) Telematic technologies have become part of our lives, and will continue to be so. They have become an object of learning themselves. Both teachers and learners have to learn them. In this regard, the difference between the teacher and the learner will be the difference between who learns first.

(3) Aspointed out by Roblyer [8], many educators now believe that the world is changing too quickly to define education in terms of specific information or skills; they believe education should focus on more general capabilities such as"learning to learn"skills that will help future citizens cope with inevitable technological change.Educators believe that knowing what questions to ask and how to ask them will be as important as, or more important than, giving the"right answers."

(4) Technologies may eventually reach the stage when they can do anything people want them to, but this capacity will be irrelevant. What will be at issue is whether it is desirable to employ them when, where, how, to whom, and for what purpose.

(5) Finally, to conclude this review, Roblyer's words are worth heeding: He urges us to make"our recognition that'technology is us'".... we must recognize the truth of Peter Drucker sstatemen:t"The best way to predict the future is to createit."Both individual teachers and teaching organizations must see themselves as enlightened shapers of our future.Each teacher must take a position as a"script writer,"helping to articulate the vision for what the future of education should look like; each should acquire skills to help work toward realizing that vision[8].

References

1. Gu, Y.: Online education in China: Problems and solutions. Innovation and Practice in China Higher Distance Education (special edition) 1, 64–75 (2005b)
2. Gu, Y.: Exploring Situated and Distributed Learnings:A Comparative Approach. Plenary speech delivered at GloCALL 2007 at Hanoi University, Vietnam (2007c)
3. Gu, Y.: Anatomizing multimedia andmultimodal learning. Computer-Assisted Foreign Language Education 2, 2–12 (2007b)
4. Gu, Y., Hall, C., Hall, E.: Using the Computerin ELT: Technology, Theory and Practice. Foreign Language Teaching and Research Press, Beijing (2006)
5. Kang, F., Song, G.: e-Learning in higher education in China: An overview. In: Spencer-Oatey (ed.), pp. 11–32 (2007)
6. Lu, Y., Gu, Y.: An ecological model of online education: An empirical investigation. Computer-Assisted Foreign Language Education 6, 17–24 (2006)
7. Mayer, R.E.: Multimedia Learning. Cambridge University Press, Cambridge (2001)

8. Roblyer, M.D.: Integrating Educational Technology into Teaching. Shaanxi Normal University Publishing Press, Xi'an (2005)
9. Skinner, B.F.: Teaching machines. Science 128, 969–977 (1958)
10. Skinner, B.F.: The Technology of Teaching. Appleton-Century-Crofts, New York (1968)
11. Targowski, A.S.: Global Information Intrastructure. Idea Group Publishing, Harrisburg (1996)
12. Tomei, L. (ed.): Integrating Information and Communications Technologies into the Classroom. Information Science Publishing, Hershey (2007)

Integration with JavaEE Framework to Build Tourism E-business System

Zhang Li[1] and Zhang Wei Xi[2]

[1] College of Computer Engineering, Jiangsu Teachers College of Technology, Changzhou, 213001, China
zhangli_3913@yahoo.com.cn
[2] Electronic & Information Engineering Department, Jiangsu Teachers College of Technology Changzhou, 213001, China
zwx@jstu.edu.cn

Abstract. Considering the disadvantages of traditional development methods for JavaEE applications, based on the system analysis, an architecture integrating Struts2, Spring and Ibatis frameworkswas proposed. A multi-tier E-commerce system for tourism was rap idly built and developed using the architecture.By practical application, it shows that new features of technique used in system improve security and stability of system. This framework technique not only make a good use of resources but also make the project developing to be compact and clear in flame. Besides good expansibility and maintainability also are provided by this kind of system.

Keywords: tourism e-business system, Struts2 framework, Spring framework, Ibatis framework, rapid build.

1 Introduction

Tourism e-commerce, is the use of the Internet, the use of computer technology, communications technology and business buying and selling network China Unicom to form a new type of commercial activity. Its features include advertising, ordering, payment, customer service and financial accounting of the use of the Internet and many other business development activities. As the complexity of the internal logic of tourism e-commerce, security, demanding, business development and changes in the form of fast, JavaEE Web application based on its level, platform independence, security and scalability of e-commerce system has become [a] major the solution. The traditional development model is to show business and high-coupling data, resulting in software is difficult to reuse. Based on the analysis of existing JavaEE technologies and the development of tourism e-commerce technology based on the status of proposed and discussed based on Struts2, Spring and Ibatis travel e-commerce platform architecture. This program addresses: ① business layer and data persistence layer logic, making operation easy while reusing the data rate is conducive to the maintenance of the code; ② reduce the network load pressure on the underlying database structure information is not exposed, can be achieved portability; ③ simplify the reuse of components, making the system reliability.

W. Zhang (Ed.): Advanced Technology in Teaching, AISC 163, pp. 197–204.
springerlink.com

2 System Architecture Design and Description of Functions

2.1 System of Business Processes

In the business process design, we follow the customer-oriented, focusing on the tourism website into a customer service platform, analysis and decision-making platform, the starting point of these three marketing platform, the overall business model shown in Figure 1: The business model reflects the corporate marketing, service and sales activities of the business relationship between the flow (flow between the data); analysts including business analysts, marketing analysts, sales analysis and service analysts; service personnel, including subscription service, distribution service, Advertising sales and service personnel; different business functions of e-commerce systems analyst in the corresponding data for analysis, the results of the analysis information to the appropriate personnel, and then by the appropriate staff for business processing activities, and will handle the information feedback to the travel e-commerce system. The first is the specific process is to provide customer service platform to provide customers with basic travel information and advice through customer service and information platform viewer (client) and the result of the interaction between tourism website feedback to the analysis and decision-making platform. Second, the analysis and decision-making platform travel through intelligent knowledge base, integrated smart travel search engine personalized content and information browsing, browsing & user-friendly anthropomorphic, and other customer contact channels to the integration of these advanced features of these results to the marketing platform. Finally, the marketing platform based on customer data was collected and the online survey, one of the ads and products, promotions, etc., to provide customers throughout the book in various ways, the cumulative consumption of the system of personalized service products.

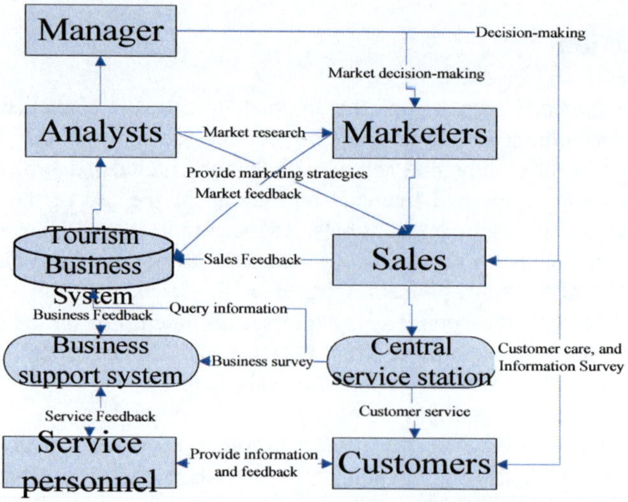

Fig. 1. Tourism E-business system business process diagrams

2.2 Tourism E-commerce Function of the System Design

Tourism E-business System[2,3] Referred to as TESystem functional structure shown in Figure 2.

2.2.1 Customer Service Platform

Tourism-oriented customer base is very broad, so it needs to mobilize the phone, web, email, portals, on-site demonstration, exhibition centers and other distribution channels of communication channels to communicate and integrate them together to achieve the information sharing between departments, in order to improve service efficiency. Through contact management, service staff can be the desired intelligence information to be collected, and the useful information stored in the database. Service through the contact management, data integration and business connections online analytical decision-making platform for customers to establish uniform standards set questions for customer service. Meanwhile, contact management business management business platform to connect to customers about our products or services, and product / service offer, etc., and one to achieve business and customer service, and to increase customer satisfaction, increased business opportunities.

2.2.2 Analysis and Decision-Making Platform

First, it will integrate a variety of database information together, and to this end have focused on the business data, clean, transfer, form a clean, consistent and comprehensive data for decision-making behavior of customers and then grouping, the use of data mining techniques, found that customer base characteristics, behavior patterns and characteristics of different types of customers, and put it into the actual content, and that the content is brand new, complete, easy to understand management and decision-making for them to make better strategic decisions, continuing to provide excellent tools to improve business support;

Finally, the various departments and employees through enterprise analysis and decision-making platform, to achieve the sharing of this data for them in the implementation of the strategic plan of the upper, improve efficiency and achieve maximum value of the individual provides a strong guarantee.

2.2.3 Sales Platform

Main features include: organizing and browsing sales information; generate reports of the sales stage, and gives businesses the stage, the time needed, the likelihood of success, history, sales evaluation, and so forth; given for sales tactics , strategy support; on the geographical (cities, zip code, region, industry, related customers, contacts, etc.) for maintenance; to a certain region and classified salesperson license; geographical re-set; based on profit, areas, priority level, time, status and other standards, the user can customize the activities to be carried on the business, customers, contacts, appointments and other aspects of the report; BBS provides similar functionality, the user can paste the secret sale on the system, can also be a the one hand, sales skills, inquiry; cost of sales management; sales commission management.

3 Overall System Architecture and Implementation

3.1 System Architecture Design

System calls through the interface between layers. Each has a clear mandate. And each exchange data through the interface, so, for each layer can be seen as an independent control. When the internal structure of each layer changes, simply modify the relevant interface implementation class and the interface remains stable, allowing the system to maintain minimum levels of inter-dependence, to achieve a loose coupling between layers, as shown in Figure 2. SSI framework described in this article direct application in the "tourism e-commerce system" is based on JavaEE architecture of distributed systems [4], using the B / S structure design, the use of tools for developers MyEclipse8.0, Oracle for the database, WebLogic is application server. Three-tier architecture using MVC, the division of the system architecture, presentation layer is responsible for the Struts2 framework of the label user interface and user interaction at the same time; business logic is in the Service by the Struts Action class to class and be responsible for specific business logic ; DAO class is responsible for interacting with the database. Specifically shown in Figure 2.

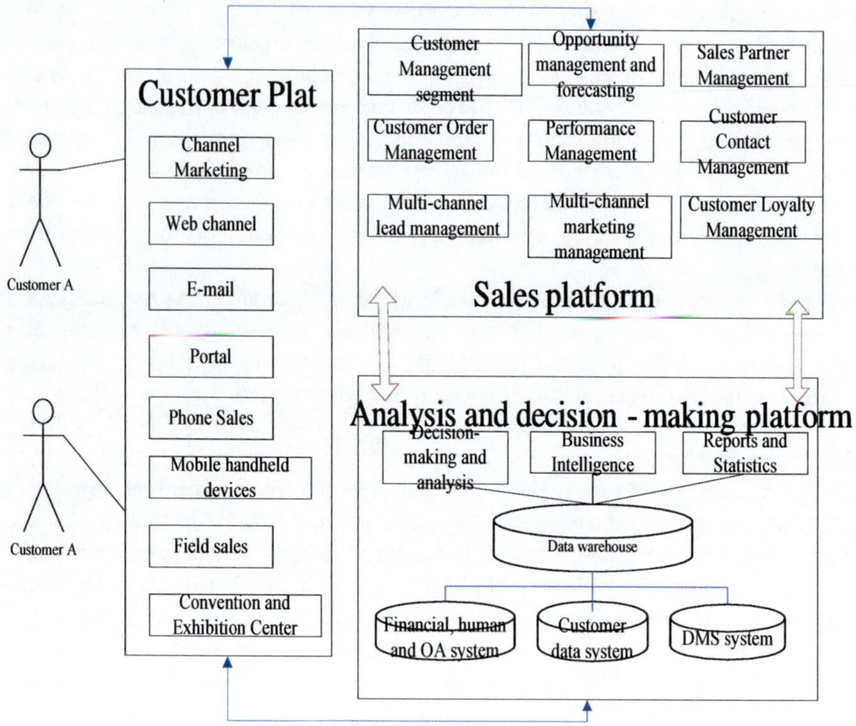

Fig. 2. Tourism E-business system Functional distribution diagrams

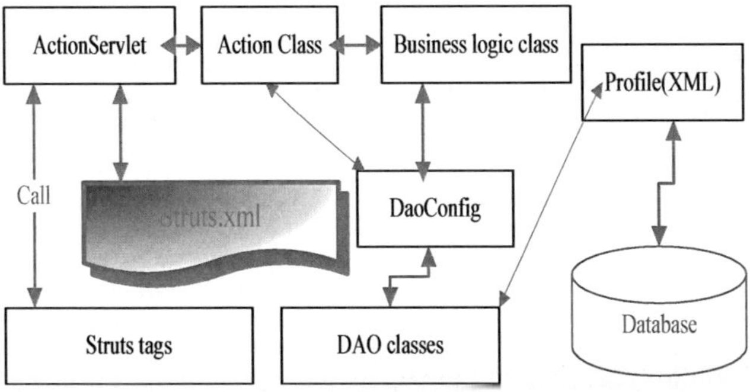

Fig. 3. Tourism E-business system architecture diagrams

3.2 System Architecture to Achieve

3.2.1 Struts2 Based on MVC Model and Display Layer to Achieve

Struts2 display layer is based on customer and service oriented interface. Responsible for the platform to interact with the system [5]. For example, a client application for registration in the system to fill a single AddCustomerForm.jsp, submitted to review the background of customer service, because Struts is the core of Struts.xml, common control components Struts filter assume the role of Controller in MVC it to check the URL Pattern match is done through the interceptor logs and other management rights, to the action class, Action class implements the business logic, action handling, steering link. Receives user requests, update the model and select the appropriate view of these components back to the user by the Struts controller components responsible, between model and view it in the schedule. Controller is based on Struts.xml configuration file to determine the traffic steering. Specific processes shown in Figure 4. The advantages are: (1), define page navigation and flow control are concentrated in an XML document, this configuration file can quickly grasp the context of the whole system; (2), web page designers are not required to traverse the change Java code to understand the application's business logic, and when business logic changes, the business logic developers are only required to Struts.xml to make appropriate adjustments and modifications. This is the development and maintenance, and upgrades are shown after the convenience and effectiveness.

3.2.2 Spring Business Layer Applications in the System

With Spring, via dependency injection, AOP applications oriented programming interface, can reduce the coupling between business components, and enhance system scalability. Spring is responsible for processing the application business logic, business checking, and transaction management; while managing the business layer object dependency; in the display layer and persistence layer adds a flexible mechanism, making them not directly linked. We are in the business layer is to complete the following two tasks. (1), in order to avoid duplication of business logic written in a large number of transaction processing code. We in the tourism e-commerce system, the transactionManager the transaction manager defined in the configuration, so that developers only need to focus on business logic, transaction management entirely to the configuration file, so the advantage is that the savings in

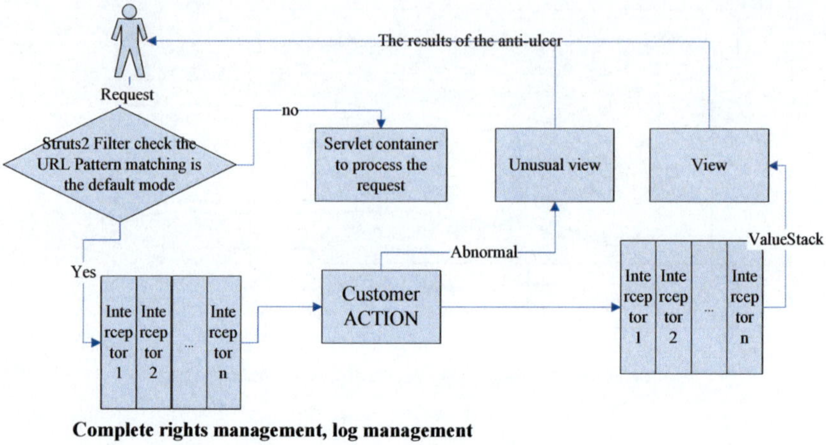

Fig. 4. Customer registration flowchart

development time and can improve development efficiency. (2), the service-dao.xml the façade class and one or more dao class configuration. The benefit of this is hard-coded does not exist between the components associated with, any component can be reused to the greatest extent. When we develop a business logic components only need to use setter method of dependency injection, between the components can be easily referenced.

3.2.3 Ibatis Data Persistence Tier Based Architecture Design and Implementation

We used to use DaoConfig class (singleton), VO model (POJO class), and ibatis framework for DAO mode and the formation of new technical data persistence layer application architecture, shown in Figure 5.

From Figure 5 we can see the data persistence layer is divided into two main layers, one of the DAO layer, the other is the SQL Map framework layer. In the DAO layer, Service layer as long as the class and by visiting DaoConfig DAO interface, you can get the resources it needs. DaoConfig class (singleton pattern to achieve) the completion of a one-time loading of the resource file dao.xml work, and dao.xml manage a large number of dao class, the benefits of doing so is to increase the efficiency of resource use, to speed up data access speed and response time; in the SQL Map framework layer, to complete three major tasks: ① complete automatic database connection and transaction management; ② SQL Map configuration, loading and optimization settings; ③ SQL Map definition files and mapping optimized. For ① and ②, mainly in the sql-map-config.xml completed, through the main configuration file can be flexible to achieve the specific configuration of the database. The following is dao.xml file in the user registration, for example, dao.xml file DAO class is responsible for providing management information to be firm and tell the manager how DAO DAO interface for the DAO implementation. For <dao> label only two properties: interface and implementation., The code is as follows:<transactionManager type="SQLMAP">

```
<property name="SqlMapConfigResource"
 value="com/tesystem/persistence/sqlmapdao/sql/sql-map-config.xml"/>
</transactionManager>
<dao interface="com.tesystem.persistence.iface.Customer"
implementation=" com. tesystem.persistence.sqlmapdao.customerSqlMapDao"/>
```

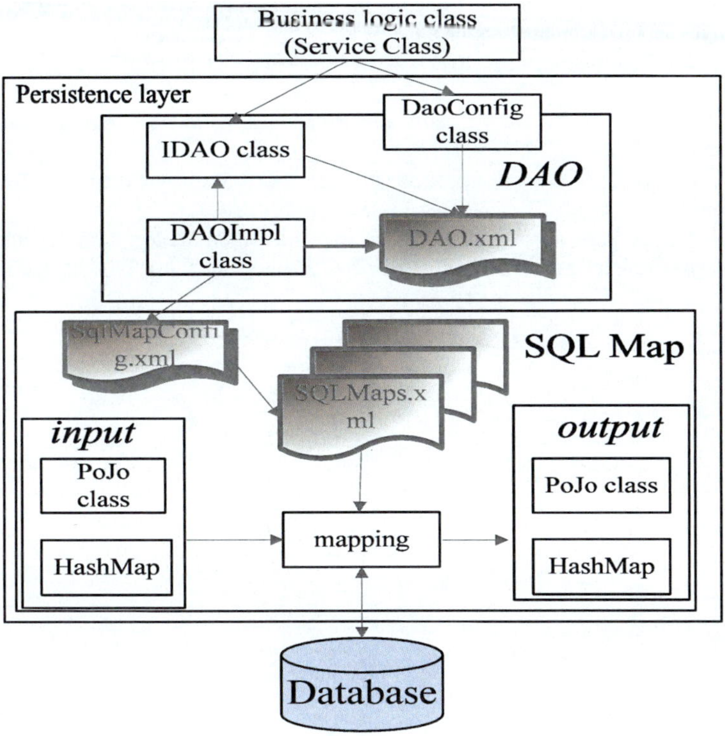

Fig. 5. Curve: persistence layer's framework design based on Ibatis

4 Conclusion

Practice shows that it solves the traditional JavaEE architecture, development is difficult, inefficient data access, it is difficult to unit test and other issues. Meanwhile, the application S2SI technical architecture for tourism e-commerce system development and shorten the system development time and improve the system's maintainability and scalability, making the system better able to adapt to changing business needs.

Acknowledgement. In this, the teachers would like to express our most sincere group thanks! You for my study and research provides a good environment, your encouragement and advice that I continue to progress. In this for you my deep gratitude. Finally, I thank my family. I thank my parents have been my deep love, I thank my friends and relatives in their lives learning to my advice and help.

References

1. Liu, T.-T., Jiang, Y.-M., Fu, J.-T.: Integration with J2EE frameworke to build reusable E-commerce system. Computer Application 26(11), 2769–2772 (2006)
2. Yan, H.: Java and patterns, pp. 10–11. Publishing House of Electronics Industry, Beijing (2002)
3. Zhang, L., Zhang, W.-X.: Design and implementation of fixed assets management system based on JavaEE 30(16), 3797–3800 (2009)
4. Yan, H.-Y., Feng, H.: Design and Realization of Social Security Comprehensive System Based on J2EE. Computer Engineering 33(23), 276–278 (2007)
5. Feng, R.-M.: Design and Implementation of High Education Student Management System Based on Struts & hibernate & Spring. Computer Engineering 35(6), 280–282 (2009)

Evaluating and Enhancing the Performance of Streaming Media Services in E-learning System*

Hu Xu-Huai, Yang Zhi-He, and Li Huan

College of Computer Science, Hunan Institute of Science and Technology
Yueyang Hunan 414006, China
huxh@163.com

Abstract. Streaming media services are powerful tools for delivering instruction content to online learners in E-learning System. However, if the capacity and efficiency of streaming server or network bandwidth is not sufficient for a lot of media files, the student have to wait the curriculum for a long time. So this paper tries to deal with streaming media communication in E-learning system, and the focus is on evaluating and enhancing the performance of streaming media services. Our experiments have demonstrated that the Disk I/O will bear the brunt of one bottleneck of server, followed by the network bandwidth while the server delivers the movies. And those problems should be solved by good services from E-learning systems, so we present a multi-tier architecture for E-learning system. In such a system, streaming server sends performance log to database server periodically, and different contents are stored in different servers, on which we can ensure the normal operation of the streaming media services to provide a stable learning environment for students.

Keywords: E-learning, Streaming media services, multi-tier architecture, Disk I/O, performance log.

1 Introduction

In recent years, E-learning technologies are being used to replace or supplement the face-to-face classroom[1]. These include videoconferencing, audio-conferencing, electronic whiteboards, application-sharing, chats, discussion tools, virtual classroom, and computer-based training[2,3,4]. A unique technology for multimedia curriculum delivery is widely used in E-learning. That is a streaming media services[5,6]. Streaming media is the simultaneous transfer of digital media (video, audio and data) so that it is received as a continuous real-time stream. Streamed data is transmitted by a server application and received and displayed in real-time by client applications. These applications can start displaying video or playing back audio as soon as enough data has been received and stored in the receiving station's buffer[7,9]. The multimedia curriculums of E-learning are delivered through streaming server, and hence the capability and efficiency of the streaming server is the key[10,15]. When a lot of files

* This work is partially supported by Scientific Research Fund of Hunan Provincial Education Department of China (No.10A045).

W. Zhang (Ed.): Advanced Technology in Teaching, AISC 163, pp. 205–211.
springerlink.com

are transmitted by the streaming server, disks must read various files simultaneously. The capability of disk I/O may not be able to support the demand. If the capacity of disk I/O can support, the bandwidth of network may not. The attitude of student learning may be affected by such situation seriously. If the capacity and efficiency of the streaming server or network bandwidth is not sufficient, the student must wait the curriculum for a long time. Once the student has waited the curriculum too long, he/she will give up the curriculum and reduce his/her learning motivation. That will make the student's learning ineffective and affect the student's learning willingness[11]. Therefore, how to provide a good network learning environment via maintaining a fast network and efficient disk I/O is an important issue[12]. In order to achieve this purpose, we must evaluate the performance of the streaming server. The result of the evaluation will be provided as the strategy for developing E-learning's structure.

2 Evaluating the Performance of Streaming Media Services

A. Hardware environment and Quality of streaming file

In the paper, the experiment environment and the quality of files for evaluating the streaming server's bottlenecks of the network bandwidth or disk drive's I/O capacity are proposed in order to put forward to investigate appropriately. As shown in Table 1.

Table 1. Experiment environment of hardware and streaming file

<table>
<tr><td rowspan="2">Streaming Server</td><td colspan="3">Hardware</td><td colspan="3">Software</td></tr>
<tr><td colspan="3">Lenovo T350 G7
2*CPU Xeon E5506
2G RAM
Giga Bit NIC</td><td colspan="3">Windows 2003
Standard Server
Media services
Media Player</td></tr>
<tr><td rowspan="2">Type of Disk Array</td><td colspan="2">RAID 0</td><td colspan="2">RAID 1</td><td colspan="2">RAID 5</td></tr>
<tr><td colspan="2">2 * 500GB
10000 rpm
Ultra320
SCSI
64MB Cache
Array
Controller : 64bit
PCI</td><td colspan="2">2 *500GB
10000 rpm
Ultra320
SCSI
64MB Cache
Array
Controller : 64bit
PCI</td><td colspan="2">3 * 500GB
10000 rpm
Ultra320 SCSI
64MB Cache
Array
Controller : 64bit
PCI</td></tr>
<tr><td rowspan="2">Stream file</td><td colspan="4">Stream file length</td><td colspan="2">Quality of file</td></tr>
<tr><td>2
M</td><td>4
M</td><td>6
M</td><td>8
M</td><td colspan="2">1280*720</td></tr>
</table>

B. Disk Cache

Disk Array offers the cache memory to increase the disk throughput and reduce the disk access times. But Disk Cache is limitation by Array controller. The utility of disk cache that we can find from Fig. 1 is useful. But when the requests from the users are quickly, Disk Cache can't support a lot of requests. Disk I/O will unexpectedly increase too fast to support user's requests. From Fig. 1, although the Disk I/O offered 25MB/s throughput near, but the network throughput did not increase, contrary it decreased. Therefore the Disk Cache is useful but can not keep from the Disk I/O increase unexpectedly.

C. The bottleneck of Disk I/O

At the present time, RAID(redundant array of independent disks) is widely used in Disk I/O of the streaming media services. RAID is a technology that provides high performance and increased storage reliability through redundancy, combining multiple low-cost, less-reliable disk drives components into a logical unit where all drives in the array are interdependent[16]. A RAID can be implemented in a number of different levels (or type of disk array). Normally, the standard RAID level varies in 0, 1, 3, and 5. We employ RAID 0, RAID 1 and RAID 5. From in Fig. 2, we know that the server delivers the streaming media files, the Disk I/O will be the bottleneck of server. And the best efficiency of Disk I/O (Output) is RAID 0 and RAID 5, the bottleneck of Disk I/O (Output) is 33.6MB/s. The lowly efficiency is RAID 1, the bottleneck of Disk I/O (Output) is 28.15MB/s ~ 29.5MB/s. But the mode of RAID 0 can support users big than RAID 5. If we hope that server can serve the largest number of clients, the mode of RAID Controller must be RAID 0, and let each server offer only one file stored in RAID 0 to serve users.

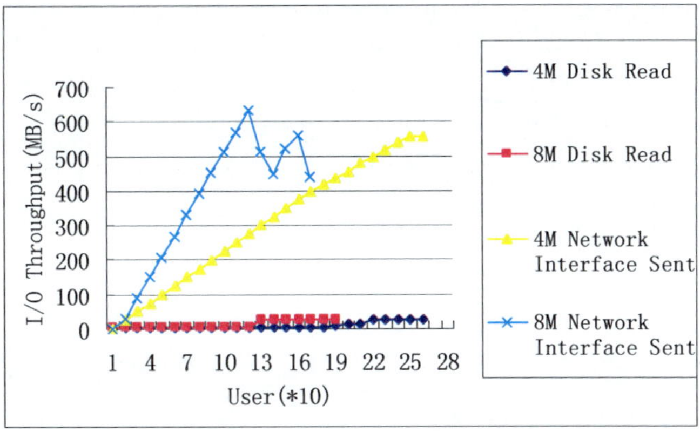

Fig. 1. The Utility of Disk Array Cache

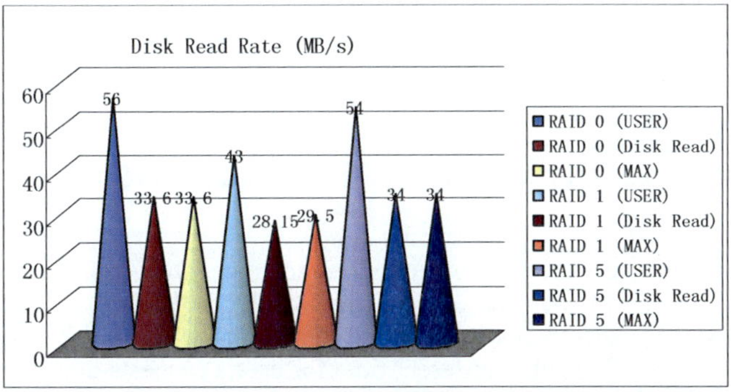

Fig. 2. The bottleneck of Disk I/O

D. Network Throughput

We know from Fig.3 and Fig.4, the bottleneck of server's gigabit NIC is about 900Mbps.If we want E-learning system to support the high-quality file to a large number of users simultaneously, it is impossible. The bandwidth of server's gigabit NIC can transmit up to 1Gbps near. Since the server can support more than two gigabit NICs, the bottleneck of the server will not be network bandwidth.

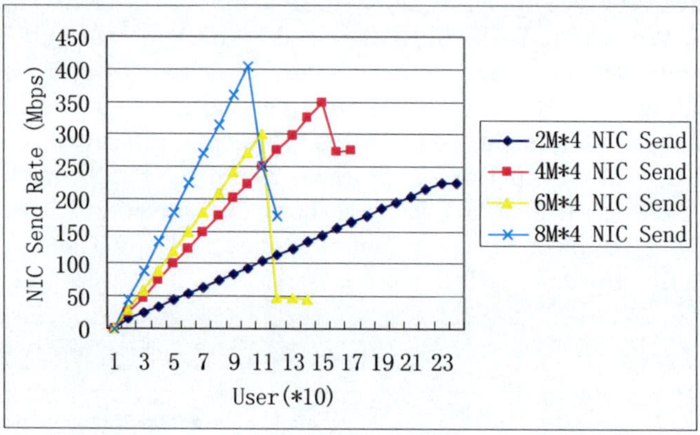

Fig. 3. Network Throughput of RAID 5 (Multiple files)

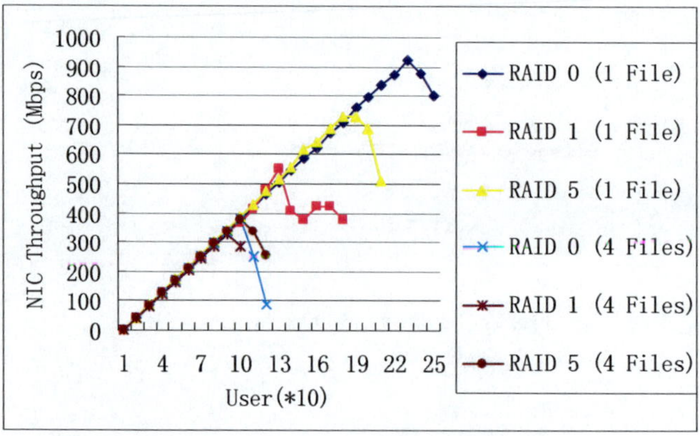

Fig. 4. NIC Throughput Contrast between different RAID Type

3 Enhancing the Performance of Streaming Media Services

A. Multi-tier architecture for E-learning system

According to the results discussed in previous sections, Disk cache can increase the capability of Disk I/O, but the size of disk cache is limited by the disk array controller[16]. If the file cannot be accessed in disk cache, it must be read from hard disk. When the files are read at different time for user's requests that cannot be found in disk cache, streaming server's disk I/O will unexpectedly become too busy to serve the users[14]. Another issue is that backbone bandwidth of Intranet is up to 1 Gbps, the server cannot offer over the bandwidth of backbone. Therefore, when a lot of streaming files put in the same streaming server to be used, the server cannot provide the most efficient services[13]. In order to overcome such problem, we propose a multi-tier architecture for E-learning system shown in Fig.5.

B. Operation of the System

In the above multi-tier architecture, the system operation is described as following:

Step1: User logon Web server and select a curriculum by client.

Step2: Web server receives the requests and query database to get the information of streaming servers.

Step3: Database sends the information to Web server. And Web server receives the information, to analyze the information and to get the streaming server with lowest loading.

Step4: Web server sends the user's request to the chosen streaming server.

Step5: The streaming server delivers content stream to user.

Step6: User receives the stream and start learning.

Step7: Web server checks the streaming server utilization rate (Disk or Network). If the streaming server's performance is approaching its capacity, the Web server will find out another streaming server with lowest loading, and send this message to file server and write this message to database.

Step8: Media file server receives the message from Web server and it will copy a curriculum file to another streaming server.

Step9: Streaming server sends performance log to database server periodically.

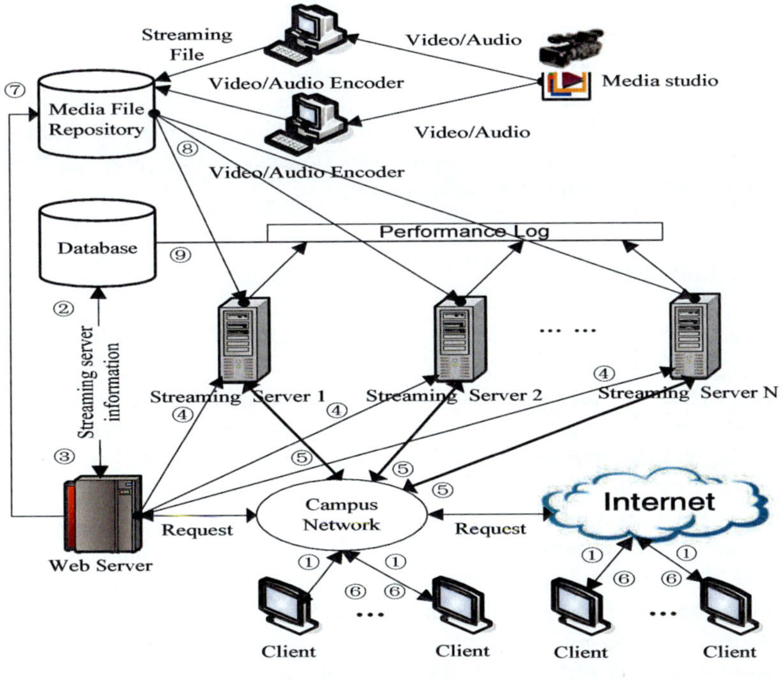

Fig. 5. Multi-Tier architecture for E-learning system

Each server, in general, only transmits one curriculum at one time in this system, but the server can offer multiple curriculums to serve users[8]. When the server cannot support the requests of clients efficiently we must create another copy to another server, and choose a threshold that the server can support the largest number of clients.

C. Threshold value of Network and Disk I/O

As shown in Fig. 1 of the previous sections, after about 130 clients come in, the disk I/O becomes busy. We can choose the threshold of Network as 650Mbps. If we hope the server can serve the requests more stable, we must choose the 600Mbps (120 clients). How can we decide the threshold of disk I/O? From Fig. 2, we could find the bottleneck of disk I/O is nearly 34MB/s (RAID 0). If the chosen threshold of disk I/O is 33.6MB/s, then when new requests come in may cause the disk effect to descent. In order to insure the disk capability, we must choose a threshold less than 33MB/s. When disk I/O is no more than 28MB/s the server can operate normally. We decide to choose the disk threshold as 28MB/s.

D. Enhancing the performance of system

As shown in Fig.5, the streaming server sends the performance log to database, and Web server reads those logs to decide the streaming server that can be chosen. First, Web server checks the disk I/O value. If it is close to 80% of the threshold of disk, then implements step 7 and step 8. If disk I/O is less than 80% of the threshold of disk, Web server will continue to check the network's loading. If it is close to 80% of the threshold of Network, then do the same operation. And the server information will be recorded in database. When the disk I/O value or network value is equal or over the threshold, Web server will choose another streaming server to serve the new user. Another problem is bandwidth of backbone. In this system, there are many servers providing services. We can put the streaming server in the different places or classrooms. Thus the students use the different files will not occupy the same bandwidth. And we can reduce the file quality to decrease the traffic of network. The file quality can be near 1.5Mbps (MPEG 1 quality), thus the server can serve more users.

4 Conclusion

In the paper, we focus our efforts on evaluating and enhancing the performance of streaming media services, and analyzing the effect of the network bandwidth and disk I/O capacity for streaming servers in E-learning systems. One bottleneck of streaming server is its disk I/O, the network is the second. Those problems must be prevented by E-learning systems to providing good services. Consequently, we present a multi-tier architecture for E-learning system. In the system, streaming server sends performance log to database server periodically, different contents are stored in different servers, so that we can ensure the normal operation of the streaming media services to provide a stable learning environment for students. The servers are distributed in the network to reduce the possibility of traffic jam in the network. This system can be used in a university or a company.

Streaming media services are powerful tools for delivering instruction content to online learners in E-learning System. In the future, the advance of the hardware and network equipment will contribute to supporting higher quality contents. And the promise of mobile computing is true “anywhere/anytime” learning[17]. With the right planning and a careful monitoring toward bandwidth consumption (network and disk I/O), students in E-learning programs can gain the benefits of streaming media with audio,video, and other forms of media on the Web.

Acknowledgment. It is a project supported by Scientific Research Fund of Hunan Provincial Education Department of China (No.10A045).

References

1. Tavangarian, D., Leypold, M., Nölting, K., Röser, M.: Is e-Learning the Solution for Individual Learning? Electronic Journal of e-Learning 2(2) (December 2004)
2. Redecker, C.: Review of Learning 2.0 Practices: Study on the Impact of Web 2.0 Innovations on Education and Training in Europe. JRC Scientific and technical report (EUR 23664 EN–2009) (February 2009), `http://ipts.jrc.ec.europa.eu/publications/pub.cfm?id=2059`
3. Allen, I.E., Seaman, J.: Staying the Course: Online Education in the United States (November 2008), `http://www.educause.edu/Resources`
4. Dorai, C., Kermani, P., Stewart, A.: ELM-N: e-learning Media navigator. In: Proceedings of the Ninth ACM International Conference on Multimedia, pp. 634–635 (October 2001)
5. LaMonica, F.L.: Streaming Media. Linux Journal (January 2001)
6. Hofmann, M., Sabnani, K.: Streaming and broadcasting over the Internet. In: IEEE Conference, June 26-29, pp. 251–256 (2000)
7. Ma, H., Shin, K.G.: Multicast Video-on-Demand Services. ACM SIGCOMM Computer Communication Review (January 2002)
8. Almeida, J.M., Krueger, J., Eager, D.L., Vernon, M.K.: Analysis of Educational Media Server Workloads, pp. 21–30. ACM (June 2001)
9. Ming, L.X.: Streaming technique and its application in distance learning system. In: IEEE International Conference, August 21-25, vol. 2, pp. 1329–1332 (2000)
10. Gary Chan, S.H.: Distributed Servers Architecture for Networked Video Servers. IEEE/ACM Transactions on Networking 9(2), 125–135 (2001)
11. Sherry, L.: Issues in Distance Learning. International Journal of Educational Telecommunications 1(4), 337–365 (2002)
12. Schocken, S.: Standardized frameworks for distributed learning. JALN 5(2), 97–110 (2001)
13. Hiltz, S.R., Wellman, B.: Asynchronous Learning Networks as a Virtual Classroom. Communications of the ACM 40(9), 44–49 (1997)
14. Hillestad, O.I.: Evaluating and Enhancing the Performance of IP-based Streaming Media Services and Applications. Norwegian University of Science and Technology (May 2007)
15. Bouthillier, L.: Streaming media delivery for higher education. Streaming Media Inc.: Innovation Series, Online Learning: Textbook Strategies for Video Education (6), 25–28 (2007)
16. Patterson, D.A., Gibson, G., Katz, R.H.: A Case for Redundant Arrays of Inexpensive Disks (RAID). In: Proceedings of the 1988 ACM SIGMOD International Conference on Management of Data, Chicago, Illinois, pp. 109–116 (June 1988)
17. You, J.: A Study of Faculty Members' Perceived Utilization of Best Practices in Distance Learning Course Design and Delivery and the Role of Instructional Designers. The University of Toledo, pp. 14–19 (August 2010)

Strategy of Reform and Recommendations of Practical Teaching of Mathematical Modeling and Mathematical Experiments Courses*

Zhaodi Xu and Xiaoyi Li

School of Mathematics and System Science, Shenyang Normal University, Shenyang, 110034, China
{xuzhd,lixy}@synu.edu.cn

Abstract. This paper shows the benefits of offering Mathematical Modeling and Mathematical Experiments courses for training the ability of students to calculate as well as to use mathematics knowledge in solving problems. By the course reforming, which are offered by the mathematics major, together with analysis and research in problems with Mathematical Contest in Modeling, we are looking for the source of the problems to make some teaching reformation and discovery to the course reforming, teaching methods and etc of Mathematical Modeling and Mathematical Experiments courses in normal universities. This paper also provides the way of teaching reformation and advices to the practice teaching processing. It can train the ability of students of Innovation and improve the ability to use knowledge of mathematics as well as computer techniques in solving problems.

Keywords: Mathematical Modeling, Mathematical Experiments, Practical teaching; Innovation.

1 Introduction

As important parts of university mathematics curriculum, Mathematical Modeling and Mathematical Experiments Courses open a new scientific field for modern mathematics. Our aim is to training the ability of students of Mathematical Modeling for practical problems and solving that by computer (Mathematical software). Mathematical Modeling and Mathematical Experiments Courses are set for Students of mathematics. It is very important for the courses setting to obtain Mathematical Contest in Modeling since the nineties of last century (since the 1990s), teaching and competition of Mathematical Modeling and Mathematical Experiments are effective ways of quality education and they not only enhance the students awareness of mathematical applications, but also improve their abilities of innovation and applying knowledge of mathematics and computer technology to analyse and solve problems.

At present, the contents of Mathematical Modeling and Mathematical Experiments courses are gradually embodied the idea of Mathematical Modeling. Based on the

* This work is supported by institution of higher education Science Foundation of Liaoning Province of China, under Grant #20060842.

W. Zhang (Ed.): Advanced Technology in Teaching, AISC 163, pp. 213–219.
springerlink.com

courses setting and improvement, the spirit of the organic content and method of Mathematical Modeling should be expressed gradually. Many universities are interested in it and the study of Mathematical Modelling and Mathematical Experiments is very important [1] and [2].

Since the beginning of this century, Mathematical Modelling courses were steadily developed in domestic universities. And the number of schools which participate in Mathematical Modelling contest increased year by year. The importance of Mathematical Modelling and Mathematical Experiments courses is widely recognized and more and more universities set up the two courses [3] and [4].

Since the course design changed from secondary to primary, as well as from elective to require courses, its course load and practice parts increase. The process of setting up Mathematical Modelling and Mathematical experiments has gone through several stages in our university. Participating in the process of mathematical modelling contest has been ten years. After several years of teaching and competition practice, problems emerged, led to consideration of the traditional courses and curriculum reform.

2 Curriculum Analysis

A. Curriculum

Mathematical Modeling courses in our university have experienced three stages: First, the Mathematical modeling is set up as an elective course in our university with the course load 52 hours. Elementary models by using the skill of problem-solving learned in middle school were taught. The content is relatively uniform. On the basis of that, the course was improved. We changed the course compulsory and the course load was increased. The content taught starts to include the advanced content of model. With elementary model counts 80% and advanced models 20%, the difficulty is slightly strengthened. Now, with the development of teaching reformation, the Mathematical Modeling course load is set to be 60 hours, elementary mathematics and mathematical counts 50% each and the ease of materials tend to be appropriate, theoretical and practical discussion and teaching hours counts 50% each.

Our Mathematical Experiments Course is open for third-year students in mathematics major, with Advanced Algebra, Mathematical Analysis, Probability and Mathematical statistics and differential equations as prerequisite courses. The course load is usually about 60 hours. Computer lessons and theory lessons count 50% equally. The reformation still continues.

B. Textbook and Content

With the popularizing of the Mathematical Modeling of teaching courses in colleges and universities, there are many textbooks of mathematical modeling [5] and [6]. But most of these materials were prepared for the science and engineering students and only a few were for advanced normal college students. Therefore, we prepared a set of suitable materials for normal college students. Compiled our university several teachers' manuscripts from years of teaching mathematical modeling, "Mathematical Modeling" was published in July 2006.

The contents in our textbook from shallow to deep, is divided into three parts, the first part is the elementary mathematical model for use of public elective courses and correspondence students and the content is relatively simple, using elementary mathematical methods and some knowledge of a small amount of advanced mathematics. The second part is the models of advanced mathematics; it is suitable for mathematics, physics, computer and biochemical students. Mathematical models in this part are complex and use a small amount of elementary mathematics method and a lot of advanced mathematics knowledge. As a part of the Mathematical Experiment, The second part is divided into basic experiments and design of experiments. By some simple practical problems, students will learn how to deal with practical problems and how to use MATLAB [8] and [9].

C. Ways and Means

The Feature of Mathematical Modeling and Mathematical Experiments Courses is that the teachers briefly explain the theory and the students take the initiative to test to experiment. Multimedia technology can be used in classroom and a Special E-mail account is also used for the Curriculum. On the other hand, website and lab were constructed by our university for the courses.

The way of practice and test of Mathematical Modeling and Mathematical Experiments Courses is different from traditional mathematics curriculum obviously. Usually practice uses the way of computer, case analysis and comprehensive assessment. Test uses the comprehensive evaluations, which includes practice, reports as well as finals, instead of a single test.

D. Modeling Contest

Our university organized students to participate in the National Mathematical Contest in Modeling for 10 years and paid more and more attention to this work. Competition results get better year after year, many people access to national second prize and other awards in Liaoning. Since 2007, our school organized students to attend international Mathematical Modeling Contest, and more people gained international second and third prizes, etc.

Although we won some achievements, we still don't have any domestic and international first prizes and many of the problems emerged in the contest: (1) Students were not interested in Mathematical Modeling. (2) The student's knowledge is limited and mathematical modeling activity is not comprehensive. Although it is related to lack of promotional dynamics, it also indicates lack of necessary teaching. (3) Although only a few junior and senior students participate, the proportion of award gained by them is large. Especially the senior students, they are not attracted by contest because they faced their graduation and employment pressure although they have brilliant mathematical basis. Freshmen and sophomores' participate in training of the competition with high enthusiasm, but gain few awards, this shows that the ability of Mathematical Modeling is closely related to the knowledge accumulated. It is a combination of theory and practical applications. Junior curriculum is not appropriate; a number of related courses are open too late. (4) In the process of solving Mathematical Modeling, the skills of computer programming of students were bad. It is hard to use mathematical software solving mathematical model problem,. Many students can't use knowledge to solve practical problems, which needs to be improved.

E. Teaching Team

In the process of teaching Mathematical Modeling and Mathematical Experiments, teachers need an ability of analyzing problems and solving problems. Our teachers have to master the core subjects of Mathematics and be good at expressing the idea of Mathematical Modeling and setting up "opening problems" to train habits and ability of students to ask questions [10]. At present, the capacity of our teachers is weak in this aspect. There are only a few opportunities for teacher-student interaction; students do not build right learning patterns.

3 Strategy and Recommendations of Practical Teaching

According to the analysis of Mathematical Modeling and Mathematical Experiments Courses, our university enforces the curriculum construction, and makes the construction and reform of Mathematical Modeling and Mathematical Experiments Courses be University key reform projects. We propose the following reform measures and recommendations.

A. Curriculum reform and Construction

At the same time of setting up Mathematical Modeling required course in mathematics, Mathematical Modeling Elective Course should be created in university. Make more students have benefit from Mathematical Modeling course. At the same time, we also make modeling contest training well. So that the students' level of Mathematical Modeling further improve. Firstly the construction of Mathematical Modeling course is basis of good results of Mathematical Modeling competition. Secondly, teachers are usually accumulating of material, rich in teaching and coaching unceasingly consummates, instilling new ideas, new methods, using modern computer modeling tools to solve problems and promote the construction of Mathematical Modeling course because the Mathematical Modeling Contest questions are mostly from the problems of the real world.

To make Mathematics Experiments course good, our school has established the computer interactive multimedia laboratory and make the Mathematical Modeling laboratory be a Mathematical Experiment center for students. To create a good environment and conditions of the experiment, Mathematics Experiment center should purchase software, such as Matlab, Maple, Lingo, Spss and so on.

The purpose of this course is not teaching some knowledge, but a request in cultivating students' interest in exploring and the ability to solve practical problems on the combination of mathematics knowledge, the ability of modeling and software.

B. Modeling contest and Extra-curricular learning

Mathematical Modeling Contest is an important way of creative ability and overall quality of training. Not only stimulate the enthusiasm of the students to study mathematics, but also improve their ability of establishing mathematical models and solving problems by computer technology. We should encourage the students to participate in that and train students' creative spirit and sense of cooperation to promote the reform of university mathematics teaching system, teaching content and methods [11] and [12].

We should encourage students to attend college Mathematical Modeling Contest and increase students' reward. For the selection of the team members of National Mathematical Contest, campus Mathematical Modeling Contest should be held. By the way of combination of Popularization and improvement, improvement is prompted by competition and Popularization is prompted by improvement.

A major feature of the reform Mathematical Modeling and Experimental teaching is organizing students study by the mode of extracurricular study group. We use group-mode to enhance Mathematical Modeling and Experimental mathematics teaching.

It is very necessary to further explore on ideas, content and methods. Mathematical modeling and mathematical experiment focused on transform the practical problems to mathematical problems and the ability of Mathematical Modeling is required. Stimulate students' interest in exploring problems. Show brilliant results in special classes and let students discuss their results.

Make students build their own Mathematical Modeling websites and build a new network platform between teachers and students. Our website covers National Collegiate Mathematical Contest in Modeling questions, award-winning paper, Mathematical Modeling applications, programming languages and application software. So, our students can have the newest information and materials and ability and competition level can be improved.

Mathematical Modeling association was established and team members can be trained according to the plan. It makes the students' innovation and overall quality is greatly improved. Mathematical Modeling Association is the largest association in our university. It belongs to Practice of entrepreneurship and innovation central, whose members are more than one thousand. In June each year, we will recruit new members and training them by special series of lectures, computer guidance, simulation exercises, experience the text flow and internal competition.

C. Case collection and Research

Collection of teaching cases and research is very important because Mathematical Modeling and Mathematical Experiments Courses involve extensive knowledge and much universities lack of teaching resources. Our teachers research and transformation the problems at home and abroad and summarize many interest examples of daily life and case with different engineering background. These examples, which reflects the basic idea of Mathematical Modeling and the basic approach, and they are simple, specific and interesting.

D. Teacher Training

Mathematical Modeling and Mathematical Experiments Courses are flexible. Therefore, teachers have to not only master certain knowledge and methods of Mathematical Modeling but also have a deep understanding about how to apply mathematics. We should organize training courses, seminars and inviting experts to enhance the professional skills of our teachers.

4 Conclusions

The construction of Mathematical Modeling and Mathematical Experiments Courses made important contributions for our university to got good grades in the National

Mathematical Contest in Modeling. Students get trained in mathematical modeling and develop the students self-learning ability and sense of innovation though these courses. It makes students have a solid foundation for greater development in future.

According to school statistics, many students who participate in mathematical modeling contest received outstanding results. There are a number of students enrolled in graduate schools and some students have also been welcomed by employers.

We hope to break the traditional teaching model through analysis of Mathematical Modeling and Mathematical Experiments Courses. Make Innovative practices and capacity-building are our goals for the teaching. Through choosing teaching contents of these courses and dealing with teaching methods, methods of scientific treatment, assessment methods and laboratory reports regally, we can improve their ability of applied Mathematics and solving problems by computer technology.

References

1. Jiang, Q.Y.: Mathematical Experiments and Mathematical Modeling. Mathematics in Practice and Theory 31(5), 613–617 (2001)
2. Ye, Q.X.: Penetrating Ideas and Methods of Mathematical Modeling and Mathematical Experimentation into the Teaching of Calculus. Journal of Engineer Mathematics 20(8), 3–13 (2003)
3. Jiang, Q.Y.: Mathematical Modeling activity and the Reform of undergraduate education. Mathematics in Practice and Theory 27(1), 92–96 (1997)
4. Li, D.Q.: China Undergraduate Mathematical Contest in Modeling, pp. 313–321. Higher Education Press (1998)
5. Li, X.Y., Xu, Z.D.: Mathematical Modeling, pp. 1–2. Liaoning University Press (2006)
6. Liu, L.F., Zen, W.Y.: Mathematical Model and Mathematical Modeling, p. 122. Beijing Normal University Press (1997)
7. Li, D.Q.: Mathematical Science and Mathematics Education. College Mathematics 20(4), 1–8 (2004)
8. Zhang, J.B., Wang, S.Z.: School mathematical modeling for students to create space for development. Journal of Mathematics Education 10(1), 11–15 (2001)
9. Hong, S.Y.: A new way to explore mathematics education. Journal of Engineer Mathematics 12(2), 91–94 (2003)
10. Li, X.Y., Xu, Z.D.: Applying the idea of Mathematical Modeling to the Course of Probability and Statistics. Journal of Shenyang Normal University 26(2), 245–247 (2008)
11. Li, X.Y., Xu, Z.D.: Mathematics Curriculum Reform from Mathematical Competition in Modeling. Journal of Shenyang Normal University 25(1), 121–123 (2007)
12. Li, S.Z., Chen, F.L.: Understanding and Practice in Constructing the Course “Mathematics Experiments”. Mathematics in Practice and Theory 31(6), 764–768 (2001)
13. Dan, Q., Zhao, J., Fu, S.L.: Inquiring in to the Contents and the Teaching Methods of the Course of Mathematics Modeling. Journal of Engineer Mathematics 18(6), 21–24 (2002)
14. Ye, Q.X.: Modeling and Analysis of the Problem "Speedily Crossing the Yangtze River". Journal of Engineer Mathematics 20(7), 123–130 (2003)
15. Ye, Q.X.: Making Activities on Applying Mathematical Knowledge at High School Level Work Effectively. Mathematics in Practice and Theory 31(5), 608–612 (2001)
16. Tan, Z.G., Jiang, Q.Y.: A Mathematical Model of Bus Scheduling. Journal of Engineer Mathematics 19, 101–106 (2002)

17. Jiang, Q.Y.: Teaching Practice and Theory of Mathematical Experiments Courses. China University Teaching (5), 18–20 (1999)
18. Li, D.Q.: Suggestions to Research in Applied Mathematics. Bulletin of National Natural Science Foundation of China (4), 223–226 (2006)
19. Li, D.Q.: Mathematical Modeling and Quality Education. China University Teaching (10), 41–43 (2002)
20. Wang, M.Z., Guo, K., Xu, W.X., Zhou, Y.: Innovative Consciousness Cultivating During Mathematical Modeling. College Mathematics 25(1), 126–129 (2009)
21. Du, X.F., Zhang, S.S., Li, X.H.: The study of Mathematical Experiments in university. College Mathematics 25(3), 10–12 (2009)
22. Jiang, Q.Y.: Machine Scheduling Problems. Models and Their Analytic Results. Mathematical Theory and Applications 19(3), 44–53 (1999); Developing the potentiality of Technical Transform Strengthening the Economic Competitiveness. Mathematics in Practice and Theory (1), 81–91 (1996)

Application of Image Simulation of Arc Characteristics in Low Voltage Miniature Circuit Breakers

Liu Jinhe, Sun Ming, and Li Zhigang

Modern Teaching Center, Hebei University of Technology, Tianjin, 300401, China

Abstract. Miniature circuit breakers (MCBs) are widely used to protect electrical facilities within commercial, domestic and light industrial buildings. Arc motion in low voltage (240 VAC) high current (103-104A.) current limiting-circuit breakers is dominated by arc root mobility. In the recent years, computer image simulation of arc characteristics has important application in the investigation and development of technical low voltage MCBs.

Keywords: Circuit Breaker, electric arc, reliability, low voltage.

1 Introduction

To improve the working reliability of low voltage electrical devices, so the core of the research is the key part, the contactor of the electrical devices for realizing the switch function with the electrical devices, as main factor decided the life-time of the low voltage electrical devices. The short-circuit current is a typical case of a transient phenomenon in the network. In the distribution network of electrical energy, electrical devices play an important part in controlling the energy supply (switching, overcurrent and surge protection etc). Its operation must be reliable not only at rated network parameters but at failure case, too. The failure is caused by short-circuit currents and overvoltages when the magnitudes achieve multiples of the rated values. Breaking the short circuit current is a task of circuit breakers and overcurrent protection devices. MCBs are ideal for the protection of power supplies, control power transformers, HVAC, refrigeration equipment, florescent lighting and many other applications requiring a primary protective device.

For many years we have dealt with problems of MCBs, mainly with short-circuit current breaking. MCBs operate as current limiting devices, able to interrupt the short-circuit currents without dangerous consequences. MCBs are widely used to protect electrical facilities within commercial, domestic and light industrial buildings. For many years we have dealt with problems of MCBs, mainly with short-circuit current breaking. MCBs operate as current limiting devices, able to interrupt the short-circuit currents without dangerous consequences. Current interruption is the core technology in low voltage circuit breakers and it is characterized by the presence of an electric arc.

The figure 1 illustrates the visualization model of the MCBs switching arc with low voltage electrical devices and the arc image gathered. Besides the function of the contactor designed is very important, the electric arc motion process played the rather important role in the switching ability, electrical life-time and failure mechanism with

W. Zhang (Ed.): Advanced Technology in Teaching, AISC 163, pp. 221–224.
springerlink.com

the contactor of the low electrical devices. Therefore, it was very necessary to adopt the proper method to investigate the arc characteristics of low voltage electrical devices and arc lighting mechanism for advanced the production designed performance of low voltage electrical devices and reliability. Image simulation has reached a significant position in the investigation and development of MCBs with electrical technical products.

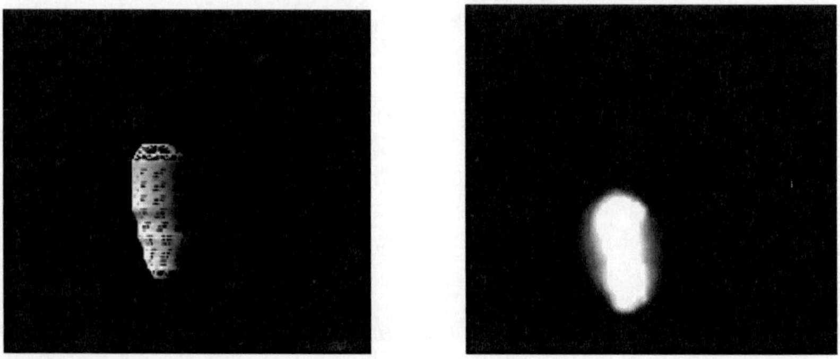

(a) (b)

Fig. 1. The visualization model of MCBs switching arc(a) and the image gathered(b)

2 Evolution of Simulation Investigations with Low Voltage MCBs

The arc simulation of the MCBs with low voltage electrical devices mainly involves the research of the 3D image forming theory and electric arc motion mechanism during electrical motion process. Although there had some progress in advance, as a new generated research method of electric arc, especially in the electric arc motion visualization was just like beginning. The research of electric arc visualization involved many technologies such as electrical devices technology, data computation, image processing and computer science. Therefore, it's too hard to measure and analyse the electric arc. It could be deeply researched with electric arc model theory, 3D finite element analysis of electromagnetism structure with electrical device, computer aided design and arc dynamic characteristics, in order to advanced understand the the complex phenomenons with MCBs contactor such as electrics, heat, magnetism during the motion process.

Various experimental and theoretical investigations have reported the study of low voltage circuit breakers. Hirofumi Takikawa[1] and Mitsuru Takeuchi[2] studied the distribution of temperature in the cross section of an arc column between separate contacts with spectroscopic detecting systems. J.W McBride investigated the influence of gas flow and gas composition on the arc root mobility in the contact region of MCBs[3]. Concerning numerical simulation, first calculations were done by Karetta[4] for arc chambers consisting of two arc runners and insulating walls using a MHD model. With a two-dimensional model, Helene Rachard[5] analyzed the influence of the magnetic forces on the shape and displacement of the arc. Swierczynski[6] developed a three dimensional model to investigate the arc motion with the influence of external magnetic field and plasma composition and transport properties. Lindmayer simulated the process of arc-splitting between metal plates in low voltage arc chutes[7].

MCBs arc models can be classified into three categories. Thermal and dielectric recovery models that describe the arc dynamical behavior considering the impact of different interrupter parameters such as nozzle size and geometry, type of quenching medium and speed of flow, pressure, etc[8-9]. Other models are based on empirical form(s) [10]. Both categories are efficient in determining the internal dimensions and quenching medium parameters of the interrupter. However, the third category of models is concerned with the arc external characteristics such as Cassie, Mayr, and modified Mayr models[11-12]. This category is recommended, as far as evaluating breaker interrupting performance in the power system is only concerned. However, H.A. Darwish and N.I. Elkalashy put forward the latest model-Universal Circuit Breaker Arc Representation with EMTP Built-in Model, and their theory has been extended to cover breaker operation in a direct test circuit and in a typical power transmission system[13].

3 Conclusions

The use of computers and simulation programs with MCBs electric arc allows solving efficiently theoretical problems, optimization of device switch reliability design, support and meet the need of improved MCBs solution technology.

References

1. Takikawa, H., Sakakibara, T.: IEEE Trans. Plasma Sci. 19, 879–884 (1991)
2. Takeuchi, M., Kubono, T.: IEEE Trans. on Comp. Pack. and 1Ianu. Tech. 21, 68–75 (1998)
3. McBride, W.J., Pechrach, K., Weaver, M.P.: IEEE Trans. on Comp. and Pack. Tech. 25, 427–433 (2002)
4. Karetta, F., Lindmayer, M.: Simulation of the gasdynamic and electromagnetic processed in low voltage switching arcs. In: Proc. 42 IEEE Holm Conf. on Electrical Contacts, Chicago, L-SX, September 16-20, pp. 35–44 (1996)
5. Rachard, H., Chevrier, P., Henry, D., et al.: International Journal of Hear and hlass Transfer 42, 1723–1734 (1999)
6. Swierczynski, Gonzalea, J., Teulet, P., et al.: J. Phys.D: Appl. Phys. 37, 595–609 (2004)
7. Lindmayer, M., Marzahn, E., Mutzke, A., et al.: The process of arc-splitting between metal plates in low voltage arc chutes. In: Proc. of the 50 IEEE Holm Conf. on Electrical Contacts, Seattle, USA, pp. 28–34 (September 2004)
8. Schoetzau, H.J., Meili, H.P., Fischer, E., Strurzenegger, C., Graf, H.P.: Dielectric Phase in an SF6 Breaker. IEEE Trans. Power Apparatus and System pas-104(7), 1897–1902 (1985)
9. Ragaller, K., Plessel, A., Herman, W., Egli, W.: Calculation Methods for the Arc Quenching System of Gas Circuit Breaker. CIGRE Report (1984)
10. Taylor, S., Wang, B., Blackburn, T.R., Jones, G.R.: Thermal Reignition Performance Limitation of a Modal SF6 Circuit Breaker under Full and Scaled Power Conditions. Univer. of Liverpool, Electrical Eng., Arc. Res. Report, ULAP-T72 (1982)

11. Habedank, U.: Application of a New Arc Model for the Evaluation of Short-Circuit Breaking Tests. IEEE Trans. Power Delivery 8(4), 1921–1925 (1993)
12. Hrabovsky, M., Mastny, V., Vostracky, Z.: Application of Mathematical Arc Model for Determination of Thermal Failure Limiting Characteristics. Cigre (1984)
13. Darwish, H.A., Elkalashy, N.I.: Comparison of Universal Circuit Breaker Arc Representation with EMTP Built-in Model. In: International Conference on Power Systems Transients, IPST 2003, USA, pp. 1–6 (2003)

A Hybrid Item-Based Recommendation Ranking Algorithm Based on User Access Patterns

Shao-Chieh Hu, Cheng-Yi Yang, and Chien-Tsai Liu*

Institute of Biomedical Informatics, Taipei Medical University, Taipei, Taiwan
ctliu@tmu.edu.tw

Abstract. Nowadays, most websites provide tremendous information organized in complex structures of web pages. Therefore, how to help users quickly find pages they are looking for is an important issue. Although a sitemap can provide navigation information *across sections of* the website, it is static and can hardly provide dynamic information based on access patterns and browsing trends. In this paper, we proposed a hybrid approach for improving recommendation ranking of the web pages for the next visit. Our raking strategy considers not only the relevance (correlation to the next page calculated by the collaborative filtering algorithm) but also the level of interest (time spent on a page) and accessibility (the distance to the next page). In order to evaluate the proposed recommendation ranking algorithm, we used the web access log (IIS log) of a website, Health 99, operated by the Bureau of Health Promotion, Taiwan. The log data was divided into training and testing sets. The measurements of the relevance, the level of interest and the distance factor were computed from the training set. The experimental results showed that the possibility of the pages in the recommendation ranking lists by our approach that were accepted by users was much higher than that proposed by the original collaborative filtering algorithm, particular in short recommendation list (< 5).

Keywords: Collaborative filtering, Accessibility, Ranking, User Access Pattern, IIS log.

1 Introduction

The purpose of personal recommendation systems is to provide recommendations to users so that they can quickly and accurately locate the information they want.

As Cline and Haynes indicated 1, web structures could become one of the obstacles for users while searching information through a website over the Internet. They summarized their viewpoints to the Internet as following. (1) The internet is like an unorganized library with plenty of materials. More worse, the volume of the materials are increasing rapidly. Thus the lack of systematic management would decrease the accessibility of information. (2) The internet is uncontrolled and unwatched, without notification, a site may disappear, change its functions or contents and move from time to time 4. Consequently, this may cause users inconvenience as they search for

* Corresponding author.

W. Zhang (Ed.): Advanced Technology in Teaching, AISC 163, pp. 225–233.
springerlink.com

information at a specific website. (3) It has been increasing difficulty in finding information for users in a website because of cognitive differences between users and website designers 5. R.

Considering web structures as a confounding factor in recommendation ranking, J. Garofalakis, P. Kappos, D. Mourloukos, M. Perkowitz and O. Etzioni [6, 7] proposed to include web page accessibility into analysis. The accessibility of a web page can be defined as the depth of the page, i.e., the number of links starting from the home page to that page.

Jhon Garofalakis & Panagiotis Kappos argued that measuring page popularity through access counts would be misleading because a page closer to home would be more easily to be seen than a page with deeper levels 6. Similarly, Mike Perkowitz & Oren Etzioni used access counts as an index for popularity measurement. However, they also took into account the distance where a page resided in order to discover the pages with high popularity but low accessibility for recommendation 7.

The visiting frequency of a web page can be measured by access counts, not by user preference. To know user preference we have to know how much a user likes or dislike an item on the website. A recommendation system needs to rank top interesting items of information for users while they are visiting. According to ways of getting preference, there are explicit and implicit feedbacks. An explicit feedback requests users actively to return rating on visiting pages. However, user may feel uncomfortable to provide rating score. This may decrease users' wiliness to provide feedbacks 8. In Order to overcome the problem, researchers proposed approaches to acquire implicit feedbacks by observing user behaviours [8, 9].

Morita & Shinoda verified the reliability of implicit feedback. The result showed there was a strong tendency to spend long time to read articles that are rated interesting 10. Besides, they also discovered there is a low correlation between the length of an article and time spent to read it. Konstan et. al. 11 discovered in the USENET group, by using reading time as preference, the same effect as those using the explicit feedback approach [12, 13].

In our research, we applied collaborative filtering algorithm and made some modification. We use reading time in calculating pages' correlation and find out correlated pages. Furthermore, consider difference in page visiting time length and steps of finding page from user behaviour. From criteria above, developed a new ranking method, to find out:

1) Page with no significant difference in correlation but with longer visiting time.
2) Page with no significant difference in correlation but shows more steps of finding page from users' behaviour.

2 Method

The goal of collaborative filtering (CF) algorithm is to predict the utility of a certain item for a particular user based on user's past behaviour and the opinion of other like-minded users 14. CF recommendation system could be divided into two main categories: User-based, comparing the activity record for a target user with the historic records of other users in order to find Top k like-minded users, through these users output the recommendation to the target user; Item-based, comparing items based on their pattern of rating across users. In item-based, it attempt to find k similar items co-rated by different users, from these similar items output the recommendation to user. Due to the lack of scalability of user-base 15, we adopted Item-based model, and calculate pages' correlation by adjusted Pearson correlation as in (1).

$$sim(i,j) = \frac{\sum_{u \in U}\left(R_{u,i} - \bar{R}_u\right) \times \left(R_{u,j} - \bar{R}_u\right)}{\sqrt{\sum_{u \in U}\left(R_{u,i} - \bar{R}_u\right)^2} \times \sqrt{\sum_{u \in U}\left(R_{u,j} - \bar{R}_u\right)^2}}$$

U is the set of all users, i and j are item, $R_{u,i}$ represent user u's visiting time on page i, $\bar{R}_u$ is the average visiting time length as before, and $sim(i,j)$ is the correlation coefficient of i and j [14, 15]. After computing we will obtain a correlation matrix, describe correlation of each page with target page, as shown in Fig. 1.

	...	*i*	*j*	...	*n*	...
...	1					
i		1				
j			1			
...				1		
n					1	
...						1

Fig. 1. Web pages correlation coefficient matrix. Row and column name represent item name.

From each row in Fig. 1 we could know, how much is other pages correlated to the target page. We obtained a list of correlated items presented $L_i = \{p_j, p_k, \ldots, p_n, \ldots\}$ sorted by $sim()$ in descending order, $p = \{sim(), T, D\}$ T is sets of users u visiting time on item, D is sets of users u steps of finding item. Next, we test the significant difference of pair $sim()$ inside L. To avoid the possibility of not satisfying the assumption, we conduct a transform as in (2) 16.

$$z = \frac{1}{2}\ln\left(\frac{1 + sim(i,j)}{1 - sim(i,j)}\right)$$

Hypothesis test of $sim()$ as in (3) 16.

$$\mathrm{H}_0: sim_1 = sim_2$$
$$\mathrm{H}_1: sim_1 \neq sim_2$$

sim_1, sim_2 represent correlation coefficient of page with target page. Transform correlation coefficient to z-value in order to conduct hypothesis testing, under the assumption of independence, the standard error between z_1, z_2 as in (4) 16.

$$\mathrm{SE}(z_1 - z_2) = \sqrt{\frac{1}{2}\left(\frac{1}{n_1 - 3} + \frac{1}{n_2 - 3}\right)}$$

The z-score of $z_1 - z_2$ z-value as in (5) 16.

$$z = \frac{z_1 - z_2}{\mathrm{SE}(z_1 - z_2)}$$

If rejected the alternative hypothesis $\mathrm{H}_1: sim_1 \neq sim_2$, p_1 and p_2 would say to have the same relation to the target page, thus we further testing user's reading time on p_1 and p_2, via wilcoxan rank sum test, hypothesis given as in (6).

$$\mathrm{H}_0: T_1 = T_2$$
$$\mathrm{H}_1: T_1 \neq T_2$$

(T_x: visiting times on item x)

If we reject $\mathrm{H}_1: T_1 \neq T_2$ this proclaim no significant differences in visiting time between p_1 and p_2. While condition established, we further test the steps of finding between p_1 and p_2 by using wilcoxan rank sum test, hypothesis as in (7).

$$\mathrm{H}_0: D_1 = D_2$$
$$\mathrm{H}_1: D_1 \neq D_2 \quad (1)$$

(D_x: steps of finding item x)

Finally, if $\mathrm{H}_1: D_1 \neq D_2$ is rejected, p_1 and p_2 will be assign the same ranking. Algorithm flow chart Fig. 2.

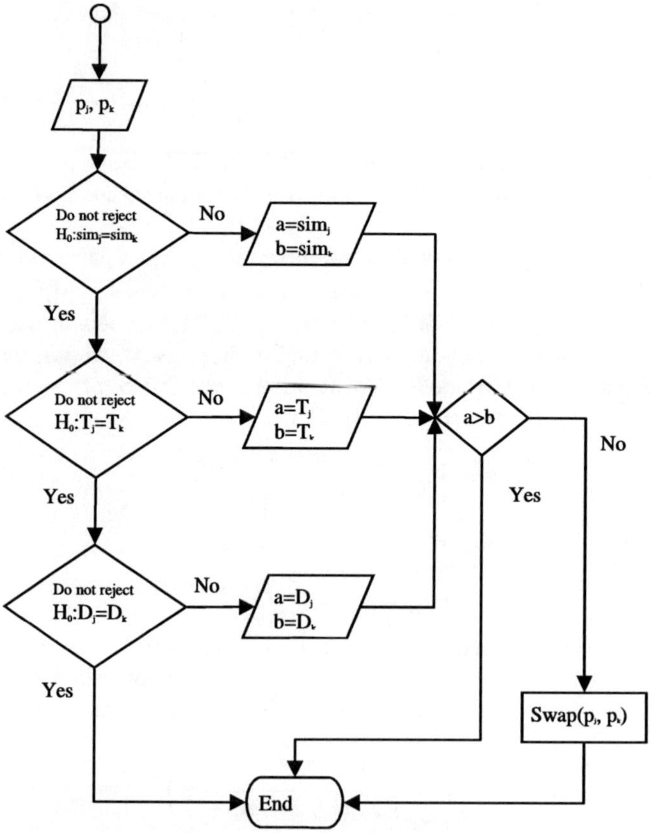

Fig. 2. The control flow of the proposed hybrid algorithm.

3 Experiment Design

A. Data Preprocessing

The IIS log contains the access paths of web pages in the web site including pictures, sounds, graphics, component objects, static web pages and other different files However, not all of the data are meaningful, therefore, to increase the analysis efficiency and accuracy, it is necessary to precede data preprocessing. We removed access paths to the graphics and component objects with file extension gif, jpeg, jpg, tif, bmp, ico, css, js…etc [17, 18]. In addition, web crawlers have been extensively used in web search engine to collect web site information. Crawlers will leave tremendous meaningless data inside the IIS log, which could greatly affect analysis on user visiting pattern, for this reason we excluded crawlers visiting information 19.

B. Data Set

The data source was the web access log (IIS log) of a website called Health 99, operated by the Bureau of Health Promotion, Taiwan. The log was archived from January 2007 to August 2009, consisting of 298,637,615 records in total, Health 99 has been a health education resource website since October 1998. It provides the public and health educators with accurate and timely health information and educational materials. Until December 2009, this web site has more than 3300 teaching material 20. After data cleaning, there were 25,071,810 records. The database was divided into testing (20%) and training (80%) sets 21. Our algorithm will be trained on each week data in training set to evaluate the recall on next coming week data in testing set.

C. Experiment Procedure

The experiment procedure is as follows:

- First, we arbitrary choose a page as a target page and a date as observation start day (2009/03/20), observing the after 12 weeks, and recall was calculated weekly.
- Second, we consider a week before each observation week as its training set, as shown in Fig. 3.
- Third, we compute the similarity between items using adjusted Pearson correlation from training set.
- Fourth, we sort the correlated items in correlation coefficient (*k* nearest) and our new method.
- Fifth, we chose top N item for the target page. In the experiment, we let N take the values 5, 10 and 20.

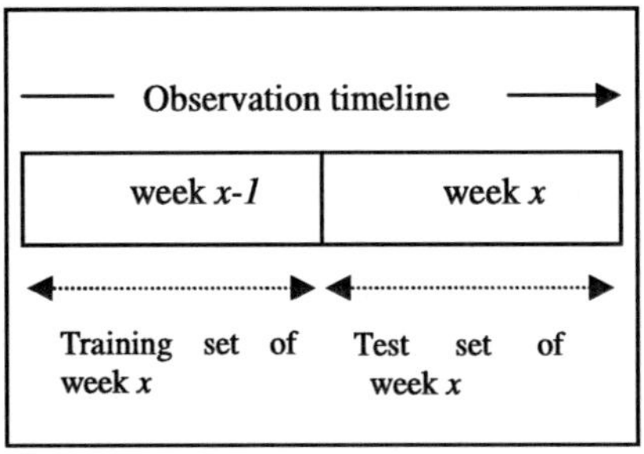

Fig. 3. Training set and testing set of week *x*.

4 Result and Conclusion

To evaluate the recall of our algorithm and the original collaborative filtering algorithm, data were divided into train set and test sets. Through each algorithm we will obtain a ranked list. Top N means the highest n web pages after ranking. Our purpose is to look into the testing set and match the items in our recommendation Top N set. The items that appear in both sets are the members of the hit set. Recall rate is defined as in (8) [21, 22].

$$recall = \frac{size\ of\ hit\ set}{size\ of\ test\ set}$$

We arbitrary choose a page as a target page and a date as observation start day. Observing 12 weeks recall under different Top N (5, 10 and 20). Each weeks' training set was based on a week before as shown in Fig. 3.

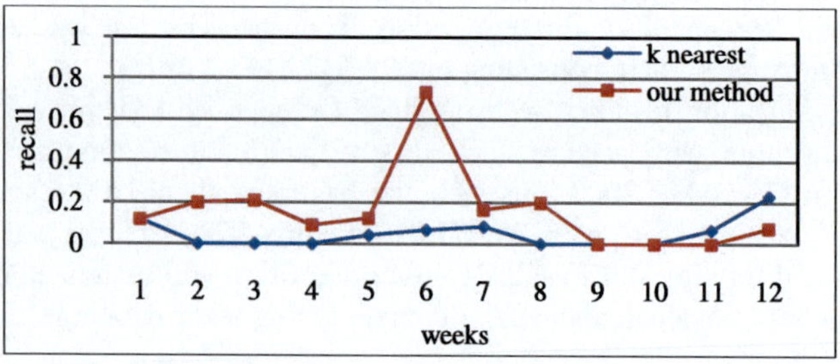

Fig. 4. Top 5 recall in 12 weeks

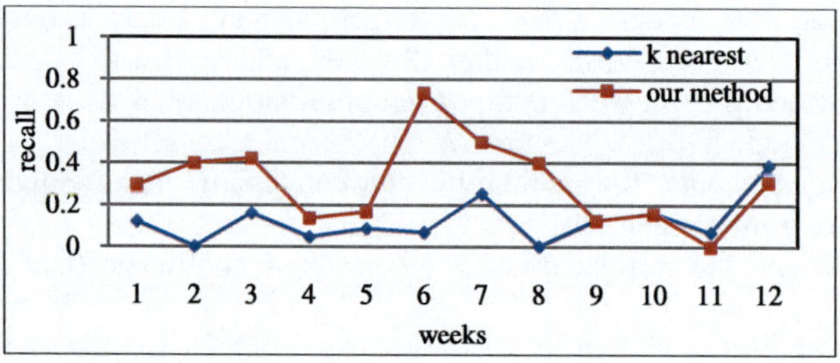

Fig. 5. Top 10 recall in 12 weeks

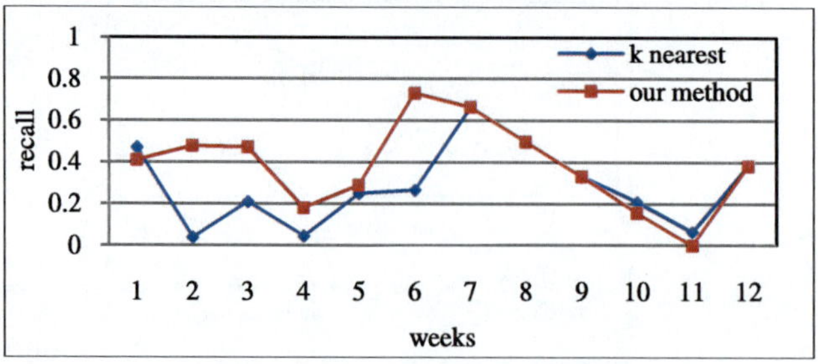

Fig. 6. Top 20 recall in 12 weeks

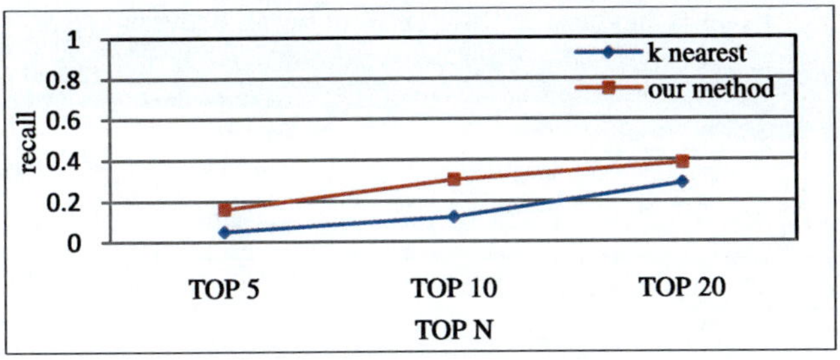

Fig. 7. Average recall for 12 weeks at different Top N

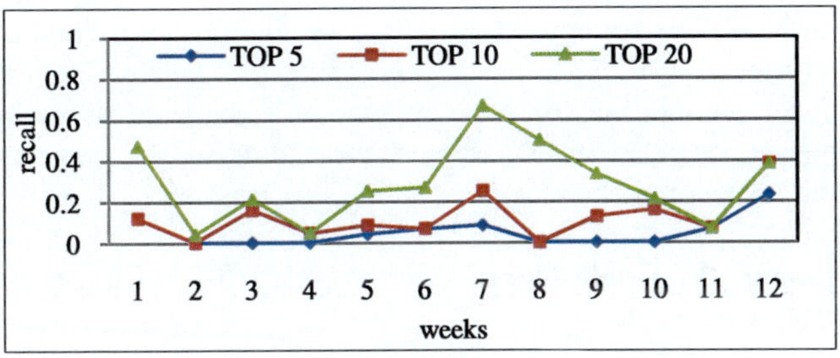

Fig. 8. k nearest recall at different Top N

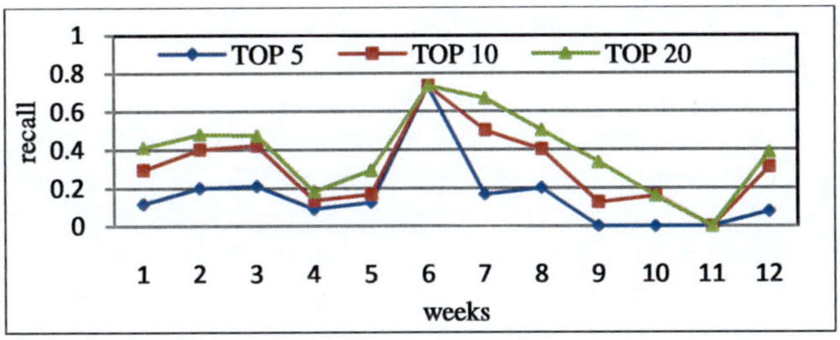

Fig. 9. Our method recall at different Top N

Table 1. Variance of Recall differences

	10–5	20–5	20–10
Our method	0.010666685	0.021474131	0.004451
k nearest	0.005522721	0.03218275	0.030594

From Fig. 4, 5, 6 and 7 shows a higher recall in our new method. Besides, as shown in Table 1, lower variation of differences between Top Ns.

Table 2. Significance Differences of Recall Between Top N

		Our method		
		Top 5	Top 10	Top 20
k nearest	Top 5	0.05019	0.6633	0.04596*
	Top 10	0.00062*	0.00920*	0.8624
	Top 20	0.00041*	0.00217*	0.2597

By using wilcoxan sign rank test on paired recall of different Top N we obtained Table 2 (cell inside represent p-value, p>0.05 no significant difference). No significant differences were found between novel Top 5 and original Top10; novel Top 10 and original Top 20; novel Top 20 and original Top 20. Thus with lower Top N, in our result re saw a higher recall. According to our experimental results, it is obviously higher probabilities of user accept the recommend pages by our method than only consider correlation, in particular more outstanding performance but with smaller number of recommend pages.

Acknowledgment. We would like to thank Professor I-Jen Chiang of TMU for his valuable advice in statistics.

References

1. Cline, R.J., Haynes, K.M.: Consumer health information seeking on the Internet: the state of the art. Health Educ. Res. 16(6), 671–692 (2001)
2. McKinley, J., Cattermole, H., Oliver, C.W.: The quality of surgical information on the Internet. Journal of the Royal College of Surgeons of Edinsburgh 44, 256–268 (1999)
3. Jacobson, R.L.: Taming the Internet. The Chronicle of Higher Education 41(32), A29–A30 (1995)
4. Pereira, J., Bruera, R.: The Internet as a resource for palliative care and hospice: a review and propasals. Journal of Pain and Symptom Management 16(1), 59–68 (1998)
5. Srikant, R., Yang, Y.: Mining web logs to improve website organization. In: World Wide Web, pp. 430–437 (2001)
6. Garofalakis, J., Kappos, P., Mourloukos, D.: Web site optimization using page popularity. IEEE Internet Computing, 22–29 (July-August 1999)
7. Perkowitz, M., Etzioni, O.: Adaptive Web sites: automatically learning from user access patterns. In: Proc. 6th Int. World Wide Web Conference, Santa Clara, CA (1997)
8. Nichols, D.M.: Implicit Rating and Filtering. In: Proceedings of the 5th DELOS Workshop on Filtering and Collaborative Filtering, pp. 31–36 (1997)
9. Wu, C.-H.: A New Approach of Auto-recommendation using Implict Feedback (unpublished)
10. Morita, M., Shinoda, Y.: Information Filtering Based on User Behavior Analysis and Best Match Text Retrieval. In: Proceeding of the 17th Annual International ACM SIGIR Conference on Research and Development in Information Retrieval, pp. 272–281 (1994)
11. Konstan, J.A., Miller, B.N., Maltz, D., Herlocker, J.L., Gordan, L.R., Riedl, J.: GroupLens: Applying Collaborative Filtering to Usenet News. Communication of ACM 40(3), 77–87 (1997)

12. Oard, D.W., Kim, J.: Implicit Feedback for Recommender System. In: Proceeding of the AAAI Workshop on Recommender System, pp. 81–83 (1998)
13. Oard, D.W., Kim, J.: Modeling information content using observable behavior. In: Proceeding of the 64th Annual Conference of the America Society for Information Science and Technology Annual Meeting, pp. 481–488 (2001)
14. Sarwar, B., Karypis, G., Konstan, J., Reidl, J.: Item-based collaborative filtering recommendation algorithms. In: Proceedings of the 10th International Conference on World Wide Web, Hong Kong, May 01-05, pp. 285–295 (2001)
15. Liu, B.: Web Data Mining, 1st edn., pp. 480–482. UOS Press, Springfield (2007)
16. Shan, M.L.: Experimental Design, 1st edn., pp. 421–423. Jeou Chou Book Co, Ltd. (1993)
17. Su, C.-P.: Apply Web Mining Techniques to Analyse the Navigation Behaviour of Visitors-Using Online Content Site as Example (unpublished)
18. Cooley, R., Mobasher, B., Srivastava, J.: Data preparation for mining World Wide Web browsing patterns. Knowledge and Information Systems 1(1) (1999)
19. Yu, J.X., Ou, Y., Zhang, C., Zhang, S.: Identifying interesting visitors through web log classification. IEEE Intelligent Systems 20, 55–59 (2005)
20. Department of Health Executive Yuan Bureau of Health Promotion Health 99 Education Resouce website, `http://health99.doh.gov.tw/en/index.aspx`
21. Sarwar, B., Karypis, G., Konstan, J., Riedl, J.: Analysis of Recommendation Algorithm for E-Commerce. In: Proc. ACM Conference on Electronic Commerce, Minnesota, pp. 158–167 (October 2000)
22. Shen, L., Zhou, Y.: A Ner User Similarity Measure for Collaborative Filtering Algorithm. In: International Conference on Computer Modeling and Simulation, pp. 375–378 (2010)

Environment Settings in College Computer Room

Feifei Yin[1] and Jing Yi[2]

[1] Information & Network Management Center, North China Electric Power University, Baoding, China
yff1021@163.com
[2] Department of Computer Science, North China Electric Power University, Baoding, China
305245040@163.com

Abstract. With the widely use of computer, to deal with more frequent computer tests and the increasingly onerous teaching tasks, and according to the requirement of the NCRE (National Computer Rank Examination) on test settings, realizing the automatic setup of system time、username and mapping network drives by applying some ways such as Visual Basic language and Dos command will meet the special needs of teaching、improve the work efficiency greatly and reduce the burden of computer rooms management teachers.

Keywords: computer room, IP change, network mapping, NCRE, Dos command.

With the widespread application of computer, computer room in university teaching takes up more and more teaching task. Up till now,there are some subjects which use computer to exam.The subjects include computer basic, computer programming language, computer graphics, multimedia, mathematics, English, AutoCAD, Matlab and so on, and involve the school. When the subject ending,there are thousands of students attending the exam. In addition, A part of National Computer Rank Examination(called NCRE) which host by test centre of the Ministry of Education use this way to exam. Certificate issued by National Computer Rank Examination come into favor of the unit. This also makes the students to attend the test. These lead to the current register rising, but the nuamber is still increasing the number of students signing up to take the examination is more than 5000,and is increasing.These exams, turnout is heavy and time is concened.All of these require a convenient and safe manner.So we adopt the mapping network drives.But the subjects is more and the number of students is large,this requires every manger is responsible for hundreds of computers' installation and maintenance. How to finish the system software of the computer in a short time, the installation of test software, and complete the relevant tests, ensure that the exam is going on smoothly, can be said to be a problem.

1 System Install

Demand on the NCRE's requirement is always changing. A few years ago, the system uses the Windows2000,but from the 29th NCRE uses the WindowsXP. In recent years,we use the WindowsXP system in the examination. The system utilizes Harddisk

W. Zhang (Ed.): Advanced Technology in Teaching, AISC 163, pp. 235–238.
springerlink.com

Protect System,and divide the computer hard disk into several sectors. Each sector install a separate system which corresponds to an examination, and according to the different test install software. After installation of the master machine which install the computer by hand in the initial stage,we use the network transmission, and copy the installed system to every computer int the room. There are several other issues you must understand before the network transmission system.

(1) The system of Master must follow the examination requirements,and put all the patch, and kill the viruses. After confirming the Safety, the anti-virus software will be uninstalled, becausesome examinations such as NCER do not require the installation of the software.

(2) In order to prevent the influence of exam, we will completely remove the races, while install the system, such as some temporary files, browser history.

After transmission, using Harddisk Protect System automatically change the IP Address and computer namem.And hidding the system after protection,this will prevent the false operation on system, and provide convenience to set the exam enviroment.

2 Test Settings

Once the system is in place, the setting of test consists of three aspects, time, user, mapping network drives. The examination involves thousands of computers' setting, and the different test needs of different environment, so it greatly increases the task of manager. For manager's convenience and little variation in every computer We can use the program to automatically set.

First, using Visual Basic language and Dos commands to generate an executable file that called set.exe and make with the system transmission. But the program can not run automatically, it Still needs manager to click manually one by one.

There are amount of work. In order to further simplify the work, we first place in system startup directory a file namedauto.exe, and make it run automatically when the system startup. The Internal procedure is as follows:

```
Dim Set_str As StringSet_str = "D:/set.exe"
Shell Set_str
'run D:/set.exe
Unload Me
```

When the computer startup, auto.exe automatically run,and its major function is to run the test setting that is set.exe.

Next, we design the test setting program that is set.exe, and the main contents are as following.

2.1 Time Synchronization

There are a lot of computer, we can not guarantee that all the computer time is accurate, But some examination is very strict on time, for example, NCRE. So before the examimation, Managers must check time, and prevent the mistake which affect the examination. Such way waste manpower.Even though, because of manual operation, it is hard to avoid the mistake.So we can use Dos command to slove the problem. It design as follows :

```
Dim Sev_Ip_Str As String
Definition with which server synchronization
Dim Shell_Time_Str As String
Definition the sentence executed by shell command
Sev_Ip_Str = "\\211.82.232.201"
'represent the IP of sever such as A_Server
Shell_Time_Str = "net time " + Chr(34) + Sev_Ip_Str + Chr(34) + " /set /yes"
Shell Shell_Time_Str
```

2.2 Change User Name

As the computer name or IP has regular pattern, that is these are in order, the Examination setting such as NCRE also has a similar rule.The program design as follows:

```
ServerName = ""
OldName = Environ("username")
'original username
UI0.usri0_name = StrPtr(NewName)
   'NewName is compute based on IP
dwLevel = 0
OldName=StrConv(OldName,vbUnicode)
ServerName = StrConv(ServerName, vbUnicode)
lRet = NetUserSetInfo(ServerName, OldName, dwLevel, UI0, 0&)
'Make use of API function to change the username
```

In order to make the change to take effect, the computer needs to restart or logout after changing the username.

```
ExitWindowsEx EWX_FORCE, 0
```

For NetUserSetInfo and ExitWindowsEx fucntions the aforesaid used must declear before use. Calling the API function in VB must pay attention to the transition betwwen Unicode code and ASCII code in a string.

2.3 Mapping Network Drives

Use of mapping network disk to setup test environment, so the server generate multiple users and folders,and the users with folders have the same name.Each user and folder corresponding to a computer in the room, and it can limit the authority, so that it can only access the corresponding folder. Mapping Network Disk use the Dos Command, take the 31st National Computer Rank Examination environment setting as an example:

```
Dim Delete_str As String
Delete_str = "net use k: /delete"
Shell Delete_str
'removed prior to mapping before the new mapping
Sev_str = "\\211.82.233.201\Ncre31\"
She_str = "net use K: " + Chr(34) + Sev_str + Chr(34) + " /user:" + Chr(34) +
"Administrator" + Chr(34) + " /persistent:yes"
'the representation of mapping is the disk of K
Shell She_st
```

2.4 Delete the Program

When the set.exe file operation and set up the test environment, but there is a file, auto.exe, in the system startup directory,in order to prevent the influence of exam, you have to delete it, so at last, auto.exe files will be deleted automatically. The program design as follows:

```
Dim Del_str As String
  Dim Path_str As String
  Path_str="C:\Documents and
Settings\Administrator\StartMenu\Program\Sartup\auto.exe"
Del_str = "cmd /c del " & Chr(34) & Path_str & Chr(34)
'deletet the file auto.exe in the directory of startup
Shell Del_str
```

Program run automatically the auto.exe when the computer startup first time, then run set.exe.It deletes the auto.exe after set.exe setting the enviroment. In formal examintion, use computer hard disk protection system to hide the disk of D.In this way, it not only greatly reduces the workload of the managers, but also have no influence on the examination system.

3 Conclusion

In this paper, We have discussed the problem in the NCRE exam,and used the Visual Basic Language, DOS command to carry out the system setting of exam automatically. This greatly improves the working efficiency. From the recent organizational processes of computer examination,we can get a conclusion that These measures have played a very significant effect.

References

1. Luo, X.: Talking about the Preparation Work of Computer Experiment Room of NCRE. Modern Computer 7, 97–98 (2009)
2. Chen, X.: Autoit Realization the Calculator Grade Examination Modification Users Name. Computer Knowledge and Technology 3, 632–633 (2010)
3. Hagler, M., Marcy, W.M.: Authentic and group learning technology using computer networks and intelligent tutors. In: Twenty-Third Annual Conference on Frontiers in Education Conference, pp. 467–470 (1993)

Preparation of Polyclonal Antibody and Expression Analysis of *Gr* in Tomato

Yuanhong Xie[1], Benzhong Zhu[2], Yunbo Luo[2], Xiangning Chen[1], and Hongxing Zhang[1,*]

[1] Beijing Key Laboratory of Agricultural Product Safety Detection and Control, Department of Food Science, Beijing University of Agriculture, Beijing, China 102206
hxzhang51@163.com
[2] College of Food Science and Nutritional Engineering, China Agricultural University, Beijing, China 100083

Abstract. The fruit ripening of Green-ripe (Gr) mutant tomato was inhibited dramatically. To determine the expression patterns of Gr in tomato, we first produced the polyclonal antibody of Gr protein. RT-PCR was used to amplify the Gr gene from green ripe tomato fruit. And the PCR product was subcloned into prokaryotic protein expression vectors pET-30a to generate recombinant plasmid. The Gr protein was induced by IPTG in BL21 (DE3) and purified by Ni-NTA agarose column. The anti-Gr serum was produced by immunizing rabbits, and the titer of the anti-Gr serum was above 5000 by ELISA analysis. Purified by the DEAE-52 ion-column, the high purification level of anti-Gr polyclonal antibody was obtained. Furthermore, RT-CPR was used in the RNA level to demonstrate that the expression of Gr gene was specialized in some cultures of tomato. For example, the expressions of Gr were higher in seed, flower and green ripe fruit than others, and the expression level were reduced by exogenous ethylene treatment in the flower and green ripe fruit. Moreover, Polyclonal antibody of Gr was used to investigate the expression pattern of Gr in protein level by the Western blotting. Our results show that the expression level of Gr in protein level was complied with the expressions in RNA. So, we suggested that the regulation of Gr was transcriptional.

Keywords: Green-ripe mutant, Prokaryotes expression, Polyclonal antibody, Ethylene.

1 Introduction

Tomato fruit ripening mutant Green-ripe (Gr) is an autosomal dominant mutant, obtained by natural mutation (Kerr. 1958). The fruit of Gr mutant was obviously immature, and the softening and lycopene accumulation of the Gr fruit was significantly inhibited during the fruit ripening (Jarret et al., 1984). Further studies showed that, during the fruit ripening, there was no obvious difference of the ethylene

* Correspondent author.

W. Zhang (Ed.): Advanced Technology in Teaching, AISC 163, pp. 239–247.
springerlink.com

production between the mutant fruit and wild type tomato fruit, the system Ⅰ ethylene and system Ⅱ ethylene were both detected (Barry et al., 2005). It indicated that the ethylene synthetic of the mutant tomato fruit was not affected, and the immaturity was due to the ethylene insensitivity of fruit. In addition, previous study found that the mutant fruit of Gr was obvious ethylene insensitive, exogenous ethylene treatment did not cause obvious ethylene reaction in Gr fruit, and the expression of ethylene relative genes were no change. Also, in the stem, leaf, organization and organs of the Gr mutant, also show different degree of ethylene insensitivity (Barry et al., 2006). And research suggested that *Gr* gene locate in the chromosome 1 of the tomato, and the length is 732bp, coding protein for 244 amino acid. The molecular weight of Gr protein is about 28KD, isoelectric point is 6.8.

Moreover, research indicated that, Gr mutant is cased by the lack of a 334 bp at the 5' sequence of *Gr* gene, which led to insensitivity to ethylene (Barry et al., 2006). The purpose of this study was to clone the tomato *Gr* gene, express the Gr protein in vitro, and prepare the polyclonal antibody of the Gr. Further, we discusses the expression patterns of *Gr* in both RNA and protein level.

2 Materials and Methods

2.1 Plant Materials

Wild-type tomato fruits (Lycopersicon esculentum cv Ailsa Craig) was grown in a heated glasshouse using standard cultural practices with regular additions of N, P, K fertilizer and supplementary lighting when required. Flowers were tagged at anthesis, and fruits were harvested at the following stages: green ripe stage, broken stage, and red ripe stage.

2.2 Cloning of *Gr* Gene and Construction of Prokaryotic Expression Vector

Total RNA was extracted from the tomato fruit (Lycopersicon esculentum cv Ailsa Craig) by the method (Xie et al., 2006), and contaminating DNA was removed using Dnase (Promega) according to the manufacturer's instructions. First-strand cDNA was synthesized with the Reverse Transcription System Kit (Promega), using equal amounts of oligo(dT), according to the manufacturer's instructions. Primers was used to amplify the *Gr* gene , sense primer 5'CGGAATTCATGGCTAGGGCACAAC 3'containing a EcoRI site, antisense primer 5'CCGCTCGAGTCACTGGATTTGGTG with a Xho Ⅰ site. The condition for amplification was at 94℃ for 3 min followed by 35 cycles at 94℃ for 30 s, at 57℃ for 30 s and at 72℃ for 30 s, plus a final extension at 72℃ for 10 min. Three independent PCR products were purified and cloned into pGEM T-easy vector (Promega) and sequenced. The PCR product of *Gr* and the expression vector pET30a were digested both with EcoRI and Xho Ⅰ, and connected with each other by T4 DNA ligase. The restruction was named as pET-Gr.

2.3 Protein Expression, Purification and Polyclonal Antibody Preparation

The recombinant was transformed into Escherichia coli BL21 (DE3) after pET-Gr was selected by restriction analysis. Gr protein expression and purification were performed as previous methods (Zhu et al., 2007) with small modification.

The preparation of polyclonal antibody was preformed according to the method (Zhu et al., 2007) and modified. Four New Zealand adult male rabbits were simultaneously immunized with a 200 μg mixture of fusion protein and an equal volume of complete Freund's adjuvant. Rabbits were re-immunized again after 2 weeks with a 100 μg mixture of fusion proteins and the same volume of incomplete Freund's adjuvant. Seven days after the final immunization, the rabbit's blood serum was harvested from the carotid artery and the polyclonal antibodies purified by salting (between 0.3 and 0.5 saturation with ammonium sulfate). The purified antibody was supplied in phosphate-buffered saline, sterile-filtered, and containing 0.02% sodium azide.

2.4 Transcript Expression of *Gr* Gene by RT-PCR

Isolation of total RNA from tomato was performed according to Xie et al. (2006). Semi-quantitative RT-PCR was performed as described by Fu et al. (2005). First-strand cDNA was synthesized using 2.5μg total RNA, 0.5μg oligo d(T)18 primer and M-MLV reverse transcriptase (Promage) to a final volume of 20 μL. RT-PCR analysis was used to investigate the expression pattern of *Gr* according to the method (Fu et al, 2007). PCR was performed in 50μL reactions using 1μL cDNA as a template. The specific primers were designed as 5'-CGGAATTCATGGCTAGGGCACAAC 3' 下游引物： 5'-CCGCTCGAGTCACTG GATTTGGTG 3'. And 129 bp fragment of endogenous tomato ubiquitin gene *Ubi3* was amplified as a control using the primers 5'-CAGGACAAGGAAGGGATT-3'and 5'-GTAGAGCAC GAGGCAGAG-3'. The condition for amplification was at 94℃ for 3 min followed by 30 cycles at 94℃ for 30 s, at 53℃ for 30 s and at 72℃ for 30 s, plus a final extension at 72℃ for 10 min. The PCR products were separated on a 1% agarose gel and compared to the amount of *Gr* transcript.

2.5 Western Blotting Analysis

5g of Wild type and ethylene treated tomato samples were ground in liquid nitrogen, and solubilized with the protein extracting buffer (400 mmol/L Tris-HCl pH7.5， 200 mmol/L NaCl， 800 mmol/L sucrose， 20 mmol/L EDTA， 10 mmol/L DTT， 2 mmol/L PMSF， 0.1% Tween-20）.

All the samples were centrifuged at 13000g for 10min. The suspension was collected as crude extraction for further analysis. Total protein was quantified by the method (Bradford, 1976). The western blotting was preformed. And the polyclonal antibody of Gr was used as primary antibody, and alkaline phosphatase marked IGg was used as detection antibody. The BCIP/NBT was used to detect the result.

3 Results

3.1 Construction of Prokaryotic Expression Vector

By RT-PCR, the *Gr* was cloned and sequenced. A 732 bp product was obtained (Fig 1). And by sequencing, we obtained the complete coding region of the *Gr*. This sequence was completely complied with the original sequence in the NCBI databases (No.DQ372895). Moreover, the *Gr* was inserted into pET-30a expression vector after digested with EcoRI and Xho I . The positive clone of the *Gr* inserted into pET-30a vector was named as pET-*Gr*, and identified by digestion (Fig 2). And the recombinant plasmids pET-*Gr* was transformed into E. coli BL21 cells.

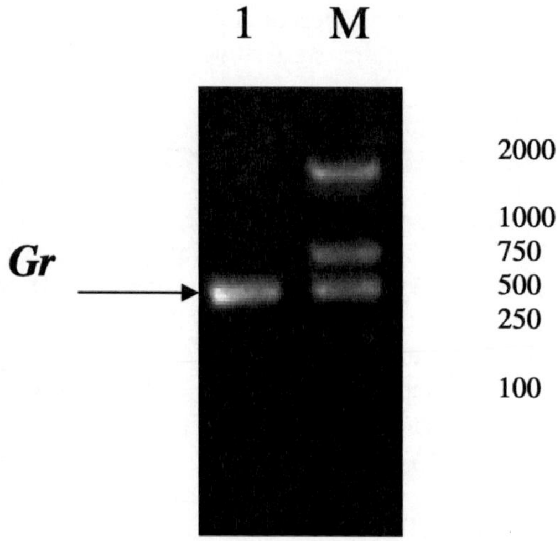

Fig. 1. RT-PCR amplication of *Gr*
M : DL2000 marker 1 : PCR production of *Gr*

3.2 Protein Expression, Purification and Polyclonal Antibody Preparation

In order to obtain recombinant protein on a large scale in the supernate, expression conditions were optimized using different IPTG concentrations (0.1, 0.5 and 1.0 mM), and no significant difference was observed, all the selected conditions could obviously induced the expression of recombinant protein (Fig 3). The soluble fraction was purified with Ni-NTA agarose column according to the manufacturer's instructions (Novagen). We obtained the purified Gr recombinant proteins as shown in Figure 4.

To prepare the polyclonal antibody of the Gr protein, New Zealand adult male rabbits were immunized. After blood, the titer of the anti-Gr serum was above 5000(Table 1).The polyclonal antibody was firstly precipitated by ammonium sulfate, and then DEAE-52 ion exchange method was used to further purify the polyclonal antibody. As shown in Fig 5, 4-7 tubes were collected and the antibody was dialyzed with deionized water. By western-blotting, the purified Gr polyclonal antibody had immunization specificity and could use for further analysis (date not show).

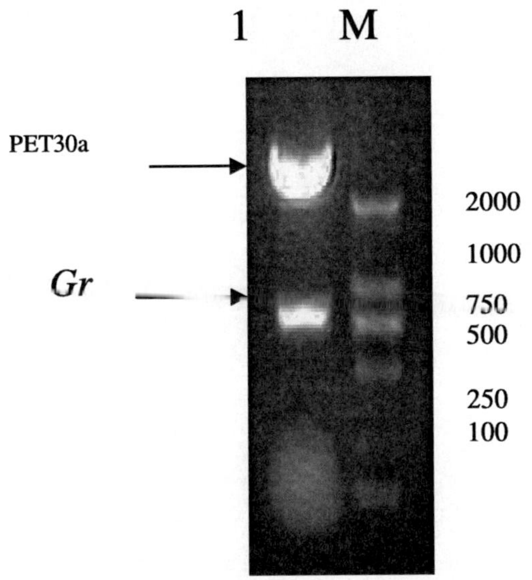

Fig. 2. Identification of recombinant pET-*Gr* plasmid
M : DL2000 Marker 1 : EcoRI / Xho I digestion product

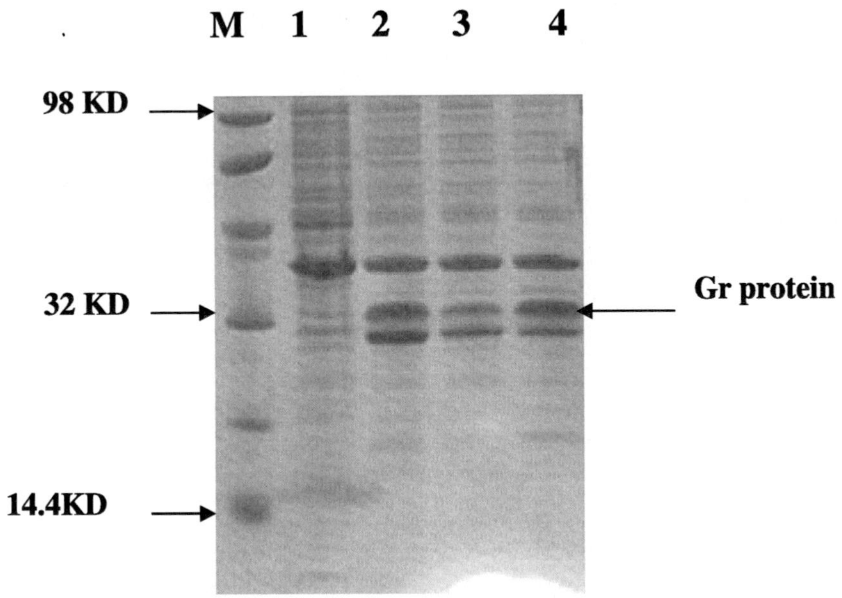

Fig. 3. Induced expression of Gr protein in E. coli
M : Marker; 1 : control; 2 : Induced by 0.1mmol/L IPTG; 3 : Induced by 0.5mmol/L IPTG ; 4 : Induced by 1mmol/L IPTG

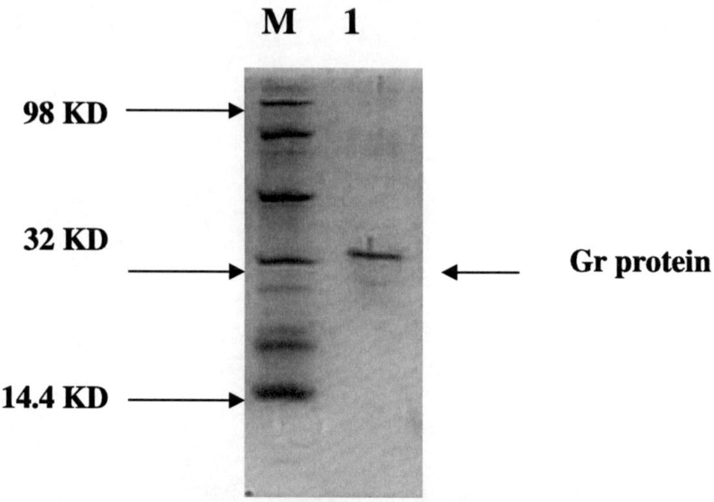

Fig. 4. Purification of Gr protein
M : Marker; 1 : Purification of Gr protein

Table 1. The titer of the anti-Gr serum

Diluted times of Serum	1/1000000	1/100000	1/5000	1/1000	1/100
OD492	0.04	0.3	1.1	1.8	>2
Negative control OD492	0	0	0.05	0.08	1.2

3.3 The Expression Patterns of *Gr* under Ethylene Treatment

To further examine the expression of Gr under ethylene treatment, both RT-PCR and Western-blotting were used. As shown in Fig 6, in different tissues and organs of wild type plant, the expression of *Gr* gene was not the same. In the tomato seeds, the *Gr* expression was significantly higher than that in stems, leaves and stems. And in the tomato fruits, the *Gr* expression in green ripe fruit was significantly higher than that in broken stage and red ripe fruit. When treated with exogenous ethylene, the transcript expressions of *Gr* had no obviously changed in seed, stem, leave and green ripe fruit of wild type tomato. However, the transcript expressions of *Gr* in flower and broken stage fruit slightly decreased compared with that in untreated tomato (Fig 6).

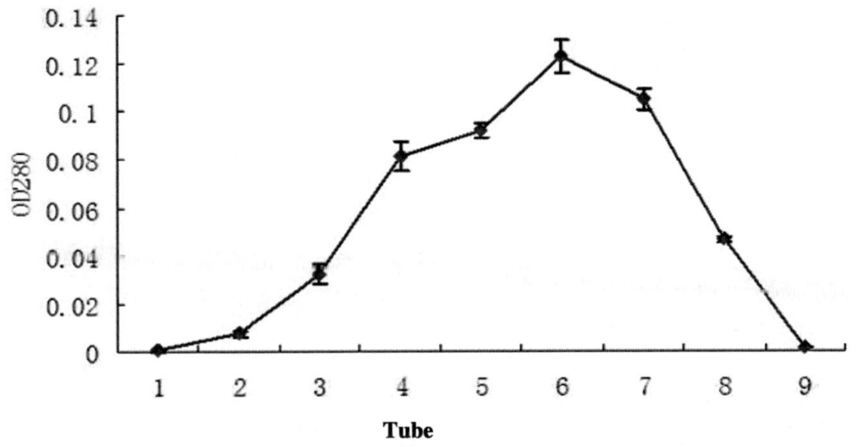

Fig. 5. Purification of Gr polyclonal antibody by DAEA-52

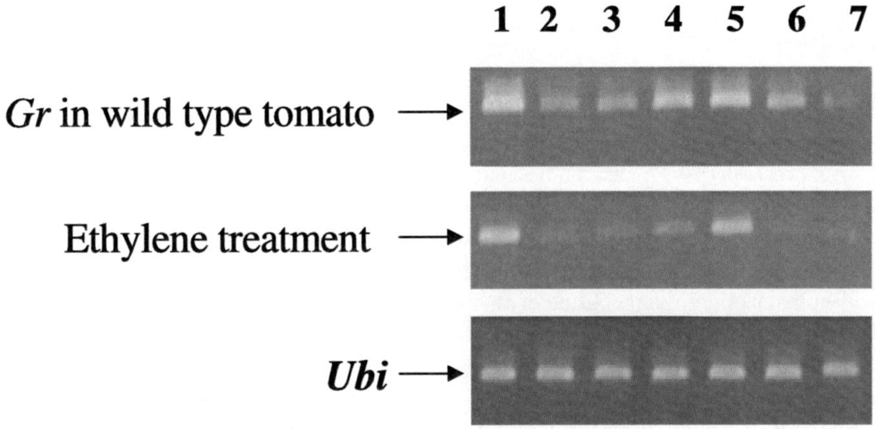

Fig. 6. RT-PCR analysis of *Gr* gene in tomato
1: Seed ; 2: Stem ; 3: Leaf ; 4: Flower ; 5: Green ripe fruit ; 6: Broken fruit ; 7: Red ripe fruit.

Further, we also investigated the Gr expression in the protein level using the Gr polyclonal antibody. Western-blotting results demonstrated that the Gr protein was low in stem, leaf and red ripe fruit of the wild type tomato, and could detect slightly. In contrast, Gr protein expression in the seeds and green ripe fruit is much higher than other samples (Fig 7). Overall, the Gr expression in protein level complied with that in RNA level in wild type tomato. When treated with exogenous ethylene, the expression of Gr protein obviously decreased in flower, green ripe fruit and broken stage fruit than control (Fig 7). All these results suggested that exogenous ethylene decreased the expression of Gr in flower, green ripe fruit and broken stage fruit both in RNA and protein level.

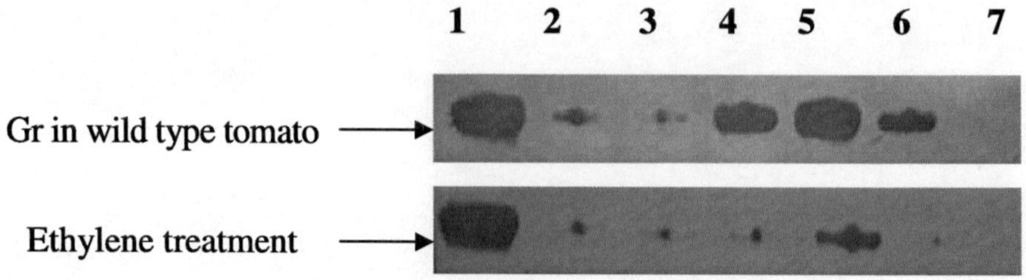

Fig. 7. Western blotting analysis of the Gr protein in tomato
1: Seed ; 2: Stem ; 3: Leaf ; 4: Flower ; 5: Green ripe fruit ; 6: Broken fruit ; 7: Red ripe fruit.

4 Discussion

Tomato is a model plant used to investigate the ripening mechanism of fruit. And the tomato fruit ripening mutants are also important materials. By using these mutants, many ripening related genes were cloned and characterized (Lanahan et al., 1994; Barry et al., 2005). Among these mutants, the Green-ripe (Gr) was also investigated. Previous study indicated that the *Gr* gene took part in the ethylene pathway of tomato fruit (Barry, et al., 2006). According to homology analysis, *Gr* gene has a high similarity with the *RTE* in Arabidopsis, and the homology reached 85%. And studies show that *RTE* involved in ethylene signal transduction in Arabidopsis. Studied suggested that the *RTE* negatively controlled the ethylene signal transduction, and over-expression of *RTE* cased obviously ethylene insensitivity in Arabidopsis (Resnick et al., 2006). Moreover, *RTE* acted at the upstream of ethylene receptors in the ethylene signal pathway (Barry, et al., 2006). So, the mechanism of fruit ripening controlled by *Gr* was especially important.

Similar as other mutant (Chao et al., 1997; Imanishi et al., 2001; Solano et al., 1998; Tieman et al., 2001) , there was a significantly ethylene insensitivity in flower and fruit of the Gr mutant. However, the ethylene induced triple response of the Gr was similar with wild type plant, which indicated that the ethylene insensitivity of the Gr was tissue specific (Barry, et al., 2005). Barry found that the Northern blotting method could not detect the transcript expression of *Gr* in wild type tomato, but could only detect the transcript expression in mutant fruit. In the Gr mutant fruit, the *Gr* expression was highest in green ripe fruit, and exogenous ethylene did not significantly affect the expression of *Gr* (Barry, et al., 2005). Among these researches, the expression of Gr in protein level was not studied. In this research, we used RT-PCR to investigate the transcript expression of *Gr* in wild type tomato. Our results suggested that the transcript expression of *Gr* was higher in seeds, flower and green ripe tomato fruit than other tissues (Fig 6). Different from the Gr mutant, exogenous ethylene decreased the *Gr* expression in flower and green ripe fruit of wild type tomato (Fig 6).

To further investigate the Gr expression in protein level, we first expressed the Gr protein in prokaryotic cells (Fig 3), and purified the recombinant protein (Fig 4). By immunized the New Zealand adult male rabbits, the polyclonal antibody of the Gr protein were prepared and purified (Fig 5). Furthermore, western-blotting was preformed to examine the expression of Gr protein. Our results suggested that the Gr protein was higher in seeds, flower, green ripe fruit and broken fruit than other tissues. And exogenous ethylene decreased the Gr protein in flower, green ripe fruit and broken fruit (Fig 7). All these results indicated that the regulation of *Gr* was transcriptional.

Acknowledgment. The item of Beijing University of Agriculture young science fund provided financial assistance.

References

1. Barry, C.S., Ryan, P.M., Andrew, J., Thompson, G.B., Seymour, D.G., Giovannoni, J.J.: Ethylene insensitivity conferred by the Green-ripe and Never-ripe 2 ripening mutants of tomato. Plant Physiol. 138(1), 267–275 (2005)
2. Barry, C.S., Giovannoni, J.J.: Ripening in the tomato Green-ripe mutant is inhibited by ectopic expression of a protein that disrupts ethylene signaling. Proc. Natl. Acad. Sci. USA 103(20), 7923–7928 (2006)
3. Bradford, M.M.: A rapid and sensitive method for the quantitation of microgram quantities of protein utilizing the principle of protein-dye binding. Analytical Biochemistry 72, 248–254 (1976)
4. Chao, Q.M., Rothenberg, M., Solano, R., Roman, G., Terzaghi, W., Echer, J.R.: Activation of the Ethylene Gas Response Pathyway in Arabisopsis by the Nuclear Protein ETHYLENE-INSENSITIVE3 and Related Proteins. Cell 89(7), 1133–1144 (1997)
5. Fu, D.Q., Zhu, B.Z., Zhu, H.L., Jiang, W.B., Luo, Y.B.: Virus-induced gene silencing in tomato fruit. The Plant Journal 43(2), 299–308 (2005)
6. Imanishi, S., Mori, H., Nagata, M.: Ethylene receptor gene homologue from tomato-ripening mutant Nr-2. Plant and Cell Physiology 42, 83–90 (2001)
7. Kerr, E.A.: Mutations of chlorophyll retention in ripe fruit. Rep. Tomato Genet. Coop. 8, 22 (1958)
8. Jarret, R.L., Tigchelaar, E.C., Handa, A.K.: Ripening behavior of the Green ripe tomato mutant. J. Am. Soc. Hortic. Sci. 109(5), 712–717 (1984)
9. Lanahan, M.B., Yen, H.C., Giovannoni, J.J., Klee, H.J.: The never ripe mutation blocks ethylene perception in tomato. The Plant Cell 6(4), 521–530 (1994)
10. Resnick, J.S., Wen, C.K., Shockey, J.A., Chang, C.: REVERSION-TO-ETHYLENE SENSITIVITY1, a conserved gene that regulates ethylene receptor function in Arabidopsis. Proc. Natl. Acad. Sci. USA 103(20), 7917–7922 (2006)
11. Solano, R., Stepanova, A., Chao, Q., Ecker, J.R.: Nuclear events in ethylene signaling: a transcriptional cascade mediated by ETHYLENE-INSENSITIVE3 and ETHYLENE-RESPONSEFACTOR1. Genes Dev. 12(23), 3703–3714 (1998)
12. Tieman, D.M., Taylor, M.G., Ciardi, J.A., Klee, H.J.: Members of the tomato LeEIL (EIN3-like) gene family are functionally redundant and regulate ethylene responses throughout plant development. The Plant Journal 26(1), 47–58 (2001)
13. Xie, Y.H., Zhu, B.Z., Yang, X.L., Zhang, H.X., Fu, D.Q., Zhu, H.L., Shao, Y., Li, Y.C., Gao, H.Y., Luo, Y.B.: Delay of postharvest ripening and senescence of tomato fruit through virus-induced LeACS2 gene silencing. Postharvest Biology and Technology 42(1), 8–15 (2006)
14. Zhu, H.L., Zhu, B.Z., Zhang, Y.L., Shao, Y., Wang, X.G., Xie, Y.H., Chen, A.J., Li, Y.C., Tian, H.Q., Luo, Y.B.: Expression of a truncated ripening inhibitor (RIN) protein from tomato and production of an anti-RIN antibody. Biotechnology Letters 29(9), 1425–1430 (2007)

Discussion and Design of Synthetically Computer Network Experiment Scheme

Kaiyu Zhu, Xuefei Gao, and Yongli An

College of Information, Hebei United University, Tangshan, Hebei, China
supzky@163.com

Abstract. According to characteristics of network experiment, experimental projects with theory and practice were put forward. With the experimental difficulty gradually deepened, we paid more attention to the design process of experiment, and cultivated students' team spirit. By multimedia of experimental guidance system, the network communication principles and experimental procedures were vividly and accurately described, and the evaluation of student experimental result was completed. Finally the synthetical experiment based on SDH experimental platform and computer networks experimental platform is introduced, so that the actual communications environment can be simulated by full use of experimental platforms with different communications.

Keywords: computer network, network laboratory, experiment management system, MSTP.

1 Introduction

The computer network is a communication major required courses, its content is to make the students master the knowledge and skills of network communication. The knowledge-transference needs to consolidate and deepen with the corresponding experiment, operational capability need to exercise in the experiment. Therefore the design of appropriate network experiment scheme is particularly important. Because the computer network experiment equipment is more expensive, so teachers should make full use of existing equipment to design diverse content. At the same time the teachers should make full use of multimedia technology to explain, demonstrate and guide experiments. The preferred solution is to complete a number of experimental and has low investment cost.

2 The Experimental Scheme Characteristics

The experimental scheme is designed in order to improve the innovation ability of students, to consolidate the existing knowledge system, arouse the students' initiative and creativity, to cultivate high - quality talents. The experimental scheme has the following characteristics:

(1) Because computer network each part of knowledge interconnected, the experiment contents should also be contextual. Only in accordance with a certain order

W. Zhang (Ed.): Advanced Technology in Teaching, AISC 163, pp. 249–254.
springerlink.com

to complete the experiment, that they can achieve the best results. First, Students complete simple experiments content, such as the production network cable, configure the network card, set up Internet services. Second, students complete a complex experimental content, such as configuring routers and switches, analysis of protocols, analysis of routing principles, etc. Finally, the various pilot projects as a whole, the students step by step to complete a complex network. When all the experimental projects are completed, the students of analysis and design capabilities have been training; by experiments, they could get greater achievement, further learning enthusiasm will be improved.

(2) Students should take the initiative to complete the experiment, rather than passive according to the instruction or the teachers to do the experiment, so the experimental program planning should be more flexible, allowing students to design and try some of the content. With the help of instruction experiment scheme and teachers heuristic questioning, students are thinking about some of the issues. Experimental project should have a comprehensive and design characteristics; experiment content and method have the characteristic of diversity; experimental results with multiple sexual characteristics; experimental results have multiple characteristics; the purpose of the experiment is in the training of students' comprehensive ability.

(3) Experimental projects is generally done in pairs, but sometimes two people do not harvest, because they are not familiar with the experimental operation or not interested in the contents of experiment. Because the network experiment has the network features, the experimental group can be composed of several students, whereas no number and range limitation. Some projects have completed in groups of 8 people, each of which have their own responsibilities. The project fully stimulates students' team spirit. Everybody into the experimental group, they develop interest in learning and sharing the joy of success. The project has the very good teaching effect.

During the experiment, such as modified configuration information, changing the access method and access objects operations will get different results. So that the experimental flexibility, students can analyze and discuss a variety of situations. By participating in the experiment, the students can deeply understand the relationship between each layer of network protocols, while students in the operation of the network hardware and software experience to lay the foundation for future work and study. Test equipment can also simulate communication failures that may occur, or skip a step in the experiment will appear what kind of problem, so as to stimulate the student interest, cause them to think. In the experiment, students can observe and study to the protocol principle and process; curriculum knowledge can be verified and supplemented; student's ability to use network software can be exercise.

3 Experimental Environment and System

3.1 Experimental Equipment

The experimental system equipment selection reflects not only meet the experimental goal and content, but also saves the experiments funds. Terminals are stably running the Windows operating system, of which 30 units for students to experiment, 2 Experimental demonstration and system files for backup. Each computer has two

100Mbps Ethernet card (for communicate with two networks), and install the Windows Server system. Packet Tracer 5 is installed on the computer for the simulation of routing and switching equipments; the Sniffer software is installed for data packet analysis Laboratory network topology as shown in Figure 1, the topology according to the actual situation may change.

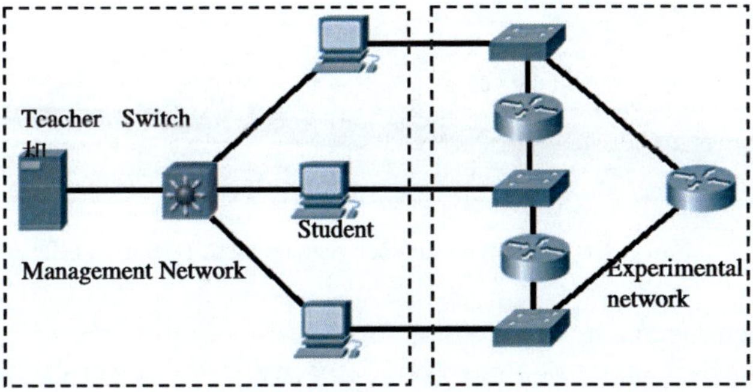

Fig. 1. Lab topology

Communication equipment is divided into three groups, each group having three H3C MSR 20-21routers, three H3C S2126 switches, and three 8 port hubs, also need a large number of unshielded twisted pair. The hub uplink port is connected with the H3C S2126, for routing protocol analysis which is needed in the shared medium to access data packets

This design has two network topologies, one for teacher management devices, and other for student experiments.

(1) Management Network. Through the management network, the teacher can monitor and guide the students. By using such as Remote Desktop or NetOp.School, the teacher can directly operate students' computers. The network is not only used for uploading and downloading files and software, but also for the operating system installation and recovery. This network segment IP is different from other network segment IP; students do not have modify permissions in this network.

(2) Experimental network. The student through the distribution frame can make the host and router or switch to any port connection and to complete the experiment configuration and mutual visits. Students have the segment network parameters modify permissions.

Comprehensive experiments also use the SDH platform to simulate multi-service communication network environment.

3.2 Experiment Management System

The experimental instructions written by VB, which not only contains the text of the experimental principle and steps, also including the use of WinCam recording teachers experimental steps and the corresponding description. Demo program supports play progress bar and pause and fast forward functions. Experimental guidance system interface shown in Figure 2. The system can send experimental data to teacher's host, for experimental operation condition evaluation of students.

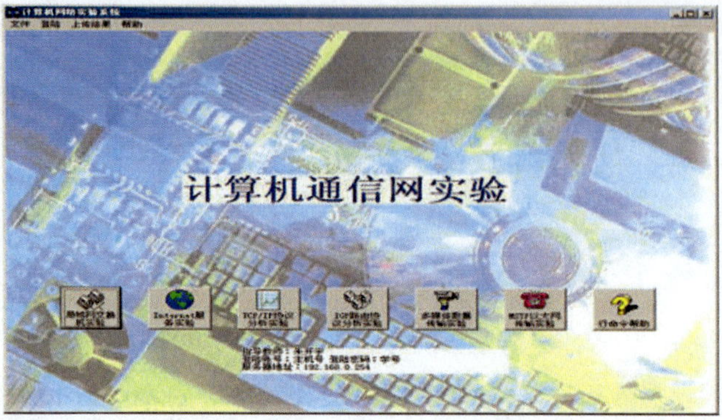

Fig. 2. Experimental guidance management system interface

The system operation is as follows: Students login using the host number as the account, student number as a password, after that in the server database, record the login IP address, the host number, student number, access time for check attendance and accomplishment evaluation. (Students can use the experimental guidance system without login). Students by clicking on different experimental icon to enter each experiment project, in project, the experiment process was shown by text and animation. If teacher need to collect experimental data, student can upload the files named by student number to the server.

The system currently contains six experimental: LAN switch experimental, Internet service experimental, TCP / IP protocol analysis experimental, IGP routing protocol analysis experiments, multimedia data transmission experiment, MSTP transmission experiment. The experiment content is from the shallower to the deeper, covering all knowledge and operation of computer network. Part of the experiment content is shown in figure 3. MSTP transmission experiment is a special experimental project, which combines telecommunications and computer data transmission network. The following will focus on the project.

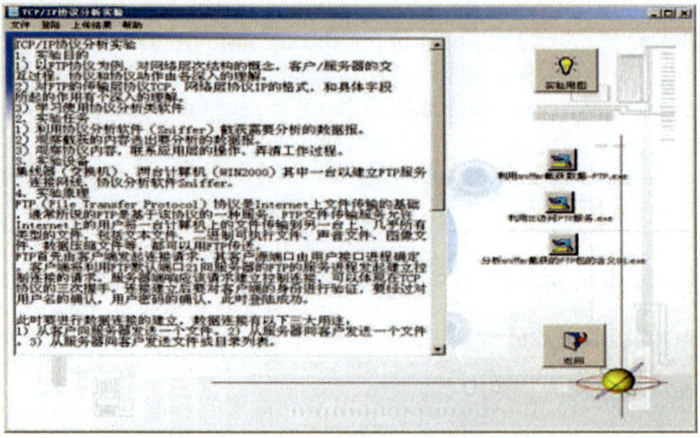

Fig. 3. TCP/IP protocol analysis interface

4 The Experimental Scheme Based on MSTP

The modern communication network is the product of a variety of integrated communications technology. On the trend of triple play (the integration of telecommunications networks, cable TV networks and the internet), the field of education and training should be more reflected a variety of network interconnection, a variety of communications technology integration. Many Chinese universities have computer networks laboratory, have been or are being built with modern communication experiment center, but these two parts are mostly independent of the training courses, so there is no concept of the overall communications network. This paper introduces a scheme can provide more comprehensive application of experimental platform, which can make full use of existing equipments simulate actual complex projects. The program is based on the SDH transport platform, which uses MSTP technology transfer Ethernet data, and uses routers compose the WAN. The scheme for the students to understand the transmission, switching, routing function is very helpful.

MSTP (Multi-Service Transfer Platform) is based on SDH multi-service transport platform, it can achieve TDM, ATM, Ethernet and other business processing and transmission, it also provides a unified network management system. MSTP supports four Ethernet services: EPL (Ethernet Private Line), EVPL (Ethernet Virtual Private Line), EPLAN (Ethernet Private LAN), EVPLAN (Ethernet Virtual Private LAN).

Routers with SDH devices are connected by 100M Ethernet; three 155M SDH equipment is configured for two-fiber unidirectional path protection ring, and the direction of the main ring is NE1 to NE2 to NE3 of the clockwise loop. Network connection uses the SDH equipment Ethernet slip board, as shown in figure 4.

In this experiment, Huawei SDH equipment Metro1000 is as the main equipment. Metro1000 supports access to a variety of business; it is used in MAN, the local transmission network access layer, dedicated access for large customers, the mobile base station access, DSLAM access and other services. Each device is equipped with OI2D the STM-4 optical interface board, EFS-4 Ethernet interface board. Ethernet access mode is the EPLAN, which has an exclusive transmission channel, and supports data exchange private network.

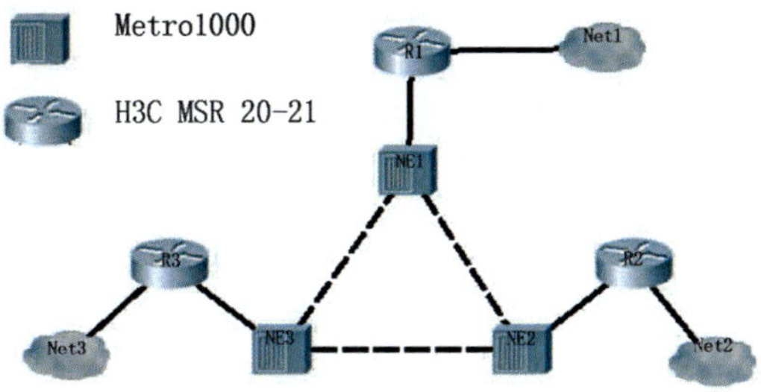

Fig. 4. Ethernet transmission experiment based on the MSTP

Each EFS-4 slip board was configured 2 VCTRUNK, which are bound to 5 VC-12 as the link between routers to provide 10M; while on the board a VB (virtual bridge) was configured for switching between PORT and VCTRUNK. Therefore, the Ethernet interfaces on three Metro1000 were equivalent to the same Ethernet switch. The configuration of each H3CMSR 20-21 must be reasonable, so that data can be transmitted through the SDH equipment. Router using dynamic routing protocols can achieve three networks interconnection. The Metro1000 was configured by T2000 NMS; H3C MSR was configured by WEB config system, also by a common form of the command line.

To complete the experiment, the students need to be very familiar with SDH network, Ethernet, routing principles. This is a very complex project, which will help students understand the concept and structure of the entire communications network.

5 Summary

The experimental content should pay attention to the combination of theory and practice. In the experiment, students can verify the theory, also can obtain network operating skills and exercise the problem-solving abilities. In this experiment, students can fully the understand knowledge; can also arouse the enthusiasm of their further learning. And the experiment possibly gives direction for students' future employment. We design communication experiments should not be limited to an experimental platform. Cross-platform experimental design is more in line with the need of multi-service and multi-network.

References

1. Tanenbaum, A.S.: Computer Networks, 4th edn. Prentice Hall RPT, USA (2003)
2. Forouzan, B.A.: Data Communications and Networking, 4th edn. McGraw-Hill (July 2006)
3. Risbood, P., Acharya, S., Gupta, B.: The Best Challenge for Next-Generation Ethernet Services. IEEE (2005)
4. TranSwitch Corporation, Data sheet of EtherMap-3 Device Ethernet into STS-3/STM-1 SONET/SDH Mapper, 25–28 (2003)

Study on Learning Performance Evaluation of Distance Continuing Education

Zhang HongYan[1,*] and Wang JiKui[2,**]

[1] College of Normal, ShiHeZi University, ShiHezi, Xinjiang, China
[2] College of Mechanical and Electrical Engineering, ShiHeZi University
ShiHezi, Xinjiang, China
zhyapple@126.com, shzwjk@126.com

Abstract. Now distance education has become an important method to realizing continuing education. It has been a key problem in continuing education how to evaluate learning performance of the learner who use distance learning as a method to take part in continuing education. This paper analyzed the characteristics of learner who was involved in distance continuing education and the relationship between learning characteristics and learning performance. At last learning performance in distance continuing education was discussed.

Keywords: Distance continuing education, learning performance, evaluation.

1 The Necessesory of Develop Distant Continuing Education

Modern society is a learning community. Some data statistics indicated that the double speed of human knowledge has been greatly lessened, which is 3-5 years now. The 10-20% knowledge of general technician is learned from school study, other 80-90% is learned from practical work or again training.

In china, the confliction promptly incremental study need with short higher education resource is more and more acute. At the present time, there are more than three millions senior high school every year who will graduate, but actual higher education recruit scale which is consisted with adult, military and party school is less than 12% recruit scale every year, which is lower than not only the scale of developed country but also developing country. To improve this condition that make everyone take in higher education and give more training chances to people who have taken part in work. Only can modern distant education spur to build really perfect and popular life-long education system.

Based on distant teaching means, modern distant education expands the time and space of continuing education. Making best of modern information technology, which can effectively bring into play the advantage of all kinds of existent education resource, is a best strategic measure to achieve civil education in present condition which is short of education resource in China. To accomplish continuing education in method of distant education, there will have more people to accept higher education.

* This work is partially supported by NPOPSS of China Grant #10XMZ018.
** Corresponding author.

W. Zhang (Ed.): Advanced Technology in Teaching, AISC 163, pp. 255–260.
springerlink.com

On 30 July 2006, Britain education and technique ministry started advisory activity whose purpose is to promote continuing education standard and performance. Education and technique ministry set up a new department named quality prompt office which put forward strategy to impel continuing education standard and performance. United Kingdom is the cradleland of distant education, and it's distant education has been focused because of it's noble development and accomplishment in distant education, in which there are lots of experience worthy of consulting. At the same time, we should wake up to the pressure of distant continuing education development.

2 Performance, Learning Performance and Distant Learning Performance

Performance refers to that work achievements, effect and benefit are acquired when people finish a task. There are lots of research on the impact factors of performance, in which *Einhorn, Hogarth* and *Libby*(1983) put forward performance and it's influence factor function:

Performance=(knowledge, ability, motivation, environment)

From this above function, we can conclude that the main effect factors on performance are knowledge, ability, motivation, and environment.

In 2004, learning performance was emphasized when American education communication and technology association gave a new definition on education technology, which are the abilities that learner use new knowledge and skill. Learning performance is not only the achievement ability of elementary knowledge and skill, but also the ability to agilely manage knowledge and skill. Study grade is quantitative outcome of learning performance, but it is not whole connotation of learning performance. Correlative researches indicate that ability and sociality should be also embodied through learning performance.

The representative form of learning performance is multifarious, which is incarnated by the quality and quantity and benefit of study efficiency. The learning performance of one person can reflect his multi-diathesis in some degree, which is the interactional result between individual diathesis and environment. Whether distant education is successful is an important index by which the learning performance of distant learner is estimated.

In network learning environment, performance function can be expressed as followed: P=K*(M*A*E).

Thereinto, P represents performance, and K represents Knowledge, and M represents motivation, and A represents ability, and E represents network learning environment.

Through this above function, we can find that in some degree several factors, such as motivation, ability and environment, are directly or indirectly decided by knowledge. It is obvious that three factors such as motivation, ability and environment of learner have influence on learning performance when the knowledge of learner is assured.

In China, scholars have given new explanation on learning performance that distant learning performance is the ability of learner to use new gained potential on the guidance of learning aim. Accordingly, Based on this new explanation, distant learning performance evaluation is researched in this paper.

3 Evaluating Ideas and Means of Learner in Distant Continuing Education

Distant education scholars have summarized the common characteristic of distant learner. *Muer* summarized three common characteristics as followed: firstly, distant learners had specific learning aim. Secondly, distant learners had strong self-managed ability. Lastly, distant learners usually lied in disadvantageous learning environment.

In 2001, Chinese distant educational scholar named *DingXingFu* put forward the common characteristic of students who take part in distant education, which are as followed: the age of learner is from 20 to 40; learners usually go to school in part time; the rate of male student is usually higher than female's; most students mainly study at home; most students have to go to work and have family; most students are not belong to privileged class and the rich; the geographical distribution of students is wider than traditional school; The individual of student is more different and so on.

In traditional education consciousness, understand of continuing education is defined sometimes as non-record education which is usually short and informal. For a certainly continuing education has many characteristic such as more kind, and quick change, short period and frequent and so on.

Distant continuing education should be accordant with distant scholastic education, so learning characteristics of distant learner which have been discussed above are also be same with the learner of distant continuing education.

3.1 The Evaluating Ideas of Distant Learning Performance

3.1.1 Trinity of Evaluation View of Teaching, Learning, Evaluating

For a long time, performance evaluation is a focal problem in education field. Teaching, learning and evaluating are research subject in education field, and the relationship of them is a focus problem which was studied by many scholars. Although many educators identified with this idea that evaluation is a part of education, but evaluation is only regarded as important means which can check up learning result of student and learning ability of teacher, and we usually pay much attention to the evaluation identification function. The evaluation of distant continuing education should break away from traditional education idea, and avoid dissevering teaching, learning and evaluation. This condition is showed in fig1.

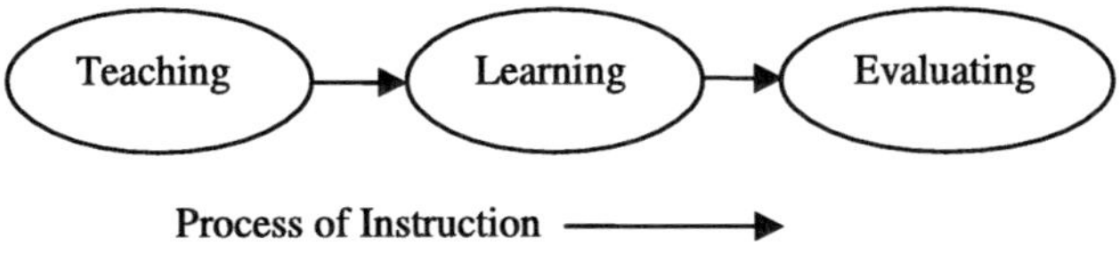

Fig. 1. Dissevering Teaching, Learning and Evaluation

The evaluation on learning in distant continuing education should be systemic whole view, which can interleave teaching, learning and evaluating. Although there are many difficult in implement of teaching and observation of student behavior and evaluation on student, which is lead because of the separation between space and time and the separation between teaching and learning, Trinity of teaching, learning and evaluating should be impenetrate one another and accelerate learning and development of distant continuing education learner. This evaluating view on teaching, learning, evaluating in distance continuing education is showed in fig2.

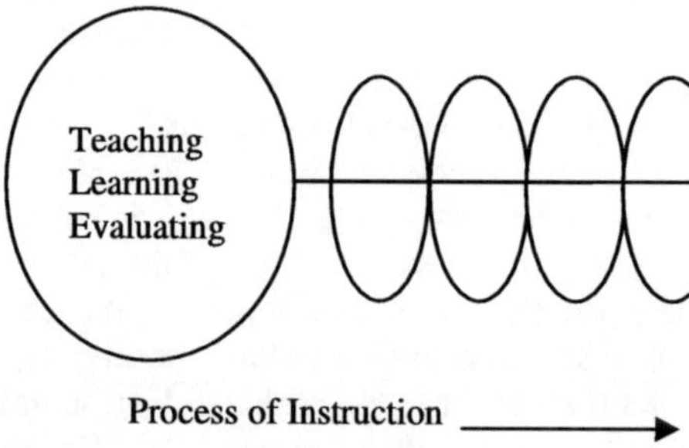

Fig. 2. Trinity of Teaching, Learning and Evaluation

3.1.2 Diversification of Evaluation View

The content of distance learning performance evaluation is very abundant. A kind of evaluation view is that the evaluation of course learning grade is not learning performance evaluation, which should include of the evaluation on skill and the evaluation on attitude. Another evaluation view is that performance evaluation should not decided completely by teacher individual, and there should have multiple subjects to take part in evaluation. These contents involve the core content of learning performance, who evaluate and which means and how to evaluation and evaluate what.

Evaluation subject is diversification. Teachers and students and group are regarded as evaluation subject, in which teacher subject is still the core of evaluation process. "Multi-agent participating in evaluation" that is advocated widely is the embodiment of diversification, which break away that only teacher do as one evaluation subject. And self-evaluation of student and mutual evaluation of student is emphasized especially.

Evaluation dimensionality is diversification. All abilities of distance learning process should be evaluated in distance continuing education. The ability of every learner is different, so the evaluation should not adopt single standard, which should have the multi-orientation.

Evaluation means is diversification. The diversification evaluation means can be embodied in the means which can be written examination or experimentation or paper or survey report and so on.

3.2 Evaluating Means of Learning Performance in Distant Continuing Education

There are three evaluating research means about distant continuing education, which are evaluations on distant transferring system, teaching scheming and learner. It is important to evaluate systemically on grade of learner, especially evaluation can insure better learning process.

Based on differently classified standard, we can evaluate learning of learner. Usually there are four kinds evaluating means, as followed.

3.2.1 Planning Evaluation

Before specifically teaching activities will be started, the preparative condition that students will do in teaching should been learned. It is must be known whether students have had knowledge and skill which is necessary to begin study.

3.2.2 Formative Evaluation

In the procedure of teaching, the progress of students should be examined to provide teachers and students with feedback information about teaching. Based on the feedback information, the behavior of teacher and student must be adjusted and improved.

3.2.3 Diagnostic Evaluation

Because there are unresolved learning difficulties in formative evaluation, which are sometimes long-term and sometimes periodical, the new evaluation methods come into being.

3.2.4 Conclusive Evaluation

After teaching is finished, the degree of teaching aim is confirmed. The whole learning process and result are assessed to estimate the development level of student and to judge if the teaching aim is fit and the validity of teaching content.

At present many overseas corporation including of *MOTOLOLA* and *INTEL* etc have been involved into the distance continuing education, but Chinese corporation involving is not more, and the study on distance continuing education is not enough. As organization who are engaged in distance continuing education must build long-term strategic planning, and establish feasible developing target, and invest designedly in distance continuing education, and have definitely market positioning, and engage professional management team and constitute interior detailed systematic manage mechanism and performance evaluation system. The views which only pay attention to hardware-facilities construction and neglect the ideas of service and evaluation should be changed. We should import market competition mechanism and gradually participate in educational market competition, and make Chinese distance continuing education be an educational means which is open to the outside world.

4 Conclusion

Based on the discussion about the evaluation on learning performance in distant continuing education, i think that formative evaluation and conclusive evaluation should be combined to evaluate learner performance. In this paper, formative evaluation should be especially emphasized, and evaluating result and keystone should not put on grade. We hope that the whole learning process should be learned through evaluation.

Acknowledgment. Thanks to my colleagues for working with us. Many thanks to my postgraduate tutor *Jiao BaoCong*, who leaded me into a completely new research field. Finally, thanks to my family for persevering with me throughout this long process.

References

1. Jiao, B.-C.: Theory and Methods for Decision Making of Educational Inforration. China Electronic Industry Publishing House (2004)
2. Addison, R.M.: Performance Technology Landscape. Performance Improvement 42(2) (2003)
3. Kaufman, R.: Why This Special Issue? Performance Improvement (2003)
4. Kaufman, R.: Thriving and Not Just Surviving: New Directions for Tomorrow's Performance Improvement Managers. Educational Technology (July- August 2000)
5. Stolovitch, H.D.: Human Performance Technology: Research and Theory to practice. Performance Improvement 39(4) (2000)
6. Marano, H.E.: A nation of wimps. Psychology Today 3584, 2–11 (2004)
7. Loveless, Tom, Brown Center Report on American Education: How Well Are American Students Learning. The Brookings Institution, `http://www.brookings.edu/gs/brown/bcreport,2004-hp-htm-63K`
8. Kaufman: Constructivist based experiential learning in teacher education. Action in Teacher Education 18(2), 40–50 (1996)
9. Goldhaber, D.D., Brewer, D.J.: Does teacher certification matter? High school teacher Certification status and student achievement. Educational Evaluation and Policy Analysis 22(2), 129–146 (2000)
10. Darling-Hammond, L., Youngs, P.: Defining "Highly Qualified Teachers":What does "Scientifically Based Research" actually tell us? Research News and Comment (13) (December 2002)

Research on the Training Mode for Applied Personnel in General Colleges and Universities

Huixin Tian[1] and Kun Li[2]

[1] Department of Electrical Engineering and Automation,Tianjin Polytechnic University
[2] Department of Management,Tianjin Polytechnic University,Tianjin, 300160, China
Tianjin, 300160, China
icedewl@163.com, lk_neu@163.com

Abstract. With the fast development of social economy, the needs for applied personnel become more and more outstanding for enterprises in China. In this paper training mode reform for applied personnel is discussed deeply especially for general colleges and universities. Firstly, the training objectives are located accurately in the site of training applied and innovative personnel for the general colleges and universities. And then, the idea and detail measures of education reform for training applied personnel are proposed in different aspects, such as basic theory teaching, practical teaching, extracurricular science and technology activities, cooperation between school and enterprise and so on.

Keywords: Higher education, applied personnel, training mode, education reform.

1 Introduction

With the fast development of economy in China, the needs of talents in enterprises become more and more strict. In the knowledge age, at the same time of rapid development of high technology, the needs of senior technical applied personnel become more urgent for contemporary society. In recent years, the higher education has stepped into a stage of popularization. The traditional undergraduate education mode focused on training academic personnel. However, this education mode can not satisfy the needs of modern social. The general colleges and universities should be located in applied undergraduate education colleges and universities. It is to say, in this kind of colleges and universities should focus on discovering a new education mode for training applied personnel based on combining theory and practice. Aiming at training applied personnel, they also should innovate in teaching philosophy and educational mechanism and become a new education system by degrees.

2 Location of Training Applied Personnel

The needs of society and enterprises for personnel have changed greatly, with the fast development of social economy. For fitting this kind of changes, the general applied undergraduate education colleges and universities have to focus on training academic

W. Zhang (Ed.): Advanced Technology in Teaching, AISC 163, pp. 261–268.
springerlink.com

personnel. The practices have proved that the location of academic personnel training objective is accurate and correct and could adapt to the development of times. The personnel being trained in this mode can satisfy the needs of society and enterprise and keep pace with the times. This kind of local general undergraduate education colleges and universities will become cradles of personnel reserve, intellectual support technology source and cultural inheritance for the developments of local industry or social economy.

It is essential difference between applied personnel and academic personnel. Different kinds of universities or colleges correspond to different types of personnel training. The academic personnel training task should be completed by teaching and researching universities (especially by 211 universities). The task of higher vocational colleges should be responsible for training the technique personnel. Its value orientation and main objective are vocational training. The applied personnel training task between above two types should be completed by general applied undergraduate education colleges and universities. However, the traditional education mode of general undergraduate education colleges and universities is the same as the one of teaching and researching universities in China. The general undergraduate education colleges and universities' training objective is to train academic and researching personnel but not to train the engineering applied personnel. But the personnel being trained by general undergraduate education colleges and universities has weak competitiveness ability compared with the one being trained by researching universities. It is because the general undergraduate education colleges and universities can not compete with researching university in several aspects, such as quality of students, quality of teacher, conditions of school running, levels of scientific research, social reputation, and so on. Therefore, it is necessary to make further discussion and reformation on training or teaching mode for general undergraduate education colleges and universities. They would pay more attention to outstanding the advantages of "applied" undergraduate education and training "applied" personnel, and keep space with the development. These general undergraduate education colleges should emphasize both basic education and training applied technique, should pay more attention to training the abilities of application and practical skill based on solid undergraduate professional knowledge. The applied personnel will have good abilities of transforming knowledge into practical engineering application.

In general undergraduate education colleges and universities, the objective of training personnel for undergraduate education will be devoted to transform basic theoretical research results into high technology and actual productivity, and the objective of graduate education will be devoted to train personnel as high level engineering and technical experts. Aiming at above objectives, in the beginning stage, teachers should let students access to professional basic knowledge. And then, on this basis, the students should be taught the knowledge and technology of practical production, such as production process, production equipment, etc. At the end, the students will have some engineering and technical abilities. They can combine theoretical knowledge and production process (or practical production problems) together to solve practical problems. At future work, these applied personnel can complete the overall program design and specific design. They also can design the details of specific measures for the specific problem and implement them into the whole production process. They can manage, monitor and maintain the technology systems. When some complex practical introduction problems happen, these personnel will find them timely, analyze and solve them through overcoming various difficulties during practical production process and management process. The applied personnel will become the applied engineers who can transform the scientific and technological achievements into actual productivity in some production areas.

3 To Lay the Good Professional Foundation of Applied Personnel by Basic Theory Teaching

The process of training applied personnel is a complex training process. During the process of training personnel's abilities of application and innovation, the basic theoretical knowledge learning is the key section especially for the following abilities forming. Furthermore, the classroom teaching is the main way to complete this section. Therefore, it is necessary to create a good classroom teaching environment form good curriculum development, curriculum enhancements to good teaching effectiveness [4,5]. This good environment can ensure the good quality of applied personnel who will be trained.

The curriculum development and curriculum enhancements will do in following aspects. Firstly, the traditional values of teaching and education have to be changed. Traditionally, theoretic teaching was more important than practice teaching; academic training was more important than technology training. In current, the professional curriculum must be closely related to the needs of practical production. Teachers should pay more attention to enhance the professional curriculum teaching section. The specific measures are increasing application theoretic courses, adding practice to required courses, and so on. These measures will be done on the premise of completing the main professional courses. These methods will let students master the professional skills corresponding to the specific course. Then the students' practice ability and passion for learning professional courses will be enhanced. Secondly, the mode of course has to be adjusted to fit the applied personnel training plan. The general undergraduate education colleges and universities might increase the percentage of integrated curriculum in the whole professional curriculum system. By this way, the basic knowledge and professional knowledge will be combined together. Then it is benefit for increasing students' overall knowledge of their professional and training students' abilities of comprehensive application. At the same time, the abilities of dealing with difficulties will also be strengthened in practical application environment.

In addition, training innovation abilities also should be paid attention. The traditional teaching mode will be changed from teacher-center to student-center. It is to say the students will become main body during the process of teaching. This transform will give full play of subjective initiative for forming an independent learning classroom atmosphere. The teachers play a leading role during the whole teaching process. They guide the students to participate in classroom teaching by different kinds of effective means. During the process of mode transforming, the reform plans will be carried out deeply firstly. And then students and teachers will be organized together to discuss the contents and values of application teaching mode.

During this process, the successful experiments of education reform will be summarized, and the disadvantages also will be found according to the students' feeling to the new teaching methods. Afterwards, the solutions of problems will be searched. During the process of teaching, teachers should pay more attention to spreading heuristic teaching and inquiry teaching actively. Let the teaching and research combined together. Teachers also should pay attention to form the classroom atmosphere of whole member participating, pay attention to how transform the knowledge into abilities. The teaching can enable the students to mastering professional knowledge and enhancing the abilities of using knowledge or skill to solve problems. This method will let students develop good habits of initiative thinking. Finally after the education reform for traditional teaching mode, the whole learning process of students will become an innovative exploration, initiative thinking, and active practice research process. Through this process, the learning and application of knowledge can be completed for students. And the students' creative ability, innovative spirits and practical application ability are also cultivated.

4 Enhancing the Application Abilities of Applied Personnel by Practice Teaching Steps

Practice teaching steps are the important links and the main way to cultivate personnel with innovative ability and manipulative ability[6,7]. Perfecting practical teaching link is necessary to take experiment teaching as main body. The experiments are independent courses involving integrated course design, synthetic experiment, and graduation design. Furthermore, the professional teaching practice and the social practice are both important complements of comprehensive practice process. Therefore, the practice teaching should be the fertile land of training applied personnel who will be adapt to the needs of society and enterprises.

Firstly, the traditional mode of the experiment teaching reform will be carried out. The traditional validation of experiment teaching after teaching by teachers must be changed. The specific measures involve changing experimental contents, improving the diversity of content and reducing the traditional verification experiments. In addition, the student-centered designing experiment and comprehensive experiments should also be strengthened because they can combine the knowledge having been studied by students and improve self-directed innovation capability of students. In the section of curriculum design should focus on increasing the practicality of designs by combining the needs of enterprise's actual production and management. The comprehensive experiments and integrated curriculum design should combine multiple experiments and professional knowledge together effectively. Teachers should pay attention to the combination of theory with practice, and provide good conditions to develop students' innovative consciousness and innovative spirit according to how the students grasp knowledge. Departments or schools should build open laboratories strongly, and program different kinds of elective experiments on the basis of existing experiment courses. These methods can provide wider practice study space for students, and make the student participate in experiments selectively according to their own hobby. Teachers also can create multiple channels to achieve practical ability training progresses by this way. At the same time, this participation way can improve the students' learning initiative and consciousness effectively to fundamentally avoid the drawbacks of traditional "spoon-feeding" teaching.

Secondly, departments or schools should strengthen the practice teaching section, dynamics. And by this way, the students will experience in the process of combining theory with practice in the conditions of actual production and management. Various practical teaching bases ill be constructed vigorously. The professional simulation laboratories will be established in school. These laboratories have the close relation of practice production. At the same time, the practical and training bases will be established outside by joining local enterprises. With these practical teaching bases as the foundation, the comprehensive training can be carried out efficiently. Students can combine a variety of professional knowledge which has been learned together in this form. Then as the various application abilities of students will be enhanced. These abilities involve professional knowledge comprehensive application ability, comprehensive practice ability, and so on. For students, the distance between school and society, enterprises will be shortened. So that, the students can adapt to the requirements of society and enterprise's in the shortest time. In addition, the existing practice section (production practice and graduation practice) should be given enough attention. It will be constructed continuously and strongly. The traditional "a visit after a walk" practice mode has to be changed. The students will step into the enterprise's actual production process really and deeply to understand the business and production activities, the operation of the enterprise and the operation mode. Combined with theoretical knowledge, the practice section should make the students understand the process of combining theory with practice. If in early production practice section, the students should clearly understand the specialized study direction and the application fields, and raise a solid foundation for the following professional knowledge learning and application ability training. If in later graduation practice section, the students enable to be familiar with production, operation and management activities. They should further study the combination of knowledge and practical production process. Through the graduation practice process, the students also can enhance the practical value of professional knowledge, clearly understand their professional direction and the corresponding requirements in actual position for knowledge and abilities. These will provide sea-mark for the students in the choice of forthcoming employment.

5 Organize Rich Extracurricular Innovation Activities

Extra-curricular innovation activities are the important means of applied personnel cultivation, and create a good growth atmosphere for students. Innovation sustains the progress of a nation, and is motive power of country developing prosperously and continuously. By adopting effective measures, extracurricular innovative activities and students' knowledge will be combined together. Teachers will ensure students to take part in scientific research and technological activities by various means. And then these activities will stimulate students' interests and passion in learning and creation.

On the basis of open laboratory students' innovation practice bases will be established. In these bases, the necessary instruments and equipment will be provided, and opened to the student's overall. Encourage students to take part in various practice activities under the guidance of experienced teachers using after school hours. These practice activities involve a variety of new product development, research and development of scientific research projects, maintenance of equipment and so on. Above practical activities are benefit to develop students' abilities in the aspects of innovation, technology applications and practical skills. Encourage students to participate in important technological competition. During these processes, the students will be guided to discover problems and solve problems in practice. Then their abilities of solving questions combining theory with practice will be improved. Meanwhile, organize the students widely participate in extracurricular science and technology activities and technological competition activities in the combination of the professional development features and students' knowledge. These kinds of technology activities not only can promote students' innovation ability but also are advantage to train students' team-work consciousness. In the meantime of competition, teachers should pay attention to create positive loose innovational atmosphere and good failure tolerated environment. These will stimulate the innovative potential of students to the greatest extent. Above competition activities will make students better understand the process from theory knowledge to practice technology, train the students' independent scientific research abilities and the abilities of communication and cooperation, and promote the students' participation and service consciousness.

In addition, schools can provide wider space for training students' innovation consciousness and broadening students' view through the way of strengthening academic communication. Schools will invite domestic and international famous experts and scholars to make special subject academic reports regularly or irregularly for students. These reports can expand students' scientific research thinking. Schools also can invite famous entrepreneurs at home and abroad for college students to do seminar that can stimulate students' innovation fervor and the pioneering consciousness, etc. Teachers and students also can carry on the different forms of the lectures and academic exchange activities. Contents can be obtained by inspired in recent, or learning a new science and technology achievements. These academic exchange activities can exercise the college students' self learning abilities and expand the science and technology horizons. These knowledge accumulations will provide the necessary knowledge reserves to college students for later innovation activities.

6 Vigorously Promote the School-Enterprise Cooperation of School-Running Model

Open convenient fast channel for training the abilities of applied personnel by using the advantages of the closely linked between local colleges with local enterprises [8].

Most of general universities and colleges are local universities. So they of undergraduate course more for local university, therefore, they should establish the practical teaching bases based on local enterprises' resource by making more links with local enterprise units towards the needs of the local economy. These bases will provide more extensive practice space for students. College teachers should have more extensive contacts with local enterprises. Encourage teachers (especially young teachers) should go deep into the local enterprises with scientific research projects transform the research achievements to productivity. And in this process, the teachers will bring and the practical problems closely related to enterprise actual production back to school as the practical problems for providing the sources of undergraduate course graduation design and graduate practical problem. During the process of solving actual production problems, the students' practical application ability and creation ability will be trained and raised. Practice proves that the practical problems issues appeal to students far more than the traditional theoretical topics. And students in higher learning interest status will completed the graduation design and research better.

Engineers and senior management personnel in enterprises will be invited to school as part-time teachers to teach some courses related applications and business practices. These part-time teachers bring the technical knowledge in enterprises into the classroom. It is benefit for students to combine the knowledge with actual production. They will also bring the actual problems, the actual needs and development requirements of enterprises into classroom to make students understand the demands of the profession and the industry trends more fully. In addition, colleges and universities can open up a new training mode combining enterprises training with school training. By this way, the student will become the full application of the coordinated development personal with good knowledge, abilities and qualities. cooperative relationship between universities and local enterprises will train the personnel as the following modes: school-enterprise cooperation training mode, enterprises training mode, customized programs training mode (according to the actual demand of enterprises). These new training mode make the applied personnel training become more pertinence and practicability.

7 Conclusion

Local general applied colleges and universities should focus on the service for local economic construction. The traditional education ideas must be transformed vigorously by constant education reform. They should locate in engineering applied innovation talents training mode. The new innovation personnel training mode will be explored in practice. The characteristic of this kind of colleges is the combination of production, study and research. They will be conveying creative application personnel with modern thinking modes and practical engineering practice abilities unceasingly for the social.

References

1. Hua, T., Hua, Z., Pin, A.: Study on Training Mode of Applied Talent. Journal of Chongqing University (Social Sciences Edition) 10(6), 170–171 (2004)
2. Pan, M.-Y., Che, R.-S.: On the positioning of application-oriented universities. Journal of Higher Education 30(5), 35–38 (2009)

3. Gu, Y., He, D.: Research on Creative Talents Training in Local Applied Universities. China Adult Education (23), 115–116 (2009)
4. Brooks, R.L.: Measuring university quality. The Review of Higher Education 29(1), 1–21 (2005)
5. Mark, M.M.: Building a Better Evidence-Base for Evaluation Theory. Fundamental Issues in Evaluation 1, 11–126 (2008)
6. Liu, Z., Yang, X., Chen, X.: Development Mode and Innovative Practice for Applied Colleges and Universities. China Higher Education (11), 34–36 (2010)
7. Pan, M.: The Problems of Production and Research Cooperative Education. China University Teaching (3), 15–17 (2008)
8. Zhu, C.: Research on the Innovative Talents Training in Research Universities. China Higher Education (11), 34–36 (2010)

Management of Using Efficiency of Large-Scale Equipment

Zhiqiang Kang, Yubo Jia, and Ying Wang

College of Mining Engineering, Hebei United University,
HeBei Province Key Laboratory of Mining Development and Safety Technique,
Tangshan, 063009 China
kzqzsh@163.com

Abstract. Management and use of large-scale research and equipment play an important role in assure its full use. It is very important how to manage and make good use of them to improve the owner's research level , teaching level and economic benefit. It notonly needs a scientific and rational management control system, but a high level, responsible and proper size management team,owning a reward system, giving full play of the teams. Only improving the management control system gradually, it can be assure that the service ability rate and using-efficiency, and can provide better service to teaching and scientific-research.

Keywords: large - scale equipment, management, equipment sharing, uses.

1 Introduction

Large-scale scientific research and teaching equipment is the school investment point, is fundamental of materials that colleges and universities engage in the teaching and scientific research, and is also the important symbol of universities strength and education levels . It is a large challenge for all the universities how to make large-scale equipment serve the discipline construction and development of universities, also can serve the scientific research of advanced level and papers published of a high level, and as far as possible create a good social benefits and economic benefits for the school.

2 The Main Problems Existing in Management and Use of the Large-Scale Equipment

Although use of large-scale equipment plays an important role in the subject construction, teaching, scientific research, achievement and so on, it is the key work of equipment management work how to optimize resource allocation of the existing equipment, coordinate the relationship between teaching and scientific research, and make the utilization of large-scale research equipment greatly increase. However it is the difficulty for equipment management work to meet the needs of undergraduate daily teaching work.

2.1 The Lack of Sufficient Feasibility before the Purchase of Equipment

Individual college only considers the college ease of use or a temporary need of the research topic before the equipment purchase, neither does not make a full investigation

W. Zhang (Ed.): Advanced Technology in Teaching, AISC 163, pp. 269–273.
springerlink.com

for the entire school and even similar equipment of this area, and also does not make fully demonstrate for the use measures, open-sharing after the equipment purchase, which causes repeat purchase, low utilization rate of equipment, make the school resources greatly wasted. Therefore, it must be strict for procurement procedures of large-scale scientific research equipment, further regulating the demonstration work of feasibility, necessity and the prediction of benefit before the large-scale equipment purchase. Such as: surveying function, index and quality, comparing the quality, price of domestic and foreign similar instrument, the implementation for operation maintenance costs each year, explaining the purpose of the equipment purchase , utility, use benefit and risk analysis, etc. What's more, strictly checks, democratic discussion, and scientific decisions must be done. To the argument report that feasibility is not full, the expert advice and signature is not sound, the equipment department shall have the right to refuse the implementation of the purchase plan.

2.2 Lack of Enough Strength for Unified Management, Common Use, and Open Sharing

Due to factors of the system, maintenance, and sources of funds, opening, resources sharing strength of large-scale scientific research equipment is poor, coordination is less. On the one hand utilization rate of a lot more valuable scientific research equipment is not high, on the other hand because of inadequacy of test funds, teachers can't use advanced equipment. Speeding up the relative rules and regulations implementation of paid for the use, and actively promoting unified management, common use, and open sharing of large-scale scientific research equipment are methods to solve the prominent contradictions between relative shortage of current investment and serious waste of the existing resources, and improve the benefit of investment.

3 The Effective Way to Improving the Benefit of the Equipment Use Is to Actively Implement the Operating Mechanism of the Centralized Management of a Large Equipment and Open Sharing

Large-scale scientific research equipment is the basic means and important material foundation of teaching, scientific research work, is the investment point of the laboratory and the discipline construction, and is also the essential condition of completing scientific research subjects of the high level. In order to strengthen the management of large-scale scientific research equipment, the enthusiasm should be fully mobilized, and we should promote internal and external opening of large-scale scientific research equipment, the sharing of resources is realized, equipment utilization rate is improved, problems in use and maintenance of large-scale scientific research equipment are solved, we should realize the purpose of self development, unified management, common use, a machine with machine of large-scale scientific research equipment. The following measures should be taken:

1) The leaders at all levels should pay attention to and set up the correct idea, and give full play to the readjustment role of the competent departments.

For any country, the purchase of a large-scale scientific research and teaching equipment whose source is the investment fund all belongs to the fixed assets, effective centralized management and good open service should be taken, and their due role should be played, which can not only make the staff room easily used, and can give more support for the instrument operation maintenance and function development teaching, will have more support for scientific research work. The competent departments of the assets have the function of supervising and managing, regulating school assets. Schools should strengthen control power to the equipment of sit idle, lower benefit, lower utilization rate, timely allocate and restructure equipment on the campus, in order to optimize the allocation of resources, revitalize the existing assets. The competent departments of assets carry on the traditional management and technical management to the whole process from purchasing equipment to discarded, in addition to the evaluation of the use efficiency of large-scale scientific research equipment, the material sharing management platform of large-scale scientific research equipment must be built, economic efficiency and to the management, to Management to Benefit.

2) Punishments and rewards should be implemented, and the enthusiasm of the staff should be fully mobilized.

Paid use and opening sharing of large-scale scientific research equipment are problems that all higher institutions have been searching for, even to this day effective management mode which is recognized commonly doesn't still form. The key reason is that the enthusiasm of the staff isn't fully mobilized, the interests relationship between the laboratory of the school and college isn't correctly handled. Therefore, when management measures of large-scale scientific research equipment paid for the use, and other supporting documents, such as benefit evaluation , are formulated in our school, the interests of the college, the school, and the individual should be fully considered, and policy tilt for the lab to the maximum limit, so that large-scale scientific research equipment paid for the use in our school can be fully implemented and can produce fruitful results as soon as possible.

3) Perfecting the operation and maintenance system of large-scale scientific research teaching equipment.

On one hand, relevant departments should urge management personnel of large-scale scientific research teaching equipment to make full use of warranty period to find that the problem and resolve it, in order to lay the foundation for the future operation. On the other hand, relevant departments revise and perfect rules and regulations, with the aid of the laboratory evaluation to promote large science and education equipment management. In addition, professional structure, age reasonable teachers and technical personnel should be equipped to manage and use equipment, in the process of use, all sorts of equipment operation rules must be strictly abided, the equipment with special requirements must be operated by the specialist, or trained personnel can use it. Equipment abnormal failure and analysis pantomime must be timely carried out, through the analysis of the fault, pooling the wisdom and efforts of everyone, and everyone recognition to the fault should be strengthened, analysis ability, maintenance ability and level are improved. It is requested that maintenance cost of a certain proportion should be leaved in production services of large-scale scientific research and teaching equipment to ensure the normal operation.

4) Increasing cost consciousness, improving efficiency concept.

With the deepening of the market economy, high school senior specialized talents as a special commodity trainers, cost consciousness and benefit concept should be rightly introduced into. No matter the experiment conditions bring subtle influence on the implementation of quality education or the development of innovative ability to students, or bring economic or social benefits to scientific research, should be are measured with input and output ratio. Especially with the further expand of universities participating in the local economic construction range, and the gradually increases the social development relying on university experiment conditions, large lateral scientific research subjects have been increasing. Relevant personnel should enhance efficiency concept, break departmental ownership, realize the open laboratory, the resource sharing, and promote open use, improve equipment utilization rate by multi-channel. This is not only the needs of strengthening equipment management, improving equipment utilization, playing the benefit of investment, and it is the demand of social development, and of the development of the situation.

5) Establishing a professional technical team with high quality.

The structure of large-scale scientific research teaching equipment is complex, of which operation needs certain skill, therefore, we should strengthen the laboratory personnel's service training, in order to improve quality, stable team. Combining features of teaching and scientific research task, we can consider that different laboratory personnel train with each other in business, which ensures that the experiment task completed, at the same time adds a wide range of skills, and avoids the drab and immobilized situation among the experiment personnel. In addition, a condition should be actively created for the technical personnel, such as attending some foreign and domestic academic activities, understanding development dynamic of advanced instrument equipment and the frontier technology, updating the knowledge structure.

6) Establishing the appraisal system of the use efficiency of expensive scientific research equipments.

In accordance with the related regulations, We should combined with the actual situation of the school, and a set of perfect, scientific, standardized, easily operated evaluation system of the large equipment using benefit. This system should include unit execution, the staff training, scientific research, function development, and comprehensively examine service to foreign, prevent one-sided. According to the assessment system, we comprehensive evaluation of large-scale scientific research should be carried out by the end of year, and according to the evaluation results, reasonable rewards and punishment system is made so that motivation, urged role can be played.

7) Establishing large-scale scientific research equipment network, promoting the sharing of resources, and improving the use benefit.

Resource sharing of large-scale scientific research equipment is a mutually beneficial relationship, along with the reform of national investment system, for the unit with a large research equipment, in order to cover their costs, it's necessary to leave no stone unturned to service to the society, and the unit which need to use large-scale scientific research equipment will make strict economy accounting, can use the somebody else's ready-made instrument, he would not have to spend money to buy. So, cooperation can optimize the allocation of resources, regulate quality services.

4 Conclusion

In short, the use benefit of large-scale scientific research equipment reflects the benefit of the school investment education and the instrument equipment's management level from a side. Only the reasonable management system, scientific management method, can manage and use well large-scale scientific research and teaching equipment, play furthest the use benefit of the equipment, better serve the teaching and scientific research. Therefore, managing and using well the equipment has very important meaning to improve the scientific research level of this unit, the teaching level and economic benefit. To have a scientific and reasonable management system, have a management team with high business level, responsibility, reasonable structure, meanwhile corresponding reward mechanism should be had, and the management should fully play the subjective initiative. Only management mechanism is gradually perfected, strict management, and a series of rules and regulations are built from equipment installation and debugging to personnel training, maintenance, etc, and these rules and regulations must be put on feet, the integrity rate and the use efficiency of the large-scale scientific research equipment can be guaranteed, and better serve the teaching and research.

Acknowledgment. The authors wish to acknowledge the funding support from HeBei Province Education Office of China (No 2011151). At the same time, acknowledge the laboratory support from The HeBei Province Key Laboratory of Mining Development and Safety Technique.

References

1. Li, Z.: Intelligent control of humanoid. National Defence Industry Press, Beijing (2001)
2. Wang, L.: Large Precision Instruments allocation by the comparative study. Research and Exploration 18(3), 112–114 (1999)
3. Zhang, Y., Liu, Z.: Introduction to Teaching Laboratory. Shandong Education Press (2002)
4. An, L., Zhang, H.: Improve efficiency in the use of large-scale scientific. Experimental Technology and Management 17(3), 123–125 (2000)
5. Lai, Y.: Full resources of large research facilities Thoughts. Research and Exploration 25(1), 134–135 (2006)

The Factors Affect Middle School Sports Curriculum Reformation and Its Countermeasures Study

Cheng Pei

College of P.E, Central China Nomal University, Hubei Wuhan 430079

Abstract. This article analysis the factors that hinder physical education curriculum reform, specific to the difficulties and problems in the implement process of "New Curriculum Standard of Sport and Health", and put forward relevant countermeasures, provide reliable basis to physical education department to guide the teaching reform of school physical education, as well as establish and implement teaching policy which accord with the characteristic of middle school sport and health.

Keywords: Curriculum Reform in Physical Education, factor, analysis, countermeasure.

1 Preface

The new curriculum reform of physical education is an important part of education curriculum reform. It gives first place to develop students' subjectivity and initiative, emphasize respect teacher and students choose content of courses, cultivate student's good habits and strong will quality, in order to develop their body, mental and social adjustment.[1]

It's the critical period of children's physical and psychological development when they are in middle school, for this reason, to make a good job of school physical education is the top priority. Implementing curriculum reform in physical education promote and develop school sport well. In the teaching process, students gain the knowledge, harvest happiness, even more accord with quality education. Nevertheless, there are still many defects, as well as many factors restrict and hinder middle implement curriculum reformation in physical education, the reform of physical education curriculum implementation should also continue to strengthen enforcement.

2 The Current Situation of Middle School PE Curriculum Reformation

2.1 Schools and Teachers Understood the Essence of the PE Curriculum Reformation Weakly

Schools and PE teachers not deep understanding the new curriculum reformation, not adapt adapt to the requirements of the new curriculum standard reformation. Schools

[1] QI You-xiang. Thinking of current high school physical education and health curriculum [J]. Journal of Adult Sports Education, 2003, (2) 10, 101.

W. Zhang (Ed.): Advanced Technology in Teaching, AISC 163, pp. 275–279.
springerlink.com

occasionally attention or pay no attention to PE. PE teachers' ideology and professional quality not adapt adapt to the requirements of the new curriculum standard reformation. It mainly great influenced and bounded by the traditional education, teaching, and the evaluation of the concept, lack of exploring spirit, there is no roots on the class education, teaching and assessment of new curriculum, even lose their bearings for curriculum reformation.

2.3 Students' Interests Not Conform to Teachers' Arrangement

According to statistics, there are 95.6 % of the students like physical exercise very much but don't like PE lesson. The reason of this problem is not that teachers teaching bad, but the studens lose interest in the content which teachers teaching. The open projects of PE lesson is limit, and most of them are not accord with students' mind.

We can know from the investigation on the part of the PE teachers: in addition to compulsory content, there also have many elective projects for the students to choose, but this can only satisfy part of students. And students' interests are different. Even part of schools teaching divided classes into boy and girl, or divide into classes on the basis of students' interests, it still hard to satisfy students' requirement.

2.4 The Lack of Sports Resource

1)The lack of sports facility resource
Because the difference of school education resourse, field and equipments are not enough that there is a great difference about opened PE projects. School scale extended indiscriminately, class is large, the field and equipments of school obviously deficiencies, lacking of money, which restrict the normal development of school physical education and health new curriculum reformation. Some devices sit idle for a long time, give rise to loss and damage. The equipments of city school is more complete than suburban school, most of suburban schools have no venue, equipments and facilities, no multimedia and other modern teaching equipments, these makes the new curriculum teaching more difficult.[2]

2)The lack of sports human resources
PE teacher plays an important role in PE curriculum reformation. To build the PE teacher team is our primary task. What's more, the shortage of teachers will also be restricted to open the sports curriculum project. Of course, we cannot only focus on the development of the PE teachers resources, school leaders and student in charge of sports activities also can be developed to be human resourses.

3)PE curriculum resources monotonously
The new teaching materials used by curriculum reformation has characteristics as follow: epochal character, openness, selectivity, it requires teacher must tapping the connotation and extension of the teaching materials according to the new curriculum concept, they must hard to develop curriculum resources and enrich the teaching content. Many teachers are only limited to inherent sports, and many teachers just

---

[2] LEI Zhi-zhao. Countermeasures about the influence of sports equipments and facilities to the middle school.[J]. Journal of sports, 1998, (6) : 75-78.

give priority to "training"but not education, steps are too meticulously, just teaching materials that makes teaching monotonous and stay in the superficial level. The school also did not fully mobilize the extracurricular activities after class, just mechanically do setting-up exercises during the break which is single and boring, student also do their job without any effect.[3]

4)Security issues make course reformation process slowly

The current family situation is that most children are the only children. The child are parents' baby and not let any fault. PE class with risks caused some resistance when school implement PE curriculum reformation, for this reason, many projects cannot carry out smoothly. School in view of safty elements, there will be less sports projects even though students like. The students have no chance to chose their interest projects that it is hard to achieve the teaching effect.

3 Advices

3.1 Strengthen the Understanding of PE Curriculum Reformation Ideals

The education department in charge, school and PE teachers should supervise and urge students recognize the importance of PE curriculum. Physical exercise can enhance body health, and also make our fatigue brain to get sufficient rest to improve the learning efficiency[4]

Change student's concept and cultivate their ability of learning by themselves. Advocate students' active participation, diligent hands and willing to explore, cultivate students' innovation consciousness and practical ability. By various education methods to improve junior middle school students' cultural knowledge level of physical and mental health, strengthen the sense of participation in physical exercise. Lead and organize the students to form a amateur sports team or sports entertainment association, develop students' initiative, make the students become the organizers and participants of activities, with the result that PE class teaching effect can be better accomplish and consolidate.

Updated PE teaching idea, establish "health first" guiding ideology. PE teacher should Deepening the reformation of physical education, updated the teaching idea and strengthen the teaching research. Proceed from the reality, insist on the integrated development educational policy, take the " health first" as the guiding ideology, face to all students, find out the model of teaching accord with actual school teaching condition which is based on the physical development law and mental development characteristics of middle school student, optimize the teaching content and process.

3.2 Increase the Strength of Teacher Training

Strengthen the PE teachers, principal training for new curriculum business, enhance the theoretical level, transform their sports view, students view, teaching material

---

[3] LI Jing-xia. Some exploration on curriculum resource exploitation and utilization[J]. Education exploration, 2003, (8) : 40-41.

[4] LU Bao-gen. Some thinking about implement course standard [J]. Sports Teaching, 2002, (1) : 25-26.

view, classroom view, teaching activities view, learning evaluation view to students, professional qualities an so on. To recognize the importance of curriculum implementation, strengthen the guidance of new curriculum implementation, make every high school in China can implement the new curriculum.

For the misunderstanding of current curriculum reformation, PE teachers should accept training, self-learning and research, combine new curriculum view with school reality, learn from each other, communication, improve and grow to maturity together. PE teachers should recognize "no change is not allowed ", determine to reinvent themselves, increase the consciousness to handle the teaching material and its scientific content gradually. Unscramble the counselling thought, education concept, reform goal of new curriculum reformation, relevant policy measures and each PE curriculum standard, on one hand, make leadership and teachers understand each subject reformation success in-depth, on the other hand, give some advice to teaching implementaion.[5]

3.3 Strengthen Extracurricular Sports Teaching

PE teachers should strengthen the study of the theory, pay more attention to teaching experience, teaching concept and teaching method, make great efforts to improve teaching ability. In addition, the way and mthod which teachers uesed to stimulate students' initiative play a crucial role in teaching process. The traditional teaching mode in which every project time is less that students cannot grasp basic skills. Only being improve by simple to difficult, step by step, from Basic skill to special skill, make students to experience their own progress and success from the constantly strengthen of general physcial quality. To improve school PE teaching effect, PE teachers must to teach extracurricular sports more and more, in this way, students can learn new method to do physical exercise as well as improve their physical quality, develop the habit of self training, being a lifelong sports practitioners.

3.4 Improve Teaching Conditions, Develop Curriculum Resources

In order to better implement the new curriculum, the education department shoud expand educational investment, improve sports field and equipment in middle school, give more money to buy necessary sports supplies and equipment. Estabilish and complete the management system to repair, maintain and use sports facility and equipment, in oder to satisfy the requirments from different students to do physical exercises. Improve teaching conditions, especially the conditions in rural middle school.

The main body to carry out the concept of the new curriculum is PE teachers. To adapt to the change of the new curriculum change and improve students' enthusiasm in learning sports knowledge and skills, there is only way that to expand the PE teachers team construction and improve their professional level and practical ability. Cultivate new style of teacher team, PE teachers respond to learn more and do some research to improve their quality and literacy. PE teacher is the key in the new

[5] CHEN Zuo-song, JI Liu.Some requirments to PE teachers of the implementation of new PE curriculum reformation [J]. Journal of Beijing Sport University, 2004, 27 (3) : 370-372.

curriculum implementation, main executives in daily PE teaching, so we must pay more attention to the PE teachers training work.[6]

According to the characteristics of various regions and schools, actively developing local PE curriculum and school-based curriculum, strengthen the curriculum resources development and application. PE teachers strive to develop the PE curriculum resources related to sports teaching activities to add teaching material and enrich the teaching content. Choose, transform or reintegrate some PE curriculum resources, combine with local and school sports characteristics, develop local sports curriculum resources, etc. Life is full of sports material, find sports curriculum resources from daily life, preferred integrate curriculum resources from teaching materials and natural geography resources around school, it will make the new sports curriculum teaching more abundant and wonderful.

3.5 Strength the Safety Consciousness

The aim of study sports curriculum is to make our physical and mental health developed. PE teachers should let the students know how to protect themselves and strength their various ability. School may add some safety lessons to give security guidance. Danger is inevitable in our life, escape is not the best way, the earthquake, fire are unexpected disasters, the best way is to learn how to get out of and protect ourselves form the disasters. Security hidden danger is there that we should face it positively and strengthening the safety management skills instead of escaping.

References

1. Estabilshed by the ministry of education of the People's Republic of China. Physical education and health curriculum standards for ordinary high school. People's Education Press (2003)
2. Qi, Y.-X.: Thinking of current high school physical education and health curriculum. Journal of Adult Sports Education (2), 10–101 (2003)
3. Lei, Z.-Z.: Countermeasures about the influence of sports equipments and facilities to the middle school. Journal of Sports (6), 75–78 (1998)
4. Li, J.-X.: Some exploration on curriculum resource exploitation and utilization. Education Exploration (8), 40–41 (2003)
5. Interpretation of physical education (and health) curriculum standard, vol. 5. Hubei Press, Wuhan (2002)
6. Lu, B.-G.: Some thinking about implement course standard. Sports Teaching (1), 25–26 (2002)
7. Chen, Z.-S., Ji, L.: Some requirments to PE teachers of the implementation of new PE curriculum reformation. Journal of Beijing Sport University 27(3), 370–372 (2004)
8. Geng, P.-X., Li, Z.-G.: Development and application of school sports curriculum resources. Physical Education (3), 4–6 (2005)

[6] GENG Pei-xin, LI Zhi-gang. Development and application of school sports curriculum resources. [J]. Physical Education, 2005, (3) :4-6

Advanced Vocational Colleges' Online Evaluation System of Teaching Quality Based on AHP

Xia Zhou[1,2], Yi Chen [1], and Zhihong Zhuo [2]

School of Computer and Control
[1] Guilin University of Electronic Technology, Guilin, 541004, China
[2] Yangjiang Vocational and Technical College, Guangzhou, 529500, China

Abstract. In the practical evaluation of teaching quality, the evaluative indexes are different at different times of different colleges. In order to rationally evaluate the teaching quality of teachers in advanced vocational colleges, this essay raised a comprehensive evaluation system based on analytic hierarchy process .According to this method, we designed a B/s online evaluation system of teaching quality. The practice indicates that this system is highly flexible and pertinent of qualitative analysis and quantitative analysis.

Keywords: AHP, Advanced vocational college, teaching quality, evaluation system, analytic hierarchy process.

1 Introduction

Teaching quality of higher vocational colleges is the life-blood of higher vocational education. It is the core embodiment of education competitiveness which determines the survival and development of schools. Nowadays, evaluating teaching quality as an important method for colleges to improve teaching quality has been paid more attention. The existing methods and evaluation index sign systems can't satisfy the requirement of teaching assessment in higher vocational colleges, influencing evaluating quality directly.

This paper proposes a new evaluating model about teaching quality and an evaluating system called B/S based on this model. In this system, the index sign system and experts will be selected by senior management according to realities. Considering human-machine mix and the principles of Meta-synthesis, experts' confidence value and weigh of each index will be determined by AHP, which will ensure fairness and reasonableness of evaluating to the greatest extent.

2 Establishment of Hierarchical Structure

In order to improve the level of teaching evaluation, we need to establish a scientific and reasonable index sign system for it which is the basement of evaluation. The index sign system in this paper is set up by Delphi Methods that obtain experts' opinions and exchange information again and again and based on considering the characteristics and laws of teaching in higher vocational education.

W. Zhang (Ed.): Advanced Technology in Teaching, AISC 163, pp. 281–288.
springerlink.com

In accordance with these principles, we set up an index sign system which is suitable for teaching evaluation of teachers. Let us take classroom teaching evaluation for example whose basic prototype is shown in Fig.1 as follows.

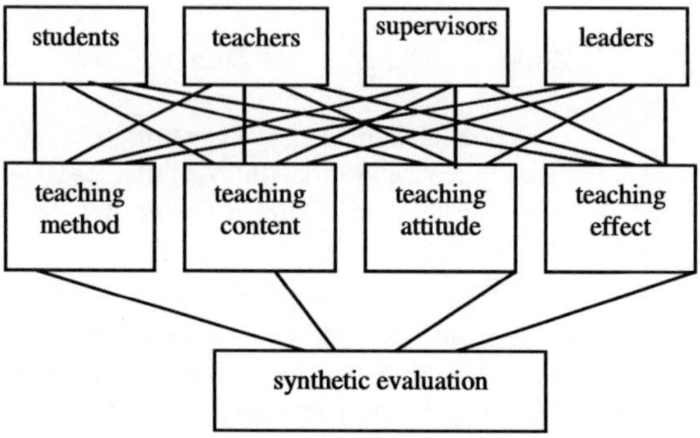

Fig. 1. Index sign system of classroom teaching quality

Hierarchy reflects relationship among elements. But in criteria layer, different criterions may have different weights when evaluating objects. They have their own proportions in the mind of decision-makers.

When ascertaining the weight of every factor that affects a certain element, the main problem we face is that these weighs are often difficult to be quantified. Additionally, if decision-makers want to ascertain the influence that each factor has on a certain element directly, they are prone to put forward some data that have different importance with practical ones and may be even contradictory from each other because of ill-considering and inability to attend everything at one time when there are many factors affecting an element.

Assume that we will compare the effects of the n factors to the factor Z, how can we provide reliable data. Saaty [3] proposes that we can create pair-compared matrix through comparing every two factors of them, which choose two factors and each time and make present the ratio of and to Z. All values of Z are put into a matrix. We say that A is the pair-compared judgment matrix.

It is easy to see that if the ratio of x_i and x_j to Z is a_{ij}, the ratio of x_j and x_i to Z is $a_{ji}=1/a_{ij}$.

Definition 1: If the matrix $A=(a_{ij})_{n\times n}$ satisfies,

(1) $a_{ij}>0$, (2) $a_{ji}=1/a_{ij}$ $(i,j=1,2,\cdots,n)$

We call it positive reciprocal matrix (obviously, $a_{ii}=1$, $i=1,\cdots,n$.

Regarding how to determine the value of a_{ij}, Saaty etc. propose citing numbers 1-9 and their reciprocal as index. Table 1 lines up meanings of indexes 1-9.

Table 1. Scales and meanings

Scale	Meaning
1	Two factors have the same importance.
3	The former factor is a little more important than the latter.
5	The former factor is obviously more important than the latter.
7	The former factor is intensely more important than the latter.
9	The former factor is extremely more important than the latter.
2,4,6,8	Intermediate values between the above adjacent judgments.
inverse of a number	If the importance ratio of factor i and factor j is a_{ij}, the importance ratio of factor j and factor i is $a_{ji}=1/a_{ij}$.

From the psychological point of view, too many classifications may exceed people's judgment, which increases difficulty in making judgments and leads to people providing false data. Saaty and others have made experiments to compare the correctness of people's judging results under different scales. The results also indicate that the scale of 1-9 is the most suitable. Considering all the elements above comprehensively, the score of evaluating will be set among 1, 3, 5, 7, 9 and 10.

3 Determining the Factor Set and Weights

3.1 Determining the Evaluation Factor Set

We assume that the factor set is S = {s1 s2 s3 ...... sn}.

3.2 Determining the Evaluation Set

We assume that the index evaluation result is divided into m grades which compose the evaluation result set F = {f1 f2 f3 ...... fm} (i = 1, 2, 3 ... m). In this paper, we assume that there are 5 grades which are respectively "excellent, good, moderate, qualified and unqualified".

3.3 Determining the Weights of Indexes

The weights of indexes is set to make synthesized evaluation reflect the effect of each factor to degree of overall impact, and the ultimate goal of teaching evaluation system is to build a kind of excitation mechanism so as to improve teaching quality and management level. Therefore, we should follow the orientation and guiding principles when we determine the weights of the teaching evaluation indexes, strengthen certain aspects of the work through controlling index weight, thereby make the system play a guiding role. We use AHP method and obtain repeated estimation and opinion of experts to form the weights of each index. The index system and weights are shown in table 2.

To ensure the objectivity of evaluation, we must choose comprehensive and reliable evaluation data and representative participation staff. Staff for evaluating classroom teaching includes specialized teachers, department leaders, students and supervisors.

Table 2. Teaching Quality Evaluation System

First-grade factor	weight	Second-grade factor	weight
teacher	0.3	teaching content	0.20
student	0.4	teaching method	0.20
supervisor	0.2	teaching attitude and preparation	0.11
leader	0.1	teaching effect	0.26
		teaching vocational skills	0.23

To ensure the objectivity of evaluation, we must choose comprehensive and reliable evaluation data and representative participation staff. Staff for evaluating classroom teaching includes specialized teachers, department leaders, students and supervisors.

4 Design of System

The construction of system for evaluating teaching quality involves the establishment of an index sign system, the selection of evaluators, the collection of evaluating information and outcome, the design and realization of system and other efforts in various aspects. The basic work of teaching evaluation is to establish an index sign system whose main task is to define evaluating indexes and the hierarchy and weighs of them. This is a complicated system engineering which belongs to the sphere of qualitative analysis.

4.1 The Basic Model of System

Evaluation index sign system directly influences the evaluating outcome. Main evaluators are school leaders, teaching supervisors and teaching managers all of which we call experts in this field. Nowadays, more and more colleges attach great importance to teacher assessments from students or from other teachers or from himself. We call these assessments generalized groups of experts which are enormous and whose opinions reflect public will and manifest democratic decision-making.

This system is designed by B/S structure. Administrators (school leaders), before evaluation, define evaluating indexes and levels, select experts, determine the demand of evaluation ,preset the weighs in sign index system and trust values for different kinds of experts. Generalized experts (students or teachers) and field experts (department leadership or teaching supervisors) log into their own interface to evaluate corresponding teachers. In this system, students can only evaluate their classroom teachers of the term, teachers can only evaluate other teachers in the same faculties, department leaders can only evaluate teachers in their faculties, but teaching supervisors have a right to evaluate all the teachers in the college. As the evaluation goes on, the system will generate evaluation results according to the established dynamic evaluation model. During evaluation, students can not see the evaluation results of teachers, teachers can only view their own evaluation information, department leaders can only see the results of their department, but teaching supervisors can view the evaluation outcome of all teachers in the college. After evaluation, the results will be published by school-level educational administrators according to situations.

The generalized experts (students and teachers) and field experts (department leaders and teaching supervisors) will make evaluation from five angles, from that of teaching attitude and preparation, of content of courses ,of teaching methods ,of teaching vocational skills and of teaching effect. But they have different emphasis. Students will evaluate teachers from four aspects, from that of teaching attitude and preparation, of content of courses, of teaching methods and of teaching effect. For example, students can make an intuitional evaluation of whether teachers have made full preparation, whether the course content is meaty, whether the teaching methods are reasonable and so on.

Compared with student evaluation to teachers, peer evaluation among teachers would be more professional and in-depth, so it will put emphasis on teaching effect, teaching vocational skills and teaching content. For example, in the aspect of teaching vocational skills, indexes of peer evaluation among teachers are whether teachers have instructed students patiently in practical operation and carried out efficient experiments while indexes are the development in students' abilities of analysis and solving problems in the aspect of teaching effect.

The emphasis of leadership assessment is the same as that of peer assessment among teachers, although evaluation indexes maybe change. For example, in the aspect of teaching vocational skills indexes are the state of participating in different kinds of vocational skill competitions and teaching reforms while in the aspect of teaching effect those are whether the form of examinations is scientific and whether teachers pay attention to the development of students' innovation ability.

Supervisors pay more attention to three aspects: teaching content, teaching methods and teaching effect. For example, the indexes of teaching content are whether it is substantial, informative and focused in important and difficult points while indexes of teaching methods are whether teachers make full use of modern teaching technique, emphasize enlightenment, bring students' initiative into play and so on

4.2 Selection of Evaluation Content

The structure of evaluation content in the system is based on improving students' competitiveness in employment. It contains two main parts: one is courses content; the other is teachers' diathesis.

1) Mind and Hand

Teaching quality evaluation should be able to enhance students' competitiveness in employment. Only by enhancing comprehensive abilities -knowing knowledge, good vocational skills and high diathesis, students are able to adapt to the social environment, develop continually, and improve their learning capacity, competitive and development forces in lifelong learning. In practice, this system takes the following items as the basic content of teaching quality assessment: stimulating interest in learning; test results with a normal distribution; acquiring knowledge and training practical skills; carrying on professional education; improving the ideological and ethical standards.

2) Valuing teachers' comprehensive development
School education should take educating people above everything, make prior development of morality and put cultivating students as workers with socialist morality and quality in the first place. Teaching evaluation system should lead teachers to lay equal stress on both educating people and teaching knowledge. In classroom, students not only learn how to behave but also how to do from teachers. "being exemplary ,imparting knowledge ,educating people and making education effective by guiding students actively" are important items when evaluating teaching quality.

5 Application Examples

5.1 Selecting Experts and Defining Evaluating Indexes and the Hierarchy

Experts who participate in evaluating include four kinds , namely students, professional teachers, department leaders and teaching supervisors. The trust values of each kind of experts are respectively 0.4, 0.2, 0.1 and 0.3 through hierarchical analysis method.

There are five first grade indexes which are teaching attitude and preparation, teaching content, teaching methods, teaching effect and teaching vocational skills. The weights of the five first grade indexes are respectively 0.11, 0.20, 0.20, 0.26 and 0.23 through hierarchical analysis method.

In this system, each first grade index includes four second grade indexes. The four second-grade indexes of teaching attitude and preparation are: selecting suitable teaching material and reference books contributed to learning, teaching discipline, preparing a lesson completely, organizing teaching courses accurately and completing teaching task on time.

The four second-grade indexes of teaching content are: plenteous, advanced, numerous-information content; precise concept; highlighting emphasis and difficulty; theory contacting with reality, integrated teaching course and systemic content.

The four second-grade indexes of teaching methods are: if methods are suitable and train of though is clear, if modern teaching means are full used and he writes on the blackboard tidily, if he pays attention to enlightening and motivating students, and if he guides students with learning methods and teaches them according to their capability.

The four second-grade indexes of teaching effect are: students' ability of analyzing and solving problems, examination, the ratio of attending a lecture and students' innovation ability.

The four second-grade indexes of teaching vocational skills are: if school assignment is reasonable and directional, and he is patient in guiding practical operation, if he participates in teaching reform positively and publishes interrelated papers, if he carry out effective experiments, and if he participates in different kinds of vocational skill competitions.

5.2 Teaching Quality Assessment

We assume that 8 experts (0022,0023, 0024, 0025, 0030, 0031, 0061, 0062) of which 0061 and 0062 present students, 0024 and 0025 are leaders in department which the teacher 00104 belongs to, 0022and 0023 are teaching supervisors of college, and 0030 and 0031 are professional teachers in department which the teacher 00104 belongs to, assess a course called Operating System which is lectured by the teacher 00104 in the first term between 2009 and 2010.

Assessment values of the second-grade indexes which 8 experts give to the teacher are shown in table 3 (Cij presents the jth second-grade index of the ith first-grade index).

Table 3. Values of second-grade indexes

Experts	*cij*	*cij*	*Cij*	*cij*	*cij*	*cij*	*cij*	*cij*	*cij*	*cij*
0022	7	7	7	9	5	7	5	7	9	7
0023	5	9	7	3	5	9	7	7	7	9
0024	7	9	7	3	5	7	3	3	7	5
0025	7	10	7	3	5	7	9	5	3	3
0030	5	7	3	3	5	10	7	9	9	7
0031	5	10	5	9	7	5	5	9	7	7
0061	5	5	5	5	7	7	5	5	5	7
0062	10	3	10	3	10	9	3	7	7	10

According to the weights of all indexes in table 3, we can obtain the synthetic course assessment value 83.32 with grade 'good' through calculating. The effect is shown in Fig.2.

Fig. 2. Effect of second-grade indexes

6 Epilogue

This paper presents a teaching quality evaluation method based on AHP, and designs an online evaluation system according to this method. This system has features of strong scalability and target. It can determine the evaluation index system and experts involved in the evaluation according to the specific circumstances of different schools, and ascertain trust values of experts and weights of evaluation indexes using AHP method. This system has been put into practice in Yangjiang Vocational and Technical College and is of good effect.

References

1. Ding, J., Ye, J.: On AHO model and fuzzy judgement in evaluation of undergraduate teaching quality. Wuhan University Journal 56(3), 241–245 (2003)
2. Wang, D., Han, Y., Dai, R.: A metasynthesis-based solution for enterprise decision making. System Engineering –Practice and Theory 23(3), 31–36 (2003)
3. Zhao, L., Wang, S., Lai, Y.: The quantitive comparative evaluation model of teaching quality estimate. Mathematics in Practice and Theory 35(1), 12–17 (2005)
4. Pan, S., Zhou, G.: The problems and improvement about evaluation of university teachers performance. Chinese Geological Education 15(2) (March 2006)
5. Huan, H., Bao, H., Zhao, J.: Study on the evaluation of practice teaching quality in higher vocational colleges. Vocational and Technical Education 30(22), 42–46 (2009)
6. Xiang, X., Yi, Z., Xu, Y.: Design and implementation of teaching quality evaluation system in colleges and universities: based on fuzzy comprehensive evaluation method and browser/server structure. Journal of Hunan Agricultural University (Social Science) 11(2), 71–78 (2010)

Development of Dual Educational System of Germany in China

Minggang Xu

Institute of Construction Method, Nanjing Technical Vocational College
Nanjing, Jiangsu Province, China
xuminggang800311@163.com

Abstract. The dual educational system of Germany is one of vocational education forms which emphasizes particularly on the skilled trainings of the students, and the study of the theory is to serve the training of the skills. The compare between the dual educational system and the traditional school vocational education of China is brought forward in this paper. The good demonstration of the dual educational system which is carried into execution in Nanjing technical vocational college is shown. The example shows that the dual educational system is a good vocational education pattern, whose sizes of running schools and the setups of specialties are adapted to the requirements of the corporations, and should be developed in Chinese vocational schools by considering of the situation of China.

Keywords: Vocational education, dual system of Germany, vocational training, innovation, educational theory.

1 Introduction

The definition of the vocational education can be comprehended by two aspects, one is the broad sense definition: the people's vocational interest are cultivated and the people's vocational ability are trained by the requirement of the society, the other is the narrow sense definition: the labourers which have different educational level are educated by different level of the vocational skills, and are well up in the basic knowledge, applied knowledge and skills of specific labor department. The seventeenth meeting of Chinese Communist Party pointed out that optimizing the educational structure, advancing the balanced development of the compulsory education, accelerating the vocational education especially are important and necessarily. At the present time, the vocational education system of China has much shortcoming which should be modified by consulting the overseas vocational education pattern such as the dual educational system of Germany.

2 Analysis of Dual System and Chinese Vocational Education

Compared to the Chinese vocational education which is traditional school vocational education, the dual educational system pays more attention to the training of the

W. Zhang (Ed.): Advanced Technology in Teaching, AISC 163, pp. 289–293.
springerlink.com

people's practice skills. The dual educational system which can bring up the practical skilled operation personnel is accepted by the corporations. Although Chinese school vocational education pattern also emphasizes on the training of the operating skills, this training pattern leaves the producing scene of the corporations, and the concentrative producing practice can not combine the theory with the practice well.

Because the students study in the given working conditions with the dual educational system, the intercommunion opportunity between the students and the corporations is much more than the school vocational education pattern, which reduces the risk of the unemployment consumedly. It is a good reference to resolve the inosculated employ of the students of the Chinese vocational school.

Chinese vocational education emphasizes on the theory study, however, the training goal of the dual educational system is the requirement of the working station, which is loved by the corporations. The training goal with the requirement of the woker- technique –examination standard, and the teaching system should be the important content of the teaching innovation of the Chinese vocational education.

In China, many small corporations have no ability to found the vocational education center, so organizing the corporations together to found the training center is an important approach to develop the vocational education.

3 Characters of Dual Educational System

3.1 Two Study Places

The corporations and the vocational schools are the two basic training places of dual educational system. The practice places are divided into practice workshop and the producing workshop. The training time of the students in the corporations is about 70% of the all training time. The corporation-training let the students master the problem of "how to do", they can make progress with the corporations at the same time, and they can master all-around shills, learn ability to solve actual problems, be familiar with the producing proceeding of the corporations, know the working criterion, establish the consciousness of self-study.

3.2 Two Statuses of the Students

Because the students study in two places (the corporations and the vocational schools), they have two statuses: students and apprentices. In the corporations, they are apprentices who sign the working bargain with the trainning corporations; in the vocational schools, they are students who keep on receiving the compulsory education.

3.3 Two Types of the Teachers

The practice teachers in the corporations and the theory teachers are the two types of the teachers in the dual educational system. The practice teachers play a key and important role in the dual educational system. They commonly are the graduates who have more than 5 years working experience after the training of the dual educational system. The theory teachers are the graduates who have completed 4 years speciality study and 2 years study in normal schools.

3.4 Two Types of Teaching Materials

In order to guarantee the standard and the quality of the training, the training books are published by the technical graduate schools. The books of the theory teaching are not uniform, so the theory teachers should increase the new information to keep up with the development of the science and the technique in the teaching.

4 Example of Dual Educational System in China

In June 2006, the Nanjing Vocational Technical College signed the agreement with the BSHHA(international Bosch and Siemens Household Appliances Group)and Hanns - Seidel foundation of Germany to run the school together. Many master-hands are trained in industrial electronic speciality, industrial mechanical speciality etc according to the reference of the dual educational system. Under the taking part in of the corporations, the Nanjing Technical Vocational College erected the practice workshop which was used not only by the BSHHA to train the client, but also the students. The equipments which can be displaced according to the renovation of the product in the practice workshop were provided by the corporations, the corporations training center also be used by the students to know the working practice.

Fig. 1. The training center of BSHHA

With the help of the Hanns - Seidel foundation of Germany, the Nanjing Vocational and Technical College have practiced and explored the soul which is emphasizing on the practice, skills, and good relationship with the corporations of the dual educational system for 20 years. The leaders also transplant the core theory of the dual educational system to the school, and make great efforts to found the teaching mechanism which the vocational service keeps close to the corporations. The Nanjing Vocational and Technical College has kept cooperation with 200 famous corporations all over the world. They carry into execution together to build practice base, train students, research the new product, and offer the high skilled experts to the corporations.

The cooperation between corporations and the school also creates the educational resource. One is that corporations and the school share in the practice base together. The building requirement of the practice base which is that corporations are set up in the vocational school, and the schools are moved to the corporations is insisted on. The BSHHA has built the corporation-training center which is 1600sq.m. and the training center of the client service in the school. These training centers can satisfy the education of the students and the workers, training of the students in the school and the training of the maintainance of the products. The other one is that corporations and the school share in the teaching resources together. The Nanjing Vocational and Technical College has erected the system which is that the experts of the corporations are invited to the school to teach the students, and the teachers of the school are assigned to study outside, on the one hand, the teachers can improve the shills themselves by joining in the training, on the other hand, the teachers can created product value and research value by joining in the employee training.

Fig. 2. The Nanjing Technical Vocational College

Fig. 3. The training scene of the students

Now, great changes have taken place of the meaning of the dual educational system. The cultivation of the technical secondary school extends to the vocational college gradually. The minority of cultivation of demonstration speciality extends to all the specialities in the school. The original edition of the dual educational system extends to the local dual educational system which accords with the situation of the society of China.

Acknowledgment. This research is helped by the teachers of the department of the civil engineering of Nanjing Technical Vocational College.

References

1. Xu, G.-Q.: Theory of vocational education. Jiaoyu Press, Shanghai (2007)
2. Ji, Z.-X.: Vocational and technical education. Fuzhou Jiaoyu Press, Fuzhou (1995)
3. Liu, J.-N.: Vocational and technical education. Shandong Jiaoyu Press, Jinan (1986)
4. Zhou, M.-X.: Explore of the essence of the Vocational education. Education and Vocation, 27–28 (2003)
5. Center of the Chinese vocational education, Dual educational system of Germany. Suzhou University Press, Suzhou (1993)
6. Jiang, D.-Y.: Study of teaching thinking of vocational education of up-to-date Germany. Tinghua University Press, Beijing (2007)
7. Cheng, G.-L.: Revelation of the innovation on the vocational education of Germany to us. Vocational Education Forum Education (2005)
8. Zhang, N.-X.: Practice and consideration of pushing the indigenous pattern of the dual system. Journal of Nanjing Technical Vacational College (2009)

Research on Nonlinear PID Control of Digital Governor for Diesel Engine*

Yong Zhou[1], Qing-jiang Zhang[1], Qi-xun Zhou[1,2], and Hong-kai Jiang[1]

[1] School of Aeronautics, Northwestern Polytechnical University, Xi'an, Shaanxi, China
yongstar@nwpu.edu.cn
[2] Department of Electrical & Control Engineering, Xi'an University of Science and Technology, Xi'an, Shaanxi, China
zhouqixun029@163.com

Abstract. In this paper, a strategy of the self-tuning control of nonlinear PID for the digital governor is presented. By analyzing the control system of the diesel engine, the basic control parameters are confirmed. Compared the digital PID with the traditional PID, a novel self-tuning control strategy is proposed, which uses the nonlinear PID algorithm. The principle of the nonlinear PID parameters is proposed, the tables of the detailed data are presented and the program with C language is made. For the better performance of this system, the Smith Prediction is used in the control strategy at the proportional process. At the end of this paper, the experiment results verified that the control strategy is reliable, the nonlinear PID parameters are rational, and the system has better dynamic response.

Keywords: Nonlinear PID, Self-Tuning, Control Strategy, Diesel Engine.

1 Introduction

The diesel engine system has been widely applied in power generation, ship and hybrid electric vehicles and other fields and the diesel itself is a quite complex transient dynamic system including the processes of gas flow, reciprocation mechanical motion, fuel injection, air fuel mixture and burning and expansion[1][2]. As for such complex system, if the traditional analog control is adopted for such complex system, the controller circuit will be complex and the system performance will be restricted; however, the ordinary digital control method failed to reach a good effect on the system servo response performance[3][4][5]. Therefore, in this paper, a kind of self-tuning control method of on-line PID parameter amendment is proposed focusing in MCU and the PID nonlinear electronic speed controller, which realizes the segment table lookup amendment of parameter dynamically and promotes the system control response.

2 Parameter Setting

The diesel generator system will change the throttle actuator expansion amount through regulation on the PWM duty ratio so as to control the diesel oil-taking amount and

* National Natural Science Foundation Project Support (grant number: 50975231).

W. Zhang (Ed.): Advanced Technology in Teaching, AISC 163, pp. 295–303.
springerlink.com

control the generator rotating speed to meet the load requirement. The system may be simplified as First Order Lag Model and in PWM speed regulation control system the smaller the PWM cycle is the closer the control effect to the continuous system. However, if the start-up and shut-down have a higher frequency, the action of actuating mechanism will be more frequently, which is sure to accelerate the abrasion of the actuating mechanism. After a comprehensive consideration, the cycle of PWM T in the paper is 2.5ms.

The system adopts the rotating speed closed loop control and the generator rotating speed is obtained through Hall sensor, that is:

$$n = 60 \cdot f / z \tag{1}$$

Where: f is the induced electromotive force frequency of sensor, z is number of gear teeth and n is the generator rotating speed.

3 Digital PID Control

Control algorithm is a key component of the electronic governor software design and PID control has such features as small computational complexity, strong real-time and easy to realize and so on as a common method adopted in the engineering filed. Therefore, the control strategy is designed on the basis of PID algorithm in this paper.

PID algorithm adopts the feedback quantity deviation proportional, integral and derivative to perform the control and the control law may be expressed in the following formula:

$$u(t) = K_P \left[e(t) + \frac{1}{T_I \int_0^t e(t)dt} + \frac{T_D de(t)}{dt} \right] \tag{2}$$

Or written to the form of transfer function:

$$G(s) = \frac{U(s)}{E(s)} = K_P(1 + \frac{1}{T_I S} + T_D s) \tag{3}$$

Where: K_P is proportionality factor ; T_I is integral time constant and T_D is derivative time constant.

In case of digital PID control, the integral item and derivative item in formulas (2) shall be subjected to dispersing processing and the following approximate conversion may be carried out as follows:

$$\begin{cases} t \approx kT & (k = 0,1,2...) \\ \int_0^t e(t)dt \approx T\sum_{j=0}^{k} e(jT) = T\sum_{j=0}^{k} e(j) \\ \frac{de(t)}{dt} \approx \frac{e(kT) - e[(k-1)T]}{T} = \frac{e(k) - e(k-1)}{T} \end{cases} \tag{4}$$

Where: T is sampling period.

Therefore, in the above-mentioned discretization process, the sampling period T must be short enough to guarantee the sufficient precision and for the purpose of convenience $e(kT)$ is simplified as $e(k)$. Substitute formulas (4) into formulas (2) to get the digital PID expression:

$$\begin{aligned} u(k) &= K_P \cdot [e(k) + \frac{1}{T_I \sum_{j=0}^{k} e(j)} + \frac{T_D}{T[e(k)-e(k-1)]}] \\ &= K_P \cdot e(k) + K_I \cdot \sum_{j=0}^{k} e(j) + K_D \cdot [e(k)-e(k-1)] \end{aligned} \tag{5}$$

Where :

k——sampling sequence number, $k = 0,1,2\ldots$;
$u(k)$——computer output value at the k sampling moment;
$e(k)$——input deviation value at the k sampling moment;
$e(k\text{-}1)$——input deviation value at the (k-1) sampling moment;
K_I——integral coefficient, $K_I = K_P T/T_I$;
K_D——derivative coefficient, $K_D = K_P T_D/T$.

4 Nonlinear PID

PID controller has a wide application, but the nonlinearity and time-varying uncertainty and other factors often exist in the practical industrial production process, and application of regular PID controller parameter often results in a poor tuning and performance and has a poor adaptability to the operation working conditions. Therefore, there points need to be improved for the traditional PID algorithm:

(1) The control parameter, once confirmed, can not be regulated on line. In case of a strong interference, the overstrike is large and the time for steady recovery time is long, which causes the dynamic and static performance of the system.

(2) In case of load changing, the integral saturation phenomenon is easy to occur and the recovery time is longer in case of load changing.

Considering that the diesel engine has a quite wide rotating speed range and there is a large span from the idling operation to stable operation, it is quite difficult to realize a single parameter PID algorithm to maintain the constant rotating speed and good dynamic response of each rotating speed point in such a wide range especially in case of sudden loading and sudden off-loading. In order to improve the control effect, the nonlinear segment PID algorithm is designed in the paper and the Smith estimation is introduced at the proportioning phase.

4.1 Subsection Setting

In this paper, the entire effective rotating speed range of diesel generator is divided into some segments and the corresponding PID parameters are calibrated respectively according to the operating characteristics of various sections, to realize the optimum control performance. P, I and D value in each section multiplies the corresponding initial value of K_P, K_I and K_D endowed by the user interface to get the proportional, integral and derivative parameters practically applied in the rotating speed closed-loop. The control process is shown as Fig. 1.

Nonlinearity segmentation principle in the system: the proportional control is applied to restrict the deviation and it depends on the deviation amount input, and the higher the deviation amount is the larger P value is; the integral regulation may realize the floating speed regulation and it depends on the total history of the deviation amount; in order to prevent integral saturation, I value is smaller when the slip is larger; differential regulation depends on deviation change rate and the higher the change rate is the higher D value is.

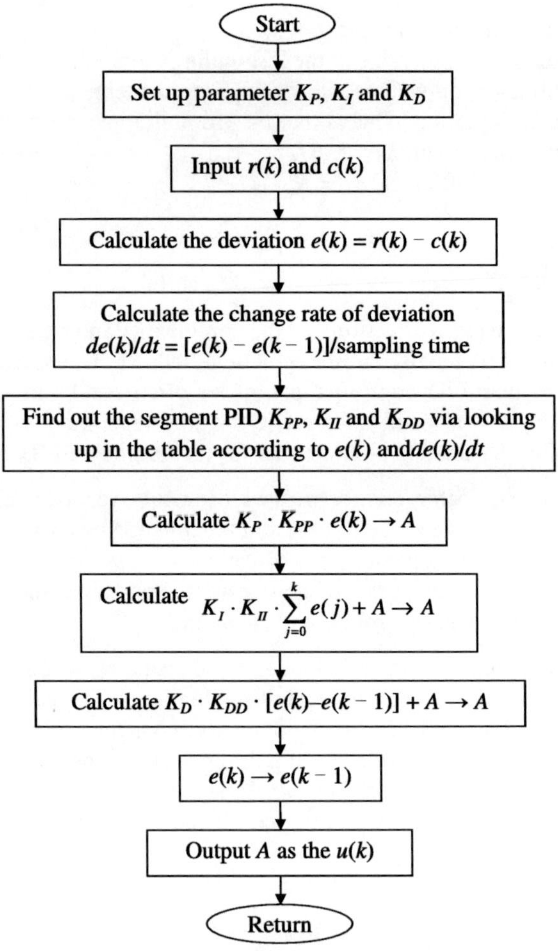

Fig. 1. Nonlinear PID control flow graph

The method for realization of nonlinear PID in this paper is: each segment rotating speed corresponds to a group of parameters and some groups of parameters compose a piece of control parameter table for real-time control system call. In case of real-time control, the rotating speed is controlled by checking up the parameter form to call the optimum PID parameter through the current rotating speed value measured. The concreter segmentation forms are shown in Table 1 - Table 3.

Table 1. Proportion segmentation table

e_k \ de_k	≥2.5	≥0.5	≥-0.5	≥-2.5	<-2.5
≥30	4.0	3.0	2.5	2.0	1.5
≥5	3.0	2.5	2.0	1.5	1.2
≥0	1.5	1.2	1.0	1.0	1.0
≥-5	1.0	1.0	1.0	1.2	1.5
≥-30	1.2	1.5	2.0	2.5	3.0
<-30	1.5	2.0	2.5	3.0	4.0

Table 2. Integration segmentation table

e_k \ de_k	≥2.5	≥0.5	≥-0.5	≥-2.5	<-2.5
≥30	0.35	0.3	0.25	0.2	0.15
≥5	0.7	0.6	0.5	0.4	0.3
≥0	0.8	0.9	1.0	0.8	0.6
≥-5	0.6	0.8	1.0	0.9	0.8
≥-30	0.3	0.4	0.5	0.6	0.7
<-30	0.15	0.2	0.25	0.3	0.35

Table 3. Differentiation segmentation table

e_k \ de_k	≥2.5	≥0.5	≥-0.5	≥-2.5	<-2.5
≥30	5.0	4.0	2.0	0.5	0.0
≥5	4.0	3.0	1.5	0.6	0.5
≥0	2.0	1.5	1.0	0.8	0.6
≥-5	0.6	0.8	1.0	1.5	2.0
≥-30	0.5	0.6	1.5	3.0	4.0
<-30	0.0	0.5	2.0	4.0	5.0

In the above list table: e_k = set value- feedback value, unit: r/min; $de_k = e_k - e_{k-1}$. According to table 1-3, the follow analysis is made:

(1) When $e_k \geq 30$ and $de_k \geq 2.5$ and $e_k < -30$ and $de_k < -2.5$ (top left corner and bottom right corner of the table), the system speed difference is quite large and the same time the difference value change direction indicates that the generator rotating speed develops towards the direction far away from the target value. Now, the coefficients of P and D reach to the maximum values and the system focuses in PD regulation and constitutes the lead correction, which rapidly restrict the system to develop towards the adverse direction and shortens the time for system recovery stability.

(2) When $e_k \geq 30$ and $de_k < -2.5$ and $e_k < -30$ and $de_k \geq 2.5$ (top right corner and bottom left corner of the table), the system speed difference is quite large but the difference value change direction indicates that the system develops towards approaching the target rotating speed. The coefficients of P and I now is small and the value of D is 0 and the system develops to the target rotating speed with a small speed which may effectively avoid overstrike and oscillation;

(3) When $-5 \leq e_k < 5$ and $-0.5 \leq de_k < 0.5$ (the middle of the table), the system speed difference is very small and the difference value is also very small and the system is in a stable state. The coefficient of I now reaches to the maximum value and the system focuses in PI regulation and guarantees the system stable precision.

4.2 Smith Estimation

It is difficult to carry out an effective control with the normal PID confronting with the control system with a large pure delayed time τ and it may result a serious overstrike and poor stability in the control process. In general, if the ratio of pure delay constant and inertia time constant is higher than 0.5, it is regarded that it is the maximum pure delay process and application of Smith estimation is a kind of quite effective measure.

Application of Smith estimation of the control strategy in this paper in ratio P, may greatly restrict the great change in rotating speed, such as sudden loading, sudden off-loading or oscillation and other conditions. In case of occurrence of the above-mentioned conditions, the oscillation control strategy may act in advance to reduce the adverse effect resulted from the system delay phenomenon, restrict the sudden change in rotating speed, improve the control performance and guarantee the system stability.

Smith pre-application of control strategy in this paper is shown as follows:

```
ek = rk - ck;              // Calculate the deviation value
dek = ek - ekp;            // Calculate the deviation change
ekNext = (1.25 * ek - 0.25 * dek);    // estimate
ekp = ek;                  // e(k)→e(k-1)
A = coefPP * ekNext;  // K_P · e(k) estimated value→ A
coefII + = coefII * ek;  // integral accumulation process
A = A + coefII;            // K_I · Σ_{j=0}^{k} e(j) + A → A
A = A + coefDD * dek;      // K_D · de(k) + A → A
```

"A" is the digital controlled variable of diesel electronic governor PWM duty ratio finically output, to complete the rotating speed PID control.

5 Experiments

In order to verify the rationality and reliability of the self-tuning parameter setting in the paper, a set of diesel generator is used as the experiment target, and the concrete parameter and control requirements are: rated speed 1500r/min, rated output power 30kw, allowable error under stable operation 0.8%, fluctuation at sudden loading or sudden off-loading is less than 100r/min, and recovery time is less than 5s.

The nonlinear PID control experiment diagram is obtained from the host computer software and the system startup, loading and off-loading waveforms are shown in Fig. 2 to Fig. 4; the vertical coordinate is the value of diesel generator rotating speed, the unit is rpm; and the horizontal ordinate is time, the unit is s/grid.

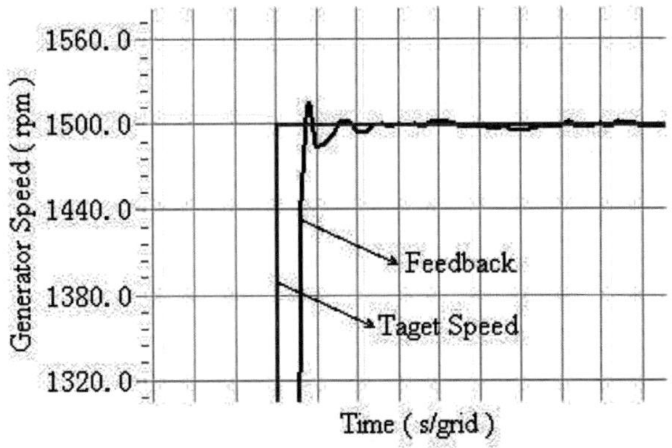

Fig. 2. Wave of startup rotating speed

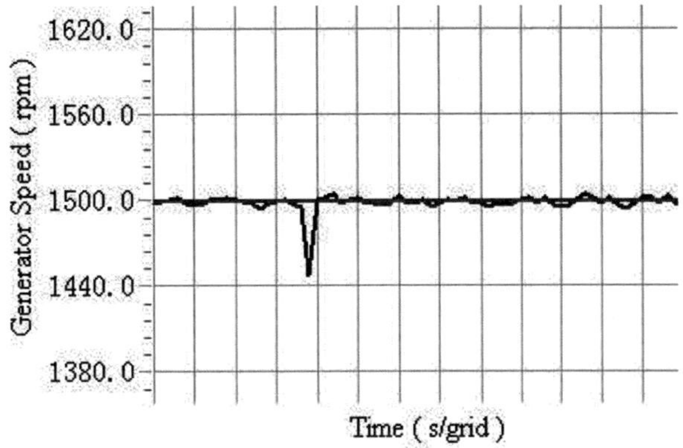

Fig. 3. Speed wave of sudden loading

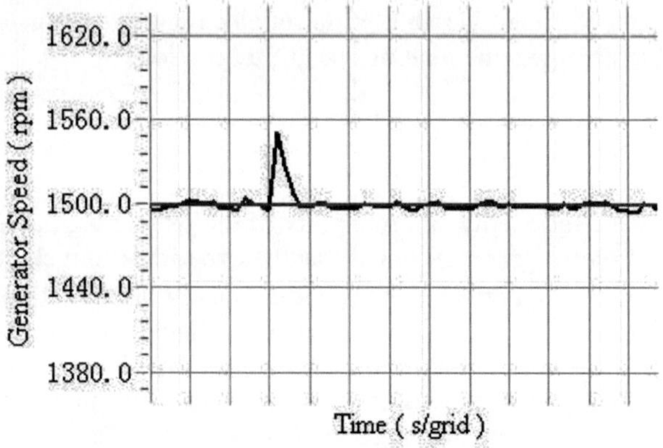

Fig. 4. Speed wave of Sudden off-loading

The experiment diagrams show that, if the system startup speed is high, the overstrike is small and the operation is stable; the fluctuations during loading and off-loading are within 40 rounds and the recovery time does not exceed 1 second and it meets the design requirement completely.

6 Conclusions

In this paper, emphasis is placed on research on diesel generator electronic governor nonlinear PID self-tuning control strategy and relevant parameter setting to verity the parameter rationality based on a large amount of experiments. The following conclusions may be drawn through analysis and research on the experiment results :

(1) The proportional regulator may make the system rotating speed approach to the target value rapidly, but the static error exists. Increase in proportionality factor K_P may reduce the static error, but too large K_P may deteriorate the system dynamic performance and result in output oscillation and failure of system to operate stably; and the too small K_P may increase the response time and influence the system stability.

(2) Introduction of integral regulation promotes the system indifference and enhances the system steady state performance at the cost of reduction in system stability. And at the same time, the appropriate integral upper limit must be set up because of integral accumulation to avoid appearance of integral saturation.

(3) In order to accelerate the control process and make a pre-judgment of the deviation change direction and control it, the differential regulation is introduced to decrease the pre-judgment and conquer the oscillation and keep the system stable.

(4) In the diesel generator regulation control system, the nonlinear PID control algorithm is applied and the self-adaptation regulation and self-adaptation are used to improve the dynamic response performance of system startup, sudden loading and sudden off-loading; and the system may look up the table at any rotating speed range to select the appropriate K_P and K_I and K_D values to realize the rapid and accurate and stable control effects.

References

1. McGowan, D.J., Morrow, D.J., McArdle, M.: A Digital PID Speed Controller for a Diesel Generating Set. In: Power Engineering Society General Meeting, July 13-17, vol. 3, pp. 1472–1477. IEEE (2003)
2. McArdle, M.G., Morrow, D.J., Calvert, P.A.J.: A Fuzzy Tuning PID Automatic Voltage Regulator for Small Salient Pole Alternators. In: Power System Technology Proceedings, pp. 103–108 (2000)
3. Augustine, D.W., Kumar, K.S.P.: A Method for Self-Tuning a PID Controller for Control of Small to Medium Sized Diesel Engines. In: IEEE International Conference on Systems Engineering, August 1-3, pp. 85–88 (1991)
4. Li, Y.-Q., Zhu, C.-Q., Huang, C.-H.: Study on Fuzzy Self-Tuning PID Parameters of Digital Governor for Diesel Engine. Transactions of CSICE 19(3), 267–269 (2001) (in Chinese)
5. Cao, H., Sun, B.-Y., Duan, J.: Self-tuning PID controller of diesel engine based on fuzzy logic. Journal of Dalian University of Technology 40(4), 465–469 (2000) (in Chinese)

A Novel Method for Modeling and Simulation of Brushless DC Motor with Kalman Filter*

Yong Zhou[1], Hong-kai Jiang[1], Qi-xun Zhou[1,2], and Qing-jiang Zhang[1]

[1] School of Aeronautics, Northwestern Polytechnical University, Xi'an, Shaanxi, China
yongstar@nwpu.edu.cn
[2] Department of Electrical & Control Engineering, Xi'an University of Science and Technology, Xi'an, Shaanxi, China
zhouqixun029@163.com

Abstract. Based on the mathematical model of the Brushless DC motor (BLDCM), a novel method of modeling and simulation for the speed control system of BLDCM with Kalman Filter is presented. In MATLAB/Simulink, the isolated functional blocks combine with the S-functions; and the model of BLDCM is established. Based on the discrete-state-space model, the recurrence formula for Kalman Filter is deduced, its block is established also. The speed feedback of BLDCM is filtered by Kalman Filter, and the PI speed controller is used. The reasonability and validity are testified by the coincidence of the simulation results and theoretical analysis. The results of experiment verified that the state space formula is correct; the correction matrix for Kalman Filter and the relevant coefficient are reasonable; and the effect of the filter is excellent. The novel method in this paper is also suitable for verifying the reasonability of other control algorithms, it offers a new thinking for designing and debugging actual motors.

Keywords: Brushless DC Motor, MATLAB/Simulink, Kalman Filter, Modeling, Simulation.

1 Introduction

The brushless DC motor, with such advantages as small volume, light weight, high efficiency, small rotational inertia and high control accuracy and so on, is widely applied in servo control, numerical control system and robot and other fields[1]. With the continuous expansion of application filed, it is required that the control system has a simple design, rational control algorithm and short development cycle, so establishment of the simulation model of brushless DC motor may effectively save the design time of control system. In this paper, a new method for establishment of brushless DC motor system simulation model based on MATLAB/Simulink is proposed on the basis of analysis of brushless DC motor mathematical model. The method, based on the motor discrete state space mathematical model, generalizes and

* National Natural Science Foundation Project Support (grant number: 50975231).

W. Zhang (Ed.): Advanced Technology in Teaching, AISC 163, pp. 305–314.
springerlink.com

deduces the Kalman Filter recursion formula and establishes the filter simulation module and the motor revolving speed is subjected to Kalman Filter for filtering and the system adopts the speed-regulation mode of PI revolving speed regulation.

2 Brushless DC Motor Mathematical Model

2.1 Model of Continuous State

The brushless DC motor operates under the state of two-phase conducting and three-phase six state, and suppose that the motor magnetic circuit is not saturated in the operation process and magnetic hysteresis loss are not calculated and the magnetic steel has a high impedance and the rotor inductive current is neglected, then the voltage balance equation of three phase winding may be expressed as[2]:

$$\begin{bmatrix} u_a \\ u_b \\ u_c \end{bmatrix} = \begin{bmatrix} r_a & 0 & 0 \\ 0 & r_b & 0 \\ 0 & 0 & r_c \end{bmatrix} \cdot \begin{bmatrix} i_a \\ i_b \\ i_c \end{bmatrix} + \begin{bmatrix} L_a & L_{ab} & L_{ac} \\ L_{ba} & L_b & L_{bc} \\ L_{ca} & L_{cb} & L_c \end{bmatrix} \times D \begin{bmatrix} i_a \\ i_b \\ i_c \end{bmatrix} + \begin{bmatrix} e_a \\ e_b \\ e_c \end{bmatrix} \quad (1)$$

Where: e is the stator winding electromotive force (V): L_a~L_c is the self-inductance (H) of each phase winding; L_{ab}~L_{cb} is mutual inductance (H) between each two phase windings; and differential operator $D = d/dt$.

If the three phase windings are completely symmetric, the back electromotive force waveform has a trapezoidal wave with the flattened width of 120° electrical angle, and then it may be regarded that the three phase winding self-inductance and mutual inductance between windings are constant, then:

$$i_a + i_b + i_c = 0 \quad (2)$$

So:

$$M \cdot i_b + M \cdot i_c = -M \cdot i_a \quad (3)$$

Formula (1) is simplified as:

$$\begin{bmatrix} u_a \\ u_b \\ u_c \end{bmatrix} = \begin{bmatrix} r & 0 & 0 \\ 0 & r & 0 \\ 0 & 0 & r \end{bmatrix} \cdot \begin{bmatrix} i_a \\ i_b \\ i_c \end{bmatrix} + \begin{bmatrix} L-M & 0 & 0 \\ 0 & L-M & 0 \\ 0 & 0 & L-M \end{bmatrix} \cdot D \begin{bmatrix} i_a \\ i_b \\ i_c \end{bmatrix} + \begin{bmatrix} e_a \\ e_b \\ e_c \end{bmatrix} \quad (4)$$

The motor electromagnetic torque equation is:

$$T_e = \frac{e_a \cdot i_a + e_b \cdot i_b + e_c \cdot i_c}{\omega} \quad (5)$$

The conversion between ω and rotor electric angle speed ω_e: $\omega_e = n_p \cdot \omega$, where: n_p is motor rotor role pair number; and the motor mechanical movement equation is:

$$T_e - T_L = J \cdot \frac{d\omega}{dt} \tag{6}$$

According to the state space theory, formula (4) may be rewritten as the space equation:

$$\begin{cases} \dot{x} = Ax + Bu \\ y = Cx + Du \end{cases} \tag{7}$$

Take the state vector $x = \begin{bmatrix} i_a & i_b & i_c \end{bmatrix}^T$, the coefficient matrix may be expressed as:

$$A = \begin{bmatrix} -r/(L-M) & 0 & 0 \\ 0 & -r/(L-M) & 0 \\ 0 & 0 & -r/(L-M) \end{bmatrix},$$

$$B = \begin{bmatrix} 1/(L-M) & 0 & 0 \\ 0 & 1/(L-M) & 0 \\ 0 & 0 & 1/(L-M) \end{bmatrix},$$

$$C = \begin{bmatrix} 1 & 0 & 0 \\ 0 & 1 & 0 \\ 0 & 0 & 1 \end{bmatrix}, D = 0;$$

The system input vector $u = \begin{bmatrix} u_a - e_a & u_b - e_b & u_c - e_c \end{bmatrix}^T$, solve the equation above and then it may obtain the motor winding current i_a, i_b and i_c.

2.2 System Transfer Function

Because the motor has only two-phase conduction at one moment, it may deduced that: $u = 2 \cdot r \cdot i + 2 \cdot (L-M) \cdot \frac{di}{dt} + 2 \cdot e$, $T_e = \frac{2 \cdot e \cdot i}{\omega}$; the winding back electromotive force waveform is the trapezoidal wave of the 120° flattened width, and when each phase winding is conduct, its back electromotive force is always in the flattened area, so $e = K_e \cdot \omega$, where K_e is the motor electromotive force coefficient, and the calculation equation above is subjected to the Laplace transform, to obtain:

$$\begin{cases} U(s) = 2 \cdot r \cdot I(s) + 2 \cdot (L-M) \cdot I(s) \cdot s + 2 \cdot E(s) \\ T_e(s) = 2 \cdot K_e \cdot I(s) \\ T_e(s) - T_L(s) = J \cdot \omega(s) \cdot s \\ E(s) = K_e \cdot \omega(s) \end{cases} \tag{8}$$

At the same time, $\omega(s) = \frac{\pi}{30} \cdot N(s)$ based on $\omega = 2 \cdot \pi \cdot n/60$; and if the load torque T_L is regarded as a kind of interference and the no load condition is considered, the motor transfer function may be obtained from equation (8):

$$\frac{N(s)}{U(s)} = \frac{\dfrac{30 \cdot K_e}{J \cdot (L-M) \cdot \pi}}{s^2 + \dfrac{r}{L-M} \cdot s + \dfrac{2 \cdot K_e^2}{J \cdot (L-M)}} = \frac{a}{s^2 + b \cdot s + c} \tag{9}$$

2.3 Model of Discrete State

In order to facilitate programming, the motor continuous transfer function is discretized. Supposed that d_1 and d_2 are two unequal roots of $s^2 + b \cdot s + c = 0$ and zero-order hold is considered and equation (10) may be obtained according to conversion of Z for equation (9) according to reference [3]:

$$\frac{N(s)}{U(s)} = Z[\frac{1-e^{-Ts}}{s} \frac{a}{s^2 + b \cdot s + c}] = \frac{A \cdot z + B}{z^2 + F \cdot z + D} \tag{10}$$

T is the system sampling cycle, in this paper $T = 0.1$ ms.

In order to obtain system state space equation, equation (10) is subjected to Z inverse transformation, to obtain:

$$n(k+2) + F \cdot n(k+1) + D \cdot n(k) = A \cdot u(k+1) + B \cdot u(k) \tag{11}$$

Take the state function: $x_1(k) = n(k)$, $x_2(k) = x_1(k+1) - A \cdot u(k)$, then it may obtain the discrete state space equation of brushless DC motor:

$$\begin{cases} \boldsymbol{x}(k+1) = \boldsymbol{G} \cdot \boldsymbol{x}(k) + \boldsymbol{H} \cdot u(k) \\ n(k) = \boldsymbol{C} \cdot \boldsymbol{x}(k) \end{cases} \tag{12}$$

Where: $u(k)$ is the motor winding terminal voltage under the control of PWM; and $n(k)$ is the motor revolving speed; $\boldsymbol{G} = \begin{bmatrix} 0 & 1 \\ -D & -F \end{bmatrix}$; $\boldsymbol{H} = \begin{bmatrix} A \\ B - A \cdot F \end{bmatrix}$; $\boldsymbol{C} = [1 \quad 0]$; $\boldsymbol{G}$, $\boldsymbol{H}$ and $\boldsymbol{C}$ are constant matrix as for the system indicated in the paper.

3 Kalman Filter

Kalman filtering method is a kind of linear minimum variance estimation and belongs to a kind of optimal estimation algorithm and it is quite easy for its discrete algorithm to realize digitization.

The system state equation is:

$$\boldsymbol{x}(k) = \Phi(k-1) \cdot \boldsymbol{x}(k-1) + w(k-1) \tag{13}$$

The system measurement equation is:

$$\boldsymbol{y}(k) = H(k) \cdot \boldsymbol{x}(k) + v(k) \tag{14}$$

Where: Φ is transfer matrix from state $\boldsymbol{x}(k\text{-}1)$ to state $\boldsymbol{x}(k)$; $\boldsymbol{y}(k)$ is the observation vector at moment $t(k)$; H is the measurement matrix; $w(k)$ and $v(k)$ are the system noise and measurement noise respectively, and their statistical properties are:

$$E[w(k), w(i)^T] = \begin{cases} \boldsymbol{Q}(k) & i = k \\ 0 & i \neq k \end{cases} \tag{15}$$

$$E[v(k), v(i)^T] = \begin{cases} \boldsymbol{R}(k) & i = k \\ 0 & i \neq k \end{cases} \tag{16}$$

$$E[w(k), v(i)^T] = 0 \tag{17}$$

Where: $\boldsymbol{Q}(k)$ is the symmetric positive semi-definite matrix and $\boldsymbol{R}(k)$ is the symmetric positive definite matrix.

The Kalman filtering process is actually a recursive process, including two steps: prediction and update.

Step 1: prediction. Under the condition that the optimum estimated value $\hat{\boldsymbol{x}}(k)$ at $t(k)$ moment is known, the system state prediction value $\tilde{\boldsymbol{x}}(k+1)$ and error covariance matrix, $\tilde{\boldsymbol{P}}(k+1)$ at the next moment of $t(k+1)$ moment, are predicted.

$$\tilde{\boldsymbol{x}}(k+1) = \Phi(k)\hat{\boldsymbol{x}}(k) \tag{18}$$

$$\tilde{\boldsymbol{P}}(k+1) = \Phi(k)\boldsymbol{P}(k)\Phi(k)^T + \boldsymbol{Q}(k) \tag{19}$$

The Kalman gain matrix $\boldsymbol{K}(k+1)$ is obtained on its basis:

$$\begin{aligned} \boldsymbol{K}(k+1) &= \tilde{\boldsymbol{P}}(k+1)H(k+1)^T \\ &\times [H(k+1)\tilde{\boldsymbol{P}}(k+1)H(k+1)^T + \boldsymbol{R}(k+1)]^{-1} \end{aligned} \tag{20}$$

Step 2: update. The estimated value is modified according to the principle of observation error and minimum variance, so as to obtain the optimum estimated value $\hat{\boldsymbol{x}}(k+1)$ of state variable at moment $t(k+1)$ and obtain the optimum estimation variance matrix $\boldsymbol{P}(k+1)$ at the same time.

$$\hat{\boldsymbol{x}}(k+1) = \tilde{\boldsymbol{x}}(k+1) + \boldsymbol{K}(k+1)[\boldsymbol{y}(k+1) - H(k+1)\tilde{\boldsymbol{x}}(k+1)] \tag{21}$$

$$\boldsymbol{P}(k+1) = \tilde{\boldsymbol{P}}(k+1) - \boldsymbol{K}(k+1)H(k+1)\tilde{\boldsymbol{P}}(k+1) \tag{22}$$

The Kalman Filter recursion formula applicable to the system is.

$$\begin{cases} \hat{\boldsymbol{x}}(k+1) = \boldsymbol{G}\cdot\hat{\boldsymbol{x}}(k) + \boldsymbol{H}\cdot\boldsymbol{u}(k) + \boldsymbol{K}(k+1) \\ \qquad\qquad \times[n(k+1) - \boldsymbol{C}\cdot\boldsymbol{G}\cdot\hat{\boldsymbol{x}}(k) - \boldsymbol{C}\cdot\boldsymbol{H}\cdot\boldsymbol{u}(k)] \\ \tilde{\boldsymbol{P}}(k+1) = \boldsymbol{G}\cdot\boldsymbol{P}(k)\cdot\boldsymbol{G}^{\mathrm{T}} + \boldsymbol{H}\cdot\boldsymbol{Q}\cdot\boldsymbol{H}^{\mathrm{T}} \\ \boldsymbol{K}(k+1) = \tilde{\boldsymbol{P}}(k+1)\cdot\boldsymbol{C}^{\mathrm{T}}\cdot[\boldsymbol{C}\cdot\tilde{\boldsymbol{P}}(k+1)\cdot\boldsymbol{C}^{\mathrm{T}} + \boldsymbol{R}]^{-1} \\ \boldsymbol{P}(k+1) = \tilde{\boldsymbol{P}}(k+1) - \boldsymbol{K}(k+1)\cdot\boldsymbol{C}\cdot\tilde{\boldsymbol{P}}(k+1) \end{cases} \tag{23}$$

Formula (23) shows that $\tilde{\boldsymbol{P}}(k)$ and $\boldsymbol{K}(k)$ and $\boldsymbol{P}(k)$ have no relationship with state and may be subjected to off-line calculation ; under the condition that the initial values of $\hat{\boldsymbol{x}}(0)$ and $\boldsymbol{P}(0)$ are known, the matrix $\boldsymbol{Q}$ and $\boldsymbol{R}$ shall be selected appropriately and they are subjected to many times of iterative computation to obtain the correction matrix $\boldsymbol{K}(k)$ steady state value of Kalman Filter; and then the optimum state estimated value $\hat{\boldsymbol{x}}(k)$ at moment k may be recursively calculated based on measured values $\boldsymbol{y}(k)$ and $\boldsymbol{K}(k)$ at moment k.

4 Model of Brushless DC Motor

Under the environment of MATLAB/Simulink, the thinking of using the module modeling is used to realize the graphical presentation of motor state space and Kalman recursion formula, and then substitute the coefficient matrix into various modules to establish the simulation model of brushless DC motor speed control system and the speed control strategy adopts the revolving speed closed loop PI control.

Fig. 1 is motor simulation mode control block diagram, mainly including: motor body module, revolving speed PI regulation module, PWM generator module, voltage inverter module and Kalman Filter module and so on, and the main modules of model are shown as in Fig. 2.

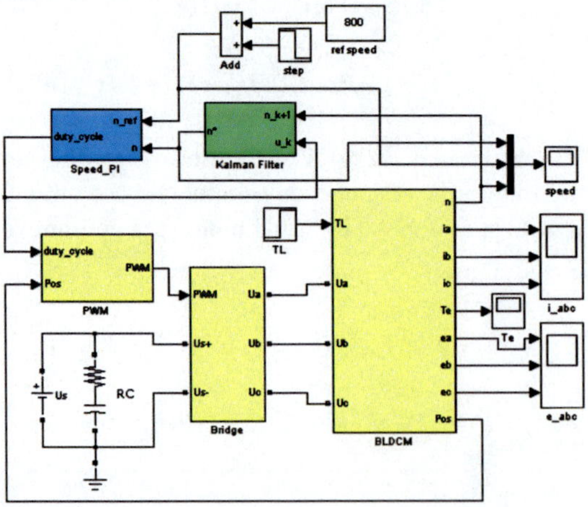

Fig. 1. Brushless DC motor simulation model

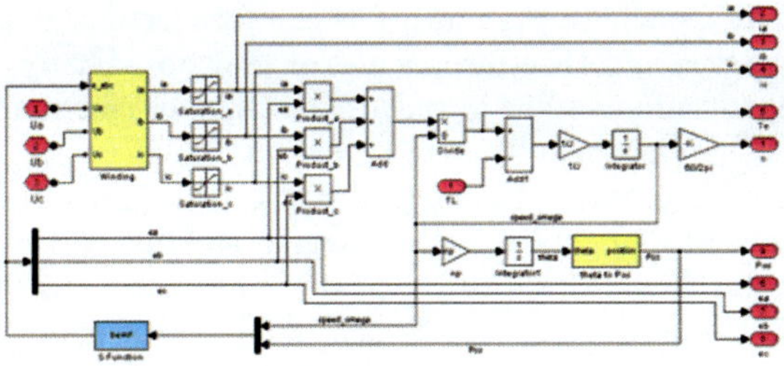

(*a*) motor block

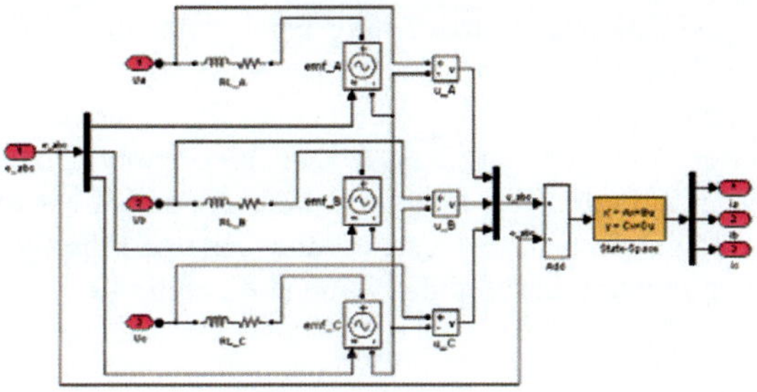

(*b*) winding module

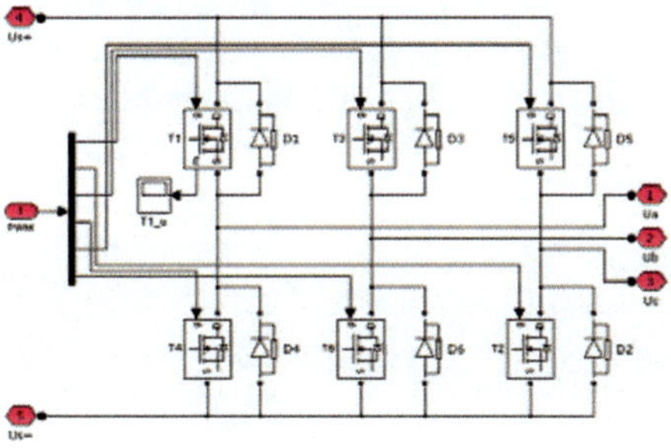

(*c*) inverter module

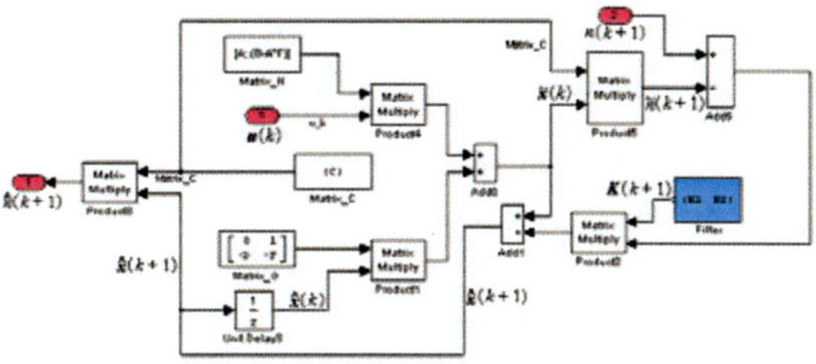

(*d*) Kalman Filter module

Fig. 2. Main function modules of motor

In the motor body module of Fig. 2(*a*), Pos is rotor electrical angle signal and the model adopts S function to realize the calculation equation. Firstly, the module shall obtain the motor phase current ia and ib and ic and obtain the electromagnetic torque Te after calculation; electromagnetic torque Te may be obtained after calculation; the motor electric angle speed ωe may be obtained after calculation of Te and then the rotor electrical angle theta may be obtained after integral; and the theta is finally converted into rotor position Pos.

Fig. 2(*b*) is the motor winding module and the paper adopts the segment linear method[4][5] to obtain winding back electromotive force and supply power and one operation cycle of motor is divided into 6 phases; each $\pi/3$ is a reversing phase and each operation phase of each phase may be indicated by a line segment; and the operation states of all the phases at one moment may be determined according to the rotor position and revolving speed signal and the back electromotive force waveform may be obtained through linear equation.

As for the inverter circuit in Fig. 2(*c*), 6 IGBTs compose three phase full bridge inverter circuit; (*d*) is Kalman filter module, where K1 and K2 included in $\boldsymbol{K}(k+1)$ is Kalman Filter correction matrix steady state value obtained through offline recursive calculation, and after calculation the finial value is K1=0.0348.

5 Simulation Analysis

The brushless DC motor parameters relevant to the paper are: $P_N = 700W$, $T_e = 1.5N{\cdot}m$, I_N = 10A, n = 800rpm, winding resistance r = 2.1Ω, L = 20mH, M = 10mH, J = $7{\times}10^{-4}kg{\cdot}m^2$, K_e = 0.1V/(rad/s), number of pole pairs np = 8, system power supply is 72V DC.

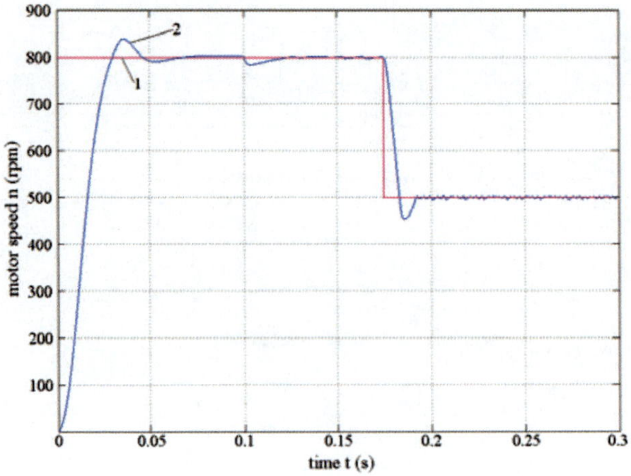

Fig. 3. Response curve of motor speed
1--target revolving speed; 2--motor revolving speed

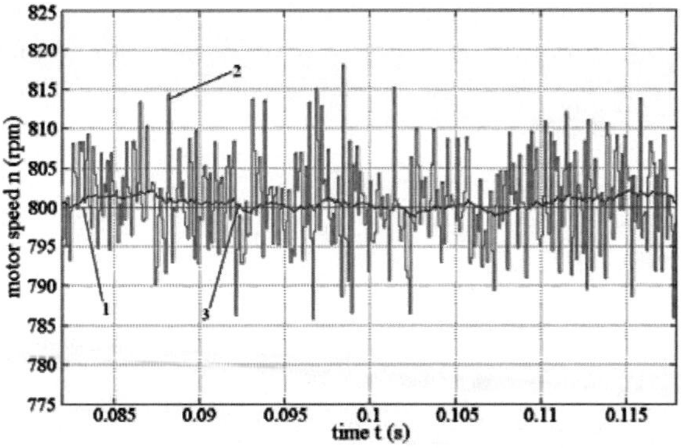

Fig. 4. Filtering effect of Kalman Filter
1--target speed; 2—motor speed without filter;
3—motor speed with filter

In order to verify the motor model static and dynamic performance, the system performs the no load startup and after it enters into the steady stat, the sudden load of $T_L = 1\text{N·m}$ shall be increased at $t = 0.1\text{s}$ and the target revolving speed shall be reduced to 500rpm at $t = 0.175\text{s}$. Fig. 3 shows that the model has a good servo performance and the steady operation is stable. In order to verify the filtering effect, the random interference is added in the system with the sampling time interval of T = 0.1 ms and diagram 4 shows the comparison of Kalman Filter filtering effect. The diagram shows that the revolving speed after filtering becomes more stable, which proves that the system state equation is correct and the steady state value obtained after recurrence is ideal; and the filtering effect is obvious.

6 Conclusion

In this paper, the brushless DC motor discrete state space mathematical model is established and in the Simulink environment the S function is combined to establish the motor simulation model and the Kalman Filter is added in the speed closed loop. The simulation result verifies the correctness and reliability of motor discrete state space model; the rationality of Kalman Filter recursion formula is proved; moreover, the model in this paper provides some simulation modules with a good versatility and portability. Therefore, the model provides a effective method and tool for analysis and design of the brushless DC motor control system and offers a new thinking for design and debugging for the motor control system.

References

1. Yin, Y.-H., Zheng, B.: A Method for Modeling and Simulation of Brushless DC Motor Control System based on Matlab. Journal of System Simulation 20(2), 293–298 (2008) (in Chinese)
2. Crnosija, P., Krishnan, R., Bjazic, T.: Transient Performance Based Design Optimization of PM Brushless DC Motor Drive Speed Controller. In: Proceedings of the IEEE International Symposium on Industrial Electronics, vol. 3, pp. 881–886 (2005)

3. Wang, L., Jiang, J.-P.: Brushless Direct-current Motor Optimal State Feedback Drive Control with Adaptive Compensation. Proceedings of the CSEE 28(4), 101–107 (2008) (in Chinese)
4. Acarnley, P.P., Watson, J.F.: Review of Position Sensorless Operation of Brushless Permanent Magnet Machines. IEEE Transactions on Industrial Electronics 53(2), 352–362 (2006)
5. Crnošija, P., Krishnan, R., Bjažić, T.: Transient Performance Based Design Optimization of PM Brushless DC Motor Drive Speed Controller. In: IEEE ISIE 2005, Dubrovnik, Croatia, June 20-23, pp. 881–886 (2005)

A Design and Applied Study of Substituted Middle School Teachers' Training Modal on the Basis of B-Learning— A Case Study of Educational Technology Training

Dehua Kong

Shanxi Normal University LinFen, Shanxi Province, China
kongdli73@126.com

Abstract. Middle School Substituted Teachers' Training is an important activity of Substitutional Teaching Practice in Shanxi Normal University. It aims at improving rural middle school teachers' teaching level and promoting the balanced development of rural elementary education. Based on the B-learning theory, the paper designs a middle school substituted teachers' training modal. Taking educational technology training as an example and combining the requirement and features of The Criterion of Educational Technology Ability in Elementary and Middle Schools (on trial), an applied study has been conducted from five aspects--- determination of training target、choice of training contents, the implementation process of training environment and the assessment of training effect, for the purpose of exploring a kind of operable and effective modal to suit rural middle school substituted teachers' training.

Keywords: Middle School Substituted Teachers, B-learning, Educational technology training, the integration of information technology and course.

1 Introduction

As one of the hot research problems in educational circles, the connotation of B-learning(Blending Learning or Blended Learning) has no precise definition. Professor He Kekang[1] thought that the so-called B-learning was the combination of the advantages of the traditional learning way and Networked Learning. That is to say, teachers must play guiding, inspiring, monitoring the teaching process in the leading role, but also fully embodies the students as the main body of learning initiative, enthusiasm and creativity. The two complement each other, in order to obtain the best learning effect. Driscoll[2] thought that B-learning meant learning process can be "a integration (or mixture) of web technology (such as virtual live classroom, collaborative learning, streaming media and the combination of text), in order to achieve a certain teaching objective; a variety of teaching methods (such as constructivism, behaviorism and cognitivism) and Instructional Technology (instructional technology or not) union, to realize the best teaching effect; any form of Instructional Technology (such as video, CD-ROM, web based training and film) and based on face-to-face teacher training; the combination of teaching technology and the detailed task, in order to form good study or work. B-learning can be thought of as a face-to-face classroom learning and online learning in two kinds of organic integration.

W. Zhang (Ed.): Advanced Technology in Teaching, AISC 163, pp. 315–324.
springerlink.com

Middle School Substituted Teachers' Training is an important activity of Substitutional Teaching Practice in Shanxi Normal University. It aims at improving rural middle school teachers' teaching level and promoting the balanced development of rural elementary education. For rural middle school teachers, it is a very scarce opportunity to receive one-term higher education. Based on the previous training experience, the paper constructed Substituted Middle School Teachers' Training Model on the Basis of B-learning and made an applied study on its theory and practice in order to improve training efficiency. A set of effective and operable training mode is expected to be explored.

2 A Design of Middle School Teachers' Training Mode Based on the B-Learning

2.1 Design Model of B-Learning Based Training Model

The core idea of B-learning is to solve problems depending on different issues, requirements. In teaching, it is to use different media and information transfer ways of learning, to achieve the minimum cost and highest learning effect. [3]

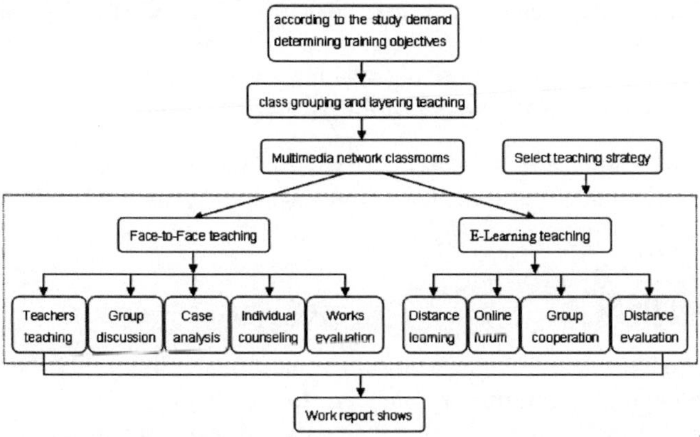

Fig. 1. The training design model based on B-Learning

In the model, the whole training process is divided into five basic steps: (a) determining training objectives according to the learning demand. Due to the diversity of learners' needs, uneven level or the learners' prior knowledge and different training target localization, therefore, before training needs analysis and diagnoses of original knowledge and skills among is not only one of the important factors to ensure the effectiveness of training, but also an important prop of the whole training process. (b) Making training content and carrying out the layer teaching. Because the original knowledge level of the learners are quite different, and the learning starting points are also not identical , the teachers should carry out the layer teaching to ensure the effectiveness of the training.(c) Customized training environment. According to B-learning learning concept, the whole training process is divided into face-to-face teaching and E-learning. In order to achieve that goal, this needs to create the material environment as well as the human resources environment, including multimedia classrooms, network learning platform, and the network teaching resources and so on, to ensure the smooth implementation of training. (d) Identify training strategy. Training strategies are keys of effective training. In the B-Learning training model, the mainly used strategies are the advance organizers teaching strategies, autonomic Learning strategies and collaborative Learning strategy, and so on. (e) Works report shows. After training, each learner should complete a teaching work and make summary report. Coaching teachers and learners evaluate and determine whether they can complete the course or not based on the evaluation criteria.

2.2 The Design Principles of the Model

(a) Goal-oriented training is to ensure the effectiveness of training; training objectives are derived from the needs of learners.

(b) Training contents as main body. For specific learners and contents, suitable content delivery and learning techniques are used to show and transmit, and appropriate methods of teaching are employed in the process of learning and teaching.

(c) Resource environment as training prop. A face-to-face teaching and network teaching are combined in the whole process to make sure related human resources and material resources effectively used.

(d) Formative evaluation and summative evaluation combined together as evaluation principle. Effective evaluation method should be able to correct learners' learning behavior and arouse learners' interest in the learning process, meanwhile motivate learners to continue learning, to produce conscious and spontaneous learners.

3 Application of Substituted Middle School Teachers' Training Model on the Basis of B-Learning

3.1 Analysis of Learning Needs, to Determine the Training Objectives

A questionnaire had been distributed to 124 Substituted teachers who are from six counties in five groups.122 valid questionnaires had been collected, and two teachers were randomly selected from each country to form 12 teachers group for individual interview. We used SPSS software to analyze the data, in order to understand the basic situation of secondary school teachers and provide basis for target location of the next training. Questionnaire is divided into four parts: The first part is the basic situation of the Substituted Teachers, including teachers' age, gender, education, job title and other basic information; second part is the mastery of information technology, the use of conventional educational software; The third part is the integration of information technology and curriculum knowledge and application status; The fourth part is past training situation and the need of this training or other investigation.

The Middle School Substituted training teachers are from Anze, Fushan, Xiangning, Pu, Liu Lin, Xi County's rural primary schools. Analysis showed the basic situation: the age ranging between 26 and 35 years old takes up 78%; undergraduate education and more than its degree account for 53.2%; 68.5% teachers' technical post is Grade two in middle schools; So, this Middle School Substituted teachers' education level is relatively high. The vast majorities are young teachers, and are the reserve force of basic education in rural schools and will play a significant role in the primary and secondary schools. Their common features are: energetic young teachers, on the accumulation and mature stage of professional development, eager to improve their education technical skills, and use of information technology for the efficiency of classroom teaching through training.

The following pie chart shows the survey results on the mastery of commonly-used educational software.

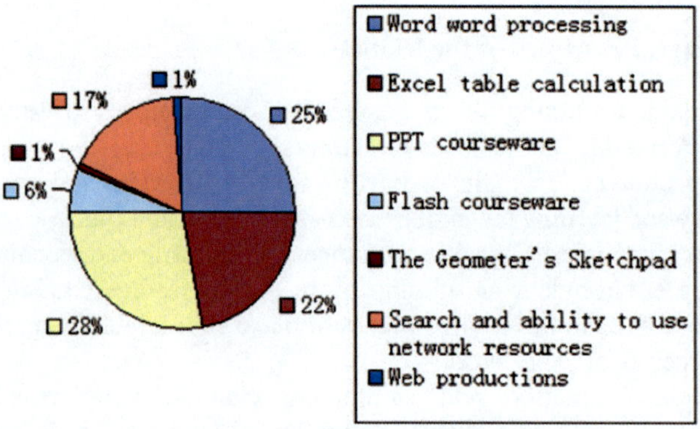

Fig. 2. The mastery of commonly-used educational software

It can be seen from the chart, the situation that these Middle School Substituted teachers master the commonly-used office software program is ideal. But we learned through interviews that most of them are dead modules operation which were mostly used to deal with computer title examination, can not be flexibility applied in teaching. PPT courseware production is the mastered best, which mainly due to the open classes carricd out in schools; A considerable part of the teachers also have ability to search and use network resources, but they mainly used to download songs, movies, cartoons and other entertainment, scarcely used in teaching; The Geometer's Sketchpad, web productions were very Rarely used. The survey indicates that most of the teachers can use information technology in the teaching process, but relatively poor in teaching methods, mainly make use of multimedia courseware platform to display, create situations. The use of information technology is still at the primary level.

66.7% of the teachers do not know the integration of information technology and curriculum. About 32.2% of the teachers know it, only 1.1% of the teachers can understand. Among them most learn the relevant information through educational magazines, books, open classes and study tours. Only 9.6% understand from teaching practice, 14.3% from a variety of training, only about 20.7% applying to teaching practice after learning. As a result, we should let the teachers know and understand the need to apply in teaching practice. In the interview, we learned that education and information for teachers under the new teaching approach is generally welcomed, but there are still vague concepts about the integration of information technology and subject knowledge. They have realized that the integration of information technology and discipline-based education reform has become an inevitable trend, willing to adjust their teaching skills to study hard and practice.

Training situation analysis about teachers' past participation in educational technology is like this. 41.5% of teachers have participated in educational technology training, 58.5% of teachers haven't participate in training. Among those who participated in training, 23.8% of teachers think that it has a great impact on their teaching practice, 56.4% of teachers think it a general effect, and 19.8% of teachers see no impact. Of the received training teachers, an investigation about the problems encountered in their teaching practice had been conducted; the results are shown in Table 1.

Table 1. Solution way of the encountered problems in Teacher's teaching practice

	Self reflection	Cooperation and exchange with other teachers	Search the Internet	Expert guidance	Set aside
Percentage(%)	18.6%	41.8%	27.1%	10.4%	2.1%

The survey showed that only 27.1% of teachers used the Internet to resolve queries encountered in the process of teaching practice, while 72.9% of the teachers did not have consciousness of using information technology to solve the problem, let alone the use of information technology for learning capability, which will be an important learning skill to constraint teachers' lifelong study.

The education technology training model demand survey results are shown below.

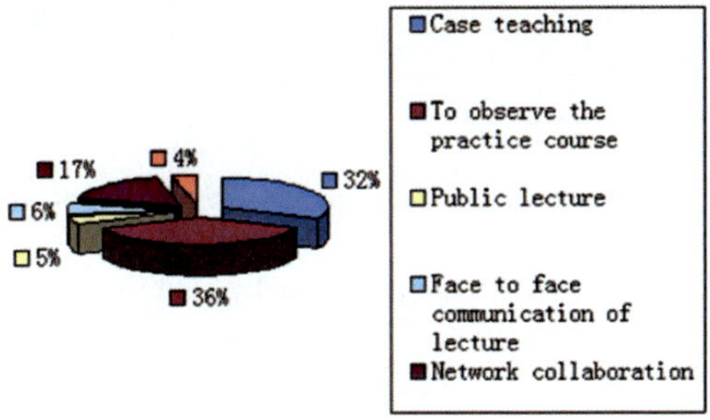

Fig. 3. Training mode

Figure 3 shows, the learners hope to study and observe the teaching ability on how to apply the new teaching ideas and methods to improve the classroom through the concrete case teaching practice. They look forward to high participation degree, initiative training; practical training in which teachers could learn through operation. Network communication and the Internet autonomous learning accounted for a large proportion, which shows teachers are aware of the importance of network learning, also want to improve the learning ability through training, thus establishing the concept of lifelong learning. It further illustrates the training scheme is consistent with the basic concept of B-learning, and combine face to face teaching with E-learning teaching together effectively, to better improve the training effect.

The survey results on the expected training content are shown in figure 4.Training needs are like these in turns: the teaching design of information technology and curriculum integration, Flash teaching courseware, cyber source search and utilization, webpage design and making, graphics and image processing technology, the network curriculum development, scientific research of educational teaching method, the use of Microsoft office and computer basic knowledge in teaching.

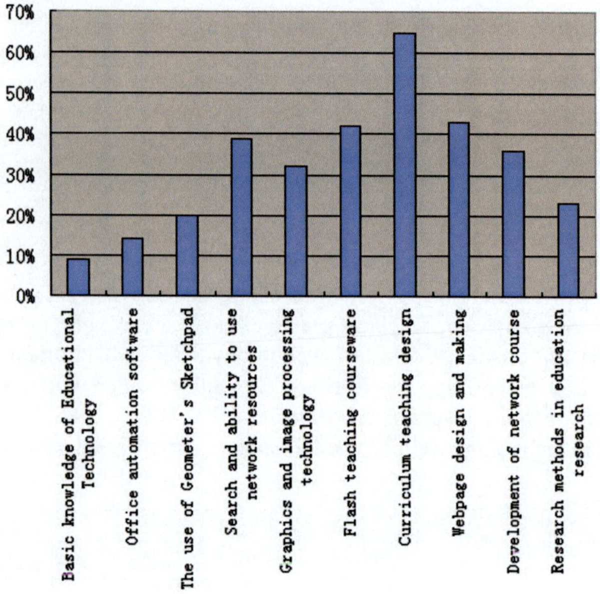

Fig. 4. The survey results on the expected training content

According to the analysis of demand for teachers and the "primary and secondary school teachers' educational technology ability standard (on trial)", we determine three respects as training objectives: The first is to master educational technology basic knowledge and basic skills; the second is able to effectively integrate information technology and discipline teaching conformity; the third is to use educational technology to carry out the lifelong learning, conduct teaching and scientific research, to promote the teachers' professional development.

3.2 Class Grouping and Layering Teaching

Divide into three levels according to three training targets: primary, intermediate and advanced, different learning levels select different training contents and the teaching models are not the same; goals are also different.

The basic training for junior class are educational technology basic knowledge, the use of commonly used office software, geometry drawing board and knowledge of network infrastructure (including the common network communication tool usage) and cyber source search. This time, among the 124 Middle School Substituted teachers, 20.4% recommended to receive primary class and they hoped that through the levels of learning, enable learners to understand the basis of information technology based on knowledge, to have a strong practical operation ability and the use of information technology in effective teaching and online collaborative learning ability.

The six elements of intermediate basic training are the use of office automation software, graphics and image processing technology, the Flash teaching courseware, integration of information technology and subject teaching and network teaching model. The training took information technology and curriculum integration as theme, combined with case teaching analysis and design, so that the trainers could truly experience the concept of integration in practice, and could apply in the future teaching, so as to better improve the efficiency of classroom teaching. 54.8% of the 124 substituted middle school teachers have a certain basis of information technology, and call for intermediate training.

The four contents of senior class are webpage design and making, network curriculum development, teachers' professional development and educational research methods. Through this level of learning, teachers are expected to accept the idea of using information technology for lifelong learning and to improve the use of information technology to conduct teaching reform and educational scientific research ability. In this training, 24.8% middle school teachers' information technology is quite good, and can accept the learning content of senior class.

In the whole training process, learners can also flexibly chose different learning content according to their given level. The ultimate goal of training is to achieve the intermediate level at the end of training, and more than half of the teachers can reach an advanced level.

3.3 The Selection of Training Environment, Strategy and the Specific Implementation Process

The training environment should be selected in the multimedia network computer room, and teaching platform is legend digital network teaching software. The platform provides a virtual learning environment to supplement classroom teaching and distance teaching platform. The platform provides real-time synchronous interaction to support a complete virtual classroom. Through the campus network and the Substitution Teaching Practice site students learn teaching resources of the network autonomously; the collaborative communications between learners, learning and training teachers, learners, experts and professors are primarily through QQ, teaching practice forum, campus network mail system and group Blogs. Learners carry out online collaborative learning through exchanging learning experience, designing subject teaching and making educational teaching research.

In the educational technology training, some teaching strategies such as advance organizer teaching strategy, autonomous learning teaching strategies and cooperative learning teaching strategy are mainly used.

The specific implementation process is:

(a) E-learning autonomous learning. After completing this level of learning content, learners could conduct autonomous learning according to their own learning needs. If they encounter problems or want to make further discussion, on one hand, learners can communicate and make mutual aid through a network; on the other hand, they can feel free to contact curriculum experts, which not only form life-long learning habits, but also can promote learners' independence and self confidence.

(b) Face to face teaching. Face to face teaching is the center of the whole training process. In the process of teaching experts' demonstration and reviews help learners improve their practical abilities. It almost can accommodate any teaching methods ranging from transfer -- accept to autonomous learning. In face-to-face process the learners' group communication, dynamic generation, systemic involvement should be a process of knowledge construction.

(c) Practice form works. By appreciating the works case learners aroused interest in learning. Teachers make concept analysis and technical analysis on the case the manufacture, and then the learners master the basic method of operation, and then complete the learning tasks through the team cooperation. In the production process, learners interact and cooperate with each other; meanwhile tutors make targeted personalized counseling. After the works, by exchange and discussion, let the learner's assess themselves and reflect their shortcomings. Reflection is a special thinking activity, which usually need to guide the learners to focus on curriculum related experience. Reflection is about not only individual, but also group. Individual reflection is likely to stay to continuously strengthen the existing views. Teachers, peer feedback, can help learners find the limitations of their own, thus to prompt a deeper reflection and active participation. [4] Through evaluation -- reflection, learners further perfect works. In this process, the learners deepen the original knowledge; develop much new knowledge; improve the ability to solve problems, and enhance team cooperation consciousness and creative consciousness.

3.4 The Selection of Evaluation Methods

The training adopted task-driven approach, emphasizing the teachers' practical experience. In order to ensure good results, the formative evaluation and summative evaluation were combined. The formative assessment was to provide feedback information to improve the practice of ongoing education quality evaluation through the analysis of the teaching plans, teaching process and the problems in the activities. [5] It emphasized the evaluation activities should take education information feedback, adjustment of the educational process, standardization of the teaching management, improvement of the quality and efficiency of education as the aims of education. [6] This training combined the learner's works with multiple evaluation principles and methods, to settle in classification, towards the learners materialized achievements in the learning process, forming electronic files (e-portfolio), based on which formative assessment is given. The last link of the training-- summative evaluation was made according to predetermined criteria. Learner knew their own change through a series of shows, quiz, namely, the degree of achieving its learning goal, a sense of success, stimulation of continuous learning.

4 Training Effect Analysis

B-Learning based teacher training model fully embodies the humanism thought, which was made according to learning tool, learning environment, learners' interaction and learning content development, learning steps and other aspects of the development. The biggest characteristics are having the commonly accepted explanation requirements and individualized learning opportunities in which the learners make diversified learn. Through the learning process evaluation and final summative evaluation we can see, by four months of training, the 89.6% teacher achieved second level target, 42.1% teachers reached the third levels target. Teachers' knowledge structure and ability structure have made great improvement.

4.1 Knowledge Structure Improvement

American educator Schulman put forward, the teachers' professional knowledge structure are made up of three kinds of knowledge: one is composed of general education teaching principle knowledge; two is the specific teaching practice professional case knowledge, three is strategic knowledge applied in solving practical problems through reflection. Earlier investigation showed, the middle school teacher had understood teaching principle rules, and now they had a deeper understanding and mastery through this training. The educational technology training process used a large number of cases teaching, which can not only show the actual teaching situation, enrich learning case knowledge, but also enrich their strategic knowledge. As teaching strategies is a tacit knowledge, it is usually demonstrated when dealing with specific problems, through expert teachers' teaching and verbal instruction. Through the experience of practice for case knowledge and strategic knowledge, at the same time through case discussion also makes the principle rules knowledge specific, vivid, it helps learners' principle rule knowledge, so as to improve the learners' knowledge structure, promoting the teacher's professional development.

4.2 Ability Structure Improvement

Through the training of educational technology, learners' ability structure has been improved, which embodied in the following aspects.

(a) The ability of using information technology and solving problems. The training teachers could make multi angle analysis to selected cases, guiding the learners to apply educational technology to fully discuss, exchange, so as to find a solution to the problem. In the process, what learners learn is not dogma in books, but a living knowledge and problem analysis and the ability to solve problems.

(b) Cooperative study ability. In the training implementation process, the learner adopted task-driving teaching method and team work to complete a teaching task, and enhance the students' awareness of cooperation, cooperation ability and research ability.

(c) Reflective ability. In the exhibition competitions, each learner actively reflected the advantages and disadvantages of their companion work, by which they deepen the original knowledge, develop new knowledge, deepen reflection and active participation.

(d) The practice innovation ability. By analyzing a large number of cases, the teaching practice, the learners were qualified with the new concept of education in actual teaching experience. Teaching experience every time promoted different innovation. After four months of training, the learners' innovative consciousness and practice ability are all improved.

(e) The classroom teaching ability. In the training process, most of the teaching process required training teachers to use information technology and curriculum integration in teaching design; learners' ability of teaching had been improved through teaching design and classroom teaching practice -- the Microteaching playback -- Reflection -- improvement.

The rural middle school teachers' ability of using information technology in teaching had been improved through the mode. It had cultivated their concept of lifelong learning and promoted the teachers' professional development. Of course, this training mode also has some deficiencies, such as the network teaching resources, evaluation system, etc. which need a further and meticulous research.

References

1. He, K.: A new look to the educational technology theory from Blending Learning. e-Education Research (3) (2004)
2. Driscoll, M.: Blended learning: let's get beyond the hype. Learning and Training Innovations (8) (2002)
3. Li, K., Zhao, J.: The theory and applied model of Blended Learning. e-Education Rescarch (7), 1–6 (2004)
4. Zhang, Z., Qi, W.: The strategies to improve the efficiency of Blended Learning. Distance Education in China (4) (2007)
5. Chen, Y.: Educational evaluation, p. 12. People' education press, Beijing (1998)
6. Zhong, Q.: The development of modern Didactics, p. 283. Educational science publishing house, Beijing (1992)

Study on the Impact of Individual Heterogeneity on Human Resource Development Needs

Yan Qin and Dan Mu

Department of Information Technology and Business Management,
Dalian Neusoft Institute of Information, Dalian, 116023, China
qinyan@neusoft.edu.cn, mudan@neusoft.edu.cn

Abstract. Heterogeneity has getting more and more attention among study on HRD (Human Resource Development). The impact dimensions of heterogeneity on organizational HRD were revealed by studying the correlation between heterogeneity and training development, heterogeneity and career development, heterogeneity and organizational development, heterogeneity and management development. This study compared the differences of HRD needs for different explicit heterogeneity by using Independent-Sample T Test, one-way ANOVA and multiple comparison analysis and found there are significant differences in HRD needs for employees with different genders, ages, working years, educational levels, majors, and growing regions.

Keywords: Training Development, Career Development, Management Development, Organizational Development.

1 Literature Review

With economic globalization, differences between employees in the future will increase, independence of employees will enhance and self-awareness will get stronger, the differences existing both in gender, race, ethnic, but also in knowledge and skills (Jin,2006) [1]. It is very important to discuss HRD (Human Resource Development) professionally, which is becoming one of the most frequent topics in the theorists.(Richard,2004; Holton,2002; Kuchinke,2000; Lynham,2000; Mclean,1998; Swanson,1997,1999,2001; Torrco,1997; Weinberger,1998） [2-6].In the theoretical research framework of HRD , heterogeneity is a growing concern （Michael,2000） . Yangjie Gu (2008) pointed out heterogeneity, also known as the diversity, was differences of the characteristics of team members such as gender, age, educational level, personality trait, work experiences. Susan, et al. (2003) thought heterogeneity could be divided into readily-detectable diversity including gender, age, race etc., underlying diversity including personality, knowledge, value etc., and the properties between the two diversities including education and working years. Heterogeneity also could be divided into task-related diversity and relations-oriented diversity [7]. Jennifer (2004) discussed the career development of women in information technology industry, studied the gender, career development and employee performance in information technology industry and pointed out that

W. Zhang (Ed.): Advanced Technology in Teaching, AISC 163, pp. 325–332.
springerlink.com

gender had far-reaching impact on career development in information technology industry [8]. Val (2006) found by depth interviews with 39 women architects that 20 of them rejected the organization's career development practice, which showed women had significant heterogeneity in career needs [9].

Many management experts and scholars focused on the heterogeneity, resulting in different perspectives on heterogeneity studies, such as capacity heterogeneity, entrepreneurs heterogeneity, knowledge employees heterogeneity, human capital heterogeneity, regional heterogeneity, team heterogeneity, R & D team heterogeneity. In addition, some scholars conducted some research on heterogeneity regarding human resource management and development, such as heterogeneity training. However, research on impact dimensions of heterogeneity in HRD activities is relatively lacking.

2 Research Framework and Research Hypotheses

Human resource in organization is heterogeneous. This paper focused on heterogeneity of human resource, studying what kind of HRD needs the employees of individual heterogeneity like different genders, different ages, different majors, different educational levels, different working years, different growing regions want , thus trying to reveal the role of heterogeneity of employee in HRD activities, and then examining the impact dimensions of heterogeneity on organizational HRD.

Heterogeneity refers to differences of the research objects. This paper mainly studied differences and diversities of human resource in organization. Heterogeneity consists of explicit heterogeneity and implicit heterogeneity. Explicit heterogeneity is obvious and can be observed easily, while implicit heterogeneity refers to potential heterogeneity difficult to be observed. Based on the literature review, this study selected explicit heterogeneity mainly from individual explicit characteristics to elaborate, including six aspects of gender, age, working years, the highest educational qualification, the major of the highest educational qualification, major geographic growing region (cultural background).

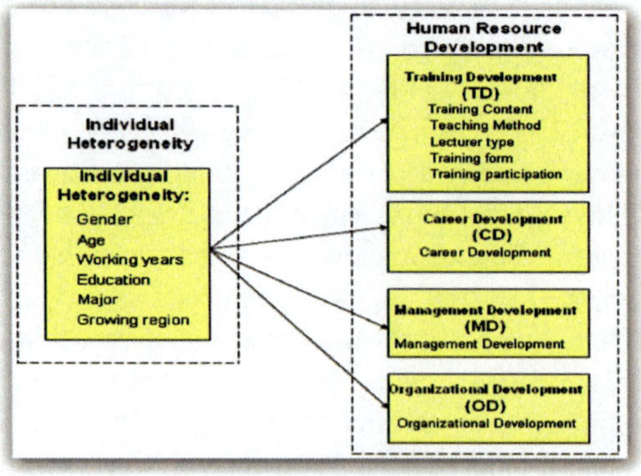

Fig. 1. Research Framework

HRD is to improve individual, group and organizational efficiency by means of training development, career development, management development and organizational development. HRD activities include Training Development(TD), Career Development (CD), Management Development (MD), Organizational Development (OD). Training development refers to educational activities about the necessary knowledge, skills, abilities, attitudes motivation, etc., for the current position and future position (after promotion or job rotation). The research framework of this paper is shown as Fig.1.

Based on the literature review, we can see that empirical research of the impact of heterogeneity on HRD needs is relatively lacking. Focusing on the framework of this paper, the following hypotheses are proposed:

Hypothesis 1: The training content which employees with individual heterogeneity need is different.

Hypothesis 2: The teaching method which employees with individual heterogeneity need is different.

Hypothesis 3: The training lecturer which employees with individual heterogeneity need is different.

Hypothesis 4: The training form which employees with individual heterogeneity need is different.

Hypothesis 5: The training participation which employees with individual heterogeneity need is different.

Hypothesis 6: The career development which employees with individual heterogeneity need is different.

Hypothesis 7: The management development which employees with individual heterogeneity need is different.

Hypothesis 8: The organizational development which employees with individual heterogeneity need is different.

3 Study Design

3.1 Questionnaire Design

This study used questionnaire survey as empirical research method. The questionnaire included preface, HRD questionnaire and personal information and organizational information of the surveyee. HRD questionnaire included four dimensions of training development, career development, management development and organizational development.

3.2 Data Collection

In this study, 1,100 questionnaires were sent, which resulted in 956 samples (87%) return, 829 (75%) valid questionnaires. In data collection statistics, the questionnaires with some missing options or with the same scale level answers to all questions due to less serious degree of the surveyee were regarded as invalid questionnaires. Questionnaires covered Dalian, Beijing, Shanghai, Shenzhen, Shijiazhuang, Guizhou, Panjin, Fuxin, Jinzhou, Taiyuan and other cities, involved over 20 organizations including HP China, Neusoft Group, Agricultural Bank of China, GE Dalian, and Accenture.

4 Data Analysis

The statistical methods used Independent-Sample T Test and one-way ANOVA, statistical software SPSS16.0.

4.1 The Descriptive Statistics of HRD Needs

The descriptive statistics of HRD needs are shown in Table 1.

Table 1. The descriptive statistics of HRD needs

Variables	Questions	Options	Number of people	Percent -age
TD : Training Develop ment	Q1 : Training Contents	1.profession development	125	15.1
		2.technical training	175	21.1
		3.management training	172	20.7
		4. foreign language training	132	15.9
		5.computer training	51	6.2
		6 .career development	69	8.3
		7. thinking development	91	11.0
		8. others	14	1.7
		Total	829	100.0
	Q2 : Teaching methods	1. lecturer teaching	236	28.5
		2. group discussion	156	18.8
		3. scenario simulation	297	35.8
		4. audio and video	100	12.1
		5. others	40	4.8
		Total	829	100.0
	Q3 : Lecturer types	1. senior manager	242	29.2
		2.professors and scholars	168	20.3
		3.professional training	378	45.6
		4. others	41	4.9
		Total	829	100.0
	Q4 : Training forms	1. on job training	390	47.0
		2. off the job training	206	24.9

Table 1. *(continued)*

		3.part time self-study	133	16.1
		4.on line learning	88	10.6
		5.others	12	1.5
		Total	817	98.6
	Q5 : Training Participation	1. good listener	264	31.8
		2.active participant	400	48.3
		3. bold questioner	137	16.5
		4. others	28	3.4
		Total	829	100.0
CD : Career Development	Q6 : Career Development	1. administration path	205	24.7
		2.professional skills path	392	47.3
		3.cross-sector path	196	23.7
		4.others	36	4.3
		Total	829	100.0
MD : Management Development	Q7 : Management Development	1.management training	303	36.6
		2. tutor training	204	24.6
		3.short-term rotation	272	32.8
		4.others	50	6.0
		Total	829	100.0
OD : Organizational Development	Q8 : Organizational Development	1. learning knowledge	84	10.1
		2. increasing capacity	304	36.7
		3.improving skills	275	33.2
		4.changing attitude	147	17.7
		5. others	19	2.3
		Total	829	100.0

4.2 Analysis of Individual Heterogeneity on HRD Needs

The paper compared the differences of HRD needs for different dimensions of individual heterogeneity including gender, age, working years, education, major, growing region, mainly using Independent-Sample T Test, ANOVA and LSD multiple comparison methods.

1) Analysis of Gender Heterogeneity
Using Independent-Sample T Test, the paper explored the differences in HRD needs of employees with gender heterogeneity.

The results showed that, in terms of training form, women are more likely to choose on job training than men; while men prefer part time self-study than women. In terms of training participation, women are more likely than men to become a good listener in training, while men are more likely than women to become an active participant in training. That showed women are relatively restrained than men in training process, more willing to be a passive listener.

In terms of career development, men are more likely than women to choose professional skills path, women are more likely than men to choose other career development than administration path, professional skills path, and cross-sector path.

The results are concordant with the fact that men have a technical preference than women. In terms of management development, men are more likely to choose management training than women; while women prefer short-term rotation than men.

2) Analysis of Age Heterogeneity

Using one-way ANOVA and LSD multiple comparison methods, the paper discussed the differences in HRD needs of employees with age heterogeneity. Samples of 46-55 years old and over 56 years old are fewer, therefore, combined with 36-45 years old as over 36 years old. Then, three groups of samples: under the age of 25, 26-35 years old, over 36 years old were analyzed.

The results showed that, in terms of training content, employees of 26-35 years old are more likely to choose professional development than those over 36 years old. Employees under 35 years old are more likely to choose technical training than those over 36 years old .Employee over 36 years old are more likely to choose management skills training than those under 35 years old. Employees of 26-35 years old are more likely to choose career development than those over 36 years old. It explained that the younger employees under 35 have relatively strong technical training needs; while more mature employees over 36 will shift the focus on management skills. This change may be related to career development phase of this age. More employees of 26-35 years old choose professional development and career development, indicating that employees of this age are in the stage of accumulation professional knowledge and career development and have a strong desire for further knowledge development and career planning.

In terms of teaching method, employees over 26 were more likely to choose lecturers teaching than employees under 25. Employees under 25 were more likely to choose group discussion than those of 26-35 years old. Employees under 25 were more likely to choose other teaching methods than lecturer teaching, group discussion, scenario simulation and audio or video teaching. It indicated that young employees under 25 hoped to have the opportunity to express their views and employee over 26 preferred lecturer teaching.

In terms of lecturer type, employees over 36 prefer professors and scholars than those under 25; while employees under 25 are more likely to choose other lecture types than senior manager, professors and scholars, professional trainer. In terms of training form, employees over 36 prefer full-time training than those under 25. Employees under 25 are more likely to choose other training form than on job training ,off the job training ,part time self-study and on line learning.

In terms of training participation, employees over 26 are more willing to become a good listener than those under 25; employees under 25 are more inclined to choose other training participation form than a good listener, active participant, bold questioner than employees of 26-35 years old . The results are concordant with the fact that employees over 26 prefer lecturer teaching and would like to be a good listener, indicating with age, employees over 26 are more sedate and hope to satisfy their thirst for knowledge by listening than those under 25.

In terms of career development, employees over 26 are more likely to choose professional skills path than employee under 25; employees under 25 are more likely to choose other career development path than administration path, professional skills path, cross-sector path. In terms of management development, employees of 26-35 years prefer management training than those under 25; employees under 25 are more likely to choose other management development form than management training, tutor training, short-term rotation compared with employees under 25. In terms of organizational development, employees under 25 and over 36 are more likely to choose to learn knowledge than employees of 26-35 years old; employees over 36 are more likely to choose to improve capabilities. Employees of 26-35 prefer to upgrade their skills than employees over 36.

3) Comparison of Working Years Heterogeneity
Using one-way ANOVA and LSD multiple comparison methods, the paper discussed the differences in HRD needs of employees with working years heterogeneity. The results showed, in training content, teaching method, training form, training participation, career development, there are significant differences for employees of different working years in HRD needs. Due to limited space, the following dimensions results were not listed in details.

4) Comparison of Education Heterogeneity
Using one-way ANOVA and LSD multiple comparison methods, the paper discussed the differences in HRD needs of employees with the highest educational qualification heterogeneity. For samples of high school or secondary specialized education were fewer, therefore combined with junior college as junior college or below. Samples of PhD degree were fewer, therefore combined with master degree as master degree or above. Then, the following three groups of junior college or below, undergraduate, master degree of above were analysed. The results showed that in training content, career development, and organizational development, HRD needs of employees with different education are significantly different.

5) Comparison of Major Heterogeneity
Using one-way ANOVA and LSD multiple comparison methods, the paper discussed the differences in HRD needs of employees with major heterogeneity. Samples of literary art, agriculture were relatively fewer, so they were combined with the others as other majors. Then, three groups of samples: science and engineer, economics and management, other majors were analyzed. The results showed that in training content, lecturer type, career development and management development, HRD needs of employees with different majors are significantly different.

6) Comparison of Growing Region Heterogeneity
Using one-way ANOVA and LSD multiple comparison methods, the paper explored the differences in HRD needs of employees with growing region heterogeneity or cultural background heterogeneity. Samples except for Northeast, North China, were relatively dispersed in other regions, so East China, Northwest China, southern China, central China, southwest China, other regions were merged into other regions. Three groups of samples data: Northeast, North China, other regions were analyzed.

The results showed that in training content, teaching method, lecturer type, training form, training participation, career development, different ways of human resource needs are brought due to different growing regions.

5 Conclusions

This paper studied the impact of individual heterogeneity of employees on HRD needs and proposed research framework and research hypotheses. The research used Independent-Sample T Test, one-way ANOVA and multiple comparison analysis. The results showed that the employees with different individual heterogeneity, whose needs for training contents, teaching methods, lecturer types, training forms, training participation, career development, management development, organizational development are different and all hypotheses are supported. Among them, employees of different genders whose HRD needs are of significant difference in training form, training participation, career development, management development. Employees with different education background whose HRD needs are of significant difference in training content, teaching method, training form , training participation, career development, management development and organizational development. Employees with different majors whose HRD needs are of significant difference in training content, lecturer type, training form, career development, management development and organizational development. Employees of different ages, working years, growing regions whose HRD needs are of significant difference in training content, teaching method, lecturer type, training form and training participation , career development, management development and organizational development. This study provides a reference for human resources managers in decision making when they organize training development, career development, management development and organizational development activities carefully catered to individual heterogeneity and implement HRD strategies. Besides individual heterogeneity, work heterogeneity and other implicit heterogeneity will be further in- depth studied in the future.

References

1. Jin, X.: Survey Ranking of Job Competencies by Perceived Employee Importance: Comparing China's Three Regions. Human Resource Development Quarterly 4(17), 371–402 (2006)
2. Richard, J., Torraco: Challenges and Choices for Theoretical Research in Human Resource Development. Human Resource Development Quarterly 2(15), 171–188 (2004)
3. Holton, E.F.: The Mandate for Theory in Human Resource Development. Human Resource Development Review 1(1), 3–8 (2002)
4. Kuchinke, K.P.: Debates over the Nature of HRD: An Institutional Theory Perspective. Human Resource Development International 3(3), 279–283 (2000)
5. Lynham, S.A.: Theory Building in the Human Resource Development Profession. Human Resource Development Quarterly 11(2), 158–178 (2000)
6. Swanson, R.A.: Human Resource Development and Its Underlying Theory. Human Resource Development International 4(3), 299–312 (2001)
7. Susan, E.J., Aparna, J., Niclas, L.E.: Recent research on team and organizational diversity: SWOT analysis and implications. Journal of Management 29(6), 801–830 (2003)
8. Jennifer, A., Kaminski, M., Reilly, A.H.: Career development of women in information technology. Sam Advanced Management Journal 6, 20–30 (2004)
9. Caven, V.: Choice, diversity and 'False Consciousness'in Women's Careers. International Journal of Training and Development 6(26), 41–54 (2006)

A Method for Permanent Magnet Synchronous Motor Control Based on Single-Loop SVPWM*

Qi-xun Zhou[1,2], Mian-hua Wang[2], Yong Zhou[1], and Yu-feng Zhang[1,2]

[1] School of Mechanical Engineering, Northwestern Polytechnical University, Xi'an, Shaanxi, China
yongstar@nwpu.edu.cn
[2] Department of Electrical & Control Engineering, Xi'an University of Science and Technology, Xi'an, Shaanxi, China
zhouqixun029@163.com

Abstract. The advantage and disadvantage of the common control methods for permanent magnet synchronous motor (PMSM) are analyzed. Combined the sine pulse-width modulation (SPWM) and field oriented control (FOC) algorithm, a novel control method of the space vector pulse-width modulation with single closed loop is presented. In MATLAB-Simulink, based on the mathematic model of PMSM, used the S-function and the algorithm of single-loop space vector pulse-width modulation (SL-SVPWM) a servo-model of PMSM is established. Via experiment, compared SPWM method to SL-SVPWM method, the different results of phase current waveform of PMSM are drawn; and the dynamic response behavior of the motor which driven by the method of SL-SVPWM is analyzed. The results of the simulation and experiment verified that the SL-SVPWM method in this paper simplified the control system of PMSM, and this control method has better control stability.

Keywords: Permanent Magnet Synchronous Motor, Sine Pulse Width Modulation, Field Oriented Control, Single Loop, Space Vector Pulse Width Modulation.

1 Introduction

The permanent magnet synchronous motor (PMSM) is widely applied in flexible manufacturing system, robot, office automation, office automation, numerical control machine and aeronautics and astronautics and other fields[1][2] with its such advantages as simple structure, small size, light weight, high power, high power factor, rotor free from heating problem, high overload capacity, small rotational inertia and small torque ripple and so on.

* Supported by the Education and Teaching Reform and Research Projects Funded of Xi'an University of Science and Technology (No. JG10069).
Supported by the Specialized Research of Education Department of Shaanxi Province(No.09JK586).

W. Zhang (Ed.): Advanced Technology in Teaching, AISC 163, pp. 333–342.
springerlink.com

In general, the PMSM has three kinds of control methods: pulse-width modulation (PWM), sine pulse width modulation (SPWM) and field oriented control (FOC) [3] [4] [5]. The PWM algorithm is simple, but the motor torque ripple is large, the winding availability is low and it is difficult to realize super-low speed operation; SPWM algorithm is easy to realize and the torque ripple is small and it may realize low speed smooth operation, but the SPWM starts from the angle of power and is pursuit of output of a three-phase symmetrical sine wave power supply of adjustable frequency and voltage, so the voltage utilization in PMSM system is low and the motor mechanical property is soft; FOC starts from the angle of motor and its objective is to make the motor generate the circular magnetic field and the flux linkage locus in motor gas-gap is close to round shape so as to reduce the torque ripple and core loss, but the control algorithm is complex and it needs to sample the motor real-time winding current and the hardware circuit is complex and the current noise is increased and the phase shift caused by filtering by use of hardware or software obviously affect the control effect[6].

In this paper, a kind of Single-Loop Space Vector Pulse-Width Modulation(SL-SVPWM)is proposed on the basis of analysis of PMSM mathematical model, and the method is based on the theory of rotor field oriented control and SPWM control is combined and the current close loop is neglected, so the hardware circuit and control algorithm are simplified, and SVPWM module is added module and third harmonic is injected after SPWM to promote the system control performance. The test proves that the method has a higher system mechanical property when compared with SPWM drive and the control performance is close to that of the traditional FOC.

2 PMSM Mathematical Model

The mathematical model of PMSM is similar to that of live excitation synchronous motor, and when the stator winding is connected to three phase symmetric current and the magnetic potential from rotating composition may form a circular rotating magnetic field in space and acts with the magnetic field generated by motor permanent magnet and forms the electromagnetic torque to push the rotor synchronous rotation [7]. PMSM mathematical model on three phase static axial system is a multivariable, nonlinear and strong coupling system, and in order to provide PMSM with the high performance control characteristic, the linearization and decoupling must be carried out through coordinate transformation to convert the time varying coefficient into constant coefficient. Coordinate transformation is the method mathematic basis for PMSM to realize magnetic field oriented control and the method and conclusion of coordinate transformation have been described in may literatures, so only the PMSM voltage balance equation [8] under the dq coordinate system is listed due to space constraints, that is:

$$\begin{bmatrix} u_d \\ u_q \end{bmatrix} = \begin{bmatrix} R_d & 0 \\ 0 & R_q \end{bmatrix} \begin{bmatrix} i_d \\ i_q \end{bmatrix} + \begin{bmatrix} L_d & 0 \\ 0 & L_q \end{bmatrix} \bullet p \begin{bmatrix} i_d \\ i_q \end{bmatrix} + \omega_r \begin{bmatrix} 0 & -L_q \\ L_d & 0 \end{bmatrix} \begin{bmatrix} i_d \\ i_q \end{bmatrix} + \omega_r \psi_f \begin{bmatrix} 0 \\ 1 \end{bmatrix} \tag{1}$$

Where: subscript d,q indicates the value of variable under dq coordinate system and $R_d = R_q = R$, L_d and L_q are the dq coordinate equivalent inductance and may be expressed as :

$$\begin{cases} L_d = L_{s\delta} + \frac{3}{2} L_{ad} \\ L_q = L_{s\delta} + \frac{3}{2} L_{aq} \end{cases} \tag{2}$$

$L_{s\delta}$ is the stator winding leakage inductance and is relevant to leakage magnetic field distribution and magnetic circuit property; L_{ad} is the stator winding armature reaction induction and is in proportional to direct axis magnetic conductance; L_{aq} is stator winding quadrature axis armature reaction induction and is in proportional to magnetic conductance.

The electromagnetic torque equation of PMSM under dq two phase revolving coordinate system is :

$$T_e = \frac{3}{2} N_p (\psi_d i_q - \psi_q i_d) = \frac{3}{2} N_p (\psi_f i_q + (L_d - L_q) i_d i_q) \tag{3}$$

The model of PMSM under dq coordinate system is obtained through coordinate transformation, such as equation (1) and (2), which is the first order linear system mathematical mode and completes the PMSM order reduction, decoupling and linearization. Equation (3) shows that the electromagnetic torque output from PMSM includes two sections: the first section is permanent magnetic torque T_m and the second section is reluctance torque T_r.The rotor magnetic flux linkage ψ_f in PMSM is constant, so it may realize the control in motor through i_q and i_d, which is the essence of field oriented control.

3 SL-SVPWM Control Principle

3.1 SVPWM Principle

Apply a certain voltage combination $U_i(S_a S_b S_c)$ on PMSM three phase symmetric armature winding and a certain magnetic flux linkage vector ψ_i will be generated. Three phase fully controlled bridge inverter may form the regular hexagon rotating magnetic field through switch of working mode according to $U_4 \rightarrow U_5 \rightarrow U_1 \rightarrow U_3 \rightarrow U_2 \rightarrow U_6 \rightarrow U_4$, as shown in diagram 3[9]. The regular hexagon rotating magnetic field may be obtained through the above working mode only and it fails to obtain circular rotating magnetic field and further analysis shall be made, and its thinking is to segment the regular hexagon to regular polygon to approach the circular magnetic field. Divide the 360° electrical angle space into six section s, that is the section $\theta(1)—\theta(6)$ shown in diagram 1; divide from the middle of rotatory orbit perimeter of magnetic flux linkage into M pieces and then there is M/6 circular arc in each section. Take any section of the M/6 circular arc, as shown $\overset{\frown}{P_1P_5}$ in section $\theta(2)$ $\overset{\frown}{P_1P_5}$ in diagram 2, approximate $\overset{\frown}{P_1P_5}$ with broken line $\overline{P_1P_2}$, $\overline{P_2P_3}$, $\overline{P_3P_4}$ and $\overline{P_4P_5}$ and then the line segments $\overline{P_1P_2}$ and $\overline{P_4P_5}$ are the forward track of stator magnetic flux linkage under the action of voltage vector U_5. P_3 is the middle point of chord $\overline{P_1P_5}$, so we may obtain :

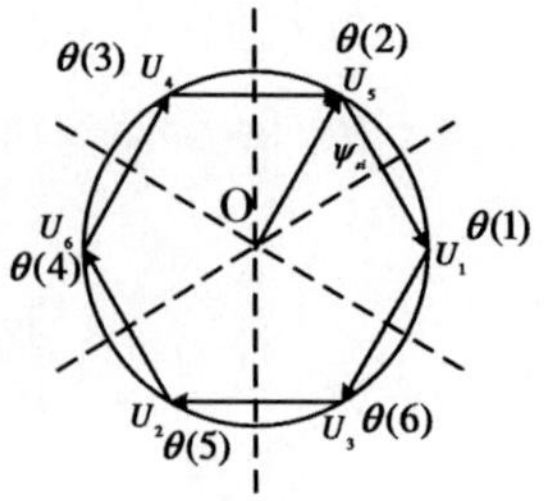

Diagram 1. Magnetic flux linkage space vector

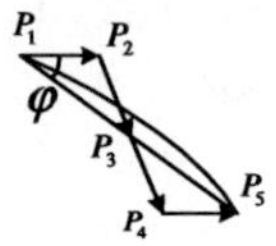

Diagram 2. Partial magnetic flux linkage space vector

$$\begin{cases} \overline{P_1P_2} = \overline{P_4P_5} = |U_4| t_4 = U_d t_4 \\ \overline{P_2P_3} = \overline{P_3P_4} = |U_5| t_5 = U_d t_5 \end{cases} \tag{4}$$

Where: t_4 a and t_5 are the action time of U_4 and U_5 respectively.

Assume that the amplitude of magnetic flux linkage space vector is R and the rotation angle frequency is ω and the time for magnetic flux linkage space vector to pass $\overline{P_1P_5}$ segment is $\overline{P_1P_5}$, then:

$$\begin{cases} \overline{P_1P_5} = \overset{\frown}{P_1P_5} = 2\pi R / M \\ 2T_0 = 2\pi / (\omega M) \end{cases} \tag{5}$$

In triangle $\Delta P_1P_2P_3$, the following may be obtained through sine theorem :

$$\frac{\overline{P_1P_2}}{\sin(\frac{\pi}{3} - \varphi)} = \frac{\overline{P_2P_3}}{\sin\varphi} = \frac{\overline{P_1P_3}}{\sin\frac{2\pi}{3}} \tag{6}$$

Simultane equation (4), (5)and(6) to obtain：

$$\begin{cases} t_4 = \dfrac{\omega R}{U_d \sin\dfrac{2\pi}{3}} T_0 \sin(\dfrac{\pi}{3} - \varphi) \\ t_5 = \dfrac{\omega R}{U_d \sin\dfrac{2\pi}{3}} T_0 \sin\varphi \end{cases} \tag{7}$$

In conclusion, SVPWM selects the appropriate voltage space vector and regulate is action sequence and action time to make the stator magnetic flux linkage space vector move on the broken line and gradually approach the circular track. The higher the modulation frequency is, the evener the average rotating speed is and the magnetic flux linkage space vector is closer to circular tract and the electromagnetic torque ripple is smaller.

3.2 SL-SVPWM Control Principle

SVPWM control system adopts the dual-loop structure of electric current loop and rotor position loop and realizes the revocation of voltage space vector through Clark and Park transformation and inverse transformation, and the system has a complex structure and it needs a complex trigonometric function operation and inverse operation, so the difficulty in realization of project is increased. The SL-SVPWM control method structure block diagram proposed in this paper is shown in diagram 3 and the current closed-loop is neglected, and only the rotor position loop is used to realize the control in voltage space vector and the forward channel SPWM module is used mainly to complete the sine check computation and the module is shown in equation 8.

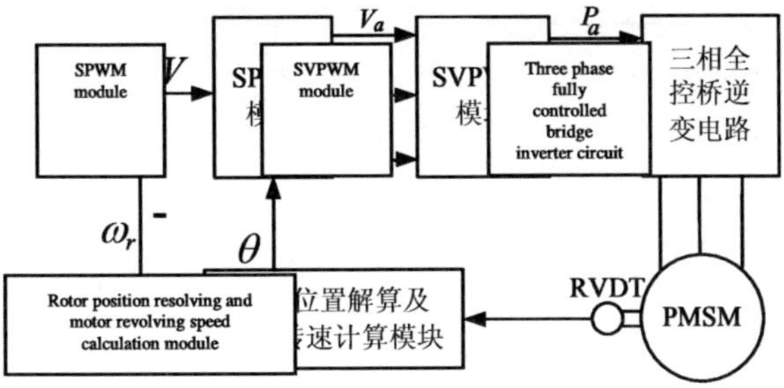

Diagram 3. SL-SVPWM control system block diagram

$$\begin{cases} U_{r1} = V \sin\theta \\ U_{r2} = V(\sin\theta + 120^o) \\ U_{r2} = V(\sin\theta + 240^o) \end{cases} \tag{8}$$

SVPWM module may be divided into six sections according to U_{r1}, U_{r2} and U_{r3}, as shown in diagram 4. The diagram shows that section judgment needs on any complex computation and may be realized through identifying the SVPWM three phase input symbol. The calculation formulas of nonzero vector action time, obtained from SVPWM principle, are shown in table 1.

Diagram 4. Relationship diagram of section and U_{r1} U_{r2} U_{r3}

Table 1. Section judgment criteria and calculation formulas

Section	Section judgment criteria	Calculation formulas
1	$U_{r1} \geq 0$, $U_{r3} \geq 0$, $U_{r2} < 0$	$t_4 = \sqrt{2}T_{PWM}U_{r3}/U_d$, $t_5 = \sqrt{2}T_{PWM}U_{r1}/U_d$
2	$U_{r1} \geq 0$, $U_{r3} < 0$, $U_{r2} < 0$	$t_4 = -\sqrt{2}T_{PWM}U_{r3}/U_d$, $t_5 = -\sqrt{2}T_{PWM}U_{r2}/U_d$
3	$U_{r1} \geq 0$, $U_{r3} < 0$, $U_{r2} \geq 0$	$t_4 = \sqrt{2}T_{PWM}U_{r1}/U_d$, $t_5 = \sqrt{2}T_{PWM}U_{r2}/U_d$
4	$U_{r1} < 0$, $U_{r3} < 0$, $U_{r2} \geq 0$	$t_4 = -\sqrt{2}T_{PWM}U_{r1}/U_d$, $t_5 = -\sqrt{2}T_{PWM}U_{r3}/U_d$
5	$U_{r1} < 0$, $U_{r3} \geq 0$, $U_{r2} \geq 0$	$t_4 = \sqrt{2}T_{PWM}U_{r2}/U_d$, $t_5 = \sqrt{2}T_{PWM}U_{r3}/U_d$
6	$U_{r1} < 0$, $U_{r3} \geq 0$, $U_{r2} < 0$	$t_4 = -\sqrt{2}T_{PWM}U_{r2}/U_d$, $t_5 = -\sqrt{2}T_{PWM}U_{r1}/U_d$

4 Simulation of Control in PMSM Servo System by SL-SVPWM

The main purpose of establishment of MSM servo system simulation model for simulation in Matlab/Simulink, is to verify the dynamic response performance of SL-SVPWM control method applied in the servo system and determine the control parameters of the revolving loop controller and provide reference for program design. The main parameters of motor are: rated power: P_e=2.1kW, rated voltageV_{dc}=270V, stator phase resistance R=0.9585Ω, direct axis electrical inductance L_d=5.25mH, pole-pairs P=2 and rotational inertia J=6.3*$10^{-3}$$Nm^2$.

Set the friction coefficient as zero and start up the motor at rated load ($T_l = 2.5N \cdot m$),and the simulation wave form in diagram 5 shows that the time for the motor to run from zero rotating speed to the rotating speed of 1000r/min is about 30ms. SVPWM module output wave form is the horseshoe curve containing third harmonic after the revolving speed reaches a stable level, as shown in diagram 6. The stator phase current wave form has a good sine property, as shown in diagram 7.

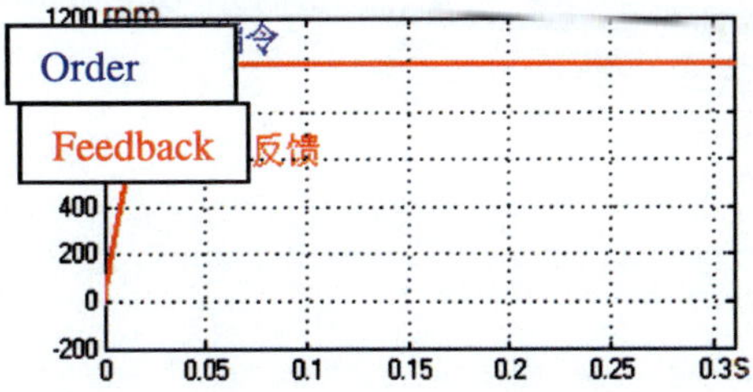

Diagram 5. Motor startup simulation curve

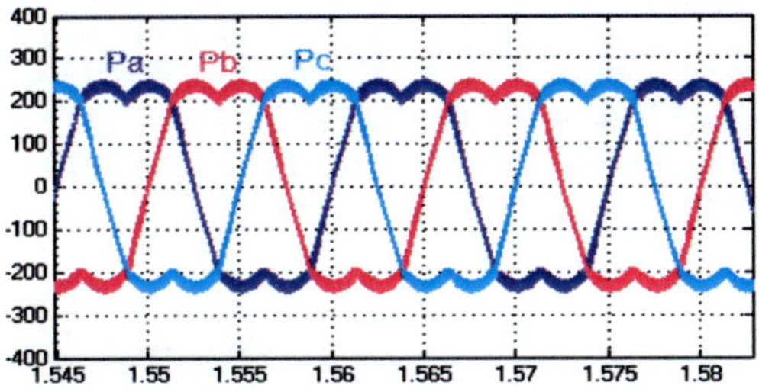

Diagram 6. SVPWM module output curve

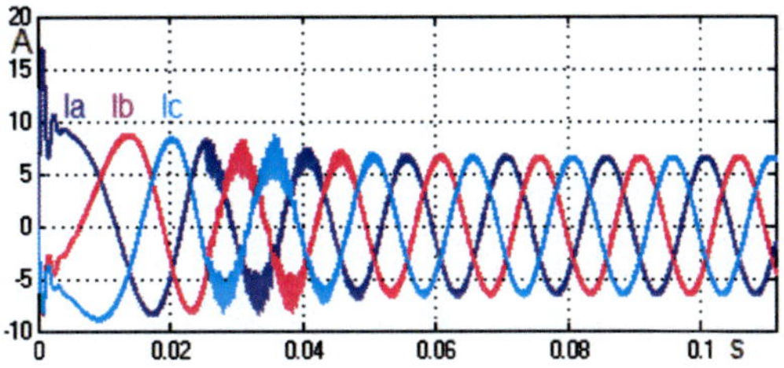

Diagram 7. Three phase winding phase current curve

In order to verify the system servo performance, the revolving speed orders are set up as triangular wave, sine wave and simulation wave form respectively, as shown in diagram 8 and diagram 9, the sine wave frequency is 15Hz and the triangular wave frequency is 20Hz, and diagram 8 and diagram 9 shows that the PMSM controlled by SL-SVPWM has a good tracking performance.

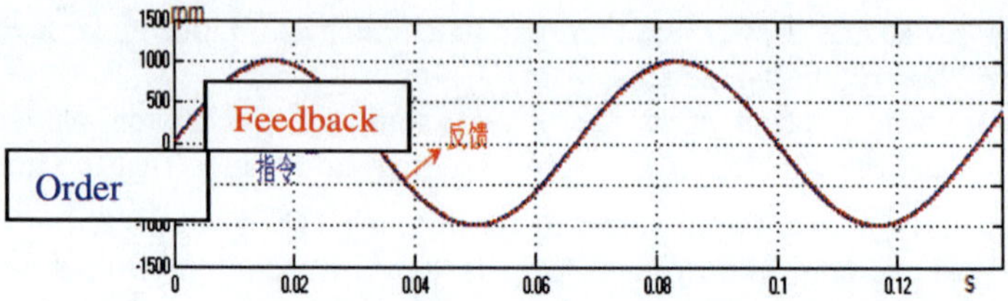

Diagram 8. PMSM sine wave revolving speed setting tracking curve

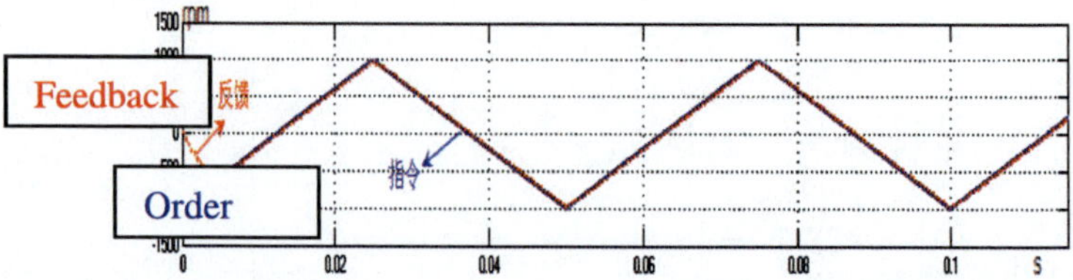

Diagram 9. PMSM triangular wave revolving speed setting tracking curve

5 Experiment Result and Analysis

PMSM servo control system consists of digital signal processor, motor and control and drive and protection and measurement circuit of motor. TMS320F2812 among which is the core of the control system and completes the revolving speed and current double closed loop PI control algorithm and generates the SL-SVPWM modulation wave.

Diagram 10 is the wave form of certain a phase current when the motor is driven; the analysis ahead shows that the start point of SPWM modulation is that the inverter output voltage is sine, but the diagram shows that the current wave form is not sine and there are concaves at the peak of wave and trough of wave and the change slope at the peak of wave and trough of wave is large. Diagram 11 is the wave form of certain a phase current when the motor is driven, and it shows that the current has a god wave form.

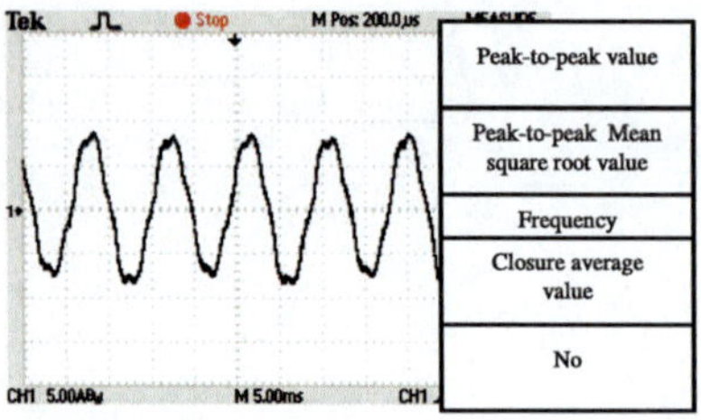

Diagram 10. SPWM phase current wave form

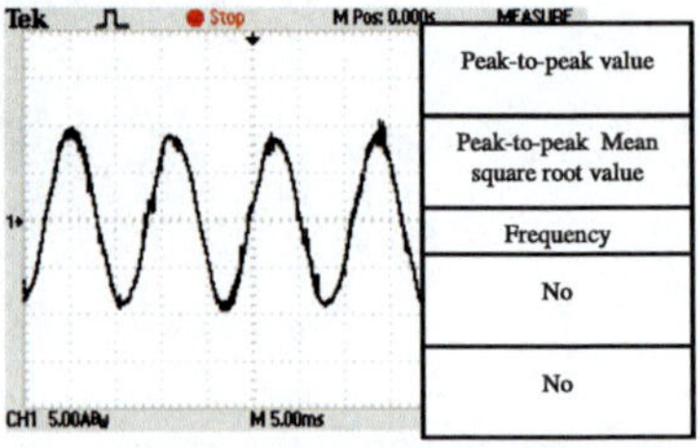

Diagram 11. SL-SVPWM phase Current wave form

Diagram 12 and 13 are PMSM servo performance test wave form when SL-SVPWM is driven; diagram 12 is the dynamic process curve when the motor start from zero speed to 5,000r/min and diagram 13 is the dynamic process curve when the motor start from zero speed to -5,000r/min and the diagram shows that the process in the transient process is about 35ms.

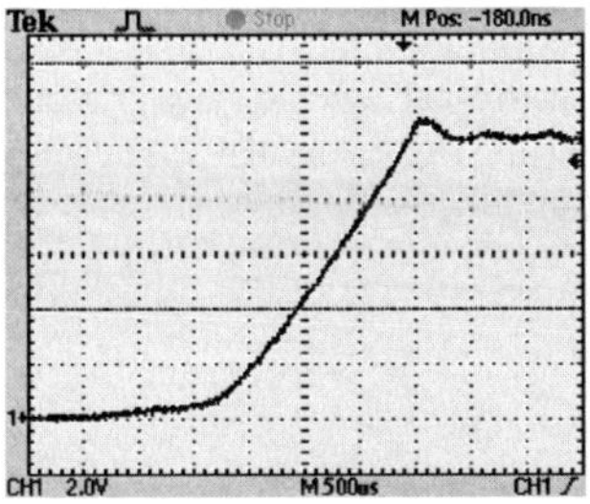

Diagram 12. positive acceleration response curve

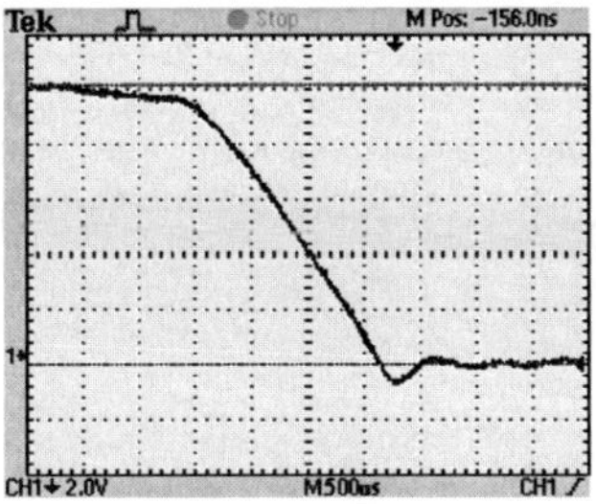

Diagram 13. negative acceleration response curve

6 Conclusion

PMSM is widely applied in various industrial fields with its superior performance, and FOC control theory has many years of history, but it is difficult to apply it in low cost system because of its complex system structure. In this paper, the single loop space vector is proposed and the electric current loop in the traditional FOC control method is omitted and the system structure is simplified and the system reliability and anti-interference performance are promoted and the system cost is reduced. The simulation and experiment results show that SL-SVPWM control method has an excellent performance and offers a new thinking for design of practical PMSM control system.

References

1. Itoh, J.I., Nomura, N., Ohsawa, H.: A comparison between V/f control and position-sensorless vector control for the permnent magnet synchronous motor. In: Proceedings of the Power Conversion Conference, Osaka, Japan (2002)
2. Perera, P.D.C., Blaabjerg, F., Pedersen, J.K., et al.: A sensorless stable V/f control method for permanent-magnet synchronous motor drives. IEEE Transactions on Industry Applications 39(3), 783–791 (2003)
3. Zhong, L., Rahman, M.F., Hu, W.Y., et al.: A direct torque controller for permanent magnet synchronous motor drives. IEEE Transactions on Energy Conversion 14(3), 637–642 (1999)
4. Yin, Y.-H., Zheng, B., Zheng, H.-X.: A Method for Modeling and Simulation of Brushless DC Motor Control System based on Matlab. Journal of System Simulation 20(2), 293–298 (2008) (in Chinese)
5. Blaschke, F.: The principle of field orientation as applied to the new transvector closed loop control system for rotating field machines. Siemens Ltd., Berlin (1972)

6. Kalman, R.E.: A New Approach to Linear Filtering and Prediction Problems. Transactions of the ASME–Journal of Basic Engineering 82, 34–45 (1960)
7. Zhou, Q., Li, S., Lu, G., Zhou, Y.: Crossed-feedback control of dual-redundancy permanent magnetic brushless dc servo system used in electro-hydrostatic actuator. In: International Conference on Electrical Machines and Systems, ICEMS 2008, pp. 1237–1241 (2008)
8. Rahman, M.A., Hoque, M.A.: On-line adaptive artificial neural network based vector control of permanent magnet synchronous motors. IEEE Transactions on Energy Conversion 13(4), 311–318 (1998)
9. Islam, R., Husain, I., Fardoun, A., et al.: Permanent-magnet synchronous motor magnet designs with skewing for torque ripple and cogging torque reduction. IEEE Transactions on Industry Applications 45(1), 152–160 (2009)

Create a Practical Teaching Environment for Applied Talents—Take the Case of Computer Specialty

Wu-Zhi Qiang[1] and Wang Yun[2]

[1] Network of Information Center of NanChang Institute of Technology, Nanchang 330099, China
[2] Computer Department of NanChang Institute of Technology, Nanchang 330099, China

Abstract. In this paper taking the case of the Computer Specialty in Newly-promoted undergraduate universities, discussed how to create a practice teaching environment for cultivating applied talents. Specifically suggested how to bulid an effectively practice teaching environment in the classroom; how to take employment as the direction to guide the training orientation of students's practical ability and how to make good use of teaching resources to create a practice platform and so on. Those thoughts and methods proposed in this paper have some references for other specialties's practice teaching.

Keywords: practice teaching, case teaching, digital means, employment orientation.

1 Introduction

Application-oriented institutions has the responsibility to cultivate applied talents with innovative capabilities, production-oriented for country, so, for an application-oriented institutions,it is a very important issue of how to achieve the tasks of developing talents with innovative, practical ability and engineering quality.

The computer practice teaching of the undergraduate colleges newly upgraded should take the market as the requirements, the employment as the guidance, to adjust the course structure. Especially to strengthen practice teaching, to develop their professional skills and practice ability as the key point, to enable students to become composite technology applied talents with solid foundation of computer basic theoretical knowledge and strong practice ability. Discusses the practice teaching system and teaching management mode, strengthen the practice teachers team construction, and active cooperation with enterprises, the rational use of school resources, construct practice base. Use of advanced teaching ideas to inspire and mobilize the students interest in learning, initiative, enthusiasm and participation, develop kinds of ability, to guide students to obtain the corresponding qualifications, to lay the foundation for the employment rate for graduates and the quality of employment.

2 Create Practical Teaching Environment from the Class

2.1 The Use of Case Teaching Method

Case teaching method began in Harvard University law school in 1910, and then gradually applicated and developed in medicine, business administration, teaching and

W. Zhang (Ed.): Advanced Technology in Teaching, AISC 163, pp. 343–348.
springerlink.com

other fields,achieved a great success. case teaching method by having learners apply the knowledge and experience to analyze andsolve real-world work situations to determine the events and issues in order to promote the migration of knowledge and skills learners, enhance their ability to solve problems. As an effective, practical and clear objective, action-oriented training.

The whole teaching process, computer application course can always focusing on a complete project cases to expand, with the advance of the progress of teaching,and gradually separate this case into anatomical sub-cases, even into smaller cases. These small case covers less knowledge, students is relatively easy to master. In the explanation of the realization of the small case, at the same time, , to inspire the student to use a variety of ways to think independently, and put forward his opinion, make a judgment and decision according to the student,s theoretical knowledge and technology level. The students gradually understand the various knowledge and their mutual links in the process. The cases was selected from the real life applications, so the students can master many of practical methods and techniques, and gain experience, get a sense of achievement, improve proficiency.

For the case teaching method can achieve better results, it has certain requirements with the materials and the teacher for the class. First, try to choose a new technology, by way of the case teaching writing teaching material; At the same time, it requires the teacher to use the case close to the practical application and the knowledge of large coverage, in the teaching process, the teacher should be able to accurately judge the student acceptance of this course, guide students' divergent thinking rationally.

2.2 Emphasis on Strengthening the Management of Curriculum Design

Curriculum design is one of the main practical teaching in every specialityof Science and engineering, it is based around a main course or specialized course, by using their knowledge of the course, and combining with practical application design to train the comprehensive analysis of the design capacity. This Course is designed to train students to master material in the context of the course when solving practical problems, think independently, comprehensive use the knowledge, gain experience, lay the foundation for the graduation design and professional work in the future.

To achieve the practical effects, curriculum design in the management need to strengthen management. First, according to computer science and technology training program development requirements of curriculum design, teachers are responsible for the implementation of the guidance system, regulate the content of curriculum design, update curriculum design exam, have a good curriculum design organization, pre-out assessment standards;During the course design, should accord to centralized training methods in computer science room, arrange regular instructor guidance, and ensure that there is enough time for communication between teachers and students and discuss between students. It have a Rigorous examination system,,and there is a uniform examination standards, if it does not meet the design requirements, need to return to modify, until completed. Course design requires communication of design content between teachers and each group or each student. so as to analyze and comment the work, helps the student to summarize the experience and the innovation in the process.

2.3 Assist Teaching with the Digital Method

The integration mode of Digital technology and teaching content is called digital learning. digital learning has the following three elements:First, the digital learning environment is the so-called information technology learning environment. Second, digital learning resources is multimedia materials that can run in multimedia computer or network environment after the digital processing. Third, the use of digital platforms and digital resources, teachers and students discuss with each other, and through to the collection and utilization of the resources to learning with e explore the knowledge, create knowledge and demonstrate knowledge, it has some ways as the resources utilization, independent found, negotiation-cooperation, practical creation.

Digital learning resources is the key of digital learning, teachers can acquire and organize by the way of exploring, students' creativity, buying and download from network and so on. through their own teaching experience and the latest application of the accumulation of technical requirements to update these resources. the teaching process of Computer courses will always face the conflict between teaching needs and existing hardware and software resources can not meet the conditions, it led to the pre-teaching objectives can not be achieved, teaching effectiveness is not intuitive. Therefore, it often need demo software to assists the teaching, to enhance students' practical skills, teachers can use the virtual software to enhance their interest in learning, it will help them understand the abstract theoretical concepts, initiative of practice, etc. Since the operation of the virtual machine will not cause damage to the physical device, the students do not have to worry about damage caused by improper operation, and then make them had the courage to play to their imagination to do personalized operation, so that it has some effects to the students creative ability.

3 Adhere to the Correct Ideological Guidance, and Actively Build a Practical Platform

3.1 Employment-Oriented, and Guide Students to Train Practical Ability

As the popularity of computer knowledge and the growing social educational channels, the advantages of current employment of college graduates in computer science is steady decline, and it is in the face of challenges and pressures, the new undergraduate colleges should increase employment guidance of computer professional. The vocational guidance and training to help students to establish a correct concept of occupation, to master the basic skills of career development and application techniques at the same time, the key is to make practical ability and market demand can have a good combination, so that it have good employment basis and competitive for graduates.

With the computer software and hardware technology and network technology develop rapidly. enterprises often need to master certain professional knowledge, has talents with strong technical skills and professional competence. Such as computer application technology, information system integration technology, computer security management, have high-quality engineering and technical talents with network engineering analysis and design capabilities, computer hardware installation and configuration capabilities, computer software installation and management and maintenance capabilities, and Web site construction and maintenance capabilities etc. Therefore, teachers should enhance the guidance of the professional related to these technical aspects for student.

Meanwhile, it is important to guide students to chooser the relevant vocational qualifications that fit them, and to help them pass the exam successfully, obtain the appropriate certificate. Most of vocational qualification certificates can reflect their level of knowledge and ability, it will be recognized by employers. there are many types of qualifying examination about Computer science, such as computer certificates, software professional and technical qualifications and so on, teachers can guide students according to their actual conditions, apply some of these, to gain employment opportunities for students, and provide a guarantee after employment.

3.2 Organize All Kinds of Practice, Build Students' Innovation Platform

Most of students emphasis on courses, however, limited by the time, they are rarely master broaden and deepen the knowledge. students are interested in these and willing to study, but they may lack the right direction, and sometimes do not know how to start.

In addition to doing an effective management in classroom learning and curriculum design, also making effective guidance in the ability to use knowledge after school.

Teachers can carry out various activities, which can enable students to use knowledge in the activity, and solve problems in practice, gain more practical experience.

For computer professional,teachers may guide student organizations to set up relevant associations, such as Network Association, Software Association, the hardware Association, etc. these associations is managed by the students. when approved, they can accept members and organize various activities by themselves. Guide students through the Association to carry out various internal communication activities to improve the common knowledge, carrying out regular outreach activities such as hardware maintenance, operating system maintenance and so on, to create opportunity to combine the knowledge and practical application, improve their related capabilities about the profession.

The school and department can carry out various competitions, such as FLASH contest, web pages contest, programming competition, and so on. This is a vivid form. it can stimulate the students' learning initiative, creativity, and inspire their positive thinking. Through access to information, books,students can use knowledge flexible, it has great help to improve the innovation ability.

3.3 Building Practice Environment by Using Effective Resources

Computer practice teaching is an important part of computer science, classroom instruction is to enable students to master the basic knowledge and basic computer skills, and computer practice teaching is aim to put the knowledge into practice, is an extension and supplement of classroom instruction. The practice teaching of computer occupies an important place whether in master the theory and principle of computer, or train student's ability of useing a computer to solve the other specialized question .

According to the actual situation and the existing conditions in school, can set up computer practice base in university relied on computer laboratory, improve the practice teaching system and teaching content; Can use mainly laboratories above the provincial level, computer basic experimental center, , computer department laboratory; you can practice the current mainstream software development methods and their applications in IT projects, practice all kind of software development tools in current, experience it's application in development process, practice mainstream network equipment's installation and commissioning.

In addition, the school's Network Information Center can also be a very good practice base for computer specialty。 Network Information Center is a unit of computer integrated application. It can provide internship opportunities in Network cabling, switching and routing equipment management, server management, software management, database management, software development, etc. part of the old equipment can be provided for students to practice.

At present, many IT companies make a further study in college's practice teaching, and set up a professional practice base. In accordance with the requirements of graduation practice, they discuss with the school and develop a detailed and thorough internship program together, provides a new way for college graduates to solve practical problems. These practice bases have modern management methods and modern business environment, and have a group of senior technical staff who worked on project development and have some experience in teaching and training. in the training process Students can fully understand the company and find a suitable opportunities and directions of development. When they are graduate and obtain employment, they can quickly familiar with the company, develop independent as soon as possible. Therefore, looking for this enterprise and strengthening the contact,, seting up graduation practice mode of School-Enterprise Joint Operations in the computer class, has an important role in Student's practical abilities.

4 Conclusion

For The teaching of computer profession in the undergraduate colleges newly upgraded, there are many problems need to be discussed, such as the condition that restricting education, how to strengthen practical ability of teachers, practice base construction program, evaluation of student's comprehensive ability, etc. but, in the teaching project, we think positive, make effort in many ways, create a good teaching environment for students. improve students' practical skills and the ability to solve practical problems, it must be able to achieve the goal that cultivate application talent with a sense of innovation, practical ability and the quality of engineering.

References

1. Wei, J.R.: A study of practical teaching reform based on computer specisity in the newly-build undergraduate universities– take employment as the direction. Jouranl of Baise University 21(3) (2008)
2. Zhang, X.M.: The Thinking of Practice Teaching Reform of the Applying Course of Tourism Undergraduate. Journal of Chengdu University (Educational Sciences Edition) 23(8) (2009)
3. Zhu, F.X., Zhi, S.-G.J.: A discussion of the personnel training mode in application-oriented Institutes. Economic Resourc.

Probe into Structuring Six-Dimensional System for Educating and Cultivating College Party Members

Liu Chaohui and Zhang Liying

Hebei Normal University of Science & Technology

Abstract. Along with the continuous growth of college student team, the number of college Party members multiplies. How to guarantee the quality of education and cultivation of college Party members has become an important task that needs much attention and in-depth research from Party-building work of colleges and universities. This article makes an analysis on the problems in process of current education and cultivation of college Party members and proposes to strengthen college Party members' education and cultivation in terms of theoretical study, moral cultivation, artistic edification, behavioral training, post dedication and system guarantee so as to actually enhance the overall quality of college Party members.

Keywords: College students, Party members, education, system.

1 Introduction

The education and cultivation of college Party members is an indispensable part of college Party-building work. Along with the continuous growth of college student team, the number of college Party members multiplies. How to guarantee the quality of education and cultivation of college Party members has become an important task that needs much attention and in-depth research from Party-building work of colleges and universities.

2 Prominent Problems in Current Education and Cultivation of College Party Members

At present, each Party committee has attached much importance to developing college Party members, specified target requirements and improved rules and regulations, so whether from quantity or from quality, the college Party members have made a great progress. However, due to characteristics of college students themselves as well as the changes and development of social situation, some problems have happened to education and cultivation of college Party members.

2.1 Theoretical Attainment Is Urgent to Be Raised

In the process of cultivating college Party members, academic achievements and real conduct usually draw much attention, while cultivation and appraisement of attainment

W. Zhang (Ed.): Advanced Technology in Teaching, AISC 163, pp. 349–354.
springerlink.com

in political theories is not strong enough and the measures are not effective, either. For example, when college students take part in Party school learning, there is usually simple training on basic knowledge about Party and Party Constitution; the way of teaching is relatively casual, lacking analysis on reality by using theoretical knowledge about Party and Marxism method so that the students hold no full enthusiasm and active attitude towards theoretical learning. The lack of political attainment in real life can be shown as not being able to reason, lack of necessary appeal and force power, even affecting college Party members' strong belief in communism.

2.2 Party Spirit Is Urgent to Be Strengthened

Party spirit is the inherent nature of a political Party. The Party spirit of the Communist Party of China is the concentrated performance of class nature of working class, is the concentrated performance of working class's nature and interests. Communist Party is a special group, people "made by special material". Each person in this group must own strong Party spirit and unique personality. However, some college Party members hold low requirements on themselves, just taking joining Party as an honorable matter, lacking necessary requirements on their own Party spirit; further more, some pose as Party members, overriding the public's interests. The consequence of college Party members' lack of Party spirit is serious, which may make a Party member lose Party principle and stand in the nick of time.

2.3 Moral Cultivation Needs Improving

Strengthening the college Party members' moral cultivation is not only necessary to cultivate high-quality college Party members, but also necessary to conduct ideological education on modern college students and cultivate qualified talents. Owing to the influence from social concept, individualism, money worship and hedonism also exist in modern college students; the motivation of joining Party is not definite; individual life is extremely materialized; take joining Party as a step to pursue material treatment; all of those seriously affect quality of developing Party members. Some Party members seek utility, attach too much importance to individual interests, pay no attention to individual attainment, and lack moral emotion and responsibility for serving people heart and soul, which seriously affect Party's impression on the young students.

2.4 Conscientiousness Needs to Be Enhanced

Each Party member must shoulder certain responsibility and carry out corresponding responsibility. But as college Party members, their main task is study and they do not shoulder corresponding social responsibility, which results in some Party members' lack of necessary responsibility sense and conscientiousness as well as more sense of honor than sense of responsibility. Because of Party member's special identity in addition that modern college students take part in relatively few social activities and mostly are the beloved in families, some Party members can not make out their own responsibility and identity after joining Party; they are not active enough when encountering difficulties and lack necessary consciousness to actively serve other people.

2.5 Exemplary Role Needs Highlighting

Some student Party members relax themselves in daily work and study after joining Party, not be able to well play a Party member's exemplary role; when encountering problems, they usually forget to develop the active influence and leading role to other students as a college Party member should do; just like general students, they can not think calmly when facing problems and do not make rational analysis; even some pose as Party members, just managing their own matters in work and study and paying no attention to matters that do not concern them, therefore, they can not play the role a Party member should do, and sometimes even let wrong opinions alone and be a capture of incorrect public opinion, which results in extremely bad influence.

3 Practice and Probe into Six-Dimensional System for Educating and Cultivating College Party Members

In recent years, the writer has made beneficial probe and practice into the cultivation education and management system of college Party members. The preliminary six-dimensional system of cultivation education and management of college Party members has been shaped, which enriches the content of cultivation education of college Party members to certain degree, specifies education management system and plays an active motivating role on guaranteeing the quality of college Party members

3.1 Theoretical Study

Theoretical study is a political task that each Party member must often carry out, especially for the college student ready to join the Communist Party of China and the new college Party members. Aiming at Party member education, the Working Regulations for Basic organization of Communist Party of China in Colleges and Universities definitely points out that: "Party organizations of colleges and universities should conduct Marxism Leninism and Maoism education on Party members, especially Deng Xiaoping's constructing education of socialist theory with Chinese characteristics and education of Party's basic line and basic knowledge; educate the Party members to try to master scientific culture knowledge and professional skill and improve political and professional quality". That is the guiding ideology for education of Party members in colleges and universities in new era. As the education of college Party members, the content must follow and carry out this guiding ideology. Since the college students submit Party membership application to Party organization, the Party organization should conduct necessary political theoretical education on them; start with education of basic knowledge about Party; rectify the motivation of joining Party; establish lofty ideal of communism and strong belief in fighting for communism lifetime, meanwhile, make them set up thought morals for serving people heart and soul. The study content includes Communist Manifesto, Party Constitution and On Communist's Spirit, important references that the predecessors of proletarian revolution like Mao Zedong and Deng Xiaoping made about communist as well as the important descriptions that comrades Jiang Zemin and Hu Jintao made about three represents. Theoretical study is endless and the communists must persist in it, but the college Party members should attach more importance to the study of basic knowledge and communist Party spirit, and then proceed in an orderly way step by step. Strengthen the assessment on study content and make out the important meaning of learning theoretical knowledge at the same time, changing from being passive to being active.

3.2 Moral Cultivation

The communist's moral cultivation is an important part of Party spirit, which refers that the communist individuals conduct the behavioral activities like self-education, self-reform, self-training and self-practice according to the requirements on communist morals in aspects of moral consciousness and moral behavior. Out Party always attaches much importance to the Party members' moral cultivation; as early as revolutionary war, in the works like Serve People and In Memory of Bethune, comrade Mao Zedong made an in-depth description on communists' moral cultivation and summoned all communists to be "a noble person, a pure person, a moral person, a person getting rid of vulgar interests and a person beneficial to people". Comrade Deng Xiaoping also warned the whole Party that: "the Party leadership must carry out communist morals and cultivation, keep calm and firmly resist the erosion from foreign ideas". Actively withstand influence of various non-proletarian ideas like money worship and hedonism in study, work and social activities; establish the value of life of the proletariat, suffering first and enjoying later; when there is conflict among individual, collective and national interests, individual interests conscientiously obey the collective and national interests.

3.3 Artistic Edification

The cultivation education of college Party members not only should pay attention to the systematization and theory of education content, but also can not ignore the artistry, elegance and interest of it. The artistic edification of college Party members include three aspects: the first is the revolutionaries' red mottoes which have educational significance. Throughout each motto full of truth, we can see the revolutionaries hold a strong belief in victory of Chinese revolution. A series of passionate voices like Hard Fortunes and Strong Nation from the revolution forerunner Li Dazhao, Prison Song from Ye Ting, Wedding on Execution Ground from Zhou Wenyong and Lovely China from Fang Zhimin explain to us the moving revolutionary spirit, red ideal and persisting pursuit. Those great voices throughout time and space should become material for educating the college students in modern times especially the college Party members. The second is the red songs with educational significance. In the sixty years since new China's birth, red songs are widely sung and spread; these songs express the vast public's endless passion for Party and the nation, so they are of special educational significance for modern college Party members. A series of classical red songs like Without Communist Party, There would be no New China and Sing a Folk Song to the Party should be taken as the important content for education of college Party members. Singing red songs widely is a kind of recollection and inheritance; the songs can rouse spirit, transmit power and fly the ideal. The red glamour is irresistible; the red memory is unforgettable and red songs should become theme of college Party members.

The third is the video works with educational significance. Red films and teleplays are more and more popular with public; therein, the bright revolutionary image and spirit inspire innumerable idealistic and ambitious young people in new times. College Party members should learn and think more about the content. Jiao Yulu, Lei Feng, Youth of Fighting and the biographical films of communist model in new times like Niu Yuru, etc. should be the content of Party member education. Artistic edification can improve the political quality of college Party members, at the same time, it can cultivate the young student Party members' human emotion to a greater extent, enrich forms and content of education and be more suitable to characteristics of young students. Content is rich and form various, easy to be absorbed and received.

3.4 Behavioral Training

The effect of cultivating and educating college Party members and improving their theoretical level and moral cultivation depends on their behavioral performance. Behavioral performance of college Party members plays an important role in establishing Party's image in young students; the college Party member's identity is first of all a student, so his or her behavior must follow the identity of an excellent student. For example, study hard, listen to the teacher carefully, obey public order, get along well with others, respect the teacher, unite classmate, etc. These excellent behavioral habits are of great significance for promoting to shape good quality of college Party members. In the process of cultivating and educating Party members, some Party members really behave inconsistently before and after joining Party, not strict enough with their own behavior and daily performance, even not as good as general students. Therefore, supervision can be strengthened through such measures as student Party member's wearing Party member sign and Party member sign being hung in the dorms of Party members; the incorrigible should be given disposition and punishment even expelled from the Party.

3.5 Post Dedication

Through shouldering cadres of student committee or mass committee in departments and classes, participating in and engaging in some concrete work in colleges, the college Party members can experience training and ordeal, improve their working ability and ideological theory level and cultivate spirit of serving people heart and soul. They can be trained consciously in college Party members to fully develop the leading exemplary role of college Party members; by shouldering learning counseling work of Party constitution group in junior grades, they can be responsible for contacting the Party activists in junior grades; through handing over tasks, entrusting task and overcoming reliance thought, cultivate and train their independent thinking and working ability to enhance comprehensive professional skills of college Party members. The exemplary role of college Party members should be omni-directional and multi-level; promote socialist moral fashion inside and outside college activities and develop exemplary role in campus civilization construction. Student Party members should play a bridge role, feedback key problems and various ideological situations from public to Party organization and provide information, conduct research and investigation and give advice to college and department student work or class work.

3.6 System Guarantee

The first is the organization development system. Cultivating college Party members starts from writing Party application; write individual ideological report; the branch determines cultivation subjects; the cultivated take part in Party school training until the branch discusses for someone to be the probationary Party member, and these procedures are compulsory. The second is theory study system. Rigorous and effective organizational life is the form and restriction mechanism of college Party members' quality optimization. Through the organizational life of standardized Party groups and Party branches, through scientific system construction like political study system, democratic life meeting system and Party lecture system as well as flexible and versatile activity modes that stress on practical results, help and educate college Party members to continuously strengthen communist conception, improve political consciousness, optimize political quality and enhance communist consciousness. The third is management evaluation system. Timely and realistically record the political performance, actual behavioral award and punishment, academic achievements, etc. of Party activists and student Party members so that the situations of Party activists and student Party members can be comprehensively reflected in order for further investigation on them. Inner-Party reports are regularly made on the performance of each Party activist and student Party member, receiving criticism and conducting broad mass appraisal.

The PE Teaching in Higher Education Based on the Integration of the Information Technology and Curriculum

Liu Daduo[1], Zhang Xianhui[2], Li Rongwei[3], Gao Chao[4], and Liu Xiangfei[5]

[1] College of Physical Education , Jilin Normal University, Siping, Jilin, China
liu_daduo@sina.com
[2] Department of Physical Education, Hebei Normal University of Science & Technology, Qin Huangdao, China
zhangxianhui@126.com
[3] Department of Physical Education, Hebei Normal University of Science & Technology, Qin Huangdao, China
sunny8026@sohu.com
[4] The No.3 Middle School of Liaoyang, Liaoning Liaoyang, China
gao_chao830515@163.com
[5] Qinhuangdao Radio and TV University, Qin Huangdao, China
liuxiangfei1980@163.com

Abstract. With the development of IT, people enter information society. The society and all kinds of jobs need IT. So it is an important task for high school to make students use computer in practice and develop students' information literacy. IITC is an efficient measure to make students have all skills. In this paper, information technology, content presentation and physical education, information technology and information technology, student learning styles and teaching methods and the overall physical education teachers, three aspects of information technology and curriculum integration for Physical Education under the discussion of information technology and aims to Integration of physical education to provide a theoretical reference.

Keywords: Information, information technology, PE Teaching in Higher Education, quality education.

1 Introduction

Integrating Information Technology into Curriculum (IITC) presents a new way of revolution on basic education. Besides its relative independence, IITC is a new pattern of education structure with close connection and succession with subject teaching. Not only as an assistant of teaching and learning, IITC stress on the using of Information technology as a tool of cognition and emotional encouraging, which is to make use of different kind of learning environment such as Self Exploration, Multiple Interacting, Collaborative and Resource Sharing to activate students initiation together with enthusiasm and train the innovation thoughts and practice capability, which is the main goal of training innovation personnel.

W. Zhang (Ed.): Advanced Technology in Teaching, AISC 163, pp. 355–360.
springerlink.com

This shows that IITC presents a effective approach of changing the traditional teaching structure and implementing the innovation personnel training. This is also the international trend of foundation education revolution. After analysis of the character and the main revolution trend of physical education in universities and colleges, this paper discuss integrating Information Technology into physical education class.

2 About IITC

Information technology (IT) is the generic terms of different kind of technologies to manage and process information with computer science and communication technology. IT was used on the design, develop, install and carry out of information system and software. It was also named(Information and Communications Technology, ICT)。

IITC means the integration between information technology and curriculum, not information technology and curriculum integration. This is the key point we understand. Integration means the process and result that two or more smaller parts like things, appearance, process, material character, relationship, information and power were integrated to a bigger integral when it follow definite objective laws or under certain condition. The definition of IITC can be understood as two definitions, Big Integratism and Small Integratism.

The concept curriculum in Big Integratism is a general concept which means that to integrate the information technology into the whole system of curriculum in order to change its content and structure. It is the revolution of the whole system.

While in Small Integratism, Curriculum was understood as teaching and the integration was between IT and subject teaching. The famous expert of educational technology in China Professor Li Kedong common that IITC is a new pattern of teaching which is to combine the course content with IT, information resource, information method and human resources dynamically to finish the class together. There are three main points:1. Informationization environment with multimedia equipment and internet.2.Information processing of the teaching content to transfer it to learning resources.3.To help the students to restructure their knowledge with information processing tools.

3 The Integrating the IT to the Present Approach of Physical Education Content

As the development of computer technology, computer multimedia become one of the important approaches of information presentation for class education for its variety functions on different kind of information. Research shows that people can remember15% of the information from language. While 65% of those information from both listening and watching can be recept and remembered. In the past teaching of PE and health knowledge, the abstract knowledge was teaching by oral explanation together with some pictures and models which was stiffness. The information technology can make up the shortage of traditional education by making the abstract information definite. In this process, those content and object which is hard to understand or observe were presented by the way of information technology which arouse students' optical function and help them to break through the difficulties. Take the teaching of Leap as the example, we used to teach by explanation and demonstration.

The traditional way can stress the important point while the demonstration finish in the twinkling of an eye and hard to be understood by the students. If the whole process was divided into run-up, takeoff, rise to the sky and all to the ground and demonstrated with the multimedia technology, the students will get better perceptual knowledge. Before the presentation, students can be given several questions such as how many steps in the process of broad jump which will be answered after the presentation. Accord with students' thought feature and help them to establish the concept of coordination, this is a better way to catch both the important point and the difficult point.

In the process of carrying out the new class, the using of multimedia can satisfy students' needs of listening and watching and active their interests of learning. Interests is the best teacher which can help the students to enjoy and looking forward to the class to meet active learning.

In the class we introduce Barrier relay, I use the ruled field, foam-rubber cushion and bench as the *cordon area* .Students hold the cap gun instead of the relay baton. The whole relay race was like battlefield with the sound of bugle call, bombing and sing of bullets. This kind of play which make students tension and happy is a good way to make the students strong and healthy.

4 Integrating the Information Technology into Study Pattern of Students

Integration between IT and the PE course in College and Universities is a revolution for the learning pattern for students. In the interactive learning environment of IT, students can choose what to learn and which level of excise to take up to their own level and interests.

4.1 Use Internet to Build the Bridge for the Communication between Students and Teacher

Internet is the biggest database of information and resource in which students can collect and search for what they need. Internet environment is the best for the initiative and enthusiasm of learner. In the research on the integration between IT and PE subject teaching, to expand the education space and make it easier for the students to learn. The training plan can be checked on the internet by the students who will do the excise by the schedule as teacher request. The students can also communicate and answer questions in the communication page to get help from the teacher. This kind of communication can develop the relationship between students and teacher by integrating IT into study.

4.2 IT Is a New Way to Develop Students' Independence of Learning

In tradition PE teaching, students learn most from teacher and teaching material. Students just following the teacher to copy and practice. Their time and space to learn by themselves is limited. The using of IT broke the simple pattern of teaching in which teacher teach and the students practice only. A new pattern of teaching in which students can learn and investigate by themselves was established to make IT the real tool for students. For example, in the design of skill class, students' gymnastics action can be presented on the wall by video and projector in the preparing part. So that the students can watch their own action to make sure the accuracy and range clearly. This is a good example of using IT to develop the practice result by feeding back the practice result to students to develop their interests and learning attitude. In the basic part, teacher can tape students' action in the former class and upload them together with teacher's demonstration and the key point of the action to the computer, so that the students can check on the computer to develop the action by themselves when they do the practice.

So IT is a good way to support the self-study by meeting students' need and make them the subject of learning. Students experience the self-control and adjustment which can optimize the teaching process and make it more efficient.

4.3 Internet Is Good for Students' Comprehensive Capability

Computer and internet provide a loose and comfortable learning environment for students which can make them more active. According to teacher's teaching plan and teaching task on internet, students can arrange their study more selectively. Variety pattern of finishing single or team work , such as searching on internet, cooperative learning and online discussion, can help to promote students' capability of information collection and analysis. The richness and interaction of material can support the self-direction in order to meet the keep enrich of the knowledge content and structure. In one of the theory courses for PE major, teacher finish the teaching task with the help of internet. The first task for students is to collect the information of Olympic Games on internet, such as the origin, the emblems and any other information of Beijing Olympic Games. Students were divided into several teams and choose one of the topics to research on the internet. They collect information on internet and discuss to screen and download the useful information to make them to webpage. With the support of teacher, students finish their task by teams and share the achievement with other teams. During one teaching unit, 45 minutes, students learn a lot not only about Olympic Games but also about the internet, including the collection, downloading, managing and application of the information. Students' information accomplishment were improved which is the best way to meet the concept and spirit of the new course for physical education and health.

4.4 Promote the Integration of Learning Pattern for the Physical Education Both in or Out of the Class

One of the teaching objectives of the PE and health course is to promote students' capability of self-exercise, self-regulation and self-evaluation. While it is difficult to pass on all the knowledge about health, nutrition and fitness to students in classroom.

The database and website set up by teachers can be visit after class and provide open and rich resources of learning which will widen the approach of learning and extend the teaching in classroom. The internet and computer integrate the classroom instruction with PE outside class, society PE and family PE, which not only developed students' interests but also means a lot for their lifelong development. Teacher can make record for students and encourage them to design the actions in two weeks. Students can show their task by team in the class and discuss with each other after class. This is a good approach for students to master the learning method and develop their capability of collecting and managing information actively.

The integration of the PE learning in and out of classroom is good for the guiding ideology of Health First and promote the harmonious development of students' body and thought.

5 Integrating IT into the Teaching Pattern of PE Teachers

The appearance of multimedia technology change the way teachers teach and the traditional dull teaching pattern. It provide a new chance for the revolution of teaching method and make learning more interesting. For example, in past, teacher need lots of time to explain the abstract part which can become visualized by the multimedia technology. So the application of multimedia information change both the teachers' quality and their teaching by push them to master the developed education concept and technology.

All kind of information such as text, sound, graph, picture and flash can be managed with multimedia technology to design different kind of courseware of different kind of sports. Those courseware can provide a vivid teaching environment and tool for teachers. which can relieve students' load efficiently and make the classroom lively. Nowadays, more and more courseware studio, e-library, digit projector classroom, multimedia language lab were opened in schools which is a good way to updating teaching pattern. How to make good use of the advantage of IT and develop the teaching efficiency and quality is the task for us.

6 Conclusion

IITC is one of the most important aspect of class revolution. It is to integrate a tool into course and help students to learn by themselves with the modern science and technology. The ability of collecting, managing and applying the information is more important for students' lifelong learning. Helping students to face the challenge of the social environment is the best proof of the effect of integration.

References

1. He, K.: Theory and method of deep integrating IT into curriculum. Education Technology (2) (2005)
2. Li, K.: Digital study-the core of integrating IT into curriculum. E-education Research (8), 49 (2001)

3. Nan, G., Li, Y., Zhu, Z.: Introduction of information education, p. 210. Higher Education Press, Beijing (2004)
4. ISTE National Educational Technology standards Projects Zhu Zhiting. National educational technology standards for students: Connecting curriculum and technology. China Central Radio & TV University Press (2002)
5. Ma, N., Yu, S.: The level of connecting curriculum and technology, vol. (l), pp. 9–13 2002(l)
6. Li, X.: Promote the well development of IT curriculum- interview Professor Dong Yuqi from NENU. Chinese E-Education (8), 5–8 (2005)
7. He, K.: The target and meaning of connecting curriculum and technology. Education Research (4), 39–43 (2002)
8. Wu, F., Fang, B.: The dimensionality and practice of connecting PE curriculum in college and Technology. An Hui PE Technology (12), 66–69 (2005)
9. Yang, P., Liao, M.: Construction of the education space for connecting PE curriculum and technology

Time Delay Dynamic System Model Identification of Hot Rolling Mill Based on an Improved PSO Neutral Network

Jianxin Zhou

Department of Electric Engineering Hebei United University
Tangshan, Hebei, China
zhoujianxin1977@126.com

Abstract. Combining improved particle swarm optimization (IPSO) with neural network, we propose a model identification method for time lag and nonlinear of hot rolling mill thickness control system. The method is used for automatic gauge control identification of hot rolling mill and the result is applied to Smith predictor. The result of simulation show that the model based on IPSO neutral network is used for Smith predictor control can avoid hypothesis of thickness control system and solution nonlinear equation. The change of thickness of hot rolling mill can be traced accurately, thus the system showing its favorable dynamic properties and precision.

Keywords: particle swarm optimization, neutral network, time lag, system identification, thickness control.

1 Introduction

Thickness control for hot strip rolling plain that there is a big lag time of the control object, the problem is the area of control has been concerned. Smith predictor control strategy is an effective method for solved the pure time delay problem. It is a control strategy based on model. Smith Predictor control is to estimate the affect that the control variables of the process output delayed. As the forecast is based on the process model of the known circumstances, therefore, Smith Predictor to achieve dynamic model must be known that is known to process the transfer function and pure lag time. Model is consistent with the real process is valid only. For most process control, process models can only approximate express the true process, and the establishment and application of Smith Predictor, there are some errors exist. If model has the greater the error, the compensation effect is worse. As the neural network can achieve a very complex non-linear mapping, and can be dynamically adjusted in real time, so it is one of the main tools in system identification.

Traditional neural network algorithm is based on gradient descent. Network convergence speed is slow in the learning process, easy to fall into local optimum. The net is extremely sensitive to change such as the initial weights of the network, learning rate and momentum parameters. This paper proposes an improved particle swarm optimization for training of BP neural network learning algorithm and neural network with optimized thickness of the hot strip mill control system to identify, avoid the problem that BP algorithm is easy to fall into local optimum, and achieved good control effect.

W. Zhang (Ed.): Advanced Technology in Teaching, AISC 163, pp. 361–367.
springerlink.com

2 Improved Particle Swarm Optimization

The basic idea of PSO algorithm randomly initializes a group of no volume and no quality of particles. Each particle as a feasible solution of optimization problems, the quality of particle is determined by a fitness function of pre-set. Each particle will be moving in the feasible solution space by a variable speed to determine their direction and distance. Usually particle follow the current best particle and obtain the optimal solution finally by each generation to search. In each generation, the particle will follow two extremes: one is the particle itself so far to find the optimal solution, the other is so far the entire group to find the optimal solution [1]

Suppose a group formed by the M particles in the search space D dimension to a certain speed flight. The state of the i particle at t time property settings are as follows:

Location: $x_i^t = (x_{i1}^t, x_{i2}^t, \cdots x_{id}^t)^T$ $x_{id}^t \in [L_d, U_d]$. L_d 、 U_d express the lower and upper search space, respectively. Speed: $v_{id}^t = (v_{i1}, v_{i2}, \cdots, v_{id})^T$, $v_{id}^t \in [v_{\min,d}, v_{\max,d}]$; $v_{\min}$ 、 $v_{\max}$ express minimum and maximum speed Respectively 。 The personal best position: $p_i^t = (p_{i1}^t, p_{i2}^t, \cdots, p_{iD}^t)^T$, global optimum position: $p_g^t = (p_{g1}^t, p_{g2}^t, \cdots, p_{gD}^t)^T$. Among, $1 \le d \le D$, $1 \le i \le M$. The position of the particle at time $t+1$ updated by the following formula:

$$v_{id}^{t+1} = \omega v_{id}^t + c_1 r_1 (p_{id}^t - x_{id}^t) + c_2 r_2 (p_{gd}^t - x_{id}^t) \tag{1}$$

$$x_{id}^{t+1} = x_{id}^t + v_{id}^{t+1} \tag{2}$$

Where, r_1 , r_2 are uniformly distributed in (0,1) interval of random numbers; C_1 、 C_2 called the learning factor. ω called the inertia weight, it's size determines particle how much the current rate of succession. To further ensure the convergence of search, Clerc [3] introduced a shrinkage factor, shrinkage factor χ is a function of parameters C_1 、 C_2 ,and the introduction of shrinkage factor, (1) becomes

$$v_{id}^{k+1} = \chi[v_{id}^k + c_1 r_1 (p_{id} - x_{id}^k) + c_2 r_2 (p_{gd} - x_{id}^k)] \tag{3}$$

$$\chi = \frac{2}{\left|2 - l - \sqrt{l^2 - 4l}\right|}, l = c_1 + c_2, l > 4 \tag{4}$$

Shrinkage factor to ensure the particle swarm algorithm converges in the search process, so this paper introduces shrinkage factor improvement PSO.

3 The Neural Network Training Based on IPSO

Most neural networks are trained using gradient descent, such as back-propagation algorithm (BP). This algorithm has a convergence for a long time, easy to fall into local minima and other defects. Particle swarm algorithm is a heuristic global optimization algorithm, so use it to train the neural network weights avoided defects of BP algorithm. Neural network training process can be seen as an optimization problem of finding a set of optimal combination of real weight, the weight has minimum error between the expected value and the output value. [3]

The particle position vectors of IPSO algorithm express the neural network connection weights and thresholds. The particle dimension value corresponding to number of the neural network connection weights and thresholds. Neural network fitness function selected the mean square error indicators. The smaller errors show that the particle has better performance in the search. Fitness function formula is as follows:

$$F=\frac{1}{N}\sum_{i=1}^{N}\sum_{j=1}^{C}(y_{i,j}-t_{i,j})^2 \tag{5}$$

Where: N is the total number of training samples; C is the number of output nodes; $y_{i,j}$ is the actual network output; $t_{i,j}$ is a network of the desired output. IPSO algorithm trained neural network algorithm is as follows [4]:

1) The IPSO-NN network structure and parameters are determined. The number of network input layer neurons is determined according to sample input vector. The number of network output layer neurons is determined according to sample output vector. Determine the number of hidden layer neurons, to determine the particle size, inertia weight minimum and maximum values, the learning factor, the number of iterations and so on. Initialize the initial location of particles according to Uniform distribution function. The initial particle velocity is set to zero.

2) The mapping is established between the particles in PSO and Parameters to be optimized. The number of input neurons by I .The number of output neurons by O. The number of hidden layer neurons by H . For the parameters to optimize in the three-layer BP network can be represented of a one-dimensional matrix. The Parameters to be optimized merge together to form the matrix size can be expressed as:

$$D=I\times H+H\times O+H+O$$

3) Calculate the fitness function value of each particle. Compare it with p_i, if it better than p_i , update p_i; compare it with p_g, if it better than p_g, update p_g .

4) The speed and location are updated according to equation (3) and formula (4), and the speed is limited.

5) If the maximum number of iterations or minimum error precision is meet, then stop the iteration and Output value and its position to adapt to the global optimum. Find a best position of particles as the final output of neural network weights and thresholds, otherwise go to step 2).

4 Thickness Model Identification Based on IPSO Neural Network

The three main features of hot strip rolling model. The performance of the mill model has uncertainty because many of the factors affecting.

The parameters of previous frame reach the next frame after a certain period of time. The controlled object model has time lag. Mill just at the end of the export the amount charged of the sensor. Control the amount of more than one rack, performance of the controlled object model features redundant control.

Neural network has a characteristic that can approximate multi-dimensional complex nonlinear function between the input and output signal. It is through the input and output samples of training to learn the function and. Any a priori knowledge of function form is not required, fitting online learning. This article will IPSO complex non-linear neural network for adaptive control of multi-variable objects. IPSO neural network simulate the rolling process and alternative Smith predictor model.The rolling process of IPSO Neural network simulation use the following functions that

$$\Delta h^N(K) = f(\Delta h(K-1), \Delta S(K), \Delta S(K-1)) \tag{6}$$

Equation (6) - $\Delta h^N(K)$ IPSO Neural Network exit thickness difference at time K; $\Delta h(K-1)$ - export thickness difference in time $K-1$; ΔS - the amount of time roll gap adjustment at time K; $\Delta S(K-1)$ he amount of time roll gap adjustment at time $K-1$. Thus, the need for input $\Delta h(K-1)$、$\Delta S(K-1)$、$\Delta S(K)$ and output $\Delta h^N(K)$ process model function learning with IPSO neural network. The learning structures that identification process model with IPSO neural network shown in Figure 1.

Which $\Delta e(K)$ is system strip exit thickness deviation and the difference between neural network output at time K. As the actual rolling process time delay, Simth system need process model that remove the delay. So a time delay joined after the neuronal network. Making the IPSO neural network simulation is to remove the delay part of the process model. In the specific implementation, we can make the same IPSO neural network input while the output goes $\Delta h(K-\tau)$ which τ is the period of delay.

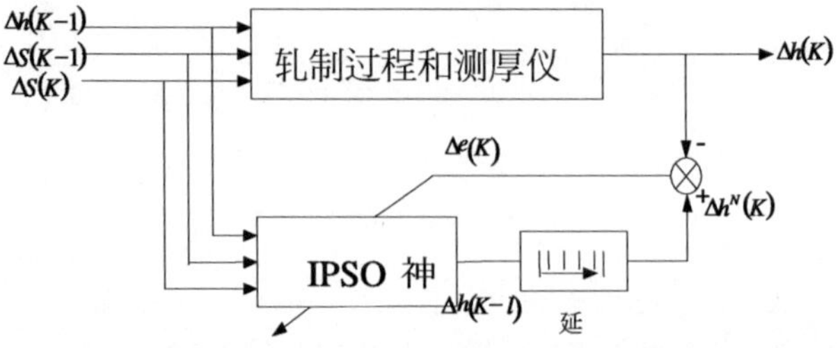

Fig. 1. IPSO neural network identification process model

This network implements a nonlinear mapping, which is a static model, while the process model is a dynamic process. So in order to build process model by IPSO neural network, the difference output of the process model and the network $\Delta e(K) = \Delta h^N(K) - \Delta h(K)$ needed feedback input to the network in order to further to amend the weights. So after a period of training process, the network becomes the system process model by learning. It enable the tracking error $\Delta e(K)$ tends to zero.

IPSO neural network model of the rolling process learning and training procedure is as follows:

(1) Initialize the IPSO neural network weights;

(2) To obtain the actual data field; a rack Smith-AGC roll gap adjustment $\Delta S(K)$ at time K . Strip exit thickness deviation $\Delta h(K)$ $(K = 1,2,3\cdots n)$, training samples $n = 400$;

(3) Normalized: to find out the amount of roll gap adjustment of the maximum ΔS_{max} and minimum ΔS_{min} , and then normalized roll gap adjustment $\Delta S'(K) = \dfrac{\Delta S(K) - \Delta S_{min}}{\Delta S_{max} - \Delta S_{min}}$. Find out strip exit thickness deviation of maximum and minimum values Δh_{min} 、 Δh_{min} , and then get normalized export bias of band steel thickness $\Delta h'(K) = \dfrac{\Delta h(K) - \Delta h_{min}}{\Delta h_{max} - \Delta h_{min}}$.

(4) The normalized volume $\Delta S'(K)$ 、 $\Delta S'(K-1)$ 、 $\Delta h'(K-1)$, as IPSO neural network input.

(5) IPSO neural network weights trained according to the steps listed in Part 3.

5 Simulation

The IPSO neural network effect of system identification is verified through the comparison of validation and GA neural network. System sampling period is 50ms, Sampling points is 200. The first 100 points is used to trained and established neural networks. After one hundred is used to verified the accuracy of the data network identification. Control volume $\Delta S(K-1)$ 、 $\Delta S(K)$ 、 $\Delta h(K-1)$ are taken as a square wave signal. Disturbance is taken as a random disturbance signal. Neural network input layer nodes are 3. The output layer node is 1.The hidden layer nodes are 4. The excitation function of hidden layer and output layer is the nonlinear sigmoid function. $c_1 = 2.5$ 、 $c_2 = 2.5$, Particle size $Size = 30$. Particle velocity range is $[0,100]$ particle dimensions $3\times4+4\times1+4+1=21$. The simulation results are shown in Figure 2 and Figure 3.

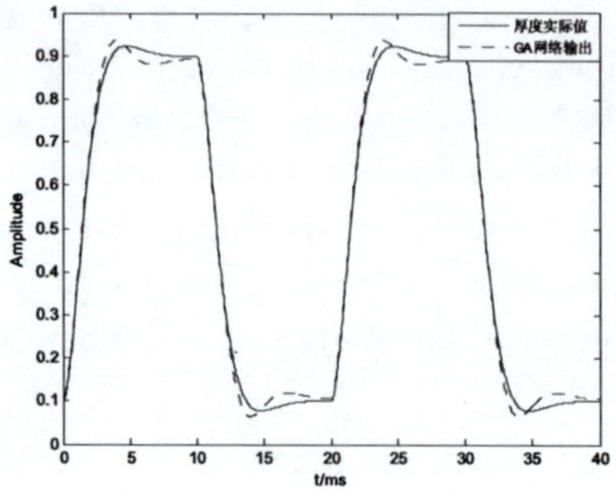

Fig. 2. Thickness of GA neural network identification results

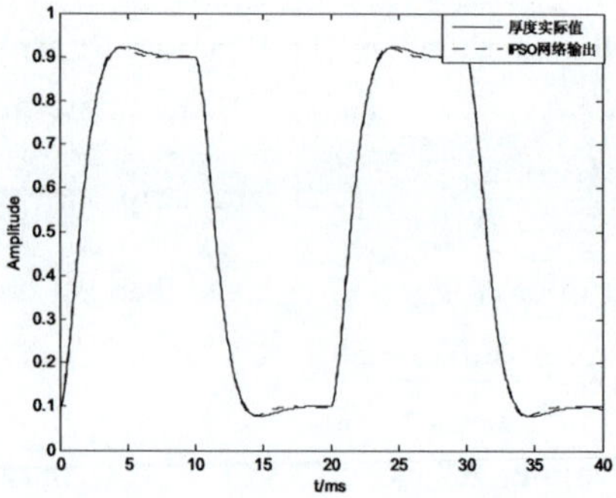

Fig. 3. Thickness of IPSO neural network identification results

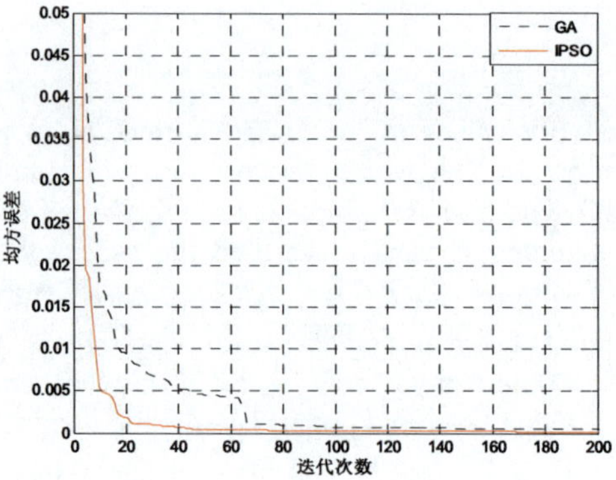

Fig. 4. MSE curve of Network training algorithm

From Figure 2 and Figure 3 shows, IPSO network tracking accuracy is higher than GA network. Simulation results show that model of the thickness system with the IPSO neural network can better reflects the actual condition. Model error is less than 3%, to meet the engineering requirements. From the control point of view, to avoid the characteristics of the controlled object for a number of assumptions and solving complex problems of nonlinear equations thickness prediction model established by use of IPSO network characteristics and faster learning ability to approximate any nonlinear mapping, and high accuracy, can replace the real thickness of the integrated control system, real-time forecasting exports to the thickness of the controlled object. It can be seen from Figure 4, IPSO neural network convergence speeder and higher precision than GA algorithm, and can be more effective for global optimization.

6 Conclusions

A new eccentricity control algorithm based on SGWT is presented in this paper. In proposed algorithm, the lifting algorithm is adopted for eliminating the influence of disturbance and noise. Furthermore, self-adaptive algorithm is also integrated to adjust the control parameters online. The simulation and comparison studies are executed and the simulation results demonstrated that the proposed algorithm has better performance than classical approach. It has less decomposing time and complicity. It is also a practical approach for identifaction and control of eccentricity.

References

1. Shi, Y., Eberhart, R.C.: A Modified Particle Swarm Optimizer. In: Proceedings of the IEEE World Congress on Computational Intelligence, pp. 69–73 (1998)
2. Clerc, M.: The swarm and the queen: towards a deterministic and adaptive particle swarm optimization. In: Proceedings of the 1999 Congress on Evolutionary Computation, Washington, USA, pp. 1951–1957 (1999)
3. Niu, B., Zhu, Y.-l., He, X.-X.: A Multi-population Cooperative Particle Swarm Optimizer for Neural Network Training. In: Wang, J., Yi, Z., Żurada, J.M., Lu, B.-L., Yin, H. (eds.) ISNN 2006. LNCS, vol. 3971, pp. 570–576. Springer, Heidelberg (2006)
4. Li, L., Niu, B.: Particle Swarm Optimization arithmetic, pp. 108–117. Metallurgical Industry Press, Beijing (2009)
5. Xiao, B.-X., Wang, X.-W.: Neural network Predictive control for superheated steam temperature based on modified particle swarm optimization. Control Theory and Application 25(3), 569–573 (2008)

Reform and Practice in Engineering Mechanics Teaching for Wood Science and Engineering Major*

Cheng Pengle[1] and Li Xiaohui[2]

[1] School of Engineering Beijing Forestry University Beijing, China
[2] 296669 troops, Beijing, China
chengpengle@yahoo.com.cn, likerle@sohu.com

Abstract. The assigned teaching period is insufficient for the huge programme of engineering mechanics teaching for wood science and engineering major, and there is no laboratory course. To reform the traditional teaching method of engineering mechanics, new teaching methods were put forward and practiced. The common things in people's daily life are introduced to the class, and relations between the common things and the engineering mechanics are established. Some simple experiments are done during the class to compensate for the lack of laboratory course. According to the characteristics of wood science and engineering major, the teaching method of treating differently and teaching accordingly was put forward. The professional courses are combined with the teachings of mechanics engineering. The integration of theory with practice inspires the students with more interest to study mechanics engineering. The students are trained for innovative consciousness and the practical ability. The students' quality is improved. Many metaphors are used in class to enhance the memory of the knowledge, and to reduce the difficulty of learning the course. The reformed methods can stimulate students' learning interest and can assure the quality of teaching. The teaching methods accord with the main orientation of teaching reform.

Keywords: Quality assurance of teaching, Engineering mechanics, Teaching reform, Teaching methods.

1 Introduction

Engineering mechanics is the combination of theoretical mechanics and materials mechanics. For non-mechanical engineering majors, the main contents of theoretical mechanics and materials mechanics are cut as necessary according with the need of different majors. The remains are engineering mechanics, the professional basic course [1,2]. Engineer mechanics not only relates to mechanics, but also broadly contacts project practice. In our daily life, we often handle something with engineer mechanics. The students of wood science and engineering major learn the basics of wood structure, how wood is produced, and how it can be combined with other

* This work was supported by Beijing Forestry University Young Scientist Fund (No. BLX2008012).

W. Zhang (Ed.): Advanced Technology in Teaching, AISC 163, pp. 369–375.
springerlink.com

materials. The students can then elect to focus on wood design and craftsmanship ranging from hand tooling and sculpture to artistic furniture design. Engineering mechanics is a basic specialized curriculum of wood science and engineering major that has strong applicability, which is an essential base for students to learn the following courses well and improve themselves further. The quality of the results of learning engineering mechanics, will influence the degree of mastering of a lot of following professional courses directly, and influence students' whole quality. So, it is of great significance to improve engineering mechanics teaching [3].

Engineering mechanics is a course with theoretical and practical application. Because the course contents principle is strong, the concept is abstract, the student is widespread to exist a study to comprehend a difficult problem. While the teaching hours are reduced to 40 hours for the reform of education and teaching, the contents of the course are rich and various. That is linked to learning difficulties of the students. The illustrations of the examples and practices demonstrated in text book are machinery components, such as turning tool, jack and belt pulley, etc. It is difficult to understand the machinery components for the students of non-mechanical engineering majors. Thus the teaching of engineering mechanics is far away from our practical application, then students consider that it is useless to learn engineering mechanics. The students' study enthusiasm is dampened and learning disabilities become more frequent. In order to solve these problems, combined with the teaching experience for years, some new methods were put forward in this article which proceeded from the features of engineering mechanics.

2 Daily Life Examples in Engineering Mechanics

Interest is one of the motive forces to improve thinking abilities, enhance learning and result in skill and knowledge. The advantages of intrinsically and interest-motivated learning appear in improved cognitive, emotional and personal outcomes as well as in the learners' identification with certain content areas of their studies [4]. Many daily life examples relate to engineering mechanics. If the examples can be analyzed and concluded their law, the study interest and the enthusiasm of students will be aroused. Some proper examples are selected according to the content of courses. Principle and concepts are introduced with heuristic method by observing and analyzing the examples. The cognition of the examples by students is improved from perceptual knowledge to rational knowledge. The rational knowledge is used to analyze and solve problems in practical engineering, so students will find all sorts of creative uses beyond just this case [5].

2.1 Torsion Test of Brittle Material

Experimental teaching is very important in cultivating the practical and scientific research ability of students, and it can train their observation ability, operate ability, analytical ability exploring ability and creation ability. There are no experimental hours in engineering mechanics of wood science and engineering major according to the curriculum standards. Students consider it difficult to understand the mechanical property of material. The mechanical property of ductile material and brittle material is often confused.

Fig. 1. Turning rupture of brittle material

Blackboard chalk is a brittle material and is very easily obtained in class. One time I asked my students what shape is the fracture surface if I twist off the chalk and students gave all kinds of answers. The chalk was twisted off and the shape of the fracture surface was a helical surface with a 45° along the axial direction (Fig. 1). The chalk was twisted off again and the shape of the fracture surface was same too. This shows that fracture surface of the brittle material follows a certain pattern. Then the reason of the phenomenon was explained to students. The maximum tensile stress resulting from the state of pure shear stress you apply occurs on planes at 45° relative to the chalk's axis. Brittle materials such as chalk are weak in tension, and you will observe that the chalk fails along a clearly defined 45° spiral line. The test is simple and easy to do, but it can make up the flaw of no test hours, and can raise the study enthusiasm of students. The test deepens understanding of the mechanical property of brittle material.

2.2 Stress Concentrations of Daily Life Examples

Either intentionally, to satisfy specific design requirements, or as a result of processing or manufacturing flaws, structural members may have geometric features that result in localized regions of high stress called stress concentrations [6]. Taking a bar of rectangular cross section containing a hole as an example, stress concentrations are introduced in textbooks. The example is not common in daily life and is not easily demonstrated, so students can hardly understand the conception of stress concentrations. A seasoning bag of instant noodles is taken as an example in class. The seasoning bag has serrated edges, so it is easy to be opened. If the seasoning bag did not have serrated edges, it would be hard to be opened. The example shows that there is stress concentration factor at the serrated edges. Students can understand the conception of stress concentrations easily through such example. The conception of stress concentrations can impress students deeply.

2.3 Application Instance of Ductile and Brittle Material in Daily Life

A notable feature of brittle materials is that their ultimate stress in compression is considerably greater in magnitude than their ultimate stress in tension. But the feature of brittle materials is that their ultimate stress in tension is considerably greater in magnitude than their ultimate stress in compression. If the two kinds of materials are combined, the ultimate stress in tension and compression of the combination materials will improved greatly. Students are shocked by the loss and casualties in the earthquake. As most of the people who die in earthquakes are killed by falling buildings, building structures must be improved so that they can withstand the power of earthquakes. I explained why these magnificent tall buildings were made of concrete and steel. Concrete is a brittle material, so its compressive strength is high, but tensile is low. Steel is a ductile material, so its tensile is high, but compressive strength is low. The combination materials of concrete and steel used in these magnificent tall buildings have high compressive strength and tensile. Ordinarily, buildings mainly load pressure. Earthquakes caused many tall buildings to sway, and the walls of the buildings loads drag force. If there was no steel in the walls, the walls would collapse by the drag force. Students on this course will acquire an in-depth knowledge and understanding of the characters and applications of the ductile and brittle materials.

According these examples in daily life, students feel that mechanics knowledge is everywhere. Students will break through the traditional thinking pattern that engineering mechanics is very abstract. They can stimulate student's interest in engineering mechanics study.

3 Treating Differently and Teaching Accordingly Based on Major Character

The teaching method of treating differently and teac Major Character hing accordingly derives from Confucius. The doctrine is developed by Zhu Xi, a famous thinker and educator in southern Song dynasty. The doctrine is the guidance of Zhu Xi summed up the experience of students reading and study. It is an important component of his educational thought [7]. It is important for teachers to transmit wisdom and impart knowledge and explain the puzzle. The students of wood science and engineering major learn the basics of wood structure, how wood is produced, and how it can be combined with other materials. Their knowledge of machinery is negligible, but many mechanical components are taken as study objects and analyzed with mechanics methods in engineering mechanics textbooks. So students are absolutely indifferent to these examples. If some examples in wood science can be introduced in class, the enthusiasm and creative thinking of the students will be aroused.

Fig. 2. Frictionless pin and hinge

The frictionless pin in hinge such as Fig. 2 is used and analyzed in the lecture on reactions equivalent to a force of unknown direction and magnitude. Students have great difficulty in understanding it. A folding fan was introduced and analyzed in my class. There is no direct interaction between each fan rib in folding fan. The fan ribs are held together by the fan pin. There is only interaction between fan rib and fan pin. From the example of folding fan, students can understand the character of pin in hinge: there is no interaction between each object on the bin and each object is only related to the pin.

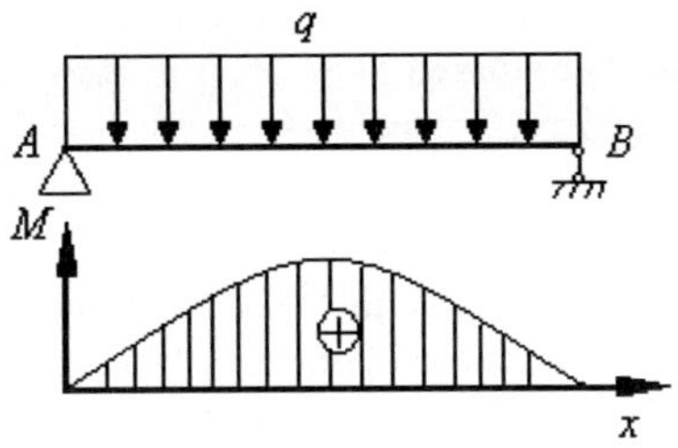

Fig. 3. Bending moment diagram

The beam in Fig. 3 is loaded by uniform distributed loads. Students have difficulty in understanding the uniform distributed loads, so bookcase is introduced in the class. Bookcase is a piece of furniture that carries static loads.

There are many loads that influence the deflections of the bookcase shelves. One of the important loads is pressure stress. The degree of deflections depends on loads, constraining factors, dimensions and character of the materials [8]. The books on the shelves are equal to the uniform distributed loads on the shelf. Through the reasonable simplification of the bookcase mechanics models, engineering mechanics and professional course are associated. The method how to use the theory is introduced, and application of the theory and method is showed.

4 Appropriate Metaphors of Engineering Mechanics in Daily Life

Memory exists in all human learning, and the cultivation of every skill is concerned with memory. There is a lot of content in engineering mechanics. There are many conceptions, theorems and formulas to be remembered. If they are not remembered, the students studying effect will be influenced. If some appropriate metaphors of engineering mechanics in daily life can be introduced in class, the memory of students will deepen and the learning confidence of students will be improved.

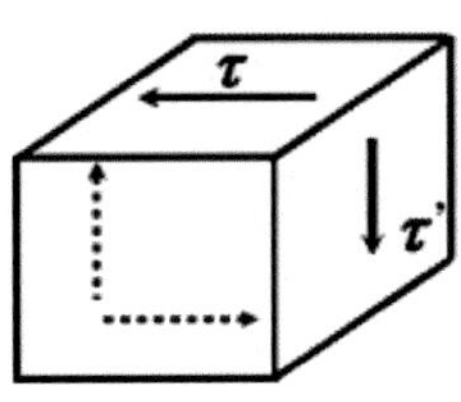

Fig. 4. State of shear stress on an element

Fig. 4 shows the state of shear stress on an element. Theorem of conjugate shear stress is "Shear stresses on perpendicular planes are equal in magnitude and have directions such that both stresses point toward, or both point away form, the line of intersection of the faces". The description of shear stress directions in textbooks is not easy memory. The arrow of the shear stress is compared to a head of a man and the other side of the shear stress is compared to a foot of a man in my class. The sleeping postures of students in their apartments are head to head or foot to foot. They are the same with the directions of shear stress. Through the metaphor, the distribution regularities are not easily forgotten by students.

Relations between the shearing force, the bending moment and the external load are very complex. The load state on the beam is shown as Fig. 3. They are uniform distributed loads and the characteristics of their moment diagram are curves. The shape of the curve in Fig. 3 is tomb-like. If the uniform distributed loads are in the opposite direction, the shape of the curve will be basin-like. Students often get basin-like and tomb-like mixed up. The uniform distributed loads are compared to rain and the shape of the curve is compared to umbrella. If the rain is top-down, the umbrella should be tomb-like to keep off rain, but if the rain is down-top, the umbrella should be basin-like to keep off rain. These vivid and effective memory methods can help to get a deeper understanding and enlighten minds. They also can reduce the difficulty in classroom teaching and stimulate students' learning interest and so on.

5 Conclusion

The assigned teaching period is insufficient for the huge programme of engineering mechanics teaching for wood science and engineering major, and there is no laboratory course, so students have difficulty in learning engineering mechanics. According to the characteristics of engineering mechanics, some reform and practice teaching methods in engineering mechanics teaching for wood science and engineering major were put forward in this paper. Some materials that can easily be obtained in daily life, such as chalks and seasoning bags of instant noodles are used in the class to make up for the inadequacy of laboratory hours. The metaphors of umbrella and sleep are introduced in the class to reduce the difficulty of learning engineering mechanics. Through the reform and practice teaching methods, the students are trained for innovative consciousness and the practical ability and their quality is improved.

References

1. Liu, S.-J., Qu, J.-L.: The study and practice of a engineering mechanics textbook. Journal of Shaanxi Normal University (Natural Science Edition) 34, 209–210 (2006)
2. Zhao, D., Lu, Z.Y.: The teaching reform of engineering mechanics according to major character. Forestry Education in China, 49–50 (May 2001)
3. Zeng, L.S.: Engineering mechanics teaching of mechanical major in forestry school. Forestry Education in China, 50–51 (January 2001)
4. Lu, K.-C., Luo, Q.-S.: Creation studies course. China Building Material Industry Publishing House, Beijing (1997)

5. Chen, T.Z.: Application of engineering case lead-in approach to the teaching of engineering mechanics in colleges. Journal of Shazhou Professional Institute of Technology 10, 43–44 (2007)
6. Bedford, A., Liechti, K.M.: Mechanics of Materials. Prentice Hall, New Jersey
7. Wang, Y.M.: The importance of treating differently and teaching accordingly. Journal of Liaoning Educational Administration Institute 26, 39–40 (2009)
8. Guo, R.-X.: The only way to develop furniture materials - fibreboard light type. Furniture & Interior Decoration, 25 (March 1996)

Psychoanalysis and Educational Countermeasures for the Fickleness of College Students Learning

Wei Yingzi[1] and Gu Kanfeng[2]

[1] School of Information Science and Engineering,
Shenyang Ligong University, Shenyang 110159, China
[2] Advanced Equipment Research and Design Center,
Shenyang Institute of Automation, Chinese Academy of Science,
Shenyang 110016, China

Abstract. Fickleness psychology is a common problem among college students recently. The main expressions are blindness, being eager for quick success and instant benefit, content with superficial understanding but ignore subsequent digestion of knowledge. Consequently, college specialty teacher should give subject courses in accordance with psychology guidance. Both the active and passive effects should be considered carefully. College education has the obligation to guide students to have a definite object in view, to teach students in accordance of their aptitude. These strategies are adopted in order to help students overcome the fickleness psychology and form a habit of active and religious learning.

Keywords: fickleness psychology, educational countermeasures, university teacher, college student.

1 Introduction

College students' learning psychology means the psychological activities of students in the learning process which mainly includes learning motivation, learning attitude, study habits, etc. Student's mental state will have positive or negative effects on course teaching. Therefore, in order to enable students to develop in the positive and favorable psychology direction, teachers should always concern about and comprehend students' learning psychology during their teaching process. This is significant for the success of teaching. It is also one of the symbols of teaching success.

Recently, fickleness psychology is a common problem among college students. They are always eager for quick success and instant benefit, content with superficial understanding but ignore subsequent digestion of knowledge. The word "fickleness" is explained for hastiness, impatience in <cihai>. The emotional expressions of college students' fickleness psychology are impatience mentality, eager for quick success and instant benefit. These students are more anxious and blind in action while compared with others. College students' learning is a complicated mental work and also an individual psychological activity. However, college students are in the transition period of psychological quick development to mature but not real mature. The college

W. Zhang (Ed.): Advanced Technology in Teaching, AISC 163, pp. 377–383.
springerlink.com

students' psychology always performs the characteristics with the contradiction and multiplicity. In the cognition, they have the intense desire for seeking knowledge, but also have the problems of ambiguous learning goal and insufficient understanding ability. They have strong sense of competition, but lack the ability of self-control. Thus, their action of study is often prone to seek quick success and instant benefits without long-term goals. They are blind arrogance, weak-willed, they either have grandiose aims but puny abilities or have strong dependence fickleness psychologies. On the other hand, little attention has been payed to psychological diathesis cultivation in primary and middle school education for a long past time. Teachers tutor their students in person and students are subject to many details. That leads to the malnutrition of students' psychology. It is hard for them to join up the strong independence and autonomy of university study. As the consequence, part of students can not adapt to it.

2 College Students' behavior Characteristics of Learning Fickleness and Its Causes Analysis

College students' learning fickleness in behavior performance basically has the following characteristics: (1) blindness (2) seeking quick success and instant benefit (3) changing about dissimilarity (4) content with superficial understanding.

2.1 Blindness

Blindness refers to the fickleness. It is the psychological states of losing learning goals, unknowing about where to go, mood uneasy, lack of the spirit of down-to-earth work. When studying, learning a foreign language is thought as the most important thing, sometimes. Mastering computer is more practical or their major subject is fundamental, sometimes. Thus, there is not a clear orientation to learning goals, so they cannot focus enough energy on studies. These factors seriously affect their development of levels and capabilities. Among current college students, a saying of "a diploma, a variety of certificates" is rampant. According to the statistics, among those who participated in obtaining certificates, the proportion of school students accounted for more than 50%. Obtaining certificates is mainly for "charge" and set foundation for future vocation re-design. It is thought that one more certificate means one more living "capabilities". It looks like a planned performance of take precautions. However, it is not truth when researching the case in detail. The costs students paid for obtaining certificates are huge. According to the statistics, during four years' college time, students take an average of 2000-3000 yuan RMB for obtaining certificates, or even near to ten thousand RMB on top. To obtain certificates increases the financial burden of college students. A more serious problem is that it interfered with students' normal learning in school. Actually, real value of some certificates is not high nowadays. Certificates don't equal to high overall quality. It is not wise for students to waste much valuable time on certificates instead of learning major subject or getting training opportunities. The upsurge of obtaining certificates reflects the college students' concerns for their own future and their confused states of mind. In the trend of appreciation of increasing knowledge, the National Postgraduate Entrance Examination continues to heat up. Additionally, as the employment difficulties of graduating students are increasing year after year, coincides with our country expand massively graduate students, it provide more competition opportunities for candidates. As a result, many college students blindly participate in the National Postgraduate Entrance Examination army without considering their own strength. Unfortunately, it may not only make them succeed, but also affect their own professional learning, or even their graduation.

2.2 Seeking for Quick Success and Instant Benefit

The students with impetuous attitude in study usually appear eager to the pursuit of the effectiveness and immediate interests. It leads to the lack of ambitious goals in life. This is actually a speculative mentality in study. The specific performances are muddling along in study, cheating in the tests and copying previous research fruits in graduation design. Its essence is “Made a fortune drilling policy loopholes”, “Lottery to get rich”. This is the refraction of social impetuous attitude of seeking for quick success and instant benefit in schools. At present, this cheating atmosphere which generated by fickleness is unfolding college campus. A lot of students seem to be no clear learning objectives after the college entrance examination. They are stumbled at the university. They do not know what to do, just dawdle away their precious youth. As a result, they lose initiative of learning, absences become their commonplace. Usually they do not study hard, do not do research. But their vanities are strong. They don’t want to fail in the exams, so the only way to pass the exams is to cheat. Many postgraduates don’t study hard, all what they want is just to pass the exams. They even showed impatience to exams.And some students said: “If you do not want to go abroad, 90 points and 60 points are not different.” This is a strong utilitarian psychology form of examination-oriented education. In addition, impetuous people are often snobby. Most of them want to get immediate benefit. They do part-time jobs just for money, which is a waste of too much learning time. Just as some students said: "it is impossible to drive a BMW car if you want to learn physics well ", that is a terrible dislocation of the value. Because of these psychological states of quick success, college students cannot really stop to study. The atmosphere of quick success has seriously affected the campus learning environment. A necessary result is that the quality of teaching in institutions of higher education has compromised.

2.3 Changing about Dissimilarities

Nowadays, few people can resist the temptation of money. College students could not withstand the temptation of the wonderful outside world. They have less self-control and lack perseverance and so on. Some students fry a stock, some students go to work, some students engage in business, a few students are willing to engage in learning on the bench. "Student Venture" is a hot topic in modern society. Many students are misled too much by some media which exaggerate the benefit of running a business. In fact, start an enterprise is not realistic for most of college students. Only a few who have special talent and special opportunities can succeed in business. Take the life sciences profession for example, it is impossible to start a business with the accumulation of two or three years learning. Not to mention that you even do not have the most basic knowledge and ability.

2.4 Content with Superficial Understanding

University study is extensive and profound, which requires students not only can realize the image of the surface and the external relations of things, even further to the information to be thinking, analysis, comparison, comprehensive, abstraction and generalization, leading to the formation of concepts. Fickleness psychology makes students cannot concentrate on what they are learning, so they cannot understand the content well. It makes one only content with superficial understanding. They just glance over hurriedly when reading material. To learning content without thorough understanding is an important performance of fickleness psychology.

3 Improve the Teaching Quality of Specialized Courses and the Education Countermeasures

The students' learning motivation, attitude and habits are formed from influence of social living conditions and education. Different social living conditions and education on students' learning have different requirements, so the reflection of learning motivation in students' mind is very complex and various. That's why students' attitudes towards study and learning habits are different. In reality, most of the students learn knowledge is mainly due to the exams for postgraduate schools or graduation test pressure. They are not really aware of professional knowledge or professional knowledge value itself. This effects students' learning enthusiasm and self-consciousness, difficulty solving ability as well.

As one of the psychological phenomena, fickleness makes student pay more attention to denotative than connotation, more attention to quantity than quality, more attention to surfer than actual and more attention to short-term than long-term. Formalism and deception expand their influence by fickleness. It's the opponent to diligent, stand on solid ground and fair competition. Fickleness psychology can make students lose themselves, make people drift with the current, and act blindly. This is extremely harmful to the individual and the society. University teachers have to try various devices to help students overcome this weakness. For this purpose, this paper gives some suggestions.

3.1 To Help the Students Establish the Correct Value Orientation

Using various teaching means to display the professional glamour and show the important status and role of the subject knowledge in the course of human development in classes. Enable students to fully understand the professional by promoting the harmonious development of man and nature to characteristics. For example, teachers can introduce the forefront of professional domain knowledge which related to students life and future practical work. Introduce the present research direction or research progress, also the difficulties in the research, and these studies of great significance to human society.Through these presentations, stimulating students' intrinsic motivation, encouraging them to foster lofty aspiration.

3.2 The Proper Use of Reward to Encourage the External Motivation

Rewards should focus on spiritual rewards than material rewards. Today, college students are independent. They need emotional communication, the encouraged words from teachers, admiring gaze, positive evaluations, equal and harmonious teaching environment and mutual cooperation between the individual. These can help the students to establish confidence, to effectively mobilize the students' enthusiasm, initiative and participation to stimulate learning motivation.

Adhere to the people-oriented and the implementation of quality education principle in the specialized teaching process. In the student's education, we should reform the content, innovation method, create a harmonious, democratic, equal relationship between teachers and students and the educational environment to make sure that students perform their subjectivity. Encourage students' active learning, active participation and independent choice. We should consider their own development as well as their needs. Provide opportunity for each student to think, to create and chances of to succeed. Guide students to participant the class activities actively and review previous course content by asking some questions. This can let the students sum up just learned points of knowledge, learn to prioritize and grasp the main points. Teachers also worked for students to create the chance for success, let them find their progress in the study activity, experience the joy of success, enhance self-confidence and ambition.

We can properly come up with some slightly difficulty questions, let the students who can solve the problem teach others on the platform. If we do it this way, in one hand, student himself can show full play of his expressive force. On the other hand, teacher can tell the positive aspect promptly, point out the insufficiency. This method can also motivate other students. We have to make sure that each college student to experience the joy and happiness of university life. Only in this way, the quality education can be implemented.

3.3 To Stimulate Interest in Learning

Ancient Chinese educators said: "the man who knows is not as well as the one who likes to know, the man who likes to know is not as well as the one who interests in knowing". Interest is the best teacher. Students' interest and love for profession is an important factor which can influence the effect of the learning and a powerful motive which can promote students to learn. Once the students lost their interest, learning will become their heavy burden. They can produce conflict or antagonism and reverse psychology. They may fall into anxiety, fear and disgust on learning. Teachers should adopt flexible teaching methods and various teaching means to display the specialized charm, to stimulate the students' interest and to attract students to participate in class activities actively. Because of the students' nature of curiosity and inquiry, they interest in the specialized experiments. Teachers should make full use of the experiment to encourage students to explore. They can find the problem and conclusion in the experiment. The experiment can meet the students' curiosity and thirst for knowledge, effectively improve their learning interest and experimental skills. For example, teachers can use MATLAB software to demonstrate simulation process of professional knowledge application in the professional course or profession basic course of the electronic information or automation directions. They should connect the specialized theories with application link. Teachers can also make the students to personally write and demonstrate software in classes. Students will have a feeling of using what they learned in classes. These can increase students' interest and help students overcome the fickleness and blind psychology.

3.4 Helping Students to Develop a Rigorous Learning Attitude

University disciplines and professional knowledge are a complete system. It includes different concepts, terminology, formulas, formulas and theorems. This system is established which always follows from simple to complex, from shallow to deep and from primary to high order. Therefore, university knowledge system itself has a tight integrity. As a college student, if he treats the knowledge with an impetuous attitude, he only learns some fragmental knowledge, he cannot step into the temple of knowledge. As China's famous scientists Lu Jiaxi said: "science has its own laws, there is no shortcut to attain the highest level in one step. We should stand on solid ground, work steadily and develop the rigorous scientific attitude. We would rather slow down our step to comprehend and make sure the truth" Therefore, college students should cultivate the rigorous learning attitude. At the same time, we should evaluate ourselves frequently to make sure if we really understand the knowledge and skills are really formed. Make sure if memory, understanding and application are firmed accurately. If we do this, we can effectively prevent the "only know the superficial knowledge, little is known about, solution is wrong" fickleness phenomenon.

4 Conclusions

University teachers should be good at thinking and active in the face of today's college students' fickleness phenomenon. Teachers should understand students' psychological state in the teaching process. Grasp the essence of the problem and give subject courses in accordance with psychology guidance to avoid students' negative factors. Both the active and passive effects should be carefully considered. College education must lead students to have a definite object in view, teach students in accordance of their aptitude. The teaching reform practice which implements people-oriented will play a positive role in promoting the teaching quality of specialized courses in universities.

References

1. Tao, G., Wang, X.: University Students' Learning Psychology. East China University of Science and Technology Press, Shanghai (2003) (in Chinese)
2. Yu, G.: Research into the college students'flippant psychology in study. Journal of XinYu College 10(3), 113–115 (2005) (in Chinese)
3. Bian, J., Zhang, G., Chen, C.: A Guide about College Students Learning Method. Harbin Institute of Technology Press, Harbin (2003) (in Chinese)
4. Li, N., Wang, X.: To master learning. Southeast University Press, Nanjing (2001) (in Chinese)

5. Brown, S.L., Eisenhardt, K.M.: The art of continuous change: linking complexity theory and time-paced evolution in relentlessly shifting organizations. Administrative Science Quarterly 42, 1–34 (1997)
6. Dong, C.: Research into the college students' flippant psychology. China Adult Education (3), 68–69 (2007)
7. Wang, W., Zhao, C.: College teacher, from scholar to great master. Modern University Education (2), 11–14 (2009) (in Chinese)
8. He, B., Peng, Z.: The causes of impetuosity psychology of the college students and preventive measures. Ata of Culture and Education (4), 29–31 (2007) (in Chinese)
9. Stephens, D.: Human impulsiveness selected for by for aging liftstyle. Future Pundit (August 2004)
10. Roger, R.: Psychoanalytic Training: Retropective Assessment by Practitioners in the Field. Master Dissertation of University of Toronto (1999)
11. Huitt, W.: Success in the information age: A paradigm shift (1999), `http://www.edpsycinteractive.org/papers/infoage.pdf`
12. Kennedy, M.: Inexact Sciences: Professional Education and the Development of Expertise. Review of Research in Education, 133–167
13. Resnick: Education and learning to Think. International Journal of Education Research 13(1), 41–46 (1989)
14. Palincsar, A.S., Stevens, D.D., Gavelek, J.R.: Collaborating with Teachers in the Interest of Student Collaboration. International Journal of Education Research 13(1), 47–53 (1989)

The Application of Mordern Information Technology in University Daily Management

He Dan[1], Xie ChengWei[2], and Yang Fang[2]

[1] Mechanical and Electrical Engineering College,
Hebei Normal University of Science and Technology, Qinhuangdao, Hebei, China
[2] Environmental Management College of China, Qinhuangdao, Hebei, China
[3] Physics and Chemistry College, Hebei Normal University of Science & Technology,
Qinhuangdao, Hebei, China

Abstract. University daily management is a very important subject, the daily management throughout the life of each student every day,if we don't manage the students's daily management, it will undoubtly affect the normal teaching and other activities. With the rise of information technology in the daily management of our colleges and universities, the application of modern information technology will save us a lot of unnecessary red tape links,so that we can establish an effective communication mechanism with sutdents and class teacher directly,and it plays a catalytic role.

Keywords: Daily management, mordern information technology, communicate.

The daily management in university is of great importance in the daily management,it's the basic of other teaching education and practice. It's a broader scope, such as student attendance, student domitory's check,the concrete statistics of every student where shall they go when holiday and when will they be back,and critisize and educate the students who debayed the school rules,plan and prepare for the graduation ceremony,arrange the students who want to work-study in school ,regist the students's card and so on. These efforts are relatively trivial, but also are essential and equally important, the daily management of colleges and universities that allows students to form a rule, they are easy to form a correct world outlook, to develop good habits,and it has a positive role in the growth of the students.

Students who just enter into college will feel unsuitable, university teacher will not meet students every day to urge them to study hard, so we should do the daily management from the college level and student level. We should establish a good communication mechanism between school with the class, the teacher with class leadership, class leadership with other students.

1 Establish an Effective Information Network

For better and more effective communication,better daily management in the university,we should establish an effective communication network, the main network can be divided into the following three levels.

W. Zhang (Ed.): Advanced Technology in Teaching, AISC 163, pp. 385–388.
springerlink.com

1.1 Information Network between the Office Serve for Students and Class Teacher

Class teacher has a a better understanding for each student, schools and colleges need to issue a notice,so we can inform the class teacher directly, and then notify the class leadership by the class teacher and students, it is convient for the message passing.

1.2 Information Network between Class Teacher and Class Leadership, the General Student

Class teacher need to achieve a comprehensive grasp of class leadership and ordinary students 's study,life,work, emotion and other information, so it's necessary to establish the network, it's benificial to the class teacher to manage daily management.

1.3 Information Network between Class Leadership and General Student

Class leadership requires the daily management of the class students, so students should establish a regular class leadership and information network.

2 The Application of Mordern Information Technology in University Daily Management

2.1 The Establishment of the Department (School) Class Public Mailbox

College need to establish the public mailbox for class teacher, we can send the emails to the mailbox when there's some recent things to apply, or appropriate form; we can send the emails including the number of application notes, a variety of forms, notices, actions, appraisal results etc to the public mailbox, students can log in the mailbox and view the letter; at the same time, each class can set up a public mailbox to send e-mail for daily management of its class, such as the use of class fees, the material and arrangements of group activities, the subject of the class class meeting .

2.2 Fection Contact

One of the biggest difference with other chat tools is that it is a software associated with the phone, it can send free instant messages from compurer fetion client to the mobile user, even if the moblephone powers off, as long as the fection is in the registry , you can receive the message, so you can inform the class teacher meeting place, the recent work to be accomplished (the identification of poor students) etc in this way, you can also inform the person who is in charge of the class specific issues. You can choose to send a group message, also choose the designated contact, can also choose to send messages in the regular time by fection software. It's convenient and eliminate the trouble calling the class teacher and class leader.We should require the person who are informed must send a text message, those who did not respond, we will contact them with the person who doesn't join fection though telephone notification. We can take the mailbox as the addition of fection, everytime after the fection contact, we can sent the concret messages and information to the department(school)'s mailbox, class teacher's mailbox, then class teacher send the email directly to the class mailbox.

2.3 QQ Group with the Nature of Discussing

QQ is a chat tool of a widespread use in contemporary society, you can create school teacher's daily student management group to exchange ideas,the teachers who are responsible for daily management in each college can join this group to discuss student problems and countermeas in the daily management and exchange experiences to each other. At the same time the school office also can be added to this group, the messages from chool level, can convey through this group, each college receives the message need to reply received words,it also saves time and money and facilitate the daily management.

QQ group can be established in all classes, it's convenient for the class teacher and class leadership to inform things, which can also be written in the group notice. General student who can use the mobile Internet can receive the messages.We need to creat a rule,you're forbidden to send unrelated pictures.I know that each class has a group, but some students who have computers always send a large number of useless pictures, so now most of the students without a computer choose to reject group messages because of monthly net traffic limited , so every class teacher is sure to inform students to develop good habits, do not send useless words and pictures in the class group,this group is only uesd to notice things and the class information communication.

2.4 The Establishment of Classes BLOG

BLOG is a popular online diary, each class can create a BLOG, each student has permission to log in,they can record the things that occur in daily life and learning,you can give them cheer before the exam,you can also record the day moving, learning experience, experience of holding activities, etc, so that each student can express his own ideas, other students can give comments, encouragement and support, class teacher can keep abreast of students thinking, learning state, also the class teacher can find the bad tendencies and resolve the promblems promptly.

2.5 The Institute 's Comprehensive Public Website

Improve the Institute 's public website, which may include the showcase of college activities, the College's publicity and competition and other important notices or arrangements, recruitment information, each time the students log in Institute website, they can see its details.

2.6 The Establishment of Hospital-Wide Student Information File

Develop a hospital-wide electronic student database including the basic personal and family information of the class students, award or punishment, duties, learning record and wheather they need the special care and other information. When we need the students's information file, a click can be easily achieved, we can learn all aspects of the student's situation.

3 The Conclusion

Through the application of modern information technology in university daily management, complex daily management of these trivial things has become organized, the relationship between college staff and the class teacher, class leadership, between class teacher and class leadership, other students become more clearer, which facilitate the daily work, a variety of information technology can complement and strengthen each other, we can convey the spirit of our college to the teacher, class leadership, students timely and accurately, thus strengthening the university's daily management and ensuring the normal education and teaching activities.

References

1. Mu, D.: The application of information technology in teaching management. Tianjin Education, 59–60 (2010)
2. Jia, L., Gao, L.: The application of information technology in university educational management. Journal of He Tian Teachers Specialist School 27(5) (2007)
3. Liu, Y., Ma, C., Gao, J.: The application of information technology in university research management. Science & Technology Information 8 (2011)

Review on Innovative Practice Teaching and Its Quality Evaluation System for Software Engineering Specialty*

Yong Zhang, Jianli Dong, Cunhua Li, Zhaohui Ji, and Junming Wu

School of Computer Engineering, Huaihai Institute of Technology,
Lianyungang, 222005, China
zhyhglyg@126.com, dongjl1019@sina.com

Abstract. Based on the analysis of training status of practice innovation talents in Chinese universities, The construction and development mechanism of high-end practice teaching environment and innovative practice teaching system are researched and discussed thoroughly. And then, the evaluation system and monitoring mechanism of practical teaching quality based on the high-end practice environment and innovative practice teaching sisyem are systematically introduced. Through actual teaching practice, thoughts and approach represented by this paper are very effective and important to train practice innovation abilities and entrepreneurial and employment awareness of the undergraduates today.

Keywords: High-end practice environment, Practice teaching system, Software engineering speciality, Practice teaching quality, Practice teaching bases.

1 Introduction

Practice shows that systematic, complete and high-quality practical teaching has an important role in cultivating the comprehensive application ability and practice innovation ability of students. With the development of quality education which focuses on training the creative spirit and practical ability in colleges and universities, the importance of practical teaching is widely recognized and valued in the current Chinese universities, especially in engineering colleges. There are two aspects which must be solved to improve practical teaching. One is to construct the practical teaching environment which consistent with the content, objectives and requirements of practical teaching, especially the high-end practical teaching environment with first-class technology and equipment. Another is to establish practical teaching

* This work is partially supported by the Strategic Cooperation Project of Guangdong Province and Chinese Science Academy Grant #2009A0091100002 and Grant #2010A090100004; Supported by Guangdong Science Plan Project Grant# 2010A040101001 and Grant# 2010A040306001.

This project is supported by Fund of Jiangsu Education Reform Research Project (Grant No. JG09004) and Chinese Association of Higher Education's Special Project (Grant No. 2010YHE009).

W. Zhang (Ed.): Advanced Technology in Teaching, AISC 163, pp. 389–397.
springerlink.com

innovation management and evaluation system to ensure and promote the steady improvement of practical teaching quality. Practical environment is the sites and base which are used for practical teaching such as experiment, practice, training and so on. High-end practical environment is the experimental sites or practice base with advanced technology, equipment and infrastructure. Practice environment constitutes the material basis of practical teaching, only constructing first-class high-end practical environment, it is possible to train first-class and high-quality innovative talents. In addition, to achieve high-quality innovative talents training in high-end environment, perfect scientific management and innovative evaluation system must be set up to ensure high-quality practical teaching [1,2].

This paper mainly researches and discusses the problems of construction mechanism of high-end practice teaching environment and innovation practice teaching system as well as practice teaching quality evaluaton system, etc. for undergraduate practice innovative ability.

2 The Significance of Practice Teaching Environment Construction

In current Higher Education in China, the based experimental environment of undergraduate engineering applied specialties already has a considerable scale after several years of construction, especially after the Ministry of Education's activities of "Evaluation of undergraduate education", the construction of basis professional laboratory have achieve a new level. However, it is the lack of university's internal and external high-end practice environment construction which restrain the development of higher education and the formation of school characteristics. What is high-end practice environment? Here we give the specific definition: high-end practice environment (also known as high-end training and practice base) is experiment, practice, training and other facilities and sites for professional talents which is in social development more cutting-edge technology and could able to meet the needs of culturing current undergraduate students in late high school comprehensive applied abilities and innovative abilities.

In the current university practice environment construction, the central and local laboratory built by government investment and construction, or the experimental center of national and provincial demonstration, played a safeguarding role in improving the practice environment in schools, enhancing the practice teaching level and forming the school characteristics. However, due to lack of government investment and long construction period, the scale of high school practice environment could unable to meet the needs of college practice teaching for a short time. Thus, in a long period, accelerating the construction and development of outside school high-end practice environment is an important model for colleges and universities to carry out practice teaching and enhance the quality of talents training. And, compared with the inner school practice environment, outside high-end practice environment construction has the inherent advantages and features for improving the quality of talent training. Practice outside school has favor in combining theory and practice, not only it can make the students a more thorough understand book knowledge, but also can allow students to think after encounter difficulties in practice, this may encourage students to learn to analyze and deal with the problem and develop their independent analysis and problem solving skills, which can improve their innovation capability and comprehensive quality. Through the outside school practice, we can also reach the purpose of making students contact social as soon as possible, understand the social and training applied innovation talents, so after graduation the students would familiar with their own work as soon as possible. Not only that, the outside school training is the link that contact universities and community, which has benefit to the combined of "production, learning and research", to promote scientific research, make the advantages of the university and practice base complement and to jointly promote the development of industries and economies. In addition, through outside school high-end environment training and practice, the teaching shortcomings which substandard economic and social development could be found, and then we can timely adjust teaching programs, improve teaching methods, enhance teaching effectiveness, as a result of giving full play to its important position and role in teaching systems[3].

Seen from the above, high-end practice environment construction in high school, especially the outside school high-end practice environment construction, has great practical significance to develop the practice creative talents training quality and promote national economic and social development.

3 Practice Teaching Environment Construction and Innovation Practice Teaching System

Practice teaching is the most effective and direct way to training the comprehensive application and practice innovation abilities. The objectives, plans, content, environment and requirements of practical teaching should be clearly developed and implemented according to professional education and the school regulations. Huaihai Institute of Jiangsu Province, located in northern Jiangsu Province, belongs to the economic, scientific and technological less developing regions compared to southern Jiangsu Province. So, because of economic constraints, not only the school should actively construct the inner school practice teaching environment, but also the construction and development of the outside school practice base. The following is the research and discuss of the outside school high-end practice environment construction program and innovative teaching management system that related to software engineering creative talents training.

3.1 Construction Programs of High-End Practice Environment

Objectively speaking, the training of software engineering applied innovative talents must require the support of high-end practice teaching and learning environment. That is, the first-class practice environment, first-class technology and equipment, first-class faculty and excellent management system is the basic conditions to ensure the first-class software engineering talents, especially for the cultivation of high-quality software professionals. So, in the process of practice environment building, we particularly emphasized large-scale integrated high-end practice environment construction. Two main directions should be take attention on, one is inner school construction including central and local governments build together, national and provincial Demonstration Laboratory and so on; another is outside school high-end practice bases construction which are built together with enterprises as well as professional build. The basic principles of internal and external laboratory and practice base construction are high-end configuration, the development of practice running and the property of binding the industry, learning and research as one. These three characteristics are the basic principles of the common property of internal and external school high-end practice environment construction. On the other hand, to the construction of external school practice environment, the enterprise feature of practice-based must be emphasized.

The principle of high-end, stress the advanced of hardware and software environment configuration and equipment of practice environment construction (including laboratory or practice base, et.al.), which could represent the current mainstream technological level existing in social and economic development and professional field. This is very important, especially for the construction of Key Laboratory of the inner school central and local governments building together, the repetitive low-quality basis laboratory must be resolutely stopped. Otherwise, it would raise the widening gap with social development. Opening principle, emphasis on laboratory and practical bases must be able to open to teachers and students, which not only meet the needs of experiment, design, training and practice and other general practice but also meet the requirement of innovation, research and development, competitions and other practical activities; opening to teachers is to become the high-end research and project development platform for teachers; open to the public is to become the practice base provided social technology services and training talents for enterprises. The combining principle stressed that the high-end practice environment must focus on the combination of practical and theoretical teaching, the combination of experimentation, training and research, the combination of practice environment and social technology service, the combination of the construction of practice teaching base and the corporation with enterprises and other construction and operation mechanism. Emphasize on the combining principle is focus on the comprehensive benefits and social potential of high-end practice environment construction and application, and play their due role in many aspects such as practice creative abilities, research, social service and so on. What must be take more attention is that the inner school practice environment must considered co-operation with the modern enterprises, this is very important and urgent not only for the university to absorb and use the advanced technology and equipment resource of enterprises, but also for students to understand the corporate culture or training as well as finding employment opportunities and developing entrepreneurial awareness [4,5].

3.2 The Construction of Innovation Practice Teaching System

To the traditional practice teaching in universities, practical teaching system establishment should focus on practice teaching programs, practice environment and management systems. Practice teaching program have systematically and clearly plan and description in general professional training programs, including course experiment, curriculum design, professional training, social practice and graduation design and so on. For each practice teaching course, its name, general outline, content, assessment methods, achievements evaluation should have a clear planning, they should be archived and issued executive in the form of documents, becoming the fundamental basis of implementing practical teaching and its management. Practice environment mainly include training plan adapting to the experiment, training, internships and other places and facilities, consisting of two types of internal and external school. Generally speaking, internal school practice environment construction give first place to professional basic laboratory, external take training and practice base construction as the principal thing. Practice teaching management system is the standards of implementing practice teaching tasks, including teaching resources management, teaching staff, teaching process, teaching effectiveness, teaching quality, assessment and evaluation, cooperation agreements and other rules and regulations [6].

With the update of professional knowledge and curriculum system, the corresponding practice teaching environment must be ensured. So, in the construction processes of the internal central and local build together lab, we have built software testing, integrated software engineering environment, embedded systems, computer application technology and other high-end training room, at the same time, we introduced and configured the advanced mainstream hardware and software facilities such as Loadrunner, Rational, ARM, .net, J2EE environment and tools. In external high-end practice environment construction, a group of entrepreneurial high-end training base, which have advanced equipment, modern software technology, practice teaching conditions, students practice and innovation direction and provided employment protection mechanism, have been built successively in Nanjing, Suzhou, Wuxi, Lianyungang, Shanghai and other areas.

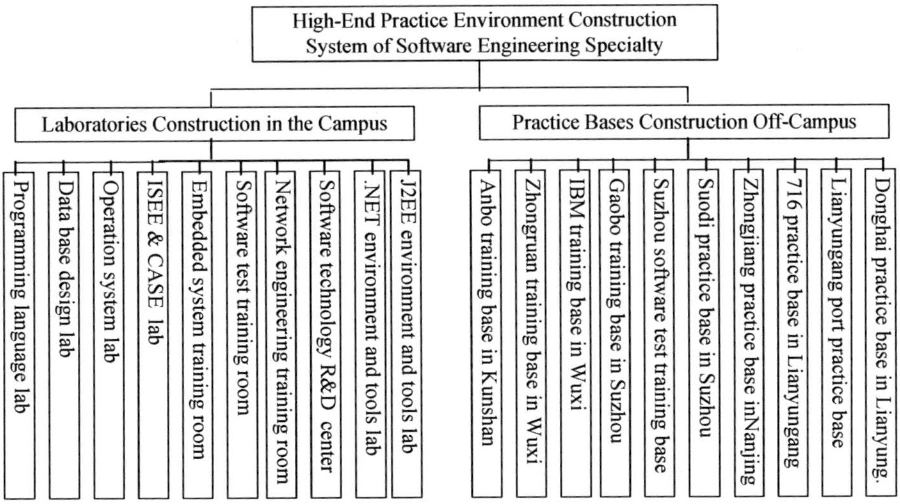

Fig. 1. Construction system of high-end practice bases and laboratries

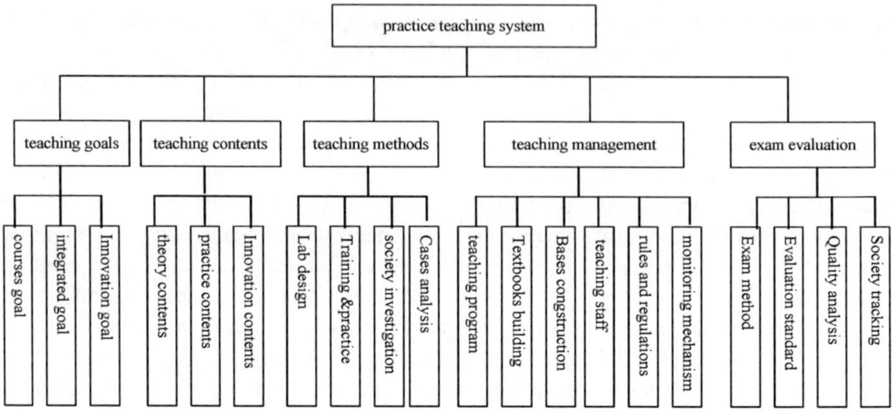

Fig. 2. Practice Teaching System

It does not mean the high-end software engineering professional talents could be trained naturally with internal and external high-end practice environment. It also needs to develop scientific and standardized practice teaching management system. The process and quality control of external practice teaching, which is the bottlenecks of current external practice and training in universities, must be give adequate attention. From overall situation of undergraduate professional talents training, our school researched and developed the management system which consisted with the practice teaching characteristics. In this respect, we firstly guaranteed the quality of external practice base. So, by critical review from nature and development potential of enterprise, infrastructure and technical equipment of base, instructor qualifications, teaching management standard, the base environment safety, quality of student life, size and number of enterprise alliance and other aspects, we choose a number of enterprises which have advanced resource and favorable conditions for running school as the software engineering high-end practice teaching base of our school, and signed long-term cooperation agreement with each practice base to clear responsibilities, rights and interests of each party. The teaching process and quality control of external practice base is an important and complex problem. So, we explore a set of new mechanism adapting to external base management in the corporation with outside base. We study and develop a series of specific management regulations and standards in all respect such as internal and external professional knowledge extension, complementary teaching resources, alternative course credit, practice process monitoring, teaching quality assessment, management system standards, quantitative employment indicators, curriculum and professional build together and so on, which make the cooperation of universities and base have rule-based, legal basis, clearly responsibility, rights and benefits, and finally achieving win-win cooperation.

The construction of high-end practice base and practice teaching system is the prerequisites for finishing practical teaching tasks. Fig 1. is the schematic diagram of high-end practice base construction system for School of Computer Engineering in Huaihai Institute of Technology, Fig 2. is the schematic diagram of practice teaching system. In Fig 1., a very important issues exists in while meet the requirements of the hardware and software equipment and technology of lab and practice base achieving advanced and mainstream, it is also must require external base having a large cooperation enterprises group to provide practice condition for post teaching practice, familiar with the corporate culture as well as employment and entrepreneurship[7].

4 The Evaluation System of Practical Teaching Quality

The evaluation of practical teaching quality, different from the signal assessment of university courses quality or teacher level, is the comprehensive strength assessment of the school conditions and educational levels. Thus, in practical teaching quality evaluation process, its evaluation index system is particularly important. Evaluation index system should not only include the hardware infrastructure construction of practical teaching, such as laboratories, practice base, technology and equipment, teaching staff, but also include software infrastructure construction, such as practical teaching system and programs, syllabus, teaching objectives, teaching management regulations and innovative mechanism. Classification from the evaluation objectives, the evaluation of the courses, teacher and teaching quality in Chinese universities can be divided into two types -- qualified and selected. Qualified is a kind of qualification evaluation in line with the society and industry standards, but selected is a kind of evaluation choosing the best one in the group. But in the actual evaluation system, it is common to use the same evaluation system to carry out the qualified and selected assessment. To evaluation methods, there are qualitative and quantitative evaluations or the combination evaluation. To the main body of evaluation, the most quality assessment in Chinese universities relies mainly on the assessments of expert, leadership and fraternity. Compared with abroad, the student-oriented and student-cantered quality assessment programs are less.

It is very important to implement universities practical teaching quality evaluation and create the evaluation index system which not only can reflects the law of higher education, teaching quality and talents training objectives in China but also could highlight the scientific, operational, practical and innovative of evaluation. In this college practical teaching quality evaluation index system construction process, we firstly developed the creative principles of evaluation index system. These principles include: the scientific development principle in accordance with universities talents training goals in new times, the incentives and innovation principle in accordance with improving practical teaching quality, people-oriented individual evaluation principle in accordance with participating by experts, leaders, colleagues and students together, the coordinated development principle in accordance with full use of social educational resources, strengthening school-enterprise cooperation and combining production and learning, the qualitative and quantitative assessment principles in accordance with easy to implement the fair, just and open evaluation.

In the teaching quality evaluation index system creation process, the factors impacting the teaching quality in teaching process can be attributed to teaching, students, teaching environment and teaching management. So, teaching quality evaluation is generally analysed from these four aspects. It is the quality of teachers, teaching attitudes, teaching content, teaching methods and tools, teaching abilities and other contents that are mainly assessed in the aspects of teachers. On the other hand, in the aspect of students, the factors that be considered mainly include the initiative, positive and enthusiasm of students learning, the improving of the basic skills and practical and innovative abilities, the training level of employment awareness venture, the innovative space and freedom of students and so on. The teaching environment refers to the construction quality, size and equipment level of laboratory and practice base and the ability of supporting practical teaching environment. The evaluation factors of teaching management mainly involve in the provisions of the rules and regulations of practical teaching management, teaching plan, contents and objectives and the development of incentives and innovative mechanisms, evaluation criteria and quality control measures and so on [8].

In the design and construction process of this evaluation index system, we use some foreign student-cantered individual evaluation ideas and indicators as the reference, while drawing on the traditional evaluation index system in China which gives priority to common evaluation factors. For example, in the determination of the subject of evaluation, in addition to reflecting take the experts, leaders and colleagues as the assessment subject, the teaching quality evaluation ideological in USA which take the students as the subjects are introduced. It is obviously embodied in students' evaluation factors and its weight distribution [9, 10]. Furthermore, in the design of evaluation index system, the integrity of teaching process, management, result and other assessment indicators as well as the complementary characteristics of qualitative and quantitative evaluation indicators and other features is fully reflected.

5 Conclusion

Through the high-end practice base construction and practice teaching innovation of our school in recent years, the software engineering professional talents training quality and practice creative abilities have been significantly improved, gradually showing the professional education characteristics of "the grassland, enough stamina". We deeply feel in teaching practice that the training of undergraduate applied innovation talents is must strengthen the high-end practice environment construction and pay attention to the training of comprehensive application and practical creative ability while keeping the advantage of systematic and integrity of traditional theory education, this is our success experience. It has been proved that it is correct and effective to emphasize enhancing the training quality of software engineering professional talents in the high-end practice environment, and the rising of the employment rate and the rate of passing postgraduate entrance exams is the best proof.

References

1. Dong, J., Wang, J.: Research on Innovative Education Thoughts and Teaching Methods for China Universities. In: 2009 International Conference on Information Management, Innovation Management and Industrial Engineering, vol. 3, pp. 379–382. IEEE Computer Society, Xi'an (2009)

2. Dong, J., Wang, Q.: Research and Practice on the Innovative Teaching Mode of Software Engineering Synthesis Design. In: The 1st International Conference on Information Science and Engineering, Track 10, IEEE Computer Society, Nanjing (2009)
3. Yang, H., Wang, Z.: Research in the Construction of Off-Campus Practice Base in the New Period. Journal of Yangzhou University (Higher Education Study Edition) 11(3), 87–89 (2007) (in Chinese)
4. Wang, W., Yao, A., Sun, W.: Research and Practice of the Construction of Off-campus Practical Teaching Base. Journal of Anhui Agri. Sci. 35(29), 9383–9385 (2007) (in Chinese)
5. Zhang, Z., Zhao, Z., Zheng, Y., Wu, J.: Construction and Practice of Practice Base Outside School. Experiment Science & Technology 7(1), 139–141 (2009) (in Chinese)
6. Huang, S., Yang, L., Zhang, Z.: Constructing Practice Bases for Engineering Specialties and Study on New Practice Modes. Journal of Guangdong University of Technology (Social Sciences Edition) 4(4), 52–55 (2004) (in Chinese)
7. Zhang, S.: The Construction and Evaluation of Modern Engineering Professinal Practice and Teaching Basees. Journal of ShanDong Institude of Business and Technology 22(5), 104–109 (2008) (in Chinese)
8. Qian, C.: New Peculiarites and development feature of Teachong Quality Assurance in Univesities. Higher-Sience Education 3(5), 37–40 (2004)
9. Lan, J., Leng, Y., Li, X., Yu, M.: On comparison of teaching quality of American and Chinese universities. Journal of Higher Education 24(2), 96–100 (2003)
10. Meng, F.: Student-oriented Teaching Quality Evaluation System of Universities and Its Construction. Higher Education Research and Evaluation, 84–89 (February 2009)

The Curricula Building for Information Management and Information System Major*

Siqing Luo, Ke Liang, and Jing Wang

Northeast Forestry University, Information and Computer Engineering College, Heilongjiang, Harbin, China

Abstract. Through the analyses of training purposes and development trends for Information Management and Information System(IMIS) major's student, and combining with the national university developmental condition and characteristics, the innovative "3+1" personnel training mode is put forward. Exemplifying the realistic development of IMIS in Northeast Forestry University(NEFU), the effective curricula system that meets the actual demand for personnel training is researched and discussed.

Keywords: Information Management and Information System(IMIS), personnel training, curricula system.

1 Introduction

Information Management and Information System is a major that closely relates to the information technology with the development of it and the increasing information demand. Nowadays a number of developed and developing countries have established the similar undergraduate degree. In China, IMIS is described first in ordinary higher school undergraduate degree catalog by ministry of education, which is composed of economical information management, information science, management information system and forestry information management, and so on. The applied and practical major is attached to management science and engineering class, which includes economics, management science and computer together. With the rapid development of social information, the major's developmental trend and personnel training purpose are established, which integrate three training modes, namely, business administration, information resource management and technology orientation, with such personnel training direction as information system design, development professional personnel, enterprise information management expert, and development and utilization of enterprise information resource etc.

In order to realize training objectives and developmental direction, the corresponding curricula are needed. Curricula are basic for teaching, which directly reflect the training measure who determines the structure and settings of it. Curricula's building is also the important and hard point for higher education reform. Curricula's settings and building

* Supported by Heilongjiang province new century higher education reform project.

W. Zhang (Ed.): Advanced Technology in Teaching, AISC 163, pp. 399–404.
springerlink.com

shall give full consideration to such factor as social needs, science and culture, the growth of new knowledge, the advancement and growth of students, and the school's own characteristics, etc..

2 Personnel Orientation and Training Objectives for IMIS

2.1 Personnel Training Objectives

The IMIS's personnel training orientations and objectives are specified by Ministry of Education, which involve training advanced professional with the theoretical basis of modern management and the knowledge and applicability to computer, as well as mastering system thinking, information system analysis and design, and information management, who can engages in levels of management at the national, industrial and commercial enterprise, financial institution and research institutes etc.. In order to further improve IMIS curricula planning, perspective and developmental requirement in China's institution of higher education, the team "China's Information System Curricula 2005(CISC2005) " is established who comes from Tsinghua University and Renmin University of China to conduct the related survey. Viewed from the findings, there is higher demand for the major's personnel who mainly serve for system development and maintenance. But there is also some shortcoming among it such as vague training orientation and shortage of effective practice session etc.. With regards to this, Information and Computer Engineering College in NEFU conducts a series of education reform for IMIS and proposes innovative "3+1" teaching mode. "3" means completing all degree-based theoretical and professional courses as the syllabus in the first three years. "1" means in the fourth year switching to specifically learn through internship, training or cooperation with the enterprise in accordance with the both-established program to targeted training. After several years of better resulting operation, the graduate students are accepted by the community. This not only adapts to the social personnel need for IMIS, but also develops a complex and innovative personnel training mode to explore and practise.

2.2 Needs-of-Community-Oriented Comprehensive Competence Training

In order to carry through the general objective of "wide caliber, solid foundation and high quality" proposed by Ministry of Education, the personnel of IMIS are oriented as the manager of management-basis, technology-survival and combination management with technology to develop, who are required to master the methods of information system analysis and design, and engage in information management and system's analysis, design, implementation and assessment. This requires the students to have the ability to organize, analyze and dispose information, and to develop and manage information system, as well as to put advanced management methods and ideas into practice.

3 IMIS's Curricula Building

The advantage of IMIS is that it focuses the study of the information composition and distributed characteristics, the information system development and design's theory, principle and means, and information's acquirement, processing, retrieval, control and usage. Simultaneously IMIS must utilize the strong practical and applicable professional advantage, which is the key to obtain the substantial development. In building curricula, not only the training objective and developmental trend are required as the schooling guide, but also both theory and practice are played equal attention to develop application-oriented technical personnel.

3.1 Knowledge System of IMIS

IMIS is a comprehensive applied discipline that includes management, system thinking and information technology. It focuses the theory and methods of the information system building and management and the theory and methods of information resource development and utilization.

The knowledge system as shown in Fig 1

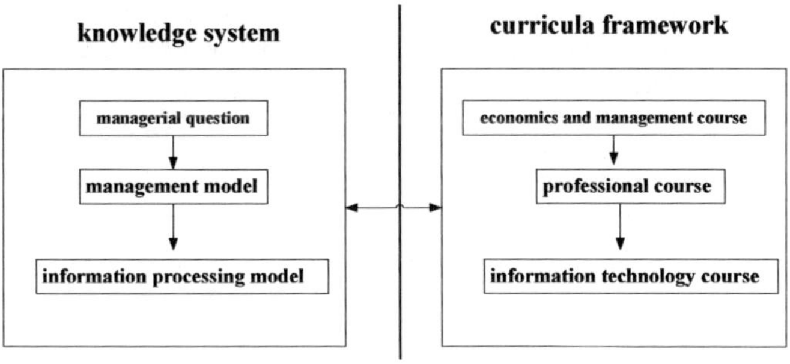

Fig. 1. IMIS knowledge system and curricula framework

The knowledge system can reflect the content of IMIS, which provide the basis for the determination of personnel training direction and curricula setting. Among it, as research background and analysis tools the theory and method of management science is not only the foundation and support of the major but also the premise of software design; system theory and thinking is widely referred for the information system discipline; economics is the background and basis, and the applied foundation to expand; computer technology is to use computer to solve the information-related management question; information technology is the tool during information system development and design. IMIS is characteristic of organically integrating management, economics and information technology rather than simply stacking knowledge. If management or information technology is placed too much emphasis, it will hard to distinguish among them. But in the survey it is founded that the society favors the IMIS personnel who are proficient in computer technology, so the information technology shall play an important role in the curricula to satisfy the social demand. Otherwise the trained student neither can use the advanced information technology tools, nor can satisfy the needs of labor market in the information ages.

3.2 Curricula Setting

The five core courses required by Higher Education Guide Committee of Management and Engineering from Ministry of Education are management science, economics, and basic application of operation research, management statistics and management information system. And the four main courses of IMIS under it are information resource management, database and data structure, basic application of computer network, system analysis and design. In order to better promote the development of IMIS, since 1998 many college and university organize scholars to research the setting of the core course for IMIS. Starting from the training direction of IMIS and referred to the United States IS2002 curricula design thinking, CISC2005 designs China's undergraduate education curricula of IMIS after taking the social demand for it into account. In setting the curriculum, guide as above and combine with the characteristics of the school and teaching model, we complete the curricula building of IMIS.

(1) Foundation courses
Include foundation of computer culture, information technology of computer, English, Humanities course, advanced mathematics, linear algebra, probability and numerical statistics, discrete mathematics, high-level language programming, western economics. Focus on the basic skill and quality and expand the knowledge of student to lay the foundation for the future learning.

(2) Professional courses
Mainly classify: ① professional skill module. Include subject-based English, webpage design and production, computer operation system, computer network technology, Java, J2ee development technology; ② theoretical basis of professional module. Include principles of management, introduction to information management, operation research, statistics, data structure, operation and management, accounting, economics, marketing, organizational behavioral science, introduction to E-commerce. ③ professional orientation module. Include management information system, database theory and application, decision support system, information economics, E-commerce website design, information system analysis and design.

(3) Practice teaching
IMIS is an applied discipline, so the simple theoretical teaching can't guarantee full access to the student's comprehensive ability to solve the practical problem and can't meet the social need of the personnel. Practice teaching plays an irreplaceable role as a key part of teaching. Through practice teaching mode, in a planned and organized manner to teach student's scientific knowledge, way of think and operational skill, which can guide student to create and design. Further improve student quality, develop their potential and personality, and stimulate the student to research actively for the actual issue in depth. At present some schools simulate the practice and practice training in the lab, and run short of the real development background so that the students neither take it seriously nor integrate the software engineering thinking into the system development, and cause the students lack of the ability of discovering and solving question, system analysis and design, and management and cooperation, which is precisely the necessary section for the personnel training. Meanwhile, different practice section are unable to been communicated effectively in many schools, which

cause that the existing regular relations among them are artificially separated, either repeat or dislocate with each other. This doesn't help the training of student's applied and innovative ability.

In order to systematically build the practical teaching system, to strengthen the contact and convergence among different courses, and ask it to dynamically adapt to the changing needs of the development, the system is composed of the below four modules.

① Professional knowledge
In the process of training student, the professional education shall be integrated. Carry on the know-how-to-manage practice in the second or third semester, that is, organize the students to Software Company or better computerized enterprise, aiming to enable students to intuitively understand what capability, knowledge and skill they must have. This allow the students to better know the discipline, and in the future learning as early as possible to determine the focus of the professional ability exercise with some guidance for the later selecting course according to its own career planning and ability characteristics.

② Course and software practice
The design of practice teaching system is close related to the course content. Among it, the professional courses necessary to open experimental teaching are high-level language programming, computer network engineering, object-oriented programming, operation system, data structure, database theory and application, statistics, application of computer technology in the network, operation research, e-commerce development and design, information system analysis and design.

In order to better complete the practice course, the students of IMIS about the system shall have a detailed understanding of the following software: C, QSB+(SAS, SPSS), object-oriented programming, SQL server, apache applied server, java, J2ee, Mysql database server. The purpose of establishing this practice teaching is to provide the basis for the later development and design of MIS and E-commerce.

③ Management system simulation development
In this part, the students will utilize what has learned, namely, the method and technology of system software analysis, design and development to accomplish the design and development of simple MIS in the appropriate context.

The process stresses the application of MIS development method and focuses on software engineering thinking, which enables the students to primary understand and master the system development process, accomplish it through comprehensively utilizing the learned knowledge of computer, management science and information organization. Combined with the specific circumstance of the development process, the teacher shall guide and evaluate.

④ Comprehensive development of the system
The process follows after finishing the first three practice teachings. Synthesize all of the professional knowledge, especially such courses as web technology, database technology, programming language and conduct the management system development based on B/S or C/S, so that the students' ability to develop MIS and solve practical question is further improved.

(5) Internship and training base
Practice teaching is the important part of higher education and the necessary process to training the ability to practice and innovate, even the shortcut to improve the social personnel quality and employability. Practice base choice is the foundational guarantee for practice teaching, which achieve the needs of school's career development and educational goal as well as the interaction with community and enterprise. At present our bases include "Neusoft group", "kingdee international software group co. ltd.", and "Beijing centergate software park" etc.. According to the actual need to arrange appropriate training, either the teacher or the business's staff guides the students into the base to participate in project development and solve the related issue.

4 Conclusion

Curricula building is a complicated set of system engineering. Because IMIS involves plenty of disciplines, during the process of building, systems thinking, integrated idea and structured approach are needed. According to different orientation, it shall be characteristic of focus on expansion towards management science, economics and IT as well as a solid foundation. Emphasis on both theory and practice to enhance the ability to deal, and build the curricula system that is fit for social need and professional development.

References

1. Wang, X., Huang, S.: Study on Curriculum of Information Management and Information System. Journal of Shenyang Normal University(Natural Science Edition) 24(3), 278–381 (2006)
2. Yang, L.: Orientation and Development of Information Management and Information System. Journal of Gansu Economic Management Institute 20(3), 63–65 (2007)
3. CISC2005 Research Group, Curriculum of Information Management and Information System of Chinese Colleges and Universities. Tsinghua University Press, Beijing (2005)
4. Liang, C., Gu, D., Li, X., Yang, S.: Exploration and Practice of the Reform of Undergraduate Education of the Specialties of Information Management. Journal of Higher Education Research 31(1), 64–66 (2008)

Assessment of Intellectual Capital Efficiency in China Transportation Equipment Manufacturing Companies: Using DEA

Yanjie Hu, Xiuli Wang, and Lanxia Yang

School of Economics and Management, Beihang University, Beijing, China
yjhu@buaa.edu.cn

Abstract. In the 21st century, knowledge is the mainstream of the era, intellectual capital has began to play an increasingly important role. Using the data envelopment analysis, the paper assessed the efficiency of intellectual capital in China transportation equipment manufacturing companies. The study shows that about two fifths of the company in the industry is located in efficiency, and there is the room for improvement in the others. The intellectual capital management efficiency of the aerospace equipment manufacturing enterprises is low.

Keywords: intellectual capital, data envelopment analysis, transportation equipment manufacturing, efficiency assessment.

1 Research Background

As the growing demand for knowledge-based products and services is shaping the global economic landscape, the role of knowledge in achieving competitive advantage has become a major consideration in management. So in enterprise the role of knowledge-based intellectual capital(IC) has become more and more important, and in achieving competitive advantage has become a major consideration too. To promote the core competitive ability of the enterprise, the management of IC and the analysis of it's efficiency has important practical significance.

Many studies have conferred upon the efficiency assessment of intellectual capital. At present, efficiency assessment of intellectual capital mainly concentrated on average intellectual capital efficiency coefficient(ICE). Representative research: Pulic's study (2004) finds that the average ICE droped from 3 to 2.5 in European 1000 companies from 2000 to 2002; Bharathi Kamath (2008) analyzes intellectual capital efficiency of the top 25 companies in pharmaceutical industry of India form 1996 to 2006, and the results show that every year the average ICE is around 3.5; Wan Xi (2006) did the descriptive statistics of the intellectual capital efficiency, human capital efficiency and structure capital efficiency of the 41 listed companies which operated best in China in 2003, and the results show that the average ICEs were 6.07、5.36 and 0.71; Yu Haizong and Deng Qian(2007) find that the ICE in high-tech industry of China was

W. Zhang (Ed.): Advanced Technology in Teaching, AISC 163, pp. 405–412.
springerlink.com

5.46 and in textile industry was 5.25; Ran Qiuhong and Deng Huiling(2009) find that different industries have different intellectual capital efficiency, and they also find that intellectual capital efficiency has difference in different enterprises of the same industry.

Although according to Pulic model, ICE can be used to measure the use efficiency of intellectual capital, ICE method is given only industry or the enterprise efficiency, and there isn't criteria to judege whether the industry or enterprise intellectual capital efficiency is in the efficiency frontier or not. Against analysis the causes of the formation of differences efficiency and improve the efficiency. While the Data Envelopment Analysis (DEA) is a kind of typical evaluation method for company performance assessment, which has become a assessment method of company efficiency in recent years. So far, however, the method seldom applied to intellectual capital management, especially in the domestic intellectual capital management research is rare.

Transportation equipment manufacturing industry is foundation industry to provide technical equipment for our national economy and our country's construction. It is the pillar of the national economy, and has important strategic significance. With the ratio of intellectual capital and enterprises' total assets is becoming more and more high, intellectual capital management has become an important topic. This research selected listed companies of China transportation equipment manufacturing as a sample, try to apply the DEA method to evaluate and manage the company's intellectual capital efficiency, so as to provide useful information for the enterprises' intellectual capital management.

This paper is divided into four parts, the first one is research background; the second part is the study design, this part introduces the model design, selection of variables and samples; the third part is the empirical analysis, analyse the intellectual capital efficiency of Chinese transportation equipment manufacturing industry and the change of efficiency. Also to analyse the intellectual capital efficiency of the aviation enterprise; the last one is conclusion.

2 Study Design

2.1 DEA Model

As previous mentioned, Data Envelopment Analysis (DEA) method is a kind of typical assement method for company performance. DEA is a linear programming-based technique, which can be used to determine the efficiency of a group of decision-making units(DMUs) relative to enenvelope(efficient frontier) by optimally weighting inputs and outputs. Additionally, DEA provides a single indicator of efficiency irrespective of the number of inputs and outputs. And the technical efficiency can directly indicate the differences of input-output projects between the DMUs and the best DMU, so it can be more convenient to point out the best way that can improve efficiency. Generally, DEA method can not only evaluate intellectual capital efficiency, but also show that the company may achieve efficiency frontier by improving the input-output. So, this paper uses CCR(Charnes, Cooper, Rhodes, 1978) model and BCC(Banker, Charnes, Cooper, 1984) model to measure the enterprise's efficiency.

2.2 Variables

DEA model is highly sensitive to the selection of input and output variables, so choose appropriate variables for the analysis is very importmant. Combine with Wann's reaearch(2006), and consider the measurable of variables and the availability of data, we select five input variables. They are number of employees, number of patents, intellectual capital stocks(beginning), number of R&D employees and management cost. Then we choose operation profits and intellectual capital stocks(end) as output variables. Among them, the unit of intellectual capital stocks(beginning) and management cost, operation profits and intellectual capital stocks(end) is million yuan.

Can't measurement can't management, the measurement of the intellectual capital is very important in this study. Considering the reseach purpose of this paper and the requirements of data, we choose Yang Manli's(2008) the current value calculation model of intellectual capital to measure intellectual capital, so we can ensure that the amount of intellectual capital is posititive and correspond with the economic meaning of intellectual capital. The measurement model:

IC(current value) = IC(current investment value) + IC(current create new value)

2.3 Sample and Data

The sample of this study is selected 68 transportation equipment manufacturing companies which are listed in the China Shanghai stock exchange and Shenzhen stock exchange. Eliminate the companies that the data is deficiency and the operation profits is negative in 2006 and 2009 years. In the end we choose 38 companies. Data is selected from CCER, CNINFO and CNIPR. The analysis of data uses DEAP (2.1) software.

3 Results and Analysis

According to the purpose of this paper, we use four steps to analyse the calculated results: the first step is relative efficiency analysis and slacks analysis. In other words, we uesed DEA to calcualte relative efficiency of 38 listed companies. Then we analyzed which companies were in efficiency frontier and could use as benchmarking companies. In the same time, we obtained slacks of input and output variables, which we used to improve intellactual capital management. The second step is that we analyzed the intellactual capital efficiency of aviation companies. The next step is to analyze how improve efficiency, use MPI to assess intellectual capital management efficiency of each company from 2006 to 2009 years.

3.1 Relative Efficiency and Slacks Analysis

Table 1 shows the relative efficiency of 38 companies of 2009 year. Using CCR model we find that technical efficiency(TE) of 8 companies is 1, other companies are less than

1, the worst number of TE is 0.159. From this we can see that technical efficiency of companies differs greatly. Accroding to BCC model, pure technical efficiency(PTE) of 16 companies is 1, 1 company is between 0.8 and 1, 6 companies are between 0.6 and 0.8, 15 companies are less than 0.6. In aspect of PTE, the efficiency of each company differs greatly too and there are two big head, but there are few companies' PTE around 0.5. If the PTE is 1, the corresponding companies are in efficiency frontier, other conditions illustrate that the corresponding companies are inefficiency. Inefficiency companies are not in efficiency frontier, so there is a big space to improve efficiency. For the inefficiency companies table 1 also offers the benchmarking companies to help improvement.

Table 1. Relative efficiency

NO.	TE	PTE	SE	RTE	Peers:	Frequency
20	1.000	1.000	1.000	-	20	19
37	1.000	1.000	1.000	-	37	9
28	1.000	1.000	1.000	-	28	6
15	1.000	1.000	1.000	-	15	4
8	1.000	1.000	1.000	-	8	3
22	1.000	1.000	1.000	-	22	3
36	1.000	1.000	1.000	-	36	3
17	1.000	1.000	1.000	-	17	2
13	0.795	1.000	0.795	irs	13	0
23	0.546	1.000	0.546	irs	23	0
25	0.485	1.000	0.485	irs	25	1
5	0.466	1.000	0.466	irs	5	1
35	0.370	1.000	0.370	irs	35	7
6	0.240	1.000	0.240	irs	6	15
14	0.223	1.000	0.223	irs	14	4
31	0.177	1.000	0.177	irs	31	1
11	0.792	0.944	0.839	irs	8,20,15, 6	0
30	0.530	0.789	0.671	irs	35,22,37	0
34	0.608	0.781	0.778	drs	20,15, 8	0
26	0.293	0.729	0.402	irs	37, 6,36,15	0
1	0.689	0.691	0.998	drs	20, 8,37	0
32	0.560	0.608	0.921	irs	20, 6,37, 5	0
4	0.602	0.602	0.999	irs	28,20, 6	0
21	0.367	0.598	0.613	irs	20,37,36, 6	0
38	0.445	0.567	0.785	irs	22,37	0
29	0.252	0.508	0.497	irs	20,14,35,37	0
3	0.251	0.472	0.531	irs	20, 6,35	0
12	0.177	0.463	0.383	irs	6,14,20,35	0
24	0.331	0.451	0.733	irs	20,28,35,22,37	0
33	0.376	0.442	0.850	irs	20, 6,25,17	0
18	0.190	0.379	0.500	irs	35,20,14, 6,37	0
27	0.314	0.333	0.942	irs	20,28, 6	0
9	0.233	0.326	0.715	irs	20, 6,35	0
10	0.272	0.319	0.855	irs	20,28, 6	0
2	0.258	0.289	0.893	irs	20,28,17, 6	0
19	0.132	0.245	0.538	irs	20,31,14	0
16	0.220	0.242	0.911	irs	28,20, 6	0
7	0.159	0.190	0.838	irs	20,15, 6,36	0

Table 2 shows the slacks of inputs and outputs variables, which stand for that each input and output variable has the room for improvement. And from table 2 we can see that 22 companies have room to improve. Some companies may locate in efficiency frontier, which show they are already very efficiency, but they still can become more efficiency. Refer to the inefficiency companies, there are big slacks of the number of employees, number of R&D employees, management cost and operation profits, so inefficiency companies can improve themselves by modulating their inputs. In other words, slacks can help inefficiency companies recognize their problems and the improving directions. For example, the company named Shanghai Areospace Automobile Electro mechanical maybe improve efficiency by decreasing the number of R&D employees and management cost, increasing operation profits.

Table 2. Input and output slacks

NO.	Input					Output	
	Number of employees	R&D employees	Number of patents	Management cost	Intellectual capital begin	Operation profit	Intellectual capital end
1	1935	417	0	0.000	13.359	0.000	310.224
2	0	258	138	45.722	0.000	0.000	0.000
3	982	0	56	14.179	0.000	0.000	43.323
4	0	1374	1406	574.949	0.000	195.269	0.000
5	0	0	0	0.000	0.000	0.000	0.000
6	0	0	0	0.000	0.000	0.000	0.000
7	279	284	0	0.000	0.000	245.063	0.000
8	0	0	0	0.000	0.000	0.000	0.000
9	1896	0	6	32.059	0.000	204.093	0.000
10	0	21	7	34.060	0.000	199.385	0.000
11	2427	0	0	9.916	0.000	0.000	387.315
12	0	52	10	0.000	0.000	16.193	0.000
13	0	0	0	0.000	0.000	0.000	0.000
14	0	0	0	0.000	0.000	0.000	0.000
15	0	0	0	0.000	0.000	0.000	0.000
16	0	127	136	54.338	0.000	224.869	0.000
17	0	0	0	0.000	0.000	0.000	0.000
18	0	285	0	0.000	0.000	8.906	0.000
19	0	0	36	7.114	38.127	353.071	0.000
20	0	0	0	0.000	0.000	0.000	0.000
21	0	115	0	39.223	0.000	53.713	0.000
22	0	0	0	0.000	0.000	0.000	0.000
23	0	0	0	0.000	0.000	0.000	0.000
24	0	14	5	0.000	0.000	0.000	0.000
25	0	0	0	0.000	0.000	0.000	0.000
26	0	55	0	8.957	0.000	0.000	8.504
27	1027	336	0	54.715	0.000	335.395	0.000
28	0	0	0	0.000	0.000	0.000	0.000
29	0	154	3	0.000	0.000	77.193	0.000
30	1832	192	6	0.000	0.000	0.000	51.560
31	0	0	0	0.000	0.000	0.000	0.000
32	2131	0	0	0.000	70.953	136.893	0.000
33	0	464	0	45.616	0.000	0.000	472.221
34	2560	383	0	0.000	179.850	278.341	0.000
35	0	0	0	0.000	0.000	0.000	0.000
36	0	0	0	0.000	0.000	0.000	0.000
37	0	0	0	0.000	0.000	0.000	0.000
38	357	274	9	0.000	27.513	106.857	0.000

3.2 Aviation Enterprises

Table 3 shows the comparison of the ratio for the aviation and non-aviation. There is only eight aviation companies in transportation equipment manufacturing industry. As is shown in table 1, in these companies only the PTE of Hubei Aviation Precision

Machinery Technology Co., Ltd is 1, others are less than 1. So there are large space to improve their intellectual capital management. And table 3 also shows that aviation companies don't have large differernce from the average level of transportation equipment manufacturing industry in the contributions of one unit patent to operation profits and the ratio for R&D employees and total employees.

Table 3. Comparison of the ratio for the aviation and non-aviation

Firm	Operation profit /employees	Operation profit /patents	Operation profit /IC begin	R&D /employees	Changing IC
Aviation	0.020	2.638	0.520	0.167	25.428
Non-aviation	0.112	2.606	2.004	0.130	513.410
All	0.097	2.607	1.831	0.136	410.677

But refer to the one staff's contributions to operation profits and one unit of beginning IC's contributions to operation profits, companies of aviation industry (0.020, 0.520) was significantly lower than the industry average (0.097, 1.831), show that the beginning IC and employees' contributions to the operation profits remain need to be further promoted in the aviation industry. The change(25.428) of intellectual capital in aviation enterprise is far lower than the industry average level(410.677), which shows that the aviation industry should strengthen the investment and management of intellectual capital.

3.3 Analyse the Efficiency Improvement from 2006 to 2009

Malmquist Production Index (MPI) aims to evaluate the performance change of a company within a period, which can be used as a kind of benchmarking management tools for checking the progress of a company. Table 4 uses MPI and PTE to divide the 38 transportation equipment manufacturing companies into three groups:

(1) High efficiency and rapid progress group(PTE = 1, MPI > 1). From 2006 to 2009, PTE of Aeolus Type Co.,Ltd, Jiangnan Mould & Plastic Technology Co.,Ltd and Shanghai Forever Co.,Ltd is 1, and MPI are 3.250, 1.075, 1.021 respectively. They have been located in efficiency frontier, and intellectual capital management efficiency also get continuous improvement. They should keep the superiority of intellectual capital management in order to maintain their competitive advantage.

(2) High and stable efficiency group(PTE = 1, MPI = 1). Although some companies was awlays in efficiency frontier, they didn't get very obvious progress in efficiency. There are 13 companies of this kind. For example Hubei Aviation Precision Machinery Technology Co., Ltd (PTE = 1, and the MPI = 1).

(3) Low efficiency and slow progress group(PTE < 1, MPI < 1). This kind of company not only lag other companies in efficiency in the years 2006 to 2009,but also have no progress in the efficiency improvement. In order to improve efficiency these companies need to change intellectual capital management strategy. There are 22 companies of this kind. For example Shanghai Areospace Automobile Electro mechanical (PTE = 0.598, and the MPI = 0.412).

Table 4. MPI

NO.	PTE	MPI
28	1.000	3.250
5	1.000	1.075
35	1.000	1.021
6	1.000	1.000
8	1.000	1.000
13	1.000	1.000
14	1.000	1.000
15	1.000	1.000
17	1.000	1.000
20	1.000	1.000
22	1.000	1.000
23	1.000	1.000
25	1.000	1.000
31	1.000	1.000
36	1.000	1.000
37	1.000	1.000
11	0.944	0.928
30	0.789	0.703
34	0.781	0.871
26	0.729	0.390
1	0.691	0.692
32	0.608	0.581
4	0.602	0.603
21	0.598	0.412
38	0.567	0.776
29	0.508	0.342
3	0.472	0.422
12	0.463	0.414
24	0.451	0.390
33	0.442	0.531
18	0.379	0.263
27	0.333	0.419
9	0.326	0.281
10	0.319	0.340
2	0.289	0.270
19	0.245	0.133
16	0.242	0.248
7	0.190	0.212

4 Conclusion

The paper uses DEA and MPI method to assess the intellectual capital management efficiency of companies in transportation equipment manufacturing industry in 2009, and analyses the progress of intellectual capital management efficiency from 2006 to 2009 years. This study finds that about 40% companies of sample have obtained the relatively satisfied efficiency, but the remaining companies still have rooms for improvement. Through the DEA efficiency analysis and slacks analysis provide the benchmarking companies for inefficiency companies and point out the improvement direction of all variables. The results of the study show that blind increase investment will not be able to raise efficiency of company in the intellectual capital management efficiency improvement process, and the key factor is to improve the effective of intellectual capital management. Particular, pay attention to strengthen the effective management of R&D employees, patents and other elements of intellectual capital. Generally, in transportation equipment manufacturing industry, aviation companies have inefficiency intelligence capital management. It is very necessary for these companies to strengthen intellectual capital management and improve creativity of intellectual capital. In addition, the aviation industry should strenghthen the management of R&D and invest enough in patent and management cost.

References

1. Jing, H.: The study of the productive forces and production efficiency in the equipment manufacturing industry listed companies. The Study of Finance and Economics Problems, 33–40 (January 2010) (in Chinese)
2. Ran, Q., Qi, F., Deng, H.: The empirical analysis of the intellectual capital efficiency in China listed companies. Accounting Communication, 61–63 (June 2009) (in Chinese)
3. Yang, M.: Research on measurement and investment benefit evaluation of intellectual capital in high-tech enterprise. Xi'an University of Technology (2008) (in Chinese)
4. Mavridis, Kyrmizogloup: Intellectual Capital Performance Drivers in The Greek Banking Sector. Management Research New 28 (2005)
5. Ante, P.: VAIC-an accounting tool for IC management. International Journal of Technology Management 20, 702–715 (2000)
6. Chang, S.-C., Chen, S.-S., Lai, J.-H.: The effect of alliance experience and intellectual capital on the value creation of international strategic alliances. Knowledge Management and Organizational Learning, 298–316 (April 2006)
7. Wu, W.-Y., Tsai, H.-J., Cheng, K.-Y., Lai, M.: Assessment of intellectual capital management in Taiwanese IC design companies: using DEA and the Malmquist productivity index. R & D Management 36, 531–545 (2006)

The Research on Anchor Ratio of Test Equating Design*

Jianhua Xiong[1], XinRong Ye[2], Shuliang Ding[1], and Fen Luo[1]

[1] Computer Information Engineering College, Jiangxi Normal University, NanChang, China
Pansy1212@sina.com
[2] Institute of Mathematics and Computer Science, Yichun University, YiChun, China
xinrong_ye@yahoo.com.cn

Abstract. Test equating is necessary for large scale item bank constructing. Before the actual test equating, not only time consuming, the consumption of wealth, but also the safety of the item may be endangered, particularly greater risk for anchor of repeated use, so the anchor ratio (R) of test equating design is expected smaller. Currently reported result in the country is: R is between 1 / 4 and 1 / 3 of the test length, test equating to be valid. So as not to affect equating accuracy ,this paper discusses R of different lengths . Monte Carlo (MC) simulation results show that, R can change with changes in test length, such as about test length for the 600,300,200,100 items, R can be reduced to 1 / 15, 1 / 12, 1 / 10, 1 / 5; when the test length is 40 items, R at least must be to 1 / 3.

Keywords: Monte Carlo Simulation, Item Response Theory, Test Equating, Equating Accuracy, Anchor Ratio.

1 Preface

Test equating is necessary for evaluating examination results, evaluating the quality of education, constructing large scale item bank, establishing and operating the CAT system [1]. Test equating is essentially a parameter conversion process, which change the ability parameter and item parameters in different forms to the same scale. Because item response theory (IRT) defined the difficulty and psychological traits (ability) in the same scale, it can also think test equating on IRT put item parameters from different test forms on the same scale to guarantee the item parameters being comparable[2, 3].

Under IRT, the scheme of test equating include equating design, measurement , answer reaction , item parameter estimation , equating coefficients estimation, equating conversion etc[2]. Equating design is the basis of test equating, in present, the nonequivalent common item test equating design is popular, anchor items is another set of items when constructing the test form, According to examinee's reaction, there are two different group examinees were measured by two different test forms, but were measured by a set of common items at the same time, this set of common items are called anchor. In practice, anchor and original test form can be putted together to

* This work is partially supported by NSF (30860084), Jiangxi province department of education science and technology research project grant to Gan Dengwen, Ministry of education in humanities and social science project(09YJCXLX012), The national education examination in the "eleventh five-year plan" scientific research planning project(2009JKS2009).

W. Zhang (Ed.): Advanced Technology in Teaching, AISC 163, pp. 413–418.
springerlink.com

measure examinees at the same time, the anchor can be independent from original test form also. When mixing anchor with the original test form in a test form, the operation is more convenient, but anchor amount must reach a certain ratio of the whole test length, the equating results will be valid, In general, the anchor ratio is [1/4, 1/3] of the test length, but for long test, such as the 600 items, 300 items, still adopt the anchor ratio, the number of anchor is very large, then lead to some problems. Such as, increasing test length, endangering the safety of the item for excessive exposure when construct item bank. This paper using Monte Carlo (MC) [4] method, in order to ensure the effectiveness of the equating results, research the minimum anchor ratio of different length of the test form, so as to save the test cost, to increase the safety of the bank.

2 The Empirical Research

It is an very important work to evaluate the result of test equating, the goal of evaluation is the reliability and accuracy on equating relationship, evaluation index is equating error, there are two kinds of error: namely random error and system error, random error comes from collecting the data of the subjects were derived from a random sample, the causes of system error is much more complex, and are difficult to quantify and find. Therefore, in order to discuss the random error, try to avoid and control system error. This paper reduce system error by selecting proper equating model and equating method.

2.1 The Choice of Equating Model

Because IRT showed more superior in equating, this paper based on the IRT equating to research, For the sake of elucidation problem but unlikely too tedious, two parameters Logistic model (2 PLM) [2] is adopted. IRT think, if there are two different test forms the X and Y, item j is the anchor of the two test forms, the item parameters of anchor in the X and Y can be different, the item discrimination note as a_{xj}, a_{yj}, the item difficulty note as b_{xj}, b_{yj} but there are some relations between them, for 2PLM, there are equating coefficient α ($\alpha \neq 0$)and β, among them

$$a_{yj} = a_{xj} / \alpha, b_{yj} = \alpha b_{xj} + \beta \tag{1}$$

$j = 1,2,\cdots M$

If the same examinee not only attends to form X but also attend to form Y, the ability of the examinee note as θ_{xi}, θ_{yi} ,there is relation:

$$\theta_{yi} = \alpha\theta_{xi} + \beta, \ i = 1,2,\cdots N \tag{2}$$

θ_{xi}, a_{xj}, b_{xj} are estimated parameters from test form X, θ_{yi}, a_{yj}, b_{yj} are estimated parameters from test form Y.

In this paper, formula (2) is adopt instead of using (1), because adopting (2) makes test equating procedure more convenient, then there is:

$$P_{xij} = \{1+\exp[-a_{xj}(\theta_{xi} - b_{xj})]\}^{-1} \tag{3}$$

$Q_{xij} = 1 - P_{xij}$

$$P_{yij} = \{1+\exp[-a_{yj}(A\theta_{xi} + B - b_{yj})]\}^{-1}\ Q_{yij} = 1 - P_{yij} \tag{4}$$

2.2 The Choice of Equating Method

In present, there are many methods of estimating the equating coefficients(equating method), Some popular equating methods, such as the Stocking-Lord criterion (SL) and the Haebara criterion(H),there are other SREcrit proposed by Xiong jianhua, Ding shuliang etc. it is well known that the equating coefficients derived from different equating methods are not the same for the same raw data.

In this paper, three equating methods under 2PLM, such as SLcrit ,Hcrit, SREcrit was measured in different test length, and repeat each experiment 30 times, calculate the equating coefficient after Wilcoxon symbols rank test [6], there is no significant difference, so in order to calculate conveniently, this paper estimating the equating coefficient all use SLcrit equating method, computation formula is as follows:

$$SLcrit = \sum_{i=1}^{N}\left[\sum_{j=1}^{M}\left(P_{xij} - P_{yij}\right)\right]^2 \tag{5}$$

2.3 Monte Carlo(MC) Simulation Process

Because it is difficult to obtain the real data, even if can get part of data, also lack representative. The paper adopt MC method, MC is essentially a statistical sampling Experiment method, because it needs Computer to generate Random observation variables, also known as the Computer Simulation Experiment method or Random Simulation Experiment method. MC simulation features are: MC simulation experiment data are simulated by computer produce or further conversion, belong to the simulation experiment; Since the data is simulation data, does not exist sampling cost and experiment price, are no effects of fatigue, etc, and it can be repeated sampling; MC method need not reason the sample to the overall sampling , it will be more direct sampling of statistics sample mean as overall statistics.

According to the choice of equating model and equating method for data simulation and calculation, mainly includes the following steps:

(1) if there are two examinee group X and Y, its ability distribution respectively $\theta_X \sim N(u_1, \sigma_1^2)$, $\theta_Y \sim N(u_2, \sigma_2^2)$, this paper generate standard normal distribution $\theta \sim N(0,1)$, then through the formula $\theta_X = \sigma_1\theta + u_1$, $\theta_Y = \sigma_2\theta + u_2$, and separately calculated two examinee group ability parameters.

(2) if there is a test, which divided into three parts [A, B, C], part test A and part test C with the same item number, all be m, the item number of part test B is n, each part of the distribution of item parameters is $\ln a \sim N(0,1)$ and $0.1 \le a \le 2.5; b \sim N(0,1)$. The two examinee group (generated by step 1) was measured by the test.

(3) if the examinee group X take part in test[A, B, C], for each examinee θ_{xi} and every item parameters (a_j, b_j), calculating P_{xij} according to the formula (3), then generate a random number r belongs to the interval [0, 1], if $r \le P_{xij}$, then $X_{ij} = 1$, otherwise $X_{ij} = 0$, get the corresponding score matrix $U_X = [u_{X1}, u_{X2}, u_{X3}]$. and the examinee group Y taking part in test [A, B, C] can also obtained the corresponding score matrix $U_Y = [u_{Y1}, u_{Y2}, u_{Y3}]$, and among them, u_{X1}, u_{Y1} is the score of part test A, u_{X2}, u_{Y2} is the score of part test B, u_{X3}, u_{Y3} is the score of part test C.

(4) estimating parameters of score matrix U_X, then get item estimated parameter $[\zeta_1, \zeta_2, \zeta_3]$; estimating parameters of score matrix U_Y,then get item estimated parameter $[\eta_1, \eta_2, \eta_3]$.

(5) if examinee group X just take part in test [A, B], examinee group Y just take part in test [B, C], so the part test B is anchor test. it is known by step (4) that the group X can get item parameters $[\zeta_1, \zeta_2]$, let X $= [\zeta_1, \zeta_2]$, the group Y can get item parameters $[\eta_2, \eta_3]$,let Y $= [\eta_2, \eta_3]$. because ζ_2 and η_2 are the item estimated parameters of part test B in different test, so if translate η_3 into $[\zeta_1, \zeta_2]$ parameters system, according to the formula (5) can calculate equating coefficient (α, β), according to the equating coefficient can convert η_3 into parameter system $[\zeta_1, \zeta_2]$ and get η_3',compare η_3' with ζ_3, In theory η_3' and ζ_3 and should be the same, if there are different, it is caused by the errors of equating , calculate the value of $\| \eta_3' - \zeta_3 \|$, let the item parameters of η_3' are (a_1, b_1) ,the item parameters of ζ_3 are (a_2, b_2), k is the number of anchor, there is:

$$Mabs = \sum_{i=1}^{k} (|a_1 - a_2| + |b_1 - b_2|) / k \quad (6)$$

(6) fix the number of part test A, C, change the number of part test B, according to the formula (6) calculate the Mabs value under different ratio anchor.

(7) statistically analyze the Mabs value calculated by step (6) (for example the Wilcoxon symbols rank inspection), observe whether there is statistical significant differences of Mabs in different anchor ratio

3 Research Conclusion

According to the above simulation program, this study rely on computer technology. Because the whole simulation calculation in process including parameter estimation error and equating error, in order to separate parameter estimation error out, underlines equating error caused by anchor ratio, this paper presents(600 + 200) item parameters and (600 + 200) * 10 ability parameters, then estimates all parameters, and extracts needed parameters from the estimated item parameters . Separately inspect test length of 600, 300, 240, 200, 160, 120, 100, 80, 60, 40 items, discuss the influence of various anchor length ratio (hereafter referred as R) on equating accuracy (see table 1), calculate the Mabs value when R is [1/4,1/3] in theory, calculate the Mabs value when R is given by this study, such as 1/5, 1/6, 1/8 ,etc, then using the Wilcoxon symbols rank inspection observe whether there is statistical significant differences of Mabs in different anchor ratio, if there is significant difference, use the theory anchor ratio still, if there is no significant difference, use the anchor ratio given by this study to save costs. Study on very large number of trials, the results show that:

(1) as the length of the test is 600 items, there is no significant difference of the Mabs value when R is 1/3, 1/4, 1/5,.....1/12. there is also no significant difference of the Mabs value when R is 1/4 and 1/25; , So 1/15 is feasible; Roughly, 1/25 is still viable.

(2) test length is 300 items, there is no significant difference of the Mabs value when R is 1/3 and 1/12,but there is significant difference of the Mabs value when R is greater than 1/12, behind neighbouring anchor ratios Mabs value of problem between there is significant difference. This shows that anchor ratio 1/12 is feasible.

(3) test length is 240 items, 200 items, there is no significant difference of the Mabs value when R is 1/3 and 1/10, but there is significant differences between 1/3 and different anchor ratio behind, so can consider anchor ratio 1/10.

(4) test length is 160 items, 120 items, 100 items, 80 items, the equating is effective when anchor ratio is 1/5.

(5) test length is 60 items, the equating is effective when anchor ratio is 1/4, and test length is 40 items, the anchor ratio must be 1/3 at least.

Table 1. The overall arrangement

Test length	Anchor ratio(R)									
600	1/3	1/4	1/5	1/6	1/8	1/10	1/12	1/15	1/20	1/25	1/30
300	1/3	1/4	1/5	1/6	1/8	1/10	1/12	1/15	1/20	1/25	
240	1/3	1/4	1/5	1/6	1/8	1/10	1/12	1/15	1/20		
200	1/3	1/4	1/5	1/6	1/8	1/10	1/12	1/15			
160	1/3	1/4	1/5	1/6	1/8	1/10	1/12	1/15			
120	1/3	1/4	1/5	1/6	1/8	1/10	1/12				
100	1/3	1/4	1/5	1/6	1/8	1/10	1/12				
80	1/3	1/4	1/5	1/6	1/8	1/10					
60	1/3	1/4	1/5	1/6	1/8	1/10					
40	1/3	1/4	1/5	1/6							

References

1. Kolen, M.J., Brennan, R.L.: Test Equating: Methods and Practices. Springer, NewYork (1995)
2. Qi, S.Q., Dai, H.Q., Ding, S.L.: Principles of modern educational and psychological measurement, pp. 201–224. Higher Education Press, Beijing (2002) (in Chinese)
3. Hambleton, R.K., Swaminathan, H.: Item Response Theory, Principles and Applications. Kluwer, Boston (1985)
4. Harwell, M.R.: Analyzing the results of Monte Carlo studies in item response theory. Educational and Psychological Measurement 57(2), 266–279 (1997)
5. Xiong, J.H., Ding, S.L., Gan, D.W.: A symmetric relative entropy method for test equating. Journal of Jiangxi Normal University (Natural Sciences Edition) 34(2), 111–116 (2010) (in Chinese)
6. Wu, X.Z., Wang, Z.P.: Nonparametric Statistical Methods, pp. 35–41. Higher Education Press, Bejing (1996) (in Chinese)

Research of the Intentional Forgetting Ability Training for Engineering Students*

Wei Zheng, Qi Lu, and Rongrong Cui

Laboratory for Optoelectronic Technology and System of the Education Ministry of China, College of Optoelectronics Engineering University of Chongqing Chongqing, China
zw3475@163.com

Abstract. Information explosion is a new challenge for college students. Research on human brain memory has shown that forgetting is an important activity of memory, thus it is one of the most important abilities for high-level cultivation of the students. Some training issues including the physiological basis for training, the training significance and the training methods are proposed in the paper. It indicates that the training process should pay more attention to the selectivity, purpose and testability.

Keywords: Intentional ability, Human brain memory, Education.

1 Introduction

One way people control the contents of their minds is intentional forgetting-voluntarily forgetting events after they have happened [1]. Forgetting the outdated information and the painful events are essential for individual's health. We stay in a social development stage with information explosion. New technologies emerge so quickly that the college students face an unprecedented situation that never appeared in the early-mid 20th century. New knowledge mixed with old knowledge and the confusions deeply affect the brain. Examining our traditional education, we stress memorization ability from birth to adult. A person with extraordinary memory is considered intelligent. There is no doubt that memory is very important during personal early education stage. It plays an important role in building comprehensive view of the world. However, the college students live in a society with information explosion. They must select the knowledge worth learning. Thus, the forgetting ability training is important to them. For the issues above, the paper do some research on the physiological basis, the significance, the training means and the precautions of the intentional forgetting ability training.

2 Physiological Basis for Intentional Forgetting Ability Training

RM Gagne, a famous psychologists and an expert in the instructional design, he proposed the information processing learning theory in which the learning process is

* This work is partially supported by the Fundamental Research Funds for the Central Universities (Project No. CDJZR10120016).

W. Zhang (Ed.): Advanced Technology in Teaching, AISC 163, pp. 419–424.
springerlink.com

divided into several stages with the requirement for different information processing [2]. The human brain's memory behaviour plays an important role in information processing. Memory is a reflection of experience in the human brain, which includes the things perceived in the past, the problems thought through, the emotions and the feelings experienced, the actions done and so on. The memory behaviour comprises memorization, retention, recognition and reproduction. In the steps of memory formation, information processing contains three ways as follows:

Coding: To obtain, process and assemble the information.
Storage: To record the information that has been assembled and rearranged.
Search: To draw out the information stored in response to a number of hints and incidents.

From the perspective of memory retention time, the memory involves sensor memory (SIS), short-term memory (STS) and long-term memory (LTS) [3]. Sensory memory corresponds approximately to the initial 200 - 500 milliseconds after an item perceived. Some of the information in sensory memory then transferred to short-term memory. Short-term memory allows one to recall something from several seconds to as long as a minute. Its capacity is also very limited. By contrast, long-term memory can store much larger quantities of information for potentially unlimited duration (sometimes a whole life span). The Working model of the human brain memory shown in Figure 1.

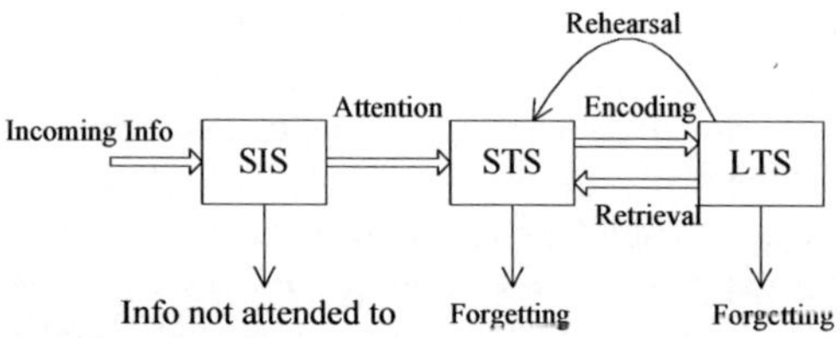

Fig. 1. The working model of the human brain memory

As Figure1 shows, forgetting is one of the important activities of the human brain memory. Therefore, it is a biased opinion to regard forgetting and remembering as opposed cognitive activities. For the information that needs no long-term attention, forgetting it naturally is the best way to develop human brain's potential ability. Ebbinghaus, a German psychologist, who put forward forgetting formula that reflects the characteristics of human memory [4]:

$$R = e^{-\frac{t}{s}} \qquad (1)$$

Where R is memory retention, S is the relative strength of memory, and t is time. Figure 2 shows the forgetting curve.

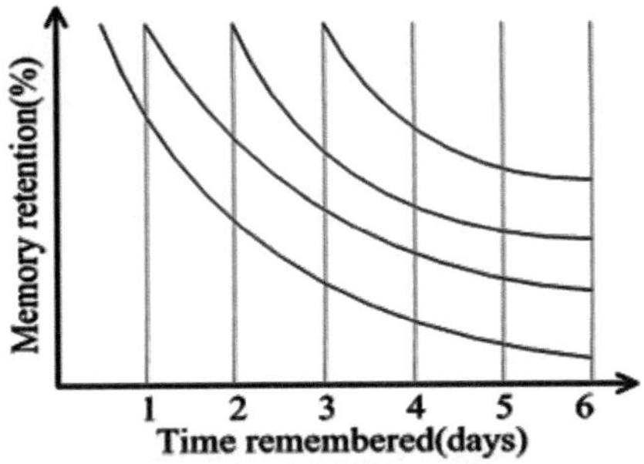

Fig. 2. Memory forgetting curve

As Figure 2 shows, memory retention in the brain decreases exponentially with the passage of time, which provides a quantitative description of the human brain's forgetting ability. However, it is evident that the memory system of human beings is very powerful. The reason is that we are good at forgetting and we trend to forget the unimportant information or the information we no longer concern. If forgetting does not exist, various contents of the memory information will bring our brain into an inefficiency situation.

3 Significance of the Training of Intentional Forgetting Ability

The intentional forgetting ability training is important for engineering students, the significance of which includes the following areas:

3.1 Requirement for Students to Adapt to New Technology Development

The learning content of engineering contains a considerable number of application technologies, such as programming languages, computer-aided design tools, applications of Single-Chip Computer and so on. College teaching process often lags behind the demand of the latest technologies as the results of the updated version of software and the development of hardware. Therefore, teaching students to forget the information that does not affect understanding of technical ideas and at the same time emphasizing the learning methods, then the students can expand their abilities maximally to accommodate new technologies.

3.2 Requirement for Improving Teaching Efficiency

Engineering students need to learn dozens of undergraduate courses and contact professional courses in upper division. Some courses are different in some contents significantly but they also have some closely relations in some contents. Take my courses as an example.

I taught two courses to the students in the same grade with the courses's names "Computer Hardware Technology" and "Application Principles of Single-Chip Computer", the contents of which are based on 80486 CPU and 8051 CPU respectively. I tried to teach knowledge by using of comparison method firstly.

However, through communication in classroom, I found that the old knowledge in students' brains often disturb their mastering process of new knowledge. I guide students to forget the memory of old knowledge in later time, and teach them to grasp the technical idea while forgetting the detail of the old knowledge. Then the students's attention is focused on the new knowledge completely which lead to a better teaching efficiency. An experiment carried out to indicate the difference between the comparison method and the forgetting method, the result of which is shown in Figure 3.

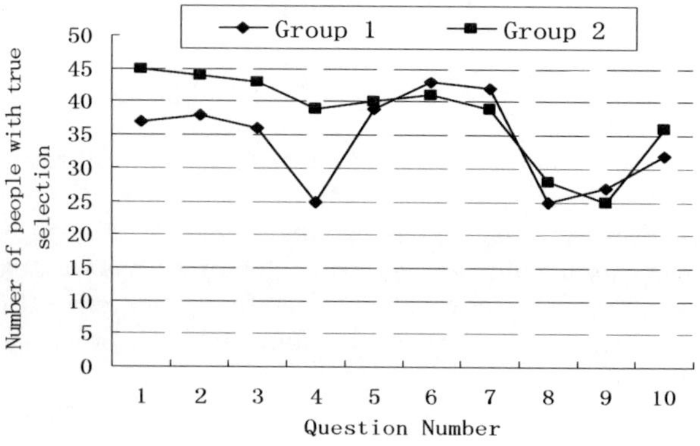

Fig. 3. Result of experiment

The students in experiment contain two groups with 50 people in each one. I applied comparison method and intentional forgetting method to Group 1 and Group 2 respectively. Then the students answered some questions. There were 10 questions totally. Question 1 to 4 is objective questions with strong confusion for old and new knowledge. Question 5 to 7 is objective questions with no association of old and new knowledge. Question 8 to 10 is subjective questions with new knowledge related with old knowledge, and they are difficult to answer. It shows that the answer accuracy of Group1 is far better than that of Group 2 for question 1 to 4. The answer accuracy is similar for question 5 to 7 because of no relationship between old knowledge and new knowledge. The answer accuracy for question 8 to 10 is similar just because the difficulties of the question have covered the differences between two methods. While after analysing the answers, we can find that group 1 is better that group 2 for question 10. The confusion between old knowledge and new knowledge is the main reason.

3.3 Requirement for Promoting the Development of Interdisciplinary

Bernard, a French physiologist, who had a famous saying: "The greatest obstacle for learning is what is known rather than unknown." It is similar with the scientific research. Sometimes a scientific problem puzzled the brain which is in a clear state. It is surprising that the ultimate resolution of the problem comes from the spiritual inspiration in sleep. In scientific history, there are many examples of the inventions in the dream. For example, Kekule, a German chemist, found the ring structure of benzene derived from the insight in his sleep. As forgetting curve in Figure 2 shows, human brain forgetting is not thorough, and there is always some information left in the memory. Some critical information may reappear in sleep or in other special occasions to create an innovative idea. In recent decades, many research results from the interdisciplinary, where the uncertain fuzzy features often present. If the researchers have a good foundation of subject knowledge, and fuzz the cognitive boundary between disciplines, then they may acquire the innovation at the intersection of disciplines. Engineering students should master the basic knowledge. However, after entering the upper division, they could acquire intentional forgetting ability training, which is beneficial for scientific ability cultivation.

4 Training Ways and Precautions

For the issues of intentional forgetting ability training in the teaching process, the author proposes several ways and precautions as following:

1) The training should be selective. It contains two aspects. First, we should select the types of the information need to forget. The engineering students in lower division learn some basic courses such as mathematics and physics. The intentional forgetting ability training is unsuitable for them. Otherwise, their basic scientific literacy may lose. On the contrary, some application courses with rapid updating rate can adopt the training. Secondly, we should select the training object. The freshman and sophomore are in a foundation building stage, thus they are inappropriate for acquiring the intentional forgetting ability training. However, the junior can attempt it. In addition, it is necessary to carry out the intentional forgetting ability training for students who have a better grasp of basic knowledge and have the desire to do research work in the future.

2) The training should be purposeful. The intentional forgetting ability training must have a distinct purpose such that the students can understand explicitly what they need to remember. So the teacher should remind the students from time to time in the classroom, teaching them which information need to remember. Then the information will come into long-term memory area of the human brain smoothly via stimulating brain repeatedly. On the contrary, the unimportant issues should be clear taught to the students with labelled in the textbook or notes. Then the brain will produce the psychological implications to the person and remind him that no longer concern about the information. The teacher should avoid mentioning the problem repeatedly in the subsequent teaching. In teaching activities, the teacher should carefully use the comparison method for old and new knowledge. Specially, we should recall little old knowledge in detail, so that to weaken the effect of old knowledge on new knowledge.

3) The effect of training should have testability. The quantitative evaluation of memory effect is easy, and it is uneasy for indication of the forgetting effect. We know that the purpose of forgetting is to get a better remembering, therefore, we can test the effect of forgetting through the experiments of forgetting the old information and remembering the new information afterwards. When the teacher prompt the students to forget old information and immediately learn the new information, the mastery effect of new information is not ideal. Let students learn new information

after isolating the old information for 3 days, 1 week and 2 weeks, the longer the separation, the better effect to master the new information. It demonstrated the forgetting curve shown in Figure 2 from another point of view. Moreover, the knowledge marked with hierarchical importance will result in different concern levels, and the knowledge marked with low importance will have low concern by students in the experiment test. Through this way, the forgetting effect is relatively good.

5 Conclusion

Traditional teaching has ignored the forgetting ability training. However, it is important for engineering students to accept intentional forgetting ability training, which has a great significance for improving student's knowledge structure, improving their cognitive efficiency and enhancing their innovation.

Acknowledgment. The authors would like to thank the Project No. CDJZR10120016 supported by the Fundamental Research Funds for the Central Universities and the Research Fund for the Doctoral Program of Higher Education of China(Grant No.20090191120010).

References

1. Payne, B.K., Corrigan, E.: Emotional constraints on intentional forgetting. Journal of Experimental Social Psychology 43, 780–786 (2007)
2. Gagne, R.M.: The Conditions of Learning and Theory of Instruction. CBS College Publishing, New York (1985)
3. Atkinson, R.C., Shiffrin, R.M.: Human memory: A proposed system and its control processes. The Psychology of Learning and Motivation (2), 89–195 (1968)
4. Ebbinghaus, H.: Memory: A Contribution to Experimental Psychology. Teachers College, Columbia University Press, New York (1885)

The Analysis and Exploration of Education Pattern on the Student's Public Morals Quality in Vocational Colleges

Chunqing Wang

Accountant Department, Zhejiang Economic & Trade Polytechnic, Hangzhou, China
wanchuqi@163.com

Abstract. Aiming at the limitation of the former research mode on public morals' meaning and the sermon form, a new research mode was presented. By collecting and analyzing some positive and negative cases of the public morals behavior, the cause why public morals quality is low was discovered and according solution was put forward. By this research, the aim is to improve presently some low public morals quality and to help to achieve socialist harmonious society. This work can also offer help for thought education work and management work of college students.

Keywords: vocational colleges, the student's public morals quality, education pattern.

1 Introduction

Public morals, which guarantee well-balanced of people's communication, refer to moral and ethical standards and obligations enforced in a society, by law or police work or social pressure, and applied to public life, to the content of the media, and to conduct in public places. Nowadays, because of relatively development of material civilization, it requests that colleague students should have not only favorable professional accomplishment, but also well spiritual civilization accomplishment. These years, some philosophies, such as constructing socialist harmonious society, education is fundamentality and morals education is first and so on, are brought forward. These philosophies' present tell us that university should pay the most attention to morals education, form education system of morals education and urge students to cultivate well and appropriate morals and behavior habit.

The Chinese idiom "It takes ten years to grow trees, but a hundred to rear people" shows that the talents' cultivation is very hard and difficult work. Most of colleague students, especially in vocational colleges, are not interesting in studying and lack attic faith and correct outlook of value. Simplex professional knowledge education is difficult to make them successful. Meanwhile, the results of tracking survey and statistics for our graduates of vocational colleges in these three years indicate that the enterprises are satisfied with our graduates' professional ability. But 70% of these enterprises also expressed their dissatisfy for graduates' only attention to their own responsible work and ignore other all things of the enterprise, even society, just like

W. Zhang (Ed.): Advanced Technology in Teaching, AISC 163, pp. 425–429.
springerlink.com

Chinese folk says "Never help it even oil bottle is falling down". All of these require that we must strengthen students' morals education and cultivation during the education of professional knowledge and skills. Through strengthening the cultivation of the students' public morals quality, not only making student's mind to avoid narrow, but also be more conductive to the cultivation of the students' psychological quality. Via these measures, the present situation of the students' public morals quality can be improved to some degree. Constructing socialist harmonious society's advancement requires enhancing all of the people's public morals quality, but it is unpractical to rapid enhancement. The youth are in the most active period of the communication and the development of the morals and ethics of the youth is an important stage of individual's socialization. Therefore, we can proceed with this group of college students and can play the role of fanning out from point to area for promoting public morals quality of the entire social.

2 Present Situation of Students' Public Morals in Vocational Colleges

In recent years, psychological quality of college students drops quickly and leads to many making suicide events and violence events. All of these sound the alarm to strengthen college students' morals and ethics education. Believing everyone has seen these phenomenon listed below. In public bus, some undergraduates don't willing to give seats to the people in need of care; in the public, some undergraduates throw waste at random, trample the flowers and grass, break the branch of the trees and so on, and there are few undergraduates to take the initiative to take good care of public properties; in campus, there are few undergraduates to turn off the light of no people's classroom and turn off the flowing tap; some undergraduates seldom take the initiative to clean public area of the dormitory and some not poor students also make falsification to obtain student loan for personal enjoyment and others.

Although for many people, all of these phenomenon are commonplace, it is extremely disharmonious and a large deviation from the requests of program for improving civic morality. As a teacher and educator, it is our responsibility to change these phenomenons. In that way, civil makings of Chinese can be improved in deed in order to build the socialist spiritual civilization and realize the aim of building socialist harmonious society.

3 The Sources of Forming Public Morals Quality of the Contemporary Undergraduate in Vocational Colleges

The current higher education has entered the stage of popularization and most of the students are the only child. With the influence of western culture and the popularity of network which has been turning the world into a "village", the communication scope and measures of people have been greatly expanded. The realization of social public morality has not only in the real society, but also in network so that somebody put forward the concept of the network morality. In this background, coupled with unique characteristics of students in Higher Vocational colleges: weak self-control ability, low learning enthusiasm and so on, the cultivation fo the students' public moral in higher vocational college confronted with hitherto unknown challenge.

Chinese idiom "Man's nature at birth is good" tells us that human nature is good. The reason why some bad things still happen should due to inappropriate acquired education and cultivation. Nowadays, our education pay more attention to students' scores so that as long as having high scores, any material requirements of these students can be meet unconditionally by their family even poor family ignoring the cultivation of thought and moral. It is just this kind of distorted education and examination system and wrong idea of students' parents, public morals quality of the undergraduates nowadays declined rapidly. Under present social and education environment, this kind of talent selection mode through the test scores is unable to change easily. Therefore in this mode it requires our educators should strengthen the students' moral quality training. At present, some schools have been taking account into moral and ethic scores during enrolling students and this is a good phenomenon. However, it still has a lot of work to do on how to establish a long-term mechanism of moral education and how to reverse the current public moral status. Corresponding solutions must be put forward timely.

4 The Research and Using for Reference of Current Domestic and Foreign Research Achievements

In practice, foreign education institutions pay more attention to the training of the practical ability and the independence ability in students' skill education and moral education. The cultivation of public morals quality is better than our work in this area.

Zhang Weiying, Chinese economist, regards that necessary condition to realize a harmonious society is some core values and ideas must be recognized and identified with by most of people, otherwise entire society will not be harmonious. Because of relatively discrete society now, some basic core values agreed by most of people are lack. In the past, some values can be accepted and complied with through coercive power in the short term in order to achieve surface harmony. However, when everyone can express their viewpoint freely, it become more and more important to have core values accepted from heart by most of people. So, all kinds of measures must be introduced to make people comply with a common core values, and this core values can be selected from Mr. Zhang Weiying's three viewpoints. These three perspectives respectively are to benefit others and no harm yourself, to benefit yourself and no harm others, and to benefit yourself and others.

During the cultivation of students' public moral, we should let everybody agree that the followings are selected sequentially: firstly, in those cases having options, we advocate to comply with the core values of benefiting others and yourself; secondly, if there are no choice, it should be advocated to comply with the core values of benefiting others and no harm yourself; thirdly, when necessary, the behavior of benefiting others regardless of yourself should be advocated.

Mr. Du Weiming, a famous educator, regards that seeing from the viewpoints of western's experience, a rational society's formation need to rely on the guide of the thinkers and the scholars, be passed to the public intellectuals, then to the public through the media. If it is reverse, i.e. the public influence on the media, the media influence on public intellectuals and public intellectuals influence on the academia, it will be very dangerous. It also shows that at the same time of bringing the democracy role into play, it must be guided positively.

Although our teachers of universities and colleges are not ideologists, they have significant responsibility and play very important role in cultivating all-round developed talents in moral, wisdom and physics for society and country. So, we should do our best in the students' ideological and political education work and resolutely implement public moral quality training on students in vocational colleges so as to promote the harmonious development of society.

The meanings and examples of enhancing the public morals of students in vocational colleges have been introduced in reference [2] [3] [4]. However, it is more important to put forward the concrete solutions and measures.

In addition, the domestic relevant researches on public moral education usually are carried out through enhancing the cultivation of psychological quality and occupation moral education. In this paper, though above analysis it is also necessary to strengthen public moral education in order to promote the psychological quality education and occupation ethics' compliance.

5 The Approach of Promoting the Students' Public Morals Quality in Vocational Colleges

In this paper, how to guide the public morals' propaganda is presented from the angle of economics and some specific measures to enhance public morals is put forward from the applicable perspective.

By analyzing and comparing some positive and negative cases of the public morals behaviors collected from three different dimensions of personal communications, public communications and network commutations, and two results can be gained. First is that the proportion of the people complying with and not complying with public morals is gained. Second, favorable public morals play a very important role and bad public morals definitely can make a negative influence for a person's development.

Through analyzing the cases, the cause why public morals quality is low was discovered and according solution was put forward. Furthermore, education pattern can be implemented via practical training and be revised and more perfected according to the feedback results so that this education pattern can better adapt to the improvement of current public morals quality and the cultivation of students' public morals quality in present vocational colleges.

We should design propaganda brochure according Parato efficiency principle of the economics and set up cultivation and education scheme through two forms of the theory and the practice feeling. Proceeding with from daily behavior, the purpose, planned and organized education and training activities are put forward to enhance students' public moral quality to some degree. The improvement of the students' public morals quality in vocational colleges can be realized through below some approaches:

First, caring out the vacation practice activities with the aim of improving public morals quality and to assign the students activities, making the parents understand our aims and means of the education in order to help us better accomplish task;

Second, making full use of weekly political study time of students in vocational colleges, let them communicate and discuss some cases about public morals in order to achieve the aim of accepting and acknowledge public morals;

Third, taking full advantage of student communists' democracy life meeting, making them to play a vanguard and exemplary role through advocating complying with public morals;

Fourth, public morals should be advocated in the public, such as the campus and the dormitory. Some NGO should be set up to make complying with and carrying forward public moral be conscious activity of most of students;

Fifth, establishing feedback mechanism of complying with public morals;

Sixth, developing humanistic education through necessary theory education methods;

Through above specific measures, present situation of low morals and ethics quality of undergraduates in vocational colleges can be improved and reformed, their consciousness of obeying public morals may be enhanced effectively and it is helpful to establish the foundation of constructing socialist harmonious society. At the same time, these measures can provide favorable education ideas and schemes for undergraduates' ideological and political education work of higher vocational education and contribute to undergraduates' management work. All of these can enhance undergraduates' level of integrity, make them adherence to occupation morals and increase undergraduates' employment quality, improve the quality of students communist and maintain the advanced characteristics of communist.

Acknowledgment. This work is supported by the Philosophy and Social Science Planning project of Zhejiang province, China (Grant No. 10CGYD101YBM), the Natural Science Foundation of Zhejiang province, China (Grant No. Y6110478).

References

1. Xu, N.: The Exploration and Practice on Improving Student Works of Vocational Colleges. China Education of Light Industry (1), 63–64 (2005)
2. Han, J.: Concerning thought and moral quality of college students. Theory Horizon (5), 182 (2005)
3. Li, P.: Citizen's Morals and Construction of Harmonious Society. Journal of Beijing Administrative College (4), 83–87 (2005)
4. Zhang, C.: Make Good Moral Be Habit. Perspectives on Education (1), 23–24 (2005)
5. Tan, D.: On Social Morality and Individual Development. Journal of Chongqing Normal University Edition of Social Sciences (1), 105–110 (2003)
6. Kai, L.: Strengthen Chinese Traditional Virtue Education, Improve undergraduate morality. Journal of Southwest University for Nationalities 26(3), 364–366 (2005)
7. Cai, X., Kang, X.: The Morality Education of College Students in the View of Promoting Citizen's Quality. Journal of Shaanxi Administration School and Shaanxi Economic Management School 19(3), 87–89 (2005)
8. Wang, W., Wang, Y.: University Morality Education Need to Put Stress on Cultivate Student's Public Consciousness. China Higher Education (7), 34–35 (2004)

A Discussion of the Prospects of the Short Pimples Out Style of Fast Attacking Penhold Grip Play*

Teng Shou Gang

Wuhan Institute of Physical Education

Abstract. World table tennis is constantly reforming and evolving year after year. The traditional penhold grip fast driving-and-blocking style using a short pimpled rubber now looks to be behind the pace of development in world table tennis today. Current Chinese table tennis team player Wang Hao has developed a unique backhand technique to achieve much success in today's game. Through comparative and logical analysis of Wang Hao's backhand technique we can find out the disadvantages of the short pimpled fast driving-and-blocking game, and techniques and tactics that can be used against it. We can also explore the urgent problems of this particular penhold grip style of play, and look at the possible future direction of the development of this style of play.

Keywords: the penhold grip, short pimpled rubber, fast driving-and-blocking game, development perspective, backhand technique, success.

1 Preface

The close to the table fast driving-and-blocking style of play is a Chinese traditional style of play that has always been China's powerful weapon against European players. Table tennis in China has gone through different periods of technological development, and China has played a positve role in promoting the development of world table tennis. Throughout the history of table tennis, fast attacking, glued up penhold players such as Zhuang Ze Dong, Jiang Jia Liang and Liu Guo Liang dominated in what was a golden age for this style of play. However, with the continued development of table tennis, and the penhold backhand's lack of innovation coupled with it's own complicated style of play, this fast driving-and-blocking style of play is currently behind the pace of development in today's game.

The traditional Chinese style of play had a brilliant record, and despite the current difficult times we should not discard it, but continue to introduce new techniques to find a new combination of this style of play, and tactics to deploy when using it.

1.1 Zhuang Ze Dong's Three Consecutive World Titles during the 1960's

In the 1960's, Chinese players played a close to the table, fast driving-and-blocking style of play, and would attack the ball left and right. Doing this, they captured the most prestigious titles the game had to offer. The playing features of this period were to take

* Wang Hao's Backhand Technique Achieves Success.

W. Zhang (Ed.): Advanced Technology in Teaching, AISC 163, pp. 431–434.
springerlink.com

the initiative to attack mainly at a fast pace. This technical style is typified by fast, accurate and relentless attacking play, moving the ball from side to side. Zhuang Ze Dong used the short pimbled rubber to play forceful shots with both his forehand and backhand. With his attacking penhold style, he won the World Table Tennis Championships men's singles gold medal three times. Together with fellow Chinese team members Xu Yin Sheng and Li Fu Rong, they defeated other teams from around the world, such as France with their unique playing style, culminating in them standing on the world championship podium several times. At that time, the world's most advanced players were the Chinese penhold players with their fast attacking style of play, and they were at the forefront of the development of the world game at the time.

1.2 Jiang Jia Liang's Double Success in the 1980's

During this period, China's fast attacking style of play kept the original features of the close to the table attacking style of the 1960's, but added to it a new shot – the loop – to deal with the evolution and development of table tennis at the highest level. The forehand loop, hit with increased velocity, together with a short pimpled backhand, was created to deal with the newest trends in world table tennis. A unique technical style was developed, and at this stage the leading Chinese exponent of this style of play was Jiang Jia Liang. He was a penholder who showed that the fast driving-and-blocking style of play was still a force in international table tennis competition. During the 38th and 39th World Championships, Jiang won the world men's singles championship twice.

1.3 Liu Guo Liang's Grand Slam in the 1990's

In the late 80's and early 90's, the Europeans seized on the fast attacking style of play by applying pressure on the penholder's backhand. The penholder was unable to take the initiative in rallies, and the bottom line was that the penholder's backhand was a major weakness. However, the Chinese table tennis team did not give up on their traditional style, and through development they came up with an innovation during these difficult times. The representative of this era was Liu Guo Liang. The innovation was the 'reverse penhold backhand' which enabled greater variety on that wing. Together with the traditional fast attacking forehand, and the backhand pushing and blocking style, this new backhand technique allowed Liu to beat many famous top players and become the men's singles champion in the Olympic Games, World Championships, and the World Cup – all at the same time. He became the first Chinese player in history to win the grand slam. ~The Chinese fast attacking penhold style of play was once again taken to a new level. This style achieved excellent results in world table tennis in the 1990's, and made outstanding contributions to the development of the game. However, the International Table Tennis Federation have carried out a series of reforms in table tennis since 2000, which has given top level players some serious challenges. One rule change, which stated that the ball should be visible at all times during the service action, required a further revision of this style. High level technical ability was required to give players the speed to win matches using the penhold fast attacking playing style, as advantages from the service became more difficult to gain.

2 The Penhold Backhand Technique

2.1 The Traditional Backhand Technique Is the Fast Block

When deploying the traditional backhand pushing and blocking style, the direction of the grip and the racket when making contact with the ball is more of a 'forward and back' movement, which doesn't allow for power to be generated from the waist. This reduces the amount of speed and spin imparted on the ball, resulting in a less threatening shot. The other technique from the past, played while at a medium distance from the table, is a backhand technique which is often played at the 'crossover point'. The player would then look for opportunities to counter attack, but this technique is usually more passive, and it is hard to find opportunities to do this. In addition, penhold players wanting to improve the quality of their backhand shots would move their body position sideways, compromising the neutral position, and making it easy to attack the space on the forehand flank. Therefore, it becomes difficult to compete when this particular fast attacking backhand method has obvious technical flaws.

2.2 Wang Hao's Backhand Technique

Stroke-making in world table tennis today are increasing in activity, speed, ferocity and power. This development requires higher quality shots with a good technique increasing in importance. Penhold players faced different types of challenges. Fast blockers who lacked an attacking backhand struggled. The play itself increased in difficulty, resulting in a slump for this style of play. However, Wang Hao's penhold attacking method gave him success in three world championships. The important point is that his bold technical innovations in playing with an open face of the bat, were based on strengthening the backhand. In fact, his improved backhand enabled him to take the initiative in rallies. It also ended the thought that the penhold grip could not succeed because the backhand was a major weakness. Wang Hao's technical innovation is to screw back his hand, and at the point of contact, propel the ball forward, using the reverse side of the bat, allowing him to play a backhand of immense speed, spin and power. Wang Hao is the world's leading exponent of this new backhand technique using the reverse side of the bat. It has brought about a technical revolution in today's table tennis. Originally, when penhold players attempted this shot, it was flawed, because the reverse side of the bat was not clearly visible, resulting in inconsistency. However, Wang Hao used a technique that had never been seen before – one where the starting position of the reverse side of the bat directly faces the opponent, so that the backhand loop is one of his main scoring areas. This is especially useful in the absence in international competition of points being won outright from serving and blocking. For a dynamic performance, penhold players now have to make full use of this new method, to cause the opponent major problems. In addition, rallying has traditionally been a relatively weak link in a penholder's game. Very often the attacks and the tactics deployed lacked variety. The penholder would often play the ball consistently from side to side without much force. However, Wang Hao sought to avoid backhand to backhand crosscourt rallies by adjusting his forehand technique. He would approach the ball from the side, and with his increased mobility, he could launch into the ball to impart greater speed and spin on the ball. Therefore, during the rally, whether Wang

Hao is near the table, a medium distance from, or far away from the table, he is in no way at a disadvantage compared with the shakehand grip players. Wang Hao's penhold style of play – while maintaining the traditional advantages of the penhold grip, has also greatly improved on it. Wang Hao's unique backhand technique makes up for the lack of a traditional penhold backhand, and brings a full range of penhold strokes against the shakehands players to create a technological revolution. The ageing Chinese tradition of the fast penhold attack has been pulled back from the brink of death, to rise to the forefront of techniques in modern day table tennis.

3 Inspiration

In table tennis today, there is greater variety with regards to ball rotation so that the penholder's fast attack style and return of serve are severely constrained.

Wang Hao was the first player to use the reverse penhold backhand technique.

This advanced technique frees him from worrying that when attacking crosscourt from the backhand wing, the opponent might return the ball to the opposite area to the one he has got ready for.

Once the penhold grip player with short pimpled rubber takes the initiative during the rally, he will be able to gain more possible opportunities to be in control of the rally with consecutive attacks right after the first aggressive shot.

The common style of being fast and accurate, and changing the direction of the ball is considered quite safe.

Instead, players must take more risks. Penhold players who use short pimpled rubber should be decisive, quick, proactive, preemptive, and further improve and develop their technique. Players will be able to rally with more success if they improve the technique of their reverse pen hold backhand. The fast attacking style assisted by short pimpled rubber will heighten the penhold player's overall awareness, and improve the quality of their forehand attack. This flourishing development has an important role to play in today's table tennis. The fast attacking style of penhold play has a very unique style of it's own, and the emergence of Wang Hao, who pioneers this style of play, shows the continuous efforts made in table tennis to constantly innovate. I believe China's trademark fast attacking penhold style of play will once again become dominant in the highest level of international table tennis.

References

1. Cai, Z.: My Penhold Attack see. Table Tennis World (October 1999)
2. Cai, Z.: Statement about table tennis made in the 2002 National Workers Conference Tianjin (2002)
3. Su, P.-J.: The Fast Blocking Penhold Style. Beijing Sport University (January 1996)
4. Wu, J.: Hold Grip Rubber Training. Table Tennis World, 7–28 (2000)
5. State Sport General Administration. Ping-Pong Training Changsheng Exploration. Beijing Sport University Press (November 2002)

Sustainability Assess of Wind Energy Project Based on Improved BSC-Hierarchy Model

Fang Fang[1] and Yu Aimin[2]

[1] Schoool of Economics and Management, North China Electric Power University, Beijing, China
[2] Logistic Services North China Electric Power University Baoding, Heibei Province, China
{orchid514,15210003909}@126.com

Abstract. As the most mature, the most development potential and the most market competitive renewable energy generation technology, wind power generation has been in a vogue. In recent years, we have seen a rapid development of China's wind power. However, there are some certain problems in wind power project design, construction and operation. The paper takes security of the wind energy project as an object of research. Considering the latest developing situation, the paper sets up comprehensive index system based on improved -BSC and matter—element analysis model for evaluating security of the wind energy project, The Practice is conductive to discover problems, ensuring a sound and stable development in our wind power industry.

Keywords: Sustainability assess, Wind energy, Improved-BSC, Hierarchy model.

1 Introduction

With current accelerating process of economic globalization, people have been consuming more and more resource of the earth, which suffers serious damage and worsening ecological situation. So the world should seek a more active and effective economic model for a greater development and living space. And under the multiple crisis situations, such as energy tension, food shortage, climate change and financial crisis, concepts of sustainable development, green economy and low-carbon economy have become a global trend and fashion.

In recent years, we have seen a rapid development of China's wind power. With newly installed capacity of 13.8 million kilowatts, surpassing the U.S., China becomes the largest newly installed capacity country in 2009. However, there are some certain problems in wind power project design, construction and operation. Sustainability assess of wind energy project is conductive to discover problems, ensuring a sound and stable development in our wind power industry.

2 Index System

This paper uses the ideas of balanced scorecard (BSC), constitutes an assessment index system for sustainable capacity in wind power project.

W. Zhang (Ed.): Advanced Technology in Teaching, AISC 163, pp. 435–441.
springerlink.com

2.1 Feasibility Analysis in Applying the Idea of Balanced Scorecard (BSC) in Constituting an Assess Index System for Sustainable Capacity in Wind Power Project

Balanced Scorecard approach is an integrated evaluation method with a set of evaluation, management and communication functions. In early 1992, Kaplan and Norton published detailed description of the benefits obtained in using the Balanced Scorecard for performance evaluation of the company. Kaplan and Norton extended the method to strategic management of the enterprise in 1993. Since then, the balanced scorecard has been widely recognized and used in the global business community, also extending to the non-profit organization.

The basic idea of the Balanced Scorecard approach is to build a complete set of comprehensive indicators system, which includes financial performance indicators, customer indicators, internal business processes, learning and growth targets and performance indicators, making systematic and comprehensively assess of sustainable capacity in wind power project.

From the above we can see, with the idea of the Balanced Scorecard, constituting an assessment index system for sustainable capacity in wind power project can not only make a more comprehensively assessment for the sustainability of wind power projects, but also provide the basis for the analysis of development status of China's wind power industry and future development, which is in favor of steady and healthy development of China's power industry.

2.2 The Construction of Sustainability Assess Index System in Wind Power Project Based on the Idea of Balanced Scorecard (BSC)

According to the ideas of Balanced Scorecard, we constitute an assessment index system for sustainable capacity in wind power project, mainly about the financial situation of wind power projects, business capabilities, resources conditions, status of enterprises management. Focusing on the above, the main four aspects are discussed and a three-level wind power project sustainability index system is established.

Indicators at all levels of sustainability assess index system in wind power project are as follows:

Table 1. Index System

	Primary index	Secondary index
Wind Energy Project	financial situation	Financial profitability
		Financial solvency
		Staying power
	business capabilities	The scale of operation
		The level of technology
		Year generating capacity
	resources conditions	Wind velocity
		Soil density
		Groundwater level
		Freezing depth
	status of enterprises management	Reserve of talents
		System sophistication
		Incentive system

3 BSC-Hierarchy Model

3.1 Compendium of Matter-Hierarchy Model

Matter-element extenics is based on matter-element theory and extension mathematics as its theoretical framework. Of these areas, matter-element is the logic cell of extenics. Being able to solve problems qualitatively and quantitatively, ordered three-dimensional group R=(N, c, v) is regarded as a basic element of describing matters. If matters N is described as n characteristics $c_1, c_2 ... c_n$ and corresponding values $v_1, v_2 ... v_n$, it can be showed that:

$$R = (N, C, V) = \begin{bmatrix} N & c_1 & v_1 \\ & c_2 & v_2 \\ & \cdots & \cdots \\ & c_n & v_n \end{bmatrix} \tag{1}$$

Extensional set extended the continuous values in closed interval [0,1] of fuzzy sets to the real axis (-∞, +∞) with correlation, to express the degree of meeting requirements when the value of matter-element is one point in the real axis. First introduced the concept of distance: suppose v is at any point in the real domain (-∞, +∞), and V= ⟨a, b⟩ is any interval in the real domain, then name is the distance of point v and interval V.

$$\rho(v, V) = \left| v - \frac{a+b}{2} \right| - \frac{b-a}{2} \tag{2}$$

The correlation function based on the distance is

$$K_i(v_i) \begin{cases} \dfrac{-\rho(v_i, V_{cji})}{\left|V_{cji}\right|}, v_i \in V_{cji} \\ \dfrac{\rho(v_i, V_{cji})}{\rho(v_i, V_{pi}) - \rho(v_i, V_{cji})}, v_i \notin V_{cji} \end{cases} \tag{3}$$

Among, $K_i(v_i)$ -- is the correlation function on the i-th matter-element to be evaluated and the j-level of matter-element;

$\rho(v_i, V_{cji})$ -- is the distance of the i-th matter-element to be evaluated and all the classical domain;

$\rho(v_i, V_{pi})$ --is the distance of the i-th matter-element to be evaluated and the corresponding section of the domain P;

$\left|V_{cji}\right|$ -- is the size of the classical field value.

The correlation of the object P to be evaluated on the level j is

$$K_j(p) = \sum_{i=1}^{n} w_i K_j(v_i), j = 1, 2, ..., m \tag{4}$$

Among, $K_j(p)$ --is the correlation of the matter-element to be evaluated and the matter-element on the level j;

w_i --is weight coefficients of the i-th characteristic to matter-element evaluation;

Matter-hierarchy method is the way of solving the rating scale problem of matter-element characteristics using the correlation of extension set.

3.2 Case Study

Choose the Taonan wind power station in jilin province for example. Take the sustainability of the Taonan wind electric field project as the appraisal object. Carry on the appraisal using the improved BSC- Matter-hierarchy Model.

The Empowering method of the sustainability of a wind electricity project evaluating indicator system is :Through analytic hierarchy process (AHP), according to the importance of the various indicators, endow them certain weight of which the sum is 1. Assignment is to give the corresponding evaluation value of the three indicators, Unifying present development situation of the wind power both our country and International, and the evaluation results are: very good, better, normal, bad, worse. Finally, endow corresponding appraisal value separately by hundred-mark system:90,70,50,30,10.

The Delphi Method was used in this paper. By putting out Index Evaluation Form and combing the data above, we can get the following comprehensive index :

Table 2. The Weight of Index

Index	U_{11}	U_{12}	U_{13}	U_{21}	U_{22}	$U_.$
weight	0.12	0.09	0.11	0.06	0.12	0.01
U_{31}	U_{32}	U_{33}	U_{34}	U_{41}	$U_.$	$U_.$
0.11	0.06	0.05	0.04	0.09	0.06	0.08

We divide the wind power project sustainability evaluation value R into five classes, that is excellent, good, normal, bad, worse. Write down the matter-element classic fields of five classes. We take the excellent classic field for Example:

$$R_1 = (R_1, C_1, V_1) = \begin{vmatrix} N_1 & c_1 & v_1 \\ & c_2 & v_2 \\ & \cdots & \cdots \\ & c_{11} & v_{11} \\ & c_{12} & v_{12} \end{vmatrix} = \begin{vmatrix} N_1 & c_1 & (a_1, b_1) \\ & c_2 & (a_2, b_2) \\ & \cdots & \cdots \\ & c_{11} & (a_{11}, b_{11}) \\ & c_{12} & (a_{12}, b_{12}) \end{vmatrix} \tag{5}$$

And we give out the other classic fields of good, normal, bad, worse, using the same method. $c_1 \sim c_{12}$ means the 12 indicators of evaluating the wind power project sustainability.

According to the evaluation method of different indicators, we regard the regional with the index data reference and the factual basis for evaluation as the classic domain-wide of wind power project sustainability evaluation matter element model. We can get that:

$$R_1 = \begin{bmatrix} N_1 & c_1 & \langle 90,100 \rangle \\ & c_2 & \langle 90,100 \rangle \\ & c_3 & \langle 90,100 \rangle \\ & c_4 & \langle 80,100 \rangle \\ & c_5 & \langle 80,100 \rangle \\ & c_6 & \langle 80,100 \rangle \\ & c_7 & \langle 90,100 \rangle \\ & c_8 & \langle 80,100 \rangle \\ & c_9 & \langle 90,100 \rangle \\ & c_{10} & \langle 0,0.5 \rangle \\ & c_{11} & \langle 95,100 \rangle \\ & c_{12} & \langle 90,100 \rangle \\ & c_{13} & \langle 80,100 \rangle \end{bmatrix}, \quad R_2 = \begin{bmatrix} N_2 & c_1 & \langle 80,90 \rangle \\ & c_2 & \langle 80,90 \rangle \\ & c_3 & \langle 80,90 \rangle \\ & c_4 & \langle 60,80 \rangle \\ & c_5 & \langle 60,80 \rangle \\ & c_6 & \langle 60,80 \rangle \\ & c_7 & \langle 80,90 \rangle \\ & c_8 & \langle 60,80 \rangle \\ & c_9 & \langle 80,90 \rangle \\ & c_{10} & \langle 0.5,1 \rangle \\ & c_{11} & \langle 90,95 \rangle \\ & c_{12} & \langle 80,90 \rangle \\ & c_{13} & \langle 60,80 \rangle \end{bmatrix},$$

$$R_3 = \begin{bmatrix} N_3 & c_1 & \langle 60,80 \rangle \\ & c_2 & \langle 60,80 \rangle \\ & c_3 & \langle 60,80 \rangle \\ & c_4 & \langle 40,60 \rangle \\ & c_5 & \langle 40,60 \rangle \\ & c_6 & \langle 40,60 \rangle \\ & c_7 & \langle 60,80 \rangle \\ & c_8 & \langle 40,60 \rangle \\ & c_9 & \langle 60,80 \rangle \\ & c_{10} & \langle 1,2 \rangle \\ & c_{11} & \langle 80,90 \rangle \\ & c_{12} & \langle 60,80 \rangle \\ & c_{13} & \langle 40,60 \rangle \end{bmatrix}, \quad R_4 = \begin{bmatrix} N_4 & c_1 & \langle 30,60 \rangle \\ & c_2 & \langle 30,60 \rangle \\ & c_3 & \langle 30,60 \rangle \\ & c_4 & \langle 20,40 \rangle \\ & c_5 & \langle 20,40 \rangle \\ & c_6 & \langle 20,40 \rangle \\ & c_7 & \langle 30,60 \rangle \\ & c_8 & \langle 20,40 \rangle \\ & c_9 & \langle 30,60 \rangle \\ & c_{10} & \langle 2,4 \rangle \\ & c_{11} & \langle 40,80 \rangle \\ & c_{12} & \langle 30,60 \rangle \\ & c_{13} & \langle 20,40 \rangle \end{bmatrix},$$

$$R_5 = \begin{bmatrix} N_5 & c_1 & \langle 0,30 \rangle \\ & c_2 & \langle 0,30 \rangle \\ & c_3 & \langle 0,30 \rangle \\ & c_4 & \langle 0,20 \rangle \\ & c_5 & \langle 0,20 \rangle \\ & c_6 & \langle 0,20 \rangle \\ & c_7 & \langle 0,30 \rangle \\ & c_8 & \langle 0,20 \rangle \\ & c_9 & \langle 0,30 \rangle \\ & c_{10} & \langle 4,18 \rangle \\ & c_{11} & \langle 0,40 \rangle \\ & c_{12} & \langle 0,30 \rangle \\ & c_{13} & \langle 0,20 \rangle \end{bmatrix}$$

Therefore, we can get the classic field:

$$R_Y = \begin{bmatrix} P & c_1 & \langle 0,100 \rangle \\ & c_2 & \langle 0,100 \rangle \\ & c_3 & \langle 0,100 \rangle \\ & c_4 & \langle 0,100 \rangle \\ & c_5 & \langle 0,100 \rangle \\ & c_6 & \langle 0,100 \rangle \\ & c_7 & \langle 0,100 \rangle \\ & c_8 & \langle 0,100 \rangle \\ & c_9 & \langle 0,100 \rangle \\ & c_{10} & \langle 0,18 \rangle \\ & c_{11} & \langle 0,100 \rangle \\ & c_{12} & \langle 0,100 \rangle \\ & c_{13} & \langle 0,100 \rangle \end{bmatrix} \qquad R_O = \begin{bmatrix} P & c_1 & 50 \\ & c_2 & 50 \\ & c_3 & 30 \\ & c_4 & 60 \\ & c_5 & 80 \\ & c_6 & 30 \\ & c_7 & 80 \\ & c_8 & 90 \\ & c_9 & 60 \\ & c_{10} & 0.5 \\ & c_{11} & 95 \\ & c_{12} & 90 \\ & c_{13} & 85 \end{bmatrix}$$

As the evaluation index value of China's wind power project sustainability within the classical domain, correlation can be calculated directly.

We can get the correlation of China's wind power projects' sustainability and rank by calculating.

$$K_1(p) = \sum_{i=1}^{5} w_i K_1(A_i) = -0.285 ,$$

$$K_2(p) = \sum_{i=1}^{5} w_i K_2(A_i) = -0.118 ,$$

$$K_3(p) = \sum_{i=1}^{5} w_i K_3(A_i) = 0.238,$$

$$K_4(p) = \sum_{i=1}^{5} w_i K_4(A_i) = -0.371,$$

$$K_5(p) = \sum_{i=1}^{5} w_i K_5(A_i) = -0.223 。$$

So we can see that the sustainability of the project in the general level, the financial situation and the general location and other index can be improved.

References

1. Shi, C.: The Balanced Scorecard; Performance evaluation system; Financial; Customer; Internal business process; Learning and growth. Nanjing Science University (2004)
2. Cai, W.: Matter-element model and its applications. Technology Literature Press, Beijing (1994)
3. Cai, W., Yang, C.-Y., Lin, W.-C.: Extension engineering technique. Technology Literature Press, Beijing (1997)
4. Chen, S.-Y.: The inaccuracy of variable fuzzy setmodel and extension engineering technique inwater resources system
5. Gao, D.-Y., Zuo, Q.-T.: Harmonious Theoretics and Practice about Human and Water, pp. 91–95. China Water Electricity Press, Beijing (2006)

The Economic Evaluation of the Wind Power Projects Based on the Cloud Model

Fang Fang[1] and Yu Aimin[2]

[1] Schoool of Economics and Management North China Electric Power University Beijing, China
[2] Logistic Services North China Electric Power University Baoding, Heibei Province, China
{orchid514,15210003909}@126.com

Abstract. The develop of wind power projects is rapid in China, but the evaluation system of wind power project economic is not perfect. This article attempts to evaluate the economics of wind power projects base on the cloud model, the transform executes between the qualitative ration and object, and the cloud parameters of the final evaluation were obtained using the competitive cloud which is included in the theory of virtual cloud.

Keywords: Cloud Model, Wind Power Projects, Economic Evaluation.

Recently, a more active and effective economic model is urgently needed to extend the development and living space for the reason that the process of economic globalization has been accelerated, whist the plunder on the resources is increasingly serious, the world is suffering serious damage and the environment stay in a worsening situation. It leads to sparking discussion and debating on the sustainable development(SD) of economic, production and society, since the concept of SD was formally proposed. Today the whole society is troubled by the issues of energy, food, climate, financial, etc., so the GE has become a trend in the global environment and development model. China is in a strategically transformation of the comprehensively implement the scientific concept of development, and the environment-friendly and resources — saving society construction The historical background and development gives new significance and mission to the green economy (GE). However, the multiple crises has given us a major opportunity to boost up the industrial restructuring and change the mode of economic development.

Since the whole world is focusing their attention on energy and environment today, developing new energy has long been recognized as a main way to address environmental and energy issues. As we all known, wind energy has a huge potential for development as a clean, non-polluting, renewable energy, especially in inaccessible remote mountainous areas, countries away from the power grid or difficult to get reached in the near future, coastal islands and grassland pastures. Also

W. Zhang (Ed.): Advanced Technology in Teaching, AISC 163, pp. 443–448.
springerlink.com

it plays a significant role in addressing these energy problems in production and life. So as the most mature, the most development potential and the most market competitive renewable energy generation technology, wind power generation has been in a vogue.

We can know that the wind power projects had a rapid development in China. And China had been the largest newly installed capacity country in 2009 all over the world. The economic evaluation of the wind power projects is conducive to understanding the development of wind power status, in order to promote the development of China's wind power industry.

1 Cloud Model

1.1 The Concept and Numerical Characteristics of the Cloud

The Cloud is a model using the language value to represent the uncertainty conversion between a qualitative concept and its quantitative representation, which mainly reflects two kinds of uncertainty of the concept in natural language, that is, the fuzziness and the randomness, and this two are completely integrated to form the mapping between the qualitative and quantitative.

Suppose U is a quantitative domain expressed in precise quantitative value, C is a qualitative concept on U, if the quantitative value x∈U is a random realization of the qualitative concept C, the certainty of c, $\mu(x)\in[0,1]$ is random numbers of a stable tendency: $\mu:U\to[0,1],\forall x\in U,x\to\mu(x)$, then the x in the distribution of the domain U is called the Cloud, each x is called a cloud droplet. If the domain corresponding the concept is n-dimensional space, then it can extend to the n-dimensional clouds.

Expressed by the Cloud Model, the overall characteristics of the concept can use the characteristics of the Cloud to reflect, the Cloud uses the its three characteristics : the Expected value E_x, Entropy E_n, Hyper entropy H_e, to overall charactcrize a concept, denoted as C (E_x,E_n,H_e). E_x is the expectation of cloud droplets in the distribution of the domain, the point that represent qualitative concept mostly, or rather, the most typical samples of the quantification of the concept; the entropy E_n represents measurable particle size of the qualitative concept, the greater the entropy is, the more macroscopic general concept is, also is the measurement of the uncertainty of the qualitative concept, decided by the randomness and fuzziness of the concept. Hyper entropy He is the measurement of the uncertainty the entropy, namely, the entropy of the entropy, decided by the randomness and fuzziness of the entropy .

1.2 Cloud Generator

Cloud Generator, CG for short, is a cloud model generation algorithm modularized by a software or cured by hardware, which establish linkages and interdependent between the qualitative and quantitative, there are quantification in the qualitative and qualitative mapping in a quantitative, there are two kinds of cloud generator: Forward Cloud Generator and Backward Cloud Generator .

Forward Cloud Generator is a mapping from qualitative to quantification, and its input is the numerical characteristic of cloud (E_x, E_n, H_e)and the number of the cloud droplets N, the output is the quantitative position of these cloud droplets in the quantitative domain space and the certainty of the concept that each cloud droplets represent.

Backward Cloud Generator is a mapping from quantification to qualitative, it will convert a certain number of precise data to the numerical characteristics (E_x, E_n, H_e) representing the qualitative concept.

2 Index System

The economic evaluation economic of wind power projects have lots of complex impact factors, it is difficult to evaluate only with the single indicator, and it must be a ulti-dimensional comprehensive evaluation. In this paper, we use the improved-BSC method to build the Index system, and use the Delphi and AHP (AHP) to determine the weigh.

Table 1. The index system and weight

	Secondary index	weight
The economic of the Wind power project	Financial profitability U_1	0.18
	Financial solvency U_2	0.15
	Staying power U_3	0.12
	construction costs U_4	0.09
	operating costs U_5	0.23
	labor costs U_6	0.10
	Price of power U_7	0.13

3 Cloud Model

3.1 The General Idea

Establishing the evaluation model of the economics of the wind power projects, first is to analyze each single factor impacting economic of the wind power projects, compose the collection of single-factor, and make sure the degree of influence of each single factor to the overall goal, namely Weight; then to confirm the collection of the remarks; then to process the collected data utilizing Backward Cloud Generator, get cloud model and its numerical characteristics of each single factor in the evaluation model of the wind farm construction process; Finally, to calculate numerical characteristics of each single factor in cloud model utilizing the integrated cloud algorithm in virtual cloud and to integrate the evaluation cloud of each single factor to form the integrative cloud of the economics of the wind power projects. Specific process is as follows:

① Determine the evaluation criteria and weights

② Determine the remarks collection, the remarks collection in this article :
V = (v1, v2, v3, v4, v5), which,
v1 (90 ~ 100 points) is the excellent;
v2 (80 ~ 90 points))is the good;
v3 (60 ~ 80 points))is the general;
v4 (40 ~ 60 points) is the poor;
v5 (0 ~ 40 minutes)is the very poor.

③ To obtain cloud model and its numerical characteristics of single factor utilizing actual data through Backward Cloud Generator. Calculating process as follows:

Input: the sample points x_i , where i = 1, 2, …, n.

Output: the numerical characteristics (E_x, E_n, H_e) reflecting qualitative concept.
Algorithm:

i utilizing x_i to calculate mean value of the sample

$$\overline{X} = \frac{1}{n}\sum_{i=1}^{n} x_i \tag{1}$$

First principal Sample canter moment

$$\frac{1}{n}\sum_{i=1}^{n} \left|x_i - \overline{X}\right| \tag{2}$$

sample variance

$$S^2 = \frac{1}{n-1}\sum_{i=1}^{n}(x_i - \overline{X})^2 \tag{3}$$

ii

$$E_x = \overline{X} \tag{4}$$

iii

$$E_n = \sqrt{\frac{\pi}{2}} \times \frac{1}{n}\sum_{i=1}^{n}\left|x_i - E_x\right| \tag{5}$$

iv

$$H_e = \sqrt{S - E_x^2} \tag{6}$$

We use the virtual cloud theory assessment of each single factor clouds the wind power project together economic assessment of the cloud, the specific formula is as follows:

$$\begin{cases} E_x = \dfrac{E_{x_1} w_1 + E_{x_2} w_2 + \ldots + E_{x_n} w_n}{w_1 + w_2 + \ldots + w_n} \\ E_n = \dfrac{w_1^2}{w_1^2 + w_2^2 + \ldots + w_3^2} E_{n_1} + \dfrac{w_2^2}{w_1^2 + w_2^2 + \ldots + w_3^2} E_{n_2} \\ + \ldots + \dfrac{w_n^2}{w_1^2 + w_2^2 + \ldots + w_3^2} E_{n_n} \\ H_e = \dfrac{w_1^2}{w_1^2 + w_2^2 + \ldots + w_3^2} H_{e_1} + \dfrac{w_2^2}{w_1^2 + w_2^2 + \ldots + w_3^2} H_{e_2} \\ + \ldots + \dfrac{w_n^2}{w_1^2 + w_2^2 + \ldots + w_3^2} H_{e_n} \end{cases} \tag{7}$$

3.2 Example Applications

Take the wind power projects as an example, in accordance with the evaluation based on cloud model. First, use backward cloud generator to process it on the basis of the data collected, get each cloud model of sub-index and its numerical characteristics in the economic evaluation of the wind power projects shown in Table 2; then use Cloud Algorithm in these virtual cloud theory to integrate assessment cloud of each single factor and get the assessment cloud of the economic evaluation of the wind power projects, the results for the C ($E_x = 57.5$, $E_n = 3.6$, $H_e = 0.46$).

Table 2. The mumerical characteristics and level

index	U_1	U_2	U_3
weight	0.18	0.15	0.12
E_x, E_n, H_e	(84.0,1.6,0.2)	(63.5,3.4,0.2)	(56.1,3.2,0.5)
level	good	general	poor
U_4	U_5	U_6	U_7
0.09	0.23	0.10	0.13
(33.2,6.5,1.3)	(23.4,6.3,0.7)	(63.2,1.6,0.3)	(88.2,1.6,0.2)
very poor.	very poor.	general	excellent

4 Conclusion

Overall, we can know that the price of power of the wind power projects is in a high level, the financial profitability is good, and Financial solvency and labour costs is general. but the Staying power is not so good , the construction costs and operating costs have had been greatly improved.

The economic evaluation of the wind power projects in China is just in a so low level, so we must pay more attention on the economics of the wind power projects. It is good for the development of power industry.

References

1. Li, D.: Uncertainty in artificial intelligence. National Defense Industry Publishing Press, Beijing (2005)
2. Shao, S., Liang, W.: Cloud model in the prediction of water supply and demand. Northwest University: Natural Science 29(71), 79–82 (2008)
3. Luo, S., Liu, S.: Based on cloud model integrated digital image quality Joint assessment. Surveying and Mapping 25(2), 123–126 (2008)
4. Hedman, K.W., Gao, F.: Overview of Transmission Expansion Planning Using Real Options Analysis. In: Power Symposium Proceedings of the 37th Annual North American, pp. 497–502 (2005)
5. Martin, O., Oliver, M.: Investment under uncertainty does completion matter. Journal of Economic Dynamics Control 31(3), 994–1014 (2007)
6. Wijnia, Y.C., Herder, P.M., Neufville, R.D.: Options for real options: dealing with uncertainty in investment decisions for electricity networks. IEEE International Conference on Systems, Man and Cybernetics 4, 3682–3688 (2005)
7. Fan, D., He, S.: A cloud model based on comprehensive evaluation model. Information Development and Economy 13(12), 157–159 (2003)
8. Li, D., Liu, C.: On the normal cloud model of. Chinese Workers Process Science 6(8), 28–34 (2004)
9. Chen, X., Duan, C., et al.: Prediction model of annual precipitation based on cloud model. In: The First International Conference on Risk Analysis and Crisis Response (2007)

Development of Shared Interactive Coursewares Based on ActiveX Controls

Jinguo He

School of Science Minzu University of China
Beijing, China
hejinguo@gsm.pku.edu.cn

Abstract. Interactive teaching means that students take part in the classroom discussion. More and more imaginative coursewares, animations as well as multimedia materials are developed to hold the attention of students. This paper discusses the development of shared interactive coursewares based on ActiveX controls. With shared interactive coursewares, different parameters or even different functions can be selected or inputted to demonstrate a complicated theory in order that interactive teaching is more effective. Furthermore, such coursewares can be inserted into PPT, word files, as well as web pages etc. They are more convenient to be shared because of the above compatibles and its flexible functions.

Keywords: Interactive teaching, courseware, PPT, animation.

1 Introduction

Interactive teaching is wildly discussed in papers. In a college math classroom, teachers can organize the class by specific methods so that students have their opportunities to think, practice and express [1-2]. Thus teachers turn to be a good advisor. Such methods are obviously good methods to improve the learning effect. However, the shortcoming is that we can not perform each lesson by such kinds of methods because such methods are time-consuming and college lessons have plenty of definitions, theorems and examples.

Modern information technologies are also used to improve interactive teaching. Web-based softwares and hardwares are developed to give the students opportunities to take part in the learning systems. Systems built by hardwares and softwares can help students do interactive experiments [3-6]. In fact, PPT files are also a product of information technologies; we use PPT files, multimedia materials and animations to improve the students' attention.

In math class and many other classes, it is very important for the students to take part in thinking. Math is gymnastics sports for brain and thinking makes students bright. The purpose of math teaching is thinking, not remembering or solving the praxes. All the praxes in the text book are already solved so that we need not the students' answer, we need they think. Students who solve the praxes by reference book or by reciting can not practice their brains; just like that students who go to the hilltop by cable car can not practice their body.

W. Zhang (Ed.): Advanced Technology in Teaching, AISC 163, pp. 449–452.
springerlink.com

In our math class, systems with hardwares and softwares are good for improving teaching effect but they need many resources, organizing a discussing class takes tool much time and may delay the schedule. In this paper, we develop shared interactive coursewares based on ActiveX controls, because such controls can be integrated into PPT files, word files as well as web pages, it is convenient to be popularized and no more harewares and softwares resources are needed. Their flexible functions include interactive parameters selecting and experiments-like interface can improve the interactive teaching effect. Its compatibles give the possibilities of sharing excellent and creative ActiveX coursewares.

2 Development of Steepest Descent Method Control

This control is developed for demonstrate the progress of steepest descent. Given the following equations:

$$\begin{cases} f_1(x, y) = 0 \\ f_2(x, y) = 0 \end{cases} \tag{1}$$

To solve the equation (1), let

$$\Phi(x, y) = [f_1(x, y)]^2 + [f_2(x, y)]^2 \tag{2}$$

Obviously $\Phi(x, y) \geq 0$, and,

$$\Phi(x, y) = 0 \Rightarrow f_1(x, y) = f_2(x, y) = 0$$

So the solutions of (2) are the solutions of (1).

Let

$$\Phi(x, y) = [f_1(x, y)\ ]^2 + [f_2(x, y)\ ]^2 = c \tag{3}$$

Thus (3) define an isoline in the $x - y$ plane, we can get a series of isolines by different c .This series of isolines shrinks to one point at the minimum, and the coordinate of this minimum point satisfy that $\Phi(x, y) = 0$.

So, we can find the minimum of (2) by steepest descent method to solve (1).

Here we simply give the steps of steepest descent method:

Step A Randomly select (x_0, y_0), calculate $\nabla\Phi(x, y)$, the gradient of $\Phi(x, y)$;

Step B Calculate the proper step size λ;

Step C Calculate the new position:

$(x_1, y_1) = (x_0, y_0) - \lambda\nabla\Phi$;

Step D If $\Phi \approx 0, (x_1, y_1)$ is the solution, else let

$(x_0, y_0) = (x_1, y_1)$, repeat step A.

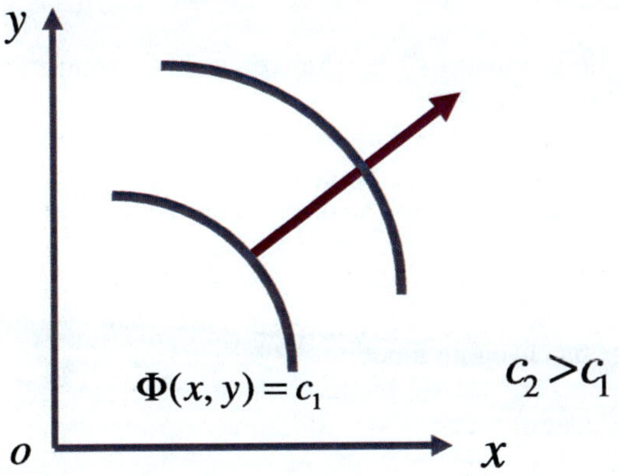

Fig. 1. Isolines of equation (3)

We use visual c++ MFC to develop steepest descent method control. We want to demonstrate the progress of descent method. Fig.2 is the interface of the control. In the interface, we can see the axes of coordinates. And given a special f_1, f_2 in equation (1), e.g., let

$$\begin{cases} f_1(x, y) = 3x - y - 2 = 0 \\ f_2(x, y) = -x + y = 0 \end{cases} \tag{4}$$

The control will draw the isoline of randomly selected point (x_0, y_0), and mark the position of (x_0, y_0). After clicking next button, the iteration begins and (x_0, y_0) moves towards the direction of $-\nabla\Phi$ and stops at (x_1, y_1). This control can give the students a visual progress that any initial point moves towards minimum point $(1,1)$.

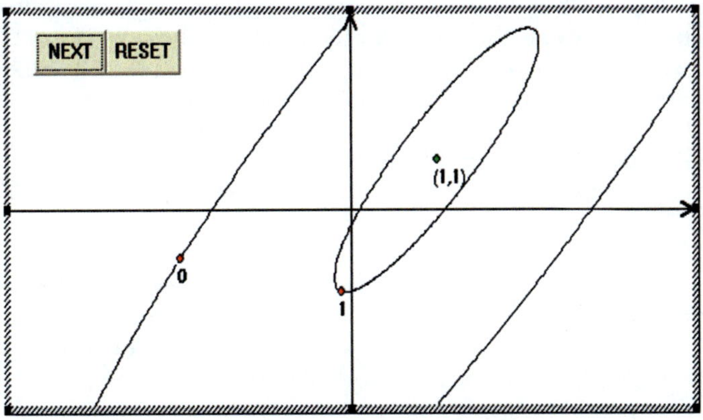

Fig. 2. Interface of the control

3 Taking Advantage of the ActiveX Control Courseware

Here we discuss taking advantage of the ActiveX control coursewares, not only the control which we developed above.

A ActiveX control are compatible components in binary mode. i.e. ActiveX control can be loaded in any containers, such as IE, word, PowerPoint, Excel etc. So we can insert our control into PPT files, thus acquire a new interactive effect without new systems with new hardware or new software. Because PPT presentation is very popular in college classroom, such control can easily be popularized.

Further more, such controls can be inserted into web pages, students can have an interactive experience by changing functions or parameters.

B Such controls can be shared by teachers in different colleges or different countries. Because controls can have different results by selecting different parameters, they can fit for every level of education. The shortcoming is that making control needs the technology of programming; however, the shortcoming can be overcome by sharing the controls.

C ActiveX control coursewares are independent components. Thus students can use them to practice after class. For example, we can make an ActiveX control which shows the $x-y$ plane curves of elementary functions, student can change the elementary functions to solve their confusions.

4 Conclusions

This paper discuss courewares made by ActiveX controls, by a real example of steepest descent control, we demonstrate that such coursewares can improve the interactive effect while being compatible with current hardwares and softwares. They can also provide experiment experiences by interactively selecting parameters. Sharing this kind of controls is also a good idea because they can provide extensive education level by different parameters.

References

1. Cheng, D., Da, L., Zhan, J.: Designing interactive education system. In: Proceedings of IEEE International Symposium on IT in Medicine and Education, pp. 450–453 (2008)
2. Li, X., Chen, J.: Study on interactive education of subject-based English. In: Proceedings of International Conference on Optics Photonics and Energy Engineering (OPEE), pp. 508–511 (2010)
3. Rahkila, M., Karjalainen, M.: An experimental architecture for interactive web-based DSP education. In: Proceedings of the 1998 IEEE International Conference on Acoustics, Speech and Signal Processing, pp. 1857–1860 (1998)
4. Zong, X.-P., Wang, P.-G.: Interaction on multimedia teaching. In: Proceedings of Second International Conference on Education Technology and Training, pp. 52–55 (2009)
5. Zhang, W., Wang, C.: Research on the Application of Computer Multimedia Technology in College Teaching Aerobics. In: Proceedings of Second International Conference on Education Technology and Training, pp. 67–70 (2009)
6. Rui, Y., Wang, Y.: Development of Web-Based University Library Teaching Reference Information System. In: Proceedings of Second International Workshop on Education Technology and Computer Science (ETCS), pp. 488–491 (2010)

The Teaching Problems in Fashion Buyer Course and Their Proposed Solutions

Junying Zhang

Fashion Design and Engineering Department
Henan Institute of Engineering
Zhengzhou, Henan Province, China
Sunny_eml@126.com

Abstract. Based on three marketing surveys data: 1.from 50 garment enterprises to understand its mode of application of fashion buyers; 2.from 100 persons who are play the role of fashion buyers to know their understanding to their work role and fashion buyer; 3.from 10 colleges which have fashion buyer course or professional to learn the settings in the professional buyers and the courses related to fashion buyers. Based On above three researches, the author found the market demands for fashion buyers and the problems in the teaching, then provided the corresponding solutions in the end of the paper.

Keywords: Fashion buyer, teaching, market, problem, solution.

1 Introduction

Based on the outsourcing process, a virtual business model was born in the clothing, department stores in European and American in the late seventies. Related to different industries such as clothing, shoes, cars, department stores etc., then fashion buyers was born as a profession. With the rapid development of apparel industry, fashion buyers, the profession has also been recognized by more and more people in the country In recent years. However, there are still many problems during in fashion buyers' working. In addition, in recent years, fashion buyer courses were set in some garment institutions to meet the market demand for talent buyers. But the buyers, as a new exploratory subject, there are also some problems in teaching. So it has far-reaching significance to understand the market demand for professional buyers and teaching problems in fashion buyer courses, then to take countermeasures, so that the garment institutions can better adjust to teaching, more better the output to the market.

2 The Current Status of the Domestic Clothing Market Buyers

From telephone survey to 50 garment enterprises in Shanghai, Zhejiang, Jiangsu, Fujian etc., and questionnaire survey to 100 persons who engiged in fashions buyers career ,we found that there were some problems below in fashion buyers situation in the apparel market At present.

W. Zhang (Ed.): Advanced Technology in Teaching, AISC 163, pp. 453–457.
springerlink.com

2.1 The Concept of Fashion Buyer Fuzzy in the Domestic Apparel Industry

At present, some foreign brands such as ZARA, NIKE etc. use buyer mode very successfully. In China the Yangtze River Delta and Pearl River Delta, 70% companies what were researched run their business in buyers business model. However, in the companies what have been used buyers business model. a half of them don't know what they use is buyers business model..

In addition, in the 100 staffs who play the role of fashion buyers in above enterprises, there are 73% persons who don't know their own just are fashion buyers; 57% persons who don't know what is fashion buyer; and they think they are apparel buyers.

Above all, the concept of fashion buyers is vague for the persons in fashion market, which is rather a buyer than a fashion buyer.

2.2 The Shortage of Highly Qualified Fashion Buyers

In the interviewed companies, the so-called fashion buyers are mostly from the fashion designers, sale staffs etc. , which make the professionalism of fashion buyers mixed. Enterprises say that the real high-quality apparel market buyers are scarce, even if the enterprises pay them high salaries. So the enterprises particularly hoped to higher quality professional fashion buyers to train in the universities.

2.3 A High Level for Buyers to Work

Professional buyer is not a simple paragraph with only the functions of procurement and copied. According to business type, the fashion buyers have been developed in three types: shop buyers, brands buyers and department store buyers. Although each type of buyers work in different scope, it is common that the least basic skills for fashion buyers can get for more than 3 years working in garment industry. That is to say,

To be a fashion buyer, you must have the following conditions:

You must be a faithful practitioner of fashion;

You must have keen insight, the predictive power and accurate sense to fashion trends;

You must own good communication skills, innate sense of curiosity and courage to explore;

You must know at least one or two foreign languages;

You have to competent the task of long-distance travel frequently.

3 The Problems in Fashion Buyers Teaching

After the research of literatures on 10 colleges what Opened fashion buyer professional or course, we found several problems in fashion buyers teaching in below.

3.1 The Lack of Teaching Book for Fashion Buyers

Currently, according to the demand of talent for fashion buyers in domestic apparel market, the fashion buyers teaching show different levels of teaching taught object. But Very few books for fashion buyers teaching written by the domestic and foreign scholars, which are mainly to translation, can not meet the needs of all levels of teaching objects.

3.2 The Fewer Relevant Curriculum for Fashion Buyers

Very It is understood that foreign university courses of professional buyers with "strategic buyers", "textiles", "popular trend analysis", "data analysis", "market strategy" and "financial management" and so on.

The domestic apparel buyer only as a course was taught to students of fashion marketing professional, which is far from enough for the students who would engage in fashion buyers career in future!

3.3 The There Is Few Practical Conditions for Fashion Buyers in Teaching

Fashion buyer is a highly practical operation course, but there is no practice field for it in common institutions, which is also a difficult obstacle for this course teaching.

3.4 The Lack of Teaching Book for Fashion Buyers

Very few books for fashion buyers teaching written by scholars in our country and aboard, which can not meet the needs of all levels fashion buyers for study.

4 The Suggestion for Fashion Buyers Teaching

4.1 The Individualized, Revised Teaching Plans and Curriculum, to Make Up the Shortage of Teaching for Different Levels of Fashion Buyers

To set up fashion buyers profession, or add several objects for marketing professional related to fashion buyers in garments institutes, is as much as possible to meet with international, so that the students can learn as much as possible in fashion buyers professional knowledge.

On the other hand, different levels learners appear in garment institutions or training institutions to meet different demands of society, so we have to render appropriate curriculum for them and revise the teaching plans to cover the corresponding teaching books shortage.

4.2 Colleges Joint with Enterprises to Solve the Difficult Problem of Training

4.2.1 Build a Training Platform to Establish and Use a Practice Base Reasonably

Fashion buyer is a very strong practical operation professional, while its practice is a required lesson before students enter into society.

Training hard for students has become a common phenomenon before students graduating from colleges. Because garment enterprises generally hope to recruit experienced buyers, in order to work immediately after recruitment. It is told by university that it is difficult to find internships, because universities have to not only pay costs, but also often need to take a human sympathy. Many college students rely on teachers, parents and private help to find internships.

Establish training bases, can completely solve the problem of students' practical difficulty. The colleges can unite with enterprises to develop practice bases in enterprises where can export order talents by using teaching orders etc.

4.2.2 Take the Form of Pre-employment to Reduce the Blindness between Work and Selecting Talents

The students usually are arrangement to practice and write paper or design in a year before graduation in fashion marketing teaching plan of garment colleges. At the stage for graduation practice, employing enterprise gives a train to students by pre-employment as much as possible.

For enterprises, it is a valid channel to find talents and reserve them from accepting students in a role of assistant buyer or training buyer. For the students' practicing can not only strengthen their practice ability, but also broaden their horizons and employment channels. So that the graduating students are familiar with the employer in advance, then they can work better. In addition, based on the demand of pre-employment, students can do some research to finish their graduate paper, which will better work future on theoretic but also ensure the paper's truth and convince by providing the original data from it.

4.3 Optimization Teaching Methods to Enable Students Deep Understanding for the Role of Fashion Buyers

4.3.1 Take Advantage of Multimedia Technology Adequately

Buyers course is highly fashionable, multi-media teaching can not be separated. Using multimedia teaching methods create a harmonious atmosphere lively can help to interest students in learning, then change it from passive to active. For example, during the teaching for popular tendency of fashion buyer forecast, It needs extensive PPT to show pictures of fashion colors and popular elements.

4.3.2 Interactive Teaching Instead of Statement Methods

We can achieve interactive teaching by interactive with the course grade to mobilize the enthusiasm of the students adequately to match up the teachers.

4.3.3 The Content Description and Analysis in More Cases

It is very beneficial to students that more cases can be decrypted and analysis by them during teaching process. Students can summarize and analyze the phenomenon, the problems in cases so that they can convey the similar problems in future works.

4.3.4 Simulation Scenarios, Role Play

We can simulate scenarios and role play to overcome the shortage of teaching conditions. For instance, we can simulate the scene of fairs, primary ordering meeting and final pre-election scenario, which the different roles can be played by students, so that the students know the corresponding work and its points. From this, the students can understand the role of fashion buyer deeply.

References

1. Diamond, J., Pintel, G.: Retail buying, Wang, Q., Gong, W.-P.(trans.), 7th edn. China Textile Press (March 2007)
2. Rick, G.: Fashion buying (International Apparel Books). China Textile Press (January 2009)
3. Wang, S.: Clothing stores "three hands" collaborative sales model. Machinery Industry Press (January 2009)
4. Wang, Y.: Apparel buyers and purchasing management, 1st edn. China Textile Press (August 2010)

The Exploration of the Mechanical Transmission Teaching in the Machine Design Course

Xinghua Li

School of Mechanical Engineering Tongji University
Shanghai, China
lixinghua@tongji.edu.cn

Abstract. In the machine design course, mechanical transmission have more knowledge, abstractive concept, complicated formula, a large number of charts, so it's difficult to understand and use. The calculation formula is divided into three types in teaching practice, such as the conclusive formulas, empirical equations and geometric relations. It needs to make some difference to them during the teaching; the chart type of data should be explained in detail, especially the selection methods, so that students can select the correct data and substituted into the formula for design and calculations. at the same time, the use of chart analysis, the basic knowledge and concepts of inductive analysis and comparison can help to develop students' ability of comprehensively researching problems. Finally, through a great work to consolidate the training, the students can be further motivated to study professional knowledge.

Keywords: Machine design, mechanical transmission, teaching method, education, chart analysis.

1 Introduction

The Machine Design course is a technical foundation course for mechanical engineering in engineering colleges [1]. This course involves many related courses, more knowledge, abstractive concept and complicated formula, so it's not easy to learn. Especially the design calculations for transmission components, in which students often reflect there are too many formulas, how to remember? They also reflect there are a lot of data in textbooks, how to correctly use them? With many years teaching practice of this part, this paper summed up some experience as following.

2 Correct Understanding of the Formulas

For the formulas of the design and calculation of transmission components in the "Machine Design" course, which was clearly pointed out in lectures: during the analysis of work conditions, this course need to do some mathematical, mechanical derivation, and formulas are indeed a lot, but there is no great number of the "formulas" of the conclusions. Thus the students called the formulas used in the process of calculus and derivation "formulas", which are purely misunderstanding. By analyzing the formulas in this part of the teaching content are divided into the following three respects, which can be treated differently:

W. Zhang (Ed.): Advanced Technology in Teaching, AISC 163, pp. 459–463.
springerlink.com

2.1 The Conclusive Formulas Derived of the Analysis of Theory

Such as standard gear drive, formulas for calculating the tooth surface contact stress σ_H, reference circle diameter d_1, Tooth root bending stress σ_F, modulus m. Such formulas will be directly used for design calculations, which is important. The formula should be familiar with the meaning of each symbol, the assumption made in the derivation should be learned, and what should pay attention to in use, in order to be able to correctly and flexibility use formulas.

In the teaching students are asked to use analysis and mapping to help remember instead of rote. For some of the commonly used (such as calculating speed V, the torque T) formulas, be sure to memorize.

2.2 Empirical Equations or Recommended Formulas

Such as V-belt drive recommended center distance a_0 best satisfy the relations (1), but in the specific design calculations, as a result of the restricted conditions, the actual center distance can not satisfy the relations (1), then you can allow the use.

$$0.7(d_{d1} + d_{d2}) \leq a_0 \leq 2(d_{d1} + d_{d2}). \tag{1}$$

Where: d_{d1} and d_{d2} are reference diameter of small pulley and large pulley in the V-belt transmission.

2.3 Geometry Relationship Formulas

For example, in the V-belt transmission, the formulas for calculation of V-belt transmission center distance and the length do not have rote, as long as graphics are able to be understood, and the formulas can be used. While in the gear drive, the geometry is relatively simple, you can use the corresponding geometric diagram to remember.

3 Correct Selection of Data

In the teaching contents of the mechanical transmission, there are also some introduction about how to select the data, how to solve problems, etc., thus giving students a clearer idea in solving problems, instead of blindly using wrong formulas.

This part of the teaching content requires selecting a lot of data, such as data in Linear graph form and data in tabular form. What factors will affect the data and how the factors impact the data should be specified.

Especially the method of selecting data through chart needs to be explained in detail. For example, when selecting the limitation of the tooth contact fatigue and to select the limitation of the tooth root bending fatigue, it should be noted that, what material the gear is made of, then according to the materials to select the chart and select the curve, so as to prevent choosing the wrong chart or curve and obtained the wrong value; Another example, when selecting the dynamic load factor and selecting the load distribution factor, the accuracy of gears and the arrangement of gears should be paid attention to, and according to the corresponding curves and tables; only based on a clear mind about the factors that affect the data can you select out the correct data; Only students learn the method of selecting data can they select the correct data and substituted into the formulas for design calculations.

4 Through Analysis and Comparison Cultivating the Ability to Develop Comprehensive Research Problems

Each completion of one kind of transmission, students can be roughly understood the contents of this drive, which is just an "individuality"; There are internal relations between the various transmission, which is called "common". If the individuality can be linked with the common, by analysis and comparison, we can deepen our understanding; and cultivate the ability of students to learn by analogy. Therefore, some forms can be prepared to achieve this purpose in teaching.

For gear failure modes, due to many causes, it's both easy to grasp, but also difficult to remember. In the lecture, table analysis (Table 1.) is used to show out all the influencing factors, analyze one by one, so that it becomes easy to understand, and easy to distinguish causes between different failure mode. In order to promote thinking, we can focus on several causes of failure mode such as contact fatigue pitting, bending fatigue fracture, scuffing, wear, etc. and make them clear. Then talk about the failure phenomenon of the tooth when the static strength is not enough, while the reason why static strength is not enough and a variety of ways to prevent or mitigate failure measures, please let the students think for themselves and fill them out.

Table 1. Failure modes of gear drive

				Contact fatigue Pitting	Bending fatigue fracture	Scuffing	Wear	Plastic deformation	Suddenly broken	
Internal cause	Tooth surface hardness HBS ≤350									
	Tooth surface hardness HBS >350									
External cause	Normal force F_n	Light load								
		Heavy load								
	Stress	Tooth surface contact stress σ_H								
		Tooth root bending stress σ_F								
	Motion	Rotational speed n	High							
			Low							
	Friction force $f F_n$									
	Working conditions	Installation								
		Lubrication	Open drive							
			Closed drive							
		Load	Smooth							
			Impact							

In another example, when the chapter about the worm drive was finished, you can list a form as shown in Table 2, and ask the students to fill it out in order to acquire the forces analysis requirements of each chapter.

When cylindrical helical gear drive was finished, you can list a form as shown in Table 3, requiring students to fill in after-school, making them seriously consider both the similarities and differences, in order to achieve the purpose of consolidation.

Table 2. Force analysis of gear drive

		Direction of peripheral force F_t	Direction of radial force F_r	Direction of axial force F_a
Cylindrical spur gear drive	Driving wheel			
	Driven wheel			
Cylindrical helical gear drive	Driving wheel			
	Driven wheel			
Straight bevel gear drive	Driving wheel			
	Driven wheel			
Worm drive	Worm			
	Worm gear			

Table 3. Comparative analysis of cylindrical gear drive

	Cylindrical spur gear drive	Cylindrical helical gear drive
Engagement principle		
Geometry relationship		
Force analysis		
Tooth surface contact fatigue strength		
Tooth root bending fatigue strength		
Design steps		
Other		

5 Through Assignment to Consolidate the Training

Through the above process of teaching, although students initially master the content and the design calculation methods for the mechanical transmission, without intensive training by examples, in the actual design calculations, students can't use freely on the purpose. Therefore, in the classroom the example is used to describe the design calculation, during the explanation of the process, the interactive teaching model is used to allow students to participate, apply the content of the theory to the actual design, and teachers do not teach too much, make students take the lead only.

After-school assignment of design for mechanical transmission [2], through the students' practical design calculations, can improve students' proficiency in design calculations, and further consolidate the knowledge of the design calculations the students learned, but also enable students to recognize their knowledge learned is useful, only to master the knowledge well can you use it to design mechanical components, in order to become a design workers, thus further motivate students to learn professional knowledge. To develop students' creation abilities and abilities of daring to deal with the challenges.

The above are my exploration in the teaching process, through the application, students' enthusiasm for learning has increased, and students' interactive participation in teaching has significantly grown up and learning initiative has also been significantly improved, the knowledge of the design and calculation of mechanical transmission has been mastered more firmly. In addition, students are able to skillfully apply the learned knowledge of mechanical transmission design calculations to the actual design calculations.

References

1. Pu, L., Ji, M.: Design of Machinery. China Higher Education Press, Beijing (2006)
2. Wang, K., He, X., Wang, X.: Course Exercise in Mechanical Design and Mechanical Design Basis. China Higher Education Press, Beijing (1995)

Research on Application of Multimedia Teaching over Literature Classroom*

Jing Guo and Jianping Zheng

School of Literature and Art
Southwest University of Science and Technology
Mianyang, Sichuan Province, China
guojing@swust.edu.cn, zhengjianping@swust.edu.cn

Abstract. Multimedia teaching has been widely used in the liberal classroom, but in process of practical application there are aspects of misunderstanding and application errors. To resolve these mistakes we should set at least the following points: clear multi-media teaching methods are as for teaching services and complementary tools, teacher is still in dominant in the teaching activities, multimedia technology should be subject to discipline under the premise of the characteristics of subjects, courseware is the core of multimedia teaching and learning activities, the desired effect of teaching activities depends on the designing level of courseware. It will be possible to practically improve effectiveness of literal classroom teaching only if the above understanding is recognized with help of multimedia technology.

Keywords: Literature classroom, multimedia teaching, misunderstanding.

1 Introduction

With the rapid development of the information technology and the advent information era, multimedia teaching has become an important part of the modern educational technology. Moreover, it has been popular and widely used in the vast majority of schools. This teaching method aided in the computer by using the comprehensive media forms, can effectively avoid and overcome the traditional teaching methods on single or one-sided shortcomings, while being able to effectively shorten the learning time and improve the quality of teaching and teaching effectiveness, to achieve the optimal teaching target. With the existence of these obvious advantages, multimedia teaching by computer education is increasingly common in China launched a sweeping storm-like. Almost all of the subject teaching the multimedia teaching method is introducing, while it is hitting the abstract communication-based language arts courses using its powerful charm. With the multimedia technology, literature teachers can stereoscopically and visually present the class contents. Moreover, with the provision of rich audio-visual perception environment, the class content is shown by the wide range of carrier which improves greatly the teaching content of the multi-dimensional perception. Additionally, the application of multimedia technology also

---

* This work is partially supported by Education Reform and Research Project of province (P09225).

W. Zhang (Ed.): Advanced Technology in Teaching, AISC 163, pp. 465–471.
springerlink.com

can dilute dull and rigid elements during the explaining of the abstract texts, thus creating a relaxed and fun learning atmosphere. However, in the courses of the practical teaching activities, there are some misunderstanding and application errors to the multimedia teaching.

2 Misunderstanding in Multimedia Teaching

2.1 Media Technology Is Used Completely, Ignoring the Teacher's Dominant Role

In the multimedia teaching activities, the media technology can dimensionally and visually present the contents which they are can not be achieved with abstract language texts. As a result, some literature teachers in the teaching process emphasize excessively the media technology, while ignoring the dominant position of teachers in the classroom. Course contents are overwhelmed completely by fancy animation and video effects, eventually leading to either information overload of multi-media teaching, or from the traditional 'people filling' to a modern 'electric irrigation'; or messy information, colorful computer screen, being similar to see a film, these can not fully embody the advantages of multi-media teaching [1]. In the traditional teaching methods, teachers can attract the attention of students with their own beautiful calligraphy writing on the blackboard, pretty cadence of reading, witty humor, body language, and explanation of the classroom language. Moreover, these can form a strong charm of magnetic field, in which the rhythm of classroom is completely controlled under the teacher's intention. In the current multi-media teaching, however, teachers take a back seat, replaced by a computer and various software tools. In this case, the teachers are habitually enslaved servants for the computer. Thus, a whole class will not leave the computer, while the pace of the classroom is not entirely under the control of teachers. The classroom teaching is evolved into fine courseware presentation. This misunderstanding leads to the teachers without careful preparation of lessons before class. We can imagine how effective this type of teaching method due to the dislocation and the loss of dominant position of teachers in the classroom. Consequently, teachers are not familiar with the content. In order to avoid complications, they only demonstrate courseware of the classes, not thinking about, having no time to attend and observe the students and exchange with the students. Thus, the classroom is lacking of flexibility and interactivity.

2.2 Incompatibility Is Existed between Multimedia Teaching Methods and Traditional Teaching Methods

Excessive emphasis on multimedia teaching under the way of media technology will lead to the abandon of traditional teaching mode. The traditional teaching modes of blackboard-writing, teacher's body language, teacher's personality charm, and the exchange between the teachers and students in classroom activities are all submerged by multimedia technology. In fact, any kind of teaching method has its own advantages and limitations, so-called “Sometimes a foot may prove short while an inch may prove long". The traditional teaching means also have the superior effect of teaching which the multimedia teaching can not achieve. For example, the use of traditional blackboard-writing has not any information of interference, which is more conducive to the teachers playing the dominant role. In addition, it is more advantageous to show the individual glamour of teachers which is real infection and inspiration to students. Moreover, the application of traditional blackboard-writing can reflect more flexibility, escaping the limitation of teaching design, adjusting measures to local conditions. Especially, when the teaching activities are not synchronized with the advance arrangements, blackboard-writing can develop more efficiency. While conducting the blackboard-writing, teachers can explain the contents, presenting completely and scientifically the knowledge how to generate and process. Thus, this can effectively deepen the understanding of the classroom content to the students, which strengthens the effect of explanation. The practice proves that the present multimedia teaching technology is not mature enough. If we, under the case of without sufficiently proficient application and operation, use completely multimedia technology and abandon the traditional teaching means, this can only make the teaching into a passive situation. Consequently, it is beneficial for the improvement of teaching quality.

2.3 The Application of Multimedia Teaching Means Does Not Reflect the Purpose of Literature Course Teaching

The important purpose of liberal arts courses teaching is to train the humanities literacy for students, in addition to the passing on knowledge. In the present practice of multimedia teaching, we often takes on knowledge points imparted, while the applications of pictures, audio, video, and animation in the courseware production are only to show the visualization of one point of knowledge or the presentation of abstract texts. However, this presentation does not reflect the purpose of humanistic quality education of liberal arts courses. Moreover, the interaction between students and teachers becomes the deduction of computers and multimedia technologies.

3 Mistaken Application in Multimedia Teaching

3.1 Abuse of Multimedia Technology

In multimedia teaching, the application of image, audio, and video, no doubt, can enhance the visualization and vividness of teaching activities. Nonetheless, during the design and utilization of classroom, partial teachers overly pursue the Rich and colorful audio-visual and animation effects. Thus it may easily overlook the rule of intentional attention and unintentional attention in the psychology. If you excessively use image, music, animation and so on, which have not direct relationship with the teaching contents, it will only make the students pay more attention to the colorful picture and sweet music. Accordingly, they can not be absorbed the teaching contents which are hided in the pictures and music. After one class, their enthusiasm on study is very high, and the class atmosphere is also perfect. The students, however, did not really learn any knowledge. Consequently, in the design of courseware, it is not the best with the more multimedia technology, yet it should appropriately select multimedia means based on the teaching content and the students' cognitive law.

3.2 Design of Courseware Is Disconnected with the Objective of Course, and the Level of Courseware Manufacture Is Lower

Because of the excessive stress on the multimedia means, teachers may pay more attention to how to make the courseware more elegant or attractive during its design. That is in a showing state of courseware, leading to lack of coherence with the teaching objective. At the same time, teaches who engage the liberal arts teaching know little about the computer software development and application. During the design of courseware, they usually paste the word texts to PowerPoint (PPT), or mechanically link texts to the related pictures, sound, and video material, etc. This type of courseware production just places the traditional blackboard writing into PowerPoint for playing. What's more, some teachers download courseware from the Internet directly, and do not make any modification before class. The quality of such courseware can be imagined. In addition, there are also two extremes in the process of multimedia courseware producing. One is the gathering of a lot of graphic and audio-visual material without selection, overmuch designing some audio-visual effects, which are dazzling, rich, and colorful. But this ignores integrity of the class and systematic of the knowledge. Therefore, the courseware is neither fish nor fowl. The other is to collect appropriate material except the text, also without design of audio, video, and animation effects. Although such courseware may is simple and practical, yet this inevitably makes people feel drab and boring.

3.3 Courseware Playing Too Fast Causes the Low Efficiency of Classroom

The advantage of multimedia is that it can instantly show lots of information, which bring about students of fresh and rich lecture content. Courseware with too much information, however, will make the teacher unconsciously speed up the switch of information, so that the students can not very well understand the knowledge, more difficult doing class notes. If there are more texts or figures, and playing with higher speed, the students could be thought passively. Thus it easily makes brain fatigue in the short time, which drops greatly the ability to receive knowledge. Therefore, after one class, the students often don't know what teacher lectures, what contents should be mastered, and what contents only should be grasped. Additionally, this makes it impossible for some students to take notes. Experiments show, the probability of what students think back the content with notes said in the class is seven times of what students do without notes said in the class [2]. Consequently, these undoubtedly and seriously affect the efficiency of lectures.

4 Primary Discussions on Solution

In the multimedia teaching of the liberal art class, to solve the above problems, we set up the following suggestions.

4.1 Multimedia Technology Is Just the Auxiliary Means, Thus the Teacher Is Still the Leading of Teaching Activities

As called "Sometimes a foot may prove short while an inch may prove long", each type of teaching has its advantage and disadvantage. The application of multimedia teaching means is to make up the shortage of the traditional teaching with "A piece of chalk, a book, a blackboard", and exert better the teacher's leading role, as the necessary technical assistance method to traditional teaching. It is not to substitute the teachers' teaching activities by the computer. In the course of design, we should deal with the relations between media teaching time, timely and appropriate lectures, blackboard writing, interactive and reflection of the students. For example, the efficiency of writing the viewpoint or conclusion into the blackboard is not certainly lower than that of playing of PowerPoint. Because the latter saves the operating time, giving an emphasis of this point to the students, making the students catch the importance of the class lectures.

Bruna, American psychologist and educator, has said "the teaching process is a kind of the constant activity with posing questions and solving the problems" [3]. So the teachers, in the teaching activities, also take an important mission of the grasp rhythm of the class, arrangement of the classroom interaction. In the classroom, teachers should avoid the "cramming", utilize the multimedia courseware to guide the students to think, discuss, answer the questions, and solve the problems. This helps them to jump out of passive and simple dilemma to accept only, making actually students become the host of the study.

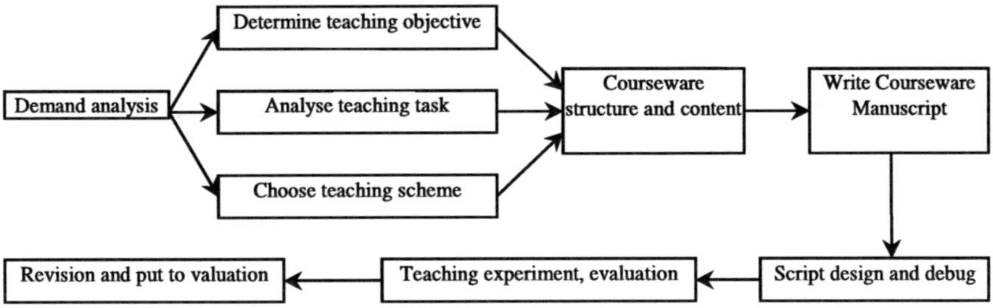

Fig. 1. Flow of courseware production [4]

4.2 Development of Multimedia Technology Should Be in Compliant with the Premise of Subject Characteristics

The multimedia technology can really better for teaching services if it is compliant with the characteristic of subject. The training of humanistic quality is one of the common features of liberal arts courses. For improving the humane quality, there must have the initiative participation. However, this can not achieve with cold computer through the sound, shape, color, and sound. For teaching activities itself, it should emphasize the role of teachers in teaching activities. In the liberal arts class, teachers should use multimedia technology to better serve the teaching purpose of humanistic quality improvement. To achieve this objective, we must emphasize the leading position of teacher in multimedia teaching. In the procedure of multimedia teaching, the standing position of teacher is not limited to the computer or the three feet platform. He can completely go to the center of students, by wireless remote controller, controlling the PPT with flip slide, full screen, demonstration, and other common control operation. Meanwhile free standing can effectively avoid the teacher to read screen as textbooks, and according to the reflection of classmates to the teaching contents teachers can add necessary expansion or instructions to some teaching contents. For doing that, teacher can also have the participation of appropriate and humor gesture language. Thus, these teachers with such teaching have affinity; such classroom atmosphere is active; such effect can achieve the best. In addition, the wide application of the photo, video, audio, animation, and media forms does not make teachers be a rigid puppet, but can completely show the personal charisma according to the special features of oneself in the classroom. For example, a teacher with magnetic voice, and good at reading, singing, or performing, in this premise, he does not need to read online audio or video of other information through the multimedia display, and should make the classroom become the stage show of teachers' personality charm. This kind of show between people has more affinity and attractiveness than cold machine. This kind of full show of personality charm can greatly active classroom atmosphere, can also greatly mobilize the enthusiasm of the students to imitate. Additionally, it can cultivate of the students' specialty, realize the purpose of educating people through teaching activity. Only in these premises the multimedia teaching can truly achieve the purpose of the humanistic quality education of the liberal arts teaching.

4.3 Courseware Production Is a Complicated Project

The production of courseware is the core of multimedia teaching activity. Moreover, the effect of teaching activity depends on the level of courseware production of teachers. "Courseware design and production involves interdisciplinary knowledge and skills, generally including the experts of teaching design, psychologists, skilled teachers, education scientific research personnel, art personnel, software design personnel, sometimes requiring music workers, camcorder staff, courseware development team, etc. So a good courseware can integrate the various factors to complete" [4]. The flow of courseware production is shown in Fig. 1. However, in the actual teaching activity of the current liberal art courses, instructors often link directly the animation or audio and video into the computer-aided instruction system (i.e. PowerPoint software) according to own understanding of teaching content and functions of computer software. Especially, for some teachers who are not familiar with computer software and language, download directly the ready-made courseware from the public network resources, even making not any changes before the class. The effect of such courseware can be imagined. We should realize that the application of multimedia technology provides a convenient to many teachers who want to slack off. But in fact, if we really want to play the great function of courseware, the teacher in the present station work actually harder than in the traditional teaching method. In fact, teachers who produce courseware are not only psychologist, experts of software technology, photographers, but also familiar with the teaching contents.

5 Conclusions

The multimedia teaching integrates audio, video, pictures, and text information, which can arouse students' vision, sense of hearing, and other various sense organs. Its utilization overcomes the single and one-sided disadvantages of the traditional teaching methods. Moreover, it can effectively reduce the study time, improving the teaching quality and teaching efficiency, optimizing the teaching goal. Because of ten years of practice time, however, the exploration of multimedia teaching in China is only just beginning, as such a service with computer language for teaching. Although in the explorative stage we did not achieved the desired results of teaching, we believe that with research of computer language and understanding it deeply, with a constantly fusion of the media technology and the characteristics of the subject, the modern information technology must bring a qualitative leap for the teaching practice.

References

1. Guo, X.L.: My opinion on multimedia teaching of university. Continue Education Research 5, 119–120 (2009)
2. Jia, G.J.: Foreign language education in psychology. Guangxi Education Publishing House (July 1996)
3. http://202.121.63.40/tsc/researchshow.asp?ShowId=828&cid=367
4. Wang, S.L.: Research on courseware production. Journal of Jiangsu Teachers University of Technology 12, 61–64 (2007)

"Post-Financial-Crisis Era" The Countermeasures for Promoting the Graduates Employment in Universities

Xingmiao Xu

Zhejiang Economic & Trade Polytechnic, Hangzhou, China
xxm2208@163.com

Abstract. Through analyzing the problems that the university employment work face in the post-financial-crisis era, two correlative countermeasures was put forward to promote employment of university graduates. Firstly, it should accelerate the "two changes" with the purpose of enhancing the employment capability of students; Secondly, it should achieve "four combinations" of changing the employment concept of graduates.

Keywords: post-financial-crisis era, graduate employment, Countermeasures.

1 Problems That the College Employment Work Face in the Post-Financial-Crisis Era

The financial-crisis in 2008 caused a great influence on our economy and employment. In recent years, the domestic economic situation presents a recovery phenomenon through the upgrade of industry and enterprise innovation. And the enterprise has not only restored but also expanded production, some places even see recruitment difficulties. However, it does not mean that our country has passed the financial crisis and the employment situation is still severe. Yuan Guiren, the ministry of education minister indicated in the 2010 national network video conference of the college employment work that the adverse impact of the international financial crisis on our country employment will continue, so the macroscopic employment situation is still serious. During post-financial-crisis period, college students face two unemployment problems, that is the influence of the economic environment and the colleges' problems.

One problem is the lag of educational reform. Recently, our higher education has fastened the speed of the reform and obtained great achievements. But on the whole, the professional setting, curriculum setting, mode of training, and other aspects of the reform still can't keep up with the pace of economic development. In the new major setting and old major transformation, we always regard the existing teachers and facilities as prerequisites, and are keen to develop majors of invests few , effective quick, easy imitation rather than plan and adjust well according to the social demands. Consequently, it causes professional convergence between different colleges or course convergence due to the carefulness of profession dividing. Simultaneously, colleges make recruiting scheme according to the current job market, and increase enrollment of students in some hot majors depending on the applied situation. When the employment market changes a few years later, without a set of effective coping mechanisms, these

W. Zhang (Ed.): Advanced Technology in Teaching, AISC 163, pp. 473–477.
springerlink.com

professional students will be passive and helpless in employment. In personnel training, cooperation between school and enterprises is strengthened recently, but reform is so timidly that the effect is not really ideal. Refer to the reasons, except for the inadequate policy and low motivation of enterprises, colleges can break the conservative teaching concepts. Attach more emphasis on theory than practice, lay more stress on classroom teaching than building practical and training base, emphasize more on imparting professional knowledge than training comprehensive quality of vocational skills. Besides, there is no effective incentive system in some aspects such as personnel training, practice of teachers, teachers and students' involvement in scientific research project research and product development.

The second is the problem of students' outdated employment ideas. China's higher education is elite education in a long time, with training goal towards city. As a "scarce resources", the employment level of college students is always higher. The concept of city values and traditional thought of "jump NongMen" are well-entrenched in many students' and parents' minds, which bring some structural problems regarding college students' employment. According to the employment survey of 2009 college students, city is the first choice for 53.2% students, less than 10% choosing to go to the county or lower level and the western, another 32.7% choosing to work in hometown due to the requirements of their parents. As for the employment unit, 64% students select the government organs, institutions, and the state-owned enterprise, which reflects the traditional complex of cadre identity and the people's employment psychology of "stable higher than all" after the financial crisis. In the Zhejiang 2009 officeholder exam, 261000 people have applied, and the enrolling proportion of the most popular position is 1770:1, both breaking the examination records. What's more, in 2010, number of applications is more than 310000.

2 Countermeasures for Promoting Employment of College Students

2.1 Accelerate "Two Change" and Enhance Students' Employment Ability

(1) Accelerate the transformation of traditional teaching model which centers on the classroom teaching, and pay more attention to the practical teaching.

Promote the whole course of educational internship mode. According to the features of all majors and the need of career development, work out the specified job skill standards and carry out a step-by-step implementation so that the cultivation of the job skill can run through the whole course of study. Strengthen the establishment of external internship bases, and try to realize the "zero-transition" from students to teachers through the co-operational school system, which integrates the training, internship and employment. Innovate the mechanism of campus and enterprise. Build up the training team and the integration of production, teaching and research. Create the opportunities for the students and teachers to improve themselves. Establish a feasible management and incentive mechanism to reach the win-win strategy between colleges and enterprises. Maintain good contacts with graduates of all previous years so as to make sure that they can offer some job vacancies to the this year's graduates. Build the internal and external job training base and job internship base, in order to enhance the students' employability.

(2) Accelerate the changes of traditional talent cultivation mode which with imparting knowledge as its center and pay more attention to students' ethical quality education.

Prime Minister Wen jiabao pointed out in the national science and technology education leadership group meeting: "students should not only learn knowledge, but also learn how to do, learn to think, learn to work, learn to survive, and learn to communicate with others. These are the whole education and the reform of teaching content." Premier Wen's explained in simple terms of the quality education under the new situation. Paying more attention to cultivating the students' good moral character is the most efficient way to improve their employability and promote the colleges' employment competition. The related questionnaire survey shows that 70.09% employers value students' moral accomplishment, in which, credibility is the prominent, amounting to 50%, followed by hard -working and dedication quality, reaching 45.65% and 32.6% respectively. In addition, qualities like humility, cooperation, etc are also important. Moral accomplishment can make up for the deficiency of the knowledge and ability, while the knowledge and ability can't make up for the defect of moral. Unfortunately, due to the historical reason, China's higher education is always emphasizing on imparting professional knowledge, but ignore the cultivation of moral quality. Thus, students that cultivated are not real talents with only profession and knowledge but neither accomplishments nor thoughts. Faces with this situation, colleges and universities should focus on the quality training, as well as promote the quality education, pay attention to moral education's penetration function in professional education, promote coordinating development of professional education and the humanities education, and assert the excellent traditional culture and humanistic spirit.

2.2 Accomplish Four Combinations and Change Students' Employment Concept

(1) Combine the career exploration and education of situation and policy.

To change the college students' employment concept, the first thing is analysis of the employment situation and publicity of employment policy. Make the students fully understand the situation of current economic development and college students' employment market supply and demand, comprehensively understand the national employment policy by organizing situation and policy symposium, career talks or studying employment publicity materials etc. Guide college graduates to combine the subjective desire and social demands, set up the correct employment concept that anyone can get success and make contributions in any area and abandon the traditional ones. Furthermore, guide the graduates to face basic level obtain employment just as a common laborer, by which they can lay a solid foundation for their long-term development.

(2) Combine the career exploration and education of outlook on life and values. Employment concept is a specific reflection of outlook on life and values, directly affecting and deciding the action and results of choosing a career. For many years, because of the social cultural environment, education orientation, objective condition

and other factors, China has faced problems of unbalanced structure in qualified personnel and considerate contradiction in college students' employment structure. On one hand, the west, countryside and grassroots level are lack of qualified personnel. On the other hand, few students choose to go to these places. Therefore, the government has set out a variety of policies to speed up the development of the western region and new rural construction, as well as to encourage college students to work in the west, rural and grassroots areas. Hence, colleges should seize the opportunities to strengthen the education of outlook on life and the values, so as to achieve the combination of ideal faith education and national education, school education and extramural education, solving students' ideological problems and practical problems. Guide and inspire students to realize their values and ideal pursuit by building up ambitious ideas and determining to strengthen themselves through grassroots work. The former education secretary Zhou jizeng said: "Elitists must develop in practice at the grassroots level. Today, no knowledge about the grassroots, no success in the work. So is the future."

(3) Combine the career exploration and mental health education. Employment has become one of the great mental pressures of college students, while good psychological quality is the strong support for students to obtain employment. In this severe employment situation, colleges should especially combine the employment education and psychological health education. Develop the employment psychological counseling and consulting services by methods of psychological lecture, team coaching or individual counseling. Guide students to fully understand the employment situation and policy, and help them to establish self-confidence, overcome fidgety and inferiority mentality. Lead the students to improve comprehensive quality, to set up the correct concepts, cherish employment opportunities, as well as to have the idea of employment first, then seeking the development.

(4) Combine the career exploration and enterprise education. Currently, under the dual effects of employment situation and favorable policy environment, trying entrepreneurship has become more and more students' choice. A survey of "College students' employment problem" conducted by Department of Education of Hebei Province shows that nearly sixty percent of college students identify entrepreneurship and also look forward to trying to start up a business. So colleges and universities should attach importantance to Enterprise Education, construct a model of teaching, training and practice. Open courses like career planning, entrepreneurship knowledge or some related courses. Hold different forms of entrepreneurship training like "SPP", "SYB". Organize business plan competition and selection activities like "Venture Star". Establish the business park in colleges and base of the enterprise education outside colleges. In project selection, establish related ancillary system in respect of project selection, capital support and business guide, to let students get exercise in the business incubation bases. And meanwhile, they can accumulate the practical experience for entrepreneurship in Venture guidance bases, which enable the entrepreneurship be a new tool in college students' employment.

References

1. Yuan, G.: In-depth study and practice of the scientific concept of development without laxation and oscilation to promote the employment of 2010 college graduates. China Education Paper (November 23, 2009)

2. Weng, H.: Over 1770:1,the application of the hottest civil service jobs in Zhejiang province reaches a record high in recent years. Today Morning (January 5, 2009)
3. Lin, J.: 31 million people apply for civil servants in Zhejiang province, most popular jobs are grass-roots positions. City Express (December 6, 2009)
4. Wen, J.: Long-term goal, based on education. China Youth Daily (January 5, 2009)
5. Tu, M.-F., et al.: Six major factors affect the employment force. Qianjiang Evening Paper (January 22, 2009)
6. Qin, S.: General Education: An important direction for the reform of undergraduate education. China Education Paper (September 25, 2008)
7. Han, M.: The chief career goal of college students quietly swerved under the financial crisis. China Youth Daily (December 11, 2008)

The Problems and Suggestions on College Teacher's Educational Technology Training in China

Yu-fang Chao

School of Business, Shandong Institute of Light Industry, Jinan, Shandong
chaochuwen@sian.com

Abstract. Based on the ICPP model, this paper analysis the problems on college teacher's educational technology training in China from four aspects the background, resource input and training course, the training result esc, applying empirical research method and literature research method. It proposed that College Teacher's Educational Technology Training in China lacks analysis to learn functions, learn-motivation and teacher's role, and proposed that resource input is too much to the hardware, lacking integration of resources. In view of this, it proposed advice that forward analysis college teacher's learn features, stimulate their motivation, hierarchical, classifying and gradually to promoting esc.

Keywords: college teacher, educational technology, training.

1 Introduction

Since 2001 China Ministry of Education carried out modern educational technology training for teachers in colleges and universities, China has invested a lot of financial, material and human resources. Proceeding from the present situation of Chinese modern education technology training for teachers in colleges and universities, it will make sense for the full utilization of training resources and significant improvement of training quality by making in-depth and comprehensive analysis of problems existed in training and exploring the optimum training measures.

This article, adopting the CIPP training effectiveness evaluation model (The following abbreviation is CIPP Model), combining the practical participation in the modern educational technology training for teachers in colleges and universities, by methods of empirical studies and literature research, conducted a comprehensive analysis of Chinese modern education technology training for teachers in colleges and universities, explored the problems and made relevant recommendations.

2 CIPP Training Effectiveness Evaluation Model

CIPP model, produced by Stufflebeam.D.L Yu in 1967, is mainly composed of such four parts as context evaluation, input evaluation, process evaluation and product evaluation, the first letter of the four activities, CIPP stands for training evaluation model. Later Stufflebeam further fined the model by breaking the product evaluation

W. Zhang (Ed.): Advanced Technology in Teaching, AISC 163, pp. 479–485.
springerlink.com

into 4 phases of impact, effectiveness, sustainability and transportability, which resulted in evaluation mode of 7 components.

The structure of CIPP model is; uses context evaluation to select targets, input evaluation to correct the plan, process evaluation to guide the implementation of the plan, product evaluation to provide a reference.

Compared with Kirkpatrick's model, four level assessment model (Ko's four level assessment model basically is for the training of the last step, with a nature of terminating evaluation, can not provides continued improvement and adjustment), CIPP model more has full process and feedback features, being more conducive in evaluation and reflection of the design of training on background, management, sustainability and application, especially at in-depth analysis of its problems in the process.

3 Analysis of Existing Problems in the Modern Educational Technology Training for Teachers in Colleges and Universities in China

According to CIPP model, from the four aspects of background, input, processes and product, this article try to make an in-deep analysis of Chinese modern education technology training for teachers in colleges and universities and explore its problematic.

3.1 Problems on Background of Modern Educational Technology Training for Teachers in Universities

CIPP Model defines the training background content as: Learning about environment-related special problems diagnosis, analysis and identification of training needs, identification of training opportunities and, among of which identification of training needs and development of training objectives are their primary tasks.

Chinese Government (leader and organizers of modern education technology training university teachers) has fully recognized the importance of modern education technology, from a macro point, being actively promoting the application of modern educational technology in education, and developing appropriate training objectives. But because of the large number of the training involved, widely distributed, vast variation students, there are so many problems in analysis of the characteristics of trainees and training needs.

1) The Lack of In-depth Analysis of the Studying Characteristics of University Teachers

The process of training is essentially such a process of guiding the trainers to learn by them. In that case, before the train design, trainers must make in-depth analysis of characteristics and laws of learning to design efficient training courses. Otherwise, ignoring those and following learning theory or hypothesis will result in unsatisfactory training.

Teacher of college and university is a special group, who has a high level knowledge, strong autonomy and independence. The learning of university teachers not only has the general characteristics of adult learning, but also has the same different characteristics. This determined that the modern education technology training for teachers in colleges and universities should not completely follow the theory of adult education, must have an in-depth analysis of college teachers studying characteristics. It was unfortunate that there are limited research on university teachers' learning characteristics and laws. You will retrieve nothing when you search the cnki (www.cnki.net) by the keyword of teacher of college and university a learning characteristics. Expanding the scope to "University teacher" and "learning" as keywords for search, there are only 22 results, only 6 articles are related with learning characteristics of university teacher, and studies are not systematical. The lack of such research results in the lack of pertinence and applicability in the design of training.

2) A lack of in-depth study of learning motivation of college teachers

Learning motivation has a direct effect in learning result, different learning will naturally result in different attitudes. For now, organizers of the modern education technology training for teachers in colleges and universities made inadequate analysis of the motivations, which is not accurate enough, in practice, even misleading their motivation, such as emphasizing that training will become one of the conditions of the titles promoted.

The lack of study of learning motivation of modern educational technology training for teachers in colleges and universities directly led to the waste of resources and will result in failure to achieve the organizer's intended target. To my actual participation in organization of modern education technology in Shandong province as an example, among a total of 100 people, randomly selected 30 persons, motivation survey was conducted to find that most (near 80%) who attend the training, regard it as an essential condition for professional promotion. An in-depth study on the motivation of teachers training in colleges and universities, inspiring and guiding proper motivation, reducing the negative impacts of improper motive, is an important topic.

3) Lack of analysis and classification on levels and functional roles of college teachers

There is a large age span and wide specialty distribution among teachers in colleges and universities, which result in a large variation in the basic learning ability. In addition, teacher of college and university is a broad group, whose functional roles have large differences, so the design of training must be adapted to these differences. Otherwise, the unified training can not satisfy different needs of those teachers resulting in waste of training resources.

For now, Beijing and Guangdong other has taken a leading step on the training level division, but there are more provinces with training model of same content. Of course, whether the training, which can be designed according to the professional nature, will work is a worthy subject of study.

3.2 Modern Education Technology Training University Teachers Output Problem

CIPP Model defines input evaluation as information collection and assessments of training resources, determination of how to make effective use of available resources to achieve the training objectives, determining the overall project planning and design of strategies, such as the need for external resources to assist.

Since 2001 year, China has invested large quantity of resources in modern educational technology for university teachers training project, establishing funding mechanisms to protect the required resources. But from the distribution of resources and effectiveness of resource use, there are questions such as:

1) More focuses are on "hardware input" instead of "software investment"

Since the implementation of modern educational technology training for teachers in colleges and universities in 2011, a large amount of resources are directed to hardware, gradually more advanced modern education technology training center are built.

Required hardware investment is one of the security conditions for achieving training goals, but if you do not have a "soft" qualified support and hardware can hardly play their appropriate role. Now, there are inadequate concerns on the soft conditions, one evidence is that the various material library has yet be established for training, which affect the outcome of the training application; the second is that there is inadequate training for trainers. Leading training needs world-class trainer, which has been neglected for many years; the last is that three is not enough resource investment in the specific training activities, which result in that most of the training is short, heavy tasks, can not guarantee the effect, which can not stimulate learning motivation of students.

2) Lacking of integration of the necessary resources

For now, different provinces in China have carried out training activities for teachers of modern educational technology in colleges and universities, municipalities also established training centers. But because of the lack of the necessary resources integration, national library does not have to establish modern education technology training resources, training resources between provinces have not joined, which led directly to a number of training resources are decentralized, independent state, unable to play its application value.

3) Insufficient use of learning resources by trainers own

Modern education technology training for teachers in colleges and universities are different from general training, university teachers are engaged in education for many years, they have a lot of learning resources. And now almost training did not take into account of college teachers' own learning resources, it is also wastage of training resources.

In the future training, how to make full use of their resources, explore their abundant practical experience to share these resources is a worthy study content.

4) Lack of planning in the process of training resources

Modern educational technique involves a number of content, which has different using frequencies and level of easiness and different requirement of mastering. Therefore, the plan of resource distribution should not equal, which should be depending on the content, the nature of students.

3.3 Problems in the Course of Modern Educational Technology Training

Process evaluation of CIPP model is responsible for providing feedback to those who are implementing training programs, and in a timely manner, amending or improving the implementation of training programs. Its main elements include: evaluation of the implementation of the programmer and related assessment on events, issues, expenses in the implementation of resource allocation.

Training process evaluation can detect problems in training timely and adjustment training content and optimize raining design to improve training effectiveness.

Assessment of China's modern education technology training for teachers in colleges and universities take Ken's theory of four-level evaluation model, mostly in the period after the training, reactions and learning assessment. Reaction layer assessments are mostly from students' impressions and feelings, mostly through training to complete the questionnaire and learning assessment is to assess knowledge, skills, knowledge, mostly through written examinations.

Although these assessments and control has a role in the training promotion, but this ex-post control can not find the training problems in time, amend the corresponding training design, this is also the current absence of modern educational technology training for teachers in colleges and universities in China.

In addition, there is inadequate necessary communication between the training, trainer, organizers and students which also caused that problems in the training can not be resolved in a timely manner.

4 Issues about the Result of Modern Education Technology Training

CIPP model's main tasks is to measure and interpret the goals achieved in the training activities, which includes student satisfaction, increased knowledge and skills, improved individual behavior and organizational performance.

Because the current assessment is limited to the reaction layer and study layer not touching evaluation of behavior and organization layers, which leads to lack of in-depth analysis of outputs from this training program, naturally which cannot be fed back to training design.

4.1 An In-depth Analysis of College Teachers Studying Characteristics to Effectively Stimulate Learning Motivation

In General, teachers in colleges and universities are more familiar with the process and teaching; they have a high level of knowledge having greater autonomy and consciousness. At the same time, they have mature physiological and psychological mechanism and very experienced teaching practice which determines that they have strong abstract generalization of thinking ability and understanding. They often do not stop on the sensibilities, but in comparison, make comprehensive judgment and abstract generalization on the basis of understanding the essence of things and laws with practice problems, by analogy and apperception. These characteristics make teachers in colleges and universities have a strong capacity to absorb knowledge compensating for mechanical memorizing capacity due to age limitations.

In addition, the teacher's character has been relatively stable, features of personality structure will makes they have strong learning self-control. These characteristics determine the learning of teachers has a clear purpose and a high level of autonomy and consciousness.

Therefore, the modern educational technology training for teachers in colleges and universities should take a more flexible design, training methods should be diversified, time span should be long term, and training should be individualized. In terms of learning motivation, we should stimulate motivations from teaching, public opinion atmosphere and practice environments in stead of title promotion.

4.2 Layering, Classifying and Promoting Training Step by Step

In terms of training organizations, training content and timing should be determined according to student's age, profession, training levels and professional categories.

In terms of training content, it is necessary to make in-depth analysis on the training content under the need of practical application in the teaching, clearing what technologies should be universal, or be on small scale use, can be used by individuals, be used through collective collaboration.

On the basis of an in-depth analysis of training content and, it is also necessary to match students with training content. At present, Beijing and Guangdong has taken rating training, achieved a certain degree of effectiveness. On this basis, educational technology training can be divided into three levels, namely, general layer, advanced layer and post-advanced layer. This allows the train to proceed step by step in depth, which helps to improve training effectiveness, being conducive to stimulate learning motivation.

4.3 Promoting the Use of Organization Mode with "Force" as a Unit

Most of the modern educational technology requires collective collaboration; especially those are difficult and need more material. To get effective conversion of knowledge and skills, in terms of student organizations, organization mode with "force" as a unit should be widely taken, even teachers who teacher similar courses within same school should take the same batch of training. This student organization model can provide a sound base for training curriculum design and practice teaching.

At the same time, in the course of training, we should take the principle of same school and same course to organize student groups in the appropriate case practice, which could cultivate application environment and habits during a training development.

4.4 Intensifying Communication in Training Processes and Focusing on Process Control

First, there must be a three-link mechanism among trainers, trainees and training organizations, establishment of communication channels.

Second, as for organizer of the training, work is not just the start of training, but should proceed throughout the entire training process. Training organizers need to made corresponding investigation in the implementation of the training, according to the implementation of the training, timely feedback to the trainers about training issues, submit proposals for amendments.

Finally, as a student, it is necessary to response problems encountered in the training process timely and correctly.

4.5 Establish the Shared "Repository"

Development of network technology makes trans-regional integration of resource is a reality. Integration of resources not only makes effective use of existing training resources, but also expands the vision of trainees and training organizers.

From the whole of modern education technology training for teachers in colleges and universities, the State is required to consolidate the training resources of different provinces. For provinces, they should make sound integration of resources of education center.

References

1. Zhang, H.-Y., Dong, Y., Chen, G.: The Current State and Recommendations of College Teacher's Educational Training in China. Modern Educational Technology (5), 51–54 (2003)
2. Stufflebeam, D.L.: The CIPP Model for Evaluation. In: Stufflebeam, D.L., Kellaghan, T. (eds.) The International Handbook of Educational Evaluation, ch. 2. Kluwer Academic Publishers, Boston (2003)
3. Heinich, R., Molenda, M., Russell, J.D., Smaldino, S.E.: Instructional Media and Technologies for Learning, July 16. Prentice Hall (2001)
4. Lu, H., Liu, X.-B., Hao, R.-F.: To Build, Modify and Validate the Model of Educational Training for Faculties. Modern Educational Technology (7), 60–62 (2009)
5. Wang, Q.-Z., Chen, H.-L.: the Applications of ICT in Classrooms in Higher Education: The Focus of Teachers. TSINGHUA Journal of Education (8), 64–68 (2004)
6. Office of Educational Technology of the US Department of Education (DB/OL) (August 5, 2008), http://www.ed.gov/technology

Applications of Modern Educational Technologies in Higher Arts Education

Zhan Yun[1] and Ma Ming-Lei[2]

[1] School of Art & Design Nanchang Hangkong University, Nanchang, China
[2] College of Computer Information Engineering
Jiangxi Normal University, Nanchang, China
thmugmcwn5@163.com, maminglei@163.com

Abstract. The new curriculum standards call for a new teaching model and implementation strategy. In recent years, the rapid developments of modern educational technologies have brought great impact on all aspects of traditional education, and will have a broad and far-reaching effect on educational reform. Modern educational technologies have posed a challenge on the existing higher arts education, and played an irreplaceable role in deepening the reform of arts education and development in the artistic potential of students, therefore, it has brought a good opportunity for the smooth implementation of new curriculum of higher arts education. Responding to this trend, this article studies how to use modern educational technologies to optimize the teaching of higher arts education, and also study the applications in the implementations of modern educational technologies in courses of higher arts education.

Keywords: Modern educational technologies, higher school, arts education, teaching model.

1 Introduction

In recent years, the rapid developments of modern educational technologies have brought great impact on all aspects of traditional education - including education ideas and concepts, educational system, and teaching content, especially brought a huge impact on teaching model and teaching methods and will have a broad and far-reaching effect on educational reform. Modern educational technologies have posed a challenge on the existing higher arts education, and at the same time, it provided technologies, methods, and cognitive tools for us to achieve quality education for arts, and played an irreplaceable role in deepening the reform of arts education and development in the artistic potential of students. Further more, because of its unique characteristics, modern educational technologies are especially suitable for the implementations of modern educational technologies. Curriculum reform and the universal access to modern educational technologies are the basic requirements and inevitable trend of our modernization development of education in the new century. Responding to this trend, this article studies the organic integration of how to use modern educational technologies and optimize the teaching of higher arts education, and also study the application value and model of the implementations of modern educational technologies in courses of higher arts education.

W. Zhang (Ed.): Advanced Technology in Teaching, AISC 163, pp. 487–494.
springerlink.com

As a culture, modern educational technologies have become an important factor in educational reform today. The recently awarded "decision of State Council on pushing the development of arts education vigorously" clearly states: "Enhance the informatizational construction of arts education and promote the application of modern educational technologies in arts education." Thus, in order to face the challenges of the new century, arts education must focus on modern educational technologies, establish the scientific concept for modern education, and continuously improve the level of hardware construction, and vigorously promote the modern methods of education, and speed up the process of informatization of arts education and campus network management, so as to further improve teaching quality and management level, promote the modernization of higher education.

2 Development of Modern Educational Technologies

Using educational technologies to engage in teaching activities is dominated by different learning theories in different historical periods. Machine-teaching (also known as programmed teaching) in early times is generally based on the learning theory of Skinner's new-behaviorism learning theory. The core of behavioral learning theory is operationally conditional reflection theory and reinforcement theory. Skinner believes that human behavior is mainly reflected by the operational behaviors composed of operational reflections, the operational behaviors are the behavior acting on the environment to produce results. In the learning environment, the operational behaviors are more representative. Skinner proposed reinforcement theory on the basis of learning a great deal of research about the learning, and put great emphasis on the importance of reinforcement in the study. Learning is also an operational behavior, the reaction rate goes up when learning, and goes down when stop. Therefore, the process of learning is a cycle of "Identify the stimulation --behavior or response--enhance the stimulation". Behavioral learning theory opens up the road for the computer-aided instruction. Programmed teaching has great influenced on the United States, Western Europe and Japan since it came out, and widely used in the teaching of all subjects.

However, as people keep on the further study of human cognitive activities, it was generally accept that the programmed teaching focuses on knowledge indoctrinization, but the communication between students and teachers are weakened, which is not conducive to creative thinking ability. Therefore, the modern educational technologies, applying multimedia technologies and network technologies as its core, is prefer to regard the students as the constructivist learning theory of information processing theme, and behavioral learning theory is isolated and criticized gradually. Constructivist learning theory holds that knowledge is not taught by teachers totally, but in certain situations, i.e. social and cultural background, learners obtain the knowledge, with the help of other people (including teachers and learning partners) and the necessary learning materials, in their process of learning via the method of constructivist theory. Since learning is in a certain situation, i.e. social and cultural background, with the help of others or through interpersonal cooperation activities, to achieve the meaning construction process, so constructivist learning theory believes, "situation", "cooperation", "communication" and "meaning construction" are the four elements or the four properties of learning environment. The learning quality is a function of the learner ability to construct meaning, rather than the learner ability to reproduce the function of teachers thought processes. In other words, how much knowledge access to learners depends on their ability to construct the meaning of relevant knowledge as per their own experience, rather than depends on learners' ability to remember and recite the content teacher taught. We believe that teaching process is a creative activity, any kind of learning theory can not conquer all over the world, cognitive differences between subject and object often requires the different learning theories in the teaching process. Behaviorism and constructivism should be equal; they were just methods applied to different aspects and different areas of teaching.

New Course Standard of Art for Universities presented the basic philosophy as "regard aesthetic as the core, develop interests and hobbies; for all students and pay attention to character development; emphasis on artistic practice, enhance the creation awareness; promote the national arts, understand multi-culture". The course is divided into six modules; attach importance to the improvement of artistic appreciation, and also concerned about the performing arts (practical) skills training. ① constructivism and behaviorism have their own merits in the implementation of these different modules in teaching. We can organically combine the two to explore a new kind of integrated teaching model.

3 University Arts Education under the Background of Modern Educational Technologies

Science and technology have made an unprecedented development during 20th century, human society transformed from the industrial era into the information age. The rapid development of science and technology promoted the great changes of social politics, economy, culture and the concept of people, and also affected the educational concept. Particularly in the new century, the constantly update of educational content, increasingly sophisticated means of education, the field of education is facing a comprehensive technical revolution. In recent years, the rapid development of modern educational technologies have brought great impact on all aspects of traditional education - including education ideas and concepts, educational system, and teaching content, especially brought a huge impact on teaching model and teaching methods and will have a broad and far-reaching effect on educational reform. The implementation of various education disciplines, based on computer network and multimedia technical support, lays out an ideal blueprint for the future education. Arts education should also focus on seeking new breakthroughs according to this new point. Modern educational technologies have posed a challenge on the existing higher arts education, and at the same time, it provided technologies, methods, and played an irreplaceable role in deepening the reform of arts education and development in the artistic potential of students. Further more, because of its unique characteristics, modern educational technologies are especially suitable for the implementations of modern educational technologies.

Modernization of education is an important component of China's socialist modernization. In the report of the Seventeenth Party Congress, Comrade Hu Jintao made clear that we will give priority to the development of education, implement quality education, develop modern distance education, and enhance the modernization level of education. With the highly developing information technologies today, we should emphasis on the use of modern educational technologies, develop the educational resources, optimize the educational process, and promote the modernization process of education vigorously in China, which are the requirements of the Party and the state to new education in the new era. This is not a figment of education experts' and policy makers' imagination, but the inevitable demand of new era and the social development. As an important component of quality education, arts education is particularly imminent in its information technologies.

4 Problems of Modern Educational Technologies in Our Higher Arts Education

Modern educational technologies use modern educational theories and modern information technologies to achieve the theory and practice of teaching optimization, through the process of teaching and learning and the design, development, utilization, evaluation and management of its resources. Currently, the modern information technologies applied in education includes analog and digital audio technologies, satellite radio and television technologies, computer multimedia technologies, digital art technologies, artificial intelligence technologies, internet communication technologies and virtual reality simulation technologies. As a new type of educational media, it plays a very important role in arts education. However, on the whole, at present there are still many practical problems of modern educational technologies in arts education in our application and popularization in the following areas:

4.1 Short of Acknowledgement and Awareness

Educational administrative departments at all levels of arts education was not able to profoundly came to realize that this is the development of arts education opportunities, because of they lack acknowledgement and awareness in the introduction and promotion of modern education technologies, which is an important tool to implement "quality education", so there is no corresponding supporting and encouraging policy to be implemented to protect the aspects of funds inputting, curriculum setting, teacher introducing and training, resulting in arts education can not brought in line with modern education, or even out of joint.

4.2 Informatization Resources Is Not Enough for Art Teaching

Design and production methods for art teaching software are obsolete, on one hand, the high-quality art teaching software are insufficient and limited, it is difficult to adapt to the urgent need for newest arts education; on the other hand, the lack of a unified and standardized implementation standards results in low-level repeated teaching and can not meet the needs of art teaching.

4.3 The Application Level of Information Technologies Is Insufficient in School Arts Education

Although some schools have recognized the importance of computer education in modern education, and paid attention to the purchase of equipment, and invested in the campus network, electronic art classrooms etc, but they ignored the "application" work, only regard these as decorations or as a "standard" for their "image projects."

4.4 The Modern Educational Technologies of Art Teachers Is in Low Level

A considerable number of art teachers lack the understanding of modern educational theory, and still stuck in the old mode of "oral teaching by examples", do not have the basic ability to use modern information technologies to carry out the art teaching, or just have the poor ability of applying computer in the initial stage, such as the "typing" and "surfing the Internet", and unfamiliar with the operation of computer. This situation not only caused some resources wasting of multimedia teaching, but also missed a lot of opportunities to improve teaching and learning.

5 Application of Modern Educational Technologies in Higher Arts Education

The traditional arts education does stifle the subject of students through teaching methods in its practice. On one hand, teachers highlight the mission of arts education, focusing on technologies, teaching, while the potential instincts of students to make creative activities can not be seen, the teachers are doing the things on behalf of the students. Finally, the nature of art won't exist any more, students are found in the same and single thinking, and that just the contrary of our expectations; on the other hand, a small number of people are engaged in a "top-style" education, rather than universal education of "education for all", therefore, students can not be individualized, and the characteristics of arts education can not be paid attention to. These issues will be improved with the introduction of modern educational technologies.

5.1 The Interaction of Computer Technologies Can Stimulate Students' Interest in Learning, Give Full Play to the Role of Cognitive Subject

Human-computer interaction (HCI) is a distinctive feature of the computer, it is important for the teaching process; it can effectively stimulate students' interest in learning, so that students will have a strong desire and motivation to learn. In art teaching activities, the human-computer interaction can train the basic art skills of students in the aspects of sight-singing, ear training, rhythm training, and also the art teaching of knowledge and theory composition. And through timely feedback to the smooth and deeper training, then their enthusiasm of learning and create can be stimulated gradually. In addition, this interaction will also help students play the main role of the cognitive. In this interactive learning environment, students can choose their own content and practice suit for their levels, based on their own ability and interest. The teaching process will be changed greatly, the traditional teaching process which is dominated by teachers will be changed, students have the possibility of active participation, rather than listening to the teachers completely, in which students can only accept passively. This creates good conditions for students to participate actively and initiatively, so that students can be the true cognitive subject.

5.2 Modern Educational Technologies Provide Diversity of External Stimulations, in Order to Exquisite and Maintain Knowledge Efficiently

The external data provided by modern educational technologies is not a single stimulation, but a comprehensive variety of sensory stimulation, it is very beneficial not only knowledge acquisition, but also to maintain the knowledge once used in the teaching process. Combination of the vision and hearing provides a colorful arts education experience, a more direct visual impact that arouses the imagination and association of students, expands their knowledge and horizons. Students will enjoy the lively and interesting music theory memory and music skills which have been boring in the past, and maintain and improve. The experiments done by experimental psychologists Treicher proved that the person who can see and hear, their knowledge, which are got through discussion and mind exchange with their own language, maintains much better than people who just look, or listen. Modern educational theory requires students to collaborate and communicate in their learning process; unique effects for the internalization maintenances of knowledge can be derived.

5.3 Use the Computer Networks to Realize Collaborative Learning, Develop the Spirit of Cooperation, and Promote the Development of Higher Cognitive Ability

The so-called collaborative teaching strategies are required to provide opportunities to the learners to observe, compare and analysis different ideas for the same problem in order to collect comprehensive wisdom. This will not only benefit for deepening understanding of issues and mastering the knowledge, but also obviously good for the development of higher cognitive ability, the cultivation of cooperation spirit and the formation of good interpersonal relationships. And therefore, the computer network-based collaborative learning attracts increasingly more and more international research and concern of educators. In our arts education, our teachers can communicate teaching experiences and methods with each other through the network, and solve the problems together to explore more effective ways and means of education to improve the art teaching. The students can learn the difficulties through the network, seek the support and help from art people in different ethnic, racial, regional, and can also enhance friendship, promote cooperation, while improving the skills to improve interpersonal relationships.

5.4 The Hypertext Features of Multimedia Technologies Achieve the Organization and Management for Most Effective Teaching Information

Hypertext is in accordance with the association thinking of human brain, employing advanced technologies of organizing and managing the information in a non-linear manner with the network structure. According to the teaching objectives and teaching content, we can use the characteristics of multimedia hypertext to integrate the relevant disciplines variety of teaching content and materials, including different media information and teaching characteristics, into an organic whole, or we can integrate the prepared knowledge and necessary expanding knowledge of the relevant disciplines for broadening their horizons, teach the students individually. This feature of multimedia technologies can combine a silent art (materials) and the vocal art (exercise) and a variety of artistic elements together, re-configure to form a constantly updating organic whole. In addition, this feature also allows different ages, different levels of students become more artistic freedom, autonomy to choose their own learning content, enhance self-learning ability and learning quality, but also enhance the students' perception and interest in art. Thus, as the core of the modern educational technologies, computer information technologies have a variety of valuable features in the aspects of optimization of arts education and teaching process. These characteristics will be able to give full play to the initiative and creativity of the students, and to create the ideal teaching environment for training students' innovative ability, this environment is the essential factor for the process of creating a new structure of teaching and learning activities. The specific application of modern technologies in arts education is not only conducive to update educational ideas and teaching methods, but has important meaning on the integration of the arts curriculum and continuity and penetration of other educations, and further promote our arts education into a comprehensive modernization and internationalization, which has extremely broad prospects.

6 Conclusion

The new curriculum standards call for a new teaching model and implementation strategy. In recent years, the rapid developments of modern educational technologies have brought great impact on all aspects of traditional education, and will have a broad and far-reaching effect on educational reform. Modern educational technologies have posed a challenge on the existing higher arts education, and played an irreplaceable role in deepening the reform of arts education and development in the artistic potential of students, therefore, it has brought a good opportunity for the smooth implementation of new curriculum of higher arts education. Responding to this trend, this article studies how to use modern educational technologies to optimize the teaching of higher arts education, and also study the applications in the implementations of modern educational technologies in courses of higher arts education.

References

1. Peng, J.: Movie and TV Appreciation. Higher Education Press (1998)
2. Zhang, R., Lu, H.: To strengthen school arts education, improve the overall quality of students. Educational Theory and Practice (6) (2004)
3. Yu, S.: On Quality Education of University Students. Journal of Guizhou Ethnic Institute (Philosophy and Social Science Edition) (3) (2005)
4. Plan for national Schools of arts education development, Ministry of Education, PRC (2001 -2010)

5. Wang, A.: From practice to policy-making. Flower City Publishing House (2005)
6. Zhang, Q., Wang, C.: Music Aesthetic Foundation. People's Music Publishing House (2004)
7. Zhu, X.: Outline of Emotional Education. Nanjing Press (1993)
8. Cardner, H.: Multiple Intelligences. Xinhua Publishing House (1999)
9. Benner: Music Education Philosophy. People's Music Publishing House (2003)
10. Qiao, S., Linfu, L., Li, J.: Modern Education Model of Talent. Guangdong Education Publishing House (1997)
11. Yin, G., et al.: Non-intellectual Factors and Their Culture. Zhejiang People's Publishing House (1996)
12. Goleman, D.: Emotional Intelligence. Shanghai Science and Technologies Publishing House (1997)

Research on the Training System of the Programming Ability for Applied Talents*

Liu Tao[1], Cheng Yusheng[2,**], Zhou Mingzheng[1], and Chen Dejun[1]

[1] School of Computer and Information, AnHui Polytechic University, WuHu, 241000, China
[2] School of Computer and Information, Anqing Teachers College, Anqing 246011, China
chengyshaq@163.com

Abstract. This paper discusses the programming ability education for students which take training applied talents as starting point according to the actual circumstances of students in local college.The system of programming ability education has been constructed based in the reserch on the teaching system, practice system, teaching methods and means, achievement evaluation, network study platform and second classroom of programming course. The practice of carrying out education proves that the individuality and characteristic of student have been given full play , the programming and applying ability has been improved.The the practical ability,autonomous learning ability and innovative spirit of student have been trained and the quality of applied talents has been improved.

Keywords: Applied Talents, Programming Ability, Training System, Innovative Spirit, Teaching Practice.

1 Introduction

In computer science research, no matter it is theoretical research and technical research results, and the final target should reflect in computer software products; And the program is the noumenon of software, quality of software need reflect by the quality of the program. Therefore, the programming ability is core ability which students in the computer and related majors must have, both basic and applied. On the one hand, and the programming ability is the foundation of studying the follow-up specialized courses; On the other hand, the programming ability is important support of applying information technology to solve practical problems, and main part of students' application ability in computer and related majors. However, through the actual situation,we know quite a number of students have certain fear emotions to design program, and even they can not grasp the basic programming method until the graduation. For adapting the transformation from higher education to the popularization and comprehensively promoting the applied talents' education, how to

* This research is supported by Anhui Polytechnic University teaching and research Foundation (ID:2011xjy03), Anhui Province teaching and research Foundation (ID:20100733), Anhui Provincial Teaching and Research Projects (20100687).

** Corresponding Author.

W. Zhang (Ed.): Advanced Technology in Teaching, AISC 163, pp. 495–500.
springerlink.com

train the students' practical ability and creative spirit is important task in teaching research nowadays. For computer and related majors' students training their programming ability is the basis of becoming applied talents [1] .

2 The Problems in Training Programming Ability

The programming ability training relies mainly on the programming language teaching, and the current programming language teaching focus on grammar teaching , using teaching model of teaching the grammar in the classroom and checking the program on the computer. This teaching model can clearly teach programming language grammar, but understanding the grammar is not equal to be able to use these statements to solve practical problems. So even if colleges offer the programming language curriculum (including the Assembly Language curriculum, several kinds of Advanced Programming Languages curriculum, Data Structure curriculum, etc.),the students' programming ability remains poor, and the reasons are as follows:

First, many colleges and universities arrange several programming language curriculum in the teaching plan, with the purpose of making the students master more programming languages to adapt to the needs of the society. But the teaching programs of these programming curricula emphasize on grammar too much, which leads to the questions that teaching key points in the programming languages appear deviation,that is, more imparted knowledge and less ability training.

Second, in the process of teaching, the teaching of programming curricula generally extends around the system of a certain course itself, and teaches knowledge of some curriculum in isolation, without considering the correlation and continuity between the differerent curricula, without making connections with related curricula, without incarnating generality and internal relations between different programming curricula in the teaching. Therefore, that can not make students master the common principle of guiding and studying other computer languages ,can not make students achieve mastery and drawing inferences about other cases from one instance, which goes against training the students' sustainable development ability [2].

Third, the practice is the weak point. Basically there are no other practice opportunities except the experiments arranged in class.The experiments arranged in class are mostly a validation style of the work. In the teaching process teachers lay stress on the theory teaching, often focus on teaching some grammar rules that students are not interested in, or focus on language itself study, and there is a question that the single classroom instructions are more and diversified practice teaching are less. Therefore, it can't conform the college students to solve practical problems,and goes against training the applied talents.

Fourth, many colleges and universities take the score of the national university computer grade examination as the score of the programming language, some teachers and students take exam standards as classroom instruction or studying standards in the programming curricula, ignoring the training of students' self-learning ability, practice ability and creative spirit, and thus this can not improve students' comprehensive quality [3].

3 The Construction of Programming Ability Training System

In view of the above problem,we should research on changing existing curriculum system, practical system, teaching methods, teaching system and evaluation methods, construct thorough and reasonable programming ability training system,make students master programming ideas and methods initiatively and quickly, and then improve students' operation ability and train students' innovative consciousness.

3.1 Constructing Basic and Applied Programming Ability Training Curriculum System

The programming ability is not only the fundament to further study other curricula for students of various kinds of computer majors, but also is the main means solve the practical problem. For this, we construct the programming ability training system constituted by technical curricula including the programming basis (C Programming Language),the object-oriented programming(C++ or Java),the enterprise-level applying and developing technique (.net or J2EE), and theory and method curricula including Data Structure and Software Engineering.In Data Structure, Software Engineering and some related curricula to programming design training contents are increased, and it is taken as the enhancement and development of basic programming ability. We combine multiple types of practical teaching link,and realize 4-year continuous line of the programming curricula.

We carry out the corresponding curricula construction around the system. According to the training goals and requirements of the applied computer talent,the overall teaching goal of programming curricula system is defined as "mastering the basic procedure-oriented programming methods, being able to program simple system program based on C; mastering the predominant object-oriented method and technology, being able to develop general application program according to specifications and criterions of the software engineering". under the overall goal the teaching aims,syllabuses and contents for every curriculum in the the programming curricula system are made. The specific teaching methods and cases are designed.

3.2 Constructing the Programming Ability Training Practice System Composed by Various Types of Practice Teaching Activities

The programming curriculum belongs to the very theoretical and practical curriculum, and it can train students not only to have solid theoretical basis, but also to have high practical ability. Only in the process of practice we can effectively train students how to use the knowledge to analysis and solve the problem. So in the process of teaching programming language, we not only need to lay stress on the cultivation of the students' basic knowledge, but also emphasize the importance of the experiment, and harmonize the relationship between different segments in teaching,composed an integral whole.

The practicality of the program design need train by a lot of practical activities and the programming design interest for students need inspire by the practical teaching. Therefore, we further intensify the status and role of the practical teaching, according to the band-aid including "interest, comprehension ,application, innovation" construct the practical system of the programming ability training composed by the segments of the curricula experiment, the curricula design, students' scientific research projects and graduation design, and design the experimental projects,topic selections and students' scientific research topics.

In the experiment of the programming design basis (C Programming Language) and object-oriented programming design (C++ or Java Language), on the basis of the students' studying psychology, we design the outline and experimental project according to imitating programming to eliminate programming fear, designing independently single structure (basic program structure or including a single class and object) program to stimulate interest in study, and programming simple application to master program design methods, among them the proportion of the designing and comprehensive experimental type is more than 60%.

In the practical segments of subsequent curriculum design, students' scientific research projects and graduation design,we combine the programming curricula with the practical applications, design various different types of topic selections, and train students' application ability and innovation ability in the program design. For example, in students' scientific research projects, we combine with information security, the embedded system and other specific application or technology field, design different topics,and further improve the students' programming ability.

3.3 Constructing the Three-Dimensional Teaching Method System Made Traditional Teaching Methods and Modern Teaching Methods

The program design is a work which also need teamwork and design personnel having some creative thinking, and it is difficult to solely rely on the traditional lectures and experimental teaching methods to ensure the programming ability. Therefore, on the basis of the specific teaching content,we reform the teaching method. For some inferential and analytical content, we comprehensively use the advantage of the traditional teaching means and modern teaching methods based on information technology, compose multimediaization and networked three-dimensional teaching system by merging many kinds of teaching methods including case driving type,discussion teaching and heuristic method with many kinds of teaching means including multimedia courseware,electronic notes and network teaching system.

The comprehensive application of many kinds of teaching methods of generates some positive effects. First of all, it can effectively arouse the students' programming interest,and play a better role in eliminating students' programming fear; Second,it can improve the efficiency of the classroom teaching and properly compress the time of the classroom teaching;Third,it can effectively improve the quality of the classroom teaching on the face of the assignments after class and experimental results; Finally,it can play a part in training the students' cooperation spirit and innovative thinking [4].

3.4 Constructing the Network Teaching System of Promoting Active Learning and Cooperative Learning

Aiming at most studetns lacking team cooperation spirit and the ability of applying existing knowledge to design and solve the practical problem,we propose that the network teaching system based on the Web technology offers the environment of achieving the active learning and cooperative learning for students, and promotes the students' active learning and cooperative learning. Therefore, we develop a network teaching system of programming curricula, under this system environment, and we develop a series of teaching resources and teaching cases of promoting active learning and cooperative learning.In them there are the background and imitativeness cases of stimulating the students' studying interest,the case of small single structure program which can strengthen students' programming confidence and promote team cooperation, and the cases which can deepen the students' application understanding, etc.

3.5 Constructing the Second Classroom Training System Focusing on Promoting Students' Employment Competence and Quickly Mastering Enterprise Mainstream Technology

In order to promote students' employment competence, we must make the students master the mainstream development technology of software. Therefore, through the second class including software training classes and technical seminars,we offer students opportunities to study mainstream development techniques of software opportunity,and lay stress on cultivating students' innovation spirit.

The specific practices include: selecting teachers having development experience commercial projects to teach,and teaching all kinds of extensive applied popular development technology and software architecture. Also, considering the vitality and the application of software technology with limited scope, for students, the most important thing is to have the ability of studying new technology and train the ability of sustainable development. For this, we combine the actual requirements of business projects,programming demonstration on the spot,tracking API source code,and help students to grasp the characteristics of new technology and regular patterns to gradually improve the learning ability. For example, students having learned object-oriented basic theory should focus on mastering the classical design mode on the second class;students having learned Java/C# mainstream programming languages should master the development framework of Struts/Hibernate/LINQ/Spring used in the enterprise largely; we have trained the AJAX technology combined with the widely appeared application pattern in recent years, such as Web2.0, SNS. In addition, we encourage students to be trained in software enterprise on their summer vacation, participate in the commercial project presided by teachers, and let students obtain some development experience and skills of commercial projects.

3.6 Reforming the Evaluation Methodology of Traditional Learning Methods and Standards and Designing a New Evaluation System

The traditional evaluation way can not assess students' ability of programming very well. Therefore, we need reform the traditional evaluation way,and design a new evaluation system composed ordinary grades and examination grades. Ordinary grades are composed of performance in class, homework, the experiment report; examination grades are composed of written examination and computer-based test. In the usual assessing we focus on students' innovation spirit.The features of ordinary grades are: making grading details public; making the score of every experiment. This not only ensures the objectivity and impartiality of the grades,and the more important thing is to arouse the students' subjective initiative of learning and meet targets of improving the quality of experiment teaching by evaluating grades. The relative scientific and perfect teaching assessment and evaluation system is established [4].

4 Conclusion

We constuct the programming ability training system on the basis of in a series of research.The quality of personnel training markedly improves according to actively carrying out the teaching practice.Recent 3 years students have taken part in the "Challenge Cup", "Robot Football", "Bi Sheng Cup" competition, and have gained gratifying achievements. From the passion of students participating in the second class and the employing units' evaluation and feedback to 08th and 09th graduates recent three years, it is successful and necessary to construct the second classroom training system with promoting students' employment competitiveness and quickly mastering enterprise mainstream technology as its heart.Employing units' are satisfied with graduates' programming ability and practice ability, and one-time employment rate is above 90%. Many colleges and universities collaboratively develop the network teaching system of resources sharing and students' self-directed learning, gradually establish faculty having a higher level scientific research,and they provide the guarantee for education teaching practice. This system not only improves the students' programming ability, but also cultivates the students' innovation spirit and ability of sustainable development.

References

1. Zhao, Q.: The position of programming language in computer science research. Journal of Taiyuan University 8(1), 129–131 (2007)
2. Qin, Y., Li, T.: The integration study about computer programming curricula in colleges and universities. Journal of Bohai University (Natural Science Edition) 28(1), 77–80 (2007)
3. Lu, X.: The training analysis to innovation ability of programming language curricula. Journal of the National University (Natural Science Edition) 16(2), 185–188 (2007)
4. Chen, Y., Sun, L.: The research on Software Engineering curricula system. The Higher Engineering Education Research (2), 140–144 (2009)

Model Design of Achievement Pre-warning in High Education Based on Data Mining

Wu Hai-feng[1,2], Cheng Yu-sheng[1,*], and Hu Xue-gang[2]

[1] School of Computer and Information, Anqing Teachers College, Anqing 246011, China
[2] School of Computer and Information, Hefei University of Technology, Hefei 230009, China

Abstract. At present, high education in china has been changed from elite education to mass education. However, a considerable number of students have academic problems. The problem that needs urgent solution in college is how to find academic problems timely and automatically which have appeared or are forthcoming, and reduce the chances of academic problems. The model of achievement pre-warning in high education based on data mining is proposed, which can find three types of academic problems-low type, landslide type and latent type, and give the appropriate pre-warming information. The information has some guidance to complete their studies for the pre-warning students.

Keywords: Data mining, academic pre-warning, model of achievement pre-warning.

1 Introduction

High education in china has gradually entered the mass education stage. The number of enrolled students has increased rapidly, poor quality of students recruited and teaching that follows naturally. Under the pressure of learning and employment, our students still face the temptation of internet. In quite a few students, indulging in the internet can result in poor academic performance. They will not finish their studies if let it be.

Academic pre-warning[1][2][3] is predictive for the academic problems and learning difficulties that has occurred or will occur in school, as well as it shows the negative consequences for students and their parents. We can take appropriate measures with actual situation to help students to finish their studies. It changes the post-processing into pre-warning which is different from traditional models of management, help students to transform the pressure into motivation and make them have clear learning goal to get through their academic career successfully.

Along with the data mining technology maturated and applied in many fields, the application of data mining technology to academic pre-warning in college is bound to benefit to discover students' problems and help teacher to improve the quality of teaching.

2 Academic Pre-warning

2.1 Background

High education in china has been changed from elite education to mass education with the enrollment expansion. A lot of problems were cropping up, such as the number of

W. Zhang (Ed.): Advanced Technology in Teaching, AISC 163, pp. 501–506.
springerlink.com

students has increased sharply, their life quality decreased obviously; students were confronted with learning and economic, employment and the psychological pressure; teachers were not enough, their teaching load was heavy, teaching quality dropped; teaching management lagged because of running across the campus.

How to improve students' autonomous learning ability, self-adjustment ability, and help them to avoid problems and overcome all difficulties in study, which is the problems need to be solved in the management of universities. Academic pre-warning mechanism is come into being with this background.

2.2 Connotation

Pre-warning, namely prior warning, can be interpreted as, "before disaster and other danger need to watch out happening, according to former summing rules or the possible presage precursor, people issued an emergency signal and reported a dangerous situation to relevant departments, in order to avoid damage happened in not informed or unprepared. So it can maximize reduce the damage."[4]

"Academic pre-warning" is one of 171 Chinese new words, which listed in article "the report of China language life situation in 2006" by the ministry of education in August 2007. It is a kind of high education management way [5].

Academic pre-warning is a education means and crisis intervention system, which can remind students and their parents timely in case of few students' academic serious problems, and adopt corresponding preventive measures to help students complete their studies.

The principle of academic pre-warning is "people first". Through the communication and cooperation between school, student and parent, guide and monitoring students in learning, which can help students complete their studies and become adults.

2.3 Object

Academic pre-warning object can be divided into three categories:

(1) Absent pre-warning

Students whose absent class hours reached a certain number in one week, one month, one term, or who don't attend the normal teaching activities certain times, need to be warned timely. Further measures will be taken if warning is invalid.

(2) Exam pre-warning

To strengthen the honesty education, students require to be abiding in the examination. Take a collective integrity exam oath on or signature activities, students who have cheating record or motivation of cheating should be warned and educated before important exam, which can reassure their fluky psychology and reduce the occurrence of cheating.

In a major, oath or signature activity, to the examination of cheating in a criminal record and motivation of the students cheating in key early and. If students violating the discipline to quickly and strict treatment, immediate effect.

(3) Achievement Pre-warning

Achievement is the inspection and final reflect of study effect. Achievement Pre-warning is a very important part of academic pre-warning, pre-warning situation mainly has the following kinds: exam failed; overall achievement appears landslide.

2.4 Difficulty

As long as the attendance and statistics on time, we can master the absent situation, the students who achieve pre-waning conditions should be warned. The key of exam pre-waning is educating before exam, handling the students who disciplined in time. The difficulty of academic pre-warning is mining the implied useful information in the achievement data.

3 Model Design of Achievement Pre-warning

3.1 Model of Achievement Pre-warning

At present, achievement pre-warning in college still stays on the stage of simple statistics and analysis, lacking of powerful tools for deep mining. Through the analysis, achievement pre-warning based on data mining is designed in this paper. Model mainly consists of three parts- low type pre-warning module, landslide type pre-warning module and latent type pre-warning module. The model is shown in figure 1.

According to the situation, pre-warning is divided into three levels: general pre-warning (★), serious pre-warning (★★), very serious pre-warning (★★★).

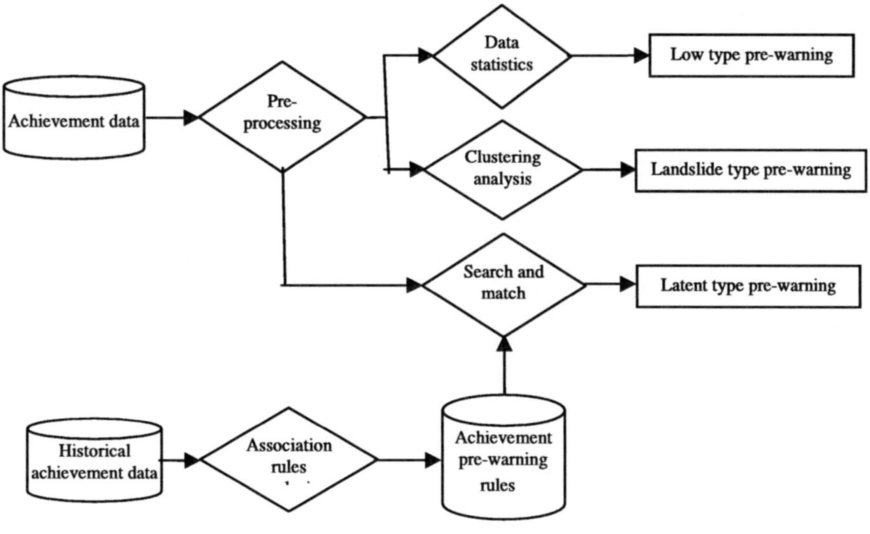

Fig. 1. Model of achievement pre-warning

3.2 Low Type Pre-warning Module

When students' achievement is low and their courses failed, they must be warned. Low type pre-warning module mainly includes the following aspects:

(1) Calculating the number of fail course

This module can calculate and output the fail course number of students for every major, every grade, and every class in one term or total.

(2) Collect the details of fail course

This module can collect and output the fail course details of students for every major, every grade, and every class in one term or total, include course name and score.

The pre-warning level is confirmed by follow standard:

General pre-warning (★) : the number of fail course is one in current term and only one in total;

Serious pre-warning (★★) : the number of fail course is one or two in current term and reach two in total;

Very serious pre-warning (★★★) : the number of fail course is one or more than one in current term and reach three in total.

3.3 Landslide Type Pre-warning Module

The traditional basis of landslide is that ranking declines than a certain criterion. Ranking is calculated with the total score or average. This method is simple, the result is intuitive. But it has certain limitation and can't accurately reflect the distribution of achievement. For example, a student's achievement is number one in last term, ranking sixth this term, but the gap between sixth and first is very small. The ranking drops five, but it is not landslide.

Clustering analysis can solve the problem effectively. For example, we can divide the achievement into five grades, set the number of clusters to five, and then the clustering result is divided into five grades: the gap in every grade is minimum; the gap between every grade is maximum. The result can accurately reflect the distribution of achievement.

The implementing steps of landslide type pre-warning module are as follows:

(1) Do clustering analysis of achievement in last term, set the cluster number to five, and then all achievement is divided into five grades (outstanding, good, medium, poor, very poor). Arrange the clustering results from large to small in order to quadratic sum of each dimension of barycenter, set cluster number to 1, 2, 3, 4, and 5.

for the semester on the results of the clustering, , then order, will the clustering and qualified each d square of heart from big to small arrangement, cluster number in turn set to 1, 2, 3, 4, and 5;

(2) Deal the achievement in this term with the same treatment.

(3) Compare the clustering results of every student in two term, if the student's grade remains in 4 or 5, then warn the student and output the information.

The pre-warning level is confirmed by follow standard:

General pre-warning (★) : the grade in this term fall a order, or the grade in two terms all stay in "poor" order;

Serious pre-warning (★★) : the grade in this term fall two order, or the grade in two terms all stay in "very poor" order;

Very serious pre-warning (★★★) : the grade in this term fall three or more than three order.

3.4 Latent Type Pre-warning Module

At present, the actual application of this module for achievement pre-warning is still blank in our country. Through the association rules mining of historical achievement, and then the association rules library of achievement can be formed. Search association rules library according to the current achievement, possible setbacks can be predicted in the next term.

The implementing steps of latent type pre-warning module are as follows:

(1) Data pretreatment

i) Screen the fail records from all records.

ii) Set the code of course to “term” + “course number” for the convenience to sort. The range of “term” is 1-8, because the four school years are 8 terms; the range of “course number” is 0-9, because the number of courses in one term is less than 10. So that, a fail record will be converted into double digits, the first digit means the term number, the second digit means the course number. Such as "higher mathematics (1)" is converted into 10.

iii) Format the database from "student id, name, fail course name" to "student id, name, fail course 1, fail course 2,......" .

(2) Mining frequent itemsets

Support reflects the universality of itemset in database. Set the right support and make the association rules have certain universality.

(3) Generate the pre-warning rules library

The association rules generated by the traditional method can't directly treat as warning rules. Because they contain many useless rules, which will reduce the accuracy and efficiency of pre-warning. In order to make the association rules have practical significance, especially do the following limits:

i) Set the right confidence. The confidence reflects the accuracy of rules. If the confidence is too small, the pre-warning rules are little credibility, and the accuracy of pre-warning is reduced. If the confidence is too big, a lot of meaningful pre-warning rules can be eliminated, and greatly reduce the effect of pre-warning.

ii) The order of antecedent itemsets, consequent itemsets, antecedent itemsets and consequent itemsets of pre-warning rules is in turn of term. The order of antecedent itemsets, consequent itemsets, antecedent itemsets and consequent itemsets of association rules is not defined. The antecedent itemsets happened before consequent itemsets in pre-warning rules. At the same time, limit the order of antecedent itemsets, consequent itemsets for the convenience of searching association rules library and matching in step (4).

iii) The scores in the same term can't appear in consequent itemsets and antecedent itemsets of one rule. Because the scores in the same term is generated at the same time.

iv) The consequent itemsets of rules only contains the course in one term. We should only predict the scores in the next term when the scores is generated. Because the prediction will be more accurate after the scores in previous terms is generated.

For a semester grades generated after, simply predict the next semester grades can, need not also forecast a few term after the performance, and the term for the first few term of generated.

(4) Search the pre-warning rules library for matching

Deal the achievement in current term with the same treatment like the step (1), search the pre-warning rules library for matching, and output the result.

The pre-warning level is confirmed by follow standard:

General pre-warning (★) : the possibility of failing a course in next term is between 40% and 59%;

Serious pre-warning (★★) : the possibility of failing a course in next term is between 60% and 79%;

Very serious pre-warning (★★★) : the possibility of failing a course in next term is more than 80%.

4 Conclusion

The implementation of academic pre-warning can detect the students who deviate from their proper development, alert and interpose promptly to help students to finish their studies.The difficulty of academic pre-warning is the grade pre-warning. The grade pre-warning model based on the data mining technology can automatically and timely detect three types of academic problems, thus the academic pre-warning has much more pertinence and efficiency.

References

1. Chi, Z.: Discussion of formation, development and improvement in pre-warning system of school work. Master thesis of Qufu Normal University (2009)
2. Hu, X., Yang, Z., Huang, W.: Exploration and practice of pre-warning system on college student management. Higher Agricultural Education
3. Ding, F., Sui, A., Cai, X.: Program Design and Optimization of student academic pre-warning system. Journal of Chengdu University of TCM(Educational Science Edition)
4. http://baike.baidu.com/view/316884.html?wtp=tt
5. http://baike.baidu.com/view/1110257.html?wtp=tt

The Research of a Cooperative Model Intrusion Detection System

Zhu Qiang[1] and Shun Yuqiang[2]

[1] Zhejiang Institute of Communication and Media, Hangzhou, P.R. China
zq.hz@163.com
[2] Changzhou University Changzhou, P.R. China
Shunyq@126.com

Abstract. The paper introduces a model of the Cooperative System of Honeypots and Intrusion Detection System (CSHIDS). The method of implementing the cooperation is: firstly, the Honeypots and HoneyWall try to record various attackers' behaviors and transfer them to Remote Log Server, secondly, the cooperative system divides the records into several classes by using the way of Unsupervised Clustering; thirdly, make the records labeled; and then extract attack-rules from the labeled records by Decision Tree; at last, add the new attack-rules to the Intrusion Detection System's rule-set in a certain time interval. The purpose is that Intrusion Detection System (IDS) can detect the new attacks. This model's effectiveness has been confirmed by the simulated experiments.

Keywords: Honeypot, Intrusion Detection System, Unsupervised Clustering, Decision Tree.

1 Introduction

With the rapid development of Internet, network security becomes a problem receiving much concern. People also pay more attention to intrusion detection and prevention. The function of Intrusion Detection System is to monitor the actions occurred in the network and system, then find out and report the intrusions or unauthorized actions for the people.

Generally, intrusion detection techniques can be divided into two groups according to the type of data they use: misuse detection and anomaly detection [1].

Misuse detection uses knowledge about known attacks and attempts to match current behavior against the attack patterns. It has the advantage that known attacks can be detected reliably with low false-positive error and economically. The shortcoming is that it cannot detect unknown attacks. [2] The introduction of Honeypot can solve this problem.

Honeypot is an information system resource whose value lies in unauthorized or illicit use of that resource[3] defined by Spitzner. A Honeypot is a program, machine, or system put on a network as bait for attackers. The Honeypot tries to capture details of the attacker's actions, and the information gained after the Honeypot is attacked can be used to learn about the attackers' tactics, motives, and tools so as to defend them for the future. [4] There are some existing Commercial or Free Honeypots: ManTrap,

W. Zhang (Ed.): Advanced Technology in Teaching, AISC 163, pp. 507–515.
springerlink.com

Specter, LaBrea Tarpit, Honeyd, Honeynets and so on.[5] The CSHIDS in this paper tends to extract attack-rules through analysis on the hack behaviors recorded in the Honeypots, and then add new attack-rules to the IDS rule-set. As a result, new attacks will be detected.

2 The Model of Cooperative System of Honeypots and Intrusion I)etection System

In order to record all kinds of attackers' behaviors, Honeypot should attract more attackers to enter into. So it needs to be designed as real as a true system by utilizing camouflage technology. The physical model of CSHIDS can be seen in figure 1.

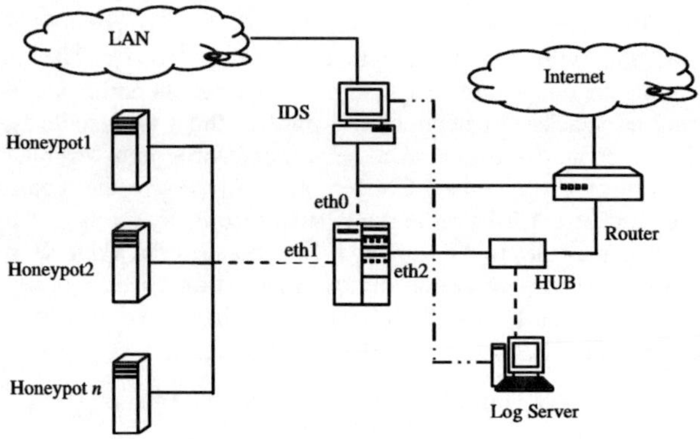

Fig. 1. Physical model of CSIIIDS

Honeypot is a typically virtual machine that emulates real machine by feigning running open ports and services which might be found on a typical machine on a network. These running services are meant to attract the attention of attackers so that they spend valuable time and resources to exploit the machine while the attackers are being monitored and recorded by the Honeypot. [4]

HoneyWall is the control center for the CSHIDS, because all traffic entering or leaving the Honeynet must go through it. The HoneyWall which separates LAN from the Honeynet network is a layer two bridge, so it is harder to be detected. The external interface of HoneyWall (eth0) is connected to the LAN. The internal interface of HoheyWall (eth1) is connected to the Honeynet systems' network. Since this is a bridge, both internal and external systems are on the same IP network, which makes an illusion that Honeynet are a part of LAN. The purpose of eth2 interface is for transferring the data gathered by Honeypots, it is connected to the Remote Log Server. Furthermore, HoneyWall allows all connections to enter into Honeypots, but restricts their leaving freely. The system strictly limits the Gangplank attack launched by the hacker to the out network. Data control includes two means: Network Intrusion Prevention System and connection counting.

The purpose of Network Intrusion Prevention System is to identify and block known attacks. It does this by inspecting each connection as it travels through the

HoneyWall. The purpose of connection counting is containing how many outbound connections an attacker can initiate from Honeypots, and when a certain limit has been met, block any more connections.

In this model, Honeypots not only attract attackers to enter passively, but also utilize IDS and HoneyWall to lead attackers to enter initiatively. Once IDS detects suspicious attack, HoneyWall will veer the attacker to Honeypots, and then the system can monitor what the attacker wants to do.

So attackers have two ways to enter the Honeypots: one is that they use attack tools to enter them when the attackers find some sensitive information or system bugs in Honeypots, the other is that the HoneyWall veers the attackers to Honeypots when IDS has detected suspicious attacks.

The logical model of CSHIDS is given in figure 2. This model includes Data Process Module. Data Store Module and Data Analyse Module.

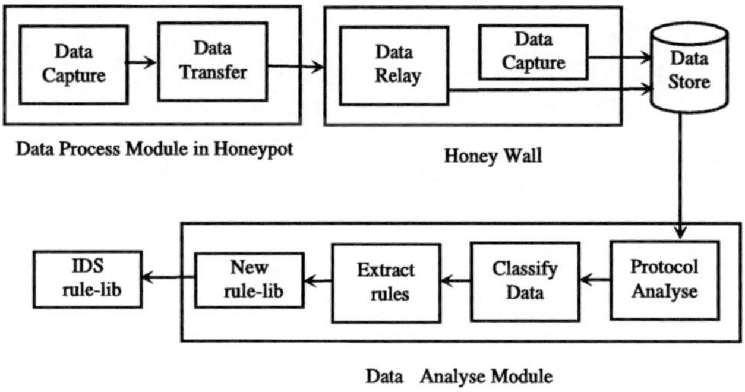

Fig. 2. The logical model of CSHIDS

In this model, the attackers' behaviors which 1-loneypots have recorded are conveyed to the Remote Log Server, and stored in real time. The system triggers Data Analyse Module to classify the records in a certain time interval or manual intervention, then extracts rules from the labeled records and adds new attack rules to IDS rule-set. So IDS can detect new attacks.

2.1 Data Process Model

Unless collecting enough information, Data Process cannot extract attackers' characters better. So, we capture data both in Honeypots and HoneyWall. Honeypot is set up into an environment without magnetic disc, which Ii changes the hacker's access to local files into the access to network files. The system action now is exposed to network, which would be recorded when we collect data.

When data has been captured by Honeypots, it will be encrypted and transferred to HoneyWall, in which it will be transferred to Remote Log Server. When data has been captured by HoneyWall, it will be encrypted and transferred to Remote Log Server directly.

Data Capture Sub-module and Data Transfer Sub-module will run in the kernel of Linux/UNIX Operating System which runs in the Honeypots, and the HoneyWall is a layer two bridge, which is to make the data gathering secluded so as to prevent attackers from finding their behaviors are monitored. Once the attackers find out their behaviors are being recorded, he would destroy the log or make some false information, which would make much more troubles in analyzing these data.

2.2 Data Store Module

Honeypot has a danger of being captured at any time, so it becomes insecure. The hacker will destroy the log after he obtains the Honeypot's supreme authority, so the system set the Remote Log Server for storing records to keep the information safe.

The Remote Log Server and Honeypot are separated by HoneyWall, communication between them adopts TCP protocol. When Honeypot transfer data to Remote Log Server, the data must pass through the HoneyWall, this deployment makes that only the log information from 1-Loneypot can reach the Remote Log Server.

2.3 Data Analyse Module

Table 1. The Set of Features of Connections

The Basic Features
Proto_type: type of the protocol;
Service: network service on the destination;
Sbytes: number of data bytes from source to destination;
Dbytes: number of data bytes from destination to source;
Features gained through analyzing higher level protocol (F1E Telnet, etc.)
num_failed_logins: number of failed login attempts;
num root: number of "root" accesses;
num_file_creations: number of file creation operations;
num access files: number of operations on access control files;
Features computed using a two-second time window:
duration: length (number of seconds) of the connection;
urgent: number of urgent packets;
count: number of connections to the same host as the current connection in the past two seconds;
serr rate: % of connections that have "SYN" errors;
rerr rate: % of connections that have "REJ";
same srv rate: % of connections to the same service.

The main function of this module is to analyze records on Remote Log Server. But the HoneyWall would veer suspicious attacks to Honeypot, which would be either attack behaviors or normal behaviors through further monitoring. Furthermore, normal users may rush into Honeypot by mistake. As a result, some normal behaviors are recorded in log. In order to extract pure attack-rules, this module should distinguish normal behaviors from attack behaviors before extracting rules. So the process of data analyzing includes four steps: Protocol Analyse, Unsupervised Clustering on data, then label cluster, finally, extract new attack rules out from labeled records.

1) Protocol Analyse: The types of records' protocol are TCP, ICMP, UDP. We should unify records of different types into a new format. Then, a connection is a sequence of IP packets starting and ending at some well defined times, between which data flows to and from a source IP address to a target IP address under some well defined protocol. We can see the set of features from table 1.

2) Unsupervised Clustering on Data: It is based on two assumptions: one is that there are obvious differences between attack behaviors and normal behaviors, which can be distinguished by their features easily. The other is that the quantity of attack behaviors is far greater than normal behaviors.

Let L be cluster radius, which is set according to experiences. Let dist(O_j, x_i) be Euclidean distance, O_j be cluster center; x_i be any record of sample. When calculating Euclidean distance, we give a weight to each feature of record. So the Euclidean distance is:

$$dist(O_j, x_i) = \sqrt{w_1 |x_{i1} - O_{j1}|^2 + w_2 |x_{i2} - O_{j2}|^2 + \ldots + w_p |x_{ip} - O_{jp}|^2}$$

Let $S=\{x_1, x_2, \ldots, x_n\}$ be a sample of records which would be classified.

Step1: select k records from S as initial cluster centers at random, and every two cluster centers satisfy the condition: dist(y_i, y_j)>L;

Step2: $O_1 \leftarrow y_1$, $O_2 \leftarrow y_2, \ldots, O_k \leftarrow y_k$; cluster number=$k$;

Step3: repeat;

Step4: choose x_i from S in turn, i=1,…,n, calculate dist(O_j, x_i), j=1,…, k; and find out O_m, which satisfy the condition dist(O_m, x_i)≤dist (O_j, x_i), j=1,...,k and $j \neq m$;

Step5: If dist(O_m, x_i) ≤L, then $C_m \leftarrow \{ x_i \}$, i=i+1; else goto Step7;

Step6: Adjust cluster center O_m, goto Step8;

Step7: add x_i to unknown cluster UKC;

Step8: until i=n.

Repeat the above operation to unknown cluster UKC, until UKC is emptied.

3) Label Cluster: For attack behaviors account for the overwhelming majority in the records, clusters can be arranged in sequence by the nUlllber of records they include. Let N be commensurable number. If tile record number of cluster is less than N, we label this cluster as noi-mal, per contra, as tile attack.

4) Extract new attack rules out from labeled records: We extract attack rules from labeled records using C4.5 algorithm, and compare these attack rules with the rules stored in tile IDS rule-set and new rule-lib. If a new extracted rule is the same as any rule in the IDS rule-set and new rule-lib, delete it. The remains are new rules. Then, we add them into the new rule-lib. The Algorithm is as follow:

Tile rules extracted by C4.5 algorithm are stored in RT[N] table, the rules from new rule-lib are stored in NR[M] table, rules from IDS are stored in IDSRT[P] table.

Step1: repeat;

Step2: select x_i from RT[N] in turn, i=1,...,N;

Step3: Compare x_i with the rules from IDSRT[P] and NR[M], if any rule is the same as x_i, delete x_i from RT[N];

Step4: until i=N;

Step5: Add the remains of RT[N] to NR[M] table of new rule-fib.

By adding the rules in new rule-lib into IDS rule-set in terms of a certain time interval or manual intervention, then IDS can detect new attacks.

3 Experiments

3.1 Datasets

KDD CUP 1999 dataset which are from 998 DARPA Intrusion Detection Evaluation Program are used to evaluate the proposed method. Every TCP connection not only includes basic features, but also includes some higher-level features that defined by Stolfo et al. to help distinguishing normal connections from attacks. For example, the "same host" features examine only the connections in the past two seconds that have the same destination host as the current connection, and calculate statistics related to protocol behavior, service, etc. Every connection has forty-one features, seven features are discrete, and thirty-four features are continuous.

3.2 Preprocessing Data

Experimental data includes seven discrete features and thirty-four continuous features. We should pre-dispose each feature before using Unsupervised Clustering.

We take protocol_type feature as an example to explain how to pre-treat discrete features. As the protocol_type is character with values of tcp, udp and icmp, it is unfavorable to further dispose, we take 00,01,10 instead of tcp, udp, icmp, and split protocol feature into two: protocol_typel and protocol_type2. When the value of protocol_type is udp, we set protocol_type l=0 and protocol_type2=l. But we do nothing with the kind of discrete features whose values are integer.

As for the continuous features, different feature has different measurement. If the values set in the features are great disparity and without disposed, it will lead to some characters of some features swallowed in the course of clustering. For example, set two vectors $x_1\{210, 0, 1\}$ and $x_2=\{70, 10, 0\}$, their Euclidean distance is:

$$\begin{aligned} dist(x_1,x_2) &= \sqrt{(x_{11}-x_{21})^2+(x_{12}-x_{22})^2+(x_{13}-x_{23})^2} \\ &= \sqrt{(210-70)^2+(0-10)^2+(1-0)^2} \end{aligned}$$

It can be seen that characters of two vectors are swallowed by the first. To solve this problem, each value of the features should be standardized. Methods we can use as follows:

Calculate mean absolute deviation s_f:

$$s_f = (1/n)\sum_{i=1}^{n}(X_{if}-m_f) \tag{1}$$

$X_{1f},.., X_{nf}$ is n features of f, m_f is the average of $X_{1f},.., X_{nf}$:

$$m_f = (1/n)\sum_{i=1}^{n}X_{if} \tag{2}$$

Standardize the values of every feature:

$$Z_f = (X_{if}-m_f)/s_f \tag{3}$$

As for isolated point, mean absolute deviation s_f more robust than standardized deviation σ_f [8].

Then, we can gain a dataset who has standardized features.

3.3 Data Classification

According to the importance of features, various weights should be given different attributes during the course of calculating dissimilarity of two records.

In order to give each feature a reasonable weight, the way we adapt is: firstly, judgments on each record's dissimilarity should be given according to experts' experiences; secondly, synthetical judgment should be carried out on the two records; at last, fuzzy relation equation is drawn out according to Synthetical decision-making model.[9] Then, we can gain every feature's weight. The weight we get by this way is more rational than that we set, which can reflect the weight of each feature more accurately in calculating the dissimilarity.

Table 2. The type and quantity of experimental records used in Unsupervised Clustering

Groups	the quantity of records
First Group probing (4232 recorders)	ipsweep(1247), nmap(231), normal(125) portswcep(1040), satan(1589)
Second Group DOS (4232 recorders)	neptune(1 636), smurf(2471), normal(125)
Third Group U2R (128 recorders)	erl(5), ps(16), rootkit(23), xterm(13), loadmodule(16), buffer_overflow(47), normal(8)
Fourth Group R2U (4232 recorders)	ftpwrit(3), guess_passwd (2626), phf(2), imap(1), multihop(18), sendmail(17), warezmaster(1402),xsnoop(4), named(17), xlock(9), normal(124)

Dataset will be divided into three classes by unsupervised clustering. They are attack set, normal set and unknown set. Here unknown set is given. As for the other two uncertain sets, the set includes more records is labeled as attack set, and the other as normal set. Repeat the operation on the unknown set until it is emptied.

3.4 Extract and Test Rules

We draw out rules from classified data by using C4.5 algorithm. We can see the rules from figure 3.

Rule 1: (1256/1, lift 3.1)
src_bytes < =1
logged_in = -1
dst_host_count>10
—>class attack [0.998]
……

Rule 9: (4049/8, lift 1.5)
src_bytes >-1
count< = -1
—>class attack [0.998]
……

Fig. 3. Rules drawn out from classified data

The quantity of test data can be seen in Table 3.

Table 3. Data used to test new rules

Groups	Total	Normal connections	Attack connections
FirstGroup	30228	26121	4107
Second Group	30228	26121	4107
Third group	2200	2080	120
Fourth Group	30228	26121	4107

It can be seen from table 4, the attack rules which are extracted from the two kinds of hack methods of probing and DOS, can obtain better result when detecting the sample-set containing the above two attack records. This indicates Unsupervised Clustering can distinguish probing and DOS invasions from normal connections better. But the test result for U2R invasions is a little worse than above two kinds, the reason is that the quantity of U2R invasions used in experiment may be too small; the result for R2U is even worse as well, the reason is that a lot of R2U attackers pretend to be legal users. The characters of attackers and legal users are so similar that Unsupervised Clustering cannot do it well.

Table 4. Test result of every group

Groups	False Positive rate	False Negative rate	Total error rate
First Group	4.3%	0.27%	3.8%
Second Group	0.20%	0.29%	022%
Third Group	74%	6.7%	7.3%
Fourth Group	6.42%	20.6%	8.34%

4 Conclusion

The CSHIDS has three characters: (1) Concealment. We utilize camouflage technology to attract attackers to enter Honeypot, and record their behaviors. (2) Capture data. Honeypot is set up into an environment without magnetic disc, which changes the access to local file into network file. When we collect information, hack action would be recorded by the system. We collect attack behaviors both in Honeypots and HoneyWall. (3) Security. Attackers' behaviors recorded in the Honeypot are conveyed to the secure Remote Log Server, and the system should make sure that attackers cannot use the Honeypot as gangplank to attack other hosts even if the Honeynet is broken down.

The process of data analyzing is: firstly, protocol analyse and classify the records stored in Remote Log Server into several classes; secondly, extract attack-rules by Decision Tree; thirdly, add the now rules to IDS rule-set, then, the cooperation of Honeypot and IDS is completed.

There are also some shortcomings, this paper has only analyzed the TCP, UDP, ICMP connections, not refer to other system logs or fusion of data from multi-resources, we will study these problems in further work.

References

1. Axeisson, S.: Intrusion detection systems: A survey and taxonomy. Technical Report 99-15, Department of Computer Engineering, Chalmers University, Mnreh (2000)
2. Han, S.-J., Cho, S.-B.: Detecting intrusion with rule-based integration of multiple models. Computers & Security 22(7), 613–623 (2003)
3. Spitzner, L.: Honeypots: Definitions and Value of Honeypots, 5 (2003), http://www.trackinghackers.com/papers/honeypots.html
4. Gubbels, K.: Hands in the Koneypot, 3 (2002), http://www.sans.org/rr/wh1tepapers/detection/365.php
5. Honeypots Solutions: So you want to build your own honeypot, http://www.tracking-hackers.com/solutions/
6. Honeynet Project: Know YoLtr Enemy: Honeynets, http://www.honeynet.org
7. Luo, M., Wang, L., Zhang, H.: An Unsupervised Clustering-Based Intrusion Detection Method. Acta Electronica Sinica 31(11), 1713–1716 (2003)
8. Jiawei, H., Kamber, M.: Data Mining Concept and Technique. China Machine Press, Beijing (2001)
9. Wang, P.: Fuzzy Sets and Application. Shanghai scientific & Technical Publishers (1983)
10. Honeynet Project. Know Your Enemy: Genll Honeynets (November 3, 2003), http://project.honeynet.org/papers/gen2/index.html
11. Richard Stevens, W.: TCP/IP illustrated Volumel: The Protocois. China Machine Press, Beijing (2000)

Particle Filtering Method for Source Localization in Wireless Sensor Network

Yong Zhang[1] and Li Wang[2]

[1] Institute of Information, Tianjin University of Commerce, Tianjin, China
[2] Institute of Information, Hebei University of Technology, Tianjin, China
zhangyong@tjcu.edu.cn

Abstract. This paper presents a new method for source localization of the gas diffusion using a wireless sensor network. The method has two steps: an initial estimation of the position of odor-source is obtained centrally by a robot and is based on particle filtering. It does not require any prior information about the position of the nodes. In the second stage, the nodes refine their position estimates employing a decentralized information filter. The paper includes several implementation aspects and experimental results.

Keywords: source localization, wireless sensor network, particle filtering.

1 Introduction

Latest advances in the low-power electronics and wireless communication systems have made the Wireless Sensor Networks (WSNs) being possible, a new generation of devices able to communicate, sense environmental variables and even process this information. These systems are aimed at various applications, one important set of applications involves monitoring diffusion phenomena gas source localization, potential applications include security, environmental and industrial monitoring, as well as pollution control [1,2].

A lot of research has been focused on detecting and localizing single or multiple odor sources with sensor networks such as [3,4] for a vapor-emitting source, [5] for a nuclear source. And [6,7] for chemical source.

In this paper proposes the use of a mobile node, a robotic vehicle with GPS, for the localization of the odor-source based on the physical models for the spatial and temporal concentration distribution of the dispersed substance from a diffusion source[1]. Particle filtering is used to process the concentration value in each node in order to localize the odor-source in a static wireless network. The method takes into account the uncertainty associated to the concentration value in order to optimally compute the mean and standard deviation of the localization of the odor-source.

The paper is structured as follows. Firstly, the approach is outlined in section 2. Section 3 will detail the proposed method to address the localization problem. Finally, the implementation and some experimental results with a real network are shown.

2 Gas Diffusion Model

The purpose of the presented technique is to localize the gas diffusion source in wireless sensor network by using the information gathered with a robot.

W. Zhang (Ed.): Advanced Technology in Teaching, AISC 163, pp. 517–523.
springerlink.com

The scheme to identify a gas diffusion source location is illustrated in [8]. Since the diffusion rate of gas molecules is usually much slower than the wind velocity, the gas molecules trail in the downwind direction as shown in Fig. 1.

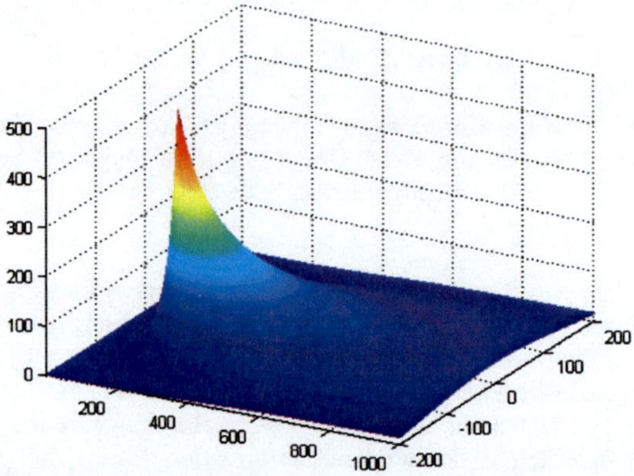

Fig. 1. Where the x-axis is taken as the downwind direction, C0 is the time-averaged gas concentration at the point (0, 100, 0).

When gas is emitted at the rate of q from the origin of a coordinate system(x, y, z) in Fig. 1:

$$C_0(x, y, z) = \frac{q}{4\pi K}\frac{1}{d}\exp(-\frac{U}{2K}(d - x)) \tag{1}$$

If a gas source is situated near the floor, its effect on the gas distribution should be considered. In this paper we suppose a gas source is placed at (0, 0, h) and a floor plane at z_0. Since there is no gas diffusion through the floor, the constraint is imposed on the equation of the gas distribution.

$$\begin{aligned} C_k(x_i, y_i) &= \frac{q}{2\pi K}\frac{1}{x_i}\exp\left[-\frac{U}{2K}(d-(x_i - x_s))\right] + \omega_i \\ d &= \sqrt{(x_i - x_s)^2 + (y_i - y_s)^2} \end{aligned} \tag{2}$$

where K is the turbulent diffusion coefficient, U is the wind speed and d is thedistance between node and the odor source . a point gas source is assumed to be set on the floor at (x_s, y_s) and the i node is assumed to be set on the floor at (x_i, y_i). ω_i is the noised which based on $N(\mu, \sigma^2)$ 。

The approach uses the sensor node recived signal to estimate the position of the emitter based on a mobile robot. The technique can be divided into two basic steps. Firstly, a Particle Filter is used to process the concentration value received from each node to compute an initial estimation of node locations in a static wireless network. The filter takes into account the uncertainty associated with the concentration value and with the robot position (it can be provided by a GPS device) in order to optimally compute the position of the node. In the second step, the initial estimation of the position of the nodes (represented by mean and standard deviation) is sent to them. A distributed Information Filter is implemented in each node in order to easily improve the localization using the signal strength received from other nodes of the network, including the mobile robot.

3 Particle Filter for Gas Diffusion Source Localization

The objective of the localization algorithm is to estimate the position of the nodes of the network from the data provided by a robot. The state to be estimated consists of the position of gas source $X_s = (x_s, y_s)$. The information about the state will be obtained from the set of measurements $Z_{1:k}$ received up to time k. This set of measurements consists of pairs of concentration C_k (the node sensor communite it to the robot) and robot position values $\{X_k^b, C_k\}$.

The method is based on Particle Filtering. In the Bayesian framework employed, the key idea is to represent the posterior density at time k $p(X_k|Z_{1:k})$, by a set of independent and identically distributed random particles $\{X_k^{(i)}\}$ according to the distribution. Each particle is accompanied by a weight $\omega_k^{(i)}$. Sequential observations and model-based predictions will be used to update the weight and particles respectively. See [9] for more details.

Particle Filter seems to be a good solution to address the localization problem. Although there are many possible implementations, in the proposed algorithm the prior probability distribution $p(X_0)$ is used as the importance (or proposal) distribution to draw the initial set of particles at time 0, i.e. $X_0^{(i)} \sim p(X_0)$. The next subsections describes the main issues in the actual implementation of the algorithm. As the likelihood function is the core of the algorithm, it is described first. Then the updating step, the prior distribution, the prediction step and the resampling procedure are detailed. Finally, some guidelines for computing the mean and standard deviation in the filter are mentioned.

3.1 Learning the Likelihood Function

The likelihood function $p(X_k|Z_k)$ plays a very important role in the estimation process. The model used here considers that $p(X_k|Z_k)$ follows a Gaussian distribution for a given value of d_k, the distance between the position of the node and the beacon node $\|X_k - X_k^b\|$:

$$C_k = \mu(d_k) + N(0, \sigma(d_k)) \tag{3}$$

This section deals with the computation of the relations $\mu(d_k)$ and $\sigma(d_k)$ that relate the received concentration in one node and the distance to the emitter, which are estimated off-line from a training data set. A couple of nodes have been distanced from 0 to 30 meters and the concentration has been recorded for each distance. This experiment has been repeated with several antenna polarizations. A least squares process was used to compute the $\mu(d_k)$ and $\sigma(d_k)$ functions that best fit the set of data.

3.2 Updating

Once learned, the functions $\mu(d_k)$ and $\sigma(d_k)$ are used online in the estimation process. Each time a new measure is received, the weights of the particles are updated considering the likelihood of the received data. The procedure is as follows. For each particle, the distance $d_k^{(i)} = \left\| X_k^{(i)} - X_k^b \right\|$ is obtained. From this distance, the mean and variance of the conditional distribution $p(Z_k | X_k^{(i)})$ are obtained, so that $p(Z_k | X_k^{(i)}) = N(\mu(d_k^{(i)}), \sigma(d_k^{(i)}))$.

The probability of the actual concentration value under this distribution is finally employed to update the weight of the particle $\omega_k^{(i)}$ .After each update stage, the weights are normalized to have a sum equal to one.

$$\omega_k^{(i)} = \frac{1}{\sigma(d_k^{(i)})\sqrt{2\pi}} \exp(-\frac{(C_k - \mu(d_k^{(i)})^2}{})\omega_{k-1}^{(i)} \tag{4}$$

3.3 Initializing the Filter

Up to now, nothing has been commented about the initial distribution from which the particles are drawn. The filter associated with a specific node is initiated when the first message is received in the mobile robot. In this case, the concentration distance functions of Figure. 1 are used inversely as in the estimation process. From the concentration values, an initial distance is estimated, and also a corresponding variance on the distance. The prior considered is then an uniform distribution on an spherical annulus, in which the inner and outer radius depend on the estimated mean and variance (see Figure. 2).

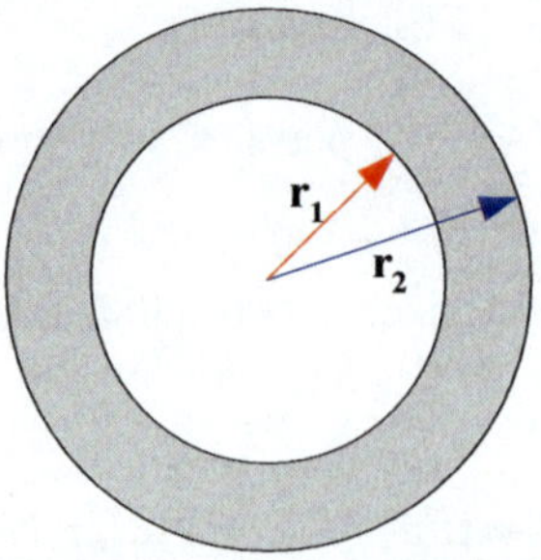

Fig. 2. Prior distribution. The estimated distance from the odor source to the sensor are a function of the inner (r1) and outer (r2) radius and its variance.

As the number of particles is limited, not all the messages received initiate the filter. Only when a concentration value corresponding to a variance below a threshold is received, the filter is initiated, in order to have a good resolution (particles per volume unit).

3.4 Prediction

The nodes of the WSN are static, so the prediction step might be ommitted (that is, with probability 1 each node is in the same position at time k and k-1). However, as the resolution of particles over the state space is limited, a random move is added to the particles, in order to search locally over the position space around the position of the previous time step. Therefore, the prediction model is: $p(X_k|X_{k-1}) = N(X_{k-1}, \Sigma_{k-1})$ .The value of Σ depends on the distribution of particles, mainly the density of particles per volume unit.

3.5 Resampling

When the filter is running, the weights of the particles with high likelihood will increase, while most of the particles will rest at places of very low likelihood on the state space. Again, as the number of particles is limited, a resampling step is included in order to increase the accuracy of the estimated position, duplicating particles with high weights and eliminating those with very low weights. In order to overcome some of the known problems with the resampling stage, two additional considerations are taken into account: first, resampling only takes place when the effective number of particles is below a threshold. The effective number is computed as follows:

$$N_{eff} = \left[\sum_{i=1}^{L} \left(\omega_k^{(i)} \right)^2 \right]^{-1} \tag{5}$$

The threshold is set to the 10% of the number of particles, so $N_{th} = 0.1$. Second, the algorithm employed for resampling is a low variance sampler. It will allow to spread the particles over the maximum likelihood areas.

3.6 Estimation of Mmean and Standard Deviation

The mean and standard deviation from the filter will be computed as follows:

$$\mu_k = \sum_{i=1}^{L} \left[X_k^{(i)} \omega_k^{(i)} \right] \tag{6}$$

$$\sigma_k = \sum_{i=1}^{L} \left(X_k^{(i)} - \mu_k \right) \omega_k^{(i)} \tag{7}$$

One of the benefits of the Particle Filter is that allows to face multi-modal or non-parametric hypothesis. While the posterior distribution will depend on the measures during the transient state, the filter approximately converge to a Normal distribution in the position of the node. It has been considered that the filter converges when σ_k is below a certain threshold during a period of time. In the implementation the threshold was set to 3m during at least 15 messages. If the filter converges at time k_0, the belief on the position of the node can be modeled as a Normal distribution such as $N(\mu_{k0}, \sigma_{k0})$.

4 Simulation and Results

In order to test the previous ideas, an simulation experimental setup was conceived. The algorithm proposed in the above section can be implemented in the gateway (laptop) of the WSN. The middleware presented in have been used.

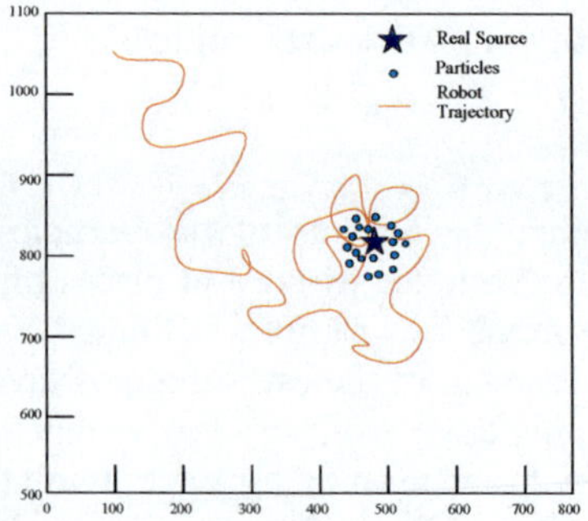

Fig. 3. Estimated position paticles of the nodes versus their real position

Then, the network can be localized as follows:

1) The mobile robot moves in the the WSN as the Figure 3 red line show..

2) The mobile robot sends a message with its position once per second.

3) When a node receives the robot message, a message with the position of the beacon and the concentration on reception is sent to the gateway of the WSN.

4) The information is stored in the gateway and the position of all the nodes are computed by using the above algorithm.

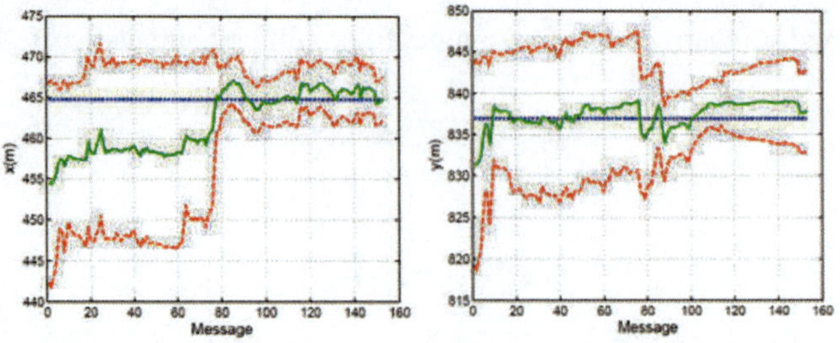

Fig. 4. Estimated gas sour position (green solid), real position (blue dash-dotted) and standard deviation confidence interval (red dashed)

Figure. 4 shows the estimated mean position of the three nodes of the network.

Figure.5 shows the estimated standard deviation from the particles compared to the error between the estimated mean and the actual position. It can be seen that the estimated deviation from the particles is consistent with the error committed.

5 Conclusion

The paper has shown a technique for estimating the position of theodor source in a WSN with a mobile node. The Particle Filter based localization has been tested online in the robot by integrateing the received concentration of the node and a distributed method for position refinement using the normal data flow of the network. Both techniques have been tested in simulation conditions and the results showed their good feasibility. Next steps we will adapt the presented approach to be implemented in real experiment.

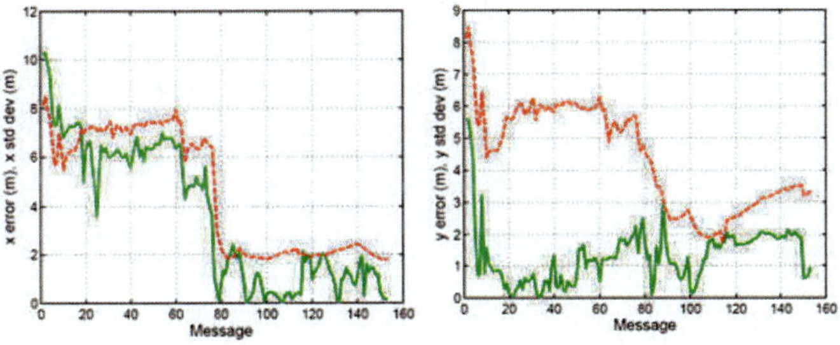

Fig. 5. Standard deviation obtained from the particles (red dashed) and error between the real position estimation and the mean position computed from the particles (green solid)

References

1. Zhao, T., Nehorai, A.: Distributed Sequential Bayesian Estimation of a Diffusive Source in Wireless Sensor Networks. IEEE Transactions on Signal Processing 55(4), 1511–1524 (2007)
2. Vijayakumaran, S., Levinbook, Y., Wong, T.F.: Maximum Likelihood Localization of a Diffusive Point Source Using Binary Observations. IEEE Transactions on Signal Processing 55(2), 665–676 (2007)
3. Nehorai, B.P., Paldi, E.: Detection and Localization of Vapor-Emitting Sources. IEEE Transactions on Signal Processing 43(1), 243–253 (1995)
4. Porat, A.N.: Localizing vapor-emitting sources by moving sensors. IEEE Trans. Signal Process. SP-44, 1018–1021 (1996)
5. Brennan, S.M., Mielke, A.M., Torney, D.C.: Radiation Detection with Distributed Sensor Networks. IEEE Computer, 57–59 (2004)
6. Vijayakumaran, S., Levinbook, Y., Wong, T.F.: Maximum Likelihood Localization of a Diffusive Point Source using Binary Observations. IEEE Transactions on Signal Processing 55(2), 665–676 (2007)
7. Sheng, X., Hu, Y.-H.: Energy Based Acoustic Source Localization. In: Zhao, F., Guibas, L.J. (eds.) IPSN 2003. LNCS, vol. 2634, pp. 285–300. Springer, Heidelberg (2003)
8. Ishida, H.: Remote sensing of gas/odor source location and concentration distribution using mobile system. Sensors and Actuators B 49, 52–57 (1998)
9. Doucet, A., de Freitas, N., Gordon, N. (eds.): Sequential Monte Carlo Methods in Practice. Springer (2001)

A Novel Reputation Model Based on Subjective Logic for Mobile Ad Hoc Networks*

Xuemai Gu, Pei Lin, and Shuo Shi

School of Electronic and Information Engineering, Harbin Institute of Technology,
Harbin, China
linpei0436102@163.com

Abstract. In mobile Ad hoc networks (MANETs), several nodes' uncooperative behaviors would significantly affect the overall performance and the security of the networks. Reputation systems have been proved to be efficient to block such behaviors. However, they failed to resist the collusion attacks of malicious nodes. A novel reputation model based on subjective logic was proposed in this paper in order to detect malicious nodes efficiently even when they colluded to interfere with the reputation assistant mechanism. The concept of opinion distance was introduced to filter malicious recommendations so that the false accusations laid by malicious nodes would have little influence on the performance of the reputation model. Simulations were conducted to evaluate the proposed model on its performance. The results indicate that it performs much better than other similar models in detecting malicious nodes, especially when malicious colluding behaviors exist in the MANETs.

Index Terms: MANET, reputation model, subjective logic, network security.

1 Introduction

A Mobile Ad hoc Network (MANET) consists of a collection of mobile nodes, which can self-organize into a multi-hop wireless network. Usually security in MANETs is achieved by preventive schemes, which actually can not prevent inner malicious nodes from acting uncooperatively [1]. Moreover, several selfish nodes may not be willing to fully cooperate for the purpose of saving battery resources. The reputation mechanism has been proved more effective to solve those problems launched by misbehaving nodes in such a MANET without a prior trust relationship. Rational nodes collect the feedback about participants' past behaviors, and share them with others to help detect the uncooperative nodes [2]. In the meanwhile, several reputation models based on subjective logic, a framework used to quantify trust and security proposed by A. Jøsang [3], have been proposed to improve the reputation mechanism in MANETs. Based on that, an Opinion Distance based Reputation (ODR) model using subjective

* This work is partially sponsored by Doctoral Program Foundation for Institutions of Higher Education of China (No.200802130023) and Postdoctoral Foundation of Heilongjiang Province (LRB09-103).

W. Zhang (Ed.): Advanced Technology in Teaching, AISC 163, pp. 525–532.
springerlink.com

logic has been proposed in this paper in order to defend against the attacks of inner nodes, and ensure the security of MANETs.

2 Related Work

In recent years, reputation systems have been proposed for a variety of applications, including the detection of misbehaving nodes in MANETs.

Buchegger and Boudec proposed a protocol called CONFIDANT [4], extending reactive routing protocols with a reputation based system. Its exclusive reliance upon negative opinions makes it act more quickly and decisively in response to selfish behaviors while treating cooperative nodes too harshly. CORE [5] was a collaborative reputation mechanism to strengthen nodes' cooperation as well. Both negative and positive behaviors were considered, but only positive reputation is shared, which makes it fail to isolate misbehaving nodes in a short time. Kane and Browne presented a reputation computation method that incorporates "uncertainty" based on subjective logic [6], which is closer to the actual network environment because of taking the networks' uncertainty into consideration. A novel reputation computation model was put forward by Y. Liu and K. Li [7]. The concept of "familiarity" is introduced, which helps utilize the recommendations more effectively. Unfortunately, this model has serious flaws in resisting collusion of malicious nodes.

3 The Opinion Distance Based Reputation Model

3.1 Subjective Logic

Subjective logic uses the term "opinion" to represent subjective trust. Node A's opinion of node B can be expressed as $\omega_{A:B} = (b_{A:B}, d_{A:B}, u_{A:B}, a_{A:B})$, where $a_{A:B} \in [0,1]$ is the base rate, $b_{A:B}$, $d_{A:B}$ and $u_{A:B} \in [0,1]$ respectively represent belief, disbelief and uncertainty satisfying

$$. \; b_{A:B} + d_{A:B} + u_{A:B} = 1$$

Let $\omega_{A:C}$ and $\omega_{B:C}$ be node *A* and *B*'s opinions of *C*. $\omega_{A\diamond B:C}$ is the consensus combination of them such that

$$\begin{cases} b_{A\diamond B:C} = (b_{A:C}u_{B:C} + b_{B:C}u_{A:C})/k \\ d_{A\diamond B:C} = (d_{A:C}u_{B:C} + d_{B:C}u_{A:C})/k \\ u_{A\diamond B:C} = (u_{A:C}u_{B:C})/k \end{cases}$$

where $k = u_{A:C} + u_{B:C} - u_{A:C}u_{B:C}$, and $k \neq 0$.

3.2 Reputation Assistant Mechanism

In the ODR model, misbehaving nodes can be rapidly detected and rational nodes need not build trust with all other nodes slowly by using reputation assistant mechanism.

Fig.1 illustrates an example of reputation assistant mechanism. Let node *S* be rational and *X* be a client requesting service. By checking the reputation table, *S* gets the local opinion $\omega_{S:X}^{loc}$. If $u_{S:X}^{loc} < \beta$ ($0 \le \beta < 1$), the local information is sufficient. $e_{S:X}$ represents *S*'s expectation of *X* being a rational node, calculated as follows:

$$e_{S:X} = b_{S:X}^{loc} + a_{S:X} u_{S:X}^{loc}$$

When $e_{S:X} \ge 0.5$, *S* treats *X* as a trustworthy node. If $u_{S:X}^{loc} \ge \beta$, *S* broadcasts to its neighbors for recommendations. *B* and *C* become recommenders and send their opinions to *S*, while *A* dose not because of its total uncertainty about *X*.

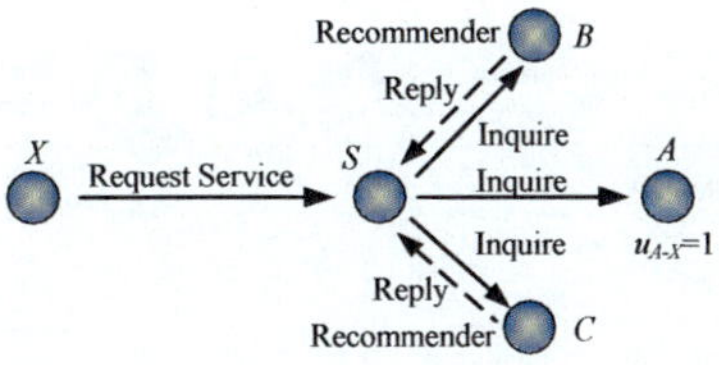

Fig. 1. Example of reputation assistant mechanism

3.3 Reputation Combination

The recommendations should be processed before combining with the local opinion. Using the average for combination would result in the loss of the limited useful information. Worse still, malicious nodes may disguise themselves as the neighbors of *S* and provide false recommendations on purpose to mislead *S*. For these reasons, a concept of "opinion distance" is brought forward to exclude malicious recommendations and restrain the malicious nodes' collusion.

Opinion distance $D_X(i, j)$ is defined as

$$D_X(i, j) = \left| \frac{b_{i:X}}{c_{i:X}} - \frac{b_{j:X}}{c_{j:X}} \right| + \left| \frac{d_{i:X}}{c_{i:X}} - \frac{d_{j:X}}{c_{j:X}} \right|$$

where $c_{m:X} = b_{m:X} + d_{m:X}$, $m = i, j$. It expresses the coherence between nodes *i* and *j*'s recommendations of *X*. The smaller the value is, the more consonant their recommendation opinions are.

The exclusion process of malicious recommendations applying the opinion distance is shown in Fig.2. Node *S* selects the most trustworthy recommender as the reference node. Then calculate the opinion distances between the reference node and every other recommenders respectively in accordance with (4). If $D_X(i, j) \le 1$, node *j*'s opinion of *X* is rational and will be accepted by S. On the contrary, it will be abandoned.

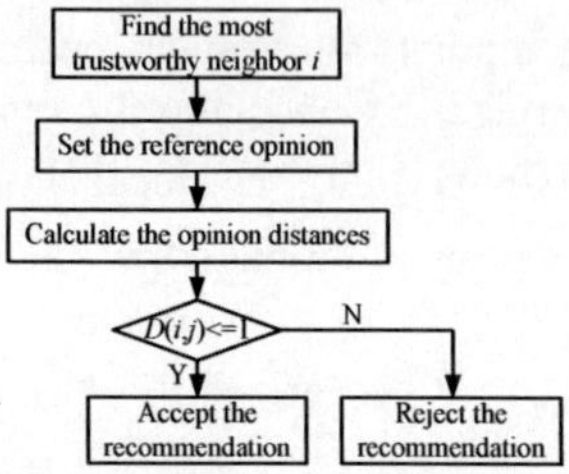

Fig. 2. The exclusion process of false recommendations

After filtering the recommendations, the ODR model weights the recommendations and then combine them with the local opinion of *X*, as shown in Fig.3

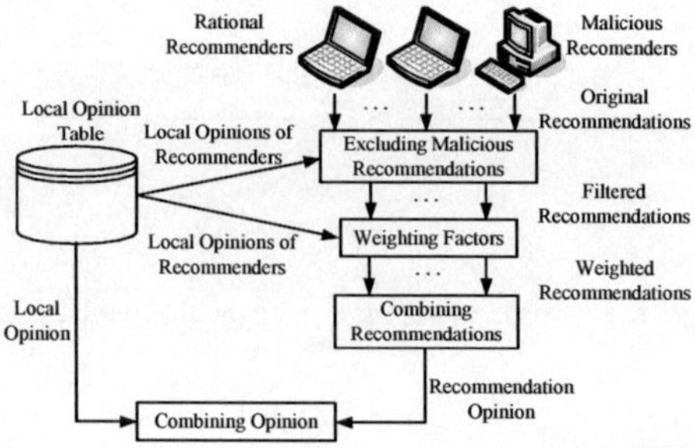

Fig. 3. The opinion combination process

Let *S* has *M* rational recommenders after the filtration, the weighting factor of *i* can be expressed as

$$v_i = b_{S:i}^{loc} f_{i:X} \Big/ \sum_{j \in M} b_{S:j}^{loc} f_{j:X}.$$

where $f_{i:X} = b_{i:X} + d_{i:X}$ stands for *i*'s familiarity with *X*. The weighted opinions are combined into recommendation opinion $\omega_{S:X}^{rec} = (b_{S:X}^{rec}, d_{S:X}^{rec}, u_{S:X}^{rec})$, whose value is given by

$$\begin{cases} b_{S:X}^{rec} = \sum_{i \in M} v_i b_{i:X} \\ d_{S:X}^{rec} = \sum_{i \in M} v_i d_{i:X} \\ u_{S:X}^{rec} = \sum_{i \in M} v_i u_{i:X} \end{cases}$$

Then node *S* gets the combination opinion $\omega_{S:X}^{con}$ by integrating $\omega_{S:X}^{loc}$ and $\omega_{S:X}^{rec}$ with the help of (2). Eventually, substituting the combination opinion into (3) yields the final trustworthiness of node *X*, according to which node *S* will decide whether to interact with *X*.

3.4 Updates on the Local Reputation Table

The local reputation table is constantly updated. Let $\delta \in [0,1]$ be a parameter that determines how much a rating changes after an interaction. Local opinions are updated following the formulas below, whose subscripts are omitted to simplify the expressions.

1) Positive Interaction: If node *S* interacts with *X*, the interaction is positive. Hence,

a) for $u \geq \delta$: $b' = b + \delta$, $d' = d$, $u' = u - \delta$;

b) for $u < \delta$: $b' = b + \delta$, $d' = d - (\delta - u)$, $u' = 0$.

In any case, the parameters should satisfy (1), so let *d*=0 and *b*=1 when *d* decreases to less than zero.

2) Negative Interaction: If node *S* refuses to interact with *X*, the interaction is negative. The update formulas resemble those of the positive interactions and can be obtained by exchanging *b* with *d*.

4 Simulaton and Analysis

4.1 Simulation Architecture

The ODR model brought forward in this paper can be applied to MANETs of any size. In order to simplify the modeling and simulation process, make the following assumptions:

- During the simulation, nodes are randomly selected, and only one couple of nodes is involved in one interaction with only one action permitted.
- The impact of complex network environment on the performance of the model is not taken into account so as to validate the model's capability to detect misbehaving nodes.

For the comparison convenience, the ODR model uses the same protocol as the common protocol in [6] and [7]. Let *X* send service request to *S*.

1) Protocol of rational nodes: If *S* is rational, it uses the reputation mechanism to decide whether to interact with *X*.

2) Protocol of malicious nodes: If *S* is malicious, the protocol is described as follows:

a) *S* flips a coin weighted by $\gamma \in [0,1]$, the probability that a malicious node would cooperate in an attempt to hide its malicious intent.

b) If the coin flip indicates to cooperate, *S* provides the service to *X*, and vice versa.

The following performance metrics are chosen to evaluate the reputation models.

1) Run length to isolate malicious nodes: One factor to evaluate the models is how long it takes to detect and throttle malicious nodes. The run length to isolate malicious nodes is the number of iterations at which the malicious nodes' success rate drops to zero for at least ten iterations.

2) Malicious nodes' success rate: Another factor of interest is how successful malicious nodes are in obtaining service. It can be measured by malicious nodes' success rate, which is the percentage of malicious nodes' success up until they are isolated.

The basic setting of the simulation parameters are shown in Table 1.

Table 1. Simulation Parameters

Parameter	**Value**
Number of rational nodes	50
Number of malicious nodes	50
The base rate $a \in [0\ 1]$	0.5
The threshold of trustworthiness $\alpha \in [0\ 1]$	0.5
The threshold of uncertainty $\beta \in [0\ 1]$	0.5
The rating change parameter $\delta \in [0, 1]$	0.1

4.2 Simulation Results

Two different simulation conditions are set up in order to evaluate the capability of the reputation models in isolating malicious nodes and resisting the collusion attack of them. The ODR model in this paper is compared with a traditional model in [8], the uncertainty-based model in [6] and the familiarity-based model in [7].

1) Collusion attack condition: In this situation, malicious nodes collude with each other and pose as recommenders to provide false recommendations, making the reputation mechanism suffer form interference.

Examine the efficiency of these models respectively with different γ. The larger γ is, the better malicious nodes mask themselves and more difficult it is for the reputation models to distinguish them. Fig.4 shows the simulation results. In Fig.4a, it is obvious that the performance in isolating malicious nodes of the uncertainly-based and the familiarity-based model deteriorates rapidly with the increase of γ, and even becomes inferior to that of the tradition one when γ becomes larger than 0.3. Conversely, the ODR model performs much better than all the others owing to adding a module to filter malicious recommendations, which indicates it much more efficient in resisting the malicious nodes' collusion. Fig.4b shows that as γ grows, malicious nodes' success rate increases likewise, but the ODR model's is still lower than the others.

2) No collusion attack condition: In this situation, the reputation assistant mechanism will not be disturbed with by malicious nodes.

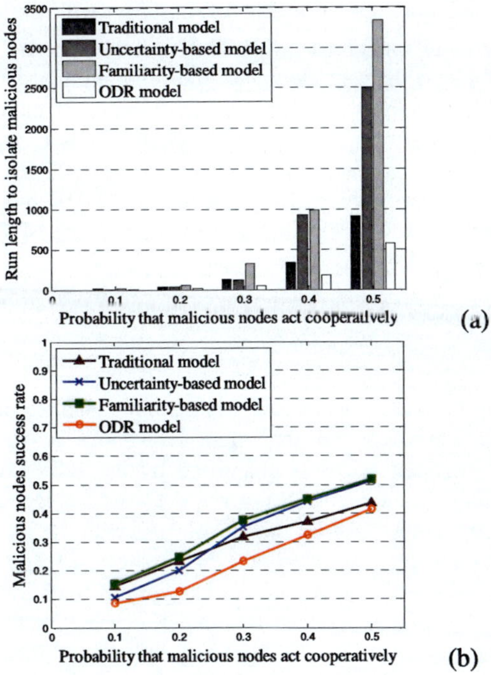

Fig. 4. The performance comparison between the four models with collusion attack.

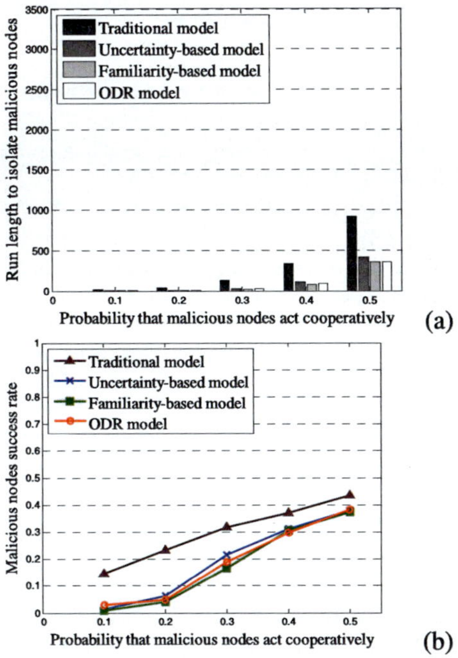

Fig. 5. The performance comparison between the four models without collusion attack.

The simulation results are shown in Fig.5. Similarly, with the increase of γ, note that run length to isolate malicious nodes and malicious nodes' success rate grow accordingly. Nevertheless, three models with the reputation assistant mechanism can isolate the malicious nodes more rapidly, especially the ODR model is still one of the best-performed model. It shows that the ODR model leads to little misjudgment of rational nodes.

In summary, the simulation results suggest that the ODR model performs better, especially when existing malicious colluding behaviors in the MANET. This model is more robust.

5 Conclusions

In this paper, an Opinion Distance based Reputation model is proposed in order to resist uncooperative behaviors and the collusion attacks of malicious nodes in a MANET. The trust between nodes is quantified by the subjective logic theory and represented by opinions. Malicious nodes can be detected and isolated quickly by using the reputation assistant mechanism. In the meanwhile, with the introduction of opinion distance's help, malicious recommendations can be excluded, and therefore the model becomes more invulnerable to the collusion behaviors of malicious nodes. The model is compared with other similar models by simulation. The results indicated that this model performs much better in dealing with malicious nodes, especially when malicious colluding behaviors exist in the MANETs.

References

1. Yang, H., Luo, H., Ye, F.: Security in mobile Ad Hoc networks: challenges and solution. IEEE Wireless Communication 2, 2–11 (2004)
2. Resnick, P., Kuwabara, K., Zeckhauser, R., Friedman, E.: Reputation systems. Communications of the ACM 43, 45–48 (2000)
3. Jøsang, A., Hayward, R., Pope, S.: Trust network analysis with subjective logic. In: Proc. of the 29th Australasian Computer Science Conference, pp. 85–94 (2006)
4. Buchegger, S., Boudec, J.: Performance Analysis of the CONFIDANT Protocol: Cooperation of Nodes-Fairness In Distributed Ad hoc Networks. MobiHOC, Lausanne (2002)
5. Michiardi, P., Molva, R.: Core: a Colaborative Reputation mechanism to enforce node cooperation in Mobile Ad Hoc Networks. In: Communication and Mutimedia Security Conference (2002)
6. Kane, P., Browne, P.: Using Uncertainty in Reputation Methods to Enforce Cooperation in Ad-hoc Networks. In: Proc. of the 5th ACM Workshop on Wireless Security, pp. 105–113 (2006)
7. Liu, Y., Li, K., Zhang, Y., Qu, W.: A novel reputation computation model based on subjective logic for mobile Ad hoc networks. In: 2009 Third International Conference on Network and System Security, pp. 249–301 (2009)
8. Wang, Y., Varadharajan, V.: Trust2: developing trust in peer-to-peer environments. In: IEEE International Conference on Services Computing, vol. 1, pp. 24–31 (2005)

A Distributed Deployment Algorithm with Mobile Sensors in Wireless Sensor Network

Fei Yang[1] and Yong Zhang[2]

[1] School of Electronic Engineering, Tianjin University of Technology and Education
Tianjin, China
[2] Institute of Information, Tianjin University of Commerce, Tianjin, China
linsio@126.com, zhangyong@tjcu.edu.cn

Abstract. The sensor networks comprise of mobile and static sensor nodes always named as hybrid sensor network, which set up for the purpose of collaboratively tasks like sensing a phenomenon. Sensor deployment is an important problem in the hybrid wireless sensor networks, which affects the performance and lifetime of the networks. This paper presents a distributed deployment algorithm for mobile sensors. The algorithm is compared with a simulated annealing based algorithm for deployment with the metrics such as coverage, uniformity, time and distance and the Simulation results are presented to demonstrate the effectiveness of the proposed approach.

Keywords: sensor networks, sensor deployment, distributed deployment algorithm.

1 Introduction

Wireless sensor networks is a novel class of computing and a new spot of information technology, most of the research on WSN has focused on the development of collaborative signal processing and power aware algorithms [1]. Sensor nodes are generally assumed to be fixed and randomly placed because of practical reasons. However, random deployment does not always lead to effective coverage, especially if the sensors are overly clustered and there is a small concentration of sensors in certain parts of the sensor field.

Our work is different from prior work on the deployment problem. The key idea of our deployment algorithm's main objective is topology improvement for longer system lifetime by utilizing mobility of robots. Modification of WSN topology in an autonomous and distributed manner using our algorithm can help in improving sensor coverage and also to prolong expected system lifetime.

Sensor deployment problems have been studied in a variety of contexts [1], [2], [3]. In the area of adaptive mobile nodes placement and spatial localization, a number of techniques have been proposed for both fine-grained and coarse-grained localization.

Recently, Meguerdichian[4] have considered the problem of location and deployment of sensors in a WSN from a coverage standpoint. Bulusu et al.'s work [5] is somewhat similar to the deployment problem that is considered here. They have

W. Zhang (Ed.): Advanced Technology in Teaching, AISC 163, pp. 533–540.
springerlink.com

investigated the problem of adaptive beacon placement for localization in a WSN. Winfield [6] considered autonomous dispersion of mobile nodes in a scenario where mobility is required to cover the entire region due to a lack of wireless network connectivity. In the incremental deployment algorithm [7], nodes are added one at a time. Loo et al. considered a system consisting of a number of cooperating mobile nodes that move toward a set of prioritized destinations under sensing and communication constraints [8]. Much research has been done about various issues related to the deployment problem. Most approaches use either a centralized solution as in the circle covering problem [9,10] and the geometric problem [11], or are restricted to a certain topology as in the coding theoretic approach [12]. Since some approaches adopt random deployment as in the set covering problem and the topology discovery problem [13], initial distribution determines the utilization of networks.

Our work is different from prior methods on the deployment problem. Our deployment algorithm's main objective is topology improvement for longer system lifetime by utilizing mobility of robots. We provide a decision and control mechanism at each robot to be used during deployment rather than random diffusion, which is used in Winfield's work[6].

The organization of this paper is as follows. In Section 2, we discuss performance metrics for a mobile WSN. Our algorithm is presented in Section 3 followed by simulation results in Section 4. Some concluding remarks are provided in Section 5.

2 Deployment Performance Metrics In

Selection of suitable measures to compare performances of different approaches is an important issue in a mobile WSN. Coverage, uniformity, time, and distance are considered as performance metrics here. In this paper, coverage is defined by the ratio of the union of covered areas of each node and the complete area of interest.

$$C = \frac{\bigcup_{i=1,2\cdots N} A_i}{A} \tag{1}$$

where A_i is the area covered by the ith node, N is the total number of node, A stands for the area of the region of interest(ROI).

In this case, the full area of that circle, i.e., πR^2 , is counted as the covered region. Because of areas covered by nodes that fall out of the ROI and the fact that overlap of covered areas between nodes should not be included while computing coverage, we need to use more nodes than simply the ratio of A and area sensed by a single node.

Uniformly distributed sensor nodes spend energy more evenly through the WSN than an irregular topology does. So, uniformity of network topology can be a good estimator for the expected system lifetime. Also, fewer nodes are required to cover an ROI when nodes are more evenly distributed. Uniformity can be defined by the average local standard deviation of the distances between nodes.

$$U = \frac{1}{N}\sum_{i=1}^{N} U_i$$
$$U_i = \sqrt{\left(\frac{1}{K_i}\sum_{j=1}^{K_i}\left(D_{i,j} - M_i\right)^2\right)} \tag{2}$$

where N is the total number of nodes, K_i is the number of neighbors of the ith node, $D_{i,j}$ is the distance between ith and jth nodes, M_i is the mean of internodal distances between the i th node and its neighbors.

In the calculation of thc local uniformity U_i at the i th node, only neighboring nodes that reside within its communication range are considered. The uniformity measure is a local measure and is computed locally because each node has access to local information only. A smaller value of U means that nodes are more uniformly distributed in the ROI.

The time spent for deployment is also important in many timecritical applications such as search and rescue and disaster recovery operations. The total time elapsed is defined here as the time elapsed until all the nodes reach their final locations. The average distance traveled by each node is related to the required energy for its movement. The variance of traveled distance is also important to determine the fairness of the deployment algorithm and for system energy utilization.

3 Distributed Self-spreading Algorithm

The algorithm is inspired by the equilibrium of molecules, which minimizes molecular electronic energy and inter-nuclear repulsion. We propose an algorithm that can cover the region of interest without any intervention from a central controller that acts remotely. We call this algorithm Distributed Self-Spreading Algorithm and discuss it in detail.

To begin with, a specified number of nodes are predeployed randomly in a given region, for instance, inside a rectangle. The sensing range (sR) and the communication range (cR) are assumed to be given. Each node can sense or detect an event within its sensing range and any pair of nodes within their communication range can communicate with each other. This communication is needed for finding neighborhood, obtaining locations of nodes in the neighborhood, and transmitting and forwarding sensed data. Neighborhood is defined here as nodes within the communication range.

The algorithm begins with the specification of cR, sR, and the initial node locations (p_0). In our algorithm, we require the quantity called expected density. This can be calculated by using

$$\mu(cR) = \frac{N\pi cR^2}{A} \tag{3}$$

where N is the number of nodes and cR is the communication range of each node, and A is the ROI. Thus, expected density is the average number of nodes required to cover the entire area when these nodes are deployed uniformly. Initial local density $D0$ of a node is equal to the number of nodes within its communication range. These densities will be used when decisions regarding positions of nodes are made.

We introduce the concept of force to define the movement of nodes during the deployment process. The force is dependent on the distance between nodes and the current local density. We define a force function that satisfies the following conditions.

(i) Inverse relation: $f(d_1) \geq f(d_2)$, when $d_1 \leq d_2$, where d_1 and d_2 are node separations from the origin.

(ii) Upper limit: $f(0^+) = f_{max}$

(iii) Lower limit: $f(d) = 0$, where $d > d_{th}$, d is the node separation and d_{th} is the threshold to define the local neighborhood.

Condition (i) is the same as in Physics, but conditions (ii) and (iii) are included to modify the model to incorporate the notion of locality. In other words, a limiting function is applied via conditions (ii) and (iii).

The partial force at time step n on the ith node from the jth node that is in the neighborhood of the ith node is calculated in this paper as

$$f_n^{i,j} = \frac{D}{\mu^2}\left(cR - \left|p_n^i - p_n^j\right|\right)\frac{p_n^j - p_n^i}{\left|p_n^j - p_n^i\right|} \tag{4}$$

where cR stands for communication range

p_n stands for the location of ith node at time step n

The density factor, which is defined as the ratio of the local density (D) and the expected density (μ) at each node, is small in sparse regions and is large in dense regions. Also internodal distance affects the partial force inversely. Closely located nodes impose larger partial forces and nodes that are far apart induce smaller partial forces.

After adding all the partial forces at the current location, each node can decide its next movement. The local information is collected from the nodes that are within the communication range and that information is used for the calculation of the local density at each node. Each node's movement is decided by the combined force at that node due to nodes in its neighborhood.

When sensor nodes are deployed in a remote and hostile region, some nodes can be affected during and after deployment period. Some nodes can lose their mobility and other nodes can lose their communication functionality. So we need a robust deployment method that can handle these difficulties. Our algorithm exhibits this kind of robustness. First, when the sensor node loses its mobility, that sensor node does not move and is considered to be an early stopped node. Neighboring nodes, if they can move, may still improve the irregular topology. Second, when a sensor node loses its communication capability, that sensor node is of no use in a sensor network. Because each node's movement is only affected by the current status of neighboring nodes, each node is adaptive to environment changes such as node failures, various terrain shapes, etc.

4 Simulation Results

We consider 100 randomly placed nodes in a region of size 100×100 to run the Distributed Self-Spreading Algorithm. We assume $sR = 2$ and $cR = 4$. Figure.1 shows the locations and coverage of the initial deployment before running the Distributed Self-Spreading Algorithm.

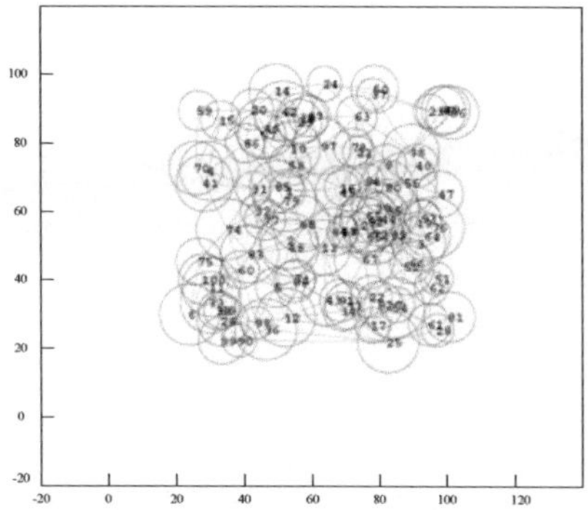

Fig. 1. Initial distribution of sensor nodes

As seen in Figure 1, some parts of the region cannot be covered by the nodes that are randomly dispersed, even though there are enough numbers of them in the given ROI.

Figure 2 shows nodes location and coverage after running the Distributed Self-Spreading Algorithm. The rectangle is fully covered after running the algorithm. Now the network is fully connected and also can cover the entire ROI. Note the spatial node distribution is more uniform than the initial random distribution in Figure 1.

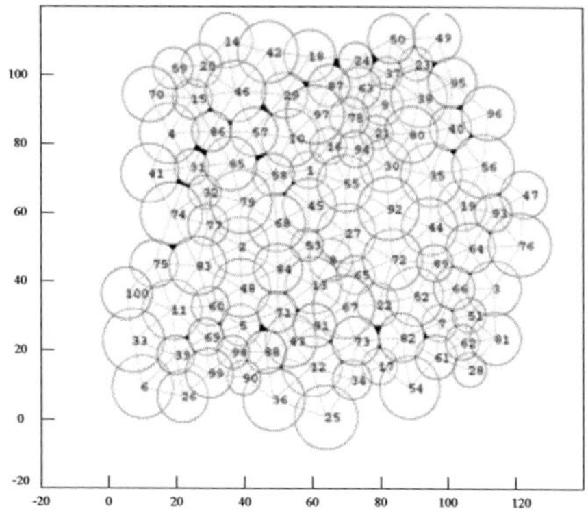

Fig. 2. Final distribution after running DSSA

For the purpose of comparison, Simulated Annealing was also used for self-spreading. To implement a Simulated Annealing algorithm, 4 main design issues should be considered. These are: the definition of the neighborhood, move operator, local energy calculation, and annealing schedule.

The performance of our algorithm is evaluated in terms of the metrics presented in Section 2. Results are presented in Figure. 3 ~6. These results are obtained for different number of nodes dispersed over a fixed ROI of size 100×100, for different node densities. Number of nodes varies from 20 to 100 and results are averaged over 100 runs.

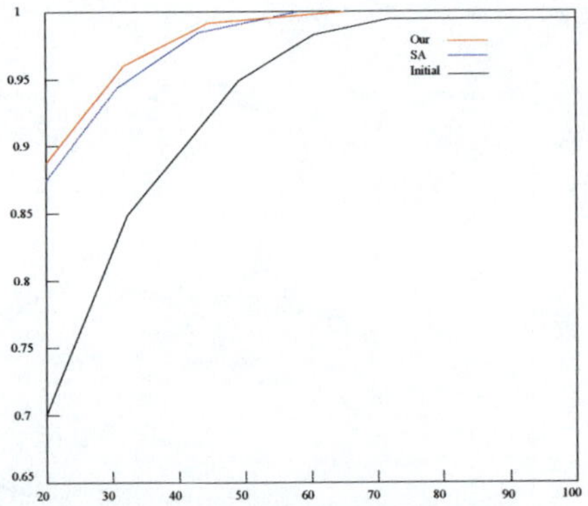

Fig. 3. Coverage versus network size

Figure. 3 shows the improvement in coverage area from the initial random deployment for both algorithms.

Figure. 4 shows the reduction in the standard deviation from the initial case. Both algorithms obtain better uniformity than the initial one and ours outperforms simulated annealing.

Figure 5 shows that our algorithm leads to faster deployment than simulated annealing.

Figure. 6 shows mean distance traveled to reach the final locations for deployment. Our algorithm requires less distance traveled than the simulated annealing algorithm.

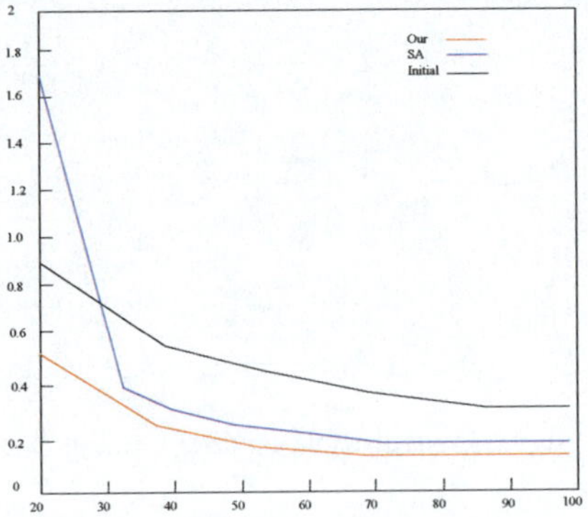

Fig. 4. Uniformity versus network size

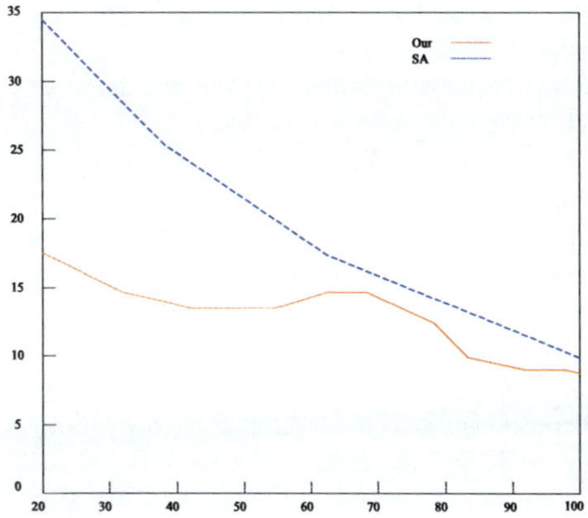

Fig. 5. Termination times versus network size

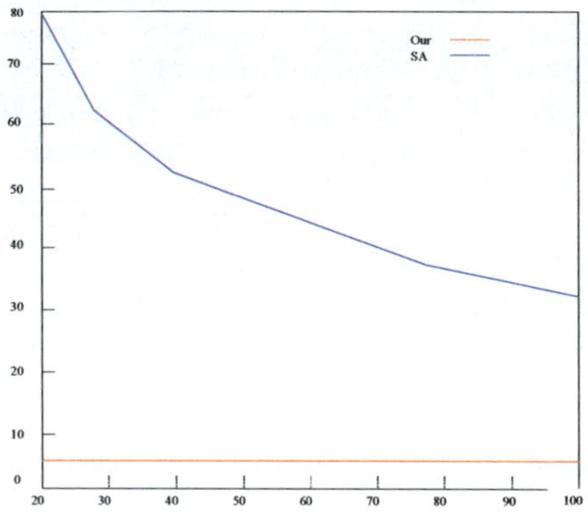

Fig. 6. Distance traveled versus network size

5 Conclusion

We considered the sensor coverage problem for the deployment of wireless sensor networks here. In this paper, we propose a distributed algorithm for the deployment of mobile nodes to improve an irregular initial deployment of nodes. Simulation results show that our algorithm successfully obtains a uniform distribution from initial uneven distribution. The performance of our algorithm is compared with the Simulated Annealing algorithm and exhibits excellent performance.

References

1. Qi, H., Iyengar, S.S., Chakrabarty, K.: Multi-resolution data integration using mobile agents in distributed sensor networks. IEEE Transactions on System, Man and Cybernetics (Part C) 31, 383–391 (2001)

2. Iyengar, S.S., Prasad, L., Min, H.: Adances in Distributed Sensor Technology. Prentice Hall, Englewood Cliffs (1995)
3. Brooks, R.R., Iyengar, S.S.: Multi-Sensor Fusion: Fundamentals and Applications with Software. Prentice Hall, Upper Saddle River (1997)
4. Meguerdichian, S., Koushanfar, F., Potkonjak, M., Srivastava, M.: Coverage Problems in Wireless Ad-hoc Sensor Networks. In: IEEE Infocom, pp. 1380–1387 (2001)
5. Bulusu, N., Heidemann, J., Estrin, D.: Adaptive Beacon Placement. In: Proceedings of the 21st International Conference on Distributed Computing Systems, Phoenix, AZ (April 2001)
6. Winfield, A.F.T.: Distributed sensing and data collection via broken ad hoc wireless connected networks of mobile robots. In: Parker, L.E., Bekey, G., Barhen, J. (eds.) Distributed Autonomous Robotic Systems 4, pp. 273–282. Springer (2000)
7. Howard, A., Mataric, M.J., Sukhatme, G.S.: An Incremental Self-Deployment Algorithm for Mobile Sensor Networks. To appear Autonomous Robots Special Issue on Intelligent Embedded Systems (2002)
8. Loo, L.-H., Lin, E., Kam, M., Varshney, P.: Cooperative Multi-Agent Constellation Formation under Sensing and Communication Constraints. In: Cooperative Control and Optimization, pp. 143–170. Kluwer Academic Press (2002)
9. Melissen, J.B.M., Schuur, P.C.: Improved coverings of a square with six and eight equal circles. Electronic Journal of Combinatorics 3(1) (1996)
10. Nurmela, K.J., Östergård, P.R.J.: Covering a square with up to 30 equal circles. Research Report A62, Laboratory for Theoretical Computer Science, Helsinki University of Technology (2000)
11. Sugihara, K., Suzuki, I.: Distributed Algorithms for Formation of Geometric Patterns with Many Mobile Robots. Journal of Robotic Systems 13(3), 127–139 (1996)
12. Chakrabarty, K., Iyengar, S.S., Qi, H., Cho, E.: Coding theory framework for target location in distributed sensor networks. In: IEEE International Conference on Information Technology: Coding and Computing (June 2001)
13. Wattenhofer, R., Li, L., Bahl, P., Wang, Y.-M.: Distributed Topology Control for Wireless Multihop Ad-hoc Networks. In: Proceedings IEEE INFOCOM 2001, pp. 1388–1397 (2001)

A Group Based Context-Aware Strategy for Mobile Collaborative Applications

Wei Wang, Junzhong Gu, Jing Yang, and Peng Chen

Department of Computer Science, East China Normal University, Shanghai, China
{wwang,jzgu,jyang,pchen}@ica.stc.sh.cn

Abstract. With significant accomplishments of wireless communication, mobile collaborative applications indispensably and inevitably become a new trend. Traditional context aware applications are not directly practicable to current mobile collaborative environments because: 1) Traditional context-aware applications mainly aim at one single user. But in a collaborative environment, the object is a group of users. 2) Traditional context-aware services are always static. In a mobile environment, however, context aware services keep on continuously changing. Fixed communication mode also changes into collaboration in motion. In this paper, a context-aware organization model for mobile collaboration and its management mechanism are proposed. We also introduce a collaborative recommend strategy based on above model. At last, the strategies we proposed are implemented in the Location Aware Mobile Cooperative System (LaMOC) system.

Keywords: mobile collaboration, context-awareness, group, LaMOC.

1 Introduction

In the recent years, we have witnessed rapid advances in the enabling technologies for mobile and ubiquitous computing, such as the increasing pervasive computing paradigm, embedded sensor technologies and wide range of wired and wireless protocols [1]. The diffusion of mobile devices in the working landscape is promoting collaboration across time and space [2]. Teams that work collaboratively can obtain greater resources, recognition and reward when facing competition for finite resources [3]. So with significant accomplishments of wireless communication, mobile collaborative applications indispensably and inevitably become a new trend.

In the mobile collaborative circumstances, traditional context-aware strategies are not appropriate for current mobile collaborative environments because:

The clients of the context-aware applications are not individual user oriented, but a group of users involved. Most traditional context-aware services and applications only support interactions of individual users. But in mobile collaborative environments, the aforementioned situation change. When group of users have common interests or undertake common activities, they share, on a regular basis, information with each other.

Traditional fixed communication oriented context-aware applications are always in a static environment. But a mobile collaborative environment allows users to move

W. Zhang (Ed.): Advanced Technology in Teaching, AISC 163, pp. 541–549.
springerlink.com

anytime anywhere and enjoy context aware services transparently. Motion is an integral part of everyday life, and ubiquitous technology must support mobility [5]. User context changes frequently subject to the user's mobility behavior and context-aware applications have to take into account user location, network resources, and device capabilities.

Current applications demands that collaboration based on static network change into collaboration in motion. Traditional collaboration relies mainly on static network. Collaborating users tend to be static, realizing simple information sharing and interaction. But in a mobile environment, collaborating users tend to be more dynamic, motivated by a shared common interest or situation. Moreover, the way a mobile user collaborates with other users tends to be more variable, asynchronous and dependent more on the user's current context.

The above characteristics of mobile collaborative environments present severe challenges to context-aware computing in the following aspects:

Which organization model should be suitable in mobile environments? How to manage it?
In mobile collaborative environments, many factors can lead to collaborative structure changes. For example: Collaboration may become meaningless as the users moving apart too far. Disconnected communication which may happen anytime may interrupt the collaboration. Some new users want to join in an existing collaboration group. So the organization model in mobile environments should be adaptive. This, in turn, requires techniques for collaboration management mechanism so as to keep the model integrated and maintain the collaboration's persistency.

Leadership challenge
Most collaboration requires leadership, although the form of leadership can be social within a decentralized and egalitarian group [11]. In some collaborative decision making strategies, leader achieve a high quality decision so as to users get more effective result. If needed, how to appoint a leader for the user group in mobile collaborative environments?

What kind of strategy a context-aware system (may be a recommend system) should take to adapt to the changing conditions and ensure the result validity?
In a traditional context aware strategy, few dynamical factors are considered so that users get invalid advice. For example, the recommendatory activity needs four persons, and someone's quit will make it impossible to continue the activity. The recommendation turns out meaningless. So the changing conditions, including changing participants, will result in invalid advices. Consequently, we require a new strategy, which is adaptive to the dynamic changes in mobile collaborative environments and also ensures the validity of results.

Indeed many services for collaboration exist today. For instance the Yahoo! Groups service allows people to discuss topics of common interest or share information. PolyLens [6] is a collaborative filtering recommender system that allows users to ask for recommendation. INTRIGUE [7] is a tourist information server that tailors the recommendation of attractions for tourists groups. However, all of this kind of services just focus on users' preferences and ignore above dynamical factors. STEAM [8] is a collaborative middleware for mobile computing. It exploits a Proximity-based Group

Communication Service [9]. But this kind of applications focusing on some dynamical factors including location does not provide enough content context-aware services for users. MobiLife [10] is a project that provides context-aware services in mobile collaborative environments. Its strategies supports group awareness and enables creation, disposal and update of simultaneous existing groups as well as their related group profiles. Differing from it, our strategy focuses on collaboration management, including the leadership voting mechanism, so as to users get more effective decision making.

In this paper, we present the architecture of Location Aware Mobile Cooperative System (LaMOC) [12], a location based service system, and outline how to solve above challenges with its context-aware strategy. In the rest of this section, we introduce the Location Aware Mobile Cooperative System (LaMOC), a location based service system where our strategies are implemented to solve the above challenges.

LaMOC is a mobile computing system, actually a hybrid environment, as shown in Fig.1. Three layers appear in LaMOC: fixed hosts (FH); mobile support stations; and mobile hosts (MH). Here FH contains the computing devices, such as PC desktops, servers, etc. Databases are active on servers, to supply reliable data services. MH means laptops, mobile phones, etc. Huge differences appear in such mobile hosts, either in computing power, or the duration of the battery. The layer connecting FH and MH is MSS. In LaMOC the IP channel based on GPRS/CDMA/WiFi or 3G is used. A web gateway works as a portal to link mobile systems and fixed systems.

LaMOC is focused on the below factors: minimizing user's efforts to access information in moving, being aware of environmental conditions and user's activities, offering natural interaction techniques, visualizing relevant information in 2D/3D maps on small screen effectively, and mobile cooperation of group users [12].

We conceive a group based adaptive context-aware strategy for mobile collaborative applications in wireless networks. First, the collaboration structure is introduced. A group is a collection of people who have a common target or similar favorites. It also has a collaboration region. When its group member left the region, he automatically withdraws the group.

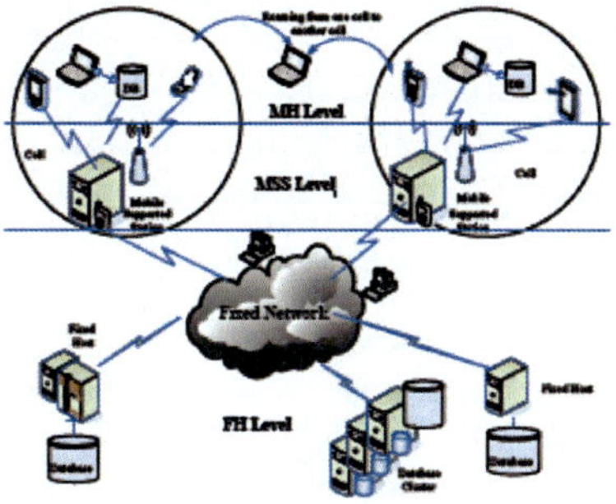

Fig. 1. LaMOC mobile computing referential model

On the contrast, anyone who moves in the region can apply to the membership. Second, we demonstrate how to adaptively maintain the functionality of the group when group structure is changing. A revised Weighted Majority Voting protocol is introduced to decide leadership for the group in dynamic environments. And then, we introduce a group feature awareness method based on group's historical activities. Despite of the changing members, an adaptive rational decision should also be provided.

2 The Context-Aware Manegament Strategy

2.1 The Group Model

In this section, we present some fundamental concept of mobile computing which are relevant to our research. Throughout our paper, we assume that cooperative work is carried out across infrastructure-based networks [13], i.e., wireless networks.

We can define the word "user" given below as a LaMOC system user equipped with GPS. In LaMOC the communication channel based on GPRS/CDMA/WiFi or 3G is used. First, we can abstract mobile users in a wireless communication circumstance as the points in a two-dimensional coordinate. Their GPS position can be abstracted as the coordinate position. So, the distance of two users can be abstracted as Euclidean distance in the coordinate. Here we must give some explanations. In real world, we can't measure the distance between two users by the length of straight-line distance. For example: It may be happened that the length of straight-line distance from one place to another is short, but we need more time to get there. Here, Using the straight-line distance is a simple case in order to explain subsequent definition.

From the abstraction above, the model of the group will be given in a two-dimensional coordinate plane below. A point set ϕdefined in a two-dimensional plane is that there must has one of the points Pi which is served as the center of a circle, whose radius is R, and all points of set ϕmust be in this circle. We call this point set as *Set Prepared for Chose* (SPC), and the circle as *Range Circle* (RC) whose radius is R.

The model allows sponsors to bound the range within which all members' messages are relevant. If some user applies to the system for establishing a group, he can define the RC whose radius is R as the range, called the *group proximity*. Therefore friends in the RC can be looked as elements of the SPC. They will receive messages, such as invitation, only if they reside inside a proximity in which messages are valid. Who accept the invitation will join in the group.

The center of the group proximity is the leader's position. We can set the position as frame of reference according to which we can get other members' position without thinking about their movement.

Fig. 2 depicts the establishment of the group G_a. User P1 was the sponsor, so he was the group leader. His friends in the RC got his invitation. P2, P3, P4, P5 got the invitation but P7 refused it. P6 was out of the RC. So the group was formed: G_a = {P1, P2, P3, P4, P5}.

If someone moves to the opposite direction, he/she will move away farther and farther from the others. It is meaningless for him/her to collaborate with the other members if he moves too far. So do the others. Another example is the change of the member's plan. Who change the plan and want to leave will not to collaborate with the others.

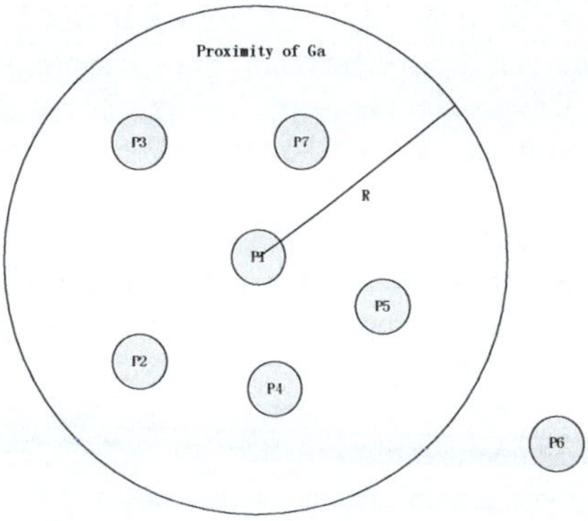

Fig. 2. The establishment of a group

Since above cases, the group must be kept normal adaptively. We can divide the changes into two classes:

- Come into the RC.
- Go out of the RC.

We can divide Class II into two sorts more explicit according to the status:

- Members
- Leader

1) . Go out of the RC

a) Members: If a group member leaves, there are two reasons for it. One reason is Active Leaving. For example, the member can send a message to quit. Another example is the client device disconnect because of the signal strength. The other reason is Passive Leaving. For example, a member moves out of the range of the RC.

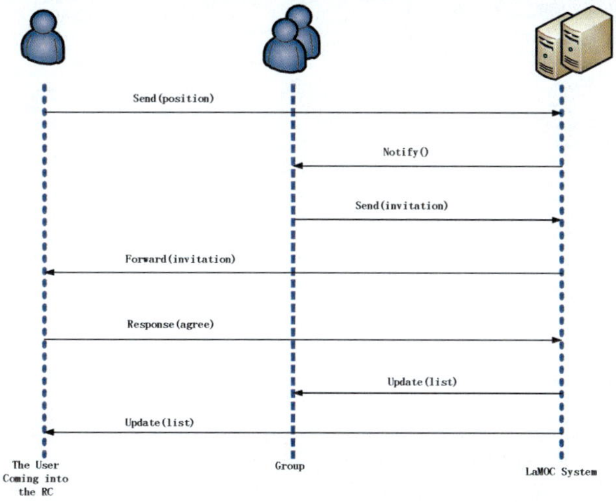

Fig. 3. The sequence of a user coming into the RC in LaMOC

b) Leader: The leave of group leader is different from the members. There is only one reason for it and it is Active Leaving. For example, the leader asks to leave or the device is offline. The problem appears. The group must have a leader and it is the most important condition for exist of the group. We can choose a new leader from the rest members.

2) Maintaining the Integrity of the Specifications

The template is used to format your paper and style the text. All margins, column widths, line spaces, and text fonts are prescribed; please do not alter them. You may note peculiarities. For example, the head margin in this template measures proportionately more than is customary. This measurement and others are deliberate, using specifications that anticipate your paper as one part of the entire proceedings, and not as an independent document. Please do not revise any of the current designations.

a) Time Mechanism

We can determine the confidence value W by the time how long the voter stays in the group. The longer the voter stays in the group, the greater the confidence value is.

b) Distance Mechanism

We can determine the confidence value W by the distance between the voter and the original leader. The shorter the distance is, the greater the confidence value is.

The above described linear choosing algorithm has several advantages. It has very little calculated amount and very small space requirements. It is easy to implement on mobile calculation.

Because the original leader leaves and the new leader is selected, the RC range depending on the leader must be changed. We must redefine it on the premise of no one out of the new RC range. The new circle range centers the new leader. So we can calculate the radius as follows:

$$Rnew = MAX(Dis1, Dis2, \ldots Disi) + Ra \tag{1}$$

Rnew is new RC's radius and Ra is added value which can be defined by the system. Disi = Dis(Pi, Pn) is the distance between any member Pi and the leader Pn. i = 1, … … , C and C is how many members in the group.

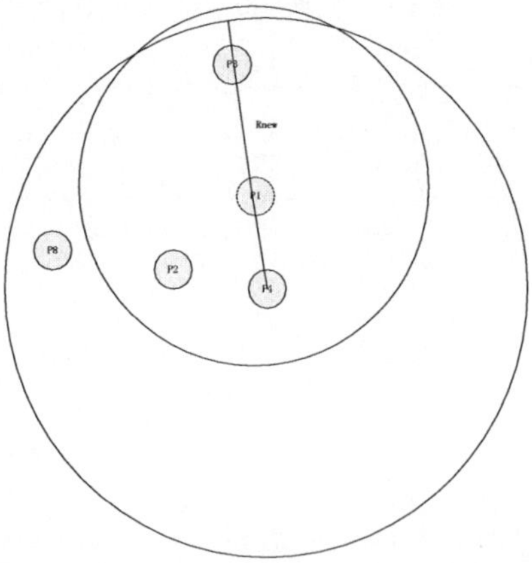

Fig. 4. Define a new circle range by distance

Fig. 4 depicts how to define a new RC by distance mechanism. The leader of the group P1 is offline. P4 is appointed to the new leader because P4 is closest to P1 in all members except P1. The radius of new circle is Rnew. It is equal to the distance which is between P4 and P3 plus additional distance. Meanwhile, P8 who is out of the range is invited to the group.

2.2 Maintaining the Integrity of the Specifications

The template is used to format your paper and style the text. All margins, column widths, line spaces, and text fonts are prescribed; please do not alter them. You may note peculiarities. For example, the head margin in this template measures proportionately more than is customary. This measurement and others are deliberate, using specifications that anticipate your paper as one part of the entire proceedings, and not as an independent document. Please do not revise any of the current designations.

3 Application Scenarios

In order to explain further above strategy, we used the following scenarios based on LaMOC. These two scenarios blow are representative examples in our daily work.

Scenario 1
One afternoon, Chen is shopping on the street. She feels dull and wants to find some friends nearby. By LaMOC system in her mobile phone, she searches for friends in 1km. In no time, she got the result that her friends Zhong and Tan are in the neighborhood. These two get message as Fig. 5 depicts: Your friend Chen invites you to the group. Agree? After their accepting the invitation, they three form a group and Chen is leader.

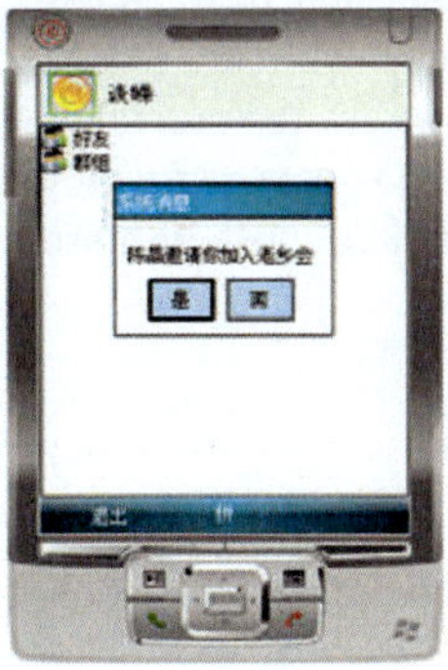

Fig. 5. The LaMOC interface of inviting

Scenario 2
What shall they do? LaMOC system recommends their favorite: playing table fun. At this moment, the system prompts: Joe is on the next street. Invite? So they invite Joe

and Joe is in the group. The system must recommend once again: drinking coffee, and then appears the position of café nearby. After consent, they go to the near café. On the way café, the group leader Chen gets a message and she must go home. So she exits the group (Fig. 6). Every member votes to choose a new leader according to the system's prompting. At last, from the leader choosing protocol, Tan is appointed to the new leader of the group by the system. The original plan is kept on.

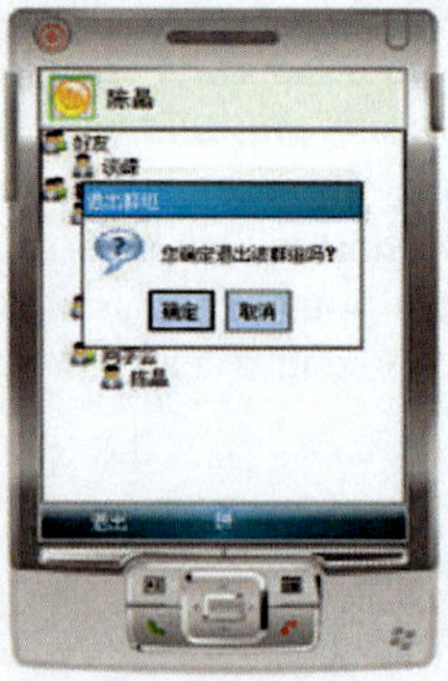

Fig. 6. The LaMOC interface of quit the group

4 Conclusions and Future Work

In this paper, we propose an adaptive context-aware strategy based on location for collaborative mobile applications. Different from traditional context-aware strategies, this strategy allows mobile users to bound a range within which their can collaborate. And we also describe how to manage group in mobile environments with solutions including a special weighted majority voting algorithm. And based on such approximate strategy we represent a way to make decision and recommendation in collaborative mobile environments. The strategies we conceived are implemented in LaMOC system which is developed as the crucial part of LaMOC engine.

In the future, we are going to consider the privacy protected problem in our strategy. It will make the system more reliable. Further, more complex scenarios in applications will be thought about.

References

1. Anagnostopoulos, C.B., Tsounis, A., Hadjiefthymiades, S.: Context awareness in mobile computing environments. Wireless Personal Communications 42, 445–464 (2007)
2. Papadopoulos, C.: Improving awareness in mobile CSCW. IEEE Trans. on Mobile Computing 5, 1331–1346 (2006)
3. Wagner, C.S., Leydesdorff, L.: The diffusion of international collaboration and the formation of a core group. Globalisation in the network of science (2005)

4. Lam, L., Suen, S.: Application of majority voting to pattern recognition: an analysis of its behavior and performance. IEEE Trans. on Systems, Man and Cybernetics 27, 553–568 (1997)
5. Satyanarayanan, M.: Pervasive computing: vision and challenges. IEEE Personal Communications 8, 10–17 (2001)
6. O'Connor, M., Cosley, D., Konstan, J.A., Riedl, J.: PolyLens: a recommender system for groups of users. In: Proceedings of ECSCW, Bonn, Germany (2001)
7. Ardissono, L., Goy, A., Petrone, G., Segnan, M., Torasso, P.: INTRIGUE: personalized recommendation of tourist attractions for desktop and handset devices. Applied Artificial Intelligence, Special Issue on Artificial Intelligence for Cultural Heritage and Digital Libraries 17, 687–714 (2003)
8. Meier, R., Cahill, V.: Location-aware event-based middleware: a paradigm for collaborative mobile applications. In: The 8th CaberNet Radicals Workshop, Ajaccio, Corsica, France (2003)
9. Killijian, M.O., Cunningham, R., Meier, R., Mazare, L., Cahill, V.: Towards group communication for mobile participants. In: Proceedings of Principles of Mobile Computing (POMC 2001), Newport, Rhode Island, USA (2001)
10. Aftelak, A., Harynen, A., Klemettinen, M., Steglich, S.: MobiLife: applications and services for the user-centric wireless world. IST Mobile and Wireless Communications Summit, Lyon, France (2004)
11. Spence, M.U.: Graphic design: collaborative processes = understanding self and others. (lecture) Art 325: Collaborative Processes. Fairbanks Hall, Oregon State University, Corvallis, Oregon (April 13 2006)
12. Gu, J., He, L., Yang, J.: LaMOC - a location aware mobile cooperative system. In: Proc. Signal Processing, Image Processing and Pattern Recognition, Jeju Island, Korea (2009)
13. Stallings, W.: Wireless Communications and Networks. Prentice-Hall, Upper Saddle River (2002)
14. Sandholm, T.: Distributed Rational Decision Making. In: Multiagent systems: a Modern Introduction to Distributed Artificial Intelligence, pp. 201–258. MIT Press (1999)

Multi-agent Negotiation with Fuzzy Logic in E-Commerce

Zuo Bao-he and Sun Yu

School of Software Engineering, South China University of Technology
Guangzhou, China
zuobh@scut.edu.cn, yusunbei@163.com

Abstract. In order to reflect the subjective preference and reduce intervention of human in E-Commerce, this work presents a model using fuzzy logic to support multi-agent negotiation. In the model, we used fuzzy logic to describe human's subjective preference, calculated evaluation function to the preference of single issue, and obtained the global preference using utility function with the differ weight of these issues. Meanwhile, we used urgent degree to represent human factors, and proposed a new dynamic concession strategy based on feedback learning. One agent should consider not only its interest but also other offer when proposing counter offer. After we analyze experimental results, the presented negotiation agent model makes negotiators achieve mutual benefits and decreases human interference effectively.

Keywords: multi-agent, agent negotiation, fuzzy logic, E-Commerce.

1 Introduction

With the applications of e-commerce becoming more and more universal, negotiation will also be more frequent. In the traditional e-commerce, buyers and sellers both negotiate with email to obtain the acceptable issues. But this method is waste of time, it's necessary to use auto negotiation which is more complex [1]. Today, multi-agent negotiation has been recognized as an important activity in E-Commerce and has become one of principal research subjects [2], [3]. The negotiation processes done by using two agents to represent the buyer and seller would be more efficient proactive, autonomous and objective to reach a mutual acceptable agreement.

The typical negotiation models have the following kinds: the Contact Net Protocol (CNP) of Davis and Smith [4], [5]; the Bayesian in Negotiation of Syeara [6]; Game Theory-based Negotiation [7]; Framework for knowledge based support [8] and debate for negotiation [8]. Based on negotiation protocols or static concession strategy, these models are overall lack of consideration for issues and human factor. However, it's difficult to evaluate satisfaction of other offer in the actual negotiation process. With the increasing number of issues, the difficulty is increased significantly. Meanwhile, there are many subjective, uncertain, and imprecise preferences of the human that can't be exactly described. In addition, one agent should consider not only its interest but also other offer when proposing counter offer.

W. Zhang (Ed.): Advanced Technology in Teaching, AISC 163, pp. 551–561.
springerlink.com

On the basis of predecessors' work [3], [10], [11], this research presents a general model of bilateral agent negotiation by fuzzy logic. Often the agents' multi-issues are not equally important in E-Commerce, so the model makes nonlinear fuzzy membership functions to get the member value of the fuzzy sets which is used to calculate evaluation function to the preference of single issue and a weighted satisfaction degree of the offer. Meanwhile, we use urgent degree to represent human factors, and propose a new dynamic concession strategy based on feedback learning. One agent should consider not only its interest but also other offer when proposing counter offer. The model makes negotiators achieve mutual benefits and effectively decreases human interference.

The rest of this article is organized as follows. Section 2 introduces the negotiation model in E-Commerce system. In section 3, we offer a brief description of the proposed intelligent negotiation agent model based on fuzzy logic presented. Section 4 shows the effectiveness of the proposed model by experiments. Finally, section 5 draw some conclusions.

2 Negotiation Model in E-Commerce System

2.1 Conditional Assumptions of Negotiation Model

In E-Commerce negotiation, agent plays two roles: buyer and seller. The roles have opposite demands, e.g. the buyer wants to obtain low price and high quality of commodities, but the seller hopes to sell ones of low quality and high price. They need negotiation for concerned multi issues, such as price, quality, and delivery time. And the preferences of issues are different. So, issues are needed to describe constraint conditions including preferences and lower limits, etc. According to certain rules of mutual communication, buyer and seller both reach acceptable issues.

In this paper, there are two agents respectively on behalf of buyer and seller who negotiate two issues (price and quality). In the negotiation model, we put forward the following hypotheses: (1) Both agents pursue maximum profits; (2) Concession strategy both agents have to adopt is the optimal one; (3) Issues are mutually independent, and their actual ranges are continuous; (4) Agents do don't know each other's preference; (5) Time is limited in negotiation, and we don't have to consider on time value; (6) Agents have the sincerity to negotiation, and cannot deceive each other; (7) Both agents consider that the failure of negotiation are the worst.

2.2 Designing Idea of Negotiation Model

The key to negotiate with Agent in E-Commerce is decision strategy and concession strategy. The decision strategy is to determine whether one agent accepts other's offer and how to make a decision including acceptance, rejection or giving a counteroffer. In this negotiation model, the satisfaction of other's offer is quantized properly with fuzzy logic because human thought is subjective, uncertainty and inaccurate to how to describe satisfaction of other offer. Besides, multi-issues are not often equally important in E-Commerce, he model makes nonlinear fuzzy membership functions to get the member value of the fuzzy sets which is used to calculate evaluation function to the preference of single issue and a weighted satisfaction degree of the offer.

Meanwhile, the concession strategy gives a description of how to propose a counteroffer. In the paper, we use urgent degree to represent human factors, and propose a new dynamic concession strategy based on feedback learning. One agent should consider not only its interest but also other offer when proposing counter offer. The model makes negotiators achieve mutual benefits and effectively decreases human interference.

2.3 Zone of Agreement

Each agent (buyer or seller) has a private retentive offer including a maximum (or minimum) limit issue that must be respected in reaching a deal as shown in Fig.1 [11], [12]. The two agents having contradictory demands must exchange their offer to reach a successful deal. In this paper, there are two issues (price and quality) in one offer. The point of S is the seller's retentive offer (the acceptable lowest price and highest quality to sell the product), the point of B is the buyer's retentive offer (the acceptable highest price and lowest quality to buy the product) and the point of X is the final offer which is accepted by the seller and the buyer. The seller hope X is close to B. On the contrary, the buyer expects X approach S. But each agent does not know its opponent's retentive offer, i.e. the retentive offer is kept hidden from its opponent. If the zone of agreement is empty, it's impossible to carry out the transaction and the next negotiation. So we must estimate whether the zone exits or not before dealing.

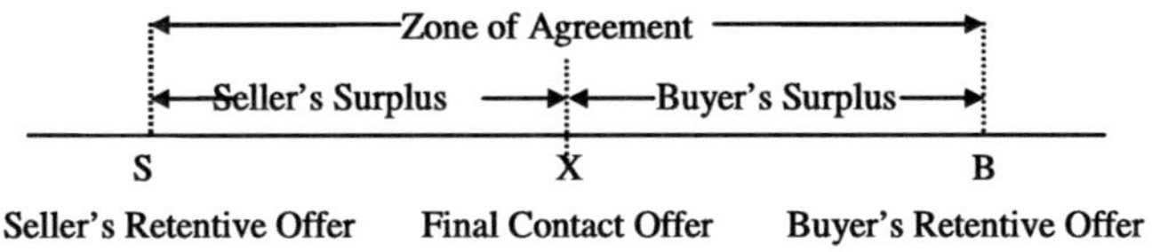

Fig. 1. Zone of agreement

A negotiation can be made possibly only if there is a zone of agreement between the retentive offers of the two agents. But the agents don't even know whether the zone exits. The rule to find a zone of agreement is the following:

if $MINP_s \leq MAXP_b$ && $MAXQ_s \geq MINQ_b$ then
    The zone of agreement exits;
    Both agents begin a negotiation.
else
    The zone of agreement doesn't exit;
    Quit the transaction.
end if

Where $MINP_s$ is the minimum price which the seller will accept to receive, $MAXP_b$ is the maximum price that the buyer is willing to pay the product, $MAXQ_s$ is the maximum quality which the seller can accept, and $MINQ_b$ is the minimum quality which the buyer can accept to buy the special commodity.

3 Modeling Agents' Negotiation with Fuzzy Logic

In this negotiation model of E-Commerce, there are three important modules: offer evaluation, decision strategy and concession strategy as shown in Fig. 2. Just as we say in the foregoing chapter, the role of them is following: The offer evaluation is to calculate the aggregated satisfaction value using utility function (U) of the given offer and then evaluate the offer whether can be accepted or not according the U which is scaled over the range from 0 to 1. The decision strategy is the evidence to make a decision including accept, reject and propose a counter offer to interact with the opponent. And the concession strategy is to generate initial offer or counter offer. So the two negotiation agents offer and respond alternatively until one agent accepts an incoming offer or quits the negotiation, or the deadline T_{max} is reached, where T_{max} is the maximum negotiation round to avoid the endless bargaining between negotiate agents.

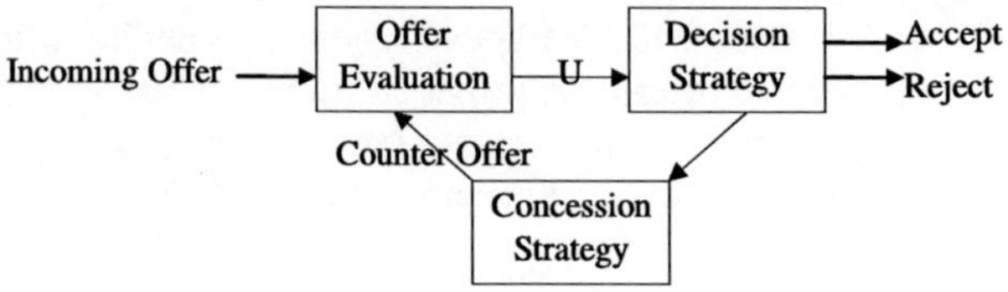

Fig. 2. The negotiation model in E-Commerce

3.1 Offer Evaluation

This module is to calculate utility function value (U) of the given offer using to evaluate the offer whether can be accepted or not. It can accept one or more input variables which are the issues of an offer and provides an output variable i.e. U based on a fuzzy logic. Here, an offer has two issues which are price and quality. In real world, it's natural and easy to use low and high to represent the utility function value of an offer. The fuzzy concepts can reflect the nature of human thoughts, which trend to be abstract and imprecise, the fuzzy logic is appropriate to express the acceptance.

To evaluate the utility function value of an offer, there are two step to be taken: (1) To get membership values of different fuzzy sets of issues according to membership functions, (2) To define fuzzy rules of issues, (3) To get a defuzzy evaluation function of one issue, and (4) To get a defuzzy utility function value (U) of an offer using different weight issues from step (1).

1) The membership functions [11]

As we mentioned earlier, price, quality and utility function have the same fuzzy sets: {*Low, High*}. We propose nonlinear fuzzy membership functions for which the reason is that the level of difficulty from low to high is different instead of being in same proportion. The corresponding membership function of low and high is shown in Fig. 3.

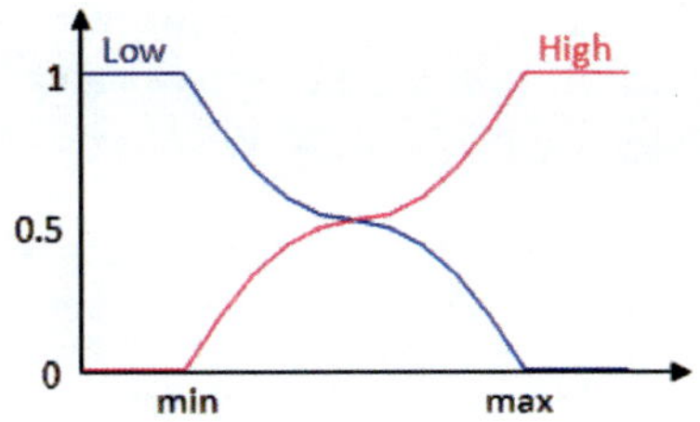

Fig. 3. The nonlinear fuzzy membership function

The nonlinear fuzzy membership function for offer price (op) is as in (1) and (2).

$$v(Low) = \begin{cases} 1, & if\ op < MINP \\ A(op-B)^2+0.5, & if\ MINP \le op < B \\ -A(op-B)^2+0.5, & if\ B \le op \le MAXP \\ 0, & if\ op > MAXP \end{cases} \tag{1}$$

$$v(High) = \begin{cases} 0, & if\ op < MINP \\ -A(op-B)^2+0.5, & if\ MINP \le op < B \\ A(op-B)^2+0.5, & if\ B \le op \le MAXP \\ 1, & if\ op > MAXP \end{cases} \tag{2}$$

Where $A = 2/(MAXP - MINP)^2$, $B = (MAXP + MINP)/2$, $v(Low)$ and $v(High)$ is a membership function defined on the input price (op). In fact, B is the intermediate value between $MAXP$ and $MINP$. When the op is from $MINP$ to B , the rate of change is decreased which show that it's more difficult to estimate a price near B to belong the Low or High. On the contrary, it's easier to estimate a price near $MAXP$ or $MINP$ to belong the Low or High concluded by the the rate of change being increased.

The two membership function as shown in (1) and (2) can alse be applied for quality (oq).

2) Fuzzy rules

According to common sense, the buyer wants to obtain low price and high quality of commodities. The higher the price and the lower quality that seller can get for his offer, the more the acceptance (utility function value) of offer. For this reason, the fuzzy rules of buyer is as follow:

if price is Low, and quality is High, then acceptance is High; if price is Low, and quality is Low, then acceptance is High; if price is High, and quality is High, then acceptance is Low; if price is High, and quality is Low, then acceptance is Low;

But the seller hopes to sell ones of low quality and high price. Consequently, the fuzzy rules of seller is is adverse to buyer's own.

3) Evaluation function

In bilateral multi-issue negotiation, when one agent receives other's offer including multi-issue, he needs to assess each of the issues as an overall assessment of the issues. Owing to opposite profit, the assessment of a same issue is opposite. The lower the price and the higher quality that buyer can get for his offer, the more the acceptance (utility function value) of an offer. But seller's acceptance of an offer is opposite. Therefore, one issue's membership value of fuzzy set can be as it's evaluation function.

For buyer, the evaluation value of price is membership value of fuzzy set "Low", i.e, $V_b(op) = v_p(Low)$; the evaluation value of quality is membership value of fuzzy set "High", i.e, $V_b(oq) = v_q(High)$. For seller, $V_s(op) = v_p(High)$, $V_s(oq) = v_q(Low)$.

4) Utility function

On our common sense, the buyer wants to buy a special product with low price and high quality. On the contrary, the seller expects to sell the product with high price and low quality to gain his profit. So for the seller, the acceptance is higher with the higher price and lowest quality. Often the agents' multi-issues are not equally important in E-Commerce, so each section will be given approximately weighting to represent the importance of issue. The method to calculate U of an offer is as follows on the assumption that an offer has the price of op and the quality of oq:

For buyer, $U(op,oq) = (v_p^{w_p}(Low) + v_q^{w_q}(High))/2$, where w_p and w_q are his own weight of price and quality, $v_p(Low)$ and $v_q(High)$ are the membership value calculated by his own maximum and minimum of price and quality according (1) and (2). In the same way, for the seller, $U(op,oq) = (v_p^{w_p}(High) + v_q^{w_q}(Low))/2$. The weights ($w_p$, w_q) of price and quality belonged to buyer and seller may be different. So their own w_p, w_q, maximum and minimum of price and quality is used to calculate their own $U(op,oq)$.

3.2 Concession Strategy Based on Feedback Learning

In this paper, a concession strategy based on feedback learning is proposed in multi-negotiation. And urgent degree (K) is used to describe human factors. One agent should consider not only its interest but also other offer when proposing counter offer and thus adjust concession range. K stands for the desire to come to an agreement. One agent with higher urgent degree has a greater willingness to reach an agreement, and will make a bigger concession. So the urgent degree of buyer and seller is opposite, and let $K_b + K_s = 1$. Besides, K_b and K_s are constants which don't change in negotiation.

For seller ($Agent_s$), at the first negotiation ($t = 1$), he first proposes his offer, i.e., $op_s^1 = MAXP^s$, $oq_s^1 = MAXQ^s$. Following the analysis mentioned above, at t round, his concession range $(\Delta op_s^t, \Delta oq_s^t)$ of buyer's offer (op_b^t, oq_b^t) is as follow:

$\Delta op_s^t = V_s(op_b^t) * K_s * op_b^t * w_p^s$, $\Delta oq_s^t = V_s(oq_b^t) * K_s * oq_b^t * w_q^s$.

After he makes concession, his counter offer (op_s^{t+1}, oq_s^{t+1}) is as in (3) and (4).

$$op_s^{t+1} = op_s^t - \Delta op_s^t = op_s^t - V_s(op_b^t) * K_s * op_b^t * w_p^s \tag{3}$$

$$oq_s^{t+1} = oq_s^t - \Delta oq_s^t = oq_s^t - V_s(oq_b^t) * K_s * oq_b^t * w_q^s \tag{4}$$

Where $V_s(op_b^t)$ and $V_s(oq_b^t)$ are seller's evaluation functions of buyer's price and quality. w_p^s and w_q^s are seller's weights of price and quality. When the counter offer's utility value is greater than seller's max utility or less than his mix utility, the seller will make no concession, i.e., $op_s^{t+1} = op_s^t$, $oq_s^{t+1} = oq_s^t$.

For buyer ($Agent_b$) , at t round, his concession range $(\Delta op_b^t, \Delta oq_b^t)$ of seller's offer (op_s^t, oq_s^t) is as follow: $\Delta op_b^t = V_b(op_s^t) * K_b * op_s^t * w_p^b$, $\Delta oq_b^t = V_b(oq_s^t) * K_b * oq_s^t * w_q^b$.

After he makes concession, his counter offer (op_b^t, oq_b^t) is as in (5) and (6).

$$op_b^t = op_b^{t-1} - \Delta op_b^t = op_b^{t-1} + V_b(op_s^t) * K_b * op_s^t * w_p^b \tag{5}$$

$$oq_b^t = oq_b^{t-1} - \Delta oq_b^t = oq_b^{t-1} - V_b(oq_s^t) * K_b * oq_s^t * w_q^b \tag{6}$$

Where $op_b^0 = MINP^b$, $oq_b^0 = MAXQ^b$. $V_b(op_s^t)$ and $V_b(oq_s^t)$ are buyer's evaluation functions of seller's price and quality. w_p^b and w_q^b are buyer's weights of price and quality. When the counter offer's utility value is greater than buyer's max utility or less than his mix utility, the buyer will make no concession, i.e., $op_b^t = op_b^{t-1}$, $oq_b^t = oq_b^{t-1}$.

Thus, it's minimum concession for the best interest that agent makes concession in other's offer. And the evaluation value of a new counter offer proposed will go down. Then, next time round, the concession range is smaller, which results in dynamic concession strategy based on feedback learning.

3.3 Decision Strategy

The module is use to make a decision including accept, reject and propose an counter offer to interact with the opponent. In the negotiation, the seller first proposes his offer to buyer. For the buyer, he receives an incoming offer from the seller in a certain round *t*. If t the *t* is more than the maximum round of negotiation (T_{max}) or the incoming offer is equal to the last one (i.e., the seller don't make a concession), he will reject the offer and quit the negotiation. If the utility of seller's offer ($U_b(O_s^t)$) is equal or greater than the utility of a counter offer prepared by the buyer in the round, also if the seller price of the incoming offer (op_s^t) is equal or less than the maximum price of the buyer ($MAXP^b$) and the quality (oq_s^t) is equal or greater than the minimum quality of the buyer ($MINQ^b$), he will accept the incoming offer from the seller to achieve agreement with the seller. Otherwise, he will send the counter offer to the seller. The decision of the seller is similar to the buyer. The module can be simplified as shown in (7) and (8).

$$Decision^{b}=\begin{cases} Quit, & if\ t>T_{max}\ \|\ (t\ge 2\ \&\&\ O_{s}^{t}=O_{s}^{t-1}) \\ Accept\ O_{s}^{t}, & if\ U_{b}(O_{s}^{t})\ge U_{b}(O_{t}^{b}) \\ & \&\&\ op_{s}^{t}\le MAXP^{b} \\ & \&\&\ oq_{s}^{t}\ge MINQ^{b} \\ Offer\ O_{b}^{t}, & otherwise \end{cases} \tag{7}$$

$$Decision^{s}=\begin{cases} Quit, & if\ t>T_{max}\ \|\ (t\ge 2\ \&\&\ O_{b}^{t}=O_{b}^{t-1}) \\ Accept\ O_{b}^{t}, & if\ U_{s}(O_{b}^{t})\ge U_{s}(O_{s}^{t+1}) \\ & \&\&\ op_{b}^{t}\ge MINP^{s} \\ & \&\&\ oq_{b}^{t}\le MAXQ^{s} \\ Offer\ O_{s}^{t+1}, & otherwise \end{cases} \tag{8}$$

4 Experiments

This paper provides experiments to illustrate the model in the solution frame work and the performance results. The application domain is applied to buy mobile phone in B2C which is a kind of E-Commerce. In this system, the buyer entered his retentive price, quality and urgent degree after choosing a seller who has your desired mobile to negotiate with him as shown in Fig. 4. Then the buyer waits the results of negotiation either acceptance on specific value of price and quality or failure of negotiation. In the experiment, the performance is obtained based on the negotiators' exchanged offers, their retentive offers, their weights of price and quality, and urgent degree.

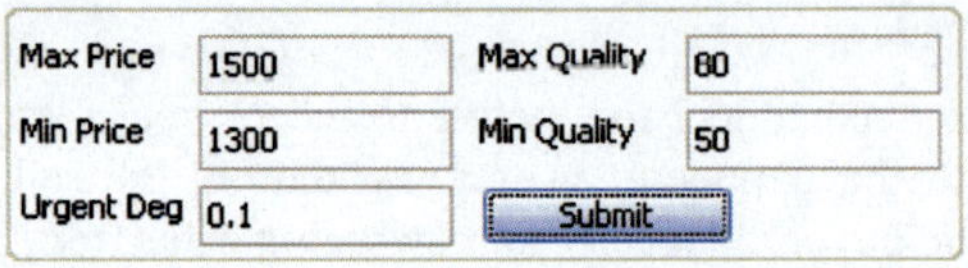

Fig. 4. Retentive offer and urgent degree inputted by the buyer

In this experiment, the $MAXP^{b}$ =1500, $MINP^{b}$ =1300, $MAXQ^{b}$ = 80, $MINQ^{b}$ = 50, w_{p}^{b} = 0.9, w_{q}^{b} = 0.1, $MAXP^{s}$ =1600, $MINP^{s}$ =1100, $MAXQ^{s}$ = 75, $MINQ^{s}$ = 50, $w_{p}^{s}=0.9$, $w_{q}^{s}=0.1, T_{max}=20$.

When adopted $K_{b}=0.1, K_{s}=0.9$, the result of negotiation is as seen in in Fig. 5. Seller's urgent to reach an agreement is very high, so his concession of offer is very huge and agree on price slightly higher than his bottom price which is is very favorable for buyer. When $K_{b}=0.5, K_{s}=0.5$, their urgent degrees are equal. Seller's concession range is smaller than the last one when $K_s = 0.9$. Finally, they reach an agreement on buyer's offer in range 6 which is both advantageous as shown in Fig. 6. When $K_b = 0.9, K_s = 0.1$, buyer's urgent to reach an agreement is very high, so his concession of offer is very huge, but the seller's concession is very small. They reach finally the agreement on buyer's offer in range 12 which is is very favorable for seller as shown in Fig. 7.

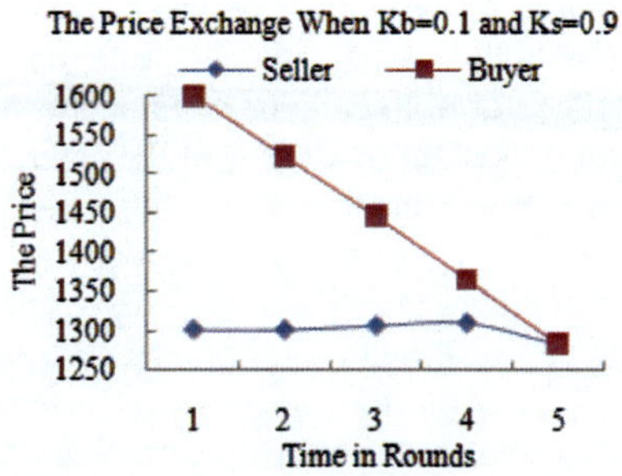

Fig. 5. The price exchange when $K_b = 0.1, K_s = 0.9$

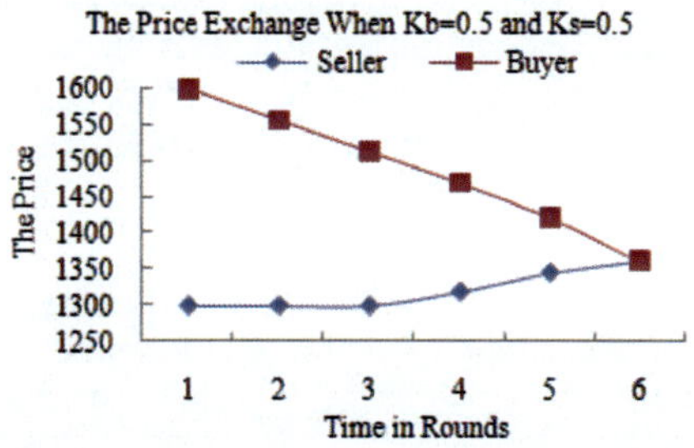

Fig. 6. The price exchange when $K_b = 0.5, K_s = 0.5$

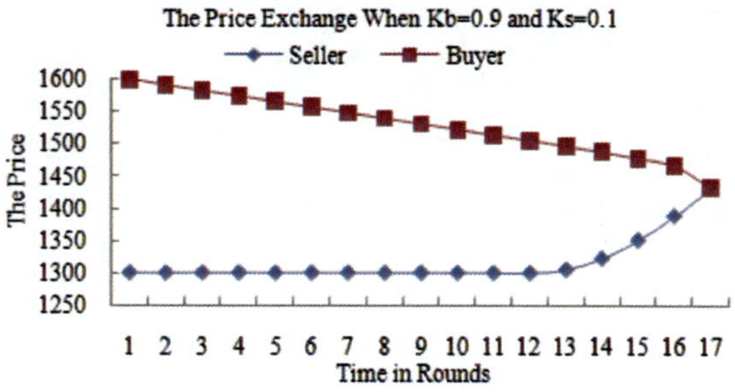

Fig. 7. The price exchange when $K_b = 0.9, K_s = 0.1$

From these above figues, it's seen that the urgent degrees of buyer and seller can affect their concession range.

5 Conclusions

This work presents a general problem-solving model to overcome the complexity of negotiation processes for automation using fuzzy logic. A negotiation can be made possibly only if there is a zone of agreement between the retentive offers of the two agents. The model makes nonlinear fuzzy membership functions to get the member value of the fuzzy sets which is used to calculate evaluation function to the preference of single issue and a weighted satisfaction degree of the offer. Meanwhile, it use urgent degree to represent human factors, and propose a new dynamic concession strategy based on feedback learning. One agent should consider not only its interest but also other offer when proposing counter offer. The model makes negotiators achieve mutual benefits and effectively decreases human interference. We can confirm that a better performance is obtained based on the negotiators' exchanged offers and their weighted retentive offer after analyzing the results obtained by changing the different parameters of the process in the experiment. Further more, the negotiation model makes negotiators achieve mutual benefits and effectively decreases human interference.

At the end of this paper, the research directions in future is that mobile agent will be introduced to E-Commerce negotiation in the ever-changing environment of the network. Buyers and seller will obtain the best benefit and the best service in E-Commerce transactions by use of mobile agent's freely moving across the entire network.

Acknowledgment. The authors thank the people making valuable comments to increasing the quality of work from the bottom of our hearts.

References

1. Wang, X., Shen, X., Georganas, N.D.: A Fuzzy Logic Based Intelligent Negotiation Agent (FINA) In E-Commerce. In: Proc. IEEE Canadian Conference on Electrical and Computer Engineering, Ottawa, ON, Canada (May 2006)
2. Feldman, S.: Electronic marketplaces. IEEE Internet Computer 4(4), 93–95 (2000)
3. Guttman, R.H., Maes, P.: Agent-Mediated Integrative Negotiation for Retail Electronic Commerce. In: Noriega, P., Sierra, C. (eds.) AMET 1998 and AMEC 1998. LNCS (LNAI), vol. 1571, pp. 70–90. Springer, Heidelberg (1999)
4. Smith, R.G.: The COntract Net PRotocol:High-Level Communication and Control in a Distributed Problem Solver. IEEE Transaction on Computers 29(12), 1104–1113 (1980)
5. Davis, R., Smith, R.G.: Negotiation as a Metaphor for Distributed Problem Solving. AI 20(1), 63–109 (1983)
6. Zeng, D., Syeara, K.: Bayesian Learning in Negotiation. In: Working Notes of the AAAI, Stanford Spring Symposium Series on Adaptation, Co—evolution and Leaming in Multiagent Systems (1995)

7. Xiao, D.: Game Theory-based Negotiation Research in Multi-agent Systems. Journal of WuHan Transportation University 24(1), 20–23 (2000)
8. Harrington, J.V., Soltan, H., Forskitt, M.: Framework for knowledge based support in a concurrent engineering environment. Knowledge-Based Systems 9(3), 207–215 (1996)
9. Jennings, N.R., Parsons, S.: Automated negotiation: prospects, methods and challenges. Group Decision and Negotiation 10(2), 199–215 (2001)
10. Ha, V.: Reasoning with Parital Preference Modles. PhD Dissertation, University of Wisconsin Milwaukee (2001)
11. Zuo, B.-H., Sun, Y.: Fuzzy Logic to Support Biliateral Agent Negotiation in E-Commerce. In: AICI 2009, vol. 4, pp. 179–193 (2009)
12. Raiffa, H.: The Art and Science of Negotiation. The Belknap Press of Harvard University Press, Mass. (1982)

Interactive Visualization of 3D Segmented Data Sets Using Simplified LH Histogram Based Transfer Function

Na Ding, Kuanquan Wang, Wangmeng Zuo, and Fei Yang

School of Computer Science and Technology
Harbin Institute of Technology
Harbin, 150001, China

Abstract. Transfer functions, whose role is essential to discriminate tissues by assigning different colors and opacities, are crucial to the quality of volume rendered images. For typical visualization applications, boundaries usually carry most of the relevant information and perform the compact abstraction of the dataset. In this paper, using the ray casting algorithm in the Visualization ToolKit (VTK), we present a simplified LH histogram transfer function design method to interactively visualize the multi-boundary of the segmented data sets. We also provide the user with interaction tools and several fundamental operations. With these interaction tools, one can choose specific tissues and boundaries of interest, and observe the rendering results at arbitrary viewpoint. The efficiency of the proposed transfer function design method is verified using the human heart data.

Keywords: LH histogram, transfer function, volume rendering, segmented data set.

1 Introduction

Scientific visualization of large 3D data sets is widely used for geographic, meteorological, and medical applications. Volume rendering can directly display the details of the volume data set, and thus has been widely used in the visualization of volume data sets. In volume rendering, transfer function should be designed to assign the optical parameters, e.g., opacity and color, to each entry of the volume data. The transfer function affects the final image to a great extent, and moreover is directly related with the user's understanding of the volume data sets [1]. A good optical transfer function should be able to isolate the regions of interest while losing the least amount of information. However, because of the complexity and diversity of the volume data, finding a universal and intuitive transfer function remains very challenging. Transfer function had been named one of the top 10 problems in volume visualization by Hanrahan in his inspiring keynote address at the 1992 Symposium on Volume Visualization [2].

An effective visualization module has to perform a problem-specific abstraction of the dataset, leading to a more efficient visual representation [3]. In most cases, boundaries carry the most important information about the volume data. Most recently, LH histogram is introduced because it can give a compact display of the boundaries in the volume data sets. Thus boundaries can be easier detected in the LH histogram.

W. Zhang (Ed.): Advanced Technology in Teaching, AISC 163, pp. 563–570.
springerlink.com

Segmented data sets are often used in computer-assisted surgical systems and diagnosis software [4]. The traditional design of transfer function for segmented data sets is only based on the scalar values. This method, however, cannot selectively visualize the objects of interest, and is inconvenient in distinguishing multiple objects of interest according to user's specific needs [5].

Interaction tools are especially important when adding dimensions to the transfer function for interactive volume exploration [6]. With the help of interaction tools, specifying transfer functions can be made more intuitive and convenient to better query or manipulation of specific objects [7].

In this paper, we propose a simplified LH histogram to design the transfer function for segmented data sets. Several interactive tools are also provided for the user to utilize the LH histogram to effectively and efficiently locate the unique boundary of particular tissues, to observe the contours of tissues, and to analyze the relative location of different tissues.

2 Related Work

Transfer function is a crucial step in volume rendering. Unfortunately, specifying a good transfer function can be a difficult and tedious task due to the lack of intuitive interface and guiding information [1]. The easiest way to design the transfer function is trial and error method. Volume data sets often require a lot of trials and the ideal transfer function may be missed because of the lacking of guiding information. When the dimension of the transfer function becomes larger than two, it would be more difficult to adjust and control the parameters.

Recent research has focused on automatic and semiautomatic techniques for transfer function design [2]. There are two main methods: image-centric and data-centric methods. Image-centric methods repeatedly adjust the parameters according to the evaluation of the rendering image, until a satisfactory rendering result is presented. The critical steps are to evaluate the image and to adjust visualization parameters. Evaluating the image requires powerful interface to effectively adjust the parameters. However, as the complexity of operation, non-related professionals without any training can not easily master the skills to render the regions of interest.

Data-centric methods design the transfer function by analyzing the characteristics of the data field. Global data attributes, such as contour line, boundary surface, and topology, are a higher level abstraction of local properties. In terms of data-centric transfer function design, a number of methods, e.g., region growing algorithm [8], image processing-based method [9] and topology-based method [10], have been proposed.

Earlier transfer functions are limited to one dimensional (1D) domains. But features of interest in volume data set are difficult to extract and visualize only with 1D transfer functions [6]. Levoy first treats the transfer function as a two dimensional function, voxel density and the corresponding gradient value are used to divide the medical data [11]. In this method the scalar data are used as the first dimension and gradient values are used as the second dimension of transfer function [12]. Subsequently, multidimensional transfer function based on scalar and gradient was proposed.

Multidimensional transfer functions based on scalar and gradient values still have some drawbacks. This can only work if all the materials in the data are present in the expected order. We cannot determine a certain boundary belong to a single arch in the histogram without ambiguity because of noise, partial volume effect and biased data field. The LH histogram method is proposed to solve this problem [13]. In the LH histogram, the compact display of the boundaries allows an easier detection of boundaries, because the LH points are easier to select compared to arches.

For multidimensional transfer functions, we may need interaction tools to resolve the potential complexities in user interface [4]. The general setup of transfer function for segmented data sets is based on the scalar values and gradient values. But users are in most cases non-experts in the field of computer graphics and image processing. Thus, the concept of gradient magnitude is not easy to be grasped for them [14].

3 Method

To use the LH histogram we first need to add two extra properties, FL and FH, for each voxel, where FL denotes the lower scalar value and FH denotes the higher value of the boundary. Voxels in the volume data having $\nabla f \leq \varepsilon$ are considered to be inside a material and are assigned $FL = FH = f(\vec{x})$. Such voxels project on the diagonal in the LH histogram. The remainders are supposed to belong to the boundaries and the second order Runge-Kutta method is used to find the lower and higher intensities of both materials that form the boundary. The lower and higher intensities are assigned to FL and FH separately. All FL and FH values are projected on the same coordinate system to generate the LH histogram. Because FL is smaller than FH, the points represented as boundaries appear only above the diagonal.

For the segmented data sets, the effect of noise, partial volume and biased data can be eliminated. Thus, we present the simplified LH histogram transfer function which only needs to search the 26 neighbors to get FL and FH. In order to speed the searching process, we first mark each voxel as exterior or interior. The current voxel is compared to its 26 neighbors, if all of the scalar values are equal, the voxel is considered inside some object, and otherwise the voxel is marked as exterior voxel. A flag array is used to record the marks before the searching process and setup of transfer function in the pipeline.

The FL and FH values of interior voxels are assigned to the scalar values of the material. As to the exterior voxels, we only need to evaluate the 26 neighbors of the current voxel. The scalar value of the neighbors which most frequently appears is the FL or FH value of the voxel, and that depends on the value is lower or higher than the scalar value of the current voxel itself. If the scalar value is higher than scalar value of the current exterior voxel, FL is assigned to the scalar value of the current voxel and FH is assigned to the statistical scalar value. On the contrary, FH is assigned to the scalar value of the current voxel and FL is assigned to the statistical scalar value. If we have more values that appear equal times, we randomly choose one value as the FL or FH value according to the comparison with the current voxel. Then all the FL and FH values are projected on the same coordinate system, and we get the simplified LH histogram where every boundary appears as one single point (Fig. 1(b)).

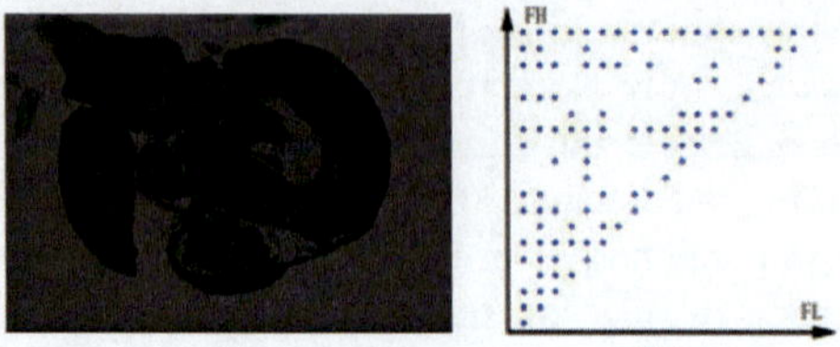

Fig. 1. The human heart data set and the corresponding LH histogram: (a) the slice data, (b) the simplified LH histogram.

4 Interaction and Tools

The Visualization Toolkit (VTK) is an open-source, freely available software system for 3D computer graphics, image processing and visualization [15]. We adopt ray casting algorithm integrated in VTK for volume rendering. VTK also has a suite of 3D interaction widgets which was adopted to implement several commonly used operations, e.g. dragging, zooming, and rotating.

A convenient and flexible setup of transfer function is very important for interactive applications. An intuitive setup of transfer function proves difficult because of a conceptual gap between the spatial and transfer function domains. In the simplified LH histogram, the boundaries are represented by points which are easier to select. We provide the user with rectangular tool to design the transfer function more naturally. One can use rectangular tool to select a series of points in the LH histogram in order to setup different colors and opacities to the contours or some corresponding boundaries of tissues. The tool can also be used to select one single point in the LH histogram, which represents a specific boundary between two tissues (Fig. 2).

In the left of Fig. 2, rectangular tool is used to select the points located outside the diagonal where FH = 80 to display the contour of the heart, and also the points located outside the diagonal where FL = 30 and FL = 31 to display contours of the right ventricle and left ventricle. The single point (30, 31) colored red in the LH histogram, represents the boundary between the right ventricle and the left ventricle. Based on the transfer function described above, the rendering result is shown in the right of Fig. 2.

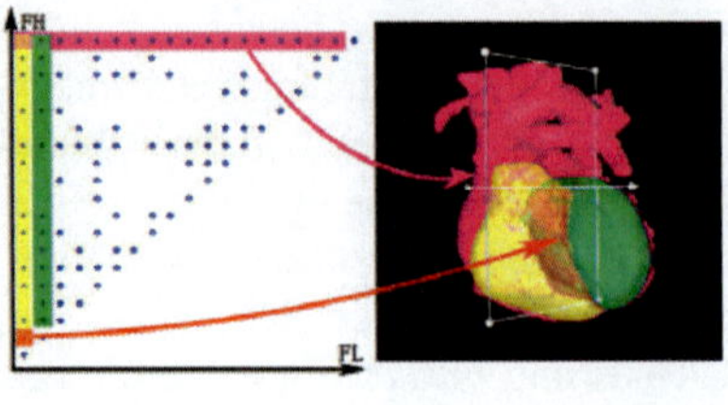

Fig. 2. Transfer function design for the visualization of the contours of heart, right ventricle and left ventricle.

After displaying the specific tissues and boundaries, the system also allows user to drag, rotate, and zoom the rendering results for better visualization by clicking and dragging the mouse. The clipping plane is also provided for the users to observe the interior structures at arbitrary viewpoint (Fig. 3).

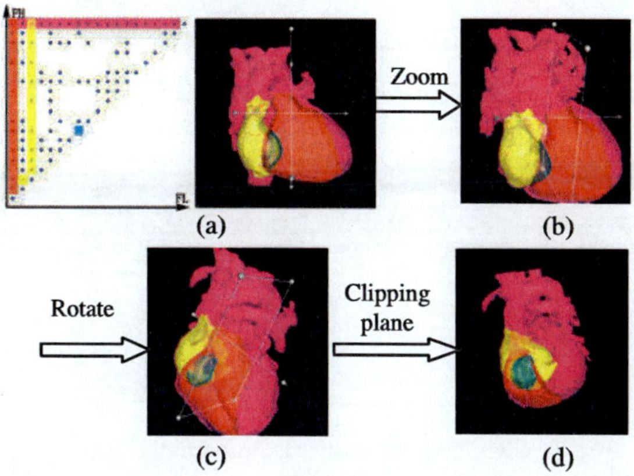

Fig. 3. The contours of the right ventricle, the right atrium and the AV valve. After zooming and rotation operation we could get the results in (b) and (c). Using the clipping plane, the AV valve inside the ventricle can be observed in (d).

5 Discussion

We use the segmented human heart data to demonstrate the performance of the proposed LH histogram transfer function design method. The size of the data set is 469 × 325 × 487. The experimental environment is: Intel(R) Pentium(R) 4 CPU, 3.00 GHz, 1.00 GB memory, NVIDIA GeForce4 MX 440 with AGP8X graphics.

In the following, we present several examples about how to use the LH histogram and interaction tools to visualize the contour of one tissue, the boundaries between tissues and one specific tissue, and the positional relationship between one tissue and the related tissues. One can also combine any of these cases together in order to get arbitrary required rendering results.

In the LH histogram, if we select all the points where FL or FH is the same value f and assign them the same color and opacity, we can get the whole contour of the tissue with the scalar value f. For the human heart data set, as shown in Fig. 4, we select all points where FL or FH is 31 and assign the red color and the contour of left ventricle is shown in the right of Fig.4.

All boundaries exposed to the air are shown in Fig.5. In the segmented human heart data set, the scalar value of the air is 80. In order to show all the tissues exposed to the air, we choose all the points with FH = 80, and assign each FL value to a specific color. The rendering result is shown in the right of the Fig. 5.

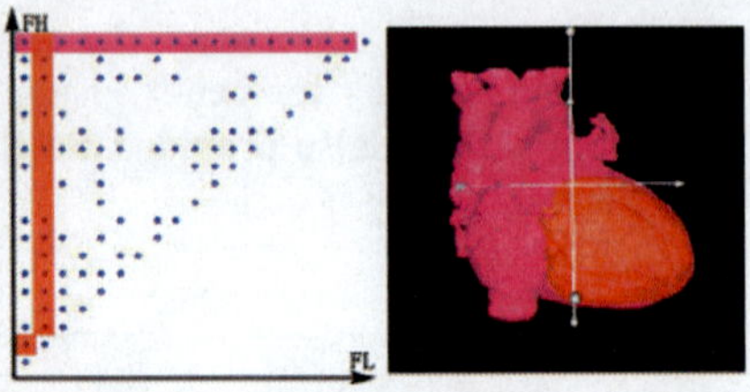

Fig. 4. The contour of the left ventricle

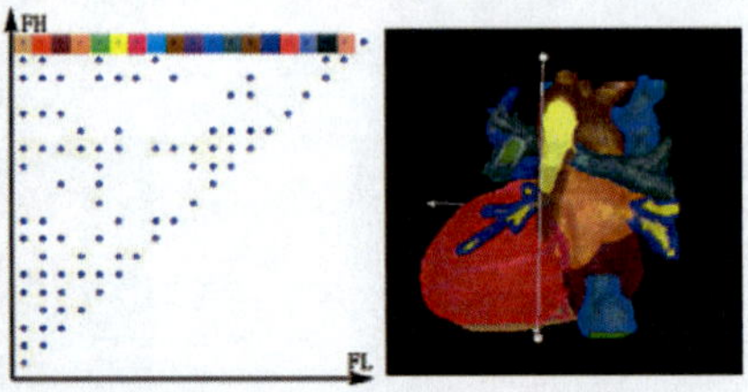

Fig. 5. All the tissues exposed to the air.

In the human heart data set, the right ventricle has many adjacent tissues. We only choose the right ventricle and one adjacent tissue, the papillaris muscles, as shown in the left of Fig. 6. In the right of the Fig. 6(a), we can visualize the whole shape of the right ventricle in red and the papillaris muscles in green. After rotating the rendering result, we get the rendering result from another viewpoint in Fig. 6(b). One can clearly see the papillaris muscles are also inside the right ventricle.

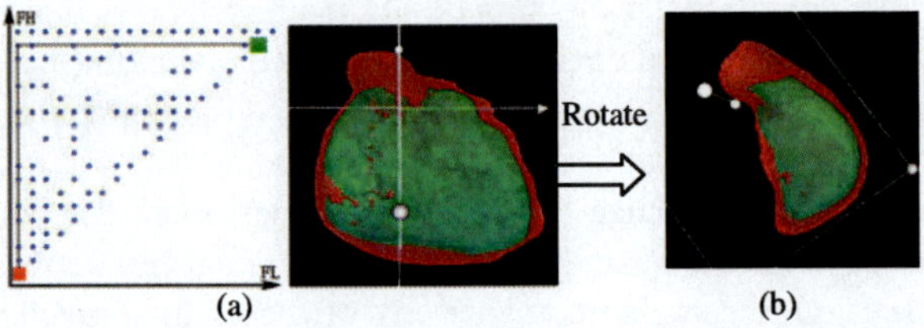

Fig. 6. The relative location of the right ventricle and the papillaris muscles.

Some combination cases are followed. In Fig. 7, five main boundaries of the aorta are selected. The interior boundary of the aorta is colored cyan, the boundary between the aorta and the right ventricle is colored red, the boundary between the aorta and the right atrium is colored green, the boundary between the aorta and Vena_cava is colored blue, the boundary between the aorta and the pulmonary trunk is colored yellow. The tissues are colored the corresponding lighter color than the boundaries, respectively.

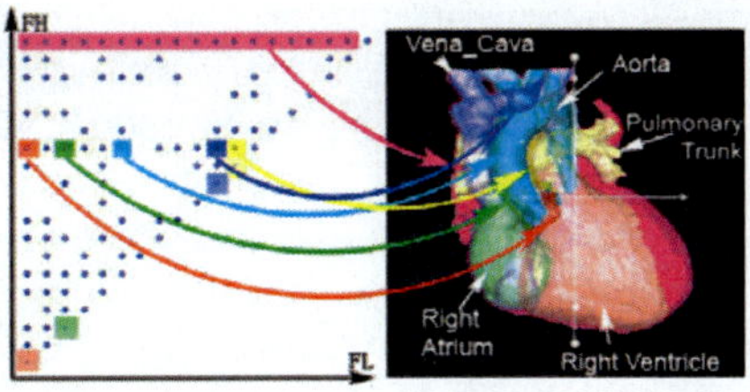

Fig. 7. Boundaries between Aorta and the related tissues.

In Fig. 8(a), we choose the two specific boundaries of the left ventricle. One is between the left ventricle and the inside material colored yellow. The other one colored red is the exterior boundary between the left ventricle and the air. In Fig. 8(b), we choose the interior tissue of the left ventricle to be visualized in green.

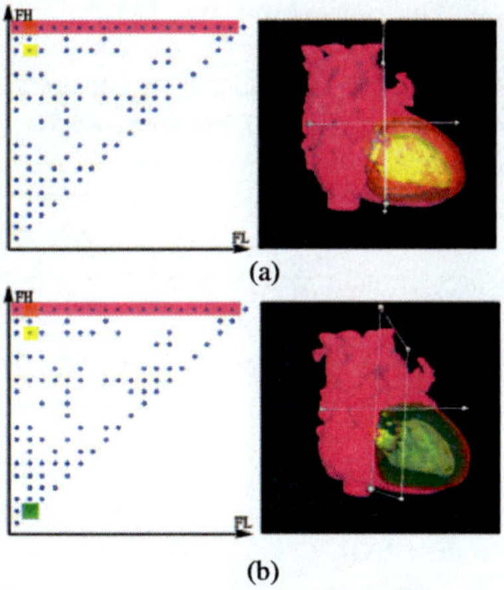

Fig. 8. The interior and outer walls of the left ventricle. (a) Only the walls to be visualized, (b) we choose the interior tissue of the left ventricle to be visualized in green.

We can use the rectangular tool to select the specific boundary of interest. In order to achieve the purpose to increase the contrast between adjacent tissues, we may assign tissues to very different colors. We can also do the dragging, rotating and zooming operations to the final rendering results to find a good viewpoint for the interior structure. The system can satisfy the specific-purpose visualization for different users by using the interaction tools.

6 Conclusion

In volume rendering, the boundaries can reveal more information and the details of the structures of the data sets, and thus have attracted considerable research attention. For the 3D segmented data sets, we proposed a simplified LH histogram transfer function design method to visualize the specific boundaries of interest, including the contours of the tissues, the boundaries between the specific two tissues, and the boundaries between one tissue and the rest tissues. We also provided the users with interaction tools to interactively select different tissues and boundaries of interest and get better visualization. Finally, we use the classified human heart data set to test the proposed method and the interactive tools. The system can provide an interactive visualization of the important tissues and the meaningful boundaries of the heart through interactive operations.

Acknowledgment. The work is partially supported by the NSFC fund under Contract No. 60872099, and the Chinese 863 fund under Contract No. 2006AA01Z308.

References

1. Zhou, F., Fan, X., Yang, B.: Prospects and Current Studies on Designing Transfer Function in Volume Rendering. Journal of Image and Graphics 13(6), 1034–1047 (2008) (in Chinese)
2. Pfister, H., Lorensen, B., Bajaj, C., Kindlmann, G., Schroeder, W., Avila, L.S., Martin, K., Machiraju, R., Lee, J.: The transfer function bake-off. IEEE Computer Graphics and Applications 21(3), 16–22 (2001)
3. Tatarchuk, N., Shopf, J., DeCoro, C.: Advanced interactive medical visualization on the GPU. Journal of Parallel and Distributed Computing 68(10), 1319–1328 (2008)
4. Bullitt, E., Aylward, S.R.: Volume rendering of segmented image objects. IEEE Transactions on Medical Imaging (2002)
5. Hadwiger, M., Berger, C., Hauser, H.: High-Quality Two-Level Volume Rendering of Segmented Data Sets on Consumer Graphics Hardware. In: Proceedings of the 14th IEEE Visualization, VIS 2003 (2003)
6. Kniss, J., Kindlmann, G., Hansen, C.: Multidimensional Transfer Functions for Interactive Volume Rendering. IEEE Transactions on Visualization and Computer Graphics 8(3), 270–285 (2002)
7. Kniss, J., Kindlmann, G., Hansen, C.: Interactive Volume Rendering Using Multi-Dimensional Transfer Functions and Direct Manipulation Widgets. In: Proceedings of the Conference on Visualization 2001, pp. 255–262 (2001)
8. Huang, R., Ma, K.L., McCormick, P.: Visualizing industrial CT volume data for nondestructive testing applications. In: Proceedings of IEEE Visualization 2003 (2003)
9. Fang, S., Biddlecome, T., Tuceryan, M.: Image-Based Transfer Function Design for Data Exploration in Volume Visualization. In: Proceedings of the Conference on Visualization 1998 (1998)
10. Fujishiro, I., Takeshima, Y., Azuma, T., Takahashi, S.: Volume data mining using 3D field topology analysis. IEEE Transactions on Computer Graphics and Applications (2000)
11. Levoy, M.: Display of surfaces from volume data. IEEE Computer Graphics and Applications 8(3), 29–37 (1988)
12. Kindlmann, G., Durkin, J.W.: Semi-Automatic Generation of Transfer Functions for Direct Volume Rendering. In: IEEE Symposium on Volume Visualization (1998)
13. Sereda, P., Bartroli, A.V., Serlie, I.W.O., Gerritsen, F.A.: Visualization of Boundaries in Volumetric Data Sets Using LH Histograms. IEEE Transactions on Visualization and Computer Graphics 12(2), 208–218 (2006)
14. Wesarg, S., Kirschner, M.: 3D Visualization of Medical Image Data Employing 2D Histograms. In: Proceedings of the 2nd International Conference in Visualization, VIZ 2009, pp. 153–158 (2009)
15. The Visualization Toolkit, `http://www.vtk.org/`

Design of Fingerprint Access Control System in Intelligent Community*

Han Cheng-hao[1], Ren-Ye[1], and Li-Min[2]

[1] Electrical and Electronic Information School Information and Engineering Jilin Architecture and Civil Engineering Institute Changchun 130021, China
[2] College of Electronic Hebei Uinversity Baoding 071002, China
hanchenghao1972@yahoo.cn, jljgyjshry@163.com,
limin118118@gmail.com

Abstract. In this paper, designing scheme of unit access control system are proposed. MO4 fingerprint identification module is basis of this scheme and single-chip machine is control center of this scheme. In the scheme design of hardware and software are introduced in detail. The structure of this system is simple and easy to operate. The error rate of fingerprint is very low and degree of intelligence is higher. The system has broad application prospects in safeguard system design of community.

Keywords: fingerprint recognition module, controller, baud rate, error rate of fingerprints.

The building of intelligent community is an important change in real estate industry in the twenty-first century. The one of important sign of intelligent communities is that safety of life and property of residents is protected reliably. The research and development of access control system unit of community is a top priority. In this paper the approach of access personnel identification and intelligent management based on combination of M04 fingerprint recognition module and the SCM is described.

1 Overall Designing Scheme of System

The controlling system is set up in residential entrances on each unit. The overall designing scheme of the system can be shown in Figure 1. The system is made up of the M04 fingerprint identification module, the microcontroller AT89S52, audio module, electromagnetic controlling lock circuit, keyboard input and display, intercom module and so on. System can identify authenticity of the user through the M04B fingerprint identification module to control open of electromagnetic lock. When the M04 fingerprint recognition module can not identify the fingerprint because of wet and dry of fingerprint or angle problem, the music module can be started to tell user the error message. Visitors can dial the user extension through the keyboard and can talkback with households.

* Foundation Project: ministry of scientific research programme funded projects(No2007K934).

W. Zhang (Ed.): Advanced Technology in Teaching, AISC 163, pp. 571–576.
springerlink.com

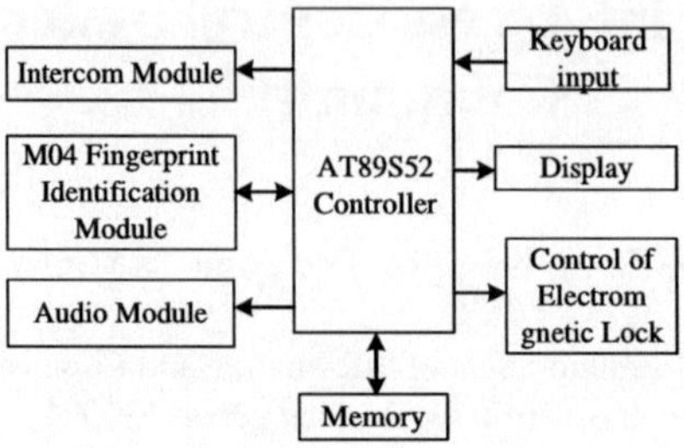

Fig. 1. Overall Designing Scheme of System

2 Design of Hardware Circuit

2.1 Connection of the Fingerprint Identification Module and the Controller

M04 fingerprint identification module is the special fingerprint module which is designed by the Changchun Hong Da photoelectronics Co. Ltd. And biometric identification technology for the second development and use. The module is made up of the fingerprint scanner and fingerprint processing plates. Fingerprint Capturing is made by advanced CMOS image sensor chip and optical technology. Fingerprint processing board is the core of the module. It can complete fingerprint image acquisition, feature extraction and matching functions under the control of sending commands. The connecting way of the M04 fingerprint recognition module and the controller is shown in Figure 2 [1-2].

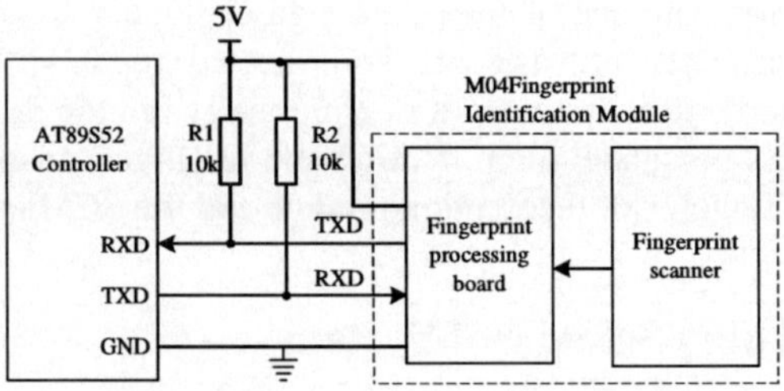

Fig. 2. Connection of Controller and M04 module

Open-drain forms are used in the input and output of fingerprint identification module. So in Figure 2 10KΩ pull-up resistor is connected TXD pins and RXD pins respectively to improve the circuit's driving capability. Mode 1 (10-bit data frame format) is used for the serial communication way between AT89S52 and fingerprint acquisition module. That is a start bit, 8 data bits and 1 stop bit. Communication baud rate is set to 9600bps. When the timer T1 is used for baud rate generator, timer mode 2 (auto reload the initial value) is selected. In the design of system crystal oscillation frequency of AT89S52 is 11.0592MHz and the initial value of X as 0FDH [3].

2.2 The Controlling Circuit of Electromagnetic Lock

The 12V power is needed for supplying electromagnetic locks. Yet working voltage of controller is 5V. So control of magnetic locks is realized by controlling solid state relays. Specific controlling circuit of electromagnetic lock is shown in Figure 3.

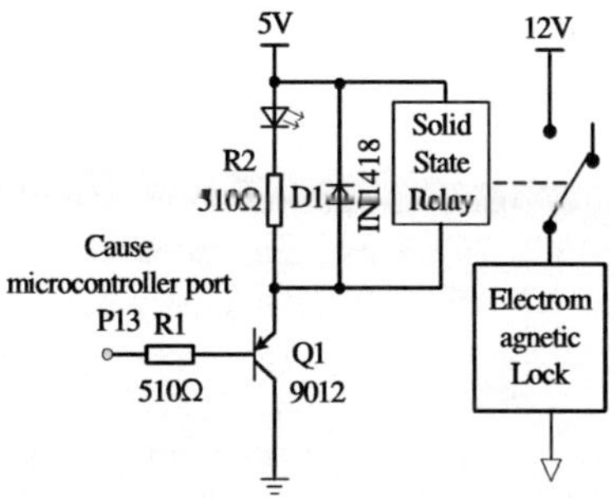

Fig. 3. Controlling Circuit of Electromagnetic Lock

2.3 Design of Memory Circuit

The CAT24WC02 memory is used in the design. CAT24WC02 is the 2048 (2K) serial E^2PROM and is organized as 256 * 8 internally. Figure 4 is the connection diagram between the controller and the memory. The A0-A2 is the address selection line and the lines can cascade up to eight memory. But the storing information of residential access control system is not a lot, a memory can be used to store information. So A0-A2 can be grounded directly. Selection line of reading and writing is connected to P2.1. Clock line is connected to P2.3 and data line is connected to the P2.2 [4].

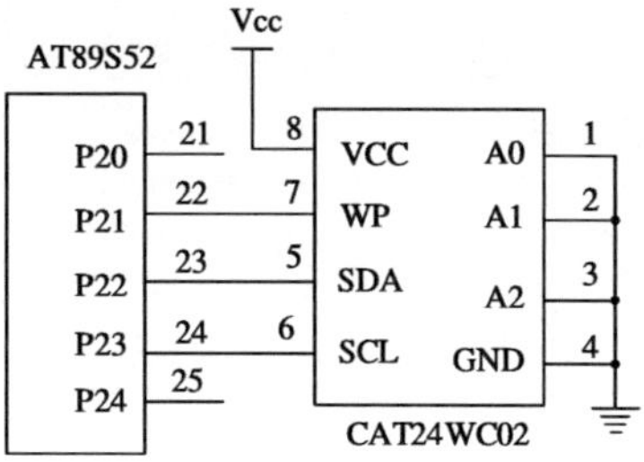

Fig. 4. Connection Diagram between controllers and memory

3 Design of Software Circuit

3.1 Communication Protocol of Fingerprint Acquisition Module

The normal data exchange is carried out between the controller and the fingerprint identification module and data exchange must follow the communication protocol. In designing software, sending data and receiving data is handled between the two sides at first.

The response means are used for the communication protocol between M04 fingerprint recognition module and controller. That is after the M04 module receives the command, it sends the response command, and then it carry out corresponding operations. Response commands include 6 bytes, i.e. 00H, 88H, 00H, 00H, 00H, 00H.

The M04 format controlled by the controller includes leading code, control command, controlling number1.the control ling number2.Control of the controlling number N. It can include:

(1) Leading codes include 7 bytes of the pilot code, that is: 00H, AAH, 00H, AAH, 55H, AAH, FFH;

(2) Control instructions include memory fingerprint, matching fingerprint, removing fingerprints, all deleting fingerprint, block deleting functions;

(3) The controlling number N obtains corresponding value according to different operation under the control instructions. The range of N is 1 to 209;

(4) M04 receives controlling instructions issued by the controller successfully or not, M04 all returns the corresponding response message. For example when the controller successfully sends stored fingerprint instructions, the return value are 00H, 66H, 00H, 00H, 00H, 00H; but when the controller failed to send stored fingerprint instructions, the return value are 00H, FFH, 00H, 00H, 00H, 00H.

3.2 Sending Data

When controller sends data to the fingerprint collection module, the maximum receiving time interval between the bytes and bytes is 4mS. But considering the stability and fast, the interval between bytes and bytes is adjusted to 4.8mS. After experimental test data errors and communications failure does not appear in 4.8mS interval operation.

3.3 Receiving Data

The wait-free interval receiving mode is used to ensure the correction of data reception. After the controller sends commands, controller waits to be receiving state immediately and it is ready to receive to data of fingerprint acquisition module. Receiving type is divided into two formats, one is fixed byte receiver, there are 12 bytes totally, i.e. 6 bytes are used as response codes and 6 bytes are used as returning codes. Another format is for the 6-byte response codes and N byte returning codes, at the same time the FFH ending sign code of two bytes is added to the format.

3.4 Storage and Comparison of Fingerprint

Procedures of the fingerprint acquisition module are divided into two parts. The two parts are storing fingerprints and comparing fingerprint. Storage part is used to input the original comparing image for the fingerprint identification module. The matching fingerprint is to compare collecting the fingerprints with the original fingerprint in order to judge which is owner of the household. Users should try to select clearer finger fingerprint for the storage fingerprint and have stored alternative fingerprint in order to use because of injury of the finger. Fingerprint storage and matching program flow chart are shown in Figure 5 and Figure 6 [5].

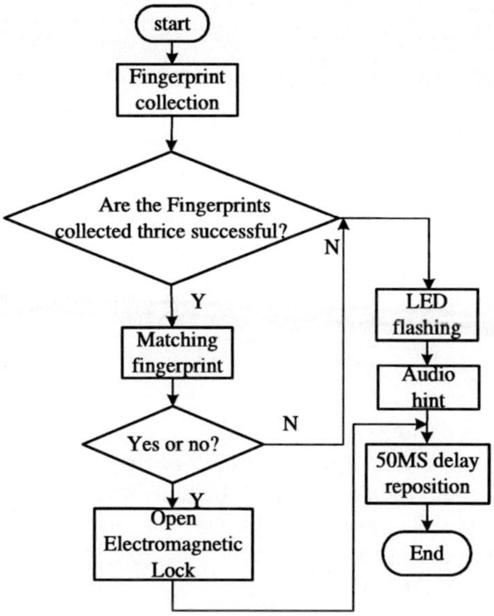

Fig. 5. Flow Chart of the Stored Fingerprint Procedures

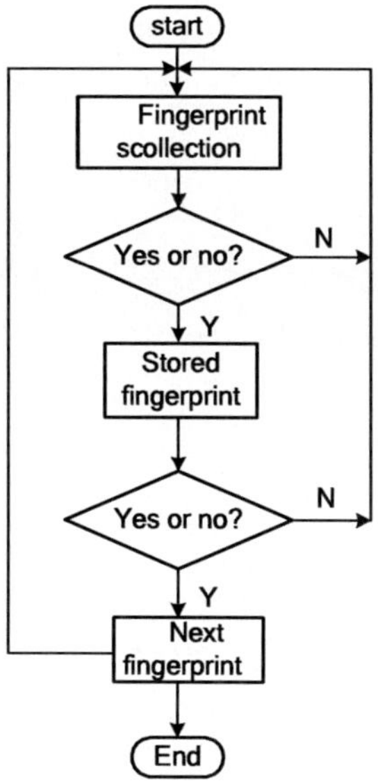

Fig. 6. Flow Chart of Fingerprint Matching

4 Conclusions

In this paper design of hardware and software are described based on M04 fingerprint access control module of the district unit. After testing, time of fingerprint collection and fingerprint matching of the system is very short (typically less than 0.5 seconds), the error rate of fingerprint is less than one ten thousandth, the whole performance is stable. There has wide application foreground in design of the intelligent community security system.

References

1. Cheng, D.: Science and Technology. In: Trojan (ed.) Principles and Applications. Beijing Hope Electronic Press, Beijing (2000)
2. Pioneer Studio. Microcontroller program design.Tsinghua University Press, Beijing (2003)
3. Xia, F.: AT89C2051 microcontroller based building directly by video intercom doorbell system. Microcomputer Information 22(1-2), 114–116 (2006)
4. Tian, P.: Researching on the Technology of Fingerprint Identification. Software School Hunan University 7(18), 4765–4768 (2007)
5. Bi, X., Zhao, W., Chen, Z.: An Embedded Fingerprint Identification System. Harbin Engineering University 26(1), 31–44 (2007)

Exploring Data Prefetching Mechanisms for Last Level Cache in Chip Multi-Processors

Mingliang Liu, Lin Qiao, Fucen Zeng, and Zhizhong Tang

Department of Computer Science and Technology, Tsinghua University
Beijing, P.R. China
liuml07@mails.tsinghua.edu.cn, qiaolin@tsinghua.edu.cn,
zengfucen@gmail.com, tzz-dcs@tsinghua.edu.cn

Abstract. Last level cache in chip multi-processors is shared by many cores; prefetchers should be fully simulated and evaluated before integrated in product. However, evaluation methodologies of proposed prefetchers vary from each other without unified simulation framework provided. This paper implements a prefetch simulation framework and evaluates seven important prefetchers, providing comprehensive evaluation methodology to newly proposed prefetchers. Moreover, we analyze the influence of storage cost and memory bandwidth quantitatively.

Keywords: cache memories, simulation, memory control and access, performance evaluation.

1 Introduction

The increasingly widen gap between processor and memory access speeds has been one of performance bottlenecks. In the multi-core era, memory wall remains a key design concern for architecture designers. Chip Multi-processors (CMPs) raise new challenges such as more deliberated on-chip memory organization, tightly-coupled core interconnect sharing last level cache (LLC) and off-chip DRAM memory bandwidth contention. Since LLC is shared by many cores, data fetched by one core may be interfered by another, especially for prefetchers and replacement policies. As a result, it is a growing need for simulation methodologies to help computer system architects with design space exploration as well as identification of performance bottlenecks in multiprocessor memory systems [1].

Data prefetcher is an important approach to bridging this expanding gap between microprocessors and memory performance, which consists a predictor determining prefetch candidates and a prefetching scheduler issuing prefetching request in time to reduce compulsory and capacity cache misses. The goals of a prefetcher are to issue enough prefetching requests to cover as many as possible cache misses, to prefetch the needed data block in time, and to reduce cache pollution and memory bandwidth competition by improving prefetching accuracy.

Unfortunately, these goals can compete with one another; for example, improving accuracy may reduce cache misses coverage. Prefetchers should be fully simulated and evaluated before integrated in product. However, evaluation methodologies of

W. Zhang (Ed.): Advanced Technology in Teaching, AISC 163, pp. 577–584.
springerlink.com

proposed prefetchers vary from each other without unified simulation framework. While managing LLC prefetchers have received considerable focus and become increasingly aggressive, a unified memory subsystem simulator can be considered as a potential means to evaluate current prefetchers quantitatively, providing comprehensive methodology to newly proposed ones.

2 Last Level Cache Data Prefetchers

In order to hide the increasingly latency between on chip cache and main memory access, architects adopt data prefetching mechanisms. Prefetchers for LLC are always more complicated than the ones for L1 or upper level cache. All of them take use of principle of locality, both spatial localities and time ones, to predict the following cache misses and prefetch them from main memory to LLC. More attention has been paid to prefetchers for LLC than L1 cache for two reasons. First, modern processors run in an out-of-order execution way and prefetchers work ineffectively in many cases. Second, load miss latency for L1 takes several cycles that can be tolerated by modern CMPs. Moreover, shared behaviors of LLC should be managed more carefully, for prefetching policies and replacement strategies.

This paper discusses hardware prefetching mechanisms. In literature, software prefetching methods have also been studied by many researchers; however, memory requests issued by a software prefetcher are not treated differently from normal ones by hardware. Prefetchers evaluated here are generic enough to be realized in all modern processors. The following are well-known hardware prefetchers to be simulated and evaluated:

- *Sequential Prefetcher.* This prefetcher is a traditional one that simply fetches next block when a cache miss happens. An effective improvement is to setup a prefetch degree, which decides how many subsequential blocks to be prefetched. For descending memory addresses, prefetcher should be smart enough to prefetch the previous ones, whose mechanism is similar to increasing cache block size and differs in influence to replacement policy.
- *Tag Sequential Prefetcher.* Tag sequential prefetcher is an improved version of traditional sequential prefetcher. It examines the hit cache blocks and issues a new related prefetching request if it is a prefetched block for the first time [2].
- *Markov Prefetcher.* This prefetcher makes use of correlations of cache miss addresses to predict the next memory access address. The Markov mode is implemented by a history table, whose size has obvious effect to prefetching correctness and coverage [3].
- *Stride Prefetcher.* Stride prefetcher finds out these memory addresses which are non-sequential but with a given step. The prefetcher is trained at beginning and issues a prefetching request if a cache miss happens at its stable state. It depends on stride distance and degree, which can be tuned automatically at running time [4].
- *Global History Buffer (GHB).* This prefetcher uses history information more effectively by putting them in a FIFO buffer, consisting pointers linked to related ones. Problems imposed by outdated entries of history table are solved and a hashed mapping table is integrated [5].

- *Adaptive CZone Delta Correlations (AC/DC).* AC/DC prefetcher is an adaptive method for prefetching data from main memory. The basic prefetch method divides the memory address space into equal-sized concentration zones (CZones), and uses a global history buffer to track and detect patterns in miss address differences between consecutive addresses within each CZone [6].
- *Tag Correlating Prefetcher (TCP).* TCP works with tags instead of cache-line addresses. The per-set tag sequences exhibit highly repetitive patterns both within a set and across different sets, as a single tag sequence can capture multiple address sequences spread over different cache sets, significant space savings can be achieved [7].

3 Simulation Methodology

Performance simulators can be either trace-driven or execution-driven. Generally, a trace-driven simulator processes pre-generated traces to determine the performance statistics by executing instructions in the traces while an execution-driven one uses instructions fed by a functional component at run time.

3.1 The TSIM Simulator

There are already many architecture performance simulators available for memory subsystems. Full system infrastructures as GEMS [8] simulate computer systems at such a level of detail including operating system. However, these simulators are usually so complicated and slow that take several months to simulate a set of workloads. At the other hand, many other excellent simulation frameworks or environments presented recently have gained little industrial interests because of lack of support for the x86 ISA, which is widely used in desktop and server computers [9].

The motivation to fill a specific niche in the simulation spectrum gives birth to TSIM, which is based on the GSIM simulation framework [9]. The TSIM is a *pin*-based trace-driven memory simulator to characterize cache performances of single threaded, multi-threaded and multi-programmed workloads. *Pin* [10] is an instrumentation tool that supports x86, x86-64 and Itanium architecture. We use it to generate workload traces and a new detailed memory model proposed to investigate access behaviors.

TSIM simulator aims at a fast and efficient tool for the research of the CMP on-chip memory subsystem. The architecture of the TSIM is shown in Figure 1, where the two pivotal components are workload traces generator and performance simulator, which are called the front-end and the back-end respectively. The front-end uses the *pin* tool to generate workload traces at run time. The back-end processes trace files with simplified pipelines, on-chip interconnect network and memory accessing model. Our simulation methodology adopts simplified pipelines and well-designed memory model causing ignorable accuracy loss, which makes it relatively simple and straightforward comparing to other memory performance counterparts. Most importantly, the statistic strategy has already been built in the top level abstraction for each module, so that the performance statistics will be completed automatically when TSIM users implement their own algorithms.

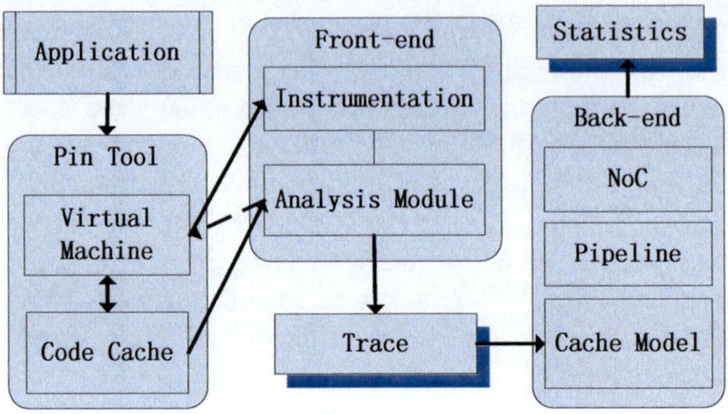

Fig. 1. Simulation framework overview.

3.2 Cache Model

Figure 2 shows the structure of the target memory subsystem modeled in TSIM. Caches or other objects on chip are connected through the on-chip interconnect network, where cache prefetcher, coherence protocol and replacement policy communicates with each other. We built an extendable prefetching framework and implemented several classic prefetchers in TSIM.

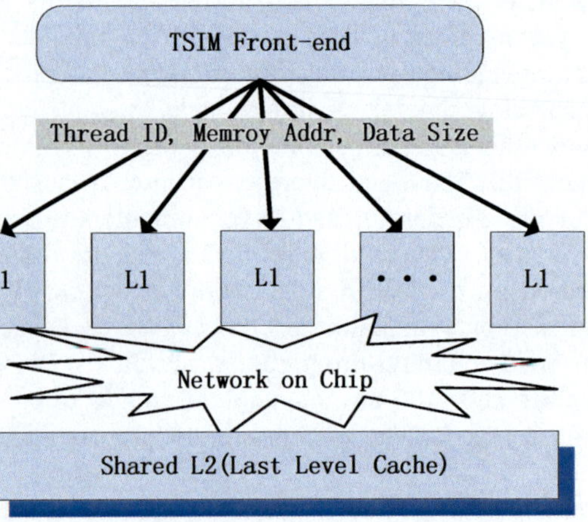

Fig. 2. Last level cache model.

Developing a detailed cache model supporting data prefetching, replacement policy and coherence is a complex endeavor, but it is immensely valuable if the modules are implemented in an extended simulator framework, say [9]. The prefetching framework of TSIM is implemented by a callback that triggers the prefetcher to handle each memory access request, miss or not. Prefetchers analyze the request's type (read or write), memory address, data size and hit/miss information to decide whether a new prefetching request is issued. New memory request should be inserted into the unified MSHRs or an independent one, assuring that the request is handled in time to cover the cache miss.

4 Results and Evaluation

To evaluate data prefetchers, traditional metrics include accuracy, coverage, and timelines. We implement prefetchers presented in Section 2 within TSIM simulator and evaluated the first two metrics in a quantitative approach. We find that the storage cost is important to memory system performance, and analyze it in detail. Meanwhile, timeliness is dependent on the first two metrics, whose simulation is complicated and remained as our future work.

4.1 Experiment Setup

Table 1 shows the experiment configuration summary. The experiment data is collected in an IA64 4-way SMP machine, running Linux OS. The benchmark is NAS Parallel Benchmark (OpenMP version 3.2) for parallel computing simulation. Each core has private L1 data cache and shares L2 as LLC. Adding private L2 to one independent core taking L3 as LLC can be simulated in the same way. We set no prefetcher for L1 cache since the goal is to evaluate the prefetchers of last level cache.

Table 1. The Experiment Configuration Summary

Configuration Name	Value
Simulated Instructions	1B (Skip 1B)
Number of Cores	16
Number of MSHRs	4
Replacement Policy	LRU
Cache Block Size	64B
Cache Associativity	4-way
L2 Hit Latency	3 cycles
L2 Cache Size	16MB
L1 Cache Size	32KB

4.2 Storage Costs of Prefetchers

Prefetchers handle memory requests in hardware logic independent on cache, and take use of some storage units. While prefetchers improve memory access performance, storage cost is not negligible, which can be tuned before implemented.

The sizes of prefetchers' storages are approximately optimal according to their papers. The sequential prefetcher has no storage cost since it simply fetches the next/previous cache block when a miss happens. Tag prefetcher is based on the sequential one and adds one bit for every cache block, indicating it is prefetched or not. The storage of tag prefetcher can be calculated as $Size_{cache}/Size_{block}$, namely number of blocks of LLC (250K here). As to GHB, storage consists two parts: a FIFO buffer and a hashed index table, the total size can be determined by $Num_{index\_table} \times Size_{index\_size} + Num_{history\_table} \times Size_{entry\_size}$, in fact, about 1MB in our simulation. The Markov prefetcher uses history table and each entry of history table cost 3 memory addresses for Markov prediction model. We can simply get the storage size by $3 \times Num_{entries\_number} \times Size_{address\_size}$, about 2M in our simulation, while the history table of stride prefetcher is $2 \times Size_{address\_size} \times Num_{entries\_number}$ since there are two addresses in a single history table entry. We simulated the stride prefetcher using about 1.8MB. Other prefetchers are similar to these ones. Most of the space on chip is contributed to caches and more storage cost of prefetchers is not suitable for modern CMPs. Prefetchers need be designed to use history buffer more effectively, not simply adding more storage units.

4.3 Memory Bandwidth Pressure

The LLC is usually shared by all the cores on chip and managing it should be carefully. Data fetched by one core may be replaced by another one. Likewise, the prefetched data may replace one useful data but contribute nothing and never be used in the future. Prefetching request issued by many different cores may compete with each other and memory bandwidth can be drained by those useless prefetching requests while useful prefetching request cannot be fulfilled in time.

Prefetchers should issue less memory request to reduce the memory bandwidth competition cases. We observed different prefetchers and record numbers of prefetching requests they issued. Figure 3 shows the numbers of requests those prefetchers issued, which reflecting the pressure to memory bus. The more prefetching requests the lower prefetching accuracy but the better memory system performance at the same time. However, Figure 3 reveals that more prefetching requests may not gain obvious performance increase (take Markov prefetcher for example). Prefetching requests are issued to MSHRs and handled when there is no normal memory request on the bus, which have higher priority than the prefetched ones. Or else memory bus may be occupied by useless prefetching requests, degrading system performance dramatically.

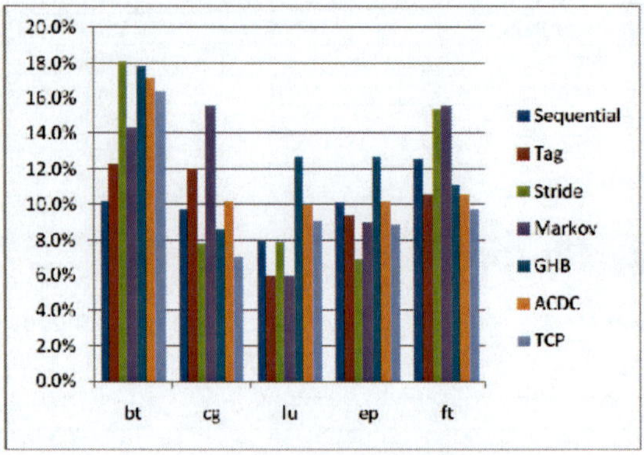

Fig. 3. Percentage of prefetching in all requests issued.

4.4 Prefetching Accuracy

The accuracy of a prefetcher is the most important metric, which consists of correctness and coverage rate.

$$Correctness = Num_{good} / Num_{prefeches} \tag{1}$$

$$Coverage = Num_{good} / Num_{misses} \tag{2}$$

The correctness reflects how many good prefetches in all prefetching requests issued by that prefetcher, while the coverage rate shows how many cache misses the prefetcher covers, which saves the load time from lower cache and improves the memory access performance. To simulate this, we use a prefetched bit to indicate whether it is prefetched or not. The results here is the total accuracy of a single shared prefetcher of LLC for all cores on chip, which does not treat different threads individual to save storage costs.

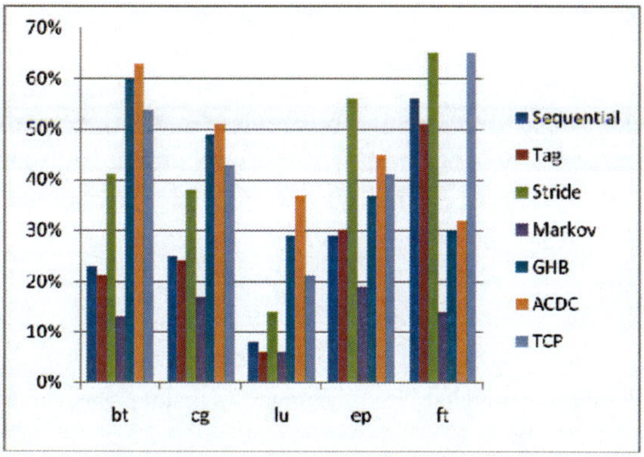

Fig. 4. Prefetching correctness for parallel applications.

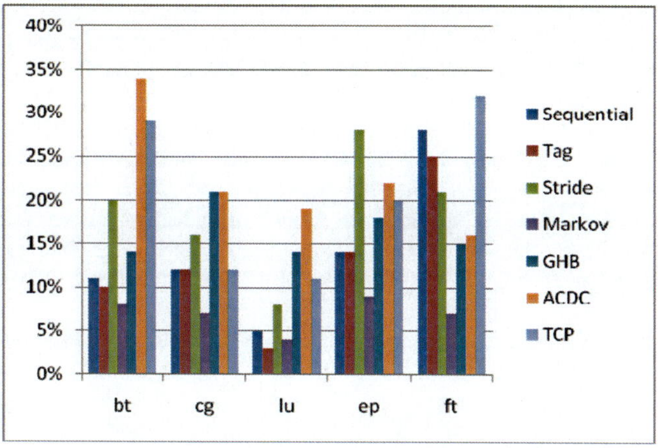

Fig. 5. Prefetching coverage in all cache misses.

Figure 4 and Figure 5 show that those prefetchers with high coverage have poor correctness which means that they are more aggressive than others. The results can also be verified according to Figure 3. This proposes that a new prefetcher for LLC should cover a degree of cache misses with a relative high prefetching accuracy. We stand on solid ground to conclude that those aggressive prefetchers should work in a constrained way so that prefetching request from individual core should be scheduled more carefully.

5 Conclusion

We implement several well-known data prefetchers for last level cache in the TSIM simulator and evaluated them in a quantitative way. The results show that these prefetchers behave differently in critical parameters, and those with high storage cost do not improve performance obviously. Considering the data share characteristic, the paper analyzes the influence of cost and memory bandwidth. Further work is going to focus on prefetching timeliness, cache pollution, and parameter sensitivities.

Acknowledgment. The authors thank Yu Chen for his pioneering work of TSIM simulator [11]. Thanks also go to anonymous reviewers for their useful feedback.

The research was partially supported by Chinese National High-tech R&D Program (863 Program) under Grant No. 2008AA01Z108 and by National Natural Science Foundation of China under Grant No. 60773149.

References

1. Lilja, D.: Simulation of Computer Architectures: Simulators, Benchmarks, Methodologies, and Recommendations. IEEE Transactions on Computers 55(3), 268–280 (2006)
2. Vanderwiel, S.P., Lilja, D.J.: Data prefetch mechanisms. ACM Computer Survey 32(2), 174–199 (2000)
3. Joseph, D., Grunwald, D.: Prefetching using markov predictors. IEEE Transactions on Computers 48(2) (February 1999)
4. Dahlgren, F., Stenstrom, P.: Effectiveness of hardware-based stride and sequential prefetching in shared-memory multiprocessors. In: Proceedings of the First IEEE Symposium on High- Performance Computer Architecture, pp. 68–77 (February 1995)
5. Nesbit, K., Smith, J.: Data cache prefetching using a global history buffer. In: Proceedings of the International Symposium on High Performance Computer Architecture, pp. 96–96, 14-18 (2004)
6. Nesbit, K.J., Dhodapkar, A.S., Smith, J.E.: Ac/dc: An adaptive data cache prefetcher. In: PACT 2004: Proceedings of the 13th International Conference on Parallel Architectures and Compilation Techniques, pp. 135–145. IEEE Computer Society, Washington, DC (2004)
7. Hu, Z., Martonosi, M., Kaxiras, S.: Tcp: tag correlating prefetchers. In: Proceedings of the Ninth International Symposium on High-Performance Computer Architecture, pp. 317–326, 8-12 (2003)
8. Martin, M., Sorin, D., Beckmann, B., Marty, M., Xu, M., Alameldeen, A., Moore, K., Hill, M., Wood, D.: Multifacet's general execution-driven multiprocessor simulator (GEMS) toolset. ACM SIGARCH Computer Architecture News 33(4), 99 (2005)
9. Liu, M., Qiao, L., Chen, Y., Zeng, F., Zhang, C.: An extensible memory simulation framework for chip multi-processors. In: International Conference on Computational Intelligence and Software Engineering, pp. 11–13 (2009)
10. Luk, C., Cohn, R., Muth, R., Patil, H., Klauser, A., Lowney, G., Wallace, S., Reddi, V., Hazelwood, K.: Pin: building customized program analysis tools with dynamic instrumentation. In: Proceedings of the 2005 ACM SIGPLAN conference on Programming language design and implementation, pp. 190–200. ACM Press (2005)
11. A memory sub-system performance simulator for chip multiprocessors (2010), online, `http://code.google.com/p/gsim-augur`

Current Research on the Fitness Coaches's Job Satisfaction in Hubei

Feng-zhen Qin and Daping Wang

School of Physical Education, Central China Normal University, Wuhan 430079, P.R. China

Abstract. Based on the questionaire survey and statistic analysis,by investigating the present situation of 30 fitness clubs aerobics instructors in Hubei,the main results are as follow: Firstly, there are mainly five factors affecting the job satisfaction: salary and welfare, work condition,development tendency, interpersonal relationship and respect. Secondly,the fitness coaches score high points in the work condition, but receive low points in the salary and welfare, even at the level of low-satisfaction. Thirdly, through the T-examination as well as multi-factor variance analysis, we know that the development tendency has significance difference in the gender; respect and work condition have significance difference in the age; salary and welfare, and work condition have significance difference in the education level.

Keywords: Hubei, Fitness Club, Fitness Coach, Job Satisfaction.

1 Introduction

With the development of the national fitness, it also will enhance the concept of fitness, for-profit health club become the best place for people to fitness, fitness coaches will no doubt demand is growing. Fitness Coaches determine the level of overall quality of customer participation in fitness activities are rational, scientific and effective. People are increasingly willing to spend money on fitness in the current, 21st century industry in the ten most popular fitness coaches will have a place.

Job satisfaction has been defined as a pleasurable emotional state resulting from the appraisal of one's job; an affective reaction to one's job; and an attitude towards one's job. Many research results show that the employee job satisfaction and organizational performance have a higher correlation between satisfaction with the organization of employees working attitude and enthusiasm more to stimulate its efficiency; the other hand, low satisfaction Employees lack the enthusiasm, and often absent, and may eventually submitted his resignation, which will undoubtedly bring to the organization adversely affect the normal operation.Satisfaction of scholars of sport organizations is increasing, study and more focused on athletes, professional team coaches, PE teachers, sports management personnel, the work in the social sports play an important role in the health club of coaches Small. The promotion of national fitness campaign will stimulate a new round of fitness craze, people will slowly come to accept the idea of spending fitness, fitness coaches can say that will play an increasingly important role.

Therefore, this study attempts to fitness coaches status of job satisfaction surveys and analysis, mainly includes three aspects: firstly, fitness coaches overall status of job

W. Zhang (Ed.): Advanced Technology in Teaching, AISC 163, pp. 585–589.
springerlink.com

satisfaction in hubei. Secondly, the fitness coaches in the job satisfaction indicators of whether there are differences. Thirdly, the demographic variables on job satisfaction of fitness coaches.

2 Methods

Method of questionnaire: Designed a questionnaire for the spectators based on the needs of this approach. In case to ensure the scientifically of this questionnaire, we accepted some suggestions from several experts and modified the questionnaire for several times. The scale consisted of 30 items: salary and welfare(5items);work condition(5items);development tendency(5 items); interpersonal relationship(5 items);respect(5 items);degree of satisfaction (5 items).Respondents were asked to rate the degree to which they satisfied or dissatisfied with each item on a 5-point Likert scale ranging from 1(strongly dissatisfied) to 5(strongly satisfied).

Method of mathematical statistics: By coordinating the received questionnaires and using SPSS 18.0 to analyze the received data, we made a conclusion.

3 Results and Analyze

Table 1 shows that in the fitness coaches in each dimension of job satisfaction, work condition dimensions of the average highest was 3.560, indicating that the fitness coaches satisfied with the work condition. The salary and welfare, and development tendency of the evaluation of the lower two dimensions, with an average lower than 3, which suggests that these two aspects of fitness coaches not satisfied. The average overall satisfaction 3.2204, at a medium level, indicating that fitness coaches has general job satisfaction.

Table 1. Analysis of fitness coaches's Job Satisfaction

Dimensions	N	Minimum	Maximum	Mean	Std.deviation
F1	186	1.40	4.60	2.6742	.9863
F2	186	1.40	5.00	3.5602	1.0512
F3	186	1.40	5.00	2.7247	1.0958
F4	186	1.20	5.00	3.2376	1.1584
F5	186	1.60	5.00	3.0892	.99294
F6	186	1.20	5.00	3.2204	1.1085

Note:F1=salary and welfare;F2=work condition;F3=development tendency;F4=interpersonal relationship;F5=respect;F6=degree of satisfaction

Gender(Table 2):Male and female fitness coaches in the job satisfaction of six dimensions, in the development tendency of the difference was very significant level, in several other dimensions of gender was no significant difference. From the scores in all dimensions, the male in salary and welfare, respect and degree of satisfaction on job satisfaction than the average female; the female in the work condition, development tendency, and interpersonal relationship on job satisfaction than male.

Table 2. A Comparion of Job Satisfaction by Sex of Participant

Dimensions	Male(N=123)		Female(N=63)		*F*	*P*
	M	SD	M	SD		
F1	2.6959	.97084	2.6317	1.0224	.228	.676
F2	3.5431	1.0674	3.5937	1.02672	.752	.757
F3	2.5707	1.0827	3.0254	1.0664	.050	.007**
F4	3.1480	1.1725	3.4127	1.1186	.864	.141
F5	3.1593	.94016	2.9524	1.0834	6.495	.200
F6	3.2553	1.1386	3.1524	1.0526	.837	.550

Note1:F1=salary and welfare;F2=work condition;F3=development tendency;F4=interpersonal relationship;F5=respect;F6=degree of satisfaction

Note2: **p*<0.05 ***p*<0.01

Age(Table 3):Age of 20 years of age,21-30,31-40 years and 41-50-year-old fitness coaches in the 4 groups, fitness coaches of different ages with respect to this dimension was a significant difference,at the level of different dimension in the work condition to achieve significant levels.

Table 3. A Comparion of Job Satisfaction by Age of Participant

	Age	N	M	SD	F	P
F1	Under 20 years old	4	2.5000	1.2701	2.153	.095
	21-30 years old	139	2.5899	.91440		
	31-40 years old	35	3.0514	1.1765		
	41-50 years old	8	2.5750	.96474		
F2	Under 20 years old	4	4.7000	.11547	3.541	.016*
	21-30 years old	139	3.4734	1.0550		
	31-40 years old	35	3.5943	1.0375		
	41-50 years old	8	4.3500	.64807		
F3	Under 20 years old	4	3.1000	.57735	.529	.663
	21-30 years old	139	2.7640	1.0698		
	31-40 years old	35	2.5543	1.2214		
	41-50 years old	8	2.6000	1.2282		
F4	Under 20 years old	4	4.4000	.00000	1.741	.160
	21-30 years old	139	3.2446	1.1881		
	31-40 years old	35	3.1714	1.0977		
	41-50 years old	8	2.8225	.89083		
F5	Under 20 years old	4	4.2000	.00000	3.912	.010**
	21-30 years old	139	3.0000	.98892		
	31-40 years old	35	3.4114	.99048		
	41-50 years old	8	2.6750	.65846		
F6	Under 20 years old	4	3.2000	.92376	2.195	.090
	21-30 years old	139	3.3180	1.0842		
	31-40 years old	35	2.7943	1.1787		
	41-50 years old	8	3.4000	1.0253		

Note1:F1=salary and welfare;F2=work condition;F3=development tendency;F4=interpersonal relationship;F5=respect;F6=degree of satisfaction

Note2: **p*<0.05 ***p*<0.01

Education Level(Table 4):Different degree of fitness coaches in the work condition,salary and welfare two dimensions of difference was very significant level. Postgraduate which the fitness coaches in the two dimensions of satisfaction were lower than the fitness coaches to obtain other qualifications. This blow to the enthusiasm of highly educated fitness coaches, they generally believe that the existing environment and conditions against themselves, can not truly reflect the value of their own, which inevitably show lower satisfaction.

Table 4. A Comparion of Job Satisfaction by Educational Levels of Participant

	Educational Levels	N	M	SD	F	P
	High school	38	3.1842	1.0928		
F1	undergraduate	77	2.5273	.90141	6.808	.001**
	Postgraduate	71	2.5606	.93785		
	High school	38	4.0789	.73306		
F2	undergraduate	77	3.3377	1.0504	6.794	.001**
	Postgraduate	71	3.5239	1.1122		
	High school	38	2.7263	1.4114		
F3	undergraduate	77	2.6312	1.0553	.577	.562
	Postgraduate	71	2.8254	.94411		
	High school	38	3.3526	.89011		
F4	Undergraduate	77	3.3481	1.1366	1.413	.246
	Postgraduate	71	3.0563	1.2927		
	High school	38	2.9842	1.1076		
F5	Undergraduate	77	2.9844	1.0181	1.694	.187
	Postgraduate	71	3.2592	.88569		
	High school	38	3.0789	1.2730		
F6	undergraduate	77	3.3662	1.0112	1.174	.311
	Postgraduate	71	3.1380	1.1133		

Note1:F1=salary and welfare;F2=work condition;F3=development tendency;F4=interpersonal relationship;F5=respect;F6=degree of satisfaction

Note2: *p<0.05 **p<0.01

4 Conclusion

Firstly, there are mainly five factors affecting the job satisfaction: salary and welfare, work condition, development tendency, interpersonal relationship and respect.

Secondly, the majority of fitness coaches are satisfied with their work. Among them, the fitness coaches score high points in the work condition ,but receive low points in the salary and welfare, even at the level of low-satisfaction.

Thirdly, through the T-examination as well as multi-factor variance analysis, we know that the development tendency has significance difference in the gender; respect and work condition have significance difference in the age; salary and welfare, and work condition have significance difference in the education level.

References

1. Liu, B.: Analysis of factors influencing satisfaction index of sport fitness and recreational industry employees. Journal of Wuhan Institute of Physical Education 41(3) (2007)
2. Xu, Q.: Research on the employee satisfaction and motivation system of Hosa fitness club in Beijing. Beijing Sport University (2007)
3. Ren, P.: Research on personal trainers of fitness clubs in wuhan. Central China Normal University (2008)
4. Robinson, Tedrick, Carpenter: Job satisfaction of NCAA Division III athletic directors: a descriptive analysis and an examination of gender differences. International Sports Journal (West Haven, CoNN 5(1), 25–32 (2001)
5. Davies, Bloom, Salmela: Job satisfaction of accomplished male university basketbaLL coaches: The Canadian context. International Journal of Sport Psychology 36(3), 173–192 (2005)

Fault Diagnosis of Ball Bearing Based on Energy Feature and Research of Intelligent Classification Method

Linxiang Lu and Yun Zeng

School of Information Engineering
Nanchang University,
Nanchang 330031, China
lulinxiang@yeah.net, zengyun1128@hotmail.com

Abstract. Roll bearing is the most important part of mechanical equipment, which has the merits of low friction resistance, convenient installment and easily realized lubrication. So it is commonly used in rotated machine. Whether bearing is normal or not affects straightly the mechanical working state, any fault or invalidation took place in the running of the machine will bring serious sequence and great economic loss. So it is necessary to check the state of the bearing and diagnose the fault. In this paper, the energy features of the vibration signal of the roll bearing are extracted by Wavelet Packet, then the method of RBF Neural Network is presented to diagnose the faults, the faults can realize intelligent classification by this method. Simulation is used to assist in the roll bearing faults diagnosis. The simulation results obtained indicated that the method of energy features extracted by Wavelet Packet is efficient and the method of intelligent classification can identify the faults well.

Keywords: roll bearing, fault diagnosis, energy features, Wavelet Packet, intelligent classification.

1 Introduction

When the roll bearing is in fault, the fault information is not easily to be extracted by a general method of time-frequency. The wavelet has the time-frequency local ability, which is suitable to analyze non-smooth signal and extract the feeble fault information of the signal. The signal can be decomposed on the random subtle frequency bands by the Wavelet Packet, different frequency bands have different information, so the energy value on different frequency bands is extracted which constructs eigenvector. The eigenvector consists of multi-elements, it is a complicated non-linear mapped relation between the eigenvector and faults type, the fault is not determinate by a general method. While RBF NN can realize the non-linear mapped relationship, it is very suitable to non-linear pattern identify and classification, the method of extracting energy characteristics based on the wavelet is combined with the method of intelligent classification of NN, intelligent fault diagnoses can be realized and the fault type can be determined conveniently and exactly.

W. Zhang (Ed.): Advanced Technology in Teaching, AISC 163, pp. 591–597.
springerlink.com

2 Wavelet Packet Theory and Energy Features Extraction

2.1 The Definition of the Wavelet Packet

The concept of the wavelet packet is put forward by M. V. Wikcer and R. R. Coifman on the basis of the wavelet transform and deduced strictly from mathematics. From the engineering the wavelet packet can be regarded as spread of the orthogonal decomposition on the function space [1].

The orthogonal scale function $\phi(t)$ and the wavelet function $\psi(t)$ is given, the two scale relationship is followed:

$$\phi(t)=\sqrt{2}\sum_{k} h_{ok}\phi(2t-k) \tag{1}$$

$$\psi(t)=\sqrt{2}\sum_{k} h_{1k}\phi(2t-k) \tag{2}$$

In the formula h_{0k}, h_{1k} is the filter coefficient of the multi-resolution. In order to spread the two scale equation, the following recursive relation is defined:

$$w_{2n}^{(j)}(t)=\sqrt{2}\sum_{k\in Z} h_{0k} w_n^{(j)}(2t-k) \tag{3}$$

$$w_{2n+1}^{(j)}=\sqrt{2}\sum_{k\in Z} h_{1k} w_n^{(j)}(2t-k) \tag{4}$$

When n=0, in (3) $\phi(t)$ is the two scale equation, so $w_0^{(j)}(t)$ is $\phi_{j0}(t)$, in (4) $\psi(t)$ is the two scale equation, $w_1^{(j)}(t)$ is $\psi_{j0}(t)$. Other $w_n^j(t)$ of the same level, when sequence number is even, it is determined by (3), when sequence number is odd, it is determined by (4). From the definition $\{w_n(t)\}_{n\in z}$ is wavelet packet decided by $w_0(t)=\phi(t)$, which are the function set including the scale function $w_0(t)$ and the wavelet function $w_1(t)$.

The wavelet packet provides a more subtle analyzed method for the signal, the frequency bands can be partitioned multilayer on the basis of the multi-resolution analysis. Not only the low frequency signal can be decomposed further, but also the high frequency signal can be decomposed further. The signal can be decomposed onto the random subtle frequency bands, the same time the resolution of the time domain and the frequency domain, the feeble fault signal can be extracted and the analysis ability is stronger.

2.2 The Extraction of Energy Characteristics

1) Based on the theory of the "energy-damage" [3]

When the signal is analyzed by the wavelet, the non-smooth component of the signal can be analyzed, especially the wavelet packet transform which can decompose the signal onto random subtle frequency bands, on these frequency bands the extracted energy build eigenvector. Yen and Lin(2000) [4]has defined the energy of the frequency bands of the wavelet packet, the signal characteristics take on the robustness rather using the energy of the frequency bands than using the wavelet packet decomposition coefficient directly.

When a signal which has rich frequency components is used as input that is the motivation of the system, the damage will restrain some frequency components and enhance some other frequency components because the restrain and enhancement of the damage aiming at different frequency components is changeable. Therefore, the output compared with the other output of the normal system, the signal energy in the same frequency band has great diversity. Damage will increase the signal energy in some frequency band and decrease the energy of the signal in some other frequency band. So the signal energy in different frequency bands contains rich damage information, the change of some energy represents a sort of damage.

When the framework received some vibrate motivation, it will generate some vibrate response, the creation of the damage will result in change of the inherent frequency, firm and damp, which will affect driving capability of the framework. The restrained and enhancement effect of the damage to different frequency components of the respond signal are different, which will result in redistribution of different frequency components of the respond signal. The wavelet packet can let the signal decomposed on different frequency bands, the energy on every frequency band is extracted, which build eigenvector, so we can use the eigenvector as the input of the NN, the fault can be diagnosed by the NN further.

2) The step of the energy characteristics extracted by the wavelet packet

Step 1: the noise of the ball bearing signal is eliminated by the wavelet packet transform, some disturb components are filtered off.

Step 2: the signal are decomposed by the wavelet packet, the decomposed layers are three, the wavelet packet function is db6, so decomposed signals of the third layer are expressed as $(x_{30}, x_{31}, x_{32}, x_{33}, x_{34}, x_{35}, x_{35}, x_{36}, x_{37})$.

Step 3: the decomposed signals of the third layer is restructured, the restructured signals are expressed as $(X_{30}, X_{31}, X_{32}, X_{33}, X_{34}, X_{35}, X_{36}, X_{37})$.

Step 4: the eigenvector of the energy is build for the reconstructed signal. The energy on every different frequency bands are $E_{3j} = \sum_{n=1}^{N} \left| X_{3j} \right|^2 \ (j = 0.,1,\ldots,7)$, N is sampling points of the signal. Then normalization, $E = (\sum_{j=0}^{7} \left| E_{3j} \right|^2)^{\frac{1}{2}}$ $\overline{E} = E_{3j} / E$ finally the eigenvector will be obtained, that is $e = [\overline{E}_{30}, \overline{E}_{31}, \overline{E}_{32}, \overline{E}_{33}, \overline{E}_{34}, \overline{E}_{35}, \overline{E}_{36}, \overline{E}_{37}]$ used as the input of the RBF NN.

3 The RBF NN

3.1 The Framework of the RBF NN[5,6]

The network of the RBF is a forward network of single latent layer, which is consist of three layers: the input layer, the latent layer and the output layer. If the number of the input sample is N, as figure 1, the number of the cells in the input layer is M, any cell is expressed by m; the number of the cells in the latent layer is I (I<N), any cell is expressed as i, the output of the any cell of the latent layer is $\varphi(X, t_i)$, which is radical function, $t_i = [t_{i1}, t_{i2}, ..., t_{iM}], (i = 1,2,..., I)$.

It is the center of the radical function; the number of the cell in the output layer is J, any cell is expressed by j. the weight value between the latent layer and the output layer is expressed by $w_{ij}, (i = 1,2,..., I, j = 1,2,..., J)$.

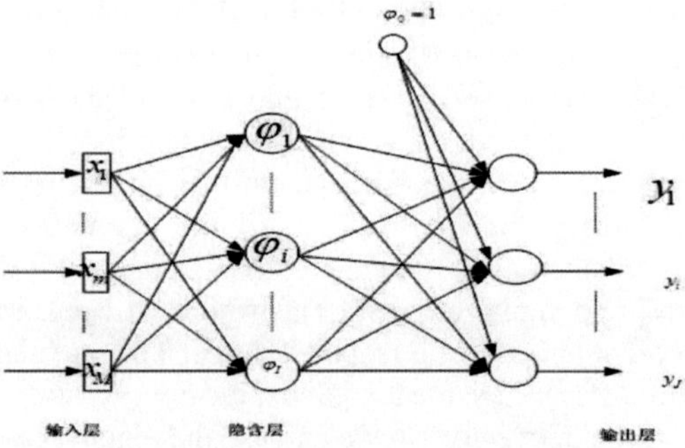

Fig. 1. The framework of the RBF NN

In Figure 1 the valve value φ_0 is set up in the output, the weight between the valve value and the output is $w_{0j} (i=1,2,...I)$.

When the input samples are X_k, the j output of the RBF NN is:

$$y_{kj}(X_k) = w_{0j} + \sum_{i=1}^{I} w_{ij}\varphi(X_k, t_i) \tag{5}$$

When the racial function is the Gauss function, it is expressed as following:

$$\varphi(X_k, t_i) = \exp(-\frac{1}{2\sigma^2}\sum_{m=1}^{M}(x_{km} - t_{im})^2) \tag{6}$$

In the formula $t_{im} = [t_{i1}, t_{i2}, ..., t_{iM}]$ is the centre of the Gauss function, σ_i is the square error of the Gauss function.

3.2 The Training of the RBF NN

The parameters of the RBF NN needed learning are: the centre of the radical function t_i, the square error σ_i and the weight value. The steps of the algorithm have two steps, the first step is: the centre and the square of the radical function, it is the learning process of non-supervise; the second step is: the learning of the weight between the latent layer and the output layer, it is the process of supervise. The non-supervise learning adopts the K-means clustering algorithm, which let input sample clustering, so the clustering centre t_i and the parameter σ_i will be found. Then the supervise learning can be carried through, when t_i and σ_i are decided, the equation of the RBF NN between the input and the output becomes linear, the weight value of the output w_{ij} can be obtained by the Least Square Method in the supervise learning process.

4 The Intelligent Diagnosis Sample

Some 406 type belling vibrating signals serve as the objects of research, the sampling frequency is 10 KHz. The signals are decomposed by the wavelet packet, the decomposed layers are three, after decomposed, the numbers of the frequency bands are eight, then the signals on the eight frequency bands are reconstructed, the energy on every frequency band is extracted, which builds eigenvector, used as the input sample of the RBF NN. The numbers of the training samples are 30, the inner raceway, outer raceway, ball fault samples are 10 respectively. The output code is followed as Table 1.

Table 1. The Expected Output of RBF

Fault type	Output code	
Outer raceway	0	0
Inner raceway	0	1
Ball	1	0

The model and error performance curve of the training network are shown as Figure2 and Figure3.

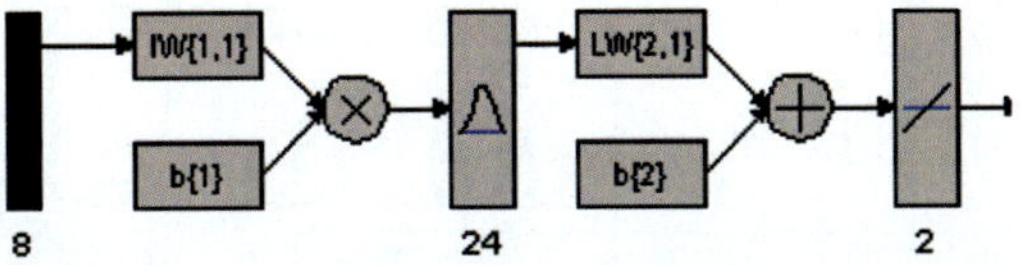

Fig. 2. The trained model of RBF

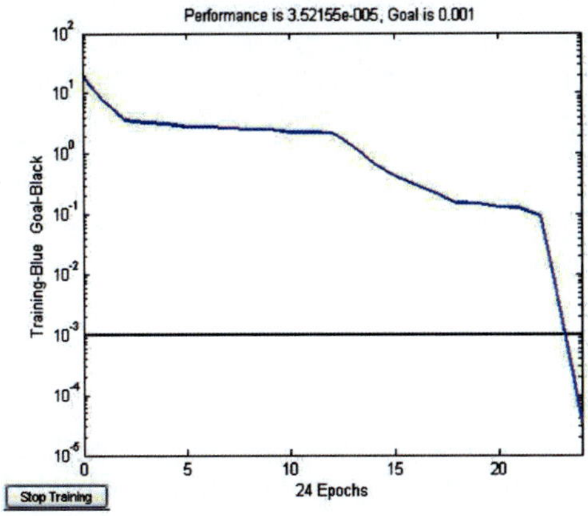

Fig. 3. The error curve of RBF

The target value of the error is set 0.001, after training, the input samples are used as test samples, which are classified by the trained RBF. The correct rate of the classification result is 100%, which shows the learning ability of the trained NN is very good. The trained time of RBF is 3.375 seconds, which shows the training speed of RBF is very fast.

The faults classification based on RBF. Another 12 testing samples selected are put into the trained RBF, so the classification results can be obtained, as shown Table 2.

From the output of the testing samples, there are 9 faults classified correct, the rate of the identifying is 75%, which shows the RBF can achieve the classification relative correctly and the wavelet packet which extracted the energy characteristics of the faults is a valid method.

Table 2. Testing Samples

Serial number	**The eigenvector $\bar{E}$**								**The out put**	**Actual fault type**
1	0.364	0.174	0.356	0.208	0.514	0.260	0.502	0.286	0.00 0.01	Outer
2	0.377	0.173	0.326	0.230	0.531	0.215	0.490	0.316	0.43 0.42	Outer
3	0.387	0.224	0.354	0.244	0.360	0.194	0.611	0.270	1.30 0.17	Outer
4	0.365	0.377	0.351	0.337	0.352	0.332	0.391	0.318	0.40 1.26	Inner
5	0.372	0.358	0.366	0.349	0.303	0.341	0.375	0.360	0.01 0.99	Inner
6	0.376	0.347	0.376	0.344	0.345	0.354	0.360	0.322	0.01 0.99	Inner
7	0.500	0.193	0.469	0.305	0.332	0.247	0.406	0.248	0.37 0.10	Ball
8	0.437	0.233	0.337	0.276	0.484	0.202	0.504	0.188	0.81 0.27	Ball
9	0.517	0.225	0.370	0.274	0.374	0.207	0.480	0.233	0.13 0.25	Ball
10	0.365	0.174	0.541	0.299	0.397	0.153	0.487	0.187	0.21 0.05	Outer
11	0.373	0.381	0.387	0.344	0.327	0.342	0.346	0.321	0.08 0.99	Inner
12	0.501	0.230	0.301	0.251	0.504	0.239	0.396	0.271	0.74 0.20	Ball

5 Conclusion

The signal can be decomposed subtle use of the wavelet packet, then the energy of every frequency band is extracted which builds eigenvector, the same time, the non-linear relation between the faults type and eigenvector is built by the RBF NN, the intelligent fault diagnosis can be realized. Based on the results of the testing samples, the faults of the fault roll bearing can be classified availably and the identify precision is very high.

References

1. Yang, F.: The engineering analysis and application of the wavelet transform. The Science Press
2. Yen, G.G., Lin, K.C.: Wavelet packet feature extraction for vibration monitoring. IEEE Trans. Industrial Electronic 47(3), 650–667
3. Hu, J., Zhang, J.: The system analysis and design based MATLAB-the wavelet analysis. Xian Electronic and Technological University Press
4. Li, H., Xun, H.: The framework damage diagnosis method based on the wavelet analysis and the NN. The Earthquake Engineering and Engineering Vibration 23(5) (2003)
5. Gao, J.: The theory of the NN and imitate sample. The Mechanical Industry Press (July 2003)
6. Zhou, K., Kang, Y.: The NN model and MATLAB imitate procedure design. Tsinghua University Press (July 2005)

Advanced Modulation Schemes Based on BICM-ID for FM IBOC

Fang Weiwei, Li Jianping, and Cai Chaoshi

School of Information Engineering
Communication University of China
Beijing, China
541764418@163.com, turbo_code@163.com, BBICAI@cuc.edu.cn

Abstract. Hybrid in-band on-channel (IBOC) broadcasting system as a scheme of digital audio radio could transmit analog FM and digital audio simultaneously. In IBOC systems, broadcasters transmit signals within the allocated channel bands, thus the bandwidth assigned for digital audio transmission is very challenging. BICM-ID can improve BER performance over Rayleigh fading channels without bandwidth expansion, thus it is used in FM IBOC and has been verified it can yield a better coding gain. However, conventional IBOC systems apply QPSK as channel modulation scheme, which can not take full advantage of constellation space compared with duobinary modulation. In this paper, various modulation schemes with the same encode rate and the same bandwidth efficiency are compared and analyzed. Simulation results show that advanced modulation scheme can effectively improve the performance of BICM-ID FM IBOC. When BER is below 10^{-6} , 32APSK modulation scheme can achieve more than 1dB coding gains over Rayleigh fading channels.

Keywords: FM IBOC, BICM-ID FM IBOC, QPSK, advanced modulation, Rayleigh fading channels.

1 Introduction

Hybrid in-band on-channel (IBOC), which is used for broadcasting digital audio signals together with analog FM programs [1], is getting more international attentions. It is designed for digital audio broadcasting (DAB) in the FM bands. As a transition to all digital DAB, IBOC is required by many broadcasters.

In hybrid IBOC system, broadcasters transmit signals within the allocated channel bands, thus the bandwidth assigned for digital audio transmission is very limited. In order to achieve better coding gains without bandwidth expansion, lots of advanced theories have been used for this system.

Bit-interleaved coded modulation (BICM), first introduced by Zehavi [2] and further studied by Caire [3], is an effective transmission scheme without bandwidth expansion. The performance of BICM can be greatly improved through iterative information exchange between the demapper and the channel decoder [4]. This system, introduced in [4] [5], is usually referred to as BICM with iterative decoding (BICM-ID).

W. Zhang (Ed.): Advanced Technology in Teaching, AISC 163, pp. 599–606.
springerlink.com

Subsequently, many literatures analyzed and improved the performance of BICM-ID [6-9]. It has been utilized in FM IBOC to achieve better performance over Rayleigh fading channels [10].

Conventional IBOC systems apply QPSK as channel modulation scheme [11] [12], which can not make full use of constellation space compared with duobinary modulation. Better BER performance can be achieved by using duobinary modulation scheme in the same bandwidth efficiency. In this paper, various modulation schemes based on BICM-ID for FM IBOC are analyzed. On one hand, various modulation schemes with the same code rate R=1/2 are compared. On the other hand, comparison among various modulation schemes with the same bandwidth efficiency is obtained.

This paper is organized as follows. In Section 2, overall system model for FM IBOC is described. BICM-ID scheme in hybrid FM IBOC is shown in Section 3. The performance of IBOC-AM system with BICM techniques are demonstrated and compared by computer simulation in Section 4, and finally Section 5 is the conclusion.

2 Background of FM IBOC System

With the progressive realization of digital television, digital radio for the CHN AM radio station may be a desirable reality. Thus the addition of a digital program signal that uses IBOC for DAB will be introduced. This system require to provide higher audio quality, more reliable reception, and additional information channels than currently available from the existing analog FM broadcasting process. And the equipment used to deliver IBOC DAB will have to meet the needs of both consumers and broadcasters in the terms of cost and performance to ensure a useful spectrum with minimum interference. Due to the existence of fading and interference in FM bands, the signal receiving for the hybrid IBOC system is a challenge.

The block diagram of the transmitter of FM IBOC system is shown in Fig. 1. To ensure appropriate energy distribution in the transmitted signal, the input binary sequences are scrambled. Subsequently, convolutional encoder codes the information bits to improve its reliability in the presence of channel impairments. The encoded bits are then fed into interleaver to mitigate the effects of burst errors. In this process, the logical channels lose their identity. A group of bits at the output of the interleaver is mapped into a mapping symbol. OFDM subcarrier mapping block assigns the interleaver matrices and system control vector to OFDM subcarriers. One row of each active interleaver matrix and one bit of the system control vector is processed each OFDM symbol to produce one output vector [13].

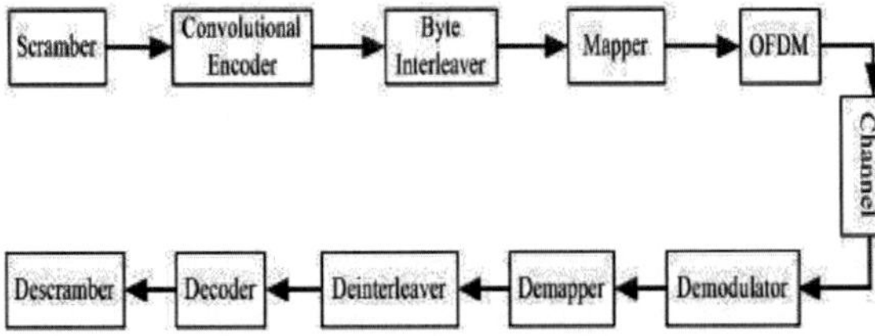

Fig. 1. Block diagram of FM IBOC system

In such HIBOC system, due to simultaneous transmission of analog FM and the digital signal, the digital signal is transmitted in sub-bands on both sides of the analog host signal at a low power level. Fig. 2 depicts the interference situation of host FM and its first adjacent [14].

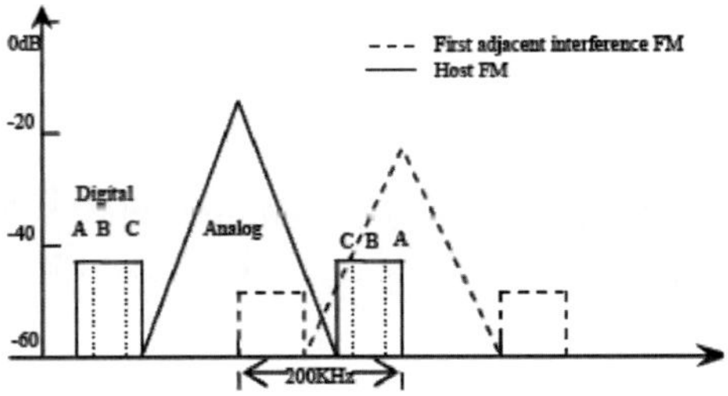

Fig. 2. Model of host FM and first adjacent interference

Broadcasters transmit signals within the allocated channel mask in FM OBOC, thus the bandwidth allocated for digital audio transmission is very limited. In order to achieve coding gain without bandwidth expansion in these systems, a coded modulation scheme which is termed as BICM-ID is used. BICM-ID employs bit-wise interleaving instead of byte-wise interleaving in IBOC. Simultaneity in BICM-ID receiver the bits are detected by iterative decoding between the demapper and the decoder.

3 FM IBOC Based on BICM-ID

BICM-ID, a bandwidth- efficient approach primarily considered for fading channels at first, can in fact be used to provide excellent performance over both Gaussian and fading channels. In this section, a brief introduction of FM IBOC based on BICM-ID (BICM-ID FM IBOC) is described.

3.1 System Model

The BICM-ID FM IBOC system is represented by the block diagram of Fig.3. Scrambered information bits are first encoded by a convolutional encoder with rate R=1/2. Then, a bit-by-bit interleaver permutes the order of encoder output bits. The bit-wise interleaving breaks the correlation of sequential fading coefficients and maximizes the diversity order of the system [2]. In addition, it removes the correlation among the sequentially coded bits, as well as the bits associated with the same channel symbol. Subsequently, the interleaved bits are broken into subsequences of m bits and then are mapped into a complex signal in S according to a certain constellation. To achieve a better performance, iterative decoder with soft output is employed. Here, signal set S is composed of $M = 2^m$ (m=2 for QPSK) complex signals.

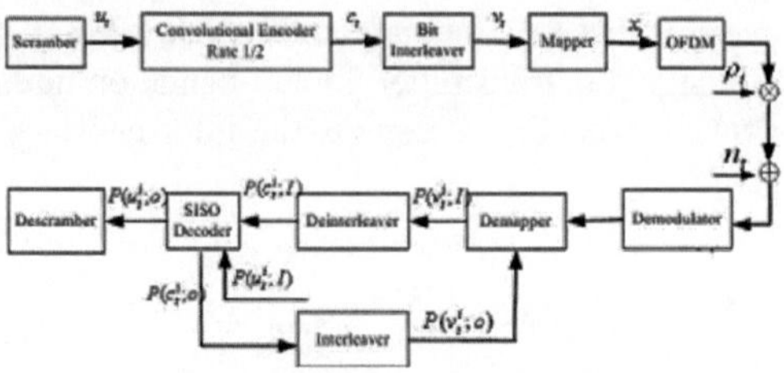

Fig. 3. Block diagram of BICM-ID FM IBOC

Denoting an interleaver output by $v_t = [v_t^1, v_t^2]$, a signal labeling map by μ, and the corresponding QPSK signal at time t by x_t,

$$x_t = \mu(v_t), x_t \in \chi \tag{1}$$

where the QPSK signal set is

$$\chi = \left\{ \sqrt{E_S} e^{j2n\pi/4}, n = 0,1,2,3 \right\} \tag{2}$$

and E_s is the energy per channel symbol. For a flat Rayleigh fading channel with coherent detection, the received discrete-time signal is

$$y_t = \rho_t x_t + n_t \tag{3}$$

where ρ_t is the Rayleigh-distributed fading amplitude with $E(\rho_t^2) = 1$ and n_t is complex white Gaussian noise with variance $\sigma_I^2 = \sigma_Q^2 = N_0 / 2$. Throughout the paper, we assume that ρ_t is known at the decoder from pilot symbols.

3.2 Iterative Decoding for BICM-ID FM IBOC

At the receiver, iterative decoding uses SISO decoder [11] and the output of the SISO decoder is fed back to the demodulator for bit metric recalculation. Suboptimum maximum log-likelihood bit metrics [2] [3] are obtained as

$$\lambda\left(v_t^i = b\right) = \log \sum_{x_t \in \chi_b^i} P\left(y_t \middle| x_t\right) \approx \max_{x_t \in \chi_b^i} \log P\left(y_t \middle| x_t\right) \tag{4}$$
$$i = 1, 2, \cdots m; b = 0,1$$

where χ_b^i is the subset of χ whose label is the binary value b at the i'th bit position. For demodulator, the maximum a posteriori probability (MAP) bit metrics are calculated as

$$\lambda\left(v_t^i = b\right) = \log P\left(v_t^i = b \middle| y_t\right) = \log \sum_{x_t \in \chi_b^i} P\left(x_t \middle| y_t\right) \tag{5}$$
$$\sim \log \sum_{x_t \in \chi_b^i} P\left(y_t \middle| x_t\right) P\left(x_t\right) \qquad i = 1, 2, \cdots m; b = 0,1$$

The priori probability $P(x_t)$ is unavailable on the first iterative of demodulation. Therefore, an equal assumption is made and (4) is used as the input of the SISO decoder, which then generates posteriori probabilities for both information and coded bits.

Here, we denote $P(q;I)$ as the priori probability for a variable q. $P(q;O)$ is the a posteriori probability. During the iterative process, the extrinsic posteriori probabilities $P(c_t^i;O)$, the output of the SISO module, are interleaved and fed back as the priori probabilities to the demodulator. Therefore, priori probabilities $P(v_t^i;I)$ for (5) can be calculated by

$$P(x_t) = P(\mu(v_t^i;I)) = \prod_{i=1}^{m} P\left(v_t^i = \tilde{v}_t^i(x_t);I\right) \tag{6}$$

where $\tilde{v}_t^i(x_t)$ is the value of the i'th bit of the label corresponding to $x_t = \mu(\tilde{v}_t)$. Using (5) and (6), the extrinsic posteriori probabilities of demodulation for the next iterative can be written as

$$\begin{aligned} P(v_t^i = b;O) &= \frac{P(v_t^i = b|y_t)}{P(v_t^i = b;I)} = \frac{\sum_{x_t \in \chi_b^i} P(y_t|x_t)P(x_t)}{P(v_t^i = b;I)} \\ &= \sum_{x_t \in \chi_b^i} \left[P(y_t|x_t)\right] \prod_{\substack{j=1 \\ j \neq 1}}^{m} P\left(v_t^j = \tilde{v}_t^j(x_t);I\right) \end{aligned} \tag{7}$$

where $i = 1, 2; \cdots m; b = 0, 1$. Therefore, we need only the priori probabilities $P(v_t^j;I)$ of the other bits on the same channel symbol v_t when recalculating the bit metrics. The receiver then uses (7) to regenerate the bit metrics and iterates demodulation and decoding. After the last iterative, the final decoded outputs are the hard decisions based on the extrinsic bit probabilities $P(u_t^i;O)$.

3.3 Modulation Scheme for BICM-ID FM IBOC

Fig.4 depicts the conventional modulation scheme for BICM-ID FM IBOC, which is used in Section 4. It is gray labeling. The shaded regions correspond to the decision regions for the first bit taking the value of 1.

It can be seen clearly that four points modulation in signal constellation is sparse and can not utilize effectively constellation space. However, duobinary modulation can make full use of the signal constellation and then improve the bandwidth efficiency, so it plays an unparalleled advantage compared with conventional QPSK modulation.

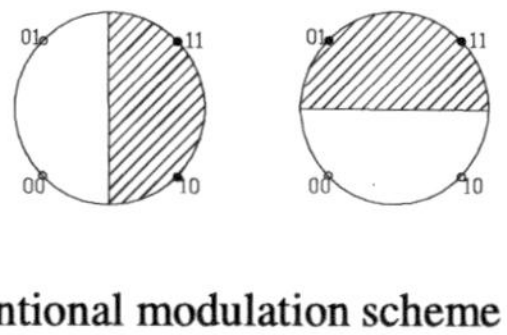

Fig. 4. Conventional modulation scheme for FM IBOC

4 Simulation Results and Analysis

In this section, the effects of BICM-ID on the performance in FM IBOC system are provided. Via Monte-Carlo simulations, we focus on the bit error rate (BER) performance over Rayleigh fading channels. Take P3 data in MP2 service mode for example, whose frame size is 4608 bits. And its encoder is built from generator polynomial (133,171,165) with the same rate-1/3 mother convolutional codes of constraint lengths 7. The iteration number is set to be 10 and Gray labeling is considered in all the simulations.

To achieve better comparison effects, our simulations consist two parts. For the one part, various modulation schemes with the same code rate R=1/2 are compared. For the other part, comparison among various modulation schemes with the same bandwidth efficiency is obtained.

4.1 Various Modulation Schemes with the Same Encode Rate R=1/2

Table 1. displays simulation parameters of various modulation schemes with the same code rate. Fig.5 describes the simulation results of various modulation schemes with the conventional code rate R=1/2 over Rayleigh fading channels.

Table 1. Simulation parameters of various modulation schemes with the same code rate

Modulation Type	Parameter				
	Generator Polynomial	Mother Code Rate	Punching Matrix	Encoder Rate	Bandwidth Efficiency
BPSK	(133,171,165)	1/3	[1 1;0 0;1 1]	1/2	1/2
QPSK	(133,171,165)	1/3	[1 1;0 0;1 1]	1/2	1
8PSK	(133,171,165)	1/3	[1 1;0 0;1 1]	1/2	3/2
16QAM	(133,171,165)	1/3	[1 1;0 0;1 1]	1/2	2
32APSK	(133,171,165)	1/3	[1 1;0 0;1 1]	1/2	5/2
64QAM	(133,171,165)	1/3	[1 1;0 0;1 1]	1/2	3

It shows that for M point modulation, the larger M value, thc worse BER performance corresponding to M point modulation under the premise of same encoder rate. And we can also observe from Table 1. that different modulation schemes possess different bandwidth efficiency. That is because, in the same signal constellation, larger M point modulation makes constellation space more crowding. Then it takes up more bandwidth and increase the density of constellation, which results in the decrease of Euclidean distance between adjacent signals. And we all know that Euclidean distance is the main factor weighing the performance of Rayleigh fading channels.

4.2 Various Modulation Schemes with the Same Bandwidth Efficiency

In order to further explore the effects of various modulations schemes for BICM-ID FM IBOC, the simulation condition with the same bandwidth efficiency are taken into consideration. Simulation parameters of various modulation schemes with the same bandwidth efficiency are revealed in Table 2. The same bandwidth efficiency is obtained by constructing encoder with different code rate. Fig.6 shows the simulation results of various modulation schemes with the same bandwidth efficiency.

Table 2. Simulation parameters of various modulation schemes with the same bandwidth efficiency

Modulation Type	Parameter				
	Generator Polynomial	Mother Code Rate	Punching Matrix	Encoder Rate	Bandwidth Efficiency
BPSK	(165)	1	[1 1]	1	1
QPSK	(133,171,165)	1/3	[1 1;0 0;1 1]	1/2	1
8PSK	(133,171,165,165)	1/4	[1 1;1 1;0 0;1 1]	1/3	1
16QAM	(133,171,165,171,165)	1/5	[1 1;1 1;0 0;1 1;1 1]	1/4	1
32APSK	(133,171,165,133, 171,165)	1/6	[1 1;1 1;1 1;0 0;1 1;1 1]	1/5	1
64QAM	(133,171,165,133,171,165,165)	1/7	[1 1;1 1;1 1;0 0;1 1;1 1;1 1]	1/6	1

From Fig. 5, it indicates when all the modulation schemes occupy the same bandwidth 32APSK modulation scheme performs the best. It can achieve 1 dB coding gains than the conventional counterpart in FM IBOC when BER is at 10^{-6}. It can be predicted from the tendency of curves, as the E_b / N_0 increases further, 32APSK modulation scheme could obtain more gains compared with original QPSK modulation.

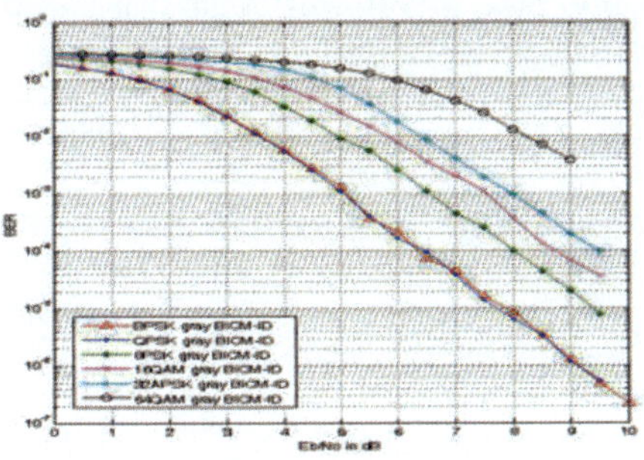

Fig. 5. Various modulation schemes with the conventional code rate

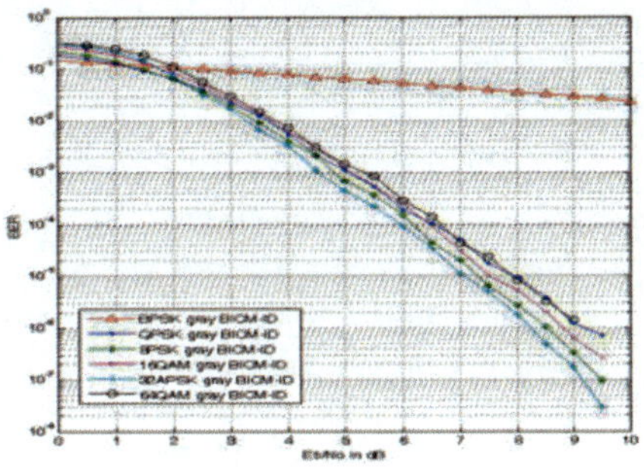

Fig. 6. Various code modulation schemes with the same bandwidth efficiency

5 Conclusions

In this paper, various modulation schemes with the same encode rate and the same bandwidth efficiency for BICM-ID FM IBOC are compared and analyzed. The advanced modulation schemes can effectively improve the performance of BICM-ID FM IBOC. When BER is at 10^{-6}, 32APSK modulation scheme can achieve 1dB coding gains over Rayleigh fading channels. As the E_b / N_0 increases further, 32APSK could obtain more gains.

Acknowledgment. The first author gratefully acknowledge the support from the fund for the key project of Chinese Ministry of Education (No. 106042) and the project sponsored by the Scientific Research Foundation for the Returned Overseas Chinese Scholars, State Education Ministry (2007[24]).

References

1. Jayant, N.S., Chen, E.Y.: Audio compression: Technology and applications. AT&T Tech. J. 74, 23–34 (1995)
2. Zehavi, E.: 8-PSK trellis codes for a rayleigh fading channel. IEEE Trans. Commun. 40, 873–883 (1992)
3. Caire, G., Taricco, G., Biglieri, E.: Bit-interleaved coded modulation. IEEE Trans. Inform. Theory 44, 927–946 (1998)
4. Li, X., Ritcey, J.A.: Bit-interleaved coded modulation with iterative decoding. IEEE Commun. Lett. 1, 169–171 (1997)
5. Li, X., Chindapol, A., Ritcey, J.A.: Bit interleaved coded modulation with iterative decoding and 8-PSK signaling. IEEE Trans. Commun. 50, 1250–1257 (2002)
6. Samahi, S.S., Goff, S., Sharif, B.S.: Comparative study for bit-interleaved coded modulation with iterative decoding. In: IEEE AICT 2009, pp. 316–318 (May 2009)
7. Valenti, M.C., Doppalapudi, R., Torrieri, D.: A genetic algorithm for designing constellations with low error floors. In: IEEE Conference on Information Sciences and Systems, pp. 1155–1160 (March 2008)
8. Tran, N.H., Nguyen, H.H., Tho, L.-N.: Performance of BICM-ID with signal space diversity. IEEE Transactions on Wireless Communications 6, 1732–1742 (2007)
9. Zou, X., He, X., Liu, W., Wang, X.: An adaptive detection scheme for iteratively decoded bit-interleaved coded modulation. In: IEEE Microwave Conference, pp. 758–762 (September 2008)
10. Feng, Y., Li, J., Dong, Y., Sha, S.: Coded modulation scheme with CPPC codes for FM IBOC broadcasting. In: IEEE Conference on Digital Object Identifier, pp. 1–4 (2009)
11. Kroeger, B.W., Peyla, P.J.: Compatibility of FM hybrid In-Band On-Channel (IBOC) system for digital audio broadcasting. IEEE Trans. Broadcasting 43(4), 421–430 (1997)
12. Scalart, P., Leclerc, M., Fortier, P., Huynh, H.T.: Performance analysis of a COFDM/FM in-band digital audio broadcasting system. IEEE Trans. Broadcasting 43(2), 191–198 (1997)
13. Cimini, L.J.: Analysis and simulation of a digital mobile channel using orthogonal frequency division multiplexing. IEEE Trans. Commun. COM-33(7), 665–675 (1985)
14. Kroeger, B.W., Brian, W.: Method and apparatus for reduction of FM interference for FM in-band on –channel digital and broadcasting system, USA, 6259893 (July 2001)
15. Benedetto, S., Divsalar, D., Montorsi, G., Pollara, F.: A soft-input soft-output APP module for iterative decoding of concatenated codes. IEEE Commun. Lett. 1, 22–24 (1997)

Research on Fault Diagnosis of CT Scanner Based on Probabilistic Neural Network

Shi Lei

College of Information Science and Engineering
Bohai University
Jinzhou 121000, China
shileiboda@163.com

Abstract. Probabilistic neural network (PNN) model is a kind of artificial neural network, which is simple in structure, easy for training and widely used. This method uses Bayes classifying and decision-making theory to constitute the mathematic model of system. With Gauss function as activating ones, it possesses the characteristics of strong nonlinear processing and anti- nterfering ability. This paper introduces a research on CT scanner fault diagnosis using PNN, and the MATLAB simulation results show that the prposed method is featured by swiftness, accuracy and easy for practical application.

Keywords: Probabilistic Neural Network (PNN), CT scanner, fault diagnosis.

1 Introduction

The CT scanner was mainly composed by radioactive source, data acquisition, mechanical drive and video graphic display etc. With the developments of computer technology, especially thanks to the slip ring technology, the integrated degree of CT scanner was higher, the arithmetic speed was faster, the physical construction was more sophisticated, the images became clearer and the competence of diagnosis and treatment enhanced unceasingly. Therefore, it was more significant to carry out a research into fault diagnosis of CT scanner for finding the trouble location and fault cause, enhancing the system movement reliability, and establishing a scientific repair system. In recent years, researches on fault diagnosis could be divided into three types: the methods based on the analysis model, such as parameters estimation, the methods based on the qualitative model and search strategy, such as the Signed Directed Graph, and the methods based on the historical data, such as the expert system and neural network. In weighing the different methods of fault diagnosis and considering many kinds of indexes on speediness, veracity and robustness, this paper presented a research on CT scanner fault diagnosis based on probabilistic neural network. The results of the simulation showed the effectiveness of this method.

2 Probabilistic Neural Network

2.1 The Network Architecture of PNN

Probabilistic neural network was a single hidden layer feed- forward network for supervised training, which was usually applied to pattern classification. Compared with

W. Zhang (Ed.): Advanced Technology in Teaching, AISC 163, pp. 607–612.
springerlink.com

other neural networks, PNN didn't need to learn and set the initial weight values, which were established by the input vectors and target values. What's more, PNN was fast in convergence, and high in accuracy, which was suited to fault diagnosis and signal classifier. We could see the architecture of PNN in Figure.1.

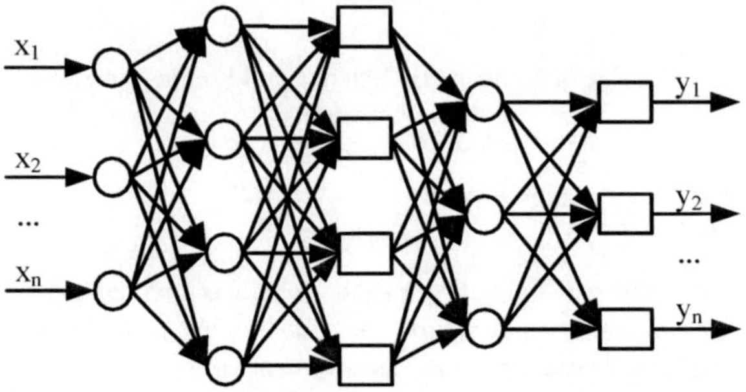

Fig. 1. Architecture of PNN

2.2 The Advantages of PNN

PNN was one kind of artificial neural network useful for pattern classification, the essence of which was the parallel algorithm based on Bayes' minimum risk rule. In comparison with conventional BP network, PNN had the following advantages:

a) The process are simples and the convergence speed is fast.

The convergence speed of BP network was slow and it's easy to be trapped into local minima value, whereas, the training of PNN was "one-step", whose training time was only a little larger than the data read time,so there was no local minima value; Furthermore, there was no rule for the choice of BP model's hidden layers and hidden nodes which could be only getted by numerical analysis method through iteration, whereas, PNN with few parameters needn't to determine the hidden note number and the number of neurons in the hidden layers.

b) PNN always quickly converged at Bayes' optimum solutions with high stability.

The conjunction weights of BP network was sensity to initial training values and varies with initial training values and then led to different and uncertain results,whereas, PNN could classify and forecast the objects on basis of Bayes' minimum risk rule, which made full use of the mutual information. That is, if there were sufficient training samples, PNN always quickly converged at Bayes' optimum solutions with high stability.

c) The training samples have strong supplemental capabilities, and tolerate particular error.

BP network needed to completely train over again after modifying the training samples, and the whole conjunction weights of BP network needed to be reassigned which is equivalent to rebuilding the whole network, whereas, PNN only needed to increase or decrease the corresponding model layers, and the new conjunction weights from input layer to model layers only needed to be assigned directly after the normalization of new samples.

d) PNN is an useful anti-jamming network with good robustness.

Fault diagnosis based on PNN had super properties and good characteristics of real-time, which could reduce the rate of positive negative and false negative. It's proved that the diagnotic rate of PNN was better than BP network with large noise.

2.3 The Classification Process of PNN

1. Hormalizd the typical samples and testing samples with the formula: $x / \|x\|$.

2. After normalization, put the typical sample x to sample layers directly.

3. After normalization, put the testing sample X to sample layers,and calculate the distance between X and x:(X-x) and add the dot product: $z_i = \sum (X-x)^T (X-x)$

4. By using kernel function, obtained high quality output: $y_k(x) = \frac{1}{N_i}\sum_{j=1}^{N_i} \exp\left[-\frac{z_i}{2\sigma^2}\right]$.

5. According to Bayes' minimum risk rule and using competitive function, it came to a conclusion that the maximum output in step 4 was the classification of X.

3 Research on CT Scanner Fault Diagnosis Based on Probabilistic Neural Network

3.1 Problem Description

For CT Scanner, the faults often occured in rack systems, high-voltage systems, artifacts in the images and software systems during operation. So owing to the complexity of the CT Scanner, it was difficult for traditional methods to find the causes and the effective solving method. This paper intro- duced a research on CT scanner fault diagnosis using PNN, and the MATLAB simulation results show that the prposed method was featured by swiftness, accuracy and easy for practical application.

During diagnosis, eight common characteristic parameters in opration were obtained firstly, such as: Location Picture Symmetry (LPS), Scanner Frame Angle (SFA), Anode Rotation Speed (ARS) , Tube Current (TC) , Data Calibration Bias(DCB), High-voltage Waveform (HVW), Imaging Deviation (ID) and Operating Temperature (OT). And then, PNN method was applied to fault diagnosis for CT Scanner and the diagnostic model was showed in Figure.2.

3.2 Establishing Probabilistic Neural Network (PNN)

During the training process, the number of the input neurons was determined adaptively based on the characteristic parameters of CT Scanner, therefore we chose 9 input neurons of PNN adding normal state(NS) to 8 common characteristic parameters of CT Scanner mentioned above, such as (LPS,SFA,ARS,TC,DCB,HVW, ID,OT and NS), which was showed in Figure.2.

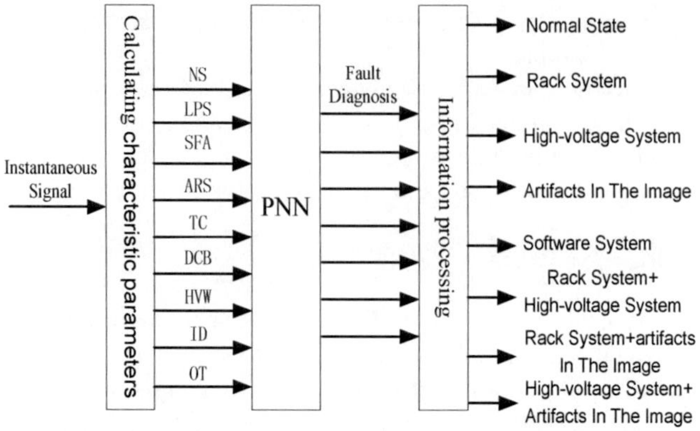

Fig. 2. Fault diagnosis model based on PNN

The ideal outputs of network had two working states: normal and fault, the faults were further divided into two levels: single fault and compound fault. Therefore, the single fault could be represented by(0,0,0,0), (1,0,0,0), (0,1,0,0), (0,0,1,0) and (0,0,0,1), corresponding to 5 fault states of CT scanner:A(normal state),B(rack system fault),C(high-voltage system fault),D(artifacts in the image) and E(software system fault), meanwhile, the compound fault could be represented by (1,1,0,0), (1,0,1,0) and (0,1,1,0), corresponding to 3 fault states of CT scanner:F(rack system+high-voltage system fault), G(rack system fault+ artifacts in the image) and H(high-voltage fault+ artifacts in the image). Consequently, we chose 8 output neurons of PNN corresponding to 8 outputs fault diagnosis system of CT scanner. At the same time, the outputs of PNN model could be translated to vector by ind2vec( ) function in MATLAB 7.0 .

Next, we chose Gaussian function radbas( ) as activation function, chose Euclidean distance function dist( ) and dotprod ( ) as weighting functions, chose netprod( ) as input function, chose compet ( ) as output function in hidden layer. Then a specialized Probabilistic Neural Network was created for fault diagnosis by using the instruction 'newpnn', the constant SPREAD was setted as 0.1, and MATLAB code was: net=newpnn(P,T,SPREAD). Where P was input sample vector, at some point now, if the inputs were very different, they should be normalized first. T was aim vector (i.e.desired output), net was the new Probabilistic Neural Network, SPREAD was the density of distribution of Radial Basis Function, which could regulate the sensitivity of Gaussian function radbas( ),i.e. the input response duration for output.

In the last place, we chose the first 8 samples of 10 as training data and the last 2 samples as testing data and during optimum searching, the standard deviation σ of Gaussian function is confirmed as 0.1. Table 1 listed the predictions and diagnosis of CT fault diagnosis system.

Table 1. The original data and forecast data of CT fault diagnosis system

No.	characteristic parameters of CT scanner								Actual Condition	Prediction of PNN	Diagnosis Result
	LPS	*SFA*	*ARS*	*TC*	*DCB*	*HVW*	*ID*	*OT*			
1	0.998	0.999	1.002	0.996	0.009	1.002	0.000	1.001	normal state	A	True
2	1.002	1.001	0.998	1.001	0.004	0.999	0.009	1.000	normal state	A	True
3	0.991	0.992	0.949	1.007	1.020	0.111	0.010	0.899	artifacts in the image	D	True
4	0.912	0.824	0.021	0.080	0.019	0.892	0.001	0.901	high-voltage system	C	True
5	0.102	0.091	0.936	0.887	0.008	0.877	0.108	0.975	rack system	B	True
6	0.919	0.884	0.953	0.999	0.015	1.013	0.890	0.941	software system	E	True
7	0.220	0.188	0.919	0.875	0.719	0.585	0.090	0.879	rack system + artifacts in the image	G	True
8	0.415	0.284	0.921	0.179	0.080	0.905	0.112	1.003	rack system+ high-voltage system	F	True
9*	0.111	0.102	0.833	0.890	0.081	0.798	0.095	0.947	rack system	B	True
10*	1.015	0.932	0.220	0.183	0.123	0.905	0.104	0.976	high-voltage system	C	True

* indicates testing sample; LPS: Location Picture Symmetry; SFA: Scanner Frame Angle; ARS: Anode Rotation Speed; TC: Tube Current; DCB: Data Calibration Bias; HVW: High-voltage Waveform; ID: Imaging Deviation; OT: Operating Temperature.

3.3 Results and Discussion

From the data summarized in Table 1, it was found that the identification probability of diagnosis reached 100% for both training samples and testing samples, which indicated that the research on CT scanner fault diagnosis based on PPN performed well for both model fitting and generalization performance. As a result, it could be applied to the active diagnosis system of CT scanner.

4 Conclusion

In weighing the different methods of fault diagnosis and considering many kinds of indexes on speediness, veracity and robustness, this paper presented a research on CT scanner fault diagnosis based on probabilistic neural network, and combined the neural network well with fault diagnosis of medical equipment. What's more, it broke through the traditional model of artificial diagnosis and maintenance, made full use of advanced computer technology in aiding medical practice, so this method was the successful practice of applying neural network in the medicine and hygiene fields. In effect, the simulation results of training samples and testing samples showed that the average precision of the method could attain 100%,which indicated that the study confirmed the feasibility and concluded that this new approach would easily be extended and expanded. And as knowledge of diagnosis increases, PNN was highly demanded to be improved under the new conditions in order to further improve diagnosis accurate rate of CT scanner.

References

1. Li, C.-F., Yang, M.-L., Xu, L., Yang, M.-Z.: A Comparative Study Of Probabilistic Neural Network And Bp Networks For Remote Sensing Image Classification. Remote Sensing For Land & Resources (04), 14–16 (2004)
2. Ji, D.-C., Song, B.-F., Yi, H.-H.: Equipment Fault Diagnosis based on Probabilistic Neural Networks and Simulation Analysis. Fire Control and Command Control (01), 84–87 (2009)
3. Zhou, Y.-J.: Rolling Bearing Condition Monitoring and Fault Diagnosis Based on Probabilistic Neural Network. Metrology & Measurement Technique (02), 33–36 (2009)
4. Sun, Z.-L., Mao, X.-C., Zhang, X.-F., Tian, W.-F.: Parameter Correction Based on Particle Filter and Probabilistic Neural Network for Pedestrian Positioning. Journal of Shanghai Jiaotong University (02), 34–38 (2009)
5. Liu, D., Cao, J., Hua, J.: Fault Diagnosis Based on Probabilistic Neural Network(PNN) for Glue Dosage of Particleboard. Process Automation Instrumentation (09), 56–58 (2009)
6. Liu, J., Xu, D., Xin, Y.: Study On Different Types Of Barium Sulfate Attenuation To X-Ray By Ct Scanner. Journal of Shandong Normal University(Natural Science) (02), 72–74 (2007)
7. Zhang, Y., Peng, X.-F.: Micro-Ct Scanning Analysis For Inner Structure of Porous Media. Journal of Engineering Thermophysics (05), 132–134 (2005)
8. Thomas, C., Krauss, B., Ketelsen, D., Tsiflikas, I., Reimann, A., Werner, M., Schilling, D., Hennenlotter, J., Claussen, C.D., Schlemmer, H.P., Heuschmid, M.: Differentiation of Urinary Calculi With Dual Energy CT: Effect of Spectral Shaping by High Energy Tin Filtration. Invest. Radiol. (30), 45–49 (2010)
9. Dybowski, R., Gant, V.: Clinical applications of artificial nerual networks, Cambridge University Press (2001)
10. Tsipouras, M.G., Exarchos, T.P., Fotiadis, D.I.: Automated creation of transparent fuzzy models based on decision trees–application to diabetes diagnosis. In: Conf. Proc. IEEE Eng. Med. Biol. Soc., pp. 3799–3802 (2008)
11. Shimansky, Y.P.: Biologically plausible learning in neural networks: a lesson from bacterial chemotaxis. Biol. Cybern. 101(5-6), 379–385 (2009)
12. Frize, M., Ennett, C.M., Stevenson, M., Trigg, H.C.: Clinical decision support systems for intensive care units: using artificial neural networks. Med. Eng. Phys. 23(3), 217–225 (2001)

An Improved Non Data Aided Carrier Frequency Offset Estimation Algorithm for OFDM Systems

Hu Dengpeng, Shi Feng, and Zhang Eryang

College of Electronic Science and Engineering
National University of Defense Technology
Changsha, China
hudengpeng@163.com

Abstract. In his paper, an improved non data aided (NDA) carrier frequency offset (CFO) estimation algorithm for Orthogonal Frequency Division Multiplexing (OFDM) systems was proposed. The proposed algorithm was based on two existed NDA algorithms which were analyzed and their channel models' assumptions were introduced at the beginning of this paper. The proposed algorithm composed the two algorithms' cost functions through a adjust factor, which could utilize data information more sufficiently than the two existed algorithms when the two channel assumptions were all satisfied at the same time. When the two channel assumptions could not be satisfied at the same time, the proposed cost function would degenerate to one of the two existed cost function automatically. At last, the simulation results confirmed that the proposed algorithm had better performance in both static channel and time varying multi-path delay channel than the two existed algorithms.

Keywords: Orthogonal Frequency Division Multiplexing, Carrier Frequency Offset, Non Data Aided, Doppler Frequency Shift.

1 Introduction

Orthogonal frequency division multiplexing (OFDM) is a promising technology for broadband transmission due to its high spectrum efficiency and its robustness to the effects of multi-path fading channels and impulse noises [1]-[3]. However, one of the major drawbacks of OFDM systems is their high sensitivity to carrier frequency offset (CFO), caused by the mismatch of the oscillators in the transmitter and the receiver. The CFO destroys the sub-carriers orthogonality and creates inter-carrier interference (ICI) [4], [5]. So it must accurately be estimated and compensated prior to the discrete Fourier transform (DFT) process.

The CFO estimation algorithm can be divided into two categories: Data Aided (DA) methods [6], [7], and Non Data Aided (NDA) methods [8]-[12]. Recently, the NDA methods become more attractive due to the high bandwidth efficiency. One NDA method has been reported in [8], which takes advantage of the modulus constellations. It could estimate the CFO in QAM modulation OFDM systems and PSK modulation OFDM systems. But its performance is tremendous affected by noise. Another NDA method is proposed in [9] based on the kurtosis metric, which measures the gaussianity

W. Zhang (Ed.): Advanced Technology in Teaching, AISC 163, pp. 613–620.
springerlink.com

of a random sequence, is used to construct a kurtosis-type cost function for fine CFO estimation. The main drawback of this technique is the modest performance in frequency selective channels. Minimizing the power difference between adjacent sub-carriers and the same sub-carrier in two consecutive OFDM symbols is used in [10] and [11] respectively to develop efficient CFO estimators. These estimators have the advantage of being given in closed form, which reduces their computational complexity. But their performances highly depend on the assumptions to the channel, and when the two channel assumption in [10] and [11] are satisfied, neither of them could use the received data sufficiently.

In this paper, we proposed a modified algorithm that composed the two cost function in [10] and [11], which could adapt to the channel by the adjust factor when channel changes and it could utilize receive data information more sufficient than the two existed algorithms. When the two channel assumptions could not be satisfied at the same time, the modified cost function would degenerate to one of the two existed cost functions automatically. The remainder of this paper is organized as follows. The signal model is presented in Section 2. Section 3 describes the proposed CFO estimator. Numerical results and conclusions are given in Sections 4 and 5, respectively.

2 Signal Model

In typical OFDM systems, a N point fast Fourier transform (FFT) is used to divide the channel spectrum into a set of N parallel sub-carriers. In order to maintain inter-carrier orthogonality and to avoid inter-symbol interference (ISI) in the presence of frequency selectivity of the channel, the last N_g samples of the IFFT output are copied and prepended at the beginning of each OFDM symbol to form the guard interval. Then the OFDM symbol transmitted over radio channel contains $N_S = N + N_g$ samples. The n th sample of l th OFDM symbol is given by

$$x_l(n) = \frac{1}{\sqrt{N}} \sum_{k=0}^{N-1} d_{l,k} e^{j2\pi nk/N}, \quad (l-1)N_S - N_g \le n \le (l-1)N_S + N - 1, \tag{1}$$

where $d_{l,k}$ denotes modulated data on the n th sub-carrier of l th OFDM symbol which is assumed to be independent of each other for different sub-carriers.

At the receiver, assuming that the sampling clock and OFDM symbol timing is ideal synchronized, the l th received symbol, which is compensated by the CFO candidate $\hat{\varepsilon}_c$, can be written as

$$y_l(n) = \sum_{i=0}^{J-1} x_l(n+i) h_l(i) \times e^{j2\pi(\varepsilon_c - \hat{\varepsilon}_c)n/N} + \omega_l(n), \tag{2}$$

where $\mathbf{h}_l = (h_l(0), h_l(1), \cdots h_l(J-1))$ denotes the channel time response during the l th OFDM block, $\varepsilon_c \in (-0.5, 0.5)$ represents the normalized CFO, and $\omega_l(n) \sim CN(0,\sigma_n^2)$ denotes the noise. Demodulated by discrete Fourier transform (DFT), the k th sub-carrier of the l th OFDM symbol can be written as

$$S_l(k) = H_l(k)d_{l,k}I_k^k + \sum_{m=0,m\neq l}^{N-1} d_{l,n}H_l(n)I_k^m + \Omega_l(k), \tag{3}$$

where $\Omega_l(k)$ denotes the DFT of $\omega_l(n)$, $\mathbf{H}_l = diag\{H_l(0), H_l(1), \cdots H_l(N-1)\} = diag\{\mathbf{F}_J\mathbf{h}_l\}$ denotes the channel frequency response during the l th OFDM block, $F_J[l,k] = \exp[-j2\pi lk/N]$ represents the first J column of the FFT matrix, where $0 \le l \le N-1$ and $0 \le k \le J-1$. The expression of I_k^m is

$$I_k^m = \frac{\sin\left(\pi\left(m-k+\varepsilon_c-\hat{\varepsilon}_c\right)\right)}{\sqrt{N}\sin\left(\frac{\pi}{N}\left(m-k+\varepsilon_c-\hat{\varepsilon}_c\right)\right)} \times \exp\left\{j\pi\left(1-\frac{1}{N}\right)\left(m-k+\varepsilon_c-\hat{\varepsilon}_c\right)\right\}. \tag{4}$$

The second sum of right hand of (3) is the ICI, which is caused by CFO, which could be thought as Gaussian distributed by the theory of Central limit theorem when large value of N. Then (3) could be farther expressed as

$$S_l(k) = H_l(k)d_{l,k}I_k^k + \Omega_l'(k), \tag{5}$$

where $\Omega_l'(k)$ is the sum of $\Omega_l(k)$ and ICI. Form (5) we can see that the CFO not only cause the ICI, but also change the amplitude and phase of current sub-carrier.

3 Proposed CFO Estimation Algorithm

Form (3) we can see that the DFT outputs would be without ICI when CFO has been completely compensated in the pre DFT stage. So in the noise free case, due to(5), the DFT outputs would be

$$S_l(k)\big|_{\hat{\varepsilon}_c=\varepsilon_c} = H_l(k)d_{l,k}. \tag{6}$$

The squared amplitude of it is

$$\left|S_l(k)\big|_{\hat{\varepsilon}_c=\varepsilon_c}\right|^2 = \left|H_l(k)d_{l,k}\right|^2. \tag{7}$$

Assuming the OFDM systems with constant modulus (CM) constellations and $\left|d_{l,k}\right|^2 = 1$, we have

$$\left|S_l(k)\big|_{\hat{\varepsilon}_c=\varepsilon_c}\right|^2 = \left|H_l(k)\right|^2. \tag{8}$$

In [10], the authors assumed that the channel frequency response changes slowly in the frequency domain; the channel frequency response on two consecutive sub-carriers is approximately the same, that is $|H_l(k)| \approx |H_l(k-1)|$. Consequently, we would have

$$\left|S_l(k)\big|_{\hat{\varepsilon}_c=\varepsilon_c}\right|^2 - \left|S_l(k-1)\big|_{\hat{\varepsilon}_c=\varepsilon_c}\right|^2 = |H_l(k)|^2 - |H_l(k-1)|^2 \approx 0. \tag{9}$$

Based on this, the authors proposed a cost function which minimizes the difference of the signal power for each pair of consecutive sub-carriers as

$$J_1(\varepsilon) = \sum_{k=0}^{N-1}\left(\left|S_l(k)\big|_{\hat{\varepsilon}_c=\varepsilon}\right|^2 - \left|S_l((k-1)_N)\big|_{\hat{\varepsilon}_c=\varepsilon}\right|^2\right)^2, \tag{10}$$

Where $(x)_N$ denotes x modulo N .Then the CFO could be estimated by

$$\hat{\varepsilon}_c = \arg\left[\min_{-0.5\le\varepsilon<0.5} J_1(\varepsilon)\right], \tag{11}$$

Where $\arg[\bullet]$ denotes getting ε off $J_1(\varepsilon)$.

In [11], the authors assumed that the channel frequency response changes slowly in the same sub-carrier of two adjacent OFDM symbols; the channel frequency response on two consecutive OFDM symbol in one sub-carrier is approximately the same, that is $|H_l(k)| \approx |H_{l-1}(k)|$. And then we could have

$$\left|S_l(k)\big|_{\hat{\varepsilon}_c=\varepsilon_c}\right|^2 - \left|S_{l-1}(k)\big|_{\hat{\varepsilon}_c=\varepsilon_c}\right|^2 = |H_l(k)|^2 - |H_{l-1}(k)|^2 \approx 0. \tag{12}$$

Based on this, the authors also proposed a cost function which minimizes the difference of the signal power for each pair of consecutive OFDM symbol in one sub-carrier as

$$J_2(\varepsilon) = \sum_{k=0}^{N-1}\left(\left|S_l(k)\big|_{\hat{\varepsilon}_c=\varepsilon}\right|^2 - \left|S_{l-1}(k)\big|_{\hat{\varepsilon}_c=\varepsilon}\right|^2\right)^2. \tag{13}$$

We can see that the principle of these two algorithms is the same that they estimate CFO the different assumption to channel frequency response. The two algorithms could work efficiently when the channel satisfy corresponding assumption, but when the channel frequency response satisfy the two assumptions all, neither of them could not use receive data information sufficiently. So we propose a modified cost function that it's a synthesis of the two cost functions, and its expression is

$$\begin{aligned} J_3(\varepsilon) = \sum_{k=0}^{N-1}\Bigg[& \mu\left(\left|S_l(k)\big|_{\hat{\varepsilon}_c=\varepsilon}\right|^2 - \left|S_l((k-1)_N)\big|_{\hat{\varepsilon}_c=\varepsilon}\right|^2\right)^2 + \\ & (1-\mu)\left(\left|S_l(k)\big|_{\hat{\varepsilon}_c=\varepsilon}\right|^2 - \left|S_{l-1}(k)\big|_{\hat{\varepsilon}_c=\varepsilon}\right|^2\right)^2\Bigg], \end{aligned} \tag{14}$$

where μ is the adjust factor, its expression is

$$\mu=\begin{cases}0.5 & ,\mu_1+\mu_2=0\\ \mu_1/(\mu_1+\mu_2) & ,\mu_1+\mu_2\neq 0\end{cases},\tag{15}$$

where $\mu_1=\frac{1}{N}\sum_{k=0}^{N-1}\left(\left|S_{l-1}(k)\right|_{\hat{\varepsilon}_c=\hat{\varepsilon}_c^{'}}\right|-\left|S_{l-2}(k)\right|_{\hat{\varepsilon}_c=\hat{\varepsilon}_c^{'}}\left|\right)^2$,

$\mu_2-\frac{1}{N-1}\sum_{k=0}^{N-1}\left(\left|S_{l-1}(k)\right|_{\varepsilon_c=\varepsilon_c^{'}}\right|-\left|S_{l-1}((k-1)_N)\right|_{\hat{\varepsilon}_c=\hat{\varepsilon}_c^{'}}\left|\right)^2$, and $\hat{\varepsilon}_c^{'}$ is the CFO candidate of last time.

Form (14) we can see that the proposed cost function could adapt to the channel by the adjust factor when channel changes and it could utilize receive data information more sufficient than the two existed algorithms. When the two channel assumptions could not be satisfied at the same time, the modified cost function would degenerate to one of the two existed cost function automatically.

From [10] and [11] we know that the equation (10) and (13) could be approximated to a sinusoid. So (14) could be approximated to a sinusoid because it's the linear summation of (10)and(13). Consequently, (14) can be expressed as

$$J_3(\varepsilon)\approx A\cos\left[2\pi(\varepsilon_c-\varepsilon)\right]+B,\tag{16}$$

where A and B are constants with real values and independent of ε_c and ε .Since the cost function described in (14) is sinusoidal, the minimization process, which is usually performed using methods such as the line search or the gradient descent, can be replaced by the curve fitting method [10], [11]. The curve-fitting method leads to a closed-form estimation of ε_c by evaluating (16) at three special trial points, namely, $\varepsilon=1/4,0,-1/4$. Then, ε_c can be obtained as

$$\hat{\varepsilon}_c=\begin{cases}\tan^{-1}(b/a)/2\pi, & a\geq 0\\ \tan^{-1}(b/a)/2\pi+0.5, & a<0 \text{ and } b\geq 0,\\ \tan^{-1}(b/a)/2\pi-0.5, & a<0 \text{ and } b<0\end{cases}\tag{17}$$

where $a=0.5\times[J_3(\varepsilon=1/4)+J_3(\varepsilon=-1/4)]-J_3(\varepsilon=0)$, $b=0.5\times[J_3(\varepsilon=1/4)-J_3(\varepsilon=-1/4)]$, $\tan^{-1}(x)$ denotes the arctangent function。

So we can conclude the steps of modified algorithm as below:

Step 1 : Initialize the adjust factor μ to 0.5;

Step 2 : Calculate $J_3(\varepsilon=1/4)$, $J_3(\varepsilon=-1/4)$ and $J_3(\varepsilon=0)$ according to(14);

Step 3 : Gets $\hat{\varepsilon}_c$ according to(17), which is the estimate of CFO in current loop;

Step 4 : Calculate μ according to(15), then go to step 2 for next estimation.

4 Numerical Results

The system performance is assessed using Monte Carlo simulation over frequency selective multi-path fading channels. In each SNR, there are 5000 times simulation runs, and 10 OFDM symbols are used in each run. The performance is evaluated by means of normalized MSE, where the MSE of the CFO estimates is normalized with respect to the sub-carriers' frequency spacing. The OFDM system considered in this paper have $N = 64$ sub-carriers modulated using quaternary phase shift keying (QPSK), and the CP length is $N_g = 16$.

For comparison purposes, the channel models used in this paper are the ones reported in [10] and [11], which consist of three fading channels. Channel 1 has five paths with delays of [0, 1, 2, 3, 4] samples, average gains of [0.35, 0.25, 0.18, 0.13, 0.09], and mean square delay spread [10] $\sigma^2(\tau) = 1.74$. Channel 2 also has five paths with delays of [0, 1, 2, 6, 11] samples, average gains of [0.34, 0.28, 0.23, 0.11, 0.04], and $\sigma^2(\tau) = 6.37$. Channel 3 has four paths with delays of [0, 4, 8, 12] samples, gains of [0.25, 0.25, 0.25, 0.25], and $\sigma^2(\tau) = 20$. The path gains are generated as complex independent stochastic variables and the fading with Jakes' Doppler spectrum. In time varying multi-path channels, the maximum Doppler shifts is 0.1 of sub-carriers frequency spacing. The MSE of the proposed algorithm is compared with the MSE of the estimators in [10] and [11], because the proposed estimator is the two estimators' modification.

Fig.1, fig.2, and fig.3 are the performances simulation results of the three algorithms in the three channels respectively. The real line is the results in static channels, and the broken line is the results in time varying multi-path channels. Form the three figure we can clearly see that the proposed algorithm have better performance than the two existed algorithms both in static channel and time varying multi-path channels.

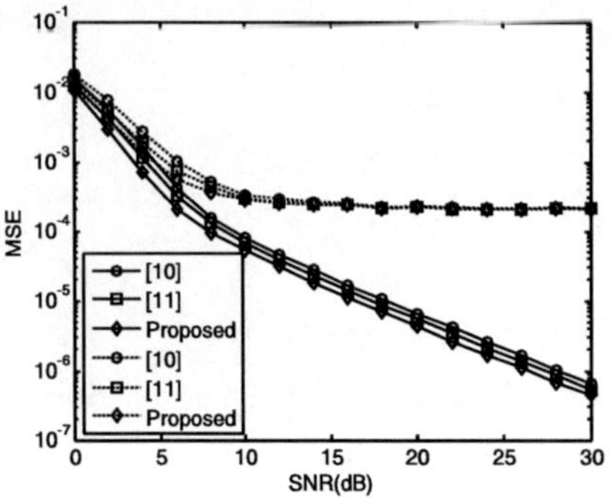

Fig. 1. Channel 1

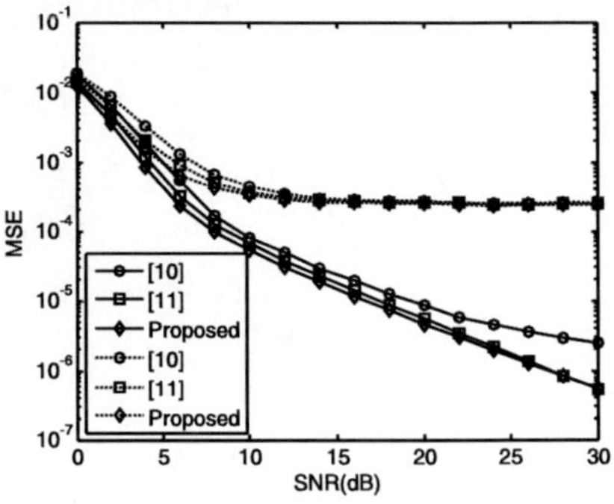

Fig. 2. Channel 2

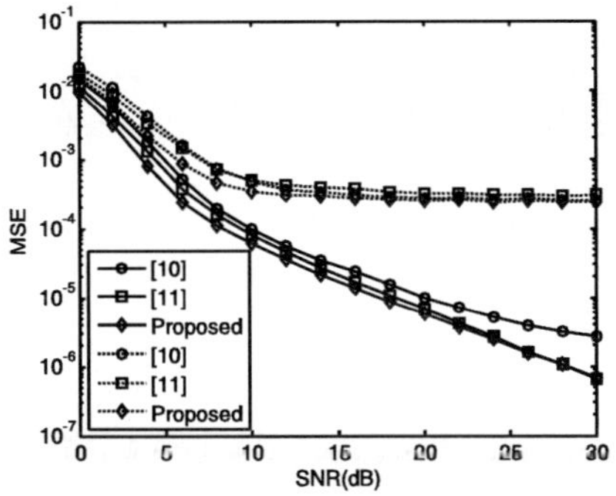

Fig. 3. Channel 3

5 Conclusion

In this paper, we have presented a modified non data aided carrier frequency estimation algorithm for OFDM systems with CM data symbols. The performance of the proposed estimator has been evaluated over various channel models, including static and time-varying frequency selective channels. Simulation results have confirmed that the proposed estimator outperforms the considered estimators over most of the considered range of SNRs. Additionally, the proposed estimator can be considered as a low complexity algorithm too because it requires only three trial values to find the CFO estimate.

References

1. Bingham, J.A.C.: Multicarrier modulation for data transmission: An idea whose time has come. IEEE Comm. Mag. 28(5), 5–14 (1990)
2. Chuang, J., Sollenberger, N.: Beyond 3G: Wideband wireless data access based on OFDM and dynamic packet assignment. IEEE Comm. Mag. 38(7), 78–87 (2000)
3. D2.10 Final Report on Identified RI Key Technologies, System Concept, and Their Assessment, IST-2003 507581 WINNER (December 2005)
4. Pollett, P.: Synchronization with DMT modulation. IEEE Comm. Mag. 37(4), 80–86 (1999)
5. Al-Dweik, A., Hamila, R., et al.: Blind estimation of large carrier frequency offsets in wireless OFDM systems. IEEE Trans. on Vehi. Tech. 56(2), 965–968 (2007)
6. Lin, H., Yamashita, K.: Subcarrier Allocation Based Compensation for Carrier Frequency Offset and I/Q Imbalances in OFDM Systems. IEEE Trans. on Wire. Comm. 8(1), 18–23 (2009)
7. Awoseyila, A.B., Kasparis, C., et al.: Robust Time-domain Timing and Frequency Synchronization for OFDM Systems. IEEE Trans. on Cons. Elec. 55(2), 391–399 (2009)
8. Al-Dweik, A., Sharif, B., et al.: Robust frequency offset estimator for OFDM with general constellation. IET Elec. Letter 44(16), 980–981 (2008)
9. Yao, Y., Giannakis, G.B.: Blind Carrier Frequency Offset Estimation in SISO, MIMO, and Multiuser OFDM Systems. IEEE Trans. on Comm. 53(1), 173–183 (2005)

10. Zeng, X.N., Ghrayeb, A.: A Blind Carrier Frequency Offset Estimation Scheme for OFDM Systems with Constant Modulus Signaling. IEEE Trans. on Comm. 56(7), 1032–1107 (2008)
11. Al-Dweik, A., Hazmi, A., et al.: Carrier Frequency Offset Estimation for OFDM Systems over Mobile Radio Channels. IEEE Trans. on Vehi. Tech. 59(2), 974–979 (2010)
12. Wu, L., Zhang, X.-D., et al.: A Closed-Form Blind CFO Estimator Based on Frequency Analysis for OFDM Systems. IEEE Trans. on Comm. 57(6), 1634–1637 (2009)

Can Strategic Trade Policies Improve Comparative Advantages of Exports? A Heterogeneous Panel Analysis of Chinese Agro-Based Manufactures

Yu Hong and Hong-wei Su

College of Humanities and Sciences
Northeast Normal University
Changchun, China
acjlhong@yahoo.com.cn, suhongwei127@sina.com

Abstract. This study tests for the effects of export oriented strategic trade policies on the comparative adv0061ntages of Chinese agro-based manufactured products, which are consisted of 39 SITC (Rev.3) 3-digit products. "Net exports ratio " is employed to capture the strength of strategic trade policies, while "revealed symmetric comparative advantage" is used to measure the comparative advantages. Using yearly data from 1992 to 2008, a heterogeneous panel cointegration test indicates a cointegrating relation across the NX and RSCA panels. No short-run Granger causal relationship of either direction can be identified. In the long-run, however, there exists a bi-directional causality. By examining the coefficients of the first-differenced lagged terms in panel vector error correction models, this study argues that the net exports of Chinese agro-based manufactures are virtually based on the comparative advantages, and the strategic trade policies which are featured by export promotion may contrarily hurt the comparative advantages in the long-run.

Keywords: strategic trade policy, comparative advantage, agro-based manufactures, panel Granger causality test.

1 Introduction

Three features can be identified for the Chinese trade since the 1990s: (1) the commodity exports have made considerable development which facilitate the miracle growth of Chinese economy, (2) the excessive trade surplus has been widely scrutinized and has led up to a rise of protectionism against Chinese goods, and (3) China has adopted strategic trade policies to support exports of a series of manufactured products, with an attempt to improve the comparative advantages and competitiveness of the relating industries [1]. These features have shaped the export-oriented economic development pattern of China and may have reinforced her dependence upon the external demands.

A strand of literatures [2]-[3] argue that for a developing country such as China, exports based on comparative advantages usually means a low-productivity

W. Zhang (Ed.): Advanced Technology in Teaching, AISC 163, pp. 621–629.
springerlink.com

specialization trap because of path-dependence [4]. In an imperfectly competitive market, trade and industrial policies of a country have crucial influences on the outcome of the oligopolistic competition between domestic and foreign firms. A government can impose export subsidies to raise profits accrued to the domestic firms and therefore to improve the social welfare [5]-[6]. Given the domestic market unchanged, an increase in exports directly means a larger economy scale for domestic firms. For increasing return industries, this expanded economy scale can be crucial to reduce costs and to obtain dynamic comparative advantages[7]-[8].

Strategic trade policies, however, usually have impacts on trade patterns by active government interference and intentional misalignment of resources, casting doubts on the effectiveness of the policies. Moreover, another question arises when exports of a country distort from the current factor endowments: can dynamic comparative advantages of a certain industry be really obtained by facilitating exports? The question can be crucial for trade in agro-based products because they have more national strategic and economic implications other than mere profit-seeking transactions. As a result, the product category is an ideal object to empirically test for the effectiveness of strategic trade policies.

2 Data and Measurements

2.1 Classification and Data

What we test for are the trade patterns of Chinese agro-based manufactures, which need to be firstly identified.

Lall (2000) proposes a classification for SITC Rev.2 products according to technological structure of exports [9]. He assigns 230 three-digit level products into six categories, among which the category of "resource based manufactures" (RB) is subdivided into "agro-based manufactures" (RB1) and "other resource based manufactures" (RB2). We follow Hong (2009) [1] to classify the agro-based manufactures using the three-digit level data of Standard International Trade Classification Revision 3. Table 1 reports the SITC Rev.3 codes for the 39 products identified.

Table 1. SITC Rev.3 Code of Agro-based Manufactures

012, 016, 017, 023, 024, 035, 037, 046, 047, 048, 056, 058, 059,
061, 062, 073, 098, 111, 112, 122, 232, 247, 248, 251, 261, 264,
265, 269, 411, 421, 422, 431, 621, 625, 629, 633, 634, 635, 641

We compile the yearly trade amount data for the period of 1992-2008 from UN Comtrade Database. Trade data of these products for each country in the database are separately summed up to obtain the world exports and imports amount. Values are in current US Dollar because our analysis is focused on the trade pattern measurements which are indexes instead of absolute values.

2.2 Measurements for Trade Patterns

The relative ability of net exports is measured by

$$NX_{ik} = (X_{ik} - M_{ik})/(X_{ik} + M_{ik}) \tag{1}$$

where NX_{ik} is the " net exports ratio" of product *k* by country *i*. *X* stands for exports and *M* represents imports. NX_{ik} measures the weight of trade balance in the total trade of a certain product. The value interval is [-1, 1] with a mean of zero. When country *i* exports product *k* without any imports, NX_{ik} =1; when there are only imports, NX_{ik} =-1; when the exports of product *k* exactly equals to the imports, NX_{ik} =0.

Comparative advantage is measured by

$$RCA_{ik} = (X_{ik} / X_i)/(X_{wk} / X_w) \tag{2}$$

where RCA_{ik} is the "revealed comparative advantage index" of product *k* exports by country *i*, X_{wk} is the world exports of product *k* and X_w is the total world exports [10]. RCA_{ik} compares the weight of product *k* in country *i*'s exports with the weight of product *k* in world exports. The value interval of RCA is [0, ∞], we normalize the index by

$$RSCA_{ik} = (RCA_{ik} - 1)/(RCA_{ik} + 1) \tag{3}$$

where $RSCA_{ik}$ is the "revealed symmetric comparative advantage index" of product *k* by country *i*, with a value interval of [-1, 1]. $RSCA_{ik}$ >0 suggests country *i* is more specialized or has "revealed comparative advantages" in product *k*; $RSCA_{ik}$ <1 implies country *i*'s dis-advantages in the specialization or exports of product *k*. The normalization enables us to compare $RSCA_{ik}$ with NX_{ik}. In equilibrium, we expect $RSCA_{ik}$ excactly equals to NX_{ik}. Figure 1 presents the mean of $RSCA_{ik}$ and NX_{ik} of the 39 agro-based manufactures.

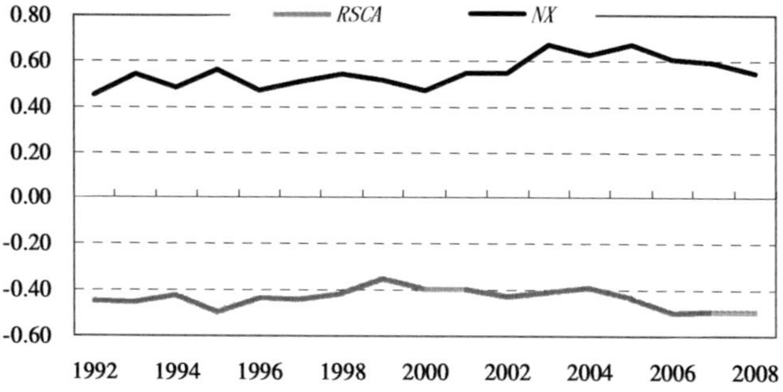

Fig. 1. NX and RSCA of Chinese agro-based manufactures (1992-2008)

The means of NX_{ik} are persistently higher than the means of $RSCA_{ik}$, strongly suggesting that China has been obtaining excessive trade surplus from the trade of agro-based products in spite of the comparative disadvantages. We view this as an evidence of the export supporting strategic trade policies.

3 Cointegration Analysis

3.1 Panel Unit Root Test

Because we focus on the Chinese trade patterns, the subscript of i is omitted for abbreviation in the following discussion. We conduct unit root tests to examine the order of integration in the NX_k and $RSCA_k$ series. We compare the results of 5 alternative methods, among which LLC and Breitung assume common unit root process; IPS, ADF-Fisher and PP-Fisher assume individual unit root process. The lags selection is based on Shwarz information criterion. Table 2 and Table 3 present the summary of the results.

Table 2. Summary of Pool Unit Root Tests for *NX*

	Level			1st difference		
Model Specification	*C*	*CT*	*N*	*C*	*CT*	*N*
Levin, Lin & Chu	-23.978 (0.000)	-6.626 (0.000)	-0.123 (0.451)	-19.872 (0.000)	-21.892 (0.000)	-21.978 (0.000)
Breitung		-0.337 (0.368)			-10.363 (0.000)	
Im, Pesaran and Shin	-7.096 (0.000)	-2.197 (0.014)		-16.372 (0.000)	-15.820 (0.000)	
ADF-Fisher	161.966 (0.000)	116.373 (0.003)	84.332 (0.292)	375.105 (0.000)	337.931 (0.000)	542.505 (0.000)
PP-Fisher	277.173 (0.000)	129.802 (0.000)	83.854 (0.305)	464.333 (0.000)	450.437 (0.000)	611.902 (0.000)

Note: C stands for individual intercept included in test equation; CT stands for individual intercept and linear trend; N stands for no exogenous variable; Probabilities are in parentheses.

Table 3. Summary of Pool Unit Root Tests for *RSCA*

	Level			1st difference		
Model Specification	*C*	*CT*	*N*	*C*	*CT*	*N*
Levin, Lin & Chu	-3.183 (0.000)	-6.865 (0.000)	-0.520 (0.302)	-18.922 (0.000)	-18.427 (0.000)	-22.093 (0.000)
Breitung		1.141 (0.873)		-7.077 (0.000)	(0.000)	
Im, Pesaran and Shin	-0.356 (0.361)	-1.902 (0.029)		-13.249 (0.000)	-15.493 (0.000)	
ADF-Fisher	90.259 (0.162)	101.715 (0.037)	85.122 (0.272)	288.972 (0.000)	356.822 (0.000)	517.475 (0.000)
PP-Fisher	105.844 (0.020)	92.362 (0.127)	89.365 (0.178)	389.817 (0.000)	407.282 (0.000)	536.136 (0.000)

Note: C stands for individual intercept included in test equation; CT stands for individual intercept and linear trend; N stands for no exogenous variable; Probabilities are in parentheses.

On one hand, 4 of the 5 methods reject the null hypothesis that the level NX_k pool is non-stationary; For the level pool of $RSCA_k$, however, the results are inconsistent. LLC rejects the null of common unit root, but no method rejects the null of individual unit roots. This implies that a level regression of NX_k on $RSCA_k$ may be spurious.

Upon taking first differences, the null of common or individual unit roots can be significantly rejected for both NX_k and $RSCA_k$. The result is robust in spite of the different assumptions for exogenous variables in test equation. This makes it possible for heterogeneous panel cointegration tests.

3.2 Panel Cointegration Test

Considering the heterogeneity across individual of the panel members, this study uses Pedroni's method [12] to test for the cointegration relationship between NX_k and $RSCA_k$ of the Chinese trade in agro-based products. The optimal lags of cointegration are 0-2 based on Shwarz information critirion. Table 4 reports the panel cointegration estimated results.

Among the 7 statistics, Panel v, Panel rho, Panel PP and Panel ADF are based on pooling the residuals of the regression along the within-dimension, while Group rho, Group PP, Group ADF are based on pooling the residuals of the regression along the between-dimension. When the model specification allows for individual intercept, 6 statistics except Group rho reject the null hypothesis of no integration at 0.001 confidence level. We conclude that there exists a long-run equilibrium cointegrating relation between NX_k and $RSCA_k$ of the Chinese agro-based manufactures.

Table 4. Panel Cointegration Test Results

Model Specification	*C*		*CT*		*N*	
Panel v	3.391	(0.000)	-0.932	(0.824)	-1.790	(0.963)
Panel rho	-3.574	(0.000)	-0.095	(0.462)	-0.058	(0.477)
Panel PP	-5.678	(0.000)	-4.213	(0.000)	-1.219	(0.111)
Panel ADF	-6.473	(0.000)	-5.463	(0.000)	-1.602	(0.055)
Group rho	0.077	(0.531)	2.307	(0.989)	2.697	(0.997)
Group PP	-4.230	(0.000)	-4.832	(0.000)	-0.837	(0.201)
Group ADF	-4.520	(0.000)	-6.116	(0.000)	-0.843	(0.120)

Note: C stands for individual intercept included in test equation; CT stands for individual intercept and a common linear trend; N stands for no exogenous variable; Probabilities are in parentheses; Automatic selection of lags (0-2) are based on Shwarz information criterion.

4 Estimation of the Panel VECM

4.1 Model Specification

Denote y_k for $RSCA_k$ and x_k for NX_k. By introducing subscript t for period t, the unrestricted regressional model is

$$y_{kt} = C_{kt} + \beta_k \cdot x_{kt} + \varepsilon_k \tag{4}$$

where C_k is the intercept and β_k is the parameter for estimation, and ε_k is the residual or the error term. (4) is a variable-coefficient model which allows for individual fixed effect. The model says that the revealed comparative advantage (y_{kt}) is linearly associated to or determined by net export ratio (x_{kt}). When $\beta_1=\beta_2=\ldots=\beta_k=\beta$, where β is a constant, (4) becomes a variable-intercept model given by

$$y_{kt} = C_k + \beta \cdot x_{kt} + \varepsilon_k \tag{5}$$

When C_1=C_2=…=C_k=C, where C and is a constant, (5) becomes a mixed-pool model given by

$$y_{kt} = C + \beta \cdot x_{kt} + \varepsilon_k \quad .(6)$$

We use F-tests to select the model by three steps.

Step 1: we estimate the three possible models and obtain the sum of squared residuals for each model, i.e. S_1 for (4), S_2 for (5) and S_3 for (6).

Step 2: we employ S_1 and S_2 to obtain F_1 statistic

$$F_1 = \frac{(S_2 - S_1)/[(n-1)K]}{S_1/[nT - n(K+1)]} \sim F[(n-1)K, n(T-K-1)] \tag{7}$$

where n is the number of cross-sections, T is the number of sample periods and K is the number of independent variables. F_1 statistic compares the unrestricted (5) and the partially restricted (6) and the null hypothesis to test for is H_0: β_k=β. A rejection of H_0 suggests the model is variable-coefficient.

Step 3: we employ S_2 and S_3 to obtain F_2 statistic

$$F_2 = \frac{(S_3 - S_1)/[(n-1)(K+1)]}{S_1/[nT - n(K+1)]} \sim F[(n-1)(K+1), n(T-K-1)] \tag{8}$$

to compare the fixed-effect (5) and the mixed-pool (6). The null hypothesis is H_0: C_k=C and β_k=β. A rejection of the null indicates that the model should be fixed-effect.

In this study, we have F_1=10.204 and F_2=55.956, both are signifcant at 0.001 level. Therefore we conclude that the optimum model is the unrestricted variable-coefficient (4).

4.2 Panel Vector Error Correction Models

NX_k and $RSCA_k$ panel data for the Chinese agro-based manufactures are heterogenous but cointegrated, making it possible and necessary to perform causality analysis using fixed-effect panel vector error correction model (VECM).

$$\Delta y_{kt} = \rho_k e_{kt-1} + \sum_{p=1}^{p} \left(\psi_{kp} \Delta y_{kt-p} + \xi_{kp} \Delta x_{kt-p}\right) + \theta_k T + c_k + \mu_k \tag{9}$$

where $\Delta(\cdot)$ stands for first differenc; the subscript p stands for lags; e_{kt-1} presents for the error correcting term, which is the residuals obtained by estimating (4); c_k is individual intercept; ρ_k, ψ_{kp}, ξ_{kp} and θ_k are the parameters to be estimated; and μ_k is the residual term of the VECM.The error correcting term, e_{kt-1}, reflects the long-run equilibrium relationship between y_{kt} and x_{kt}. The lags of the first differenced terms, $\Delta(\cdot)$, include the information of short-run shocks. Considering that each product may have an individual trend, (9) allows for individual time effect (θ_k).

The lagging period (p) of (9) is determined by following procedure: firstly, we assume p=1,2,3 and 4 and estimate (9) seperately; secondly, we compare the AIC and SIC statistics of the VECM estimations, and decide the lagging period by Akaike and

Schwarz information criteria. Consistent to the results of Pedroni panel cointegration test, we have p=2.

By interchanging the dependent variable by Δx_{kt}, we can obtain another vector error correction model to examine the effects of y_{kt} on x_{kt}.

$$\Delta x_{kt} = \rho_k^* e_{kt-1} + \sum_{p=1}^{p} \left(\psi_{kp}^* \Delta y_{kt-p} + \xi_{kp}^* \Delta x_{kt-p}\right) + \theta_k^* T + c_k^* + \mu_k^* \tag{10}$$

5 Panel Granger Causality Tests

When the inclusion of the lags of an independent variable helps to improve the explanatory power of the model, we say that the independent variable Granger causes the dependent variable [13]. Following [14], we perform the short-run and long-run Granger non-causality tests based on VECM.

Hong (2009) argues that the key to identify short-run and long-run effects is to examine the assumption of "other conditions keeping unchanged" [1]. If focusing on the coefficients of the lagged difference terms in the VECMs, the effect revealed is short-run because of the neglection of the long-run equilibrium relationship (e_{kt-1} in our model). Specifically, we test for the Wald restriction of the null of $\xi_{k1}=\xi_{k2}=0$ for (9) and $\psi^*_{k1}=\psi^*_{k2}=0$ for (10) to examine the short-run Granger non-causality.

There are two dimensions of understanding of long-run Granger non-causality: the first dimension is whether the long-run equilibrium relation (as represented by the error correcting term in the VECM) exerts significant effects on the dependent variables. We examine this by testing for the Wald restriction of the null of $\rho^*_k=0$ for (9) and $\rho^*_k=0$ for (10); the second dimension of understanding is whether the lagged differences of the independent variable exert significant effects upon the dependent variables via the long-run equilibrium relation, We examine this by testing for the Wald restriction of the null of $\rho^*_k=\xi_{k1}=\xi_{k2}=0$ for (9) and $\rho^*_k=\psi^*_{k1}=\psi^*_{k2}=0$ for (10). The results are shown in Table 5.

Table 5. Panel Granger Causality Tests Based on VECM

Independent / Dependent	Δy_{kt}		Δx_{kt}	
Δy_{kt-1}, Δy_{kt-2}			83.664	(0.311)
Δx_{kt-1}, Δx_{kt-2}	68.418	(0.773)		
e_{kt-1}	10.889	(0.000)	3.143	(0.000)
Δy_{kt-1}, Δy_{kt-2} and e_{kt-1}			1.637	(0.000)
Δx_{kt-1}, Δx_{kt-2} and e_{kt-1}	4.799	(0.000)		

Note: chi-square statistics are reported for short-run Granger causality tests, when the dependent variables are Δy_{kt-1}, Δy_{kt-2} and Δx_{kt-1}, Δx_{kt-2}; F statistics are reported for long-run Granger causality tests, when the dependent variables are e_{kt-1} and the joint of Δy_{kt-1}, Δy_{kt-2} and e_{kt-1} as well as the joint of Δx_{kt-1}, Δx_{kt-2} and e_{kt-1}; Probabilities are in parentheses.

- In the short-run, we find no evidence for Granger causality of any direction. In other words, the Chinese strategic trade policies, being featured with export promotion, distort the Chinese trade patterns in agro-based manufactures, making the net export ratios (y_{kt}) unassociated with the revealed symmetric comparative advantages (x_{kt}) in the short-run.
- In the long-run, there are strong evidences for Granger causal relations across the panels. Firstly, the long-run equilibrium relation (e_{kt-1}) significantly Granger causes both net export ratios (y_{kt}) and revealed symmetric comparative advantages (x_{kt}) at 0.001 confidence level; Secondly, there exists bi-directional Granger causalities between y_{kt} and x_{kt}.
- After restricting $\xi_{kp}=\xi_p$ to (9) and $\psi^*_{kp}=\psi^*_p$ to (10), the panel VECMs become variable-intercept and enable us to examine the common coefficients of (9) and (10). We have ξ_{k1}=0.059 and ξ_{k2}=0.003 for (9) and ψ^*_{k1}= -0.008 and ψ^*_{k2}= -0.002 for (10). Simply put it, an increase of x_{kt} has positive effects on y_{kt}, but an increase of y_{kt} has negative effects on x_{kt}, implying that the Chinese strategic trade policies unexpectedly hurt the comparative advantages of the agro-based manufactures in the long-run.

6 Conclusion

This study explores the Chinese trade patterns of 39 SITC 3-digit agro-based manufactures. Using heterogenous panel data of 1992-2008, we find an equilibrium long-run cointegrating relation between net export ratios and revealed symmetric comparative advantages.

Granger causality tests based on panel vector error correction models reveal that in the short-run, there exists no significant Granger causal relationship between the two measurements for trade patterns; in the long-run, revealed symmetric comparative advantages exert positive effects on the relative ability of the Chinese net exports. The strategic trade policies do Granger cause the comparative advantages, but the long-run effects are negative. The poor long-run performance of the export promotion casts doubts on the effectiveness of strategic trade policies.

Acknowledgments. This work is supported by grant 2006JDXM100 and 09JZD00037 from Ministry of Education of People's Republic of China.

References

1. Hong, Y.: A study of changes and distortions of Chinese commodity trade pattern. Ph.D. dissertation, Northeast Asian Studies Academy, Jilin University, Changchun, Jilin, pp. 30–34 (December 2009)
2. Cimoli, M.: Lock-in specialization (Dis)advantages in a structuralist growth model. In: Fagerberg, J., Verspagen, B., von Tunzelmann, N. (eds.) The Dynamics of Technology Trade and Growth. Edward Elgar, Aldershot (1994)
3. Damuri, Y.R., Atje, R., Gaduh, A.B.: Iintegration and trade specialization in east Asia. CSIS Working Paper, No. WPE094 (2006)

4. D'Costa, A.P.: Export growth and path dependence: the locking-in of innovations in the software industry. Science Technology and Society 7(1), 51–89 (2002)
5. Spencer, B.J., Brander, J.A.: International R&D rivalry and industrial strategy. Review of Economic Studies 50(4), 707–722 (1983)
6. Brander, J.A., Spencer, B.J.: Export subsidies and international market share rivalry. Journal of International Economics 18(1-2), 83–100 (1985)
7. Helpman, E., Krugman, P.R.: Market structure and foreign trade: increasing returns, imperfect competition, and the international economy. MIT Press, Cambridge (1985)
8. Krugman, P.R.: Rethinking international trade. MIT Press, Cambridge (1994)
9. Lall, S.: The technological structure and performance of developing Country Manufactured Exports, 1985-1998. Oxford Development Studies 28(3), 337–369 (2000)
10. Balassa, B.: The purchasing-power parity doctrine: a reappraisal. The Journal of Political Economy 72(6), 584–596 (1964)
11. Dalum, B., Laursen, K., Villumsen, G.: Structural change in OECD export specialisation patterns: de-specialisation and 'stickiness'. International Review of Applied Economics 12(3) (July 1998)
12. Pedroni, P.: Critical values for cointegration tests in heterogeneous panels with multiple regressors. Oxford Bulletin of Economic and Statistics 61(S1), 653–678 (1999)
13. Engle, R., Granger, C.: Cointegration and error correction: representation, estimation, and testing. Econometrica 55(2), 257–276 (1987)
14. Lee, C.C., Chang, C.P.: Energy consumption and economic growth in Asian economies: a more comprehensive analysis using panel data. Resource and Energy Economics 30(1), 50–65 (2008)

Adaptive Equalization Using a New RLS Algorithm Combined with the Coordinate Descent Method

Xiangyuan Bu and Yaoyao Yang

School of Information and Electronics, Beijing Institute of Technology,
Beijing, China
bxy@bit.edu.cn, yangyaoyao1010@163.com

Abstract. The adaptive equalization technique is an effective approach to eliminate the inter-symbol interference (ISI). It is known that the RLS algorithm has been widely used to update the tap weights of equalizers in most cases. However, it is restricted to being implemented on hardware platforms because of its large computational complexity. Considering this, we have proposed an improved algorithm combined with the coordinate descent method in this paper. The simulation results confirm that the novel algorithm achieves faster convergence rate and lower steady-state errors than the traditional RLS algorithm.

Keywords: adaptive equalization, RLS, coordinate descent method.

1 Introduction

It is well known that the inter-symbol interference (ISI) is the main obstacle in wireless digital communication systems, and things become worse with speed increasing. In mobile channels, the inter-symbol interference (ISI) is introduced by multi-path effects, which will result in the high error rate in receivers. The equalization technique can be an effective approach to mitigate or eliminate the ISI. As mobile radio channels are stochastic and time-varying, the equalizer should track variations of the channel, which is called the adaptive equalizer [1].

Generally speaking, the adaptive equalization technique can be divided into two categories--linear and non-linear equalization. Furthermore, for each structure there is a class of algorithms to adaptively adjust the tap weights of the equalizer. With respect to non-blind equalization, LMS (least mean squares) and RLS (recursive least squares) are two basic algorithms.

In practical applications, the decision feedback equalizer (DFE), which is belong to non-linear equalizers, has been used mostly in fading channels. Due to the faster convergence rate and lower tracking errors of the RLS algorithm compared with the LMS algorithm, the former may be preferred in conditions, where short training sequence is needed, such as TDMA systems [2]. However, a very important consideration in the usage of RLS is the high computational complexity. That is the reason why some improved algorithms based on it have been developed over the past years [3]. In this paper, we will also investigate a new algorithm combined with the idea of the coordinate descent method.

W. Zhang (Ed.): Advanced Technology in Teaching, AISC 163, pp. 631–638.
springerlink.com

The rest of the paper is organized as follows. In Section 2, the principles of traditional RLS algorithm and coordinate descent method are presented respectively, and in Section 3 we propose the combined algorithm for adaptive equalization. In Section 4, we simulate a communication system on the platform of MATLAB and compare the performance between the traditional and novel algorithms. In the end, we draw a conclusion in Section 5.

2 Preliminary Knowledge

2.1 Traditional RLS Algorithm

In order to present the algorithm, it is more convenient to use vectors and matrices. So the input and tap weights of the equalizer can be defined as follows respectively.

$$\boldsymbol{x}(n) = [x(n), x(n-1), x(n-2), \cdots, x(n-N+1)]^T \tag{1}$$

$$\boldsymbol{w}(n) = [w_1(n), w_2(n), w_3(n), \cdots, w_N(n)]^T \tag{2}$$

where N is the number of tap weights. Then, the output signal of the equalizer $y(n)$ and the error signal $e(n)$ are denoted in

$$y(n) = \sum_{i=1}^{N} w_i(n)x(n-i+1) = \boldsymbol{w}^H(n)\boldsymbol{x}(n) \tag{3}$$

$$e(n) = d(n) - y(n) = d(n) - \boldsymbol{w}^H(n)\boldsymbol{x}(n) \tag{4}$$

where $d(n)$ is the reference signal or the desired signal.

The traditional RLS algorithm can be regarded as a special case of the Kalman filter. Its cost function is defined in accordance with the exponentially weighted least squares criterion, and the expression is

$$J(n) = \sum_{i=1}^{n} \lambda^{n-i} | e(i) |^2 \tag{5}$$

where λ is called the exponential-weighted factor or the forgetting factor, a positive number less than 1. From the equation above, we can see that the old data decay exponentially, which is useful in a non-stationary environment.

To minimize $J(n)$, set the gradient of it to zero, then we get the normal equation [4]

$$\boldsymbol{R}(n)\boldsymbol{w}(n) = \boldsymbol{p}(n) \tag{6}$$

where $\boldsymbol{R}(n)$ and $\boldsymbol{p}(n)$ are defined as

$$\boldsymbol{R}(n) = \sum_{i=1}^{n} \lambda^{n-i} \boldsymbol{x}(i)\boldsymbol{x}^H(i) = \lambda \boldsymbol{R}(n-1) + \boldsymbol{x}(n)\boldsymbol{x}^H(n) \tag{7}$$

$$\boldsymbol{p}(n) = \sum_{i=1}^{n} \lambda^{n-i} \boldsymbol{x}(i)d^*(i) = \lambda \boldsymbol{p}(n-1) + \boldsymbol{x}(n)d^*(n) \tag{8}$$

Note that $\boldsymbol{R}(n)$ and $\boldsymbol{p}(n)$ stand for the auto-correlation matrix and cross-correlation matrix between the input signal and the desired signal of the equalizer respectively.

If the matrix $\boldsymbol{R}(n)$ is nonsingular, then $\boldsymbol{w}(n)$ can be calculated by

$$\boldsymbol{w}(n) = \boldsymbol{R}^{-1}(n)\boldsymbol{p}(n) \tag{9}$$

According to the matrix inversion lemma, and defining $\boldsymbol{P}(n) = \boldsymbol{R}^{-1}(n)$, we derive

$$\boldsymbol{P}(n) = \lambda^{-1}\boldsymbol{P}(n-1) - \lambda^{-1}\boldsymbol{k}(n)\boldsymbol{x}^H(n)\boldsymbol{P}(n-1) \tag{10}$$

where, $\boldsymbol{k}(n) = \dfrac{\boldsymbol{P}(n-1)\boldsymbol{x}(n)}{\lambda + \boldsymbol{x}^H(n)\boldsymbol{P}(n-1)\boldsymbol{x}(n)}$, is the gain vector.

Substitute (10) into (9), it follows that [4]

$$\boldsymbol{w}(n) = \boldsymbol{w}(n-1) + \boldsymbol{k}(n)\xi^*(n) \tag{11}$$

where, $\xi(n) = d(n) - \boldsymbol{w}^H(n-1)\boldsymbol{x}(n)$, is the prior estimation error. Then the traditional RLS algorithm can be summarized below.

a) Initialization: $\boldsymbol{w}(0) = \boldsymbol{w}_{initial}$; $\boldsymbol{R}(0) = \delta^{-1}\boldsymbol{I}$ (δ is a small positive constant)

b) For the $k-th$ *iteration,* $k = 1, 2, \cdots$:

- Calculate the output signal, $y(k) = \boldsymbol{w}^H(k-1)\boldsymbol{x}(k)$
- Calculate the error, $\xi(k) = d(k) - y(k)$
- Calculate the gain vector, $\boldsymbol{k}(k) = \dfrac{\boldsymbol{P}(k-1)\boldsymbol{x}(k)}{\lambda + \boldsymbol{x}^H(k)\boldsymbol{P}(k-1)\boldsymbol{x}(k)}$
- Update the matrix $\boldsymbol{P}(k)$, $\boldsymbol{P}(k) = \lambda^{-1}\boldsymbol{P}(k-1) - \lambda^{-1}\boldsymbol{k}(k)\boldsymbol{x}^H(k)\boldsymbol{P}(k-1)$
- Update the tap weights, $\boldsymbol{w}(k) = \boldsymbol{w}(k-1) + \boldsymbol{k}(k)\xi^*(k)$

2.2 Coordinate Descent Method

As we know, numeric iterative algorithms can be used to solve the normal equation. Here we will adopt line search methods including CG (conjugate gradient) [5] and CD (coordinate descent) algorithms to complete iterations. The CD algorithm is described specially below because of its lower computational complexity than the CG algorithm.

The principles of the numeric iterative solution indicate that solving the equation, $\boldsymbol{R}\boldsymbol{w} = \boldsymbol{p}$, is equivalent to minimizing the quadratic function

$$f(\boldsymbol{w}) = \frac{1}{2}\boldsymbol{w}^H\boldsymbol{R}\boldsymbol{w} - \boldsymbol{w}^H\boldsymbol{p} \tag{12}$$

In order to minimize it, search direction vectors must be determined before each iterative operation. In the coordinate descent algorithm, we choose columns of the $N \times N$ identity matrix, i.e. $\boldsymbol{p}_1 = [1, 0, 0, \cdots, 0]^T$, $\boldsymbol{p}_2 = [0, 1, 0, \cdots, 0]^T$, ..., $\boldsymbol{p}_N = [0, 0, 0, \cdots, 1]^T$, as search directions. Then, use $\boldsymbol{p}_i$ ($i = 1, 2, \cdots, N$) in turn at every iteration. According to searching methods, the step-size on the search direction can be calculated by

$$\alpha^{(k)} = \boldsymbol{r}_i^{(k-1)} / \boldsymbol{R}_{i,i} \tag{13}$$

where $\boldsymbol{r}$, whose definition at the time instant k is $\boldsymbol{r}^{(k)} = \boldsymbol{p}(k) - \boldsymbol{R}(k)\boldsymbol{w}(k)$, is the residual vector. If $\boldsymbol{R}_i$ denotes the $i-th$ column of the matrix $\boldsymbol{R}$, then $\boldsymbol{r}$ is updated through [6]

$$\boldsymbol{r}^{(k)} = \boldsymbol{r}^{(k-1)} - \alpha^{(k)} \boldsymbol{R}_i \tag{14}$$

The algorithm is clearly presented as follows.

c) *Initialization:* $\boldsymbol{w}(0) = \boldsymbol{w}_{initial}$; $\boldsymbol{r}^{(0)} = \boldsymbol{p} - \boldsymbol{R}\boldsymbol{w}(0)$

d) *For the* $k-th$ *iteration,* $k = 1, 2, \cdots$:

- Calculate the step-size on the direction $\boldsymbol{p}_i$, $\alpha^{(k)} = \boldsymbol{r}_i^{(k-1)} / \boldsymbol{R}_{i,i}$
- Update the solution vector, $\boldsymbol{w}(k) = \boldsymbol{w}(k-1) + \alpha^{(k)}$
- Calculate the residual vector, $\boldsymbol{r}^{(k)} = \boldsymbol{r}^{(k-1)} - \alpha^{(k)} \boldsymbol{R}_i$

3 Proposed Adaptive Equalization Algorithm

Considering the implementation of adaptive equalizers on the hardware platform, we hope that the computational complexity should be as low as possible in the condition of good performance. As the complexity of the traditional RLS algorithm is quite high, a new algorithm combined with the coordinate descent method, has been proposed in our adaptive equalization.

According to the coordinate descent method described in Section 2, we calculate the step-size on the certain search direction. Instead of using $\boldsymbol{p}_i$ ($i = 1, 2, \cdots, N$) in turn at every iteration, the index i in $\boldsymbol{p}_i$ is determined by the maximum of the vector $\boldsymbol{r}$, which will improve the convergence rate obviously. The equation can be expressed as

$$i = \arg\max\{abs(\boldsymbol{r})\} \tag{15}$$

Applying the method above to the normal equation of RLS, we can give the steps of the combined algorithm as follows.

e) *Initialization:* $\boldsymbol{w}(0) = \boldsymbol{w}_{initial}$; $\boldsymbol{R}(0) = \boldsymbol{I}$; $\boldsymbol{p}(0) = \boldsymbol{0}$; $\boldsymbol{r}(0) = \boldsymbol{p}(0) - \boldsymbol{R}(0)\boldsymbol{w}(0)$

f) *For the* $k-th$ *iteration,* $k = 1, 2, \cdots$:

- Update the auto-correlation matrix $\boldsymbol{R}(k)$, $\boldsymbol{R}(k) = \lambda \boldsymbol{R}(k-1) + \boldsymbol{x}(k)\boldsymbol{x}^H(k)$
- Update the cross-correlation matrix $\boldsymbol{p}(k)$, $\boldsymbol{p}(k) = \lambda \boldsymbol{p}(k-1) + \boldsymbol{x}(k) d^*(k)$
- Choose the maximum index i, $i = \arg\max\{abs(\boldsymbol{r})\}$
- Calculate the step-size on the direction $\boldsymbol{p}_i$, $\alpha_i(k) = \boldsymbol{r}_i(k-1) / \boldsymbol{R}_{i,i}$
- Update the tap weights, $\boldsymbol{w}(k) = \boldsymbol{w}(k-1) + \alpha(k)$
- Calculate the residual vector, $\boldsymbol{r}(k) = \boldsymbol{p}(k) - \boldsymbol{R}(k)\boldsymbol{w}(k)$

Of course, the output signal at the time instant k is also achieved by $y(k) = \boldsymbol{w}^H(k-1)\boldsymbol{x}(k)$.

Based on the analysis of the new algorithm, we can see that it employs less multiplications and additions between matrices to complete updating tap weights. In fact, the usage of (15) roots in the essence of the selective partial updating, which means that only parts of tap weights are updated at every iterative operation [7]. So, we can call the proposed algorithm SPURLS-CD algorithm.

4 Simulation Results

In this section, we simulate a communication system on MATLAB, which is shown in Fig.1.

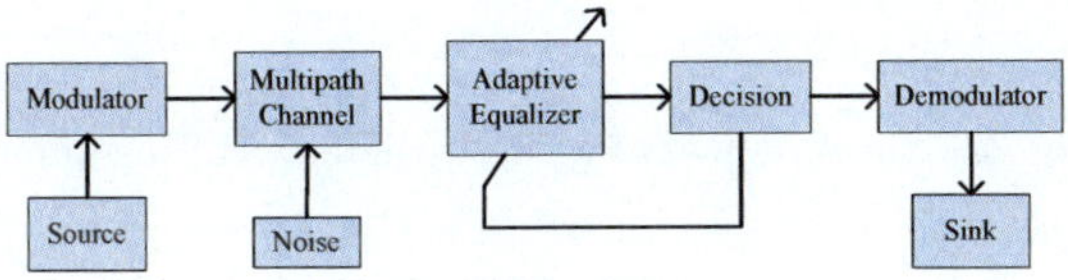

Fig. 1. The system model

The corresponding parameters of each module are:

- Modulation type: QPSK
- Channel models: COST 207TU and COST 207RA [8]; the sampling time and the maximum Doppler shift are set to be 2.6us and 100Hz, respectively.
- The structure of the adaptive equalizer: DFE, in which the orders of feedforward and feedback filters are specified as eight and four.
- The algorithm of the adaptive equalizer: Traditional RLS algorithm and the proposed one in which forgetting factors are both 0.96; the length of the training signal is 25.

4.1 Convergence Performance

The number of symbols that the transmitter sends is 500. After QPSK modulation, the signal passes through the COST 207TU channel model, to which add the Gaussian white noise at the same time. We choose SNR=15dB here. The constellations of the signal before and after the equalizers are shown below.

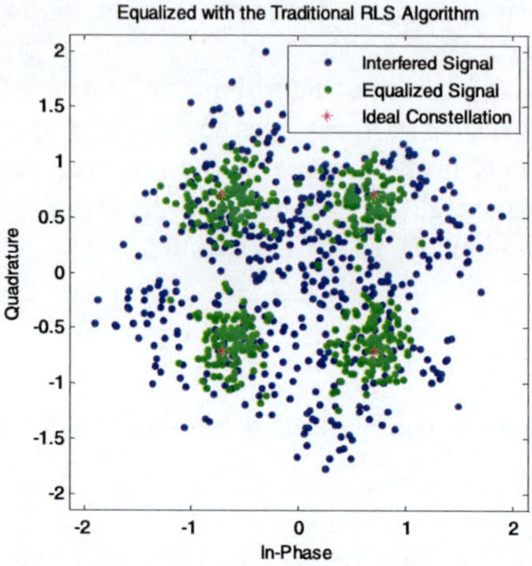

Fig. 2. Constellation of using the traditional RLS algorithm

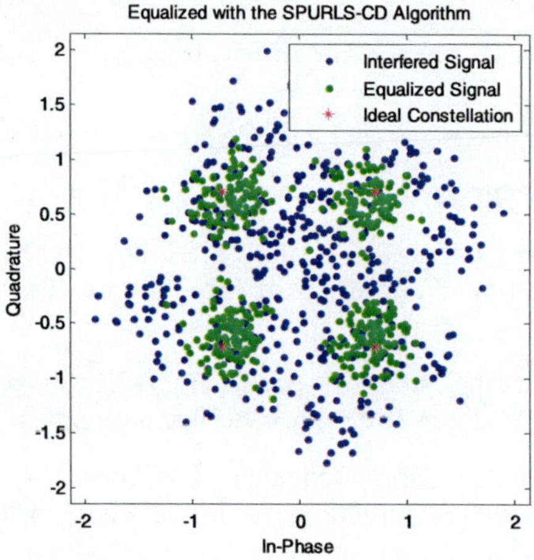

Fig. 3. Constellation of using the SPURLS-CD algorithm

The results in Fig.2 and Fig.3 illustrate that the constellation of the equalized signal marked in green is inclined to the ideal one, which is much more convenient to be demodulated, decreasing the error ratio in the receiver.

However, it seems that there is no obvious difference between two algorithms if we only compare the constellations. In order to manifest the advantages of the SPURLS-CD algorithm over the traditional one, the convergence curves are obtained by ensemble averaging over 200 independent experiments, which are displayed in Fig.4.

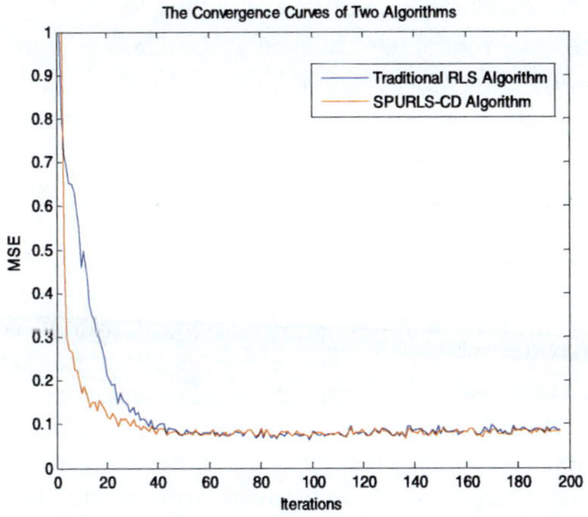

Fig. 4. The convergence curves of two algorithms

Fig.4 shows the MSE performance of the SPURLS-CD algorithm versus the traditional one. It is clear that the former speeds up the convergence rate and goes into the steady state quickly. In addition, the MSE of the traditional RLS algorithm still keeps a little bigger even after the training sequence, whose length is set to be 25. Thus it can be seen that the novel equalization algorithm is more suitable for conditions, in which the training signal is as short as possible.

4.2 Tracking Capability

Good tracking performance is necessary when the channel environment varies dramatically. Typically, we start with the COST 207TU channel model, and then change it to the COST 207RA channel model. The length of the transmitted signal is 350, and other parameters are the same as the above case. Then, the curves can be achieved in Fig.5.

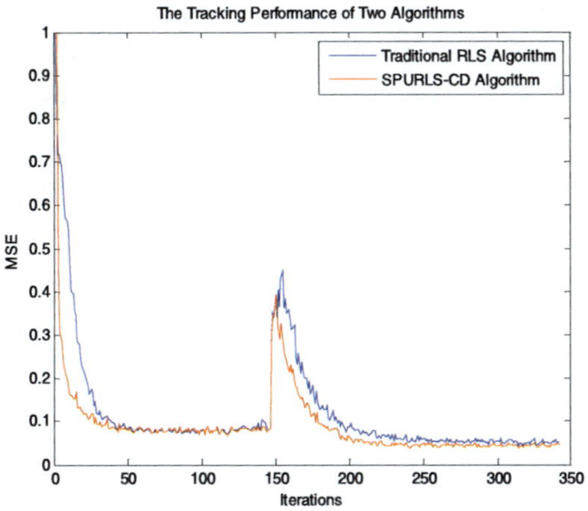

Fig. 5. The tracking performance of two algorithms

Fig.5 shows that the proposed algorithm can adjust to the variation through fewer iterations since the channel model has changed at the $150-th$ iteration. Moreover, the steady-state errors after adjustment are relatively lower than the traditional RLS algorithm. In other words, the novel algorithm is superior in tracking the variations in channels.

5 Conclusions

We have developed a new adaptive equalization algorithm combined with the coordinate descent method, which is called SPURLS-CD algorithm. As expected, the proposed algorithm achieves lower computational complexity comparing with the traditional RLS algorithm. Furthermore, it performs much better in convergence rate and tracking capability, especially when the SNR in the channel is low. So, it provides a good reference for the practical implementation of the adaptive equalization technique.

References

1. Rappaport, T.S.: Wireless Communications Principles and Practice, 2nd edn., pp. 249–266. Publishing House of Electronics Industry (July 2006)
2. Li, J.-D., Guo, T.-Y., Wu, G.-Y.: Mobile Communications, 4th edn. Publishing House of Xidian University (December 2006)
3. Diniz, P.S.R.: Adaptive Filtering: Algorithms and Practical Implementation, 2nd edn. Publishing House of Electronics Industry (July 2004)
4. Haykin, S.: Adaptive Filter Theory, 4th edn., pp. 121–143. Publishing House of Electronics Industry (July 2006)
5. Zhao, S., Man, Z., Khoo, S.: Conjugate Gradient Alogrithm Design with RLS Normal Equation. In: 2007 6th International Conference on Information, Communications & Signal Processing, pp. 1–5 (2007)
6. Zakharov, Y.V., White, G.P., Liu, J.: Low-Complexity RLS Algorithms Using Dichotomous Coordinate Descent Iterations. IEEE Transactions on Signal Processing 56(7), 3150–3161 (2008)
7. Dogancay, K., Tanrikulu, O.: Adaptive Filtering Algorithms with Selective Partial Updates. IEEE Transactions on Circuits and Systems-II: Analog and Digital Signal Processing 48(8), 762–769 (2001)
8. Patzold, M.: Mobile Fading Channels, pp. 208–246. Publishing House of Electronics Industry (January 2009)

An Improved CUSUM Algorithm in Cognitive Radio

Song Xiao-ou, Xiang Xin, Bi Du-yan, and Wang Xing-hua

Department of Avionic Electric Engineering, Air Force Engineering University
Xi'an China
e_miracle@163.com, xxisdn2002@sina.com

Abstract. In this paper, the conventional primary user signal detection algorithms' common weakness and advantages of the primary user signal detection scheme based on CUSUM algorithm are analyzed. To avoid the statistic's dithering which increases the detection delay, an improved CUSUM algorithm with a slipped pretreatment window is proposed. Simulation shows that the detection delay of the improved algorithm is decreased obviously.

Keywords: cognitive radio, CUSUM, detection delay, slipped pretreatment window.

1 Introduction

In recent years, cognitive radio has attracted intensive research because of pressing demand of efficient frequency spectrum usage. In a cognitive radio system, secondary radio users try to find 'blank spaces', in which the primary user is not working. Energy detection, matched filtering and feature detection are representative primary user signal detection methods [1]. These methods are to minimize the miss probability subject to constrain on false alarm probability. A common feature of these methods is that the detection is block-based detection. That is the cognitive radio takes a block of samples, computes a statistic and compares it with a threshold. Block-based detection can not achieve the minimal average detection delay. And there is a signal to noise ratio (SNR) wall phenomenon [2] in block-based detection, which refers to the situation that if the SNR of received signal is smaller than a threshold, these detection algorithms will not work, no matter how many samples are in each block. To conquer the challenges above, primary user signal detection schemes based on sequential detection [3] are proposed in [4] and [5]. Sequential detection makes decision at every sample time without a fixed block length. And after enough observed samples, the final determination can be made without SNR Wall phenomenon [3]. Among these sequential detection schemes, CUSUM algorithm is adopted in the situation that the signal distributions are known [4]. As observed, the statistic's dithering in CUSUM algorithm increases the detection delay. To avoid the dithering, an improved algorithm is proposed with a slipped pretreatment window. Simulation shows that the detection delay of the improved algorithm is decreased obviously.

The rest of the paper is organized as follows. In Section 2, the system model is introduced. In Section 3, CUSUM algorithm is illustrated. In Section 4, an improved CUSUM algorithm is proposed. In Section 5, simulation is carried out. Finally, in Section 6, some conclusions are offered.

W. Zhang (Ed.): Advanced Technology in Teaching, AISC 163, pp. 639–644.
springerlink.com

2 System Model

Assuming that a cognitive user tunes to a frequency band and begins taking samples, primary user signal detection is a distinction between two hypothesizes:

(1) Hypothesis H_0: there is only the noise without primary user signal in the detected frequency band.

(2) Hypothesis H_1: there are primary user signal and noise in the detected frequency band. The discrete time model is:

$$H_0 : X[i] = N[i] \qquad i=1,2,\ldots \tag{1}$$

$$H_1 : X[i] = S[i] + N[i] \quad i=1,2,\ldots \tag{2}$$

where $X[i]$ is the received signal sample, $N[i]$ is white Gaussian noise with variance σ^2, $S[i]$ is the received primary user signal assumed to submit to the distribution $N(0,W)$[4]. Hence under the hypothesis H_0, $X[i]$ submits to the distribution $N(0, \sigma^2)$. And under the hypothesis H_1, $X[i]$ submits to the distribution $N(0,W+\sigma^2)$. As shown in Figure1, first $X[i]$ submits to distribution P_0 with probability density p_0. Then at an unknown time τ, $X[i]$ changes to distribution P_1 with probability density p_1. Only the case of detecting the primary user emergence is considered in this paper. Hence P_0 is distribution $N(0, \sigma^2)$, and P_1 is distribution $N(0,W+\sigma^2)$. If H_0 is true, no primary user signal is detected in the observed frequency band, then continue observing. If H_1 is true, primary user emerges in the observed frequency band, then raise alarm. We assume alarm time is t_a.

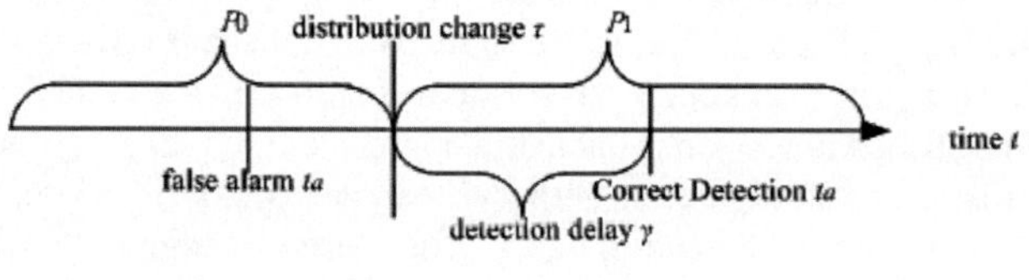

Fig. 1. Course of primary user emergence

The event $\{t_a<\tau\}$ is a false alarm event. And the mean time to false alarms is defined as:

$$\overline{T_0} = E_{p0}\{t_a\} = \sum_{t_a=1}^{\infty} t_a p_0(X(t_a)) \tag{3}$$

The value of $\overline{T_0}$ depends on the detection strategy and p_0. The event $\{t_a \geq \tau\}$ is a correct detection. The detection delay $\gamma=t_a-\tau$ is a random variable, whose value depends on the samples after τ. The conditional mean delay is defined as :

$$\overline{T}_{1,c} = E_{p_1}\{\gamma = t_a - \tau \mid t_a \geq \tau, X_1^{\tau}\} \tag{4}$$

Once a prior distribution is assigned to τ, the conditional mean delay can be fixed. However, in real applications, it is difficult to assign a suitable prior distribution. Then we consider the worst case delay [6]

$$\overline{T}1 = \sup_{\tau\geq 1} ess \quad \sup E_{p_1}\{\gamma = t_a - \tau \mid t_a \geq \tau, Y_1^{\tau}\} \tag{5}$$

Obviously, a larger mean time to false alarms and a smaller supreme detection delay are preferred.

The strategy in this paper is to minimize $\overline{T}_1$, while maintaining $\overline{T}_0$ to be larger than a conveniently set threshold T_{th}. Hence the cognitive detection problem is formulated as [7]:

$$\min_{\Gamma} \quad \overline{T}_1 = \sup_{\tau\geq 1} ess \quad \sup E_{f_1}\{\gamma = t_a - \tau \mid t_a \geq \tau, X_1^{\tau}\}$$
$$s.t. \ \overline{T}_0 \geq T_{th} \tag{6}$$

3 Cusum Algorithm

CUSUM algorithm is briefly introduced in this section. With the known W and σ^2, the log-likelihood ratio for $X[i]$ is:

$$l(X[i]) = \ln\left\{\frac{p_1(X[i])}{p_0(X[i])}\right\} = \frac{WX^2[i]}{2(W+\sigma^2)\sigma^2} + \frac{1}{2}\ln\left\{\frac{\sigma^2}{W+\sigma^2}\right\} \tag{7}$$

Before the primary user emergence, the mean value for $l(X[i])$ is:

$$E_{p_0}\{l(X[i])\} = \int p_0(x)\ln\left\{\frac{p_1(x)}{p_0(x)}\right\}dx = -D(p_0 \parallel p_1) \leq 0 \tag{8}$$

After the primary user emergence, the mean value for $l(X[i])$ is:

$$E_{p_1}\{l(X[i])\} = \int p_1(x)\ln\left\{\frac{p_1(x)}{p_0(x)}\right\}dx = D(p_1 \| p_0) \geq 0 \tag{9}$$

in which $D(p_0 \parallel p_1) = -\frac{W}{2(W+\sigma^2)} - \frac{1}{2}\ln\left\{\frac{\sigma^2}{W+\sigma^2}\right\}$ is the Kullback-Leibler divergence of p_0 from p_1. $D(p_1\|p_0)$ is defined similarly. Thus, after the primary user emergency, the mean value for $l(X[i])$ has a positive drift. Hence we define the detection statistic as:

$$\begin{cases} Z_t = \{Z_{t-1} + l(X[t])\}^+ \\ Z_0 = 0 \end{cases} \quad t=1,2,\ldots \tag{10}$$

where, $\{a\}^+=\max\{a,0\}$ which means that the accumulation of the log-likelihood ratio is set "0" while it is negative, and maintains its value while it is positive. Alarm time t_a is:

$$t_a = \inf\{t : Z_t > h\} \tag{11}$$

where, h is the threshold. Referring to the sequential analysis [8], the relationships between h, $\overline{T}_1$ and $\overline{T}_0$ is

$$\overline{T}_1 \leq \frac{h+\zeta}{D(p_1 \parallel p_0)} \tag{12}$$

$$\overline{T}_0 \geq e^h \tag{13}$$

where, $\zeta = \dfrac{\int_{c_1}^{\infty} l^2(x)p_1(x)dx + \int_{-\infty}^{-c_1} l^2(x)p_0(x)dx}{D(p_1 \parallel p_0)}$, $c_1 = \sqrt{\dfrac{(W+\sigma^2)\sigma^2}{W}\log\left(1+\dfrac{W}{\sigma^2}\right)}$. Hence the corresponding threshold h is computed by (13). And the worst case delay is thus bounded by (12).

4 Improved Cusum Algorithm

According to equation (10), with the system parameters σ^2=1, W=3, τ=100 and σ^2=1, W=1, τ=100, statistics changing courses of CUSUM algorithm are shown in Figure2. The value of Z_t increases quickly after primary user emerges. But there is dithering in the Z_t increasing course, which slows down the statistic increasing speed and hence increases the detection delay. A small detection delay will allow the design of a spectrum reuse scheme that has minimal impact on the primary users. So an improved CUSUM algorithm is proposed to conquer the dithering and moreover decreases the detection delay.

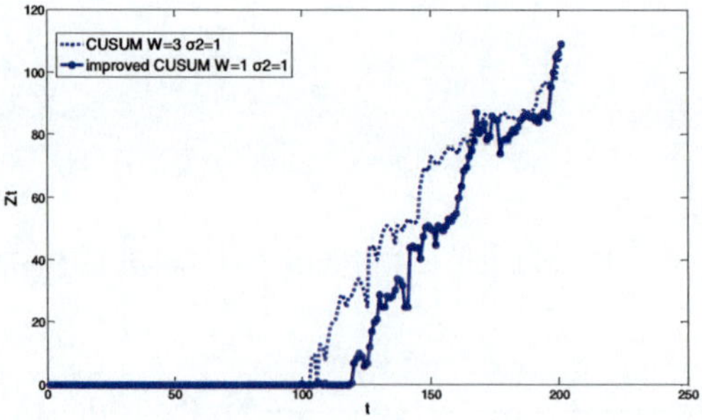

Fig. 2. Changing course of CUSUM algorithm statistics

As observed, in some time after primary user emergence, the log-likelihood ratios are mostly positive values, among which there is single negative value with run length 1. And the statistic's dithering is caused by the negative log-likelihood ratios. So a slipped pretreatment window is adopted in the improved CUSUM algorithm, in order to eliminate the single negative value. A simple pretreatment algorithm is proposed to illustrate the advantage of the pretreatment.

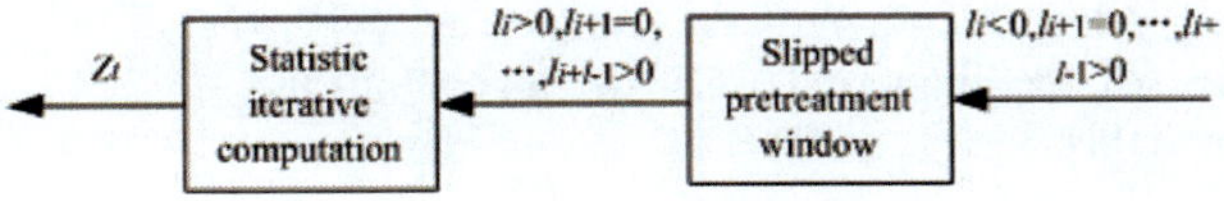

Fig. 3. Pretreatment algorithm

As shown in Figure3, before the iterative computation for Z_t, the log-likelihood ratios are first input into a slipped pretreatment window with window length L. If the i th log-likelihood value is negative, and after it, every value of the continuous L-1 log-likelihood ratios is positive or "0", we set the i th value as "0".

The improved CUSUM algorithm steps are described as follows:

Step1: Referring to the system requirement, T_{th} is set. And h is determined by equation (13).

Step2: Observing sequential signal samples $X[i]$, compute log-likelihood ratios by equation (7).

Step3: Input the log-likelihood ratios into the pretreatment window to be pretreated, eliminating the single negative value.

Step4: Iteratively compute the detection statistic Z_t by equation (10). And the current Z_t is stored.

Step5: Compare Z_t with h. If $Z_t>h$, primary user is detected to emerge, and raise alarm. If $Z_t<h$, there is no primary user, then back to step2 to continue observing.

5 Simulation

With the system parameters L=4 and L=6, σ^2=1, W=3, τ=100, the statistics changing courses of the improved CUSUM algorithm are simulated and compared with that of CUSUM algorithm.

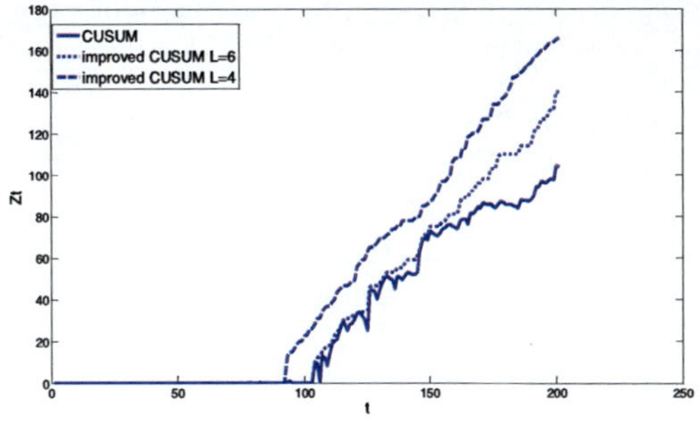

Fig. 4. Statistics changing courses

Figure4 shows that Z_t of the improved CUSUM algorithm increases more quickly and smoothly than Z_t of CUSUM algorithm. And Z_t of the improved algorithm with the window length L=4 increases more quickly than Z_t of the improved algorithm with the window length L=6. But it is also noted that when L=4, Z_t already has obvious positive drift before the primary user emergence, which may lead to false alarm. Hence we choose the window length L=6. With the parameters T_h=1000, σ^2=1, W=3, τ=100, L=6, the performance of the improved CUSUM algorithm and CUSUM algorithm are simulated comparably.

As shown in Figure5, the detection delay probability distribution peak value is around γ=11 in CUSUM algorithm, and around γ=6 in the improved CUSUM algorithm. Hence the improved algorithm can achieve smaller detection delay than CUSUM algorithm.

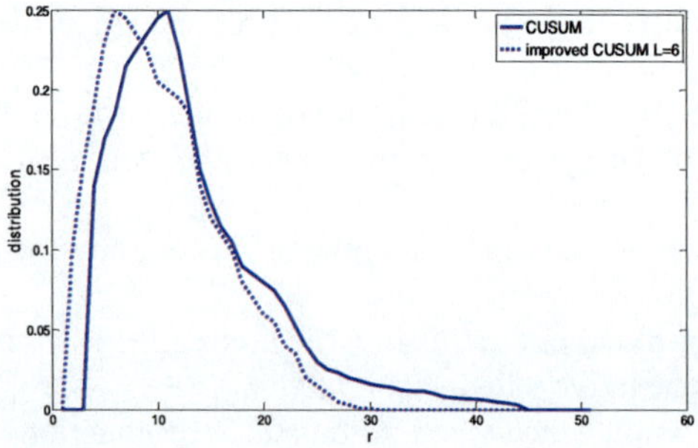

Fig. 5. Detection delay probability distribution.

6 Conclusion

Base on analysis of the advantages and weakness of CUSUM algorithm in cognitive radio, an improved CUSUM algorithm is proposed to avoid the statistic dithering, which decreases the detection delay. Slipped pretreatment window is adopted in the improved algorithm, to eliminate the single negative value. Simulation shows that the detection delay of the improved CUSUM algorithm is smaller than that of the traditional CUSUM algorithm. The pretreatment algorithm is easily designed in this paper, and it is proved to be useful. So exploiting more efficient pretreatment algorithm is of interest.

References

1. Proakis, J.G.: Digital Communications. MacGraw-Hill Higher Eduation (2001)
2. Tandra, R., Sahai, A.: SNR walls for signal detection. IEEE Journal on Special Topics in Signal Processing 2, 4–17 (2008)
3. Wald, A.: Sequential Analysis. Wiley, New York (1947)
4. Lai, L.F., Fan, Y.J., Vincent Poor, H.: Quickest Detection in Cognitive Radio: A Sequential Change Detection Framework. In: IEEE GLOBECOM (August 2008)
5. Li, H., Li, C., Dai, H.: Quickest spectrum sensing in cognitive radio. In: Pro. Conf. on Information Science and Systems, Princeton, NJ (March 2008)
6. Page, E.: Continuous inspection schemes. Biometrika 41, 100–115 (1954)
7. Moustakides, G.: Optimal procedures for detecting changes in distributions. Annals of Statistics 14, 1379–1387 (1986)
8. Lorden, G.: On excess over the boundary. Annals of Mathematical Statistics 41, 520–527 (1970)

The Research on Subject Database Division Based on Entity Similarity

Hongjun Zheng[1], Jinli Peng[1], and Duo Wang[2]

[1] College of Information Science and Engineering, Yanshan University, Qinhuangdao, China
[2] Party School of CPC Shijiazhuang Municipal Committee, Shijiazhuang, China
wcw@ysu.edu.cn, {pengjinli_001,Duow2000}@163.com

Abstract. The division of subject database, which is a major task of information resource planning, is the important way to realize the information sharing by solving the "Information Isolated Island" problem. The entity similarity, which makes the relations among entities more clearly, is presented by researching the existing methods of subject database division. Meanwhile, the subject database is divided in terms of the cluster analysis methods, which can conquer the bi-direction problem appeared in the process of aggregation analysis. Finally, an evaluation function is put forward to assess the results of division so that the division of subject database is more reasonable.

Keywords: subject database, entity similarity, cluster analysis, evaluation function.

1 Introduction

At present, the most serious challenge of enterprise informationization construction is transformation and reconstruction of the data environment. Information systems in many enterprises are separately developed and lack common, unified data standards. In the early 1980s, in order to improve the data sharing and reduce data redundancy, developed countries proposed a "strategic data planning" theory and method, reconstructed high-level data environment by using holistic view and integrity actions. A main purpose of the information resource planning is to design a series of subject databases in order to resolve the "information isolated island" problem, which can not be solved by the error idea for a long time, to realize the integration of application systems and information resources sharing. James Martin put forward the concept of subject database and the theoretical methods in order to solve these problems[1]. He has made great contributions, which has a great reference meaning for us. Subsequently, Gao Fu-xian[1], Wang Yu-shu, Dong Pei-ming[2] and others studied the subject database and achieved certain achievements. Up to now, the division of subject database is seldom given specific formal methods, most of which are dependent on some written language steps given by the systems analysts and sales' management experiences.

In this paper, we conduct a deep technical analysis and research about the division of subject database and then propose the concept of entity similarity. The subject database is divided by using cluster analysis methods, which can effectively

W. Zhang (Ed.): Advanced Technology in Teaching, AISC 163, pp. 645–651.
springerlink.com

overcome the bi-direction problem appeared in integration analytic process. Finally, we introduce the application of this method in detail through the application example.

2 Overview of Subject Database

The subject database[3] is known as intensive data environment. It is relevant to a variety of business subjects, rather than specific computer application procedures. The basic features of subject database are the following four points:

- Face business-oriented theme, rather than document-oriented statements.
- Share information, and stress that each different application system shares a common subject database.
- Data is collected and processed on the spot, the same data must enter system one time and place and be used many times and places.
- The data structure of subject database is composited by standard basic forms and follows the database design specifications.

Definition 1: Ent = {BM, A, E} is an enterprise, BM is enterprise business model; A is a set of activities in which entities participated; E is a set of all entities of enterprises which can be defined as: $E=\{e_1,e_2,..,e_n\}$, $E_1,E_2,..,E_k$ are entity groups constituted by elements of E, if $E= E_1 \oplus E_2 \oplus .. \oplus E_k$($\oplus$is direct sum), $SDB=\{E_1,E_2,..., E_k\}$ is a planning of subject database, each E_i is called a subject database of Ent.

Each subject database is a set of some entities, reflecting the relationship between entities. Although the system analysts can divide the subject database based on the functions of the enterprise activities, different system analysts can obtain a division of different subject database. The division of subject database is determined solely by persons, but also should be automatically generated by the algorithm. It is a precondition of determining subject database division using calculation method to do qualitative and quantitative description aiming at the relationships of entities and activities.

3 Entity Similarity

In 2005, Zhou Yan-tao[3] proposed an integration analytic method to divide subject database based on the entity affinity. But entity integration analytic method has the defect of bi-direction. In order to overcome the problem of bi-direction and make the technology of subject database division more widespread, in this paper the concept of entity similarity and a calculation formula of similarity between any two entities borrowing idea form Sorenson Coefficient are given.

Definition 2: Suppose that $E = \{e_1,e_2,..,e_m\}$, $A = \{a_1,a_2,..,a_n\}$, $T_{m\times n}$ is matrix of entity and activity, defined as follows: Table 1 is a simple example of matrix.

$$t_{ij} = \begin{cases} 0, & if \quad e_i \quad isn't \quad involved \quad in \quad a_j \\ 1, & if \quad e_i \quad is \quad involved \quad in \quad a_j \end{cases}$$

Table 1. Matrix of entity and activity

	a_1	a_2	a_3	a_4	a_5	a_6
e_1	0	0	1	1	0	0
e_2	1	1	1	0	0	0
e_3	0	0	0	0	1	0

Definition 3: the set constituted by all activities in which entity e participated is recorded as Sa(e), the set constituted by all entities involved by activity a is recorded as Se(a).

Definition 4: The number of activities in which any two entities simultaneously participate, that is the number of 1-1 match is recorded as *a*; the number of activities in which only one entity participated, that is the number 1-0 or 0-1 match is recorded as *b*, then similarity formula of any two entities is:

1) $s\left(e_i, e_j\right) = 2a/\left(2a + b\right)$

2) $s\left(e_i, e_j\right) = s\left(e_j, e_i\right)$

3) In general, $0 < s\left(e_i, e_j\right) < 1$

Definition 5: Let $R = \begin{pmatrix} r11, \cdots, r1q \\ \vdots \\ rp1, \cdots, rpq \end{pmatrix}_{p\times q}$ indicates the similarity matrix between any two entities, in which $r_{ij} = s\left(e_i, e_j\right)$.

Definition 6: Suppose that $E=\{e_1,e_2,\ldots,e_k\}$is a set of entities, e is an arbitrary entity, so that $GS\left(e, E_i\right) = \sum_{i=1}^{k} s\left(e, e_i\right)/k$, as the generalized similarity between entity e and entity set E_i.

The generalized similarity between entity and entity set reflects the affiliation degree of an entity with an entity set. The higher the value is, the more closely they link, and it is used to reflect the affiliation degree of an entity with the subject database.

4 Algorithm of Subject Database Division

4.1 Algorithm Idea of Subject Database Division

The division of subject database, that is the overall top-down planning of enterprise, all data sources of which can be divided into a number of management units-the subject database. Cluster analysis is a method of studying "things of one kind come together" in mathematical statistics[4]. In this paper, we cluster all the entities in enterprises using cluster analysis method based on entity similarity to get the initial division results of subject database. The basic idea is as follows:

Step1: Entities needed to be classed are recorded as $X = \{X_l\}$ (l= 1, 2 ... n), each entity in set X is one class.

Step2: According to entity similarity formula to compute entities similarity, get the matrix R of entity similarity.

Step3: Traverse R, find a greater value of similarity r_{ij} corresponding entity (e_i, e_j), as core entity groups of subject database, the remaining entities are non-core entities.

Step4: Establish the matrix Mij of core entity group and non-core entity ($m_{ij} =(s(e,e_i)+s(e,e_j))/2)$).

Step5: Cluster entities in accordance with the M_{ij}: allocate non-core entities to core entity group with the greatest value of similarity. If the similarity values of non-core entity with all core entity groups are very small, put the non-core entity as a separate subject database.

4.2 The Realization of Subject Database Division Algorithm

According to entity similarity and cluster analysis algorithm proposed in this paper, we can get the initial division of subject database, and then standardize the results to get the ultimate subject database of information resource planning. The specific algorithm is described as follows:

Line 1: For i =1 to n-1 do // n is the number of entity

Line 2: For j=i+1 to n

Line 3: Compute similarity of e_i and e_j

Line 4: $R_{ij}=S(e_i,e_j)$ // establish matrix R of entity similarity.

Line 5: max{S_1, S_2,... Sn} //identify the larger value of entity similarity from R as core entity groups.

Line 6: End For

Line 7: End For

Line 8: Compute the generalized similarity of ek and (e_i, e_j)

Line 9: $M_{ij}=GS(e_k,(e_i, e_j))$ //generate the matrix M of core entity group and non-core entity.

Line 10: For each ({e_1, e_2,...,e_k }) / /cluster the non-core entities as following steps.

Line 11: IF $GS(e_k, (e_i, e_j)) > max(m_{ij})$

Line 12: Cluster e_k to core entity group (e_i,e_j)

Line 13: Else

Line 14: The similarity values of non-core entity with core entity groups are very small, separate the non-core entity into a subject database.

Line 15: Fetch next non-core entity and repeat line 11 until all of the non-core entities are completely grouped.

Line 16: End IF

Line 17: End For

4.3 The Evaluate Function of Subject Database Division

After division, the subject database needs to be evaluated. This paper presents a method of evaluation function to evaluate the subject database division, which can make the division results more science and reasonable.

Definition 7: Suppose that Ei={e_1,e_2,..,e_k}is a set of all different entities. Ai is activity set in which Ei participated, Ai=Sa(e_1) ∪ Sa(e_2) ∪ ...Sa(e_k). Supposing Ai={a_1,a_2,...,a_t}, then all entities Ai participated are recorded as $E_i^{'}$={Se(a_1) ∪ Se(a_2) ∪ ...Se(a_t)}. The activity relevant number of Ei is recorded as $\lambda_a(E_i) = |E_i| / |E_i^{'}|$.

Definition 8: Suppose that G={E_1,E_2,E_3,..,E_k} is a division of subject database of Ent, Let RG = $\sum_{i=1}^{n} \lambda_a(E_j) \times |E_j| \times GS(E_j) / |E|$,in which, $|E| = |E_1| + |E_2| .... + |E_n|$. Function RG is called as evaluation function of subject database division, RG∈[0, 1]. Function value is called reasonable degree of subject database division. The greater the value of RG is, the more reasonable the division is.

5 Application

In order to verify the algorithm described in this article, some entities planning of Marine Office is given. e_1 ~ e_{11}, respectively represent sailing plan of domestic ships, information of key-tracking ship, check of key-tracking ship, information of deviating track ship, statistics of deviating track ship, stranded information of ship, information of key-tracking reason, PSC safety inspection information, import and export visa, import and export inspection form and the sailing plan of international ship. According to similarity formula of entity, we calculate the similarity of entities to obtain matrix R, as shown in Table 2.

Table 2. Matrix of entity similarity

	e_1	e_2	e_3	e_4	e_5	e_6	e_7	e_8	e_9	e_{10}	e_{11}
e_1		0.5	0.2	0.5	0.5	0.5	0.6	0.5	0.0	0.5	0.91
e_2			0.2	0.9	0.7	0.6	0.6	0.5	0.3	0.4	0.6
e_3				0.9	0.9	0.6	0.6	0.5	0.23	0.2	0.6
e_4					0.92	0.2	0.7	0.5	0.42	0.0	0.6
e_5						0.2	0.8	0.3	0.6	0.6	0.6
e_6							0.4	0.0	0.0	0.0	0.0
e_7								0.2	0.0	0.0	0.0
e_8									0.5	0.0	0.0
e_9										0.9	0.0
e_{10}											0.0
e_{11}											

Divide subject database using cluster analysis method:

1) Traverse similarity matrix R, each pair of entities with higher similarity constitutes a core entity group. Because the similarity value of e_9 and e_{10} with other entities is small, but the similarity value of e_{10} and e_9 is high, so (e_9, e_{10}) also can be used as a core entity group, therefore we can obtain three core entity groups:

$E1=\{(e_4,e_5)|r4,5=0.92\}$

$E2=\{(e_1,e_{11})|r1,11=0.91\}$

$E3=\{(e_9,e_{10})|r9,10=0.90\}$

2) Establish the matrix M_{ij} of core entity group and non-core entity in accordance with the formula of similarity, as shown in Table 3.

Table 3. Similarity matrix of core entity group and non-core entity

	e_2	e_3	e_6	e_7	e_8
(e_1, e_{11})	0.55	0.4	0.3	0.3	0.25
(e_4, e_5)	0.8	0.9	0.2	0.75	0.4
(e_9, e_{10})	0.35	0.21	0	0	0.25

3) Cluster entities based on the cluster algorithm.

- S (e_2, (e_4, e_5)) = 0.8, cluster entity e_2 to core entity group (e_4, e_5), constitute entity group (e_2, e_4, e_5).
- S (e_3, (e_4, e_5)) =0.9, cluster entity e_3 to core entity group (e_4, e_5), constitute entity group (e_2, e_3, e_4, e_5).
- S (e_7, (e_4, e_5)) = 0.75, cluster entity e_7 to core entity group (e_4, e_5), constitute entity group (e_2, e_3, e_4, e_5, e_7).
- The value of similarity between non-core entity e_6, e_8 and the core entity group is very small, so e_6 and e_8 separately form the subject database.

4) Through the cluster analysis, we can get the following subject databases.

- $E_1 = \{e_1, e_{11}\}$ form the subject database of ship plan.
- $E_2 = \{e_2, e_3, e_4, e_5, e_7\}$ form the subject database of tracking ship information.
- $E_3 = \{e_9, e_{10}\}$ as visa test subject database.
- $E_4 = \{e_6\}$ as the subject database of ship stranded.
- $E_5 = \{e_8\}$ as the subject database of security checks.

The cluster results get through cluster analysis should be reviewed by the planning group and the sector representatives based on the evaluation function, and then combined with the data storage locations to further adjust the content and structure of subject database in order to obtain the subject database we need. But it is the conceptual model of subject database, using standardized methods and theories of data structure, format each entity of subject database into the third paradigm that is the basic form, determine the link between the basic tables, we can get the logic model of subject database. This method of subject database division has been applied to a Maritime Office and achieved good results. It proves that this method has the feasibility.

6 Conclusions

The establishment of subject database provides a standard, specification, unified data access mode and data for overall planning of information resources. In this paper, the subject database is divided by using cluster analysis algorithm with the combination of entity similarity, which can overcame the bi-direction problem appearing in integration analytic process, Therefore, subject database division is more orderly, standardized and efficient, especially for large number of entities and activities with the complex relationships between them and the staff with little experience.

References

1. Gao, F.-X.: Information Resource Planning - the Basis of Information Technology Projects, pp. 221–235. Tsinghua University Press, Beijing (2002)
2. Wang, Y.-S., Dong, P.-M.: The Mathematical Formula of Subject Database Planning Reasonable Estimate. Journal of Software 8, 93–98 (1997)
3. Zhou, Y.-T.: Research on Integration Analysis Methodology of Subject Database. Computer Engineering and Applications 41, 168–170 (2005)
4. Chen, X.-G., Li, K.-L.: Application in Personalized Service Based on Cluster Analysis of Association. Journal of Intelligence 28, 196–197 (2009)

A Extension to Triangular Projected Iterated Function System

Huichun Hua and Bo Xiong

School of Mathematics and Physics North China Electric Power University
Baoding, China
{huahuichun,xiongbo0728}@126.com

Abstract. Triangular rational projected IFS model is presented in the paper. This model is generalized from the triangular projected IFS model we proposed before. The properties of triangular rational projected IFS model and the multi-resolution mesh generation algorithm for generating triangular surfaces rapidly are also presented. The main applications are surface modeling, shape description and geometric surface compression.

Keywords: Triangular surfaces, approximating, multi-resolution, surface modeling.

1 Introduction

Surface representation plays an important role in a variety of disciplines, including computer graphics, computer-aided design, computer vision, and geographic data processing, etc. A wide variety of representation methods have been proposed for modeling these surfaces, such as NURBS or B-splines, these models only can represent smooth surfaces.

In order to propose an efficient solution to the problem of rough surfaces approximation, the fractal theory is introduced and many good results have been obtained [1, 2]. Unlike [1, 2], in [3] and [4], C. E. Zair and E. Tosan have proposed a model for fractal curve and surfaces. This model combines the classical iterated function system (IFS) model and the free form approximation theory based on a set of control points. These points allow an easy and flexible control of the fractal shape generated by the IFS model and provide a high quality fitting, even for surfaces with sharp transitions. This model is called the projected IFS model. In [5], E. Guérin, E. Tosan and A. Baskurt have proposed an approximation method for curves based on this model. In [6], E. Guérin, E. Tosan and A. Baskurt presented the extension of this method to surfaces.

Considering the applications of surfaces defined in the triangular domain, we proposed the triangular projected iterated function system to approximate triangular surfaces in [7,8] and we also studied its applications in surfaces morphing in [9], In these papers, we have given some properties and illustrations, finding that triangular projected iterated function system is less flexible.

In order to get a more flexible model to represent the triangular surfaces, we propose an extension model by introducing the weight function. The new model is called triangular rational projected iterated function system, which has higher free degrees and has better representative capability.

W. Zhang (Ed.): Advanced Technology in Teaching, AISC 163, pp. 653–659.
springerlink.com

2 Approximation Model

In this section, the fractal approximation model called triangular rational projected IFS model is proposed, which is the generalized form of triangular projected iterated function system(TPIFS) we proposed in [7,8].

Similar to TPIFS model, the iterative space is defined by B^J :

$$B^J = \left\{ (\lambda_j)_{j\in J} \mid \sum_{j\in J} \lambda_j = 1 \right\},\quad J = \{1,...,m\} \tag{1}$$

And the iteration semi-group is constituted of matrices as follows:

$$S_J = \left\{ \boldsymbol{T}_i \mid \sum_{j=1}^{m} \boldsymbol{T}_i(j,k) = 1, \forall k = 1,...,m \right\} \tag{2}$$

Where $\boldsymbol{T}_i (i = 1,...,n)$ is the subdivision matrix. The attractor of S_J iterated on B^J is called weight function, denoted as follows:

$$\boldsymbol{\Phi}(u,v,w) = [\phi_1(u,v,w), \phi_2(u,v,w),...,\phi_m(u,v,w)]^T \tag{3}$$

Now denote the control points as homogeneous coordinates:

$$\boldsymbol{c}_i^{\omega} = (\omega_i \boldsymbol{c}_i, \omega_i) \tag{4}$$

Where $\omega_i \neq 0$. Triangular rational projected IFS model can be defined as:

$$F(u,v,w) = \frac{\sum_{j\in J} \phi_j(u,v,w)\,\omega_j \boldsymbol{c}_j}{\sum_{j\in J} \phi_j(u,v,w)\,\omega_j},\quad J = \{1,...,m\} \tag{5}$$

The sum of weight function $\boldsymbol{\Phi}(u,v,w)$ equals to 1, i.e.

$$\sum_{j\in J} \phi_j(u,v,w) = 1 \tag{6}$$

3 Properties of Triangular Rational Projected IFS Model

3.1 Affine Invariance

Triangular rational projected IFS model is invariant under affine maps, which means that the following two procedures yield the same result: a) first, compute the model and then apply an affine map to it; b) first, apply an affine map to the control points and the model.

Affine invariance is a direct consequence of the property of weight function showed as (6).

Let us discuss a practical aspect of affine invariance. Suppose we plot the surfaces of 100 points under rotation. We can apply the rotation to each of them and solve the model. Or, we can apply the rotation to the six control points, and then get the model. The first method needs one hundred applications of the rotation, while the second needs only six.

3.2 Degenerated Model

Set $\omega_i (i=0,...,m)$ equal to 1, triangular rational projected IFS model will degenerate into triangular projected iterated function system.

Deduced from (5), if $\omega_i (i=0,...,m)$ equal to 1, the denominator will equals to $\sum_{j\in J}\phi_j(u,v,w)$. Considering (6), triangular rational projected IFS model will equals to $\sum_{j\in J}\phi_j(u,v,w)\boldsymbol{c}_j$, which is the triangular projected iterated function system.

From this property, we can clearly see that triangular rational projected IFS model is the extension form of triangular projected iterated function system.

3.3 Endpoint Interpolation

Given a triangular rational projected IFS model:

$$\left\{B^J, T_l, P_{i,j,k}, \omega_{i,j,k}, l=1,...,N, i+j+k=n\right\}$$

Where $B^J=\left\{(\lambda_j)_{j\in J} \mid \sum_{j\in J}\lambda_j=1\right\}$, T_l is the subdivision matrix, $P_{i,j,k}$ is the control points, $\omega_{i,j,k}$ is weight. Denoting the control points as:

$$\boldsymbol{c}_i^{\omega}=\left(\omega_i\boldsymbol{c}_i, \omega_i\right)$$

If existing one subdivision matrix T_m with the $m-th$ column equals to:

$$e_m=\begin{pmatrix}\underbrace{0 \;\; \ldots \;\; 0}_{m-1} & 1 & 0 & \ldots & 0\end{pmatrix}^T$$

Then the attractor of triangular rational projected IFS model must pass through the point of c_m.

Proof: As all the subdivision matrix is compressive, according to the fix-point theorem, the attractor the TPIFS model is existing and unique, which can be gotten by a special iterative process as follows:

Set the initial points as :

$$B^0 = \begin{bmatrix} \dots & 0 & \dots & \dots \\ \dots & \dots & \dots & \dots \\ \dots & 1 & \dots & \dots \\ \dots & \dots & \dots & \dots \\ \dots & 0 & \dots & \dots \end{bmatrix}_{h\times M} \tag{7}$$

Where every column in B^0 is an element $\left(\lambda_j\right)_{j=1,\dots,h}$ of B^J , M is the size of initial iterative points set. The iterative process is scheduled to:

$$\begin{aligned} &B^1 = \left[T_k B^0 \quad T_m B^0 \quad T_{m+1} B^0 \quad \dots \quad T_{m+N-1} B^0\right] \\ &\dots \\ &B^{n+1} = \left[T_k B^n \quad T_m B^n \quad T_{m+1} B^n \quad \dots \quad T_{m+N-1} B^n\right] \\ &\dots \end{aligned} \tag{8}$$

Because the $m-th$ column vector in T_m is $e_m = \begin{bmatrix} \underbrace{0 \quad \dots \quad 0}_{m-1} & 1 & 0 & \dots & 0 \end{bmatrix}_h^T$, then T_m can be written as:

$$T_m = \begin{bmatrix} \dots & 0 & \dots & \dots \\ \dots & \dots & \dots & \dots \\ \dots & 1 & \dots & \dots \\ \dots & \dots & \dots & \dots \\ \dots & 0 & \dots & \dots \end{bmatrix}_{h\times h} \tag{9}$$

So:

$$T_m B^0 = \begin{bmatrix} \dots & 0 & \dots & \dots \\ \dots & \dots & \dots & \dots \\ \dots & 1 & \dots & \dots \\ \dots & \dots & \dots & \dots \\ \dots & 0 & \dots & \dots \end{bmatrix}_{h\times M} \tag{10}$$

Thus, the $m-th$ column vector of B^1 equals to vector e_m , the $m-th$ column vector of B^{n+1} also equals to vector e_m , when $n \to \infty$, the $m-th$ column vector of B^∞ equals to vector e_m . The attractor can be gotten by projecting B^∞ on the control points.

Considering that an element $\left(\lambda_j\right)_{j=1,\ldots,h}$ of B^{∞} equals e_m ,we can get :

$$P\left(\lambda_j\right)_j = c_m \tag{11}$$

As a result, the attractor of triangular rational projected IFS model passes the point of c_m .

4 Multiresolution Mesh Generation

In order to generate the triangular rational projected IFS model rapidly, we can consider only the value of the grid surfaces. The algorithm can be shown as follows:

Step1: compute the eigenvector of matrix $B_{001}, B_{010}, B_{100}$.

Step2: divide u, v, w into 2^p pieces, and then compute the values at these $2^p - 1$ points.

1) no-boundary points

$$for \;\; i = 1, 2, \ldots, 2^p - 1$$
$$for \;\; j = 1, 2, \ldots 2^p - i - 1$$
$$solve \;\; \phi(i, j), \; then \; compute \; F(i, j)$$
$$\text{end}$$
$$end$$

2) boundary points

$$while \;\; u = 0 \text{ or } v = 0 \text{ or } w = 0$$
$$for \;\; i = 1, 2, \ldots, 2^p - 1$$
$$solve \;\; \phi(i, j), \; then \; compute \; F(i, j)$$
$$end$$

2) corner points

$$F(1,0,0) = P\phi(1,0,0)$$
$$F(0,1,0) = P\phi(0,1,0)$$
$$F(0,0,1) = P\phi(0,0,1)$$
$$then \; converting \; into \; F(i, j)$$

5 Approximation Results

The multi-resolution generation algorithm of triangular rational projected IFS model has been tested by two surfaces: a smooth triangular surface and a rough triangular surface.

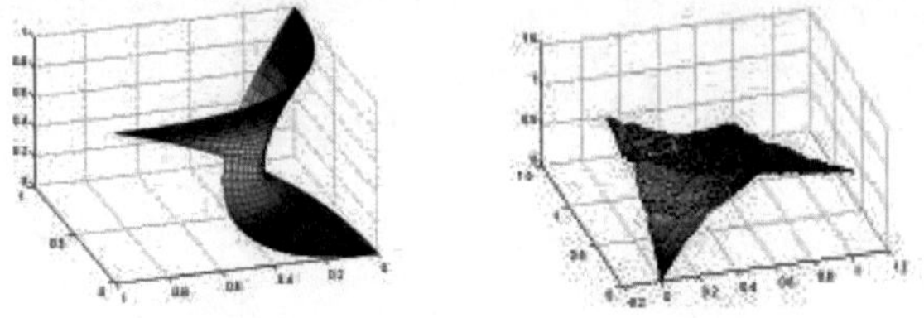

Fig. 1. Multi-resolution generation of smooth and surfaces

Fig. 1(left picture) shows the result of smooth surface, we can see that it is the same as the result of TPIFS model we presented in [7,8],and Figure 1(right picture) shows the result of rough surface.

Because the weight function is introduced into triangular rational projected IFS model, model has more free degrees. Obviously, triangular rational projected IFS model has more advantages over TPIFS model in shape editing. We also give some examples of shape editing represented by triangular rational projected IFS model in Fig. 2.

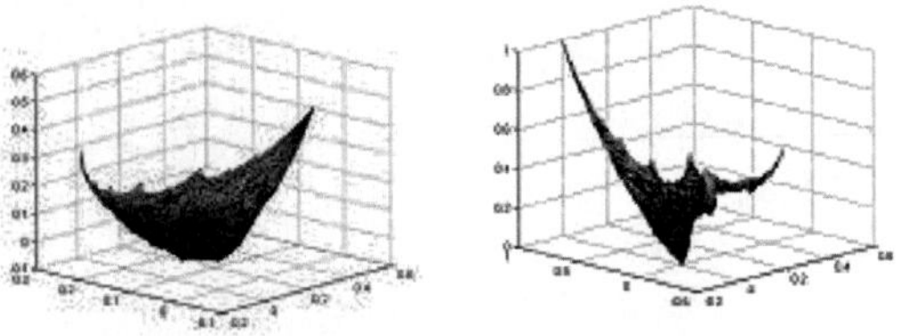

(a) original surface (b) adjust ω

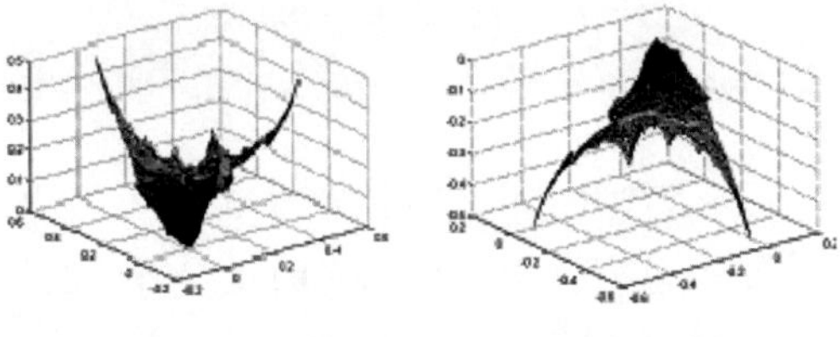

(c) adjust more factors (d) minus weight factors

Fig. 2. Shape editing by weight factor adjustment: (a)~(d)

6 Conclusion

A generalized model from TPIFS model proposed in this paper. Triangular rational projected IFS model has more advantages over TPIFS model: 1)it can represent all the surfaces which can be approximated by TPIFS model owing that TPIFS model is its Degenerated form. 2) it has more free degrees, that makes it possible to edit the shape more easily. Moreover, the properties of this model were discussed in detail. We also gave the multi-resolution mesh generation algorithm to generate the triangular surfaces rapidly. At the end, illustrations showed the approximation performance of triangular rational projected IFS model.

Acknowledgments. The work was partially supported by NCEPU science foundation for young scholars (93209705).

References

1. Zheng, H., Ye, Z., Lei, Y., Liu, X.: Fractal properties of interpolatory subdivision schemes and their application in fractal generation. Chaos, Solitons & Fractals (December 2005) (in Press, Corrected Proof)
2. Zai, Z., Chen, D., Lu, S.: Reconstruction of a fractal rough surface. Physica D: Nonlinear Phenomena 213, 25–30 (2006)
3. Zair, C.E., Tosan, E.: Fractal modeling using free form techniques. Computer Graphics Forum 15, 269–278 (1996)
4. Zair, C.E., Tosan, E.: Unified IFS-based model to generate smooth or fractal forms. In: Le Mehaute, A., Rabut, C., Schumaker, L.L. (eds.) Surface Fitting and Multiresolution Methods, pp. 344–355. Vanderbilt University Press, Nashville (1997)
5. Guérin, E., Tosan, E., Baskurt, A.: Fractal coding of shapes based on a projected IFS model. In: ICIP 2000, vol. II, pp. 203–206 (2000)
6. Guérin, E., Baskurt, A., Tosan, E.: Fractal approximation of surfaces based on projected IFS attractors (2001), `http://www710.univ-lyon1.fr/~et/WLIGIM/textes/equipes/IMAGE/publi/guerin_EG2001.pdf`
7. Hua, H., Jiang, D., Song, W., Ao, B.: Fractal approximation of triangular surfaces based on TPIFS model. Journal of Computer-Aided Design and Computer Graphics 17(1), 174–178 (2005)
8. Hua, H., Zhao, M.: Multiresolution fractal approximation of smooth or rough surfaces defined in the triangular field. In: Proceeding of ICMLC 2008, Kunming, pp. 2957–2963 (July 2008)
9. Hua, H., Gu, G., Wang, T.: A new approach of 3D shape blending. In: Proceeding of Fourth International Conference on Image and Graphics, Chengdu, pp. 863–867 (August 2007)

An Evaluation Approach to the Normativity of Handwritten Chinese Character Based on Vector Cross-Product

Si Huayou[1,2], Qi Hengnian[1], Deng Fei[1], and Zhuang Li[1]

[1] School of Information Engineering, Zhejiang Agriculture and Forestry University, Hangzhou 311300, P.R. China
[2] Key Laboratory of High Confidence Software Technologies, Ministry of Education, Beijing 100871, China

Abstract. With the wide use of information technology, the ability of young people to write Chinese characters is declining. In order to improve the ability, in the paper, we propose an on-line evaluation approach to the normativity of a handwritten character, which is based on the determination of the relationships between writing sub-strokes by vector cross-product. By comparing the relationships between sub-strokes of a handwritten character with the relevant relationships of a standard character, the method can easily determine whether or not a handwritten character is normative. This approach can be used to automatically evaluate handwritten characters in real time and allow paperless teaching character writing.

Keywords: Handwritten Chinese Character, Normativity Evaluation, Stroke, Vector Cross-Product.

1 Introduction

With the wide application of information technology, more and more Chinese writing was replaced by keyboard input. The ability of young people to write Chinese characters is declining, which has become a social problem that can not be overlooked [1]. In order to enhance awareness of the standard Chinese characters writing, improve the capacity in writing and foster interest in writing of young people, in particular primary and secondary students, Ministry of Education of the People's Republic of China has developed a corresponding writing specifications and evaluation criteria.

As a matter of fact, Chinese writing assessment is an essential prerequisite to promote the ability to write Chinese character. It mainly involves evaluation to the normativity of handwritten character. Furthermore, the key issue of the evaluation is how to judge the normativity of the intersection, joint and separation relationships between strokes or sub-strokes of handwritten characters. For example, as far as the character "工" is concerned, if the middle vertical line of the character is beyond the top horizontal line, namely, the vertical line intersects with top horizontal line; the writing is non-standard and become another character, i.e., "土". In fact, strokes and relationships among them are important presentation forms of Chinese characters. They are the basis to determine whether character writing is normative.

W. Zhang (Ed.): Advanced Technology in Teaching, AISC 163, pp. 661–668.
springerlink.com

The traditional methods just focus exclusively on the final result of a writing and usually based on the teacher's marking. Out of question, the artificial methods of evaluation is subjective, high cost and inefficient. Although, the downward trend in writing ability is due to the increasing application of information technology, as a wise idea, the issue to improve the capacity in writing should be resolved through the development of information technology. Accordingly, it is necessary to develop the technology to let computer aid to evaluate the normativity of writing in real-time.

In the paper, we provide an on-line and real-time Chinese writing normative evaluation method, which is based on determining the relationships between sub-strokes of handwritten character. By comparing the relationships of a handwritten character with the relevant relationships of a standard character, the method is used to determine whether or not a handwritten character is normative. Thereby, we can achieve the quantitative normativity evaluation and assessment to handwriting character. By the method, we can also realize real-time computer guidance to characters standard writing.

The rest of the paper is organized as follows: In Section 2, we describe the main process of the normativity evaluation to handwritten character. In Section 3, we analyze the key technique in our approach. In Section 4, we discuss the implementation and properties of the approach. In Section 5, we present the related work. In Section 6, we draw a conclusion and point out our future work.

2 The Main Process of Normativity Evaluation Approach

When using our approach to evaluate the normativity of writing, a main process framework should be adopted, which can be divided into five activities, shown in figure 1.

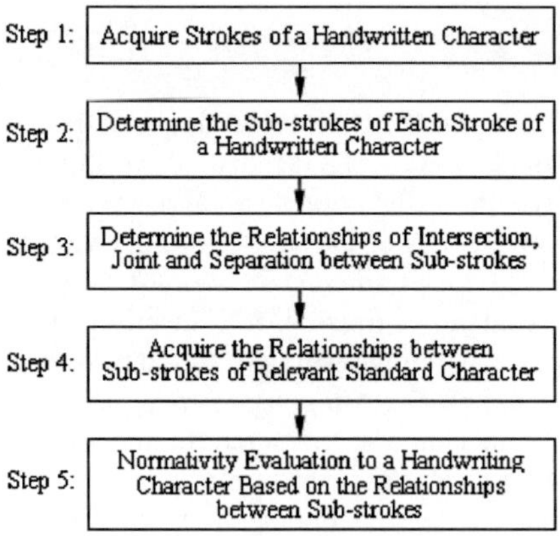

Fig. 1. The main process of normativity evaluation to a writing

In order to facilitate the detailed description of the normativity evaluation approach, here, we first explain each step in the process as follows:

Step 1: It is a unit to acquire strokes of a handwritten character. When a user writes character on a handwriting input device, the device would converted touch pressure to a serial of pixels and record the pixels as strokes of handwritten Chinese character.

In the paper, in order to facilitate to calculate the cross-product of vectors, we create rectangular coordinate system and assign each pixel acquired a corresponding coordinate. When we create the coordinate system, we first get the most up, most down, most left and most right borders of a whole character writing, and then based the borders create a smallest cube that contains the writing, shown in Fig. 2. Naturally, we can view the left-bottom corner of the cube as the coordinate origin of the coordinate system. Based on the coordinate system, we can calculate the coordinate of each pixel acquired on a stroke of a handwritten character.

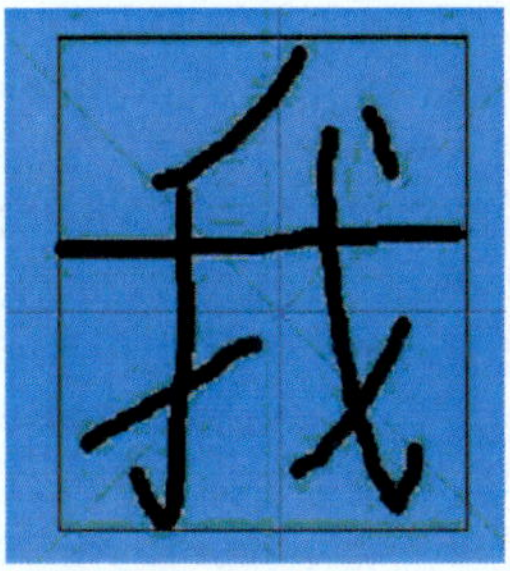

Fig. 2. The smallest cube containing writing

Step 2: The step is to determine the sub-strokes of each stroke of a handwritten character. The stroke of a character refers to a trace of a pen from down to up when a character is written. Specifically, based on the standard in GB13000.1, five basic stroke forms are defined in Chinese character—Horizontal Line, Top-Down Vertical Line, Left-Downward Slope Line, Dot and Twist stroke—which are called 'heng', 'shu', 'pie', 'dian' and 'zhe' in Chinese pinyin respectively. Among these stroke forms, 'zhe' always is composed of several sub-strokes, while 'heng', 'shu', 'pie' and 'dian' can be viewed as the strokes which are composed of one sub-stroke. "sub-stroke" can be named here as 'biduan' in Chinese pinyin. Biduan is an intermediate unit without an inflexion point. Therefore, biduan can be represented with two coordinate. One refers to the starting point of a biduan and the other to its end point.

Since strokes are a fragment trace of a pen, it is easy to separate strokes from a handwritten character and record its order in chronological sequence. However, when a stroke is separated from a handwritten character, it is not so easy to determine which sub-strokes it contains. In this paper, we use a method to identify every sub-stroke from a stroke. The method is described in subsequent section.

Step 3: In the step, we determine the relationships between sub-strokes, which is identified in step 2.

The relationships among strokes in a character are a kind of important information. A character is presented only based on its strokes and the relationships among them. The stroke order of a character is just a simple relationship among strokes. It can be used to initially recognize Chinese character or determine whether a handwritten character is normative. In fact, the relationship only based on stroke order of character is insufficient. For instance, when we practice writing Chinese characters and want to write a character '甲', if the middle vertical line in the character is too long and beyond the top horizontal line, the handwriting is non-standard and it become another character '申', although the strokes and stroke order are all normative.

Accordingly, we need further to consider the relationships of sub-strokes in a handwritten character. We should determine the relationships of intersection, joint and separation between any pair of sub-strokes in a handwritten character. Based on the relationships of sub-strokes we calculate, we can further determine whether or not the writing character is normative. Therefore, in the later section, we focus on how to determine three kinds of relationship between sub-strokes.

In the paper, we adopt the triple (From, To, State) to represent a relationship between a sub-stroke and another sub-stroke. The "From" and "To" in the triple indicate the relevant sub-strokes of a character. The "State" in the triple indicates the type of relationship between the two sub-strokes. When our method is implemented, if the type of relationship between the two sub-strokes is intersection, the "State" is assigned as 0. And the joint as 1, separation as 2.

Step 4: The task of the step is to acquire the relationships between sub-strokes of relevant character in standard font. The relevant character is the character that user try to write. The relationships between sub-strokes also refer to intersection, joint and separation.

For a character in standard font, the intersection, joint and separation relationships between sub-strokes can be determined and stored in advance. When a handwritten character need be determined whether it is normative, we directly retrieve out the relationships among sub-strokes of the relevant standard character.

Step 5: By the step, we finally evaluate the normativity of a handwriting character.

As mentioned above, the standard in GB13000.1 prescribes five basic stroke forms, which are called 'heng', 'shu', 'pie', 'dian' and 'zhe' and numbered 1, 2, 3, 4 and 5 respectively. It also takes stroke order of a character as an index in standard font. For example, the code of stroke order of the character '我' is '3121534'. Based on our previous work [2], we can easily identify each stroke of a handwritten character and encode the handwritten character according to its stroke order.

If the stroke order code of a handwritten character is not coincident with the stroke order code of the corresponding standard character, we can determine that the handwritten character is not normative. Otherwise, we preliminary determine that the handwritten character is normative. We should further determine its normativity based on the relationships between sub-strokes. We compare each triple (From, To, State) of handwritten character with the corresponding triple (From, To, State) of standard character. If all the results of comparison are acceptable, we believe that the handwritten character is normative. Otherwise, it is not normative.

3 The Key Techniques of Our Approach

3.1 Sub-stroke Recognition from a Stroke

As a part of stroke, according to the standard of character stroke, sub-stroke can be viewed as an intermediate unit without an inflexion point. Therefore, after extracting a stroke, we can adopt the distance method to recognize sub-stroke from the stroke. Following the direction of the handwritten stroke, select three consecutive pixels and acquire their coordinates, i.e., B, C and D, in the stroke.

If the result of formula $(BC + CD)\text{-}BD\times\sigma$is larger than zero, we believe that a cut-off point exits between point B and D on the stroke. Hereinto, BC, CD and represent the distance between relevant pixels and σ represents threshold value in the range of $1.00\leq\sigma\leq1.11$. The value of σ is used to adjust the bending degree of the three points.

The process, as above, which is used to determine a cut-off point of a stroke, should be an overlay detection process. For example, if there are 10 pixels in a stroke of handwritten character, first, we get the pixels 1, 2 and 3 to determine whether there is an inflection point, then, get the pixels 2, 3 and 4, until get the pixels 8, 9 and 10. This is to say, turn the focus from starting point to end point and find out all the cut-off points of the stroke.

Besides, in the high-resolution, pixel-intensive cases, we can also get the pixels according to intervals. For example, if there are 100 pixels in a stroke in direction of the handwriting, we can first get the pixels 1, 5 and 10, and then get the pixels 5, 10 and 15, and so on.

If we find out all the cut-off points of the stroke, we can easily extracting every sub-strokes of the stroke. At the same time, it is necessary and feasible to obtain the coordinate of the pixels of each sub-stroke starting point and end point.

In addition to method which adopts three pixels to determine the inflection point, we can also determine the inflection point based on the changes of pixel coordinate. If there is no inflection point in a stroke, the x-axis and y-axis coordinate of continuous pixels changes regularly. Otherwise, the regular change will be broken.

3.2 The Method to Determine the Relationships between Sub-strokes

Based on the standard in GB13000.1, in a character, only three kinds of relationships may exist between any two sub-strokes, i.e., intersection, joint and separation. If there is not any pixel point shared by two sub-strokes, the relationship of the two sub-strokes is separation. If there is a pixel point shared by two sub-strokes and the point is starting point of one sub-stroke or end point of the other sub-stroke, the relationship of the two sub-strokes is joint; else if the point is not end point or starting point of the two sub-strokes, the relationship is intersection.

In the paper, the method to adopt to determine the relationships between two sub-strokes is described as follows:

First, we obtain the starting point P1 and end point P2 of sub-stroke P and obtain the starting point Q1 and end point Q2 of sub-stroke Q. Then, we determine the relationship based on the vector cross product.

If ((P1, Q1)×(Q2, Q1))×((P2, Q1)×(Q2, Q1))<0, and ((Q1, P1)×(P2, P1))×((Q2, P1)×(P2, P1))<0, we can believe that the relationship between sub-stokes P and Q is intersection.

Hereinto, (P1, Q1)×(Q2, Q1) represents the cross product of vector (P1, Q1) and (Q2, Q1); ((P1, Q1)×(Q2, Q1))×((P2, Q1)×(Q2, Q1)) represent the product of vector cross product A and B, where A is the vector cross product of vector (P1, Q1) and vector (Q2, Q1), and B is the vector cross product of vector (P2,Q1)and vector (Q2, Q1).

If ((Q1, P1)×(P2, P1))×((Q2, P1)×(P2, P1)) = 0, or ((P1, Q1)×(Q2, Q1))×((P2, Q1)×(Q2, Q1)) = 0, the relationship between sub-stokes P and Q is joint.

Otherwise, the relationship between sub-stokes P and Q is separation.

In fact, a property of the vector cross-product is applied to the method to determine the relationship between the two vectors. The property can be described as follows:

If P×Q>0, P is in a clockwise direction of Q; if P×Q<0, P is in the anti-clockwise of Q; if P×Q=0, P and Q are in a line.

For the approach, an optimization can be used: Before we determine the relationship between sub-strokes P and Q based on vector cross product, we can create two rectangles. The first rectangle takes sub-stroke P as its diagonal line, while the second rectangle takes sub-stroke Q as its diagonal line. If the two rectangles are disjoint, the two sub-strokes must be disjoint. Therefore, the relationship of the two sub-strokes is separation. If the two rectangles are intersecting, we must further determine the relationship between the two sub-strokes based on vector cross product.

Since it is very easy to determine whether two rectangles are disjoint, if we can determine two rectangles are disjoint, we do not need to judge relationship of the two relevant sub-strokes by vector cross-product. Our experiments prove that this optimization indeed reduce a large number of calculations.

4 Implementation and Discussion of the Approach

As an evaluation approach to normativity of Chinese character writing, we have implemented it based on Stroke order specifications on GB13000.1 Chinese character set [3]. It proves to be feasible and efficient method to evaluate the normativity of writing.

This approach can be used to evaluate handwritten Chinese characters and allow paperless teaching character writing. When the user, adopting mouse, tablet, and touch screens as handwriting input device, writes Chinese character, the method can records the process information of writing and analyzes the relationships between any two sub-strokes of the writing. Then, based on the relationships among sub-strokes, the normativity of writing can be evaluated. As we all know that the relationship among strokes in a character is a kind of important information, for a character is presented only based on its strokes and the relationships among them. Therefore, the evaluation approach to normativity of writing is reliable and feasible.

In fact, the approach overcomes the shortcomings of subjectivity of manual marking. The normativity of writing Chinese characters can be objectively evaluated in real-time, with advantages of low cost and high efficiency. Therefore, the approach can be used to evaluate Chinese writing to achieve intellectually teaching and counseling.

5 Related Work

The research work [4-9] on Chinese Character Recognition has been done for many years, which has yielded fruitful results. The goal is to do its best to identify handwritten Chinese character. Generally, it does not care about whether or not a handwritten character is ill-formed or incorrect. In fact, the current handwriting recognition technology records the coordinates of pixels and the vector-value of handwriting direction in chronological order when the user writes a character. According to strokes, the technology splits them into different sequence set. Then, it compares handwriting with font sequence of the standard character set and takes the highest similarity of Chinese characters as the recognition results to recommend to the user. The work seldom cares about normativity of writing.

It is different from our work in the paper. On the contrary, we are mainly concerned with the normativity of a handwritten character and try to determine whether or not a Chinese character is written correctly and normatively. Therefore, we pay attention on chronological process and writing information of a Chinese character. Our purpose is to provide a method which is computer-based and real-time evaluation approach to the normativity of writing according to standard font, by which order, location and relationship of strokes or sub-strokes and the overall structure of writing evaluated. Our method aims to assess the writing level of the recognizable handwritten character.

In addition, our method is different from the general calligraphy assessment [10] too. Although calligraphy assessment also evaluates the quality of character writing, it is not from a normative point of view of writing, but from the perspective of the art of Chinese character calligraphy.

6 Conclusion and Future Work

In the paper, we provide an approach to evaluate the normativity of handwritten character based on the recognition of the relationships between any two sub-strokes of a handwritten character. The approach firstly acquire strokes of a handwritten character and determine the sub-strokes of each stroke in a handwritten character; next, determine the relationships between any two sub-strokes; then, acquire the relationships between sub-strokes of relevant standard character; and finally, evaluate the normativity of a handwriting character based on the relationships between sub-strokes.

In particular, we emphasize on how to separate sub-strokes from a stroke and determine the type of relationship between two sub-strokes. In order to recognize sub-strokes, we apply the method of distance detection and discuss it in detail. For the issue to determine the type of relationship between sub-strokes, according to the coordinates of sub-stroke, we propose a solution which is based on vector cross-product and present it in the paper in detail too.

Through its implementation and our experiment, the evaluation approach can be considered feasible and effective. It can be used to evaluate Chinese writing.

Although the method can determine whether or not writing is recognizable and normative, it can not determine the aesthetic level of writing, this is, it can not score the merits of the overall structure of the degree. This is our further work.

Acknowledgment. This work has been supported by the Natural Science Foundation of Zhejiang Province of China under Grant No. 2009C33087 and the Natural Science Foundation of China under Grant No. 60970082.

References

1. Qi, H.-N.: A Size-independ Ent method FOR Chinese Character Writing Structure assessment. In: Proceedings of the 7th International Conference on Machine Learning and Cybernetics, Kunming, China, July 12-15 (2008)
2. Qi, H., Si, H., Deng, F., Zhuang, L.: A Robust and On-Line Evaluation Approach to Normativity of Chinese Character Writing. Technology report. School of Information Engineering, Zhejiang Agriculture and Forestry University (March 2010)
3. Fanao, T.: Stroke order specifications on GB13000.1 Chinese character set. Shanghai Education Publishing House, Shanghai (1999)
4. Stallings, W.: Approaches to Chinese Character Recognition. Pattern Recognition 8(2), 87–98 (1976)
5. Cheok, A.D., Jian, Z., Chng, E.S.: Efficient mobile phone Chinese optical character recognition systems by use of heuristic fuzzy rules and bigram Markov language models. Applied Soft Computing Journal 8(2), 1005–1017 (2008)
6. Liu, C.-S., Ding, X.-Q.: Study of character recognition using writing style consistent. Zidonghua Xuebao/Acta Automatica Sinica 33(11), 1121–1127 (2007) (Language: Chinese)
7. Liu, C., Jaeger, S., Nakagawa, M.: Online recognition of Chinese characters: the state-of-the-art. IEEE Transactions on Pattern Analysis and Machine Intelligence 26, 198–213 (2004)
8. Wei, M., Meng, M., Wu, Z., Ge, Y., Shen, F.: Online Chinese Character Recognition Based on Force Vector. In: The Sixth World Congress on Intelligent Control and Automation, WCICA 2006, vol. 2, pp. 9609–9612 (2006)
9. Li, Y., Xie, M.: Chinese character recognition based on character reconstruction. In: International Conference on Communications, Circuits and Systems, ICCCAS 2009, pp. 460–463 (2009)
10. Lv, X., Huang, D., Song, E., Li, P., Wu, C.: One Radical-Based On-Line Chinese Character Recognition (OLCCR) System Using Support Vector Machine for Recognition of Radicals Bioinformatics and Biomedical Engineering. In: The 1st International Conference on ICBBE 2007, pp. 558–561 (2007)

Study on Inherent Mechanism of Strategic Transformation Capability

Tang Jianxiong[1] and Huang Jianbo[2]

[1] Business School of Central South University
Tourism College of Hunan Normal University
Changsha, P.R. China
[2] School of Business Central South University
Changsha, P.R. China
hntjx168@163.com, jbhuang@mail.csu.edu.cn

Abstract. This paper begins with the four basic propositions of strategic transformation capability, exploring the internal mechanism of strategic transformation capability, and puts forward to strategic transformation capability is interacted to form by strategic transformation and the capability, also it is the capability integration system which composed of four capabilities including environment distinguish, resource integration, management control and continuance innovation, and it evolves as a self-organizing system relying on the continuing synergetic relation among above four capabilities and becomes a source of competitive advantage.

Keywords: Strategic Transformation Capability, Internal Mechanism, Competitive Advantage.

1 Introduction

Strategic transformation has become an important part of strategic management of the enterprises in recent years. However, with the intense competition, uncertainties and intangibles of the market environment, the strategic transformation has become more and more risky, which results in nearly 40% transformation enterprises ended in failure. Surprisingly, in the high failure rate of the strategic transformation, some enterprises have made a continuous success of the transformation, such as IBM, Lenovo, Galanz. Considering of the successful transformation results, there should be a question why some companies can make a successful transformation while others can not? The fact shows that the firms focus on getting more capabilities through successful transformation than their competitors (Ginsberg, 1988) [1]. Strategic transformation capability influences the company performance, which helps to enhance the competitiveness and ensure the company's sustainable growth. Therefore, it is important to understand the inherent mechanism of strategic transformation from a theoretical and practical view. However, it is still in the preliminary stage for strategic transformation capability research. Today, people know a little about strategic transformation capability, even the relationship between strategic transformation capability and competitive advantage.

W. Zhang (Ed.): Advanced Technology in Teaching, AISC 163, pp. 669–674.
springerlink.com

Thus, we will make a research about the inherent mechanism according to strategic transformation capability's four basic questions: How to form strategic transformation capability? What is the characteristic as a special capability of enterprise? Dose the evolving rule of strategic transformation capability reflect a self-organizing system? What is the relationship between strategic transformation capability and competitive advantage?

According to the logic, all things are judged by proposition. So what we talked above just constructs four basic propositions of strategic transformation capability. Later, we will discuss the transformation with lots of theories of the economics, management, Organization Behavior and so on. We are trying to help the academia and the enterprise to penetrate the transformation through a new perspective, in order to recognize the inherent mechanism of transformation and offer some theoretical direction in practice.

2 The Alternation of Strategic Transformation and Capabilities Forms Strategic Transformation Capability

Capability is a complex concept, there have been many scholars (Richardson, 1972 [2]; Prahalad and Hamel, 1990 [3]; Foss, 1998 [4], etc.) from different perspectives to research and define the company capability. Looking at the literatures about capability, we believe that the capability is not only an rational resource distribution activity process, on behalf of the resources reciprocity in the organization, but also has significant inertia and corporate monopoly, and to be changed according to the solution of the practical problems. Otherwise, capability emphasizes that strategy management should match the requirements of external changeable environment through integrating and re-allocating organizational skills, resources (Teece et al. 1997) [5]. Strategic transformation capability as a special capability, its formation is closely related to strategic transformation, the essence of strategic transformation is a process which changes the original orientation to form a new pattern through firm resources re-allocation, integration and management (Bossidy and Charan, 2005 [6]; Zhou faming, 2006 [7] etc.). Therefore, strategic transformation capability is a kind of capability based on strategic transformation, a complex whole of "strategic transformation" and "capability" (as shown in Figure 1). Actually, strategic transformation is a process of the capability accumulation and conversion, and it provides a development direction, velocity and important point for capability. Obviously, this capability is dynamic, positive feedback and growing which focuses on capability promotion, expansion and span. Enterprises should do their best to foster and devote to strategic transformation capability, because it exceeds the growing upper limit to ensure continuous growth by the implementation of strategic transformation (Figure 1).

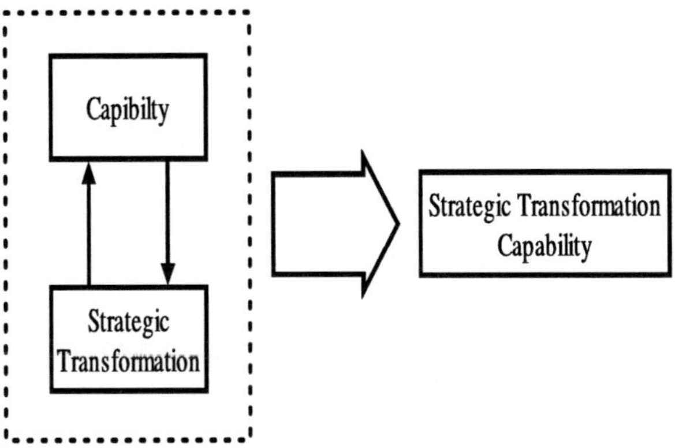

Fig. 1. The relationship betweens strategic transformation and capability

3 Strategic Transformation Capability Is a Source of Competitive Advantage

3.1 Co-ordination in Strategic Transformation Capability

Strategic transformation capability is a self-organization system which composes by the collaboration of environmental identifying, resource integrating, management controlling and continuous innovating. Each capability sub-system accumulates the energy by adapting the changeable environment and constantly internal organizational learning, then they transform strategic transformation capability from low-level to high level structural change in the process of continuous interaction and co-evolution. It furtherly explains strategic transformation capability is a kind of dynamic capability which integrates, establishes and re-allocates internal and external resources, then reflects the integrity, evolutionary, open and dynamic(Teece et al., 1997) [5]. Strategic transformation capability is an order parameter of sustainable development for the enterprises, and its formation is not a simple addition of each capability sub-system, but a synergetic result of the self-organization. The order parameter comes from the internal system rather than the external role. The capability sub-systems acts independently without excellent collaboration before the forming of strategic transformation Capability, even some capabilities own a temporary advantage but also unsustainable in the long term. Only all capabilities including of environmental identification, resource integration, management control and continuous innovation achieve the critical state of collaboration, they will form the relationship of cooperation, then leads to the emergence of the order parameters, and at last forms strategic transformation capability. Once the order parameter forms, it will dominate all the capability sub-systems to act with the parameter's order (such as the selection and allocation of the resources). That is the existing condition of opposite side for each other. In a word, the dynamic evolution of strategic transformation capability is embodied in the collaboration process of the environmental identification capability, resource integration capability, management control capability and constantly innovation capability, which is an important source of the competitive advantage (Figure 2).

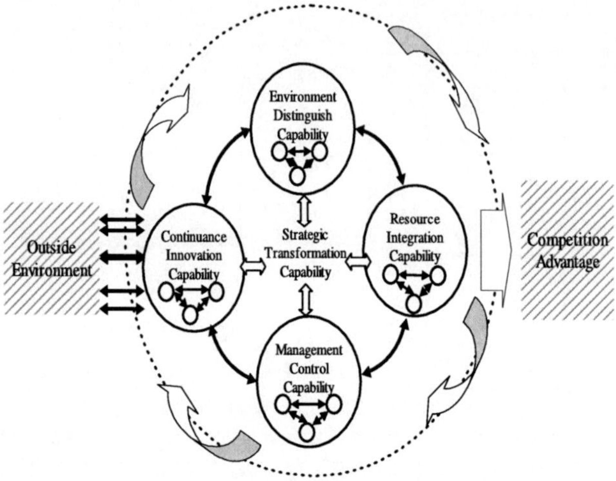

Fig. 2. The collaboration of the sub-systems of strategic transformation capability

Secondly, the collaboration of the capabilities including environmental identification, resource integration, management control and continuously innovation is mainly reflected in the interdependence, interaction and mutual influence. Among them, the environment distinguishing capability is a premise of strategic transformation capability, a window that reflects strategic transformation capability from outside to inside and forms the combining characteristics. It provides a prerequisite for the resource integration capability's operation by determining the strategic transformation direction and adjusting the transformation content when it anticipates the environmental changeable factors. The resource integration capability is the foundation of strategic transformation capability, and it provides an operation base for the management control capability through the internal and external resources selection, configuration and integration. The management control capability is the security factor of strategic transformation capability, and it can ensure the smoothly implementation of the target and the direction of strategic transformation. The continuous innovation capability is the dominant factor of strategic transformation capability, and it creates a new business model by revolution and innovation of the routines for enterprises to keep a sustained competitive advantage. Therefore, strategic transformation capability comes true self-improvement and evolvement in the constantly collaboration and circulation process of the four capabilities, so the four capabilities play synergies at the same time and no clear priorities in a complete strategic transformation process.

3.2 Strategic Transformation Capability Decides the Competitive Advantages

The heterogeneity of enterprises which mainly reflects the capability differences determines the competitive advantages [8]. Strategic transformation capability as a dynamic capability effects on the cycle of enterprise competitive advantage. According to the viewpoint of George S. Day [9], the competitive advantage exists in a cycle process of dynamic environment, the enterprise will transform and invest a new strategic resource to enhance or rebuild the new competitive advantage when the original competitive advantage becomes weaken in the changeable environment. Any single capability existed in the strategic resource can only be a basis of the sustainable competitive advantage, but can not ensure the advantage for a long term (Teece et al., 1997) [5]. It is no exception in the field of the strategic transformation Accordingly, strategic transformation capability plays an important role in every link of the competitive advantages cycle. The strategic transformation begins with the environmental change, the enterprises rely on the environment distinguishing capability to change the strategic content according to the environmental change, and foresee the lots of complex factors of environment which impacts on the competitive advantage, and then determine the strategic transformation direction and adjust the transformation content, and further builds the strategic competitive advantages by integrating and re-allocating the internal and external resources in the plenty of strategic-related activities. The competitive advantage of enterprises comes from not only the factor endowment, but also the using and linking means of the factors (Penrose, 1959) [10]. These use and linking means of the factors to be a solidifying routine by the enterprises to integrate inside and outside resources (Nelson & Winter, 1982) [11], and at last relying on the management control capability to ensure strategy implementation smoothly after the integration of resources. Rumelt (1984) [12] believes that there exists some uncertainty in the enterprises which is fundamental and closely related to the special resources (hidden resources) that can lead to the success of the enterprises. Hidden resources include resources of entrepreneurs, inimitable corporate culture, intangible assets such as reputation and successful management experience of problem-solving, its common ground is that they are accumulated with an uncertain, inimitable process in a very long time by the enterprise. Under normal circumstances, it is very difficult for a company to copy or imitation competitive advantage of others, such as management control, innovation and so on. The Enterprise continually innovate to create new routines, and to produce the innovation capability, then innovative form a new business model in the integrating resources process, and to adapt the changeable environment by establishing the new competitive advantage through the successful transformation (Tang Jianxiong, Wang Guoshun, 2008) [13]. In this way, the difference of the innovation capability leads to the heterogeneity among the enterprises, which constitute the sustainable competition advantage furtherly. Therefore, the heterogeneity means the probability for the enterprises to obtain to the economic "rent".

In a word, strategic transformation capability which plays a durative effectiveness is not depending on some single capability but the collaboration of the sub-capability in the integration process. That is, every capability formed in the strategic transformation has a positive impact on the firm performance, and this impact is specific and inimitable by other competitors. The sustainable competitive advantage embodied in strategic transformation capability is a combined, dynamic competitive advantage based on the synergy of the capabilities, which brings a constantly growth for the enterprise.

4 Conclusion

In this paper, we begin with the four basic propositions of strategic transformation capability, and then do some researches about inherent mechanism of strategic transformation capability. We systematically analysis some issues around strategic transformation capability about its formation, characteristics, the capability evolution rule as well as the relationship with the competitive advantages and so on. It comes out that strategic transformation capability is a result of the interaction by the strategic transformation and the capability, and is a capability integration system which composed of environmental identification, resource integration, management control and constantly innovation. Its evolution displays a self-organizing system, which relies on the interdependence, interaction and mutual influence of the capabilities including of environmental identification, resources integration, management control and self-innovation to become the competitive advantage source of enterprises.

Without doubt, strategic transformation capability is a very new research area in itself which is full of practical value, and it is still in the preliminary stage both at home and abroad. There still are many issues worthy for us to research constantly and deeply. This paper only discussed some the basic question in this field, so the forward-looking and practice guidance of the strategic transformation theory are worth seeking furtherly.

Acknowledgment. This paper is supported by Central South University Postdoctoral Science Foundation Project.

References

1. Ginsberg: Measuring & modeling changes in strategy: theoretical foundation & empirical direction. Strategic Management Journal 9, 559–575 (1988)
2. Richardson, G.B.: The organization of industry. Economic Journal 82, 883–896 (1972)
3. Prahalad, C.K., Hamel, G.: The core competence of the corporation. Harvard Business Review 68(5/6), 79–93 (1990)
4. Foss, N.J.: The classical theory of production and the capabilities view of the firm. Journal of Economic Studies, Glasgow 24, 307–314 (1997)
5. Teece, D.J., Pisano, G., Shuen, A.: Dynamic capabilities and strategic management. Strategic Management Journal 18, 509–533 (1997)
6. Bossidy, L., Charan, R.: Confronting Reality–Doing What Matters to Get Things Right. Publishing House of CITIC, Beijing (2005)
7. Zhou, F.: On the growth of small and medium-sized strategic transformation process. Journal of Economic Freely 6, 72–74 (2006) (in Chinese)
8. Zhang, W.: Study on Firm Strategic Capabilities. Publishing House of Science, Beijing (2005) (in Chinese)
9. Day, G.S., Reibstein, D.J., Gunther, R.E.: Dynamic Competition Strategy. Publishing House of Shanghai Traffic University, Shanghai (2003) (in Chinese)
10. Penrose, E.: The Theory of the Growth of the Firm. Oxford University Press (1959)
11. Nelson, R.R., Winter, S.G.: An Evolutionary Theory of Economic Change. Havard University Press (1982)
12. Rumelt, R.P.: Towards a Strategic Theory of the Firm. In: Lamb, R.B. (ed.) Competitive Strategic Management. Prentice-Hall, Englewood Cliffs (1984)
13. Tang, J., Wang, G.: An Empirical Study on the relationship between Chinese Enterprise Strategic Transformation Capability and its Performance. Journal of Systems Engineering 1, 15–19 (2008) (in Chinese)

Adaptive Neural Network Applications in Ship Motion Control

Hu Guanshan and Pan Weigang

Shandong Jiaotong University
Jinan, China
hugshan@126.com, panweigang1980@163.com

Abstract. For the purpose of improving the performance of ship motion control, this paper proposes an intelligent reference modeling adaptive controller for ship steering. The controller is based on artificial intelligence. We used fuzzy logic and neural networks to design the feedback controller, used multilayer perceptron neural network to design the reference model and the identification network. Based on the fuzzy control and neural network, an intelligent adaptive control algorithm was presented in the paper. In order to enhance adaptive characteristics of the controller, the parameters of membership functions and connection weights were revised online by neural network learning algorithm. The simulating result indicates that the performance of the ship controller is valuable and easy to implement.

Keywords: Adaptive control, neural network, ship motion.

1 Introduction

The automatic instrument of ship maneuvering is a important equipment in shipping, it is used to maintain a given direction track, its performance directly affects the navigation of the economy and security. Traditional autopilots adopt PID controller, with the development of control theory, many control algorithms were used in autopilot for achieving high performance. Holzhuter has tested the LQG optimal control scheme. Unar and Murray-Smith have studied the neural networks. Jia X has also presented the sea trial results of a prototype of a tracking autopilot for a full-scale ship, which based on the artificial neural network. Fossen has developed a nonlinear passive observer for marine vessel control. It may be seen that whenever a new control algorithm is introduced, it must be used in the field of vessel controlling. [1]

Recent advances in computing technology have prompted the so-called "Intelligent Control" theory, in which control algorithms are developed by emulating certain characteristics of intelligent biological systems. Adaptation and learning are the most interesting features that can be used for the design of more sophisticated and reliable controllers. Neural networks and fuzzy logic paradigms represent the most popular of such algorithm that can be used in order to provide the control systems with the ability to learn and adapt to different operating conditions.

A neural network is a massive system of parallel distributed processing elements connected in a graph topology. Neural network deals with cognitive tasks, such as

W. Zhang (Ed.): Advanced Technology in Teaching, AISC 163, pp. 675–683.
springerlink.com

learning, adaptation, generalization and optimization. Indeed, recognition, learning, decision-making and action constitute the principal navigation problems. Neural controllers learn, may cause systems to be unstable. In addition, sufficient training patterns are usually difficult to obtain, and training time for the whole dynamic range is very long. So fuzzy logic and neural networks are used.

The rest of the paper is organized as follow. First we discuss adaptive neural network application for ship in section 2. In section 3, we discuss our method of designing a reference modeling adaptive control based on neural network of ship motion, which depends in using two multilayer neural networks for NNRM and NNI. We design a feedback controller for the system using fuzzy and neural network in section 4. Simulation results are included in section 5. The text is summarized our conclusions and gives the notes for our further research in this area.

2 Adaptive Neural Network in Steering

2.1 Automatic Steering of Ship

The control of ship steering, is essentially achieved by controlling the rudder. An automatic steering mode, for example, measures the ship's course and navigation position information to automatically maintain a ship sailing in a given course. As shown in Figure 1.

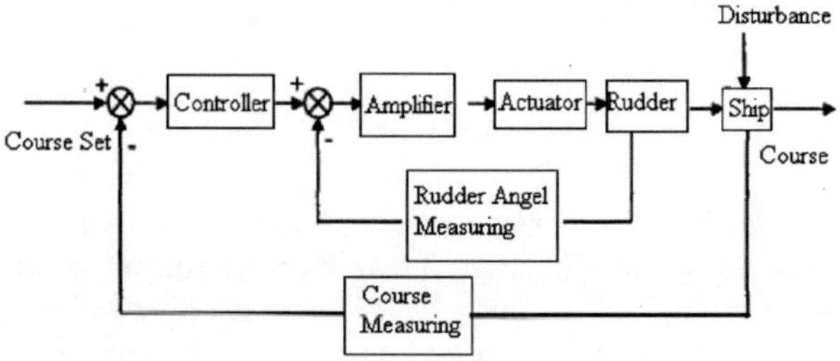

Fig. 1. The structure of ship automatic steering.

Automatic steering device, first, make the ship as required stability on the scheduled course. Second, significantly reduced the rudder deflection angle and the number of fixed rudder and, therefore, reduce the bias caused by the sailing rudder resistance. But the control of ship is nonlinearity, time-varying, and uncertainty complex system, traditional control way cannot to meet the exact requirements.

2.2 Neural Networks for Ship Control Applications

The neural networks for control applications have witnessed a growing increase in the use and study. The advantage of the neural network is the fact that under certain conditions, the parameters of the network can be adjusted from input to output data. A learning system relies on the emphasis in the approximation of the functional mapping representing the controller, but not an adaptive system. In fact, using the approximation theory, the controller design reduces in finding an appropriate functional mapping from the measured and desired system's output to a control action that will produce a satisfactory behavior of the control system. A control system that, during the functional approximation, treats every distinct operating situation as a novel one is limited to adaptive operation, whereas a system that correlates past experiences with the current situation, and that can recall and exploit those past experience, is capable of learning.

Sailing ship, the subject of the disturbance is more complex. To ensure the ship is at sea in accordance with the setting course, it must be timely given the appropriate steering signal to rudder deflection, correcting the deviation caused by disturbance. As the ship parameters of the nonlinear, time variability and uncertainty, etc., it is difficult to establish accurate mathematical model of the ship. Traditionally, adaptive control can not adapt to this situation. Since fuzzy control does not depend on the precise mathematical model, self-learning ability of neural networks, fuzzy control and neural networks combined intelligent control technology into the ship's steering control system design, may solve the issue. [3]

Two different methods exist for the adjustment of the controller parameters, namely direct and indirect methods. In the former approach, the control parameters are directly adjusted based on an observed output error, while in the latter one the parameters of the plant to be controlled are estimated and the parameters of the controller are adjusted based on these estimates. It is important to note that the choice of the particular control structure will determine the properties of the adaptation algorithm to be used. In the following sections an indirect adaptive neural controller is proposed for the steering problem of sailing ship, a proper choice of the identifier and reference model is implemented by function neural networks.

3 The Adaptive Neural Network Controller Design

Reference model based on neural network intelligent adaptive ship course control system is shown in fig2. The input of control object is the rudder angle, the output is sailing course. NNRM is the structure of neural network reference model, its input is a setting course and its output is the desired course. NNI that the structure of neural network identifier, is as the ship's identification model, the algorithm by error back-propagation learning was, and to provide gradient information for fuzzy neural network controller online learning. FNNC, used as a feedback controller, its input is deviation of course, output is rudder angle. The weights and parameters of the FNNC is adjusted according to deviation and $\frac{\partial y_1}{\partial u}$, the purpose is to enable the ship course can follow the reference model output.

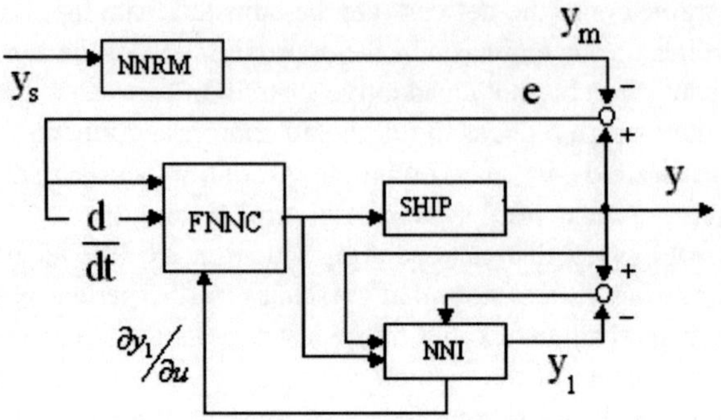

Fig. 2. A reference modeling adaptive control based on neural network of ship sterring

3.1 Neural Network of NNRM and NNI

Here we used a multilayer perceptron neural network to design the NNRM and NNI. Multilayer perceptron is a layered feed-forward network. These networks have found their way into countless applications requiring static pattern classification. The main advantage is that they are easy to use, and that they can approximate any input/output map. [3] The multilayer perceptron network is shown in Figure 3.

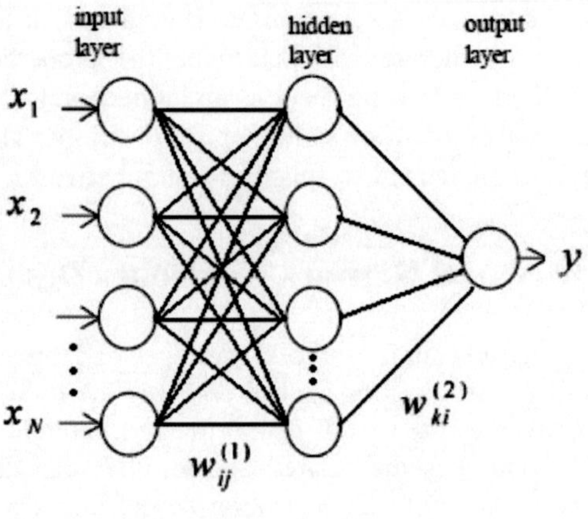

Fig. 3. The multilayer perceptron neural network

The back-propagation learning algorithm is based on the error-correction principle. The learning process is as follows. On the first layer the input data are given and then are calculating the response of the network. The error ei between an actual output and a desired output is acquiring by formula (1).

$$e_i(n) = d_i(n) - y_i(n) \qquad (1)$$

Where di is desired output and yi is the actual output. During the back-propagation compute the δ i by propagating the errors backwards. Update the weights using formula (2).

$$w_{ij}(n+1) = w_{ij}(n) + \eta\delta_i(n)x_i(n) \tag{2}$$

Where ω_{ij} is the weight between ith neuron from last layer and jth neuron in the next layer, η is the learning rate. This process is repeated for the next input-ouput pattern until the error is below a specified threshold or a maximum number of iterations are reached. We used the minimization of the average squared error as evaluate function. Eavg is given by formula (3).

$$E_{avg} = \frac{1}{N}\sum_{n=1}^{N}\frac{1}{2}\sum_{j=1}^{V}(d_i(n) - y_j(n))^2 \tag{3}$$

3.2 The Neural Network Identification of Reference Model

By the course changing of the ideal ship for online identification, we can get the NNRM model. It is shown in fig 4. Yr is the setting course, and y is the ideal steering angle. The deviation of the two signals Y and Ym is as adjust signal for reference model NNRM. Learning algorithm is shown as in Section A.

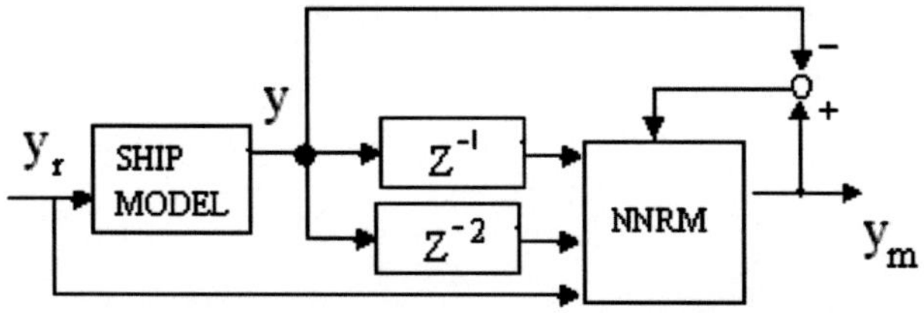

Fig. 4. The neural network identification of reference model

The learning structure of NNI identifier neural network is similar as NNRM except the input is rudder angle. It can proved gradient information for the feedback control FNNC.

4 The Fuzzy Neural Network Controller

4.1 The Fuzzy Neural Network of FNNC

Here we used a fuzzy neural network to design the feedback control. The structure of neural network controller is shown in Figure 5. [4]

There are two input parameters in the first layer. The second layer has fourteen neurons. In the third layer, there have forty nine neurons as the fourth layer.

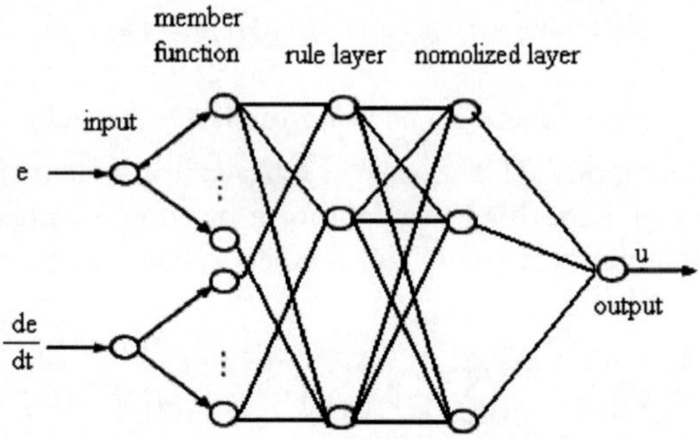

Fig. 5. The structure of neural network feedback controller

The first layer is the network input. The nodes directly connect with the input vector of the various components, and the value is sent to the next layer.

$$f_i^{(1)} = x_i^0 = x_i,\ x_i^{(1)} = g_i^{(1)} = f_i^{(1)},$$

In the second layer, each node represents a linguistic variable value. It is a vague process. That is, calculating the weight of the input value which belongs to the fuzzy set membership function.

$$f_{is_i}^{(2)} = -\frac{(x_i^{(1)} - c_{is_i})^2}{\sigma_{is_i}^2},\ x_{is_i}^{(2)} = \mu_i^{s_i} = g_{is_i}^{(2)} = e^{f_{is_i}^{(2)}},$$

$\mu_i^{s_i}$ is a membership function.

A node in the third layer is a fuzzy rule, which is to be used to match the premise of fuzzy rules.

The nodes of the fourth layer are as the third layer. It is the realization of normalized computing.

$$f_j^{(4)} = x_j^{(3)} / \sum_{j=1}^{m} x_j^{(3)} = \alpha_j / \sum_{j=1}^{m} \alpha_j$$

The out layer realizes clarity for the calculation.

$$f_k^{(5)} = \sum_{i=1}^{49} \bar{\alpha}_j \omega_{ij},\ x_1^{(5)} = u$$

In the rule layer of NNC, there are 49 rules. It is shown the fuzzy rules as in Table 1.

Table 1. The rules of the ship steering control

EC / E	NB	NM	NS	ZE	PS	PM	PB
NB	NB	NB	NB	NB	NM	NS	ZE
NM	NB	NB	NB	NM	NS	ZE	PS
NS	NB	NB	NM	NS	ZE	PS	PM
ZE	NB	NM	NS	ZE	PS	PM	PB
PS	NM	NS	ZE	PS	PM	PB	PB
PM	NS	ZE	PS	PM	PB	PB	PB
PB	ZE	PS	PM	PB	PB	PB	PB

4.2 Parameter Learning of FNNC

During dynamic learning process, learning algorithm is the indirect back learning of FNNC. As the control object is not to know in fact, FNNC study can not directly be required gradient $\partial y / \partial u$, so it often get gradient information instead by $\partial y_1 / \partial u$ from NNI.

Learning parameters are the connection weights of the last layer and the central value and width of membership functions of the second layer. We use the BP algorithm to adjust the parameters of FNNC. Take the variance of ship out and NNRM out as object function.

$$E_C = \frac{1}{2}(y_m - y)^2 .$$

So we can obtain from the equation that

$$\frac{\partial E}{\partial net^{(5)}} = -e_c \frac{\partial y1}{\partial u} .$$

A gradient is:

$$\frac{\partial E}{\partial c_{ij}} = -\delta_{ij}^{(2)} \frac{2(x_i - c_{ij})}{\sigma_{ij}^{\ 2}} ; \delta_{ij}^{(2)} = -\frac{\partial E}{\partial net2}$$

$$\frac{\partial E}{\partial \sigma_{ij}} = -\delta_{ij}^{(2)} \frac{2(x_i - c_{ij})^2}{\sigma_{ij}^3} ;$$

The learning algorithms of connection weights, the central value and width of membership functions are shown below formula.

$$\omega_{ij}(k+1) = \omega_{ij}(k) - \beta \frac{\partial E}{\partial \omega_{ij}}, i = 1; j = 1, ..., 49$$

Where, $\frac{\partial E}{\partial \omega_{ij}} = -e_c \frac{\partial y_1}{\partial u}$.

$$c_{ij}(k+1) = c_{ij}(k) - \beta \frac{\partial E}{\partial c_{ij}}, i = 1; j = 1, ..., 49$$

$$\sigma_{ij}(k+1) = \sigma_{ij}(k) - \beta \frac{\partial E}{\partial \sigma_{ij}}, i = 1; j = 1, ..., 49$$

5 Simulation Results

In neural networks or fuzzy controller, nonlinear mathematical model of ship motion often is used to provide training and learning data. There we use the Nomoto model of ship to describe nonlinear equation.

$$\dddot{\psi} + (T_1 + T_2)\ddot{\psi} + T_1 T_2 \psi = K T_1 T_2 [\delta(t) + T_3 \dot{\delta}(t)]$$

Where, ψ is the course of ship, $\delta(t)$ is the rudder angel. In our experiment, we set T_1=0.008, T_2=0.143, T_3=0.054, and K=0.185 by calculating one ship model.

First, we can get the ideal steering course of ship by changing the rudder angel. Through learning as shown in section 3, we get the neural network reference model. Then, based on the neural network reference model, we use FNNC described in Section 4 to do the steering control simulation of ship. The result is shown as Fig.6.

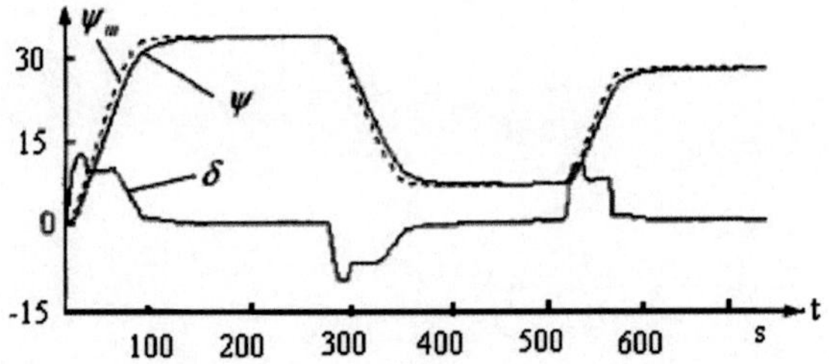

Fig. 6. The control simulation result

Where, ψ is the value of the ship course, ψ_m is the out value of NNRM, δ is the rudder angel. We can seen from simulation results that the ship steering control without overshoot, steering a reasonable and fast tracking, get a better control effect.

6 Conclusion

In this paper, the problem of designing a course-keeping autopilot for ship is considered. To achieve the energy efficiency for ship course keeping, an adaptive model neural network intelligent control strategy is proposed for ship steering control. To test the method the ship control simulation has been built. Via simulating, it is illustrated that the system can adjust the parameters online to improve the adaptability. In future we suppose to use neural algorithm not only for a ship motion in a theoretical model but for a complex actual ship steering.

Acknowledgment. This work was supported by Natural Science Foundation of Shandong Province (ZR2009FQ024). It is a pleasure to acknowledge our thanks on the subject.

References

1. Jia, B., Guang, R., Zhang, H.: Marine An Approach of Signal Detecting on Ship Control Test-bed. In: Proceedings of 6th International Symposium on Test and Measurement, vol. 4, pp. 3407–3410 (June 2005)
2. Sui, J., Lin, Y., Guang, R.: A Genetically Optimized Fuzzy Neural Network For Ship Controller. In: Proceedings of the 2006 IEEE International Conference Mechatronics and Automation, pp. 1367–2371 (June 2006)
3. Tiano, A., Zirill, A., Yang, C., Xiao, C.: A Neural Autopibt for Sailing Yachts. In: IEEE MED (August 2001)
4. Jiang, C., Wang, C.: Intelligence control and application. Science Press, Beijing (2007)

Research and Practice of Bilingual Teaching for Electrical and Electronic Technology

Chen Peijiang

School of Engineering Linyi Normal University
Linyi, Shandong, 276000, China
chenpeijiang@163.com

Abstract. The competition of international market and talent propose more demand on higher engineering education, and cultivating the professional technicians with strong ability of using English in practice is one of the goals of higher engineering education. Then, bilingual teaching has attracted more and more attention. Bilingual teaching in China is in the early stage and faces many problems. Based on exploring the meaning of bilingual education, with the bilingual teaching for electrical and electronic technology, the ideas and objectives of establishment of bilingual teaching system are introduced, the characteristics of the construction of the course materials is described, the bilingual teaching methods facing the ordinary college students are discussed, the bilingual teaching practice is carried out. The result shows that the implementation of the bilingual teaching for electrical and electronic technology is feasible with the concerted efforts of both sides of teaching and studying, and it can lay a good foundation to cultivate bilingual professionals with extensive engineering knowledge.

Keywords: electrical and electronic technology, bilingual teaching, teaching method, teaching practice.

1 Introduction

Along with the integration in the global economy, the higher education of China is geared to international standards. English is the most common language in the world, and it is the main carrier of all kinds of information in the current age. Bilingual teaching is to teach non-English language courses in English and understand them in mother language auxiliary. Bilingual teaching is an important part of quality education in China. Higher education institutions should advance with the times, promote bilingual teaching vigorously, train professionals who can speak English, write English and communicate scientific and technical research in English, and take measures to encourage and promote bilingual teaching[1].

Based on the bilingual teaching course of electrical and electronic technology, this paper explores the laws and methods of bilingual teaching, establishes a relatively complete bilingual teaching system, including the bilingual teaching programs, teaching materials, evaluation methods, website building and extra-curricular teaching activities, and forms the corresponding bilingual teaching models and teaching

W. Zhang (Ed.): Advanced Technology in Teaching, AISC 163, pp. 685–691.
springerlink.com

methods, so that students can master the professional knowledge and English formulation of course of electrical and electronic technology, make students combine learning of technical course with ability of English application well, and progressively train ability of thinking, analyzing and solving problems in English.

2 Meaning of Bilingual Teaching

2.1 Bilingual Teaching

Bilingual teaching refers to use both languages—mother language and the second language (here means Chinese and English) to organize, arrange, implement teaching activities so that students can describe the same knowledge using both languages to understand, think and then grasp the knowledge. As the globalization of the world economy, culture and technology, international exchange is deepening. English, as the main carrier, becomes universally accepted world language. Bilingual teaching creates environment and opportunities of applying English. The interest of professional learning and different ways of thinking brought by different culture of English language will benefit and help students learn cultural quintessence in different professional courses[2].

2.2 Meanings of Bilingual Teaching for Electrical and Electronic Technology

College courses can be divided into basic courses, professional basic courses and specialized courses. Basic course is the course that all professional should study. Professional basic courses lay some foundation for the various professions. They have close relationship with specialized courses. Electric and electronic technology, as the specialized basic course of mechanical, electrical, computer, communications, is the introductory course brick for students to master this course. It is important for electric and electronic technology to carry out bilingual teaching.

1) The fine original materials in the developed country can be used in the process of bilingual teaching, which provide students an international education and learning environment. Foreign materials pay more attention to update and develop the knowledge. They are edited by experienced teachers or the forefront of academic disciplines. The authors often continuously add new content, absorb new results and publish new textbooks. Using this material, students can master new information and advanced technical knowledge in this field[3].

2) Bilingual teaching can enable students to experience and feel the education concept, education system, teaching methods and thinking ways. China's traditional education is exam-oriented, focus on transferring theoretical knowledge systematically and completely, and neglects training ability of thinking independently. While developed countries focus more on people's overall ability and quality, attach importance to develop creativity and skills of analyzing and solving problems independently. It spreads knowledge through easy ways, and mobilizes the students' interest in specialized courses.

3) Bilingual education can improve student's comprehensive ability to use English. Foreign language is a full range of acquisition process, and run through every stage of learning. Bilingual education enables students to use this international academic

language into research and work, so as to grasp the latest international expertise and technical skills. They constantly enrich and improve their qualities and capabilities in the increasingly intense international competition[4].

In addition, bilingual teaching combines the different characteristics of Chinese and Western education, not only highlights solid foundation and reasoning rigorously of Chinese education, but also introduces developing students' creativity, imagination, lively classroom teaching, actively participation in Western education. It is useful to improve the standard of teaching.

3 Existing Problem of Bilingual Teaching

Bilingual teaching is still in its infancy in China, and it is a novelty in the field of education. There are several problems in many aspects.

3.1 Teaching Level Needs to Be Raised

Teachers are the most critical key in the process of bilingual teaching. Teachers not only require proficiency in professional knowledge, but also have considerable level of spoken English skills. There are many problems in reading, such as, oral expression of some symbols and special formula will be the obstacle of accurate description, may be neglected. In addition, how do we to translate some specialized vocabulary authentically also need profound professional knowledge. Therefore, qualified teachers are seldom, and it is impossible to educate a large number of qualified teachers in a short time.

3.2 Selection of Teaching Material Is Difficult

In general, reading levels of Chinese students are usually better than their hearing. During the early stages of bilingual teaching, students understand and consolidate classroom knowledge by materials from not understanding to understanding. Therefore, appropriate bilingual teaching material is a key factor of success[5]. There are many problems in the selection of foreign original materials. As the original materials in the domestic distribution are very lack, the selection is limited. The contents of teaching materials are jumbled, so we can find suitable materials only through carefully selection.

3.3 English Levels of Students Are Various

A small part of students have higher level English, so they can adapt quickly to bilingual education. But most students are poor in reading and hearing in English, and seldom can learn the terminology related to the professional. In this environment, bilingual teaching can not achieve the expected goals, but dampen the student enthusiasm, and affect the normal course. In addition, the students come from different regions and schools, the English level are different. All these can form gap between English proficiency and academic objectives. Therefore, bilingual teaching should choose students with a certain level of English, so that we can learn the professional, at the same time, continue learning English.

4 Practice of Bilingual Teaching

In the process of bilingual education, teachers, student language level and materials play important roles. Colleges and universities should attach great importance to bilingual education, enhance student quality as an important task, actively carry out the reform, and promote bilingual education in a planned and systematic way.

4.1 Strengthen the Construction of Teachers

We must firstly cultivate a group of teachers with fluent English and adapting to bilingual education teachers when set up bilingual education programs. Currently we can choose young teachers with higher standard of English, give full play to the backbone of the young teachers. Qualified university should organize teachers to undertake short-term English language training, so that they can understand the subject abroad, but also learn the advanced teaching methods, and can improve their English.

4.2 Build the Teaching Material Actively

Starting from the teaching content and English level, drawing on reasonable framework of knowledge, knowledge unit and time allocation of the Chinese language syllabus, we develop the course syllabus of electrical and electronic technology in English.

When we carry out the bilingual teaching of course of electrical and electronic technology, curriculum group first encountered two problems. First, we could not find a suitable original material matching with the syllabus. Second, it is difficult for non-electric students to read the original electronic books. Therefore, we compiled the teaching materials of the course to meet the demand of bilingual teaching.

The main features of the teaching materials are as follows: it use the original material of Circuit Analysis, "Electrical Engineering" as the references, borrow the core contents of the original materials such as the accurate expression of technical terms, the basic concepts and laws and syntax habits of technological English.

4.3 Adopt Different Teaching Strategies

1) Build good bilingual teaching environment, rationally use English and Chinese to teach. It is necessary to build good English environment in line with the implementation of bilingual education. English interactive is important in the classroom teaching, so as to culture students' ability of academic communication in English. According to the student's acceptance, we select the English proportion, and explain the concept in bilingual, which help students understand the concept. When explaining the difficulties of knowledge, we mainly use Chinese to help students to understand.

2) Adopts heuristic teaching method. Classroom teaching adopts heuristic methods, and carries out discussion, so as to arouse the students' attention to the stress and difficulties and thinking of some difficult problems. For example, when speaking the transition process of the circuit, we introduce the theme by the starting and stopping of the train. We can tell students that there are many similar phenomena in nature, there is transition process from a steady state to another steady state. The circuit is the same. We make the students has an initial perceptual knowledge of the circuit transition process, so that help the students learn and master. We should be good at summarizing to improve the learning efficiency. Succinctly speak and much train, repeatedly deepen understanding the basic concepts, basic principles and basic method, make full use of various types of questions, exercises, self exercises to achieve skillful master.

3) Combine theoretical teaching with experimental teaching. For example, when teaching the theory of Kirchhoff's laws and Thevenin theorem, we should verify the conclusion is correct or not, we can let the student certify in the experiment, which can deepen the understanding of the law and theorem.

4) Make use of modern technology. Combining the traditional method with modern methods, we use advanced multimedia technology to display some difficult so as to make students pay more attention to the classroom teaching. Vivid courseware facilitates students learning, and increases the capacity of classroom teaching. Additionally, we show students complex circuit analysis through modern circuit analysis software to achieve good teaching.

5) Expand the second class. To achieve the desired teaching objectives, classroom teaching is just not enough. Therefore, it is necessary to construct a second classroom. In the past few years, the program group had invited experts at home and abroad to conduct seminars to make students know electrical new technology and stimulate enthusiasm for learning. Students carry out a variety of extra-curricular activities such as science and technology competition works to improve the interest in new discoveries. Take full use of the campus network, carry out on-line tutor and student exchange to improve learning all around.

6) Class combines with race. The course has strong engineering nature, therefore, it is particularly important to train engineering practical ability. Course group explores a practice teaching mode. Students choose instruments and tools by themselves, learn with task, complete the task in the study, show results in the competition and teachers guide in the entire learning process. Meanwhile, according to the requirements of bilingual teaching, the students should write technical documentation in English.

In the teaching process, we strive to create the environment of mutual exchange, mutual inspiration and heart to heart communication, activate students to tap reserves of self-education resources, improve the education radiation. We always guided by modern education concepts and have played a very good role in improving the attractiveness of the course, the effectiveness of education, student participation and sense of identity.

4.4 Diversified Evaluation Methods

According to the characteristics of course nature and syllabus requirements of electric and electronic technology bilingual teaching, the examination should be basic, theoretical and practical. It can not only examine basic theory, basic knowledge, but also test the capacity of comprehensive application of each knowledge point and analyzing problems using basic principle.

In the examination we focus on evaluating students' comprehensive quality and innovation, emphasis on assessing learning ability of research study, exploratory learning and collaborative learning to reform simple assessment methods and overcome gap between results and ability. In the experiment, we do not write papers, but do experiment operation by oneself. We test the ability of finding and analyzing problems by practical hands-on methods. This test format increases coverage of the circuit knowledge point, highlights the importance of basic knowledge.

Evaluation methods are various. In addition to the final assessment, we pay more attention to usual evaluation, including homework, experiment, skill test, quizzes in the class, stage assessment, writing scientific papers, data translation, classroom performance. We make the assessment run through the whole process of learning. We inspect full range of learning, including the basic knowledge the students understand, ability of analyzing and solving problems, and professional literature search capabilities.

The results show that it is useful to improve attention in class, focus on learning and practice ability, which improve the overall quality of students.

4.5 Build Bilingual Teaching Curriculum Website

The dedicated website of the course of electrical and electronic technology contain all required contents, provide a good platform for interactive teaching. Students can download a full set of the complete syllabus, electronic courseware, and course-related technical terms and key points, download a bilingual experiment notes, Chinese and English Exercises, the information with strong engineering practical, so that they can enhance the ability to apply knowledge. Students can communicate with teachers on line, bring forward course-related requirements, opinions and recommendations. Thus it can greatly facilitate the students to preview and review, encourage them to suggest improvements and recommendations of the courseware, and continuously improve the courseware.

5 Practice and Effect Analysis

On the basis of the material construction of textbook, courseware, laboratory classes and discussion sessions, the bilingual teaching of the course of electrical and electronic technology is carried out to the undergraduates in Linyi normal university since 2004. The progress situation is very good nowadays. By comparing the students scores of the with the examination with same difficulty before and after the bilingual teaching as shown in Tab. 1, we can say that the student capacities of mastering knowledge and applying knowledge have not been affected, and the ability of using English proficiency is greatly improved.

Table 1. Score comparison

ExamType	Number	[90-100)	[80-90)	[70-80)	[60-70)	[0-60]
Bilingual Exam	35	5(14.3%)	11(31.4%)	12(34.3%)	5(14.3%)	2(5.7%)
Chinese Exam	38	6(15.8%)	14(36.8%)	10(26.3%)	7(18.4%)	1(2.6%)

In 2004, the beginning of bilingual education, many students did not support bilingual teaching because of heavy studying task. But, after the improvement of teaching methods and many efforts, most students have been able to adapt and support for the bilingual teaching of the course by 2007. The survey results show that the 75 percent students are very glad to accept bilingual education, and have good evaluation to the course construction of bilingual education of the course team and teaching effect. We believe that we can get better effect by the efforts of the teachers and students.

6 Conclusion

The course team has established a complete set of bilingual teaching system for the course of electrical and electronic technology after 6 years efforts. We have construct teaching materials both in and out of classes, and gradually grope the methods of bilingual teaching, and form our own characteristics. Appropriate teaching materials and teaching aids, vivid and emotional courseware, a variety of teaching activities, and practical teaching of bilingual teaching methods, ensure the prosperous development of the bilingual teaching for *Electrical and Electronic Technology*. The teaching can achieve the common objective. The students can not only master the professional curriculums, but also improve the English proficiency of undergraduates. Then, carrying out bilingual teaching for electrical and electronic technology is very important. Bilingual teaching is not only the conformity requirement of higher education to international norms, but also a new idea for basic education innovation of higher education. We believe that bilingual teaching has good development prospects.

References

1. Hu, T.: Study on Bilingual Teaching of Engineering University. Journal of North University of China(Social Science Edition) 24(S1), 44–46 (2008)
2. Ge, H.-J., Liu, H.-C.: Research and Practice of Bilingual Teaching System for Electrotechnics and Electronics. Journal of Electrical & Electronic Education 28(1), 51–53 (2006)
3. Ai, Y.-L., Li, D.: Practice and Experience of the Bilingual Teaching about the Electronics Technology Course. Higher Education Forum (2), 26–28 (2009)
4. Li, M.: Research and practice on bilingual teaching based on electric circuit in the institutions of higher learning. Journal of Shenyang Jianzhu University(Social Science) 10(4), 510–512 (2008)
5. Liu, D.C., Hu, P.: Exploration and Conception of Compilation of Electrical Textbooks in English for Teaching of the Circuit Course. Journal of Electrical & Electronic Engineering Education 25(6), 76–78 (2003)

Research on How to Optimize the Structure of China's Foreign Reserve

Ma Jie and Li Lei

School of Economics and Management,
Bei Hang University,
Beijing, China
majie_buaa_sem@126.com, lilei_bh@126.com

Abstract. To improve the operation efficiency of the Chinese foreign reserve, the true difficulty is to set scientific standards for the dynamic adjustment of the foreign reserve. This paper incorporates the CVaR model into the framework of "Risk-Return", and constructs a new programming model on the currency composition of China's foreign reserves. Based on the model we found and applying the monthly data during 2005.1~2009.7, we analyze the currency structure of the China's foreign reserve, and derive the optimal currency weights in different scenarios. According to the empirical analysis results, along with the international finance theory and the factual status, we confirm the availability of the new model.

Keywords: Foreign reserve, Structure management, Conditional Value-at-Risk, Convex optimization.

1 Introduction

By the end of 2009, the China's foreign reserve amounted to 2.4 trillion dollars, which brings some negative economic effects. It is estimated that 70% of reserves was invested to the assets marked via dollar (Jianhuai Shi, 2007). The single currency structure of Chinese foreign exchange is dangerous, whose value is easily threatened by the scale of US's debt and the Sino-US relations. Without regard to the continuous depreciation of dollar since 2000, the market value of China's foreign reserve has still shrunk over 1.2 trillion yuan according to our rough estimation for the sake of the reform of RMB exchange rate.

In order to enhance the operation efficiency, China founded the national investment company in 2007. However, the inept performance of China's investment company proves that it is just a small step to improve the efficiency of the Chinese reserve's operation to constitute a system framework, under which the authority entrusts a part of foreign reserve to the special investment entity. To fulfill the guideline of IMF (2004), the true difficulty is to set scientific standards for the dynamic adjustment of the foreign reserve. Facing the changing international financial environment, it is a serious theoretic issue and a great challenge for China to keep her huge foreign assets from depreciating.

W. Zhang (Ed.): Advanced Technology in Teaching, AISC 163, pp. 693–701.
springerlink.com

2 Literature Review

Through logistic regression, Chinn and Frankel (2005) found several factors that affect the currency weight in foreign reserve and proved there exists non-linear relation among them. Reid W. Click (2006) found the new added reserves in China, Japan and other Asian countries were marked by euro assets at least 50%, which indicated the substitution role of euro for dollar. Ewe-Ghee Lim (2007) founded a SEM model including five kinds of currencies, estimated it by SUR method and found that trend tracing and portfolio adjustment is a better strategy, relative to purchasing and holding. However, these researches are based on the aggregated data of IMF, which do not think about the difference of diverse countries.

Papaioannou, Portes and Siourounis (2006) thought the central bank has special responsibility on the whole economy and the common "mean-variance" analysis do not fit to analyze the foreign reserve structure, and they reconstructed a "dynamic mean-variance model"; their empirical research showed the importance of euro might be overestimated. According to the diversification demand of global bond investors, Y. Campbell, Medeiros and M.Viceira (2007) constructed another model that permits overselling, which is also based on the "mean-variance framework"; their results showed when the financial situation is in turbulence, capital will flee to the so-called "security currency", such as dollar, euro and Swiss franc. Though there is no special research on the structure of China's foreign reserve by foreign scholars, their methods and thoughts provide the meaningful references for us.

Based on the Markowitz model, Shuzhen Zhu (2002) derived the effective frontier of the foreign exchange portfolio, but their optimal result did not include dollar, which is unreasonable. Daoji Shi, et al (2005) used the simulation method to compute the Value-at-Risk (VaR) of the portfolio which included yuan, yen and Korean won, and their work focused on how to compute the accurate VaR without regard to the optimal structure of foreign reserve. Based on fuzzy mathematic method, Shenggang Yang, et al (2009) founded a multi-goal and multi-restriction model to analyze the reserve structure, however the hypotheses of this model on the currencies membership functions are kind of simple.

In sum, most of researches employed the "return-risk analysis framework", and the difference lied in how to measure the risk or return. When gauging risk, lots of documents used VaR to replace variance for variance can not depict the negative deviation and the concrete loss figure. However, compared to VaR, conditional value-at-risk (CVaR) can measure the risk more accurately even when the return has a fat tail distribution, and can treat the situation that the financial market is in the extreme scenario. Based on the "mean-variance framework" of the former research and considering the drawbacks of variance or VaR to measure risk, this paper takes CVaR to gauge risk, incorporate it into the "return-risk framework" on the reserve structure, and construct a new model to analyze how to optimize the structure of China's foreign reserve in light of the change of international financial situation.

3 Constructing and Solving the Model

3.1 The Depiction and Simplification of the Method

Suppose $f(x, y)$ is the loss function about the decision vector x, and regard x as the position or weight of every asset in the foreign reserve and y as the uncertainty that affects the loss, such as the price or return. If set $p(y)$ as the density function of y, then the probability that $f(x, y)$ is no more than a given threshold η can be shown as below:

$$\varphi(x,\eta) = \int_{f(x,y)\le\eta} p(y)dy \tag{1}$$

$\varphi(x,\eta)$ is the cumulative distribution function of η when set x fixed. And then we define:

$$\tau_\alpha(x) = \min\{\eta \in R : \varphi(x,\eta) \ge \alpha\} \tag{2}$$

$$\phi_\alpha(x) = (1-\alpha)^{-1}\int_{f(x,y)\ge\tau_\alpha(x)} f(x, y)p(y)dy \tag{3}$$

Where $\tau_\alpha(x)$ is VaR and $\phi_\alpha(x)$ is CVaR when set the significance level as $\alpha \in (0,1)$.

It is difficult to compute CVaR directly for VaR occurs in its definition equation. To resolve this problem, define a function as below [10]:

$$F_\alpha(x,\eta) = \eta + (1-\alpha)^{-1}\int_{y\in R^n} [f(x, y)-\eta]^+ p(y)dy \tag{4}$$

where $[t]^+ = \max\{t,0\}$. It can be prove that $F_\alpha(x,\eta)$ has the features as follows[11]:

1) $F_\alpha(x,\eta)$ is convex and differentiable on η;
2) VaR equals to the value of η which makes this function minimum;
3) as $F_\alpha(x,\eta)$ gets minimized on η, we can get CVaR as below :

$$\phi_\alpha(x) = \min_\eta F_\alpha(x,\eta) \tag{5}$$

Set (x^*,η^*) is the solution of the optimization problem (5), then $F_\alpha(x^*,\eta^*)$ is the optimal CVaR we seek and the optimal portfolio weight is x^*. When $f(x, y)$ is convex on x, $F_\alpha(x,\eta)$ is also convex on (x,η) and $\phi_\alpha(x)$ is convex on x, so the minimum problem of equation (5) is the issue of convex programming.

As the analytical formula of $p(y)$ is difficult to obtain, we took the scenario method and got J scenarios of $y_1, y_2, ...y_J$. So, the $F_\alpha(x,\eta)$ can be approximately expressed by the discrete method as below:

$$\hat{F}_\alpha(x,\eta) = \eta + (1-\alpha)^{-1}\sum_{j=1}^{J}\pi_j[f(x,y_j)-\eta]^+ \tag{6}$$

where π_j is the probability that scenario y_j occurs. If the loss function $f(x,y)$ is the linear function of x, $\hat{F}_\alpha(x,\eta)$ should be convex and piecewise linear. And if we introduce the dummy variable z_j, the $\hat{F}_\alpha(x,\eta)$ can be expressed by the linear function $\eta + (1-\alpha)^{-1}\sum_{j=1}^{J}\pi_j z_j$, and the linear restriction sets are showed as follows:

$$z_j \ge f(x,y_j) - \eta, \qquad z_j \ge 0, j = 1, ...J \tag{7}$$

By the transformation above, we can transform the complicated CVaR defined originally into the linear function and restrictions, which can be solved through linear programming (LP) method eventually.

3.2 Loss Function and Constraint Condition

Suppose there are n kinds of currency assets, whose original weight vector is $x_0 = (x_{1,0}, x_{2,0}, ...x_{n,0})^T$, denote $x_t = (x_{1,t}, x_{2,t}, ...x_{n,t})^T$ as the optimal currency weight at period t we seek and denote $y_t = (y_{1,t}, y_{2,t}, ...y_{n,t})^T$ as the return vector of reserve portfolio, which is independent on x_t. Since the loss means the negative return, the loss function at period t can be defined as below:

$$f(x_t, y_t) = -y_t^T x_t = -(x_{1,t}y_{1,t} + x_{2,t}y_{2,t} + ... + x_{n,t}y_{n,t}) \tag{8}$$

According to the expectation theory, the currency return can be split as:

$$E_t(y_{i,t+1}) = b_{i,t} + E_t(s_{i,t+1} - s_{i,t}) \tag{9}$$

where s denote the logarithm of exchange rate and b_i is the interest rate of currency i. For simplicity, suppose the central bank can foresee the fluctuation of exchange rate, so equation (9) can be rewritten as (10):

$$y_{i,t} = b_{i,t} + (s_{i,t+1} - s_{i,t}), \quad i = 1, ...n \tag{10}$$

VaR has been the standard tools to gauge and control risk, which is applied widely in practice nowadays, after the new Basel capital accord introduced it as risk measurement tool to financial institutions 10 years ago. However, as we mentioned above, VaR has some pitfalls to measure the risk and this paper will take CVaR to replace it, and if we set ω as the risk tolerance level, the restriction can be expressed as $\phi_\alpha(x) \le \omega$. As the loss function in equation (8) is linear and convex, the $\phi_\alpha(x)$ should be also linear and convex on x, and so the linear constraint sets based on CVaR function can be expressed as follows:

$$\eta + (1-\alpha)^{-1} \sum_{j=1}^{J} \pi_j z_j \le \omega ,$$

$$f(x, y_j) - \eta \le z_j, \qquad z_j \ge 0, j = 1, \ldots\ldots J \tag{11}$$

3.3 Model Construction on the Structure Optimization of China's Foreign Reserve

Summarize the definition and analysis above, the optimization problem that makes the loss of foreign reserve minimum based on the CVaR risk constraint and the framework of "Return-Risk" can be expressed as follows:

$$\min_{x,\eta} \quad \sum_{i=1}^{n} -E[y_i]x_i \tag{12}$$

$$s.t \begin{cases} \eta + (1-\alpha)^{-1} \sum_{j=1}^{J} \pi_j z_j \le \omega & (13) \\ \sum_{i=1}^{n} (-y_{ij} x_i) - \eta \le z_j, j = 1, \ldots\ldots J & (14) \\ lb \le x_i \le ub, i = 1, \cdots n & (15) \\ \sum_{i=1}^{n} x_i = 1 & (16) \\ x_i \ge 0, i = 1, \cdots n & (17) \\ z_j \ge 0, j = 1, \cdots J & (18) \end{cases}$$

where equation (12) is the objective function, which ask the loss of reserve portfolio minimized; equation (13) is the CVaR constraint defined in equation (11); equation (14) and (18) derive from the definition equation $z_j = \max\{f(x, y_j) - \eta, 0\}$; equation (15) preliminarily set the lower and upper bounds for the weight of currency *i* based on the prior information; equation (17) forbids the overselling in reserve operation. Via solving the linear programming problem defined here, we can derive the optimal weight vector x^*, as well as the η^* and $F_\alpha(x^*, \eta^*)$.

4 Empirical Research

4.1 The Choice of Reserve Currency

The principles to choose reserve currencies include economic power, stability of currency value and demand of trade match. Starting from the first two rules, obviously the dollar, euro, yen and pond might be the possible candidate. Meanwhile, Europe, US and Japan are the largest three trade and investment partners of China. Considering the sterling is still strong and London is the largest international financial center in the world, we set pound as a candidate also. Switzerland often takes a neutral stance in international affairs and many international businesses proceed by Swiss franc. Based on the practical status of international finance, this paper chooses five kinds of currency as the reserve currency, including dollar, euro, yen, pound and Swiss franc.

4.2 Preliminary Constraint on the Weight

Considering China takes an export-oriented economic growth strategy, the trade structure plays an important effect on the currency weight. So we investigate the distribution of the main trade partner of China through the export and import data, which is shown in table 1.

Table 1. Area Distribution of China'S Main Trade Partner and Related Weight （Unit：%）

share	2001	2002	2003	2004	2005	2006	2007	2008	Avg
US	15.79	15.66	14.85	14.69	14.88	14.92	13.9	13.03	14.72
Japan	17.21	16.42	15.69	14.54	12.97	11.78	10.85	10.42	13.74
HK	10.97	11.15	10.27	9.76	9.61	9.43	9.07	7.95	9.78
Taiwan	6.34	7.19	6.86	6.78	6.42	6.13	5.73	5.04	6.31
German	4.61	4.48	4.9	4.69	4.45	4.44	4.32	4.49	4.55
England	2.02	1.84	1.69	1.71	1.72	1.74	1.81	1.78	1.79
France	1.53	1.34	1.57	1.52	1.45	1.43	1.55	1.52	1.49
Swiss	0.47	0.44	0.41	0.41	0.41	0.38	0.43	0.44	0.42

Since HK dollar pegs dollar, its weight can be emerged into dollar. Dollar is still the main settling currency in the international business, and the trades of China with Asian, African and Latin American countries are usually settled by dollar also. So excluding the weight of euro and yen in table 1 and considering the dominance position of dollar in China's reserve (Jianhuai Shi, 2007), this paper sets the lower bound of dollar as 50%.

After euro occurs, most of the independent currency of the European countries are replaced by euro, and the trades of China with German, France and other Euro union countries are mainly settled by euro; so, we roughly set the lower bound of euro as 15%. Although the weight of Japan in China's trade structure is within the 0.42~17.21%, and the average weight is 13.74%; however, the trade between China and Japan is not settled through yen entirely and Japan has taken a zero interest rate policy for a long time, so we adjust the weight of yen downwards and set the lower bound of yen as 10%.

4.3 The Empirical Results and Analysis

To consider the balance between avoiding the reserve value deprecating and maintaining a relatively stable currency structure to save the trade commission, this paper chooses month as the adjustment period[1]: select the monthly data that include 57 sample points during 2005.1~2009.9, and obtain 56 scenarios which is supposed as occurring with equal possibility, so $\pi_j = 1/56$。 Set the original currency weight vector as $x_0 = (0.2, 0.2, 0.2, 0.2, 0.2)^T$, and the confidence level is $\alpha = 0.90$. Based on the data and hypotheses above, we take the software of Matlab to compile the program and at first do a tentative computation under the condition without any preliminary constraint coming from trade structure, and the computation result is shown in table 2.

Table 2. Optimal Currency Weight without Trade Structure Constraint (Unit: %)

weight	dollar	pound	euro	Swiss franc	yen	CVaR
ω=0.005	86.41	7.59	6.01	0	0	0.0264
ω=0.01	77.38	11.19	11.43	0	0	0.0279
ω=0.015	68.33	14.87	16.80	0	0	0.0294
ω=0.02	58.83	29.79	11.38	0	0	0.0308

The tentative computation result shows that dollar assets should be the main part of China's foreign reserve, no matter how the risk constraint value ω is given. The newest IMF report issued in Mar 2010 shows that 62.1% of the global foreign reserve is still dominated in dollar, which coincides with our computation result in table 2. However, with ω increasing, the corresponding CVaR increases too and the optimal dollar weight decreases, accompanied with the weight of pound and euro increasing.

The balance of payment shows the European Union is the largest economic and trade partner of China, and we have set the lower bound of the weight of euro as 15%. In the results in table 2, the optimal euro weight is less than 15% in 3 kinds of scenario, which is unreasonable and shows that the risk constraint value of ω =0.015 might be an available choice.

Japan is the second largest economy and the economic ties of Japan and China is very close, however, the tentative result on yen weight is 0%, which is obviously unpractical. Seeing from the position of pound in China's trade structure, it is unreasonable that its optimal weight in reserve is over 10% and even 29.79%. To revise the tentative results, according to the preliminary constraint on the weight of currency asset above and the model we constructed in part III: set the scenario of ω =0.015 as the reference, and set the lower bound of yen is 10% and the upper bound of pound is 5%. Based on these hypotheses, we derive the optimal currency structure of China's foreign reserve, and the new result is shown in table 3.

1 The trade data come from the official website: ww.mofcom.gov.cn; the data about exchange rate and interest rate come from the website of British bank association: www.bbalibor.com

Table 3. Optimal Currency Weight Under Trade Structure Constraint (Unit: %)

weight	dollar	pound	euro	Swiss franc	yen	CVaR
ω=0.015	57.42	4.77	23.34	0	14.48	0.0273

Compared with the scenario of ω =0.015 in table 2, the result in table 3 shows: the weight of dollar decreases obviously while the weight of euro increases largely, which implicates euro can indeed replace dollar in some extent under the practical constraint condition, and this coincide with the researches by many foreign scholars (Galati and Wooldridge,2006; Click,2006; Ewe-Ghee Lim, 2007)。 The pound weight in the optimal reserve is 4.77%, which is less than the upper bound we set above but is more than the average trade weight of pound in China's trade structure that is only 1.79%, and this result reflects the strong sterling and the relative higher interest rate of pound in the recent 4 years.

It is surprising to find each weight of Swiss franc is 0% whether in table 2 or in table 3, which indicates that China need not hold assets marked by Swiss franc, and this result is different from the people's intuition. Furthermore, the weight of yen increases to 14.48% from 0% and is a bit higher than the average weight of yen in China's trade structure, which reveals the fact that Japan is the main capital inflow country of China and the interest rate of yen is close to zero for a long time. Simultaneously, the corresponding value of CVaR in table 3 decreases compared to that in table 2, which shows that the risk of foreign reserve portfolio reduces also. In other words, given the equal risk constraint condition, the optimal reserve portfolio obtained in table 3 is more stable than that derived in table 2 and can resist the market risk much better.

5 Conclusion

Using CVaR other than VaR to gauge risk can make the optimal structure of foreign reserve more stable, and let central bank easy to adjust the currency composition with the change of international finance status. This paper incorporates the risk measurement method of CVaR into the model construction on China's optimal foreign reserve structure. First of all, we take CVaR to gauge the risk of reserve portfolio. Referring to the trade structure, we determine 5 kind of currency as the reserve assets and preliminarily set the lower or upper bound on their weight. Based the "return-risk analysis framework" and series of hypotheses, we eventually construct a new model to optimize the China's foreign reserve structure.

Based on the theoretical model and applying the monthly data during 2005.1~2009.7, we analyze the currency structure of China's foreign reserve, and derive the optimal currency weights in different cases and different policy preference of the central bank. The empirical results show that euro can replace dollar in some extent, and China should increase the weight of euro when dollar appreciates in the short term and depreciates in the long-run, so as to prevent the huge foreign reserve shrinking continuously. At the same time, our research shows pound might be a useful choice when China intends to diversify its composition of foreign reserve; however, it seems that there is no need to hold the unpopular currency, such as Swiss franc, and we can not take it for granted on this issue. In addition, the empirical research shows yen has no any attraction to China as the value reserve, however, when considering the trade constraint and the fact that Japan is the main capital inflow resource of China, it is still necessary for China to hold some yen reserves.

Acknowledgment. This paper is supported by the National Natural Science Foundation of China (NSFC No: 70803003) and by the Social Sciences Fund of the Chinese Education Ministry (No: 07JC790053) .

References

1. Shi, J.: Reflection on China's Foreign Exchange Reserves. International Economic Review 2, 51–53 (2007)
2. International Monetary Fund, Guidelines for Foreign Exchange Reserve Management. IMF Research Report (2004)
3. Chinn, M., Frankel, J.: Will the Euro Eventually Surpass the Dollar as Leading International Reserve Currency? NBER Working Paper, No 11510 (July 2005)
4. Click, R.W.: On the Composition of Asian Central Bank Reserves: Will the Euro Replace the Dollar? Journal of Asian Economics 17, 417–426 (2006)
5. Lim, E.-G.: Do Reserve Portfolios Respond to Exchange Rate Changes Using a Portfolio Strategy?: An Econometric Study Using COFER Data Rebalancing. IMF Working Paper, No 293 (December 2007)
6. Papaioannou, E., Portes, R., Siourounis, G.: Optimal Currency Shares in International Reserves: The Impact of the Euro and the Prospects for the Dollar. Japanese Int. Economies 20, 508–547 (2006)
7. Zhu, S.: The Analysis on the Return and Risk of Chinese Foreign Reserve Portfolio. Shanghai Finance 7, 26–28 (2002)
8. Shi, D., Di, N.: An Analysis on the Risk Correlation of the Exchange Rate Portfolio. Systems Engineering 23, 90–94 (2005)
9. Yang, S., Long, Z.: Analysis of Currency Composition of China's Foreign Exchange Reserves Based on Fuzzy Decision Theory. The Theory and Practice of Finance and Economics 30, 8–13 (2009)
10. Uryascv, S.: Conditional Value at Risk: Optimization Algorithms and Applications. In: Proceedings of the IEEE Conference on Computational Intelligence for Financial Engineering, pp. 49–57 (August 2000)
11. Rockafellar, R.T., Uryasev, S.: Optimization of Conditional Value at—Risk. Journal of Risk 2, 21–41 (2000)

Model Checking Web Applications Based on Web Navigation

Mingyue Jiang and Zuohua Ding

Center of Math Computing and Software Engineering Zhejiang SciTech University
Hangzhou, China
jiang_my@126.com, zouhuading@hotmail.com

Abstract. With the development of web technologies and the increasing size of web sites, the structure of the web application becomes more complicated, and it becomes a difficult task to model and analyze it. Web navigation, which is useful for clarifying requirements and specifying implementation behavior, can be helpful in the analysis of web application. The structure of a web application can be checked based on the web navigation model. In this paper, we present a method to model and check a web application based on web navigation. We first model the web navigation by statechart, and then translate the statechart model to Promela description (the input language of SPIN), meanwhile, the properties to be checked are specified by LTL (Lineal Temporal Logic); finally the web navigation can be checked by SPIN. A case study demonstrates the advantages of our method.

Keywords: Web navigation, statechart model, LTL, SPIN.

1 Introduction

With the rapid growth of the World Wide Web, more and more new technologies have been employed to construct web site, such as JavaScript, VBScript, Java, ActiveX, etc. These technologies provide lots of convenient services to us, however, they also bring difficulties to web application analysis and maintenance due to the increasing size and complicated structure.

Navigation is a critical aspect in web systems. As [1] indicated, navigation models are useful for clarifying requirements and specifying implementation behavior. Navigation model describes the navigation capabilities, i.e., the path on which users can traverse to explore the information required, and it is considered as from dynamic point of view. Many models have been build proposed to model web navigation[2,3,4,5,6], based on which the analysis for web applications can be conducted. Moreover, some models are used for web engineering[7,8]. In this paper, we will check the web application based on the navigation model.

The model used in our paper is statechart model proposed by Leung et al[9]. The reason is that this model can model both static and dynamic web pages. Based on this model, we develop a method to check web navigation. We first model the web navigation by statechart, and then translate this statechart model to Promela language,

W. Zhang (Ed.): Advanced Technology in Teaching, AISC 163, pp. 703–711.
springerlink.com

meanwhile, the properties to be checked are specified by LTL, finally web navigation can be checked by SPIN.

The remaining part of the paper is organized as follows. In Section 2, the general idea of web navigation modeling will be introduced. The verifier and the constructing of its input model will be discussed in section 3. In section 4, a case study is presented to show our method. The conclusion of this paper will be given in section 5.

2 Modeling Web Navigation

The basic web navigation is accomplished by clicking Hyperlinks. However, the development of web technologies makes web navigation more complex. Since various client side scripts and programs are embedded in web pages, the navigation between web pages not only depends on clicking hyperlinks, but also depends on dynamic HTML pages.

Finally, complete content and organizational editing before In order to analyze web navigation, we need first to model the navigation. In this paper, we adopt the navigation model proposed by Leung et al[9], which models the web navigation with statechart.

The method proposed by Leung et al can be generalized as follows:

- Window.

A window is a statechart.

- Web page.

A web page is modeled as an composed state, and its sub-states represent possible browsing positions.

AND state is introduced to describe a web page with frames.

- Hyperlink.

A hyperlink is mapped to a state transition, and the hyperlink activation is the event that triggers the transition.

The select connective is used to group all hyperlinks target of a page.

- Synchronization.

Synchronization between windows, frames is achieved by event broadcast.

- Selection.

The select connective is used to group all hyperlink targets of a page.

Parameterized-OR is used to model dynamic server side.

With the proposed model, the behavior of a web application can be depicted by statechart, thus the analysis can be conducted by checking this model.

3 Moving from Statechar to Promela

In Leung et al's work, only statechart model is presented, and the verification is not provided. In this section, we will solve this issue. By defining some rules, we can translate statechart model to Promela, and then use SPIN to check the model. The properties to be checked are described by LTL.

3.1 Transform Statechart to Promela

1. Statechart

If a system has multiple windows, each window is modeled as a statechart, otherwise, the whole structure of a system is described by a statechart.

An *init* process represents a statechart that denotes a whole system, however, the statechart that denotes a window will be described by a common process.

2. State

There are two kinds of states in statechart model: composite state and atomic state.

A composite state is represented by a process in Promela, and the content belong to this state (sub-states and transitions) will be contained in this process.

An atomic state is modeled as a variable, here we assume it is a bool variable. Because the content of an atomic state is hidden, we only need to record its status, for example, fired or unfired.

3. Description of activated state

For a composite state, when it is activated, it means that it is time for the state to execute. So the activating of a composite state can accomplished by starting corresponding process, which is described as *run P()* in Promela(P is the name of a process). Particularly, related info will be passed as parameter to the process.

For an atomic state, it is modeled by a bool variable. We will set its initial value as false, and when the state is activated, the variable's value will be set as true.

4. Transition

Transitions link states together, and describe transformations between states. For the sake of simplicity, we use HState to denote the state which activates the transition, and EState to represent the state which is activated by the transition.

Suppose a transition links HState to EState with event e. If HState is activated and event e happens, that is, the transition is fired. Then EState will be fired. There are some cases when using Promela to describe a transition and its related events.

Assume S_1 and S_2 are two composite states, and S_{11} and S_{21} are two states. There is a transition T between S_{11} and S_{22}, and the related event is jp_1. S_{11} is HState and S_{21} is EState(In Promela, we use capital letter to denote composite state, and small letter for atomic state).

a) HState and EState are both atomic state, and they belong to the same composite state.

In this case, S_{11} and S_{21} are atomic, and they belong to the same composite state. The transition T and its related activity will be described as 's11->s21', where '->' represents the event generation, and s11 and s21 denote the status before and after the event respectively. Because S_{11} and S_{21} are both atomic state, s11 is the expression describes the status of S_{11}, and s21 is the expression describes the status of S_{21}.

b) HState and EState are both atomic state, but they belong to different composite states.

In this case, S_{11} and S_{21} are atomic, but they belong to different composite states. Suppose S_{11} is the sub-state of S_1, and S_{21} is the sub-state of S_2.

Channel will be introduced to passing message between different composite states. Channel is a data type with two operations: send and receive. We can send a message into a channel with send operation and get a message from the channel with receive operation.

The description of the transition and its related event will be divided into two parts, each in one composite state, and message passing will connect these two parts together.

```
chan ch1=[1] of {bool};//the channel can be defined
locally or globally, which depend on its effect, i.e., if it is
used inside a process, it is defined as local variable, or, it
should be defined globally.
proctype S1(){
    ...
    s11->ch1!true; //Message is sent to activate EState.
    ...
}
proctype S2() {
    ...
    ch1?[true]->atomic{s22=true;ch1?t};//Once message
is receive, the EState is activated.
    ...
}
```

c) Either HState or EState is composite state, and they belong to the same composite state.

In this case, S_{11} and S_{21} belong to the same composite state. Assume that they all belong to S_1.

The transition description is placed in process S1, and there are two cases.

If S_{11} is atomic (S_{21} is composite), then transition description is 's11->run S21()',where the activation of S_{21} is depicted by starting corresponding process.

If S_{11} is composite, then the transition description is as follows:

```
proctype S1(){
chan ch1=[1] of {bool}; //passing message between S1
and S11.
run S11(ch1); //ch is passed as a parameter to S11.
'ch1!true' will be placed at the end of S11.
ch1?[true]->atomic{s21=true;ch1?t}; //ch1 receives the
required message, and then S21 is activated.
}
```

d) Either Hstate or EState is composite state, but they belong to different composite state.

S_{11} and S_{22} belong to different composite state, suppose they belong to S_1 and S_2 respectively.

If S_{11} is an atomic state, then the transition description is as follows:

```
chan ch1=[1] of {bool};
proctype S1() {
   ...
   s11->ch1!true;
   ...
}
proctype S2() {
   ...
   ch1?[true]->run s22();
   ...
}
```

If S_{11} is composite, then the transition description is different. Two channels are employed, one is for message passing between S_1 and S_2, another one is used for message passing between S_1 and S_{11}.

```
chan ch1=[1] of   bool};
proctype S1() {
   ...
   chan ch=[1] of {bool};
   run S11(ch); //s11 is activated, 'ch!true' will be
placed at the    end of process S11.
   ch?[true]->atomic{ch1!true;ch?t} //s11 has executed,
and the transition is   fired.
   ...
}
 proctype S2() {
   ...
   ch1?[true]->atomic{s22=true;ch1?t}
 //Estate is    activated.
   ...
}
```

e) HState and EState are both composite state, and they belong to he same composite state.

In this case, transition description is similar with the second case of c), the difference is that the EState activation is represented by running a process after the transition is fired.

f) HState and EState are both composite state, but they belong to different composite state.

In this case, transition description is similar with the second case of d).

5. Selection.

The selection structure will be modeled with selection statement in Promela.

6. Concurrency and synchronization.

The concurrency between frames is described by atomic statement, and synchronization between frames, windows will be achieved by channel.

3.2 System Property Specification

System properties can also be described by Promela, so that we can check if the obtained model matches the properties of the system. If the properties are violated, they will be checked by SPIN and the violation trace will be given. Some statements are used to assist the checking, such as assert and never. In this paper, LTL (Linear Temporal Logic) is applied to specify system properties. LTL is the formal logic used for verification in SPIN, and it is based upon the propositional calculus. LTL formulas can be translated into never claim (a Promela construct) automatically in SPIN, and then the never claim is checked instead of LTL formulas.

In the aspect of structure, there are kinds of properties can be checked. In this paper, we mainly focus on following properties.

1) Page reachability.

For a web application, it is necessary to make sure that every web page can be visited. That is, the reachability of pages should be guaranteed. The reachability of all pages of a web application (pages in one statechart model) specified as

$$\Box(start \rightarrow \Diamond P_1 \wedge \ldots \wedge \Diamond P_n),$$

where *start* denotes the starting point of a statechart model, which refers to the first displayed page of a web application. P_1, P_2, ...,P_n denotes reach status of each page respectively. This property is hold if every page is eventually reached (P_i is true) after the first page of a web site is displayed.

2) Dead end

If there is a page, in which no links can leads to others page, what's more, the back action is forbidden. We call this page as a dead end. In other word, we can not go to anywhere from a dead end page.

Suppose *P* is a page, the number of links on this page is recorded as PN, and the state of back button is *B*(browser effect is not considered in this paper, and buttons, such as back, refresh and so on, are regarded as invalid),the properties that P is not a dead end is described as follows.

$$\Box(PN>0 \vee B)$$

For a web navigation, if the visited pages are $P_1,P_2,...,P_n$, then the dead end property can be checked by following formula.

$$\Box((PN_1>0 \vee B_1) \ \&\& \ (PN_2>0 \vee B_2)\&\&...\&\&(PN_n>0 \vee B_n))$$

If these properties are violated during model checking, the relevant information of violation will be obtained by analyzing guided simulation output after verification. What's more, we also can find where the problem is. So the structure defects of a web application can be determined.

4 Case Study

In this section, the Online Magazine System (OMS) is employed to illustrate our method. OMS is composed by four modules: login, book info, book reading and download. The statechart model of OMS is displayed in figure 1.

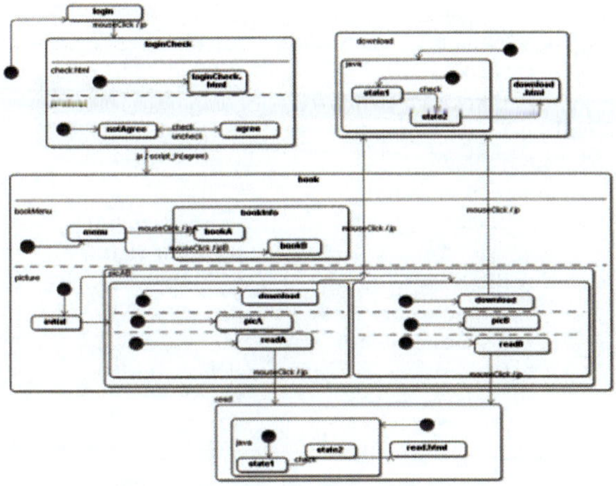

Fig. 1. The statechart model of Online Magazine System

By translating the statechart model of OMS to Promela, page reachability and dead pages are checked. These two properties are checked together (each one has its corresponding LTL specification), and *&&* is applied to connect these two LTL formulas to construct the result LTL formula. In figure 2, the simulation output of SPIN is given.

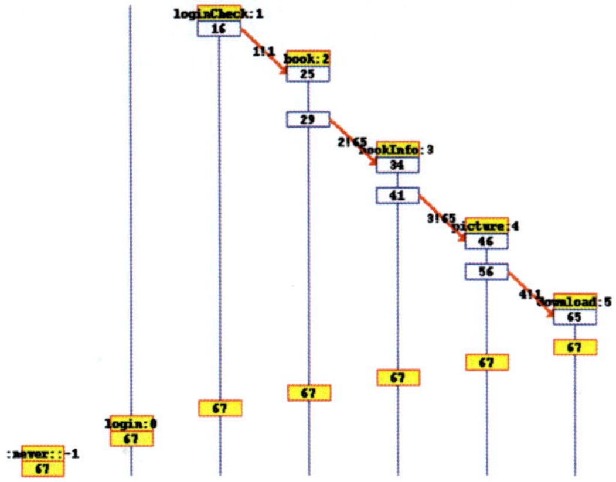

Fig. 2. The simulation output of SPIN

As a result, the verification output is displayed in figure 3.

The verification output tells us that there exits an error, which is caused by the violation of never claim, and a guided simulation is advised. With the help of guided simulation, we find the never claim is violated because *download* page is a dead end. Actually, *read* page is also a dead end, but it is not found by verification, because the verification stopped when the first violation is found.

```
(Spin Version 5.1.5 -- 25 April 2008) Warning: Search
not completed
    + Partial Order Reduction
Full statespace search for:
    never claim              +
    assertion violations     + (if within scope of claim)
    acceptance    cycles     + (fairness disabled)
    invalid end states   - (disabled by never claim)
State-vector 80 byte, depth reached 72, errors: 1
      197 states, stored (358 visited)
      261 states, matched
      619 transitions (= visited+matched)
       91 atomic steps
hash conflicts:          0 (resolved)
```

Fig. 3. The verification output of SPIN

5 Conclusion and Discussion

In this paper, we have presented a method to check web applications based on a statechart model. The transformation rules from state chart model to Promela have been defined, meanwhile, the properties related to system behavior are specified by LTL. The verification can be automated with SPIN.

Many aspects of web application can be reflected by web navigation model. In the future, we will focus on the completeness and effectiveness of the navigation model, and kinds of system properties will be considered and verified.

Acknowledgment. This work is partially supported by the National Key Research program under Grant No. 90818013.

References

1. Han, M., Hofmeister, C.: Modeling and verification of adaptive navigation in web applications. In: International Conference on Web Engineering, ICWE 2006, pp. 329–336 (2006)
2. Conallen, J.: Buiding web applications with UML. Addison-Wesley (2002)
3. Oliveira, M., Turine, M., Masiero, P.: Statechart based model for hypermedia applications. ACM Transactions on Information Systems 19(1), 28–52 (2001)

4. Ding, Z., Jiang, M., Pu, G., Sanders, J.W.: Modelling and Verification of Web Navigation . In: Gaedke, M., Grossniklaus, M., Díaz, O. (eds.) ICWE 2009. LNCS, vol. 5648, pp. 181–188. Springer, Heidelberg (2009)
5. Paulo, F.B., Masiero, P.C., Liveria, M.C.F.: Hypercharts:extended statechart to support hypermedia specification. IEEE Transactions on Software Engineering 25(1), 33–49 (1999)
6. Turine, M.A.S., Oliveira, M.C.F., Masiero, P.C.: A navigation-oriented hypertext model based on statecharts. In: Proceedings of the 8th ACM Conference on Hypertext, pp. 102–111 (1997)
7. Ceri, S., Fratemali, P., Bongio, A.: Web modeling language(WebML):a modeling language for designing web sites. In: Proceddings of the 9th International World Wide Web Conference, pp. 137–157 (2000)
8. Escalona, M.J., Aragón, G.: NDT:A model-driven approach for web requiremetns. IEEE Transactions on Software Engineering 34(3), 377–390 (2008)
9. Leung, K.R.P.H., Huia, L.C.K., Yiu, S., Tang, R.W.M.: Modeling web navigation by statechart. In: Proceedings of the 24th International Computer Software and Applications Conference, pp. 41–47 (2000)

ESPRIT-Based Joint Angle and Frequency Estimation Using Multiple Delay Output

Sun Zhongwei, Li Jianfeng, Liu Jian, and Zhang Xiaofei

Department of Electronic Engineering, Nanjing University of Aeronautics & Astronautics, Nanjing, China, 210016
{szwsg,lijianfengtin,lj0405205}@126.com

Abstract. This paper presents a novel ESPRIT algorithm based joint angle and frequency estimation using multiple delay output. The algorithm can estimate the joint angles and frequencies, as using multiple output, the estimation accuracy is greatly improved compared with conventional algorithm. The useful behavior of the proposed algorithm is verified by simulations.

Keywords: ESPRIT, angle estimation, frequency estimation, Uniform linear array.

1 Introduction

Antenna array has been used in many fields such as radar, sonar, communications and seismic data processing. The direction-of-arrival (DOA) estimation of signals impinging on an array of sensors is a fundamental problem in array processing. The problem of joint DOA and frequency estimation arises in the applications of radar, wireless communications. For example, these parameters can be applied to locate the mobiles and to locate pilot tones in space division multiple access systems. Furthermore, a precise estimation of these parameters is helpful to attain a better channel estimate and thus enhances the system performance. Optimal techniques based on maximum likelihood [1] are often applicable but might be computationally prohibitive. Some ESPRIT-based joint angle and frequency estimation methods have been proposed in [2-8]. Zoltowski [2] discusses this problem in the context of radar applications. Pro-ESPRIT is proposed to estimate angle and frequency. Haardt [3] discusses the problem in the context of mobile communications for space division multiple access applications. Their method is based on Unitary-ESPRIT, which involves a certain transformation of the data to real valued matrices. Multi-resolution ESPRIT is used for joint angle frequency estimation in [4]. ESPRIT method is used for frequency and angle estimation under uniform circular array in [6-7]. The others joint angle and frequency estimation method is proposed in [9-10].

This paper uses multiple delay output, so as to achieve the purpose of improving estimation accuracy. This algorithm is the improved algorithm of conventional method. The proposed algorithm is applicable to uniform linear array.

W. Zhang (Ed.): Advanced Technology in Teaching, AISC 163, pp. 713–719.
springerlink.com

Denote: We denote by $(.)^T$ the matrix transpose, and by $(.)^H$ the matrix conjugate transpose. The notation $(.)^+$ refers to the Moore–Penrose inverse (pseudo inverse).

2 The Data Model

There are K sources to reach uniform linear array with M elements. Suppose that the *i*th source has a carrier frequency of f_i. The signal received at the *m*th antenna is

$$x_m(t) = \sum_{i=1}^{K} e^{j2\pi(m-1)df_i \sin(\theta_i)/c} s_i(t) + n_i(t) \tag{1}$$

where θ_i is direction of arrival (DOA) of the *i*th signal, *d* is array spacing. $s_i(t)$ is the narrow-band signal of the *i*th source. $n_i(t)$ is the white noise, which has zero mean and variance of σ^2. In order to estimate frequency, we add the delayed outputs τ for the received signal of array antenna, as shown in Fig.1. We suppose that $0 < \tau < 1/\max(f_i)$.

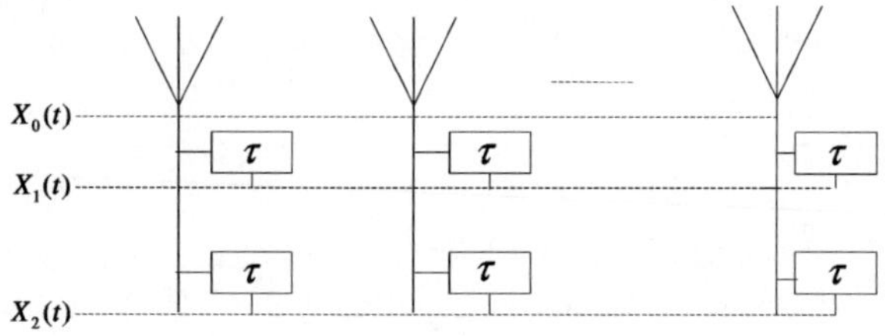

Fig. 1. The received signal with delayed output

The delayed signal for Eq.(1) with delay τ is

$$\begin{aligned} x_m(t-\tau) &= \sum_{i=1}^{K} e^{j2\pi(m-1)df_i \sin(\theta_i)/c} s_i(t-\tau) + n_i'(t) \\ &= \sum_{i=1}^{K} e^{j2\pi(m-1)df_i \sin(\theta_i)/c} s_i(t) e^{-j2\pi f_i \tau} + n_i'(t) \end{aligned} \tag{2}$$

where c is velocity of light. We assume that channel state information is constant for N symbols. The received signal of array antennas can be denoted as

$$X_0 = AS + N_1 \tag{3}$$

where the source matrix **S**, the direction matrix **A** and the noise matrix N_1 are shown as follows

$$S = [S_1 \quad S_2 \quad \cdots \quad S_K]^T \tag{4}$$

$$N_1 = [n_1(t), n_2(t), \cdots, n_K(t)]^T \tag{5}$$

$$A=\begin{bmatrix} 1 & 1 & \cdots & 1 \\ e^{-j\alpha_1} & e^{-j\alpha_2} & \cdots & e^{-j\alpha_K} \\ \vdots & \vdots & \ddots & \vdots \\ e^{-j(M-1)\alpha_1} & e^{-j(M-1)\alpha_2} & \cdots & e^{-j(M-1)\alpha_K} \end{bmatrix} \tag{6}$$

where $\alpha_k = 2\pi d f_k \sin\theta_k / c$,k= 1,2,…,K. According to Eq.(6),we define

$$A=\begin{bmatrix} A_1 \\ a_M \end{bmatrix}=\begin{bmatrix} a_1 \\ A_2 \end{bmatrix} \tag{7}$$

where A_1 is the first (M-1) row of A , a_M is the last row of A . A_2 is the second to Mth rows of A , a_1 is the first row of A .The delayed signal for Eq.(2) with τ can be denoted as

$$X_1 = A\Phi S + N_2 \tag{8}$$

where

$$\Phi = diag\{e^{-j\beta_1}, e^{-j\beta_2}, \cdots, e^{-j\beta_K}\} \tag{9}$$

where $\beta_k = 2\pi f_k \tau$, k= 1,2,…,K.

The delayed signal for Eq.(2) with 2τ can be denoted as

$$X_2 = A\Phi^2 S + N_3 \tag{10}$$

According to Eq. (3), Eq. (8) and Eq. (10), we define

$$X=\begin{bmatrix} X_0 \\ X_1 \\ X_2 \end{bmatrix}=\begin{bmatrix} A \\ A\Phi \\ A\Phi^2 \end{bmatrix} S + N \tag{11}$$

where $N=[N_1, N_2, N_3]^T$.

3 Joint Angle and Frequency Estimation

We can use received signal to attain the direction matrix **A** and the delay matrix $\mathbf{\Phi}$, and then estimate angle and frequency. The covariance matrix of the received signal can be reconstruct via $R_X = XX^H$. Using eigenvalue decomposition of R_X , we can get the signal subspace E_s . In the free-noise case, E_s can be denoted as

$$E_s=\begin{bmatrix} A \\ A\Phi \\ A\Phi^2 \end{bmatrix} T \tag{12}$$

where **T** is a K×K full-rank matrix.

3.1 Frequency Estimation

According to Eq. (12), we define $\mathbf{E}_1$ and $\mathbf{E}_2$

$$E_1 = \begin{bmatrix} A \\ A\phi \end{bmatrix} T \; ; E_2 = \begin{bmatrix} A\phi \\ A\phi^2 \end{bmatrix} T \tag{13}$$

According to Eq. (13),

$$E_2 = \begin{bmatrix} A\phi \\ A\phi^2 \end{bmatrix} T = \begin{bmatrix} A \\ A\phi \end{bmatrix} TT^{-1}\phi T = E_1 T^{-1}\phi T \tag{14}$$

Let $\Psi = T^{-1}\phi T$, so $\Psi = E_1^{+}E_2$. Because $\mathbf{\Psi}$ has the same eigenvalues as $\mathbf{\Phi}$, we use eigenvalue decomposition on $\mathbf{\Psi}$ to get β_k, k=1,2,…,K. And then estimate frequency f_k , k= 1,2,…,K. Eigen decomposition of Ψ ,We can get the eigenvalues ψ_k ,k=1,2,…,K.

$$f_k = \frac{1}{2\pi\tau} angle(\psi_k) \tag{15}$$

where angle (.) denotes taking the phase angles.

3.2 Angle Estimation

According to Eq. (12), Take first (M-1)th rows of E_s to get E_3, which is shown as follows

$$E_3 = A_1 T \in \mathbb{C}^{(M-1)\times K} \tag{16}$$

Take second to Mth rows of E_s to get E_4,

$$E_4 = A_2 T \in \mathbb{C}^{(M-1)\times K} \tag{17}$$

We can get

$$E_4 = E_3 T^{-1}\varphi T \tag{18}$$

where

$$\varphi = diag\{e^{-j\alpha_1}, e^{-j\alpha_2}, \cdots, e^{-j\alpha_K}\} \in \mathbb{C}^{K\times K} \tag{19}$$

where $\alpha_k = 2\pi d f_k \sin\theta_k / c$, k=1,2,..,K.

Let $\Omega = T^{-1}\varphi T$, so $\Omega = E_3^{+}E_4$ .Because Ω has the same eigenvalues as φ, we use eigenvalue decomposition on Ω to get α_k, k=1,2,..,K.and then estimate θ_k, k=1, 2, …,K. Eigen decomposition of Ω, We can get the eigenvalues ξ_k, k=1,2,…,K.

$$\theta_k = \arcsin\left(\frac{c}{2\pi f_k \tau} angle(\xi_k)\right) \tag{20}$$

4 Simulation Results

We present Monte Carlo simulations that are to assess the angle and frequency estimation performance of JAFE algorithm. The number of Monte Carlo trials is 1000. 8-element-uniform linear array is used in the simulations. Note that N is the number of snapshots; K is the number of the sources. Define $RMSE = \sqrt{\frac{1}{1000}\sum_{m=1}^{1000}\left[a_m - a_0\right]^2}$, where a_m is the estimated angle/frequency, and a_0 is the perfect angle/frequency.

Simulation 1: The performance of our proposed algorithm is investigated. K=3 and N=400 in this simulation. Their DOAs are 10°, 20° and 30°, and their carrier frequencies are 500 kHz, 700 kHz and 900 kHz. Fig.2 shows the performance of our proposed algorithm with SNR=20dB, 35dB. From Fig.2 and Fig.3 we find that our proposed algorithm works well.

Simulation 2: We compare our proposed algorithm with single delay output method. K=3 and N=400 in this simulation. From Fig.4 we find that our proposed algorithm has better angle-frequency estimation performance than single delay output method.

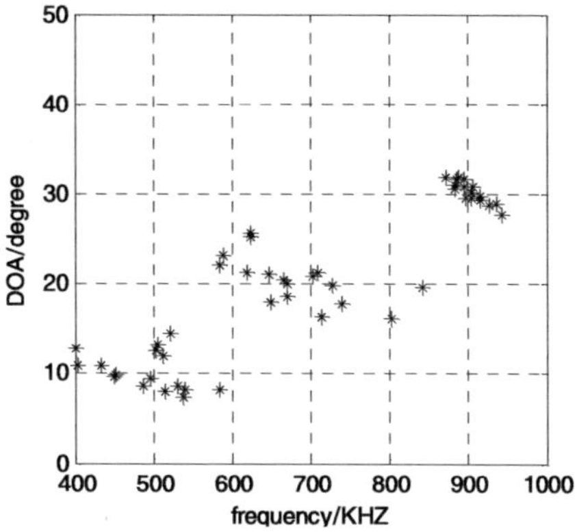

Fig. 2. Angle-frequency scatter, SNR=20dB

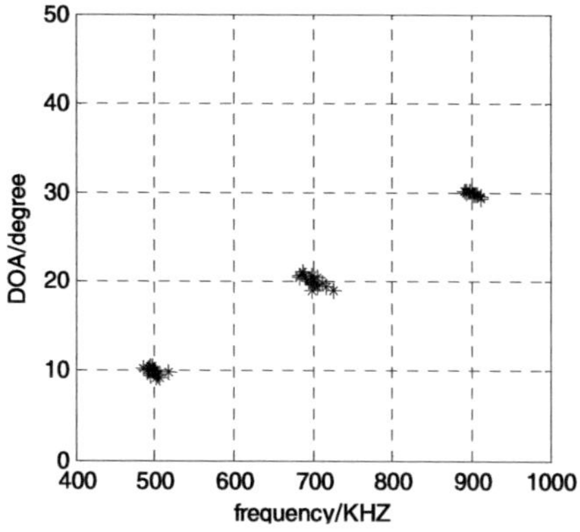

Fig. 3. Angle-frequency scatter, SNR=35dB

Simulation 3: JAFE algorithm performance under different snapshots N is investigated in this simulation. K=3 in this simulation.Fig.5 shows the angle-frequency estimation performance under different N. We find that the angle-frequency estimation performance of JAFE algorithm is improved with N increasing.

Simulation 4: The performance of our algorithm under different source number K is investigated in the simulation. N=400 in this simulation. The source number K is set 2, 3 and 4. JAFE algorithm has the different performance under different source number, as shown in Fig.6. From Fig.6, we find that angle and frequency estimation performance of JAFE algorithm degrades with the increasing of the source number K.

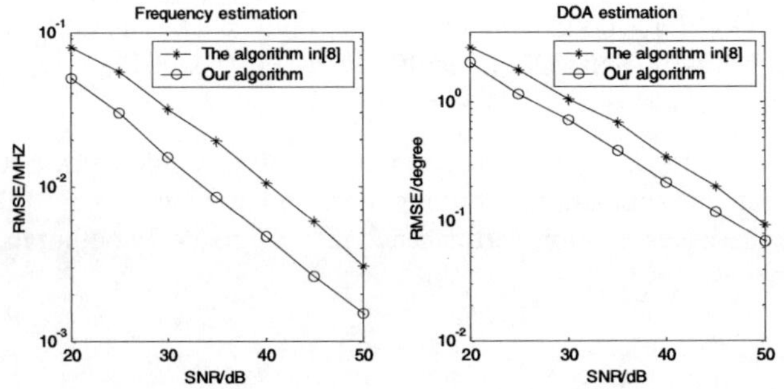

Fig. 4. Angle-frequency estimation performance comparison

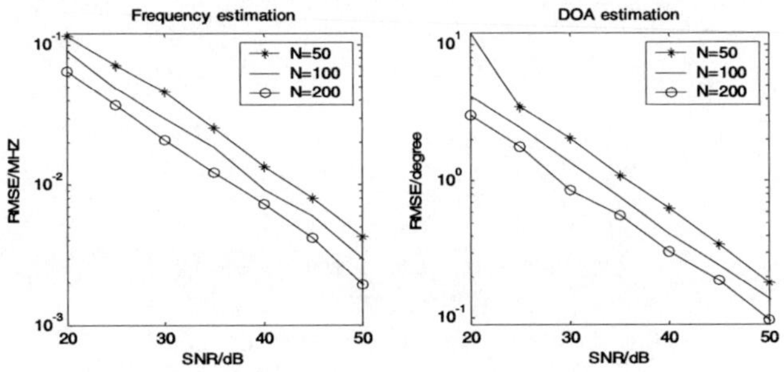

Fig. 5. Angle-frequency estimation with different snapshot N

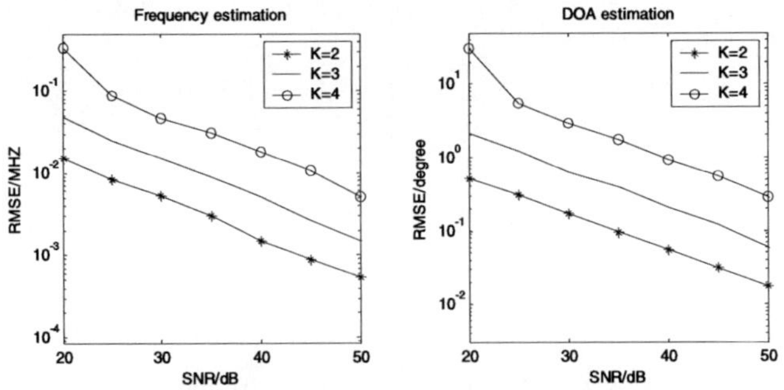

Fig. 6. Angle-frequency estimation with different sources

5 Conclusion

Our work links the uniform linear array parameter estimation problem and we presents a new ESPRIT algorithm-based joint angle and frequency estimation using multiple delay output. Our proposed algorithm is thought of as a generalization of ESPRIT. The advantage of our proposed algorithm using the multiple delay output over the conventional algorithm is that the estimation accuracy has been greatly improved.

References

1. Djeddou, M., Belouchrani, A., Aouada, S.: Maximum likelihood angle-frequency estimation in partially known correlated noise for Low-Elevation Targets. IEEE Trans. Signal Processing 53(8), 3057–3064 (2005)
2. Zoltowski, M.D., Mathews, C.P.: Real-time frequency and 2-D angle estimation with sub-nyquist spatio-temporal sampling. IEEE Trans. Signal Processing 42, 2781–2794 (1994)
3. Haardt, M., Nossek, J.A.: 3-D unitary ESPRIT for joint angle and carrier estimation. In: Proc. ICASSP, Munich, Germany, pp. 255–258 (1997)
4. Lemma, A.N., van der Veen, A.-J., Deprettere, E.F.: Joint angle frequency estimation using multi-resolution ESPRIT. In: Proc. ICASSP, Seattle, WA, vol. 4, pp. 1957–1960 (1998)
5. Lemma, A.N., van der Veen, A.J., Deorettere, E.F.: Analysis of joint angle-frequency estimation using ESPRIT. IEEE Trans. on Signal Processing 51(5), 1264–1283 (2003)
6. Chen, H., Wang, Y., Wu, Z.: Frequency and 2-D Angle Estimation Based on Uniform Circular Array. In: 2003 IEEE International Symposium on Phased Array Systems and Technology, pp. 547–552 (2003)
7. Fu, T., Jin, S., Gao, X.: Joint 2-D Angle and Frequency Estimation for Uniform Circular Array. In: 2006 International Conference on Communications, Circuits and Systems Proceedings, vol. 1, pp. 230–233 (2006)
8. Wang, S., Zhou, X.: Direction-of-Arrival and Frequency Estimation in Array Signal Processing. Journal of Shanghai Jiaotong University 33(1), 40–42 (1999)
9. Jia, W., Yao, M., Song, J.: Joint frequency, two dimensional arrival angles estimations via marked signal subspace. In: The 8th International Conference on Signal Processing, vol. 1, pp. 16–20 (2006)
10. Lin, C., Fang, W., Wu, K., Lin, J.: Fast Algorithm for Joint Azimuth and Elevation Angles, and Frequency Estimation via Hierarchical Space-Time Decomposition. In: IEEE International Conference on Acoustics, Speech and Signal Processing, vol. 2, pp. 1061–1064 (2007)

A GAF Algorithm of Dislocated Grid for Wireless Sensor Network*

Chen Zu-jue, Ouyang Ye-long, and Liu Shao-qing

College of Computer Science and Communication Engineering of Jiangsu University
Zhenjiang Jiangsu, 212013, China
chenzujue@126.com, oyangyelong10@163.com, king8411twins@yeah.net

Abstract. For the problem of limited energy in wireless sensor networks, a GAF algorithm of Dislocated Grid (GAFDG) is proposed, based on the traditional GAF algorithm of Aligned Grid (GAFAG). The algorithm divides monitoring areas into several dislocated grids, each node belong to corresponding grid according to its coordinate and the side length of grid. The nodes within the same grid compose a cluster. Selecting a node as cluster head, whose residual energy above the average residual energy and close to intra-cluster nodes' center. The backbone network of wireless sensor networks is composed by each cluster head, and forwarding information according to GeRaF routing protocol. Theoretically proofing the single-hop coverage area of GAFDG is increased by 12% than GAFAG. Emulational results show that both energy consumption and network lifetime Of GAFDG are better than GAFAG.

Keywords: wireless sensor networks, dislocated grid, GeRaF, routing protocol, GAF algorithm.

1 Introduction

Wireless Sensor Networks (WSN) is a self-organizing wireless communication networks, formed by a large number of micro-sensors[1]. With the development of microelectronic technology, sensor technology and com- munication technology, WSN has broad applications and developing prospects in science, medicine, business and national defense[2].

Sensor nodes' energy is limited, and sending and receiving data consume most of it, so energy efficient routing strategies to prolong network lifetime is essential. Seeing from the structure of network topology, WSN routing protocols can be divided into flat routing protocols and cluster routing protocols., Because of convenient topology management, efficient energy utilization, simple data fusion, as well as good scalability, cluster routing protocols have became the focus of the current routing technology.

Currently, many cluster routing protocols are complex in cluster head(CH) selection algorithm and cluster division algorithm, and the number of nodes in each cluster is

* National High Technology Research and Development Program (863 Program) (2006AA10Z258); Zhenjiang Social Development and Technology Support Projects (SH2009002).

W. Zhang (Ed.): Advanced Technology in Teaching, AISC 163, pp. 721–728.
springerlink.com

uneven, so causing uneven energy consumption (EC), therefore, they are difficult to be practical applications. However, the positioning technology of sensor node mature gradually, so using nodes' location information to divide clusters that becomes one of the methods to solve the problem of uneven division[3]. Take literature [4] for example, in this paper, the optimal grid size is studied based on link model reachable with probability in realistic environment. Reaching probabilities of neighboring grids and path are analyzed, and the effects of node density and grid size on reaching probabilities is studied, giving the best grid selection criteria at last. In literature [5], a grid-based routing protocol in WSN was proposed on the basis of node location. The protocol uses grids to avoid the complicated CHs selection and cluster division algorithm, also restricts the routing flooding data packages in single grid. In literature [6], a cluster routing protocol based on virtual grid is proposed. The Sink node dynamically and randomly builds virtual grid, meanwhile forms cluster structure in the proposed protocol. It selects CHs according to the residual energy and the EC of all nodes in each cluster, not only reduces control message and redundancy message, but also makes the energy load distributed more evenly. In literature [7], Z. Yuan et al proposed a novel geographic routing algorithm based on virtual force on the basis of GAF. The forces between a certain node and other nodes or boundary were calculated. The effect forces were utilized to control the movement of mobile nodes to a new location. The improved node sleep-scheduling could duly awaken sleep nodes. In literature [8], an improved GAF algorithm with hexagon-based virtual infrastructure is proposed. The goal of this algorithm is to lengthen the side of the grid, to enlarge the area of grid and to use the nodes in the overlap areas as transfer nodes among the CHs. HGAF-h[9] is the improvement of the GAF algorithm of honeycomb structure. Firstly, each cell divides into several sub-cells, and sequentially labels the sub-cells. Making each active sub-cell always in the center of corresponding cell by mobiling cells' boundaries, that is, making CHs are always in the cluster's central location. Literature [10] proposes a topology control algorithm based on honeycomb structure. Then with addition of greedy forwarding algorithm and perimeter forwarding algorithm in GPSR, the routing void or obstacles could be easily bypassed, thus the network lifetime be prolonged, and the energy saving be realized.

Using the location information of nodes, the paper proposes a GAF algorithm of Dislocated Grid for WSN.

2 The Fundamental of GAF Algorithm

GAF (geographical adaptive fidelity) is location-based routing algorithm, proposed by Xu et al[11]. Initially, it was primarily design for mobile Ad hoc network application, and also suitable for WSN. The basic idea is: the network region is divided into several virtual grids in equal size, and each grid will select a node to maintain an active state in a certain period of time, while other nodes go to sleep. The selected node on behalf of other nodes within the same grid to monitor and report data to base station or Sink node. Each node in the network area obtain its location information by GPS or other positioning methods to determine which grid is it belongs to and other nodes within the same grid. Multiple nodes within a grid is equivalent to the cost of packet routing, taking turns to sleep. GAF routing algorithm balances nodes' EC, extends network lifetime without affecting the routing fidelity. With the increasing number of nodes in the area, network lifetime can be greatly extended.

3 The GAF Algorithm of Dislocated Grid

The traditional GAF algorithm divides monitoring areas into several square grids of equal side length, and each square grid be aligned. But the GAF algorithm of this paper, the square grids between the upper and the lower are dislocated arrangement. Fig.1 and Fig.2 respectively describe two structures.

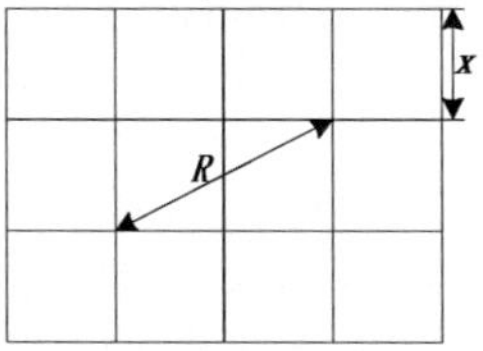

Fig. 1. Aligned Grid Model

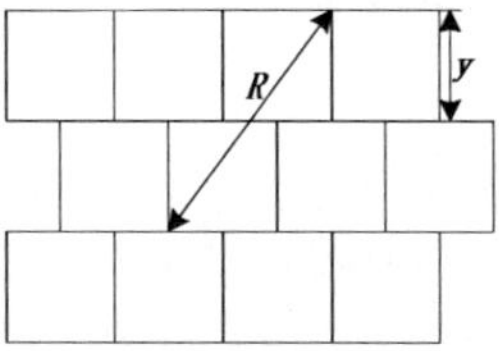

Fig. 2. Dislocated Grid Model

3.1 The Establishment of Virtual Grid and Clusters Formation

Using the similar method of LEACH to circularly establish clusters. Before establishing virtual grids, Sink node needs to send a reference coordinate to the whole network, which consumes a large amount of energy. To reduce the control signal, reconstructing virtual grid once every M rounds. Therefore, Sink node needs to send reference coordinate, parameter M and grid duration at the same time.

Sink node randomly selects a reference coordinate (x_s, y_s) within the network. And establishing virtual grid with r_{grid} for the unit. The size of r_{grid} according to the node density and clusters' scale, The nodes within the same grid compose a cluster. Node (x_i, y_i) joins the corresponding identified cluster according to its coordinate and follow the formula (1), (2).

1) When $\left\lceil \frac{y_i - y_s}{r_{grid}} \right\rceil$ is a even number, the grid which node (x_i, y_i) belongs to is calculated as:

$$\left(\left\lceil \frac{x_i - x_s}{r_{grid}} \right\rceil, \left\lceil \frac{y_i - y_s}{r_{grid}} \right\rceil \right) \tag{1}$$

2) When $\left\lceil \frac{y_i - y_s}{r_{grid}} \right\rceil$ is a odd number, grids rightward shift $0.5r_{grid}$ units, and the grid which node (x_i, y_i) belongs to is calculated as:

$$\left(\left\lceil \frac{x_i - x_s}{r_{grid}} + 0.5 \right\rceil - 0.5, \left\lceil \frac{y_i - y_s}{r_{grid}} \right\rceil \right) \tag{2}$$

3.2 Cluster Head Selection

In this M rounds cycle, the CH helps select the next CH, sends the message of CH selection. The time-slot of new CH will transfer to the old one, but the time-slot of the remaining nodes do not change. Due to the new CH doesn't know intra-cluster nodes' information, the remaining nodes send their residual energy and coordinates when they send data to the new CH. Using "the principle of minimum energy consumption " to select CHs, namely, choose a node as a CH, whose residual energy above the average residual energy and close to the center of the inter-cluster nodes.

$$\min \| n_i - cg \| \quad (E_{ni} \geq \overline{E}_{cls}) \tag{3}$$

Among them, n_i is node's coordinate, cg is intra-cluster nodes' center coordinate, E_{ni} is the residual energy of node n_i, $\overline{E}_{cls}$ is the average residual energy of the cluster which node n_i belongs to. cg and $\overline{E}_{cls}$ are calculated by CH.

Selecting the node that is the closest to intra-cluster nodes' center as CH can minimize the total EC, consumed by each node send data to CH, and make the distribution of CHs is more uniform. Without loss of generality, scattering N nodes in a square region, and each node's coordinate is (x_i, y_i), $i = 1, 2, \cdots, N$. Looking for a point (x, y) which can minimize the sum of the square of distance to other N nodes. Namely:

$$\min \sum_{i=1}^{N} [(x - x_i)^2 + (y - y_i)^2] \tag{4}$$

Solutions are: $x = \frac{1}{N}\sum_{i=1}^{N} x_i$, $y = \frac{1}{N}\sum_{i=1}^{N} y_i$

The solutions are the coordinate of center. The solutions of the quadratic function can only be selected from the N nodes' coordinates, and the node which is the closest to the center, its coordinate can satisfy formula (4). The distinction of sum EC among each round is determined by the sum of the square of distance, which between intra-cluster nodes and CH. If the sum of the square of distances is the least, it can ensure sum EC is the least.

3.3 Data Transmission

Data transmission of intra-cluster nodes adopt TDMA method. Have received data, CHs will fuse data and forward the fused results to Sink node through neighboring CHs. The backbone network, formed by CHs, is equivalent to grids network. The backbone network can transmit data by location-based routing protocols, for example, GeRaF (Geographic random forwarding) routing protocol which was proposed by Zorzi et al[12-13]. Data transmission among CHs use grids' *ID* to mark the source and the destination , thus it needn't know neighboring CHs' *ID*; When the CHs selection are not complete in synchronous, it can send data directly to the grid which is closer to Sink node.

4 Single-Hop Coverage Area

Supposing nodes' communicated radius is R, the area of the square grid in Fig.1 and Fig.2 are S_X and S_Y. To ensure the connectivity of network, x and y must satisfy following inequalities:

$$x^2+(2x)^2 \le R^2 \tag{5}$$

$$(3y/2)^2+(2y)^2 \le R^2 \tag{6}$$

Solutions of inequality (5) and inequality (6) are:

$$x \le \frac{R}{\sqrt{5}},\ y \le \frac{2}{5}R$$

$$S_X = x^2 \tag{7}$$

$$S_Y = y^2 \tag{8}$$

So,

$$S_X(\max) = (\frac{R}{\sqrt{5}})^2 = \frac{R^2}{5} \tag{9}$$

$$S_Y(\max) = (\frac{2}{5}R)^2 = \frac{4}{25}R^2 \tag{10}$$

The structures of aligned grid and dislocated grid, assuming the former and the latter's single-hop coverage area are S_1 and S_2, so they are corresponding to the $5S_X$ and $7S_Y$, namely:

$$S_1 = 5S_X \tag{11}$$

$$S_2 = 7S_Y \tag{12}$$

Thus,

$$S_1(\max) = 5S_X(\max) = R^2 \tag{13}$$

$$S_2(\max) = 7S_Y(\max) = \frac{28}{25}R^2 \tag{14}$$

$$\frac{S_2(\max)}{S_1(\max)} = 1.12 \tag{15}$$

After using dislocated grid structure, single-hop coverage area is increased by 12%, it is should be able to reduce routing hops and save nodes' energy.

5 Simulation Testing and Analysis

The experiment compare the performance between GAFDG and GAFAG (which was mentioned in literature [5]).

5.1 The Construction of Experimental Platform

This experiment is simulated in network simulation software NS2. Using the same EC model as literature [6]. In this model, receiving and sending data' EC is calculated as follows:

$$E_R = E_{elec} \times k \tag{16}$$

$$E_T = E_{elec} \times k + E_{amp} \times k \times d^2 \tag{17}$$

Formula (16) shows the EC of receiving *Kbit* data, and formula (17) shows the EC of sending *Kbit* data to a node which d meters away from. The experimental parameters are showed in Table.1.

Table 1. Experimental Parameters

Parameter	Values	Parameter	Values
Area	600*600m^2	Fusion EC	5nJ/bit/message
Nodes number	2000	Sending data	2Kbit
Initial energy	0.125J/node	Control signal	200bit
Eelec	50nJ/bit	Grid	50*50m^2
Eamp	100pJ/bit/m	M	2

5.2 Energy Consumption Testing and Analysis

Fig.3 shows the percentage of the nodes' average EC of the two algorithms. Seeing from the figure, the EC performance of GAFDG is superior to GAFAG. With the increasing number of rounds, the node's average EC also increases in two algorithms. However, at the same number of rounds, GAFAG's EC is greater than GAFDG. When nodes' average EC is about 85% in GAFDG and 100% in GAFAG are having same number of rounds. The reasons are: firstly, the former's single-hop coverage area bigger than the latter, be bound to reducing the routing hops and reducing end-to-end EC among nodes; Secondly, the former use "the principle of minimum energy consum- ption" to select CHs, can minimize the total EC which each node send data to CH consumed, and the distribution of CHs is more uniform; Furthermore, reconstructing virtual grid once every *M* rounds, so reduces the EC which is used for spreading control signals.

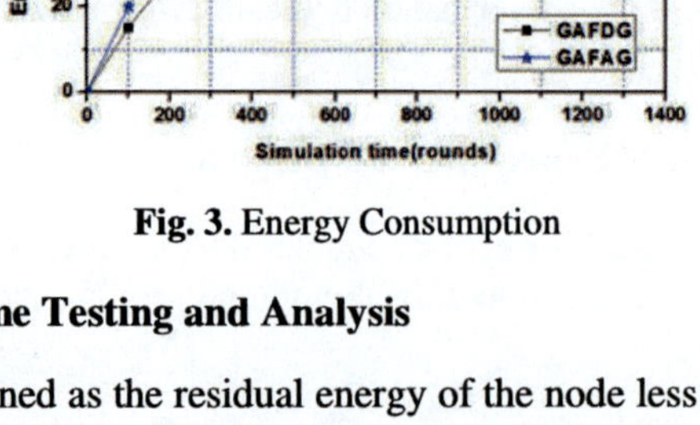

Fig. 3. Energy Consumption

5.3 Network Lifetime Testing and Analysis

The death node is defined as the residual energy of the node less than 80% of its initial energy. Network lifetime is defined as the number of rounds when the death nodes account for 40% of the total number of nodes. The death nodes appear at 450 rounds in GAFDG, while 300 rounds in GAFAG. There are 800 death nodes at 750 rounds in GAFAG, and according to the definition of network lifetime (800/2000=40%), knowing the latter's network lifetime is 750 rounds. However, the death nodes is less than 800 when at 900 rounds in GAFDG, so the former' network lifetime should be longer than 900 rounds. The main reason is that the reference coordinate (x_s, y_s) is randomly generated by Sink node. For the principle of minimum energy consumption, most of the CHs close to the center of the intra-cluster nodes. Reference coordinate's transformation directly affects CHs selection, and the more frequently transfor, the better to the uniformity of CHs selection, thereby, it is good for balance energy load.

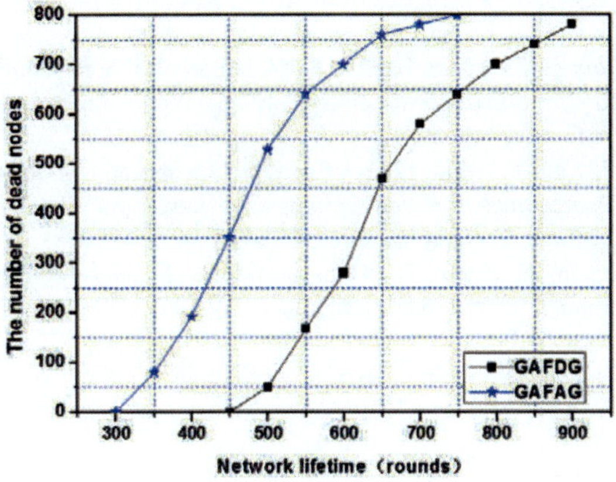

Fig. 4. Network Lifetime

6 Conclusion

The algorithm divides monitoring areas into several dislocated grids, and the nodes within the same grid compose a cluster. Selecting CHs according to the principle of minimum energy consumption, therefore, the EC is evenly distributed to the whole network, and prolonging the network lifetime. Emulational results show that the algorithm has good performance, it is a energy efficient and load-balanced cluster routing algorithm.

References

1. Hana, K., Reza, S.: A New tree-based Routing Algorithm for Energy Reduction in Wireless Sensor Networks. In: 2009 International Conference on Signal Processing Systems, pp. 116–120 (2009)
2. Shen, L., Jiang, Y.: Study on the influence of data fusion on clustering energy in wireless sensor networks. In: IEEE/INFORMS International Conference on Service Operations, Logistics and Informatics, pp. 481–486 (2009)

3. Sakhaee, E., Wakamiya, N., Murata, M.: GPS-free disaster-scale mapping and energy-efficient alerting scheme in a wireless sensor network. In: Second International Conference on Sensor Technologies and Applications, pp. 73–80 (2008)
4. Liu, X., Yu, H., Hou, H., Hu, H.: Grid Routing Based Oil Link Reachable with Probability in Wireless Sensor Networks. Journal of Electronics & Information Technology 30(9), 2259–2262 (2008)
5. Shen, Y., Xu, Q., Pei, Q., Ma, J.: Grid-based routing protocol in wireless sensor networks. Journal on Communications 30(11A), 96–100 (2009)
6. Yun, C., Wang, P.: Virtual Grid based Routing Protocol for wireless sensor network. Computer Applications and Software 6(9), 200–203 (2009)
7. Yuan, Z., Li, J.: Geographic routing algorithm based on virtual force for WSN. Application Research of Computer 26(6), 113–116 (2009)
8. Liu, S., Liu, L., Tao, J.: Improved GAF Algorithm with Hexagon-Based Virtual Infrastructure. Complwer Technology and Development 19(1), 39–43 (2009)
9. Takashi, O., Tokuya, I., Susumu, I.: HGAF-h:A hierarchical honeycomb cooperative power saving architecture for sensor networks. In: IEEE International Conference on Mobile Data Management, pp. 542–547 (2009)
10. Zhao, Y., Cao, P.: Grid-based for Energy Conservation around Obstacles in Wireless Sensor Networks. In: Proceedings 5th International Conference on Wireless Communications, Networking and Mobile Computing, WiCOM 2009, pp. 1–4 (2009)
11. Xu, Y., Heidemann, J., Estrin, D.: Geogrph-informed energy conservation for Ad hoc routing. In: 7th Annual International Conference on Mobile Computing and Networking, pp. 70–84 (2001)
12. Zorzi, M., Rao, R.: Geographic random forwarding (GeRaF) for ad hoc and sensor networks: Multihop performance. IEEE Transactions on Mobile Computing 2(4), 337–348 (2003)
13. Zorzi, M., Rao, R.: Geographic random forwarding (GeRaF) for ad hoc and sensor networks: Energy and latency performance. IEEE Transactions on Mobile Computing 2(4), 349–365 (2003)

An Improved Code Modulation Scheme Based on BICM for IBOC-FM

Song Yang[1], Li Jianping[1], Cai Chaoshi[1], and Yang Wei[2]

[1] School of Information Engineering Communication University of China
Beijing, China
[2] Tianjin Sino-German Vocational and Technical College
icysongyang@eyou.com, turbo_code@163.com, BBICAI@cuc.edu.cn,
rosebush_y@126.com

Abstract. In-Band On-Channel(IBOC) digital audio broadcasting(DAB) is a method of digital audio radio, which could transmit analog FM and digital audio simultaneously. However, due to the existence of fading and interference in FM bands, and transmit capacity has reached its limitations for the bandwidth, thus, adding digital transmission availability in the crowded FM band is a challenging proposition. Diversity is one of the crucial techniques for digital transmission. Bit-interleaved coded modulation (BICM) can provide spatial diversity without occupying high bandwidth. Meanwhile, Low Density Parity Check(LDPC) code with low density scheme can be a better choice as IBOC FM system which allows a certain degree of freedom in design, In this paper, a type of coded modulation scheme was proposed in Hybrid in-band on-channel system, by use of BICM embedded into LDPC code is instead of BICM embedded convolutional codes for IBOC FM system, would result in coding gains without bandwidth expansion. The simulation results show that the proposed scheme can effectively improve the receiving performance of an FM IBOC system over Rayleigh fading channels. When BER is below 10^{-4} can achieve more than 2dB coding gains.

Keywords: IBOC, FM, LDPC, diversity, BICM, Rayleigh fading channels.

1 Introduction

Hybrid In-band On-channel have been proposed in FM band for digital audio broadcasting [1]. Digital audio is transmitted simultaneously with an analog host signal which can be transmitted simultaneously within licensed band. However, the bandwidth allocated for digital audio transmission is limited. Since the system uses convolutional encoding, which adds redundancy to the digital data, to achieve coding gains without bandwidth expansion in each logical channel. But the wireless transmission environment presents a fundamental technical restriction challenge for reliable high-speed communications.

Low Density Parity Check (LDPC) Code was first introduced by G.G.Gallager in 1962[2] and specified by a parity-check matrix with very low density. The meaning of low density is that the ones in the matrix are very sparse and most elements of the

W. Zhang (Ed.): Advanced Technology in Teaching, AISC 163, pp. 729–735.
springerlink.com

matrix are zeros. LDPC can achieve performance closer to he Shannon's limit. It has been shown to perform extremely well even with computationally efficient iterative decoding [9]. On one hand, at a high code rate, the LDPC codes achieve good bit error rate (BER) performances [7]. On the other hand, there is no bit-interleaver needed as the random structure of LDPC which can establish statistically in dependence of the received bits itself. Therefore, it is more suitable for commercial applications.

Bit-interleaved coded modulation (BICM), first suggested by Zehavi [4] and further studied by Caire [5], is a new channel coding scheme using bit interleaver to maximum the code diversity to achieve better reliability over Rayleigh fading channels.

In this paper, an improved channel coding scheme based on BICM embedded LDPC for IBOC FM system is proposed. This paper is organized as follows. In section 2 some principles on coded modulation and LDPC encoder computation for FM IBOC DAB are given. The improved channel coding scheme for IBOC is presented in section 3. Simulation results are given in section 4, and finally conclusion is made in section 5.

2 IBOC FM System Model

The block diagram of IBOC in FM system is shown in Fig.1. This system delivers digital audio and data services to mobile, portable, and fixed receivers from terrestrial transmitters in the existing frequency radio bands. The input binary sequences are scrambled in order to ensure appropriate energy dispersal in the transmitted signal. The information bits are then coded by convolutional encoder. The latter is fed into independent ideal interleaves. Finally, the decision on the transmitted sequence is taken with the aid of the Viterbi algorithm. Broadcasters may continue to transmit an analog FM. Simultaneously the improved, higher-quality, and more robust digital signals allow themselves and their listeners to convert from analog to digital radio while maintaining their current frequency allocations. IBOC using Puncturing Convolutional encoder [6] consists of three primary operations: mother code generation, puncturing and parallel-to-serial conversion [8].

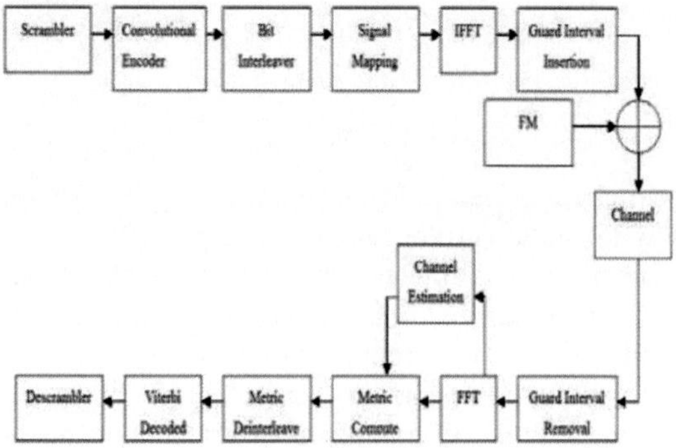

Fig. 1. A Block diagram of the BICM- IBOC in FM System

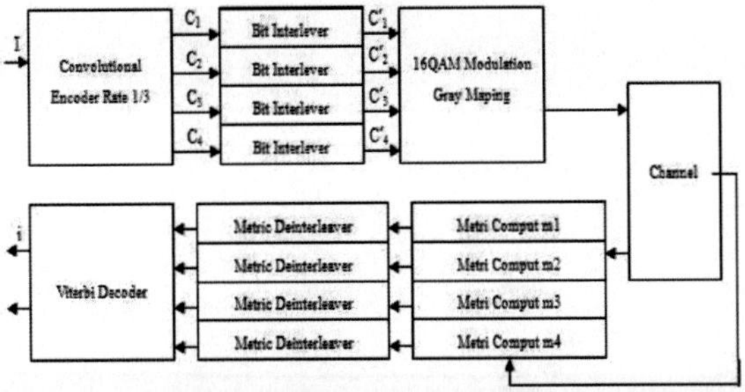

Fig. 2. BICM system diagram

3 The BICM-LDPC Scheme for IBOC FM

3.1 Bit-Interleaving Coded Modulation

In [4], Zehavi showed that the performance of coded modulation can be improved over Rayleigh fading channels by bit-wise interleaving at the encoder output, and by using an appropriate soft-decision metric for a Viterbi decoder at the receiver. Then, BICM was suggested [5]. BICM is a new channel coding scheme which has a better performance over Rayleigh fading channels and AWGN channels. It has been proved to be a very power-efficient approach providing those state-of-the-art codes. An example of a BICM scheme in the hybrid IBOC-FM system is shown in Fig. 2.

3.2 LDPC Code

LDPC is a kind of linear codes. It is so named because it is specified by a parity-check matrix with very low density. Low density means the ones in the matrix is very sparse and most elements of the matrix are zeros. An (N,j,k) low-density code is a code of block length N with a *M* by *N* matrix which gives the rate *(N-M)/N*. Each column contains a small fixed number $j \geq 3$ of 1's and each row contains a small fixed number k>j of 1's as in Fig.3(j=3,k=4)[2].We call this type of codes as regular codes when the number of ones in each column and each row is identical. If not, it is irregular codes. There is no bit-interleaver needed as the random structure of LDPC. LDPC can establish independence of the received bits itself, yield better coding gains with low complexity. The belief propagation decoding algorithm of the LDPC codes is based on a parity-check matrix with very low density. Therefore, the implementation in hardware is kept in low cost. If the system allows a certain degree of freedom in design, then the rate of LDPC codes may flexibly be chosen for variety communication systems [10].

1	1	1	1	0	0	0	0	0	0	0	0	0	0	0	0	0	0	0	0
0	0	0	0	1	1	1	1	0	0	0	0	0	0	0	0	0	0	0	0
0	0	0	0	0	0	0	0	1	1	1	1	0	0	0	0	0	0	0	0
0	0	0	0	0	0	0	0	0	0	0	0	1	1	1	1	0	0	0	0
0	0	0	0	0	0	0	0	0	0	0	0	0	0	0	0	1	1	1	1
1	0	0	0	1	0	0	0	1	0	0	0	1	0	0	0	0	0	0	0
0	1	0	0	0	1	0	0	0	1	0	0	0	0	0	0	1	0	0	0
0	0	1	0	0	0	1	0	0	0	0	0	0	0	1	0	0	1	0	0
0	0	0	1	0	0	0	0	0	0	1	0	0	0	0	1	0	0	1	0
0	0	0	0	0	0	0	1	0	0	0	1	0	0	0	1	0	0	0	1
1	0	0	0	0	1	0	0	0	0	0	1	0	0	0	0	0	1	0	0
0	1	0	0	0	0	1	0	0	0	1	0	0	0	0	1	0	0	0	0
0	0	1	0	0	0	0	1	0	0	0	0	1	0	0	0	0	0	1	0
0	0	0	1	0	0	0	0	1	0	0	0	0	1	0	0	1	0	0	0
0	0	0	0	1	0	0	0	0	1	0	0	0	0	1	0	0	0	0	1

Fig. 3. Low-density matrix N=20,j=3,k=4.

3.3 BICM-LDPC

In addition to presenting LDPC codes, Gallager also provided a decoding algorithm that is effectively optimal. At first a hard-decision algorithm, whose idea is to "flip" bits until all parity checks are satisfied. He also gave a soft-decision algorithm that is exactly the sum-product or belief-probability algorithm now. The theorem to compute the posterior probability of each transmitted bit being 1 as follows:

$$\frac{\Pr\left[x_d=0\middle|\{y\},S\right]}{\Pr\left[x_d=1\middle|\{y\},S\right]}=\frac{1-\mathrm{P}_d}{\mathrm{P}_d}\prod_{i=1}^{j}\left[\frac{1+\prod_{i=1}^{k-1}(1-2\mathrm{P}_{il})}{1-\prod_{i=1}^{k-1}(1-2\mathrm{P}_{il})}\right] \tag{1}$$

In this paper, an iterative algorithm proposed by D.J Mackay[11] is adopted of the above complex algorithm and introduced: R_j is the set of column locations of 1's in the *j* row, $R_{j\backslash i}$ is the set of column locations of 1's in the *j* row excluding location *i* ; C_i is the set of row location of the 1's in the *i* column; $C_{i/j}$ is the set of row location of the 1's in the *i* column excluding location *j*; $r_{ji}(b)$ is the probability of the *j*th check equation being satisfied given bit c_i=b and the other bits have distribution given by $\{q_{ij}(b)\}$ is probability that c_i=b given extrinsic information from all check nodes except the *j*th node. The algorithm iterates back and forth between $\{q_{ij}\}$ and $\{r_{ji}\}$ using formulas as follows:

$$r_{ij}(0)=\frac{1}{2}+\frac{1}{2}\prod_{i\in R_{j/i}}(1-2q_{ij}(1)) \tag{2}$$

$$r_{ji}(0)=1-r_{ji}(0) \tag{3}$$

$$q_{ij}(0)=(1-P_i)\prod_{j\in c_{i\backslash j}}r_{ji}(0) \tag{4}$$

$$q_{ij}(1)=P_i\prod_{j\in c_{i\backslash j}}r_{ji}(1) \tag{5}$$

Owing to those advantages, BICM-LDPC has been proved to be a very power-efficient approach provided in [3], The LDPC code used in the simulation of this section has the codeword length of 576 and the code rate of 1/2. M=10.20.80.100 are chosen as the iterative number and the performance curves are plotted respectively over Rayleigh fading channels. It is clearly shown that, the larger iterative number would result in the better BER performances, however it also brings in more complexity calculation. Considering the computational complexity, iterative number 20 is a better selection for BICM-LDPC, with a great performance which gains approached to iterative number=100.

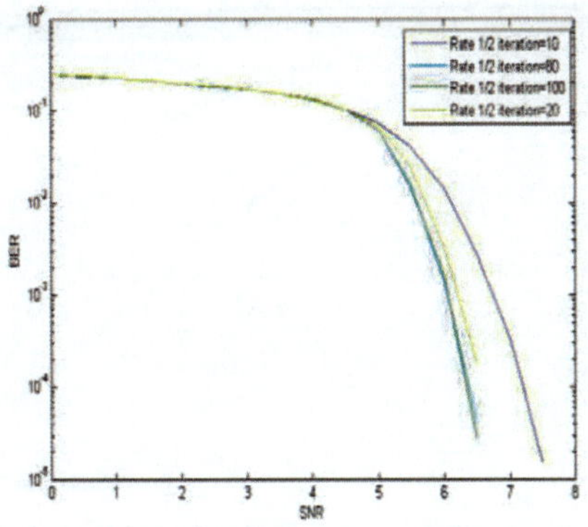

Fig. 4. LDPC decoding performance with different iterative number

4 Simulation Results

In the improved scheme, BICM embedded LDPC code for instant of BICM embedded convolutional codes for IBOC FM system. We focus on the bit error rate(BER) performance over Rayleigh fading channels and AWGN channels transmission environment. All the code rates (1/2) supported by IBOC standard are simulated. Here we consider 16QAM and QBSK modulation, Gary mapping and the maximum number of decoding iteration is set to100.

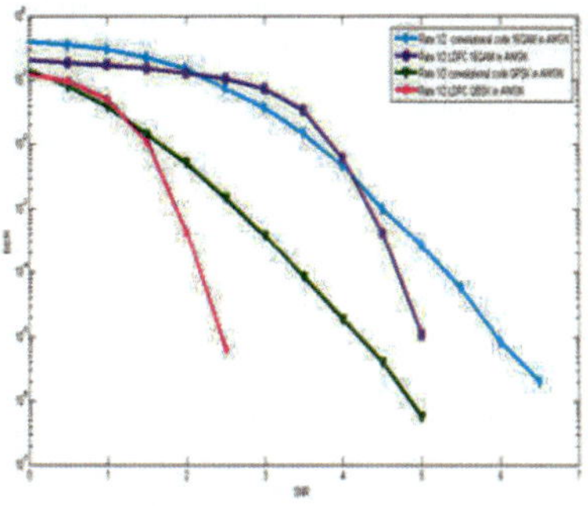

Fig. 5. The BER performance of conventional system of IBOC and the improved scheme with 16QAM and QBSK modulation over AWGN channels

Fig.5 shows the BER curves of the improved BICM-LDPC scheme and the conventional scheme of IBOC over AWGN channels. The improved scheme is kept about 1-2dB code gains when BER is below 10^{-3} with QBSK modulation and about 0-1dB code gains with 16QAM modulation. The next shows over Rayleigh fading channels when BER is below 10^{-4} can achieve more than 2dB coding gains at the Fig.6.Simulation results show that a good coding modulation scheme can improve coding gains without bandwidth expansion and can be a good optional application for specific bandwidth demands in FM IBOC system.

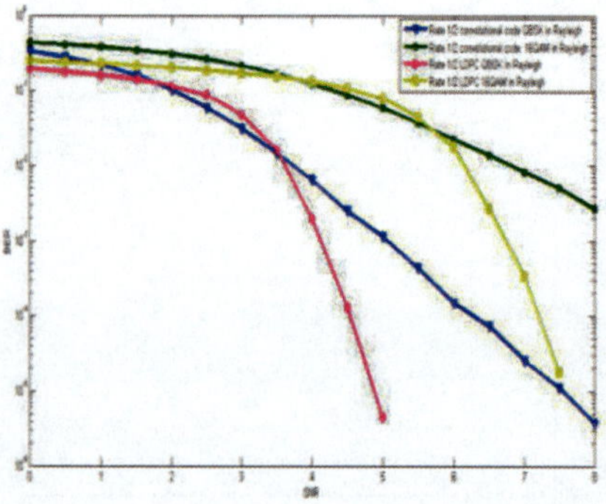

Fig. 6. The BER performance of conventional system of IBOC and the improved scheme with 16QAM and QBSK modulation over Rayleigh fading channels

5 Conclusion

An improved BICM-LDPC scheme, which can get coding gains compared with BICM-Convolutional code in IBOC FM system, is considered in this paper. Simulation results confirm that, the BER performances of the proposed improved coding scheme based on BICM-LDPC for IBOC outperforms the traditional scheme. Compared with the traditional scheme, an improved performance with BICM-LDPC at 16QAM and QBSK modulation can get better coding gains without bandwidth expansion over Rayleigh fading channels and AWGN channels in IBOC FM system.

Acknowledgment. This paper is supported by the key project of Chinese Ministry of Education(No. 106042) and the project sponsored by the Scientific Research Foundation for the Returned Overeas Chinese Scholars, State Education Ministry(2007[24]).

References

1. HD Radio Air Interface Design Description-Layer 1 AM. Rev. E, pp. 9–14, March 22 (2005)
2. Gallager, R.G.: Low-density parity-check codes. MIT Press, Cambridge (1963)
3. Caire, G., Taricco, G., Biglieri, E.: Bit-interleaved coded modulation. IEEE Trans. Inform. Theory 44, 927–946 (1998)

4. Zehavi, E.: 8-PSK trellis codes for a rayleigh fading channel. IEEE Trans. Commun. 40, 873–883 (1992)
5. Caire, G., Taricco, G., Biglieri, E.: Bit-interleaved coded modulation. IEEE Trans. Inform. Theory 44, 927–946 (1998)
6. Chung, S.-Y., Lou, H.: Multilevel rs/convolutional concatenated coded qam for hybrid IBOC-AM broadcasting. IEEE Transactions on Broadcasting 46(3), 49–59 (2000)
7. Abematsu, D., Ohtsuki, T., Jarot, S.P., Kashima, T.: Size Compatible (SC)-Array LDPC Codes. In: VTC 2007, pp. 1147–1151 (October 2007)
8. Feng, Y., Li, J., Dong, Y., Sha, S.: Coded Modulation Scheme with CPPC Codes for FM IBOC Broadcasting. IEEE Trans. Inf. Theory, 1–4 (September 2009)
9. Awano, T., Kasai, K., Shibuya, T., Sakaniwa, K.: Three-Edge type LDPC Code Ensembles with Exponentially Few Codewords with Linear Small Weight. In: ISITA 2008 (December 2008)
10. Zhang, H.-G., Yuan, D.-F., Ma, P.-M., Yang, X.-M.: Low-density party-check code(LDPC) schemes with BICM. In: ICCT (April 2003)
11. Mackay, D.: Good emr correcting codes based on very sparse matrices. IEEE Trans. Information Theory (March 1999)

An Improved Spatial Division Multiplexing of STBC Scheme Based on BICM

Song Yang, Li Jianping, and Cai Chaoshi

School of Information Engineering
Communication University of China
Beijing, China
icysongyang@eyou.com, turbo_code@163.com, BBICAI@cuc.edu.cn

Abstract. Spatial Division Multiplexing (SDM) technique combined with Space-Time Block Coding(STBC) from multiple antennas can achieve high data rates for digital transmission in Multiple Input Multiple Output(MIMO) system. However, the transmit capacity has reached its limitations for the bandwidth, thus, enhancing digital transmission efficiency becomes reality. Bit-interleaved coded modulation (BICM) can provide spatial diversity which is one of the key techniques for transmission, and can get better coding gains without increasing the bandwidth. Meanwhile, STBC which is the technique of temporal is applied to raise transmission reliability. In this paper, BICM is utilized to be embedded into spatial division multiplexing of two Alamouti codes, which would lead to coding gains and not increase the bandwidth. The simulation results show when BER is below 10^{-4}, the improved channel coding scheme can achieve about 1~2 coding gains over Rayleigh fading channels.

Keywords: MIMO, SDM, STBC, BICM, diversity, Alamouti code.

1 Introduction

In recent years, the use of multiple antennas at transmitters and receivers has gained popularity due to potential of such systems to increase capacity [1]. The increased spectral efficiency of such MIMO systems is important and the spectral efficiency can be drastically increased, boosting the throughput of the communications link while requiring the same amount of bandwidth. Since bandwidth is to the limit, improving effectively becomes important. Diversity technique by sending signals that carry the same information through different paths can get better coding gains without increasing the bandwidth. It is expected that maximal diversity gain can improve channels utilization. However, increasing the diversity gain comes at the expense of decreasing the multiplexing gain. Traditionally, MIMO systems have been designed to implement maximal diversity gain to increase transmission reliability. The first spatial diversity scheme has been proposed by Alamouti [2] to transmit data from two antennas to implement either maximal diversity gain to increase transmission reliability. And then extended by Tarokh et al. in [3] for any number of transmit

W. Zhang (Ed.): Advanced Technology in Teaching, AISC 163, pp. 737–743.
springerlink.com

antennas. However, the symbol rates of OSTBC were shown in [4]-[5] to be upper bounded by 3/4 for more than 2 transmit antennas in [6]. Another transmit diversity scheme, the so-called ABBA code, has been proposed first by O.Tirkkonen et al [7] for 4 transmit antennas. This rate 1 scheme is derived from the permutation of two Alamouti codes, and provides spatial diversity to the detriment of the orthogonality of the code. Thus, due to the self-interference terms in the detection matrix, the performance of the ABBA code is inferior to the one of the orthogonal designs. Some schemes called hybrid MIMO schemes consisting of STBC and spatial multiplexing schemes combine transmit diversity and spatial multiplexing, which can achieve the diversity gains and multiplex gains at the same time [10]-[15]. In this paper,we consider temporal diversity gains combined with spatial diversity gains.

Space-Time Block Code (STBC) provided diversity by multiple antennas. Meanwhile, bit-interleaved coded modulation (BICM), which can provide spatial diversity, was first introduced by Zehavi [8] and further studied by Caire [9] to improve system performance under a variety of channel conditions. This is done by increasing the diversity order by use of bit interleaving at the transmitter. At this scheme, employment of STBC provides temporal diversity and BICM provides spatial diversity.

In this paper, an improved channel coding scheme based on BICM embedded Spatial Division Multiplexing(SDM) of STBC system [16] provides spatial and temporal diversity and result in a robust system under a variety of channel conditions. The modulation is with 16 QAM and 64 QAM and mapping with Gray and the channel is over Rayleigh fading channels.

This paper is organized as follows. In section 2 some principles on coded modulation and hybrid MIMO system model are given. The improved channel coding scheme for BICM with hybrid MIMO is presented in section 3. Simulation results are given in section 4, and finally conclusion is made in section 5.

2 System Model

Consider a MIMO system in fig.1 with m transmit antennas and n receive antennas which can be modeled by the linear relationship:

$$y=\mathrm{H}x+n \tag{1}$$

Let H be the $n\times m$ channel gain matrix. $n=[n_0,n_1,.....,n_{N-1}]^T$ is the zero-mean additive white Gaussian noise(AWGN) complex vector which has i.i.d entries $n_i\sim CN\ (0, N_{0/2})$, $x=[x_1,x_2,.....x_{N-1}]^T$ is the transmitted vector, $y=[y_0,y_1,.....,y_{N-1}]^T$ is the received vector and $(.)^T$ denotes transposition. The element hij of H is the complex path gain from transmit antenna j to receive antenna i. The receiver is assumed to have ideal channel estimates so it can separate and decode the symbols transmitted from each antenna. The ability to separate the symbols is due to the fact that in a scattering environment, the signals received at each receive antenna from each transmit antenna appear to be uncorrelated:

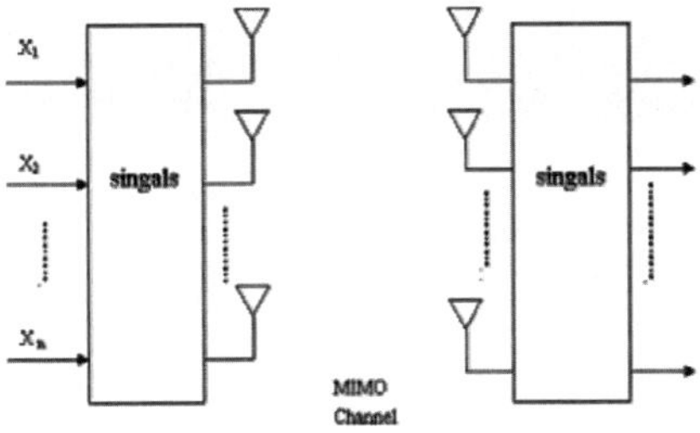

Fig. 1. MIMO System Model

2.1 Alamouti Code

The Alamouti Scheme[2], which is for two transmit antennas and one receive antenna, is one of the most elegant space-time codes for the transmit diversity system. And it can be extended to two receive antennas. The channel at time t can be modeled by a complex multiplicative distortion $h_0(t)$ for transmit antenna 0 and $h_1(t)$ for transmit antenna 1. Assuming that fading is constant across two consecutive symbols, these two complex multiplicative distortions can be written as follows:

$$h_0(t) = h_0(t+T) = h_0 = \alpha_0 e^{j\theta_0} \tag{2}$$

$$h_1(t) = h_1(t+T) = h_1 = \alpha_1 e^{j\theta_1} \tag{3}$$

where T is the symbol duration. The received signals can then be expressed as follows:

$$r_0 = r(t) = h_0 c_0 + h_1 c_1 + n_0 \tag{4}$$

$$r_1 = r(t+T) = -h_0 c_1^* + h_1 c_0^* + n_1 \tag{5}$$

where r_0 and r_1 are the received signals at time t and $t+T$ and n_0 and n_1 are complex random variables representing receiver noise and interference.

The combiner shown in Fig. 4 builds the following two combined signals that are sent to the maximum likelihood detector:

$$\tilde{c}_0 = h_0^* r_0 + h_1 r_1^* \tag{6}$$

$$\tilde{c}_1 = h_1^* r_0 - h_0 r_1^* \tag{7}$$

Substituting (2) ~ (5) into (6) and (7) we get:

$$\tilde{c}_0 = (a_0^2 + a_1^2) c_0 + h_0^* n_0 + h_1 n_1^* \tag{8}$$

$$\tilde{c}_1 = (a_0^2 + a_1^2) c_1 - h_0 n_1^* + h_1^* n_0 \tag{9}$$

After the signal combining, the combined signals are sent to the maximum likelihood detector which, for each of the signals c_0 and c_1, uses the decision rule expressed as follows:

Choose c_i if :

$$(a_0^2 + a_1^2 - 1)|c_i|^2 + d^2(\tilde{c}_0, c_i) \le (a_0^2 + a_1^2 - 1)|c_k|^2 + d^2(\tilde{c}_0, c_k),$$
$$\forall i \neq k \quad (10)$$

The Alamouti code with two transmit antennas and one receive antenna is used.

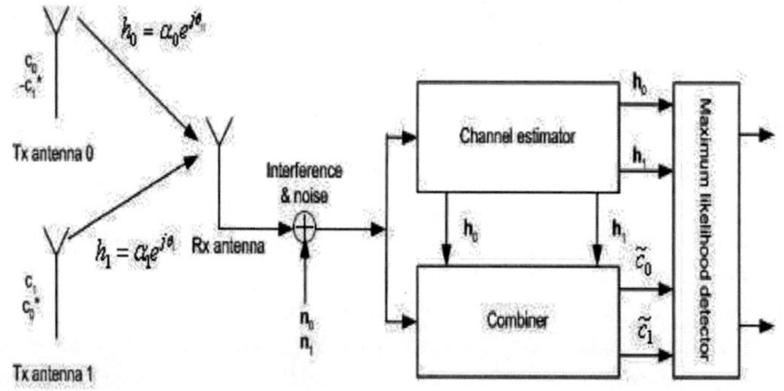

Fig. 2. The Alamouti coding system with 2 tx antennas and 1 rx antenna

2.2 Bit Interleave Code Modulation

In [8], Zehavi showed that the performance of coded modulation can be improved over Rayleigh fading channels by bit-wise interleaving at the encoder output, and by using an appropriate soft-decision metric for a Viterbi decoder at the receiver. Then, BICM was suggested [9]. BICM is a new channel coding scheme which has a better performance over Rayleigh fading channels. It has been proved to be a very power-efficient approach providing those state-of-the-art codes. An example of a BICM scheme is show in Fig.3, two binary sequences at the encoded into four binary sequences, and then fed into four independent ideal interleavers. At the receiver, a fade noisy version of the transmitted signal is passed through a demodulator, four metric computation units and four metric deinterleavers. Finally, the decision on the transmitted sequence is taken with the aid of Viterbi decoder.

3 BICM Embedded into SDM of STBC

In this configuration, a hybrid system with N antennas at the emitter can transmit simultaneously N/2 streams, and each stream is transmitted over two anrennas and with bit interleave code modulation, than though mapping and according to Alamouti coding. The system with four transmit elements is depicted on Fig.3.This results in transmitting four complex symbols s_1,s_2,s_3 and s_4 according to the following code matrix:

$$C = \begin{bmatrix} s_1 & s_2 & s_3 & s_4 \\ -s_2^* & s_1^* & -s_4^* & s_3^* \end{bmatrix} \quad (11)$$

Where the symbols of the nth column of C are sent from the nth antenna.

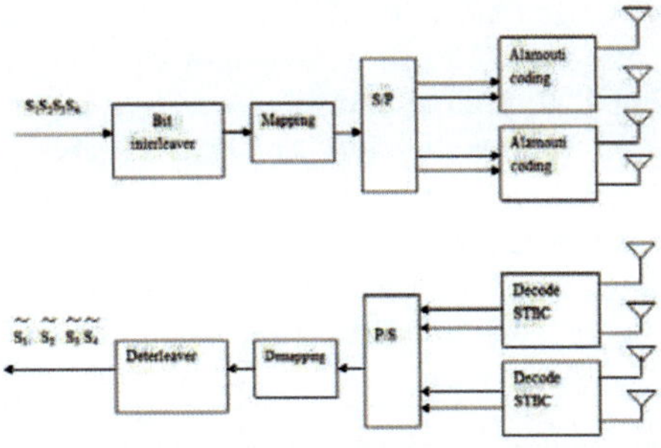

Fig. 3. Two Alamouti codes

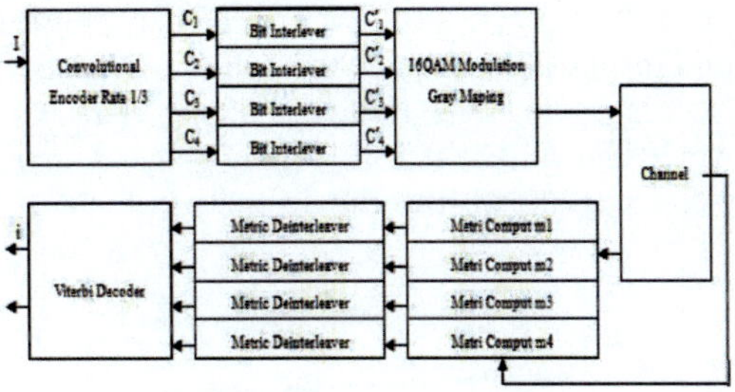

Fig. 4. BICM System Diagram

4 Simulation Results

To show the performance embedded BICM into SDM of STBC, simulation results present the two Alamouti codes in a MIMO system with 4 transmit antennas and 4 receiver antennas over flat Rayleigh fading channels. The maximum number of decoding iteration is set to 20 10 and 5. Here we only consider 16QAM, 64QAM and Gary mapping. It is clearly shown that, the larger iterative number would result in the better BER performances.

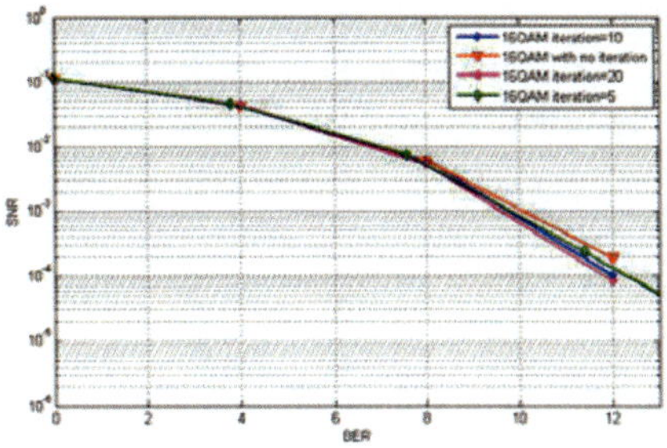

Fig. 5. Compare the BER performance of SDM of STBC scheme with the improved scheme with 16QAM and Gray mapping over Rayleigh fading channels

Fig.5 shows the BER curves of the SDC of STBC with different iterative over Rayleigh fading channels and 16QAM Gray mapping. When BER is below 10^{-4}, iterative 20 can get about 1 dB coding gains and iterative 10 and 5 can get about 0~1 dB coding gains compared with no iterative. And the next shows over Rayleigh fading channels when BER is below 10^{-5} the improved channel coding scheme can achieve more than 2dB gains with iterative 20 and iterative 10 and 5 can get about 1~2 dB coding gains at the Fig.6 .Compared with the traditional scheme, the new scheme embedded into BICM can get a better BER performance.

5 Conclusion

An improved STBC scheme, which can get coding gains due to spatial diversity, is considered in this paper. The proposed SDM of STBC is constructed by embedding into BICM. Simulation results demonstrated that when BER is below 10^{-4}, iterative 20 can get about 1 dB coding gains over Rayleigh fading channels with 16QAM. And with 64QAM iterative 20 can get about 2 dB coding gains. A good coding modulation scheme can improve coding gains with spatial diversity in SDM of STBC.

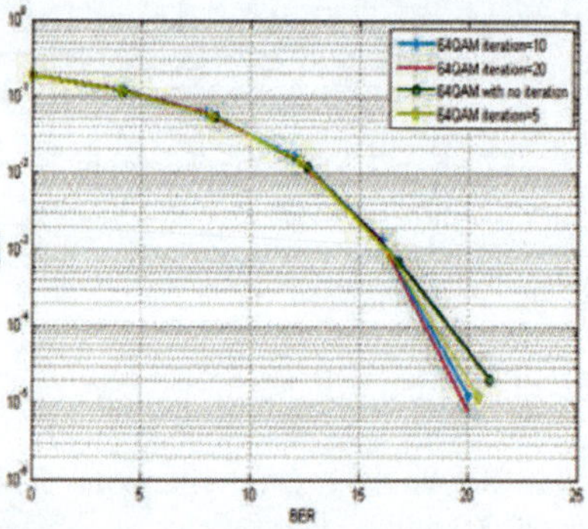

Fig. 6. Compare the BER performance of SDM of STBC scheme with the improved scheme with 64QAM and Gray mapping over Rayleigh fading channels

Acknowledgment. This paper is supported by the key project of Chinese Ministry of Education(No. 106042) and the project sponsored by the Scientific Research Foundation for the Returned Overeas Chinese Scholars, State Education Ministry(2007[24])

References

1. Foschini, G.J., Gans, M.J.: On limits of wireless Communications in a Fading Environment when Using Multiple Antennas. Wireless Personal Communications 6, 311–335 (1998)
2. Alamouti, S.M.: A Simple Transmit Diversity Technique for Wireless Communications. IEEE Journal on Selected Areas in Communications 16(8), 1451–1458 (1998)

3. Tarokh, V., Jafarkhani, H., Calderbank, A.R.: Space-time block codes from orthogoal designs. IEEE Trans. Inform. Theory 45(5), 1456–1467 (1999)
4. Wang, H., Xia, X.-G.: Upper bound s of rates of complex orthogonal space-time block codes. IEEE Trans. Inf. Theory 49, 2788–2796 (2003)
5. Lu, K., Fu, S., Xia, X.-G.: Closed form designs of complex orthogonal space-time block codes of rates (k + 1) / (2 k) for 2k - lor 2k transmit antennas. IEEE Trans. Inf. Theory 51, 4340–4347 (2005)
6. Sandhu, S., Paulraj, A.J.: Space-time block coding: A capacity perspective. IEEE Commun. Letters 4, 384–386 (2000)
7. Tirkkonen, O., Boariu, A., Hottinen, A.: Minimal nonorthogonality rate one space time block codes for 3+Tx antennas. In: Proc. IEEE Int. Symp. on Spread Spectrum Techniques and Applications 2002, pp. 429–432 (September 2000)
8. Zehavi, E.: 8-PSK trellis codes for a rayleigh fading channel. IEEE Trans. Commun. 40, 873–883 (1992)
9. Caire, G., Taricco, G., Biglieri, E.: Bit-interleaved coded modulation. IEEE Trans. Inform. Theory 44, 927–946 (1998)
10. Thompson, J.S., Tan, H.S., Sun, Y.: Investigation of hybrid MIMO techniques. In: Fifth IEE International Conference on 3G Mobile Communication Technologies, pp. 1–5 (2004)
11. Aragon, J.R.C., Fernandez, E.M.G.: Combining diversity and multiplexing in wireless communications systems. In: 2006 International Telecommunications Symposium, pp. 540–544 (September 2006)
12. Do, M.V., Chin, W.H., Wu, Y.: Performance study of a hybrid space time block coded system. In: 2006 1st International Symposium on Wireless Pervasive Computing (January 2006)
13. Freitas, W.C., Cavalcanti, F.R.P., Lopes, R.R.: Hybrid MIMO Transceiver Scheme with Antenna Allocation and Partial CSI at Transmitter Side. In: IEEE 17th International Symposium on Personal, Indoor Personal, Indoor and Mobile Radio Communications, pp. 1–5 (September 2006)
14. Zhou, L., Oishi, Y.: Improved Performance for Hybrid STBC and Spatial Multiplexing OFDM Systems with Linear Receivers. In: IEEE 18th International Symposium on Personal, Indoor and Mobile Radio Communications, pp. 1–5 (September 2007)
15. Xu, N., Zhong, X., Sun, Y.: An Adaptive Hybrid MIMO Scheme. IEEE Trans. Inform. Theory, 562–565 (2009)
16. Rouquette-Leveil, Gosse, K., Zhuang, X., Vook, F.W.: Spatial Division Multiplexing of Space-time Block Codes. In: IEEE Communications Theory, ICCT 2009, pp. 1343–1347 (2009)

A Traffic Light Detection Method

Chunhe Yu and Ying Bai

Department of Electronic and Information Engineering, Shenyang Aerospace University
Shenyang, 110034 P.R. China
chunhe_yu@tom.com

Abstract. In order to reduce accident at traffic intersections, the algorithm of traffic lights detection which is applied in a vehicle driver assistance system is designed by using the image processing technology. The system of traffic light detection includes three parts: a CCD camera, an image acquisition card, and a PC. Based on RGB color space, the algorithm extracts red, green, and yellow objects in the image firstly; For the purpose of eliminating disturbance in the environment, the features of traffic lights are used to verify the object identity, and then the types of traffic signals are judged.. The results of experiments show that the algorithm is stable and reliable.

Keywords: traffic intersection, traffic lights detection, vehicle driver assistance, image recognition.

1 Introduction

With the development of IVS (Intelligent Vehicle System), many achievements have been gained in the field of vehicle driver assistance system, such as the technology of car anti-collision, the technology of road curb detection and the technology of pedestrian detection [1~4], etc. Because many traffic accidents often take place at intersections, IVS at intersections has been studied in recent years [5~8]. For example, the system of traffic light recognition can be applied in the driver assistance system [9, 10], which can remind drivers to pay attention to their soundings when they reach an intersection and help him make a decision.

The information of traffic signal can be obtained through wireless communication technology, but it is high cost for installing wireless communication equipment and is difficult to construct the related hardware facilities. In order to simplify the hardware, the image processing technology can be used to recognize traffic lights. This method has several advantages, such as low price, high performance, and easy upgrade. Applying image processing method, Park [11] studied round traffic lights detection by judging the shape and size of an object, in which the arrow-shaped traffic light was not discussed. Chung [12] adopted HIS model to recognize traffic lights, in which the road scene was simple and some interference problems such as vehicle lamps, street lamps were not considered. So in this paper traffic lights including round and arrow-shaped signal light detection

W. Zhang (Ed.): Advanced Technology in Teaching, AISC 163, pp. 745–751.
springerlink.com

problems were discussed in a complex environment. The rest parts are organized as follows: The first is hardware structure, the next is traffic lights detection algorithm, and then is the experiment, the last is the conclusion.

2 Hardware System Desige

Our traffic lights detection system includes a CCD camera, an image acquisition card, a PC (main frequency: 1.7GHz, DRAM: 256M). The CCD camera whose the frame frequency is 30FPS and the resolution is 680×480 is mounted on the vehicle front windshield. The color image sequences can be acquired by the image acquisition card. The PC is in charge of the tasks of traffic lights detection, road curbs detection and pedestrian detection, and the results are displayed on the PC LCD. The work process is as follows: As a vehicle travels at non-intersection (no traffic lights in the scene), the program of road curb detection operates, while it travels at intersection (traffic lights in the scene), the program of pedestrian detection operates, and if there is a hazardous event occurring, the assistance driver system give a warning signal to the driver.

3 Traffic Lights Detection Algorithm

3.1 Object Extraction Based on Color

Based on experience, when the traffic light is on, its corresponding color component is much bigger than other color in RGB color space. According to this feature, red, green, and yellow color pixels of traffic lights can be extracted by subtraction between two color components. That is: when the red traffic light is on, R component is bigger than G & B component; when the green traffic light is on, G component is bigger than R & B component; when the yellow traffic light is on, R & G component is bigger than B component. The object extraction algorithm is shown as follows: if $x_R - x_G \geq T_R$ & $x_R - x_B \geq T_R$,then $x_R = 255$, $x_G = 0$, $x_B = 0$;if $x_G - x_R \geq T_G$ & $x_G - x_B \geq T_G$,then $x_R = 0$, $x_G = 255$, $x_B = 0$;if $x_G - x_B \geq T_{YG}$ and $x_R - x_B \geq T_{YR}$, then $x_R = 255$, $x_G = 255$, $x_B = 0$;(T_R,T_G, T_{YG} ,and T_{YR} are the thresholds, if the conditions are met, the correspond pixel is purified for color segmentation.); and other pixels are set to white color.

As shown in Figure 1, there is an intersection scene in figure 1(a); the extracted pixels are shown in figure 1(b). Because there are two types of the traffic lights, one is round and the other is arrow-shaped. In order to detect the two types of the traffic lights, the image is segmented by image region growing method firstly, and then the minimum out-connected rectangles are made to represent the segmentations. Because the traffic lights are often set at the top of an image, and in order to quicken the detection program, only the 1/3 region of the image is segmented in the algorithm. The results of the segmentation are also shown in Fig. 1(b).

Because there are often some color objects (such as red, green or yellow object) disturb the traffic lights detection in the complex environment, the verification method is needed to judge whether the traffic light exists or not.

(a) An intersection scene

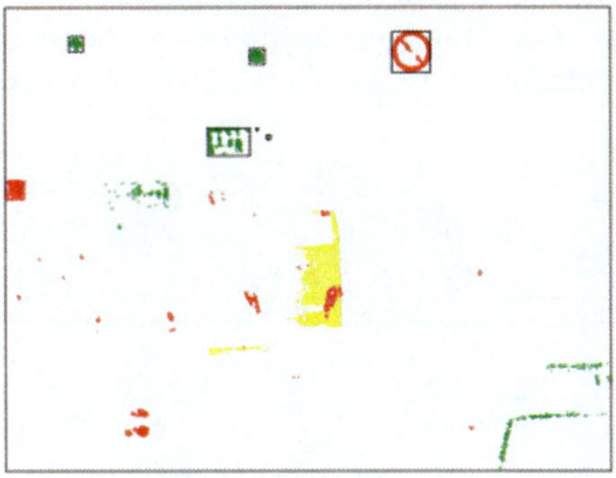

(b) The results of extracted pixel and segmentation

Fig. 1. Traffic lights extraction based on color

3.2 Rule-Based Traffic Light Verification

There are many kinds of disturbance in the object extraction, such as vehicle lamps, vehicle bodies and road signs. In order to extract the true traffic lights, the disturbance must be eliminated.

1) Traffic light verification based on the shape of an object
Traffic lights have some shape features: the minimum out-connected rectangles are approximately square; and the rectangles have certain area when a vehicle is close to the intersection; in addition, a traffic light rectangle density of color pixel point is much more than a road sign rectangle. Those features can help to extract traffic lights from complex environment. The judgments are listed in details as follows: (given L(i), W(i), S(i) as a rectangle's Length, width and area)

(1) Judge if the minimum out-connected rectangle of an object is reasonable or not. If it is not reasonable, the object rectangle is eliminated as a disturbance. (Supposing *Th*1 is the threshold of length-width ratio,*Th*2, *Th*3 is the area threshold of traffic lights)

$$1/Th1 \le L(i)/W(i) \le Th1 \tag{1}$$

$$Th3 \le S(i) \le Th2 \tag{2}$$

(2) Judge whether the density of the color pixel points in the minimum out-connected rectangle is reasonable or not. If it is not reasonable, the object rectangle is regarded as a disturbance and should be eliminated, (Supposing Z stands for the total of the color pixels in the segmentation; $Th4$ is the density threshold)

$$Z(i)/S(i) \geq Th4 \tag{3}$$

2) Traffic light verification based on position relationship
In a complex scene, there are often many vehicle lamps which may be judged as traffic lights because they can meet the condition of the traffic lights. According to the CCD camera installing angle, the lamps often appears below the 2/3 image and the traffic lights often appear at the top of an image. As shown in figure2, the traffic lights usually lie in a horizontal line. If there is a pedestrian traffic light in the scene, whose position is lower than the vehicle traffic lights, and by the position features of traffic lights, the segmentations of traffic lights can be verified precisely.

Fig. 2. The positions of traffic lights

3.3 Signals of Traffic Lights Extraction

According to the extraction rule of an image pixel points, the type of the traffic signals is decided by the color in the region of segmentation. There are three signal types, which include a red traffic light signal, a green traffic lights signal and a yellow traffic lights signal.

In order to distinguish an arrow-shaped traffic light from objects of rectangle segmentations, a central point of the segmentation rectangle and a gravity point of the pixels in the rectangle are calculated. The arrow-shaped traffic lights are shown in figure 3. If the distance of the two points is larger than the threshold Th6, the traffic light is judged as an arrow-shaped traffic light, otherwise, it is judged as a round traffic light. Once the arrow-shaped traffic light is judged, its direction can be easily identified by the relationship between the central point and gravity point. If the gravity point is on the left, its direction is left; if on the right, its direction is right; if on the top, its direction is forward.

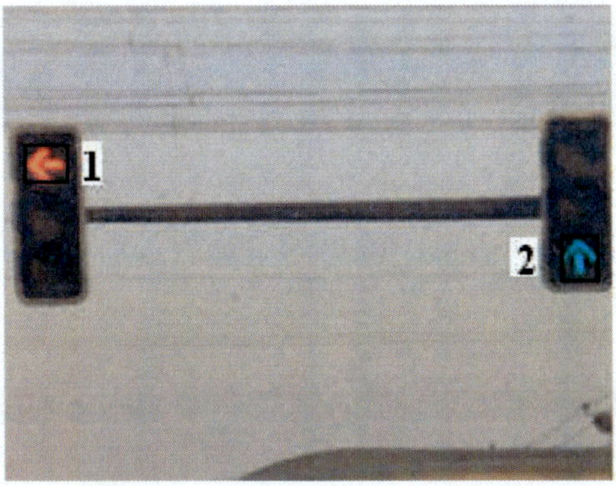

Fig. 3. Arrow-shaped traffic lights

4 Experiment

The experiments were carried out in the scene as shown in figure4. In fig.4 (a) there were one round and one forward arrow-shaped traffic light, which were in green signal; In fig.4 (b) there were one arrow-shaped traffic light, one round traffic light, and one pedestrian traffic light, which were in red signal; in fig.4(c) There were three round traffic lights, which were in allow green signal. The results of object extraction and segmentation about traffic lights are shown in figure 5, in which the thresholds of color point extraction were: T_R=100, T_G=100, T_{YR}=100, T_{YG}=100. The detection results were shown in figure 6.

In figure 5 (a), the right round roadside sign "1" was filtered by its area size and density of red pixel points. In fig. 5 (b), segment "3" was filtered for its low position (which was a pedestrian traffic light) than segment "1"& "2" , and "1"& "2" were basically in the horizontal direction. There were many non-target objects, such as vehicle lamps, road signs, etc. which are shown in figure 5 (c). Those non-target objects were well eliminated according to the features of the traffic lights' position and alignment. From the results of the detection, we can see that the traffic lights were extracted accurately, and can give the right direction signal as well.

(a) (b) (c)

Fig. 4. Traffic light in the intersection scene

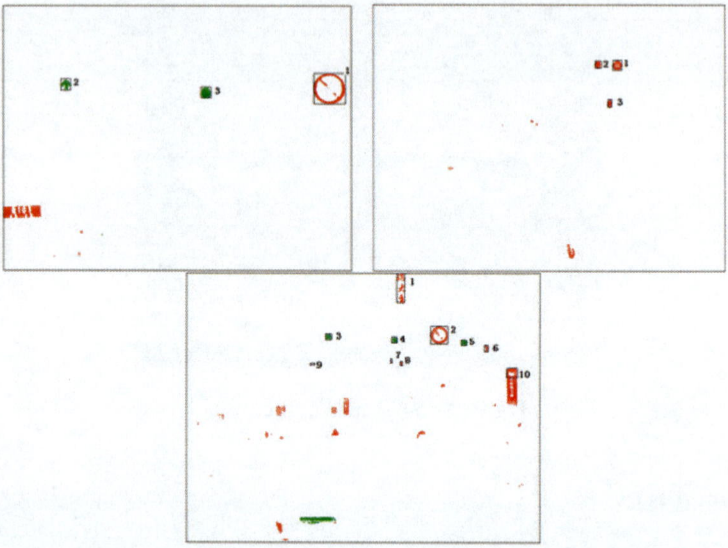

(a) (b) (c)

Fig. 5. Object extraction and segmentation

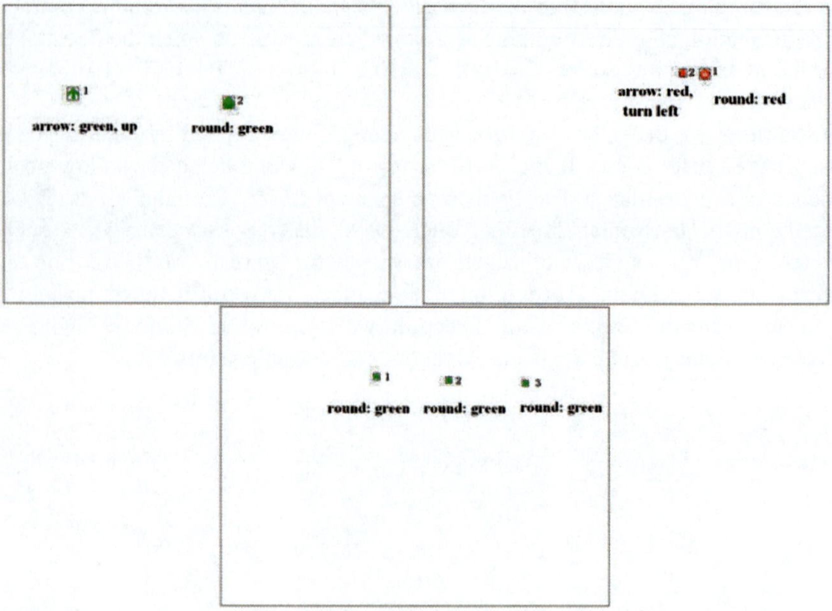

(a) (b) (c)

Fig. 6. The results of traffic light detection

5 Conclusion

By using the object extraction in RGB color space and the rule verification, the traffic lights can be well extracted in the complex environment. The algorithm can obtain the round traffic lights as well as arrow-shaped traffic lights, and can give traffic signals accurately. Experiments show that the algorithm has the characteristics of stability and reliability. In order to strengthen the detection effect, the tracking algorithm will be taken into account in future study.

References

1. Masoud, O., Papanikolopoulos, N.P.: A novel method for tracking and counting pedestrians in real-time using a single camera. IEEE Transactions on Vehicular Technology 5(50), 1267–1278 (2001)
2. Cristovbal, C., Johann, E., Thomas, K.: Walking pedestrian recognition. IEEE Transactions on Intelligent Transportation Systems 3(1), 155–163 (2000)
3. Jansson, J., Gustafsson, F., Ekmark, J.: Decision making for collision avoidance system. Society of Automotive Engineers, World Headquarters, Pennsylvania, pp. 1–8 (2002)
4. Chen, Y., Huang, X.Y., Yang, S.G.: Research and Development of Automotive Collision Avoidance System. Computer Simulation 23(12), 239–243 (2006)
5. Benmimoun, A., Chen, J., Neunzig, D., et al.: Communication-based intersection assistance. In: IEEE Intelligent Vehicles Symposium, pp. 1–5 (June 2005)
6. Fuerstenberg, K.: A new European approach for intersection safety - the EC-Project INTERSAFE. In: IEEE Conference on Intelligent Transportation Systems, pp. 493–504 (September 2005)
7. Long, L., Andreas, F., Roberto, B., Zhang, W.H.: Communication-Based Intersection Safety: Motivation, Challenges and State-of-the-Art. In: International Workshop on Intelligent Transportation, Hamburg, Germany, pp. 1–5 (2009)
8. Le, L., Festag, A., Baldessari, R., Zhang, W.: V2X Communication and Intersection safety. In: 13th Int. Forum on Advanced Microsystems for Automotive Applications, Berlin, Germany, pp. 1–14 (2009)
9. Yelal, M.R., Sasi, S., Shaffer, G.R., et al.: Color-based Signal light Tracking in Real-time Video. In: Proceedings of the IEEE International Conference on Video and Signal Based Surveillance, Washington, USA, pp. 67–76 (2006)
10. Kamarudin, A., Riza, A., Rozmi, I.: Intelligent Transport System for Motorcycle Safety and Issues European. Journal of Scientific Research 4(28), 600–611 (2009)
11. Park, J.H., Jeong, C.S.: Real-time Signal Light Detection. International Journal of Signal Processing and Pattern Recognition 2(2), 1–10 (2009)
12. Chung, Y.C., Wang, J.M., Chen, S.W.: A Vision-Based Traffic Light Detection System At Intersections. Journal of Taiwan Normal University: Mathematics, Science & Technology 47(1), 67–86 (2002)

VRLA Battery Management System Based on LIN Bus for Electric Vehicle*

Changhao Piao[1,2], Qifeng Liu[2], Zhiyu Huang[2], Chongdu Cho[3], and Xiaonong Shu[2]

[1] Key Laboratory of Network Control & Intelligent Instrument
Chongqing University of Posts and Communications, CQUPT Chongqing, China
[2] Research Center of Energy Electronics, College of Automation
Chongqing University of Posts and Communications, CQUPT
Chongqing, China
[3] Department of Mechanical Engineering
INHA University of Korea, Incheon Korea
changhaopark@hotmail.com, ltiger2007@126.com

Abstract. A battery management system (BMS) based on the LIN-Bus was designed for the Valve Regulated Lead Acid Battery (VRLA). The VRLA BMS plays an important role as a vehicle power supply system which is part of the entire vehicle system. In order to improve the stability of various sub-systems inside the vehicle, VRLA parameters estimation, including current, voltage and temperature, are very important. However, due to lead-acid battery charging and discharging are a complex process of the electrochemical process. The VRLA BMS is consisted of a VRLA battery, a NI USB-8476 LIN transceiver manufactured by National Instruments(NI), a battery test system(BTS) and a high stability, high reliability ADuC7033 for which is high precision and rapid data acquisition ARM7 chip. The hardware design of sampling circuit, the whole VRLA BMS system structure, the software design of keil ARM7 and LabVIEW8.5 and LIN Bus communication were introduced for realtime on-line monitoring and controlling. Experimental results show that measurement is high credible and the error is within 1%. The BMS that was proposed by this paper can be used in Electric Vehicle(EV).

Keywords: VRLA, BMS, LIN communication, EV, DAQ.

1 Introduction

In recent years, as mankind's increasing attention to environmental problems, major domestic auto manufacturers, such as Chang'an, BYD, also stepped up their electric car (EV) research. As part of the electric vehicles, lead-acid storage battery plays an important role in popularizing and promoting EV[1-4].

ADuC7033 chip which is produced by Analog Devices Inc. (ADI) has 96Kbytes of read-only memory unit (ROM), 6Kbytes of static random access memory cell (SRAM),

* This work is supported by Natural Science Foundation Project of CQ CSTC (2008BB6084).

W. Zhang (Ed.): Advanced Technology in Teaching, AISC 163, pp. 753–763.
springerlink.com

dual-channel high-precision ADC, LIN interface peripherals and it is one of ARM7 family. Through accessing the chip, we can acquire the high-precision parameters of lead-acid battery, such as voltage, current, temperature of battery.

LIN-bus technology is an emerging technology as assistant of a wide range CAN bus communication network. When the performance, bandwidth and complexity of CAN are not needed, LIN has been widely used in low speed system. Compared with the CAN bus, LIN bus based master-slave structure, the use of one-way communication, a lot of wiring harness to reduce the weight and cost, is widely used in automotive, industrial automation, aerospace, medical and other fields.

NI-USB-8476LIN device is LIN bus data transceiver which is produced by National Instruments (NI) Co. Ltd. It is mainly used for the development of the host computer, when carried out a certain configuration, such as channel selection, baud rate settings, can be true to receive data on the LIN bus, even if the bus load of 100% of the cases, it does not appear the phenomenon of missing frames.

In this paper, we developed a distributed battery management system based on the LIN-bus considering the structure and function requirement of the VRLA battery used for the EV[5]. The BMS we developed can sample the battery's voltage, current, temperature information, and communicate with the electronic control unit (ECU).

2 Lin Bus Technology

LIN network consists of a master node which includes a primary task and secondary task, and multiple slave nodes which include secondary task only. LIN-frame formed by the frame header and response messages. LIN standard frame format of the data is shown in Figure 1.

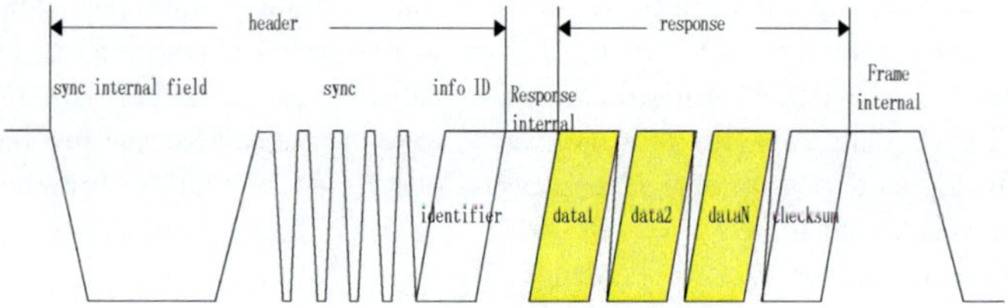

Fig. 1. LIN data frame format

Frame header including the synchronization interval field, synchronization, and information field identifier field. Frame header sent by the primary task. After sending the synchronization interval field, the primary task sends synchronization field (0x55). Secondary node uses sync field to adjust its baud rate to match with the received signal's baud rate. After sending the sync filed, the primary task sends a byte of information identifier, where 0 to 3 bits indicate information category, 4 to 5 bits indicate message length, and 7 to 8 bits indicate parity. It is information category rather than information's destination address that information identifier sends. This allows multiple nodes receive the same information, and the data can be exchanged in a variety ways. Secondary task of the secondary node judges whether the data relating to itself according to information identifier, and than to determine how they handle the data. Response message consisted of 2, 4 or 8 bytes length and a checksum byte is sent from secondary task of secondary node. And check sum byte indicates the end of data frame, which calculated all the bytes of data obtained (do not contain the information identifier byte and synchronization field byte)[6].

3 System Structure

The VRLA battery is equipped on the foreside of EV. The nominal voltage of the battery is 12V and the nominal capacity is 45Ah. Figure 2 shows the structure of the developed BMS based on LIN-bus. As part of the host computer system, PC displays real-time data received from the LIN bus. NI USB-8476 LIN is master node which can transmit information identifier of secondary node to the LIN network. ADuC7033 development board is slave node which is responsible for sending sampling data to the LIN bus, for example lead-acid battery voltage, current, temperature and estimated state of charge (SOC). The voltage, current, temperature in the battery should be sampled because of its rigorous demands on the safe working. In order to reduce the wire harness, a distributed type of BMS based on the LIN-bus is developed. The BMS is made up of one master module (NI USB-8476 LIN) which is located in foreside of the automobile and one slave module (ADuC7033 Slave Board) which is located in the VRLA battery. The sampling module samples the voltage, current and temperature of the battery. And then these data are transmitted to the LIN bus. The NI USB-8476 LIN master module receives the battery's information from the LIN bus. Moreover the master module communicates with the ECU according to the LIN protocol of the EV. The master module can also communicate with PC through the USB interface so that the battery's information can be displayed on the screen of the PC and the BMS can be debugged on line.

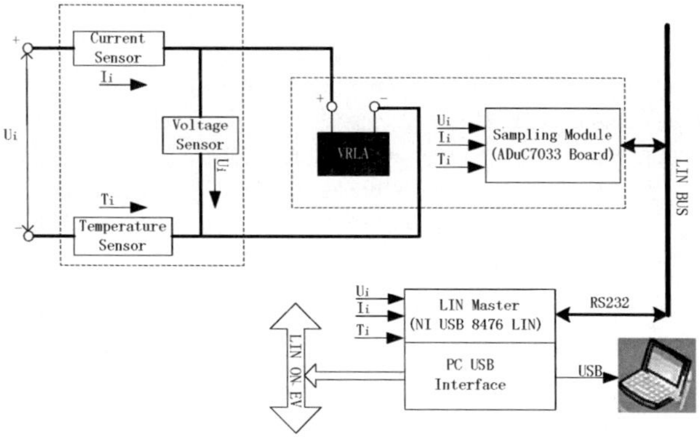

Fig. 2. The structure of the developed BMS based on LIN-bus

4 Hardware Design

4.1 Sampling Module

This paper selects ADuC7033 which is automotive battery data collection dedicated chip as the chip has a high reliability and strong anti-interference ability. This chip contains internal integrated LIN controller, thus we can connect it to the LIN network conveniently.

The ADuC7033 development board slave module samples the VRLA battery's voltage, current, temperature through its' high precision A/D converter. In order to achieve precision measurement, I channel should be isolated with V/T channel. Current measurement channel is distinguished from voltage and temperature measurement channel which shares the same V/T channel of the chip.

Figure 3 shows the ADuC7033 data acquisition circuit of the hardware implementation.

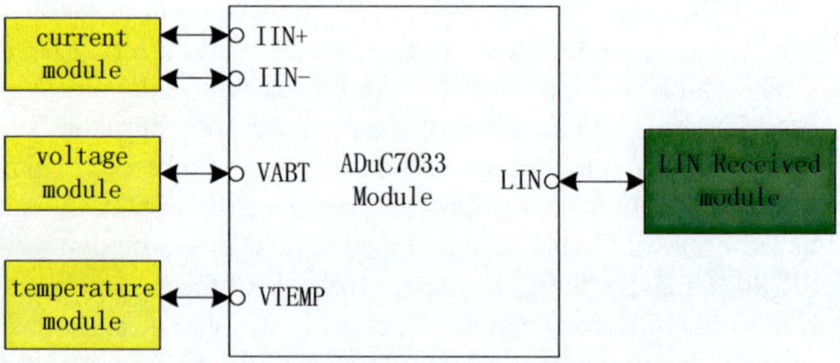

Fig. 3. ADuC7033 data acquisition circuit

4.2 Master Module

NI USB-8476 LIN manufactured by NI Co. Ltd is single LIN data bus transceiver used to communicate with other sampling modules. And NI USB-8476 LIN interfaces, featuring the Atmel ATA6620 LIN transceiver, are compliant with the LIN 1.3/2.0 and J2602 specifications and offer software-selectable master/slave termination. The ATA6620 features baud rates up to 20 kb/s and offers advanced power management with a low-power sleep mode. In this paper, we use NI USB-8476 LIN as master module through programmable LabVIEW software in the PC. Further more, the sampling module can adjust itself baud rates. How the NI USB-8476 module communicate to sampling module is shown Figure 4.

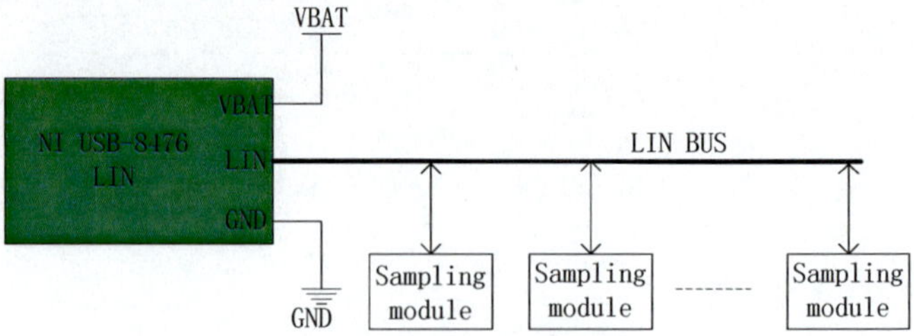

Fig. 4. NI USB-8476 LIN communication circuit

4.3 VRLA Data Acquisition System

Battery management system is playing an important role in hybrid electric vehicle. Data acquisition, including current, voltage, temperature, as one of vital part of the BMS, is achieved by Figure 5. The left of the picture acquires the data of the VRLA battery by battery test system (BTS), which is one of NEWARE corporation battery test products and it has high acquisition precision. In this paper, we use those data as practical data of the VRLA. The right of the picture acquires those same data using ADuC7033 data acquisition board made by Zhi-yu Huang, who is one of authors of this paper. In order to verify the data acquisition precision, we compare the data curve acquired by BTS and ADuC7033.

Fig. 5. Data acquisition circuit by BTS and ADuC7033

The all devices connection is shown in Figure 5. It is Ethernet interface that connects BTS and PC1, which processes the data acquired by BTS. It is USB interface that connects USB-8476 LIN device and PC2, which processes the data acquired by ADuC7033 board. From bottom explore board to PC2, the data are processed via decimal to hexadecimal, then to decimal, and then to hexadecimal, finally those data transferred to PC2 through LIN bus.

For example, the current voltage of the VRLA battery in some time is 12V. In bottom layer, the slave converts this decimal data to hexadecimal data as follow Equation (1).

$$ADCOUT = \left[\frac{VIN * PGA}{VREF} - ADCOF\right] * \frac{ADCGN}{ADCGNNOM} \quad (1)$$

5 Software Design

Figure 6 describes the keil ARM7 process of programming. Firstly configuring the LIN interface and LIN baud rate, in this paper, we use NI USB-8476 LIN No. 0 channel as LIN interface and the LIN baud rate is set 19.2kbps. However, the LIN baud rate can be set arbitrarily among certain range. And slave can adjust itself baud rate through keil bottom programs to match the master's. LIN master node sends information identifier 0x25, and then slave node starts secondary task to determine the information sent by the master node whether identify with itself. If the same, LIN communication slave module starts the data acquisition task, and sends the collected data to the LIN bus. And then NI LabVIEW collects current, voltage, temperature information of the VRLA battery and writes them to excel files. We can program using MATLAB7.0 to load those files containing current, voltage, and temperature data and describe the curve in PC2 which processes the data acquired by ADuC7033. As the same time, BTS has also completed acquisition task. Through comparing those two curves, we can judge whether the acquisition precision is lower than expected value, in this paper is set to 1%, if it does, continuing the program, if not, starts LIN bootloader to revise the programs on-line in the PC2.

From this figure, we can know the program starts from the LIN initialization. When the master sends the identifier that is unique for the slave. Through compare with the true value of battery, judge whether the error is satisfied the requirement. If yes, continue, else, return the main function and revise the functions.

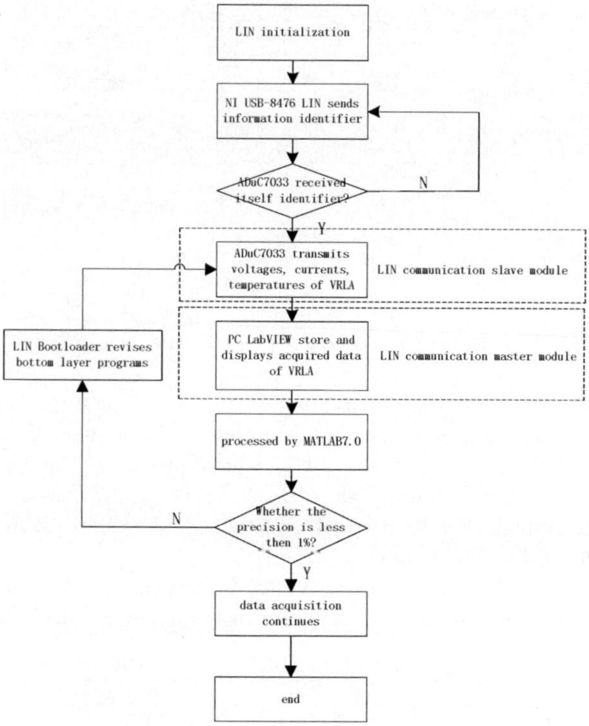

Fig. 6. Program block diagram structure of data acquisition

6 Experimental Results

This experiment is carried out in Research Center of Energy Electronics (RCOEE). At the beginning, we must set work condition of the VRLA in the PC1 through Battery Test System (BTS), in this paper, the work condition is set as follow table1.

Table 1. Battery Test System work condition set

work condition	Battery Test System VRLA discharge current (mA)	terminal voltage(V)
NO.1	9500	11.5
NO.2	9000	11.0
NO.3	10000	10.5
NO.4	8000	10.0
NO.5	7000	9.5
NO.6	9000	9.0
NO.7	8000	8.0
NO.8	10000	5.0

The BTS data acquisition internal time is set 1 second. In order to match the speed of the acquisition, the process time in PC2 is also set 1 second.

6.1 Current Measurement

The voltage of shunt resistor is amplified by ADuC7033 chip programmable gain amplifier (PGA), and then goes into the I channel AD converter. In this paper, the gain is set 64. We can get the current of circuit through the value of voltage and shunt resistor by using the law of ohm. In this paper, real-time current measurement is achieved by using Keil ARM and LabVIEW programming. In order to reduce errors and improve accuracy, we use the differential signals input. The experimental result shows the effect is very well and the accuracy is very high. The current acquisition curve is shown in Figure 7 and measure error is below 1%.

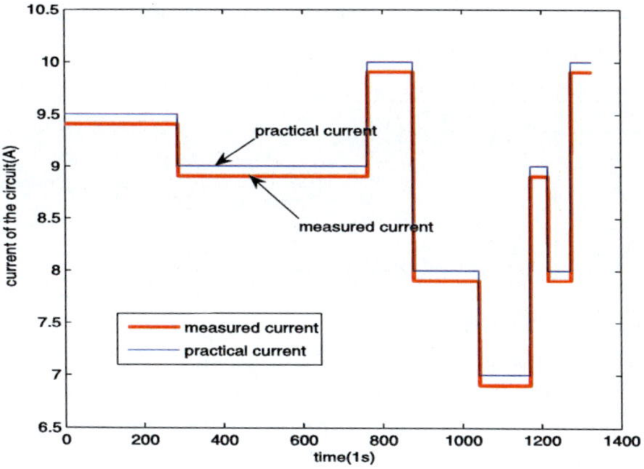

Fig. 7. Current acquisition experimental result

6.2 Voltage Measurement

In this paper, the open circuit voltage of valve regulated lead acid battery is measured by using the V/T channel AD converter of ADuC7033 chip when the input voltage value attenuated by 24 times.

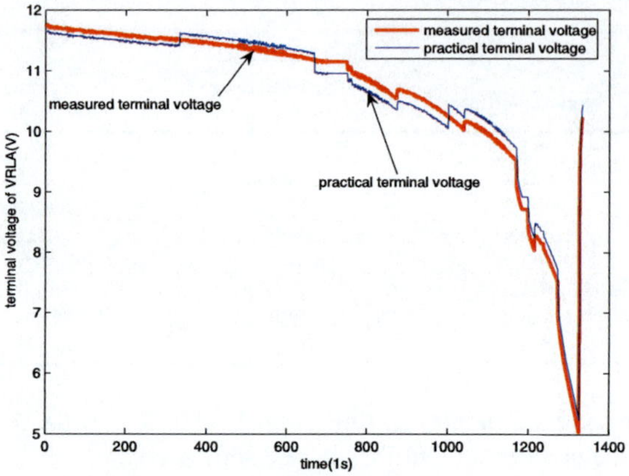

Fig. 8. Voltage acquisition experimental result

The voltage acquisition experimental result illustrates data precision is very high and the effect is perfect. The voltage experimental result curve is shown in Figure 8 and the results show the acquisition error is less than 1%.

6.3 Temperature Measurement

In this paper, the vicinity temperature of lead-acid battery is measured by using NTCMF3700 which is one kind of NTC thermal resistor and its' thermal resistance is 10kohm under 25 cent degree and its' thermal coefficient is 3700. The thermal resistor with high precision, high reliability and fast response characteristics can be described by the follow Equation (2) and Equation (3).

$$R_T = R_0 e^{B(\frac{1}{T} - \frac{1}{T_0})} \tag{2}$$

$$= R_0 e^{B(\frac{1}{273.15+t} - \frac{1}{273.15+t_0})} \tag{3}$$

where all of variables are defined as follows,

t_0: Room temperature(25 centidegree)
t: The current ambient temperature (unit: centidegree)
T_0: Kelvin corresponding temperature as the ambient temperature is 25 centidegree
T: Kelvin corresponding temperature as the ambient temperature is t centidegree
B: Thermal coefficient of thermal resistance
R_0: Thermal resistance as the ambient temperature is 25 centidegree
R_T: The thermal resistance of the current temperature

In order to guarantee the reliability and high precision measurement of temperature of VRLA battery's surface, we do thousands of measurements to NTCMF3700, and finally we can get the relationship between resistor and temperature of NTCMF3700. Table2 shows this relationship among the 0 cent degree to 50 centidegree.

Table 1. The relationship between T and R of NTCMF3700

T	R	T	R	T	R
0	25.116	1	24.130	2	23.189
3	22.291	4	21.434	5	20.616
6	1.9835	7	1.9088	8	1.8375
9	1.7693	10	1.7041	11	1.6417
12	1.5821	13	1.5250	14	1.4703
15	1.4179	16	1.3678	17	1.3197
18	1.2737	19	1.2296	20	1.1872
21	1.1466	22	1.1077	23	1.0703
24	1.0344	25	1.0000	26	0.9669
27	0.9352	28	0.9046	29	0.8753
30	0.8471	31	0.8200	32	0.7939
33	0.7688	34	0.7447	35	0.7214
36	0.6991	37	0.6775	38	0.6568
39	0.6368	40	0.6176	41	0.5990
42	0.5811	43	0.5639	44	0.5473
45	0.5312	46	0.5158	47	0.5009
48	0.4865	49	0.4725	50	0.4591

The temperature acquisition experimental result is shown in Figure 9. The blue wire is measured by BTS's DS18B20 temperature sensor, while the red wire is measured by ADuC7033 explore board's NTCMF3700 thermal sensor. From the curve, we can know the relatively acquisition error is 0.57%, which is lower than 1%.

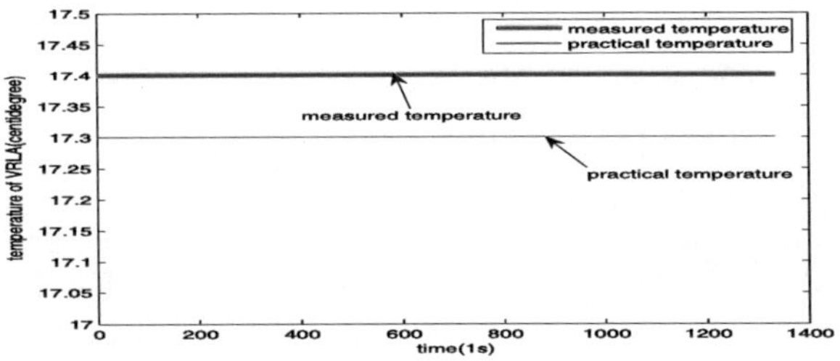

Fig. 9. Temperature acquisition experimental result

6.4 ADuC7033 LIN Interface Wave

In the paper, the information identifier is 0x25. Once the ARM7 DAQ experimental board receives this information identifier, ADuC7033 slave module starts data acquisition task and returns the voltage, current and temperature of VRLA battery. One of LIN data frame is shown in Figure 10. From the graph, we can know the time to transmit a bit is about 100us, so the LIN communication baudrate has been set to 9600bps. The waveform was captured by RIGOL DS5022ME oscilloscope, and from left to right of signal represents synchronization interval field which holds on 13 bits, sync byte 0x55, information identifier field, data field and checksum. From the returned data, we can know the data format is the same as Figure 1.

Each byte included synchronization byte, information identifier, data field and checksum starts with the failing edge and ends with the rising edge. According to this principle, the synchronization field is 0x55, the information identifier is 0x25, the data field is 0x00, 0x1f, 0x2f, 0x7f, 0x01 and 0x63, and the checksum is 0xea.

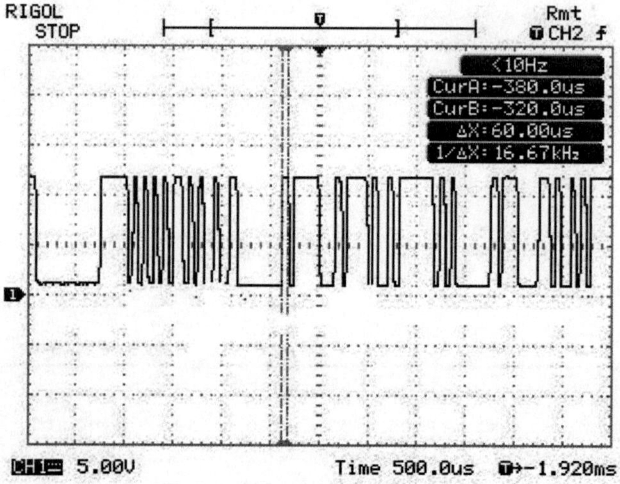

Fig. 10. LIN bus data frame

7 Conculsion and Summary

Valve regulated Lead-acid battery real-time data collection and transmission plays an important role in vehicle safety and operating reliability. In this paper, we have completed a more accurate measurement of battery parameters through LIN bus technology and the design of circuit and measurement error is less than 1%. The BMS (Battery Management System) is a bridge between the VRLA Battery and the EV (Electric Vehicle), which can be used to improve the performance and the reliability of the VRLA battery. This paper has developed a BMS based on the LIN-bus which can reduce the wire harness and is convenient to install and debug. The developed BMS can sample the voltage, current, temperature of VRLA battery with a reliable sampling circuit. The developed BMS has succeeded in working on VRLA in Research Center of Energy Electronics. It will benefit for the electric car.

References

1. Cao, L., Wen, W., Zhou, S.: The Design and Realization of a Distributed Measure Control System for the Performance Test of Engine. In: Proceeding of the International Conference on Sensing, Computing and Automation, ISCCA 2006 (May 2006)
2. Yan, Z.: Influence of the charging mode on the performances of lead / acid batteries used for electrical vehicles. Batteries (6), 261–263 (2001)

3. Chiu, H.-J., Pan, P.-L., Yeh, W.-H.: A novel rapid charger of lead-acid batteries with energy recovery. In: Eighteenth Annual IEEE Applied Power Electronics Conference and Exposition, APEC 2003, February 9-13, vol. 2, pp. 756–759 (2003)
4. Chan, C.C.C., Sun, F., Zhu, J.: Technology on Modern Electric Vehicles, pp. 238–239. Beijing Institute of Technology Press, Beijing (2002)
5. Wu, K.: Theory and Application on the CAN-bus. Beijing University of Aeronautics and Astronautics Press, Beijing (1996)
6. Singh, P., Reisner, D.: Fuzzy logic-based state-of-health determination of lead acid batteries. In: The 24th Annual International Telecommunication Energy Conference, pp. 583–590 (2002)
7. Cai, C.H., Du, D., Liu, Z.Y.: Battery state-of-charge(SOC) estimation using adaptive neuro-fuzzy inference system(ASFIS). In: The 12th IEEE International Conference on Fuzzy System, St. Louis, pp. 1068–1073 (2003)
8. Plett, G.: Extended Kalman filtering for battery management systems of LiPB-based HEV battery packs: Part 3. State and parameter estimation. Journal of Power Sources 134, 277–292 (2004)
9. Zhu, Y., Han, X., Tian, G.: Research on estimation technology of traction battery SOC for electirc vehicle. Chinese Journal of Power Sources 24(3), 29–32 (2000)
10. Deweck, O.L.: Multiobjective Optimization: History and Promise. In: The Third China-Japan-Korea Joint Symposium on Optimization of Structural and Mechanical Systems, Invited Keynote Paper, GL2-2 (2004)
11. Miller, J.F., Thomson, P.: Cartesian Genetic Programming. In: Poli, R., Banzhaf, W., Langdon, W.B., Miller, J., Nordin, P., Fogarty, T.C. (eds.) EuroGP 2000. LNCS, vol. 1802, pp. 121–132. Springer, Heidelberg (2000)
12. Liu, R., Zheng, S.Y., et al.: An Efficient Multi-Objective Evolutionary Algorithm for Combinational Circuit Design. In: Proceedings of the First NASA/ESA Conference on Adaptive Hardware and Systems, pp. 215–221 (2000)
13. Felder, R.M., Rousseau, R.W.: Elementary principles of chemical processes, 3rd edn., pp. 501–503. John Wiley & Sons press, New York (1978)
14. Lin, C., Qiu, B., Chen, Q.: A comparative study on power input equivalent circuit model for electric vehicle battery. Automotive Engineering 28(3), 229–234 (2006)
15. Buchmann, I.: Batteries in a portable world, 2nd edn., pp. 81–84. Cadex Electronics Inc. Press, Richmond (2001)

Analysis of a Ka-Band Phase-Shifter Using Distributed MEMS Transmission Line Structure

Zhen Kelong, Lv Shanwei, and Zhang Yan

School of Electronic and Information Engineering
Beihang University
Beijing, China
zhenkelong@163.com

Abstract. The design of 4-bit Ka-band DMTL (distributed MEMS transmission line) Phase-Shifter consisting of the CPW (coplanar-waveguide) transmission line was constructed on the 100 μ m GaAs substrate, with MEMS bridge capacitors loaded periodically on the transmission line. By changing the height of the MEMS Bridge, the phase velocity of the CPW line can be varied to yield a true time delay phase shift. The full wave simulation results demonstrate that the return loss can be lower than -10dB, and the insertion loss can be lower than -4dB respectively, with smaller than 3° phase shifter error in the frequency range below 38GHz.

Keywords: MEMS, Phase Shifter, Ka-band, Insertion loss.

1 Introduction

Because of the advantages of high speed and intellectualization, the phased array is widely applied in radar and communication fields. The phase shifter is an important component part of phased array, but the insertion loss and power consumption of phase shifter increase with the increase of working frequency.

The MEMS(micro-electromechanical system) devices have many excellent performances, such as low power consumption, high isolation, low insertion loss, overlarge band and low cost [1]-[5], so the phase shifter based on MEMS technology has a good application prospects especially in millimeter frequency range [6]-[8].

Base on the analysis of the distributed MEMS transmission line phase-shifter principle, the models of shunt MEMS switch and 4-bit Ka-band DMTL phase-shifter were created and simulated with the HFSS and ADS software in this paper. The study proves that the DMTL phase-shifter has excellent millimeter wave characteristics.

2 Principle of the DMTL Phase-Shifter

RF (Radio frequency) MEMS switch is the basic element of DMTL phase-shifter, it controls the "on" or "off" states of RF signals by mechanical actions. It has the

W. Zhang (Ed.): Advanced Technology in Teaching, AISC 163, pp. 765–772.
springerlink.com

performance of low insertion loss and high linearity, and be widely applied in the frequency band from 0.1GHz to 100GHz.

2.1 RF MEMS Switch Working Priciple

The structure of RF MEMS switch is shown in Fig.1, including CPW, dielectric film and MEMS bridge which is made of metal film.

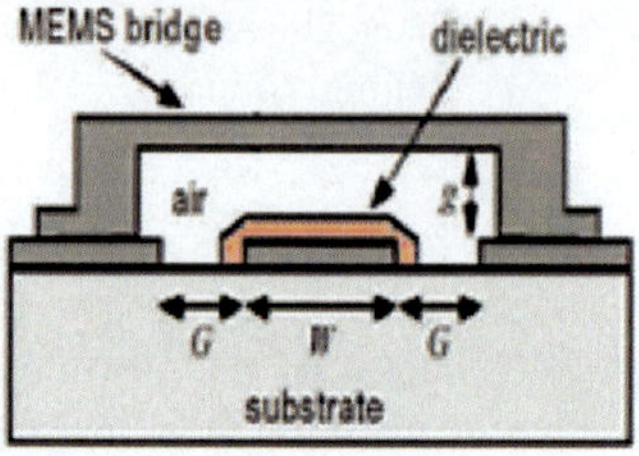

Fig. 1. The structure of MEMS switch

When the driving voltage between the bridge and CPW conduction band was added, the thin metal bridge will be bent down because of electrostatic attraction. The electrostatic attraction can be given by (1).

$$f_{down} = \frac{1}{2}\varepsilon_o A E^2 = \frac{\varepsilon_o A V^2}{2(g + {t_d}/{\varepsilon_r})^2} \tag{1}$$

Where A is area of bridge overlapping the conduction band; V is the driving voltage; g is the distance between the CPW central conductor and MEMS Bridge; t_d is the dielectric film thickness; ε_r is the dielectric film relative dielectric constant; ε_o is the air dielectric constant.

The electrostatic attraction causes the MEMS Bridge bend downward, and the change of the MEMS bridge may cause the opposite restoring force. When the restoring forces is equal with the electrostatic attraction, as shown in (2), the MEMS bridge will maintain in the state of equilibrium [9].

$$kx = \frac{\varepsilon_o A V^2}{2(g + {t_d}/{\varepsilon_r})^2} \tag{2}$$

Where k is the elasticity coefficient of MEMS Bridge; x is the downward displacement of the MEMS Bridge.

The equivalent circuit model of the switch includes shunt capacitance C, resistance R and inductance L, as shown in Fig.2.

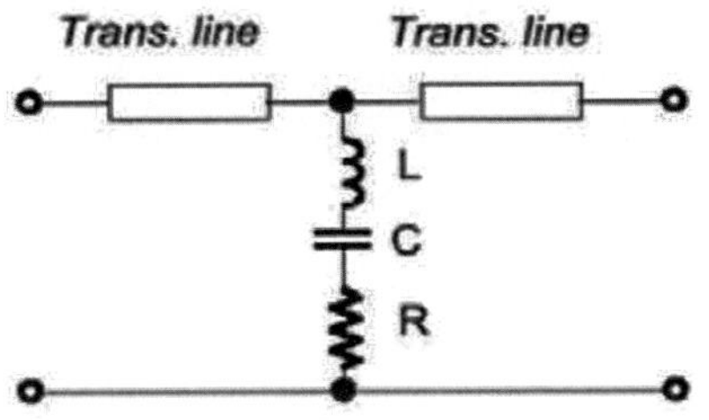

Fig. 2. The equivalence circuits of MEMS switch

The shunt capacitance plays an important role in MEMS switch because of its sensitivity to frequency, and it can be regard as a parallel plate capacitor which is composed with MEMS Bridge and the central conduction band. The deformation of the MEMS bridge can change the capacitance value of the parallel plate capacitor, then the transmission performance of the CPW is changed too. The simulation results in ADS are shown in Fig. 3 and 4.

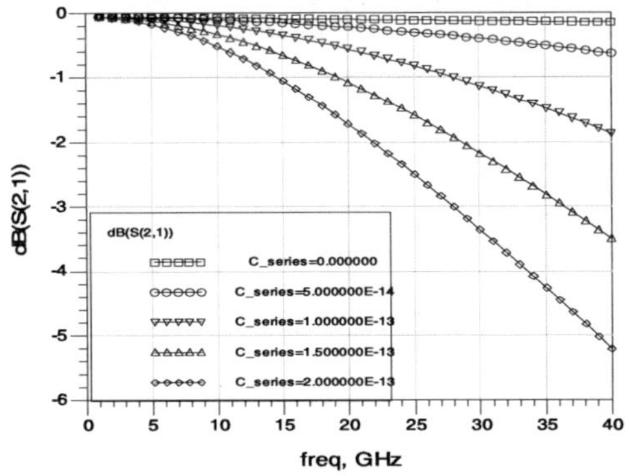

Fig. 3. Transmission performance of the CPW with different shunt capacitance

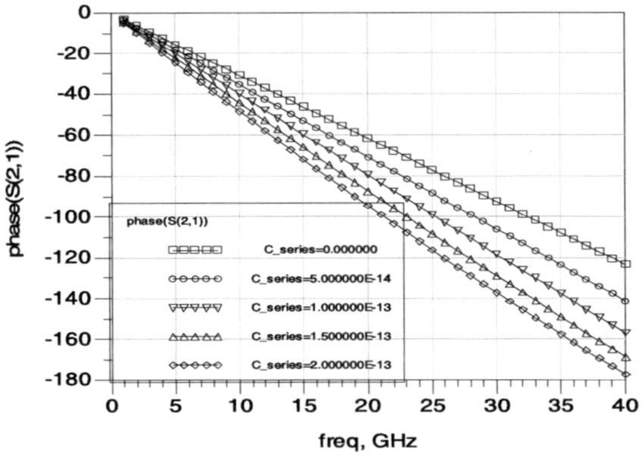

Fig. 4. Phase-shift performance of the CPW with different shunt capacitance

It can be concluded: the phase shift and insertion loss increase with the increase of the working frequency when the capacitance is unchanged; the phase shift and insertion loss increase with the increase of the capacitance when the working frequency is unchanged. So different phase shifts can be given by changing the MEMS Bridge's height.

2.2 DMTL Phase-Shifter Working Principle

The DMTL Phase-Shifter is consisting of CPW transmission line with MEMs bridge capacitors loaded on, it is shown in Fig.5, and its Equivalent circuit is shown in Fig.6.

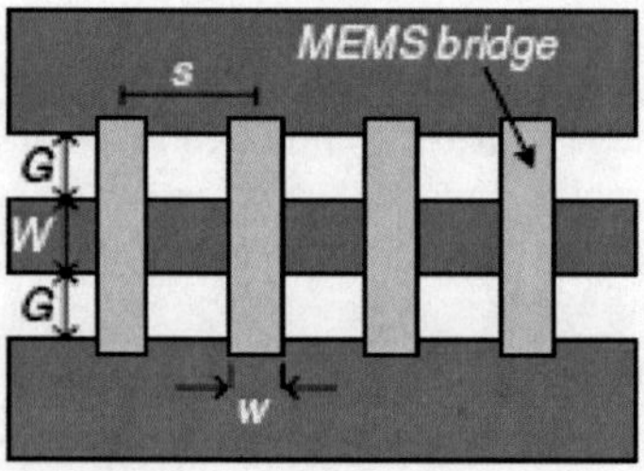

Fig. 5. Structure of the DMTL Phase-Shifter

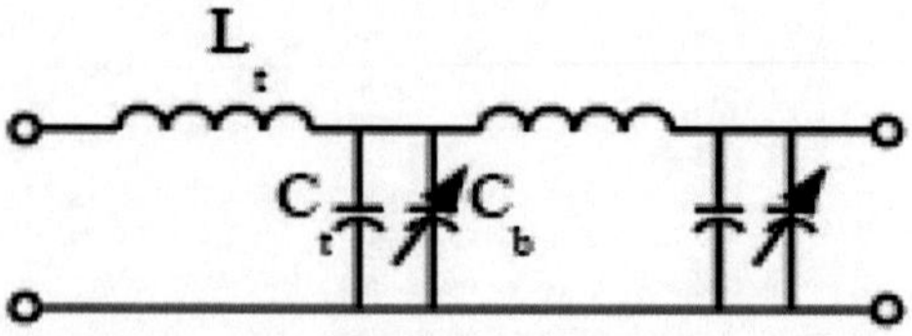

Fig. 6. Equivalent circuit of the CPW with switch loaded

In Fig.6, C_b is the shunt capacitance of the MEMS switch loaded, C_t is the capacitance parameter of the ideal CPW, L_t is the inductance parameter of the ideal CPW, the value of C_t and L_t can be given by (3).

$$C_t = \sqrt{\varepsilon_{eff}} \Big/ cZ_o \quad ; \quad L_t = C_t Z_o^{\,2} \tag{3}$$

Where ε_{eff} is the equivalent dielectric constant; Z_o is the characteristic impedance of ideal CPW.

When the MEMS switch was loaded, the characteristic impedance of CPW transmission line Z_l is given by (4).

$$Z_l = \sqrt{\frac{L_t}{C_t + C_b/s}} \tag{4}$$

The phase velocity V_l is given by (5).

$$V_l = \frac{1}{\sqrt{L_t(C_t + C_b / s)}} \tag{5}$$

The s is the periodic interval of loaded MEMS switch.

As the driving voltage increases, electrostatic attraction makes the MEMS bridge curved down, then C_b varies from C_{bup} to C_{bdown}. The ratio between the C_{bup} and C_{bdown} is named capacitance rate, it cannot be too big, or the insertion loss will increase. The phase difference between the states "up" and "down" can be given by (6).

$$\Delta\phi = w\sqrt{L_t C_t}\left(\sqrt{1+\frac{C_{bup}}{sC_t}} - \sqrt{1+\frac{C_{bdown}}{sC_t}}\right)$$
$$= wZ_o\left(\frac{1}{Z_{bup}} - \frac{1}{Z_{bdown}}\right) \tag{6}$$

Select the CPW whose characteristic impedance is greater than 50Ω, load the MEMs bridges capacitors periodically over the transmission line, and you can get the slow-wave structure not only whose characteristic impedance is 50Ω but also which can meet the requirement of phase-shift. Because of the bragg reflection, the upper limit of the DMTL structure working frequency f_{bragg} can be given by (7).

$$f_{Bragg} = \frac{1}{\pi s\sqrt{L_t(C_t + C_b / s)}} \tag{7}$$

The insertion loss will increase rapidly if the working frequency f_o is much greater than f_{bragg}. It is necessary that the working frequency should be fulfilled the following requirement in actual application, as shown by (8).

$$f_{Bragg} > 2f_o \tag{8}$$

3 Modelling and Simulation of the DMTL Phase-Shifter

The DMTL phase shifter was fabricated on the GaAs substrate whose thickness is 100μm, the central conductor of CPW and the MEMS bridge were all made of Au, and the key parameters of the phase shifter were given by Table 1 as follows.

Table 1. The parameters of phase-shifeter

Parameters	Values (μm)	
Width of the CPW central conductor band (W)	100	
Distance between the central conductor and the Ground (G)	150	
MEMS switch loaded periodic interval (s)	280	
Width of the MEMS bridge (w)	50	
Thickness of the CPW conductor band and MEMS bridge (t)	1	
Thickness of the GaAs substrate (H)	100	
Height of the MEMS bridge (g)	Up	Down
	2.5	1.0

The DMTL phase shifter model with double MEMS switches loaded on the CPW can realize 22.5° phase-shift. The simulation results of the phase-shifter insertion loss by HFSS software are shown in Fig.7.

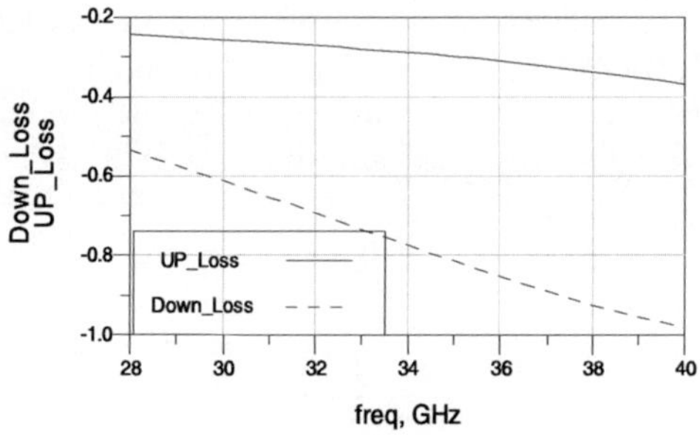

Fig. 7. Insertion loss characteristic curve

The simulation results of the phase-shift are shown in Fig.8

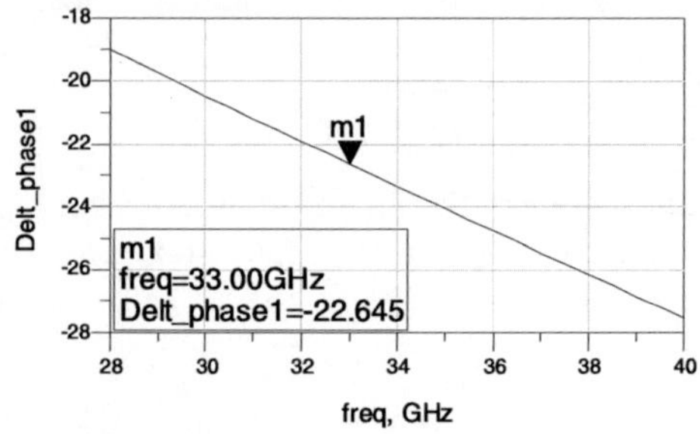

Fig. 8. Phase-shift characteristic curve

It can be concluded that the return loss can be lower than -10dB, and the insertion loss can be lower than -4dB respectively, with smaller than 3° phase shifter error in the frequency range below 38GHz.

In the same way, the phase-shift structure of 45° can be obtained with four MEMS switches loaded on the CPW; the phase-shift structure of 90° can be obtained with eight MEMS switches loaded on the CPW; the phase-shift structure of 180° can be obtained with sixteen MEMS switches loaded on the CPW, then the 4-bit Ka-band DMTL Phase-Shifter can be obtained by connecting all the phase-shift structures in series. The simulation results of different phase shifts by the ADS software are shown in Fig.9 to 11.

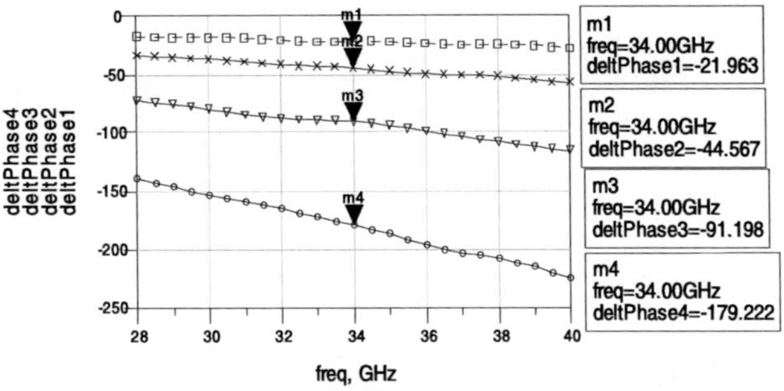

Fig. 9. Phase-shift characteristic curves

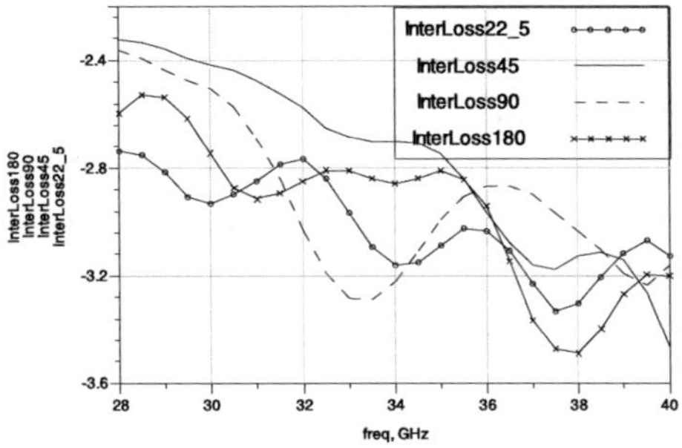

Fig. 10. Insertion loss characteristic curve

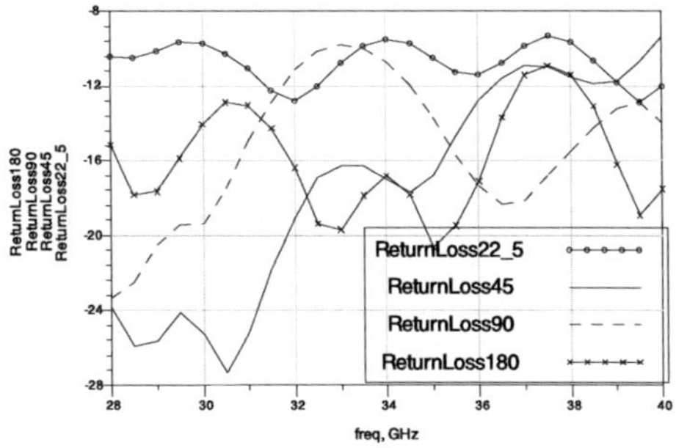

Fig. 11. Return loss characteristic curve

The simulation shows that, under the conditions of meeting the phase-shift error restraint, the insertion loss of the DMTL phase shifter is far lower than the loss of the ferrite phase shifter and the PIN diode phase-shifter. Moreover, the DMTL phase shifter has excellent millimeter wave feature of small size and low DC power consumption. It is beneficial to the miniaturization of the millimeter-wave transceiver front-end.

4 Conclusion

Base on the study of the MEMS switch structure and equivalent circuit, the DMTL phase-shifter principle was analyzed. By simulation with HFSS and ADS software, it was proved that the DMTL phase-shifter could ensure not only the phase-shifter accuracy but also low insertion loss. Because the MEMS technology is compatible with the monolithic microwave integrated circuit (MMIC) technology, the DMTL phase-shifter is conductive to integration. It is important to research of millimeter wave phased-array antenna miniaturization.

Acknowledgment. This work was supported in part by the National Natural Science Foundation of China under Grant 60901001 and in part by the National High Technology Research and Development Program of China under Grant 2008AA12A216.

References

1. Scardelletti, M.C., Ponchak, G.E., Zaman, A.J., Lee, R.Q.: RF MEMS phase shifters and their application in phase array antennas. In: The 2005 IEEE Annual Conference on Wireless and Microwave Technology, WAMICON 2005, pp. 191–194 (2005)
2. Rock, J.C., Hudson, T., Wolfson, B., Lawrence, D., Pillans, B., Brown, A.R., Coryell, L.: A MEMS-based Ka-band 16-element sub-array. In: Aerospace Conference, pp. 1–11. IEEE, Big Sky (2009)
3. Constant, S.B., Nicole, P., Menager, L., Labeyrie, M., Fourdin, C.: MEMS technology for Radar front end modules. In: Radar Conference - Surveillance for a Safer World, Bordeaux, pp. 1–6 (2009)
4. Rebeiz, G.M., Tan, G.-L., Hayden, J.S.: RF MEMS phased shifters: Design and applications. IEEE Microwave Magazine 3(2), 72–81 (2002)
5. Jin, B.-S., Wu, Q., She, H.-Y., Fu, J.-H., Li, L.-W.: Modeling and Design of the Novel MEMS Phase Shifter for Ka Band Applications. In: IEEE International Symposium on Antennas and Propagation and USNC/URSI National Radio Science Meeting (2005)
6. Kim, M., Hacker, J.B., Mihailovich, R.E., et al.: A DC-to-40GHz four-bit RF MEMS true-timedelay network. IEEE Microwave Wireless Cop. Lett. 11(2), 56–58 (2001)
7. Barker, N.S., Rebeiz, G.M.: Distributed MEMS true-time delay phase shifters and wideband switches. IEEE Trans. on MTT 46(11), 1881–1890 (1998)
8. Kim, H.-T., Park, J.-H., Yim, J., et al.: A compact V-band 2-bit reflection-type MEMS phase shifter. IEEE Microwave and Wireless Components Lett. 12(9), 324–326 (2002)
9. Rebeiz, G.M., Muldavin, J.B.: RF MEMS switches and switch circuits. IEEE Microwave Magazine 2(4), 59–71 (2001)

File Protection System Based on Driver

Li Wei, Chen Hao, Liu Peng, and Ye Shichao

Control and Computer Engineering School
North China Electri Power University
Beijing, China
epulw@126.com, chenhao321123@163.com

Abstract. Aiming at finding a solution to protect the security of file system which contains core information, this paper presents a file protection system model based on driver , and has implemented the file protection system based on the model. The System blocks the process which accesses to the protected files, while detecting the existence of corresponding local hardware signals and the key, to decide whether to allow the access or not. Iteliminates malicious manipulation of external invasion or internal Trojan and realizes the protection of core information.

Keywords: file protection, mandatory hardware confirming, transparent encryption, filter driver.

1 Introduction

With the rapid development of information technology, informatization has become an inevitable trend of development of enterprises. While improving production and management efficiency, enterprises' information security requirements are put to an unprecedented level. In today's electronic office, internal documents, electronic data is regarded as one of the most important assets of enterprises. its security requirements are self-evident. Leakage of important data, whether on business or personal, will bring huge losses. Therefore, the problem of protecting the security of enterprise information, especially the security of the file system, is needed to be solved urgently.

2 Related Research

Generally Speaking, two ways of protecting files are available [1]: the one based on access-control and the other based on data encryption.

Access-control, which based on priority, is normally implemented by Kernel-mode driver. Once hackers get enough permission, they can access confidential data unscrupulously. In addition, due to the fact that the file is stored on disk in plain-text, there are potential safety problems. So, a single access-control solution is not strong enough.

Data encryption is dependent on the encryption algorithm and key management. Even if hackers get the chance to access confidential data after break through various barriers, they can only receive a pile of useless data, since the data cannot be

W. Zhang (Ed.): Advanced Technology in Teaching, AISC 163, pp. 773–780.
springerlink.com

decrypted currently. Considering that traditional encryption software requires users to encrypt and decrypt manually, which is not easy to operate, while relying on Cryptography solely, it is still not a satisfactory solution.

3 The Technique of File Protection System

The system in this paper uses the file system filter driver technology and has implemented the combination of the access control and data encryption. Simultaneously, it uses the force hardware recognition technology to ensure the reliable operation on the document.

3.1 The Model of File Protection System

To build the model of the file protection system, it should describe a file accessing operation abstractly at first. The file accessing operation is constituted by process set and file set, and it is defined as followed:

Set $P=(p_1,p_2,\cdots,p_m)$ as process set,and $p_i \in P, i=(1,2,\cdots,m)$ is a process in the system.

Set $D=(d_1,d_2,\cdots,d_m)$ as file set,and $d_i \in D, i=(1,2,\cdots,n)$ is a file in the system.

Operation set $A=\{a(i,j)|a(i,j)=a(p_i,d_j), p_i \in P, d_j \in D\}$ which is a accessing operation of process p_i to file d_i.

$A=P\times D$ is accessing operation of every process to every file.

Rule set $R=\{r(i,j)|r(i,j)=r(p_i,d_j), p_i \in P, d_j \in D\}$, $R\in A$ is a permission for accessing operation of process to file.

At a certain time, set process t to the file accessing operation as $a^{(t)}(i,j), a(i,j)\in A$.

Hardware signal set $S=\{0,1\}, s\in S$. If $s=1$,there has signal from local hardware device; else $s=0$, there is no signal from local hardware device.

Key state set $K=\{0,1\}$, $k\in K$. If $k=1$, there is a right key; else $k=0$,there is a wrong key or no key.

At t' time, any local hardware whether send operation signal can be denoted by $s^t, s\in S$.If $s^t=1$,the operation signal is sent by the local hardware; else $s^t=0$, the operation signal is not sent by the local hardware.

Result set $V=\{0,1\}$, $v\in V$. If $v=1$, there is a permission for process accessing operation to data; else $v=0$, there is a prohibition for process accessing operation to data.

File protection system processing procedure:

$$\text{v}=\begin{cases}1 & a^{(t)}{}_{(i,j)}\in R,\ s^{(t)}=1\,and\,k=1\\ 0 & otherwise\end{cases}$$

When a process tries to operate on a file, and this file accessing operation is accord with the rule. At the same t time if hardware signal exits and right key exist, operation is permitted; else, the operation is forbidden.

3.2 System Modules

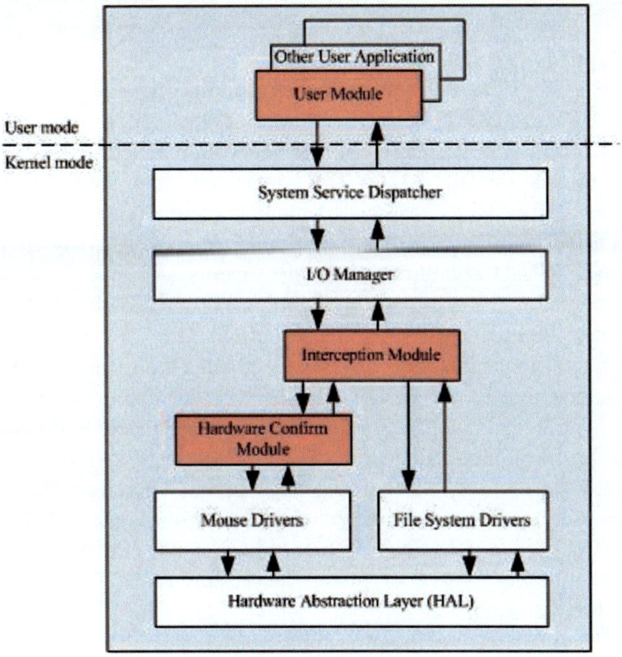

Fig. 1. System model

The model is consists of three modules: the user module, the interception module and the hardware confirm module. The relationship among the modules is shown in Figure 1.

The functions of these modules are as follows:

(1) User module: provide configuration functions of the file protection; start or stop the kernel driver; provide a friendly graphical user interface.

(2) Interception module: provide file transparent encryption and decryption, access control, logging and other functions; meanwhile supply the user module with control interface.

(3) Hardware confirm module: judge whether the user input is sent from the local hardware or not, completely eliminate the Trojan horse remote accessing, and ensure the safety of the core files; furthermore ensure critical operations are issued from THIS machine to realize the operation of non-repudiation.

3.3 Key Technology

1) Mandatory Hardware Confirming

Mandatory hardware confirming technology is based on the operating system kernel. By detecting devices such as keyboard or mouse port registers and the state of these registers, it determines whether the operation is performed locally or not (e.g. operation from local mouse or keyboard).Since most of the Trojans are hidden in the user's computer, and act stealthily, mandatory hardware confirming technology can effectively prevent Trojan horses and other unauthorized operations from accessing to the protected files.

2) Transparent Encryption and Decryption

Through the file system filter driver, we realize the interception of file access-related IRP (I/O REQUEST PACKET) operation. And then through custom message dispatch routine we realize transparent encryption and decryption functions [2]: Before the file data are written to disk, the IRP_MJ_WRITE operation of the document is intercepted, in its dispatching routine, the data carried by IRP are encrypted and then written to disk. So as to complete the transparent encryption; when the file data are read from the disk, the IRP_MJ_READ operation of the documents is intercepted. After the read requisition is completed, the data carried by IRP are decrypted and then returned to the top level. So as to complete transparent decryption. The data on the disk is always stored in the form of cipher text, thus effectively ensures the security of data.

It is worth noting that in part of the Windows applications, files reading / writing use a memory mapping mechanism [4], thus the process's access to the memory cannot be captured by the file system filter driver. At this point, we should deal with it when the file goes into the memory through the page faults. Therefore, we must intercept page read/write requests IPR which has a special tag (irp->Flags has IRP_PAGING_IO or IRP_SYNCHRONOUS_PAGING_IO). In addition, due to Windows file caching mechanism, it is unnecessary to encrypt and decrypt the cache data. So the type of the IRP is required to determine.

3) File Access Control

Access control is to limit access to critical resources by some way,to prevent the damage caused by the intrusion of unauthorized users, or inadvertent operation of legitimate users. File access control do real-time monitoring to the operations of the processes on the file. Legal operation will be executed, and the illegal operation will be prohibited. So that the file will be well protected.

In filter driver, it's necessary to obtain the name of the current access process and the Full Path of the file being accessed. Here's the method used to get the name of the current access process:First use the PsGetCurrentProcess() function to get the EPROCESS structure of the process,then get the name of current process by the offset address in EPROCESS structure. But Microsoft did not publish the structure. In order to ensure that the correct process name in the different versions of Windows can be obtain, we can tcompute the offset address of the SYSTEM process in the EPROCESS structure to get the names of other processes by the offset address. The way of obtaining the full path of the file being accessed is as follows: get the file path of FileObject after the file open IRP processing, and save the correspondence of pointer and the path of FileObject in a hash table. When a request of file operation comes, the corresponding path of FileObject can be obtained by searching the hash table.

4) The Choice of Encryption Algorithm

Cryptographic system can be divided into two parts: symmetric encryption and asymmetric encryption. Features of symmetric encryption are: fast encryption , but has the difficulties in key management; asymmetric encryption does not have key management problem, but the encryption algorithm is complex and slow. The file protection system uses hybrid encryption, which means it uses symmetric encryption algorithm to handle confidential data to be protected, meanwhile it uses asymmetric encryption algorithm to protect the symmetric encryption key. This method ensures the efficiency of data encryption, as well as the data security.

4 The Implementation of File Protection System

4.1 User Module

The user module offers an interface of system configuration, and offers bidirectional communication with the interception module as well.

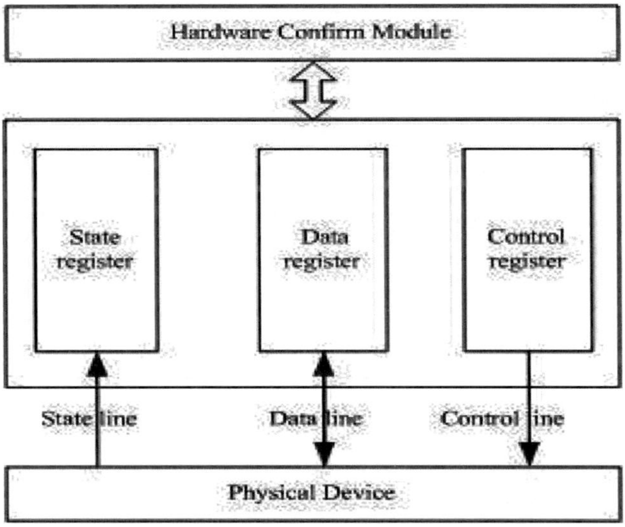

Fig. 2. Communication procedure

By means of the DeviceIoControl in WDM [8], the user moudle can communicate with the interception module. The user module send the I/O Control Code to the interception module through the DeviceIOControl() fuction to finish the communication. Meanwhile, the event method is used in the communication between the interception module and the user module as is showed in figure 2. The idea of this method is that the user module creates an event object though the function CreateEvent(), and uses the DeviceIOControl() function to transmit the event handle to the driver, and then the user module process call the waitForSingleObject() function to wait the trigger event. After that the driver call the ObReferenceObjectByHandle() function to obtain a pointer which point to the event object. When a special device event occurred, the driver will trigger the event, and the event will call the WaitForSingleObject() function to suspend the user process.

4.2 Interception Module

Based on filter driver, the filter moudle realized the access control and lucency encryption of the file system by means of IRP leaching, as well as comunication with the user moudle and the hardware confirming moudle. The core arithmetic are as follows:

Step 1: Interception module monitor the file access operations:

a. If the interception module has examined that a process tried to operated on a file which is protected, then go to Step 2
b. Return to Step1

Step 2: Interception module get the process ID to determine the legality of the file access operations

a. If the operation is legal, then go to Step 3
b. Otherwise, go to Step 8.

Step 3: Interception module issues hardware confirmation request and Notify the user module to show hardware confirmation window,then go to Step 4

Step 4:The Interception module detects hardware signal

a. On receiving the notices from the Interception module, then go to Step 5
b. Otherwise, go to Step 8.

Step 5:The Interception module try to get user's key

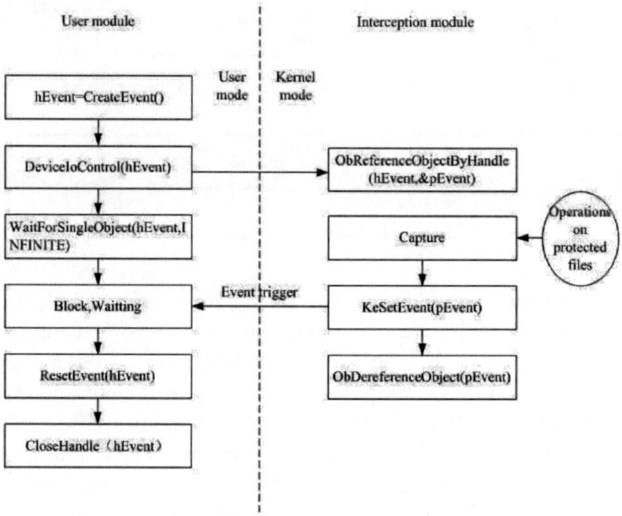

Fig. 3. Hardware confirm module detects the port registers

a. Correct key got, then go to Step 6
b. Otherwise, go to Step 8.

Step 6: the interception module encrypt or decrypt files which is protected, then go to Step 7

Step 7: If the operation on the protected file is allowed, go to Step 1 Continue to monitor

Step 8: The operation on the protected file is denied,go to Step 1 Continue to monitor

4.3 Hardware Confirm Module

The hardware confirm module detects the Hardware signal through the drivers running in kennel mode.

When requests of hardware confirming from the interception module are received, the hardware confirm module modifies the IOAPIC relocation table, and meanwhile inserts a new interruption into the interrupt descriptor table(IDT) to handle the keyboard or mouse interrupt. If any hardware signal exists either in the keyboard interface or the mouse interface, he interrupt handler will be invocated which will receive the signal and then inform the interception module to determine whether to discharge this operation or not. Keyboard, mouse and other hardware devices connect through the interface components with the system bus, and then communication by means of data and signal with CPU. Each interface component includes a group of registers [10]. By testing the state and the data of the port registers, the hardware confirm module can identify whether the hardware signal exists or not, just as the picture shows below:

4.4 System Analysis

This system is realized as a driver. Because drivers run in kernel mode, which are protected strictly by the operating system, this program is much more superior in safety and efficiency. Meanwhile, since drivers run in the lower layers of the operating system, this system can perform encryption and decryption transparently, without changing users' habits. Even if the file on disk is copied illegitimately, he or she receives only "junk files", because the file contains only cipher-text.This enhances the safety of the data.

After this program is installed, all the files in assigned folder will be encrypted forcibly with access-control and hardware confirm operations. When the protected files are accessed legally, this system will check out the hardware signal to ensure the effectiveness of the accession, which ensures the access comes from local users rather than Trojans.

5 Conclusions

In this paper, we analyzed the requirement of guaranteeing the kernel information in enterprise firstly, and then proposed a method to protect the file system based on driver, which focus on key files in protecting the enterprise's information. This system ensures authorized access to the protected files by access-control, mandatorily hardware confirmation and transparent encryption and decryption, which protects the key information successfully. The practice shows that it is a sound solution to meet the requirement of data security and preventing hackings. Besides, it can also serve as a competent guard to protect private information. Overall,this system makes higher commercial value and better social benefits.

References

1. Zhao, M.-W., Mao, R., Jiang, R.-A.: Transparent Encryption File System Model Based on Filter Driver. Computer Engineering 35(1), 150–152 (2009)
2. Gu, Z., Zhou, L., Lv, G.: The access control technology of spatial data files based on file system filter driver. In: 11th IEEE International Conference on Digital Object Identifier, pp. 734–737 (2008)
3. Takizawa, Y., Kourai, K., Chiba, S., Yanagisawa, Y.: A Secure File Access Control System for Desktop PC. In: Computer System Symposium, IPSJ, pp. 79–86 (2007)
4. Tan, W., Yang, X., Shao, J.-L.: Windows Kernel Security Programming, pp. 223–273. Electronic Industry Press (2009)
5. Liu, H.-Y., Fan, J.-L., Ma, J.-F.: Research Advances on Access Control. Mini-Micro Systems (2004)
6. Ohmiya, M., Shinagawa, T., Kato, K.: Simple Access Control using Filename Prefixes. In: Computer System Symposium, IPSJ, pp. 97–105 (2007)
7. Forouzan, B.A.: Cryptography and Network Security. Tsinghua University Press (2009)
8. Wu, A.-H., Tai, M., Yu, H.-T.: Windows 2000/XP WDM Device Driver Development. Electronic Industry Press (2003)
9. Dai, M.-E., Shi, J.-Q.: Micro-computer Technology and Application. Tsinghua University Press (2006)
10. Mao, D.-C.: Windows Kernel Scenario Analysis, pp. 729–946. Electronic Industry Press (2009)

A Miniaturized UHF RFID Reader Antenna

Zhongbao Wang[1], Shaojun Fang[1], Shiqiang Fu[1], and Jun Zheng[2]

[1] School of Information Science and Technology, Dalian Maritime University, Liaoning, 116026, China
[2] Beijing Electro-mechanical Engineering Institute, Beijing 100074, China
{wangzb,fangshj,fushq}@dlmu.edu.cn, junzheng999@yahoo.com.cn

Abstract. A miniaturized circularly polarized (CP) patch antenna is proposed for ultra high frequency (UHF) radio frequency identification (RFID) reader applications. The antenna is composed of a vertical ground, a U-shaped strip, and a square patch with truncated corners. A novel U-shaped strip feed structure is proposed for good impedance matching. Broadband CP operation is achieved by using a thick air-layer substrate (about $0.08\lambda_0$) and selecting an optimal size of the truncated corners. The antenna has an impedance bandwidth (S_{11} < -20 dB) of about 141 MHz (861-1002 MHz or 15.1%), a 3-dB axial ratio (AR) bandwidth of about 30 MHz (899-929 MHz or 3.3%), and a gain level of about 7.2 dBi or larger within the 3dB AR bandwidth. The dimension of the antenna is 13 cm × 13 cm × 2.8 cm. The small size and high gain make this element suitable for portable UHF RFID reader applications.

Keywords: Vertical ground, u-shaped strip, circularly polarized, patch antenna, RFID reader.

1 Introduction

Radio frequency identification (RFID) is a technology that provides wireless identification and tracking capability. In recent years, RFID technology has been rapidly developed and applied to supply chain management, animal tagging, and electronic payment [1-3]. In this regard, a great demand of UHF RFID system is anticipated to replace the barcode system. Globally, each country has its own frequency allocation for UHF RFID applications [4]. For example, RFID UHF bands are: 840.5-844.5 and 920.5-924.5 MHz in China, 866-869 MHz in Europe, 902-928 MHz in America, and so on.

Reader antenna is one of the important components in RFID systems, which is used to transmit or receive RF signal from a tag. Usually, circularly polarized (CP) antenna is used in readers for ensuring the reliability of communications between readers and tags [5], [6]. Many types of antennas have been used in UHF RFID systems, wherein patch antenna is one of the most commonly used [7-9]. The size reduction and gain enhancement of UHF RFID reader antenna have been key issues to the system developer. Some CP antennas have been reported in literatures [10-12]. Generally, to enhance antenna gain, a large ground plane has been used for RFID

W. Zhang (Ed.): Advanced Technology in Teaching, AISC 163, pp. 781–787.
springerlink.com

reader antenna [13]. However, a big size antenna cannot be applied to portable UHF RFID applications. It is also reported that a CP patch antenna with a high-impedance ground plane (HIGP) introduces higher gain effects for operation [14]. However, HIGP increases the complexity of the antenna configuration. To reduce size and maintain high gain, a corner truncated square-ring patch, which is surrounded by a vertical ground, is reported in [15]. Similar to the other single-fed single-patch CP antenna, the antenna suffers from narrow axial ratio (AR) bandwidth (less than 1%). It is not suitable for broadband system applications, such as 902-928MHz RFID system.

To solve the problem, a novel single-fed CP antenna is proposed in this paper. The antenna is composed of a vertical ground, a U-shaped strip, and a square patch with truncated corners. The U-shaped strip feed structure is used for good impedance matching; the vertical ground is also used for size reduction. Broadband CP operation is achieved by using a thick air-layer substrate (about $0.08\lambda_0$) and selecting an optimal size of the truncated corners. The antenna is designed to operate at 902-928 MHz for RFID reader applications. In the next section, design concepts of the proposed CP antenna are introduced, followed by simulation results and analysis.

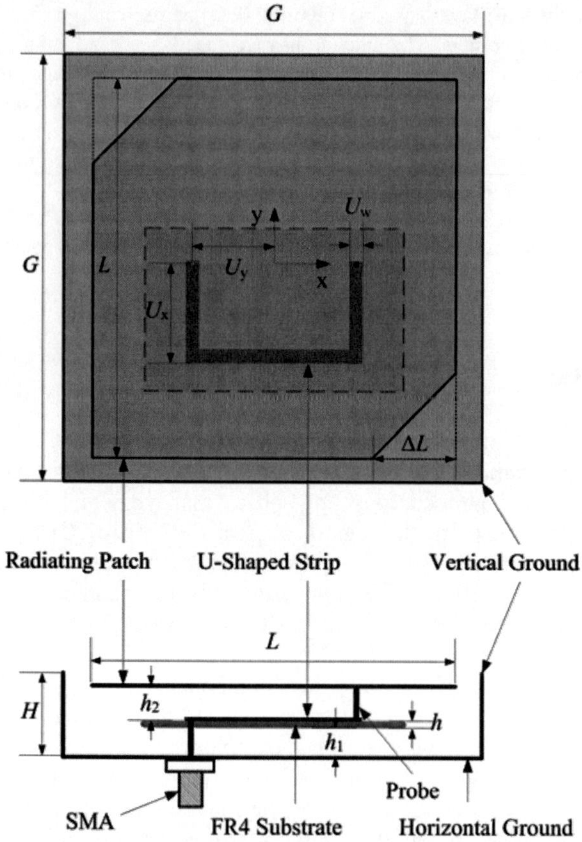

Fig. 1. Geometry of the proposed antenna.

2 Antenna Structure

The structure of the proposed antenna is shown in Fig. 1. The centre frequency of the antenna is chosen at f_0= 915MHz (λ_0=328mm). The antenna is composed of a U-shaped strip, a FR4 substrate, a square patch with truncated corners, a vertical ground, and a horizontal ground. The U-shaped strip of a width of U_w (5 mm) is printed on the upper side of the FR4 substrate with thickness h = 1 mm and relative permittivity ε_r = 4.4. One end of the U-shaped strip is connected to the radiating patch by the probe of diameter of d =1.27 mm, while the other one is connected to an SMA connector. By properly selecting U_x (36.9 mm) and U_y (distance between the feed point and the centre of the main patch, 26.5 mm), the input reactance and resistance can be easily determined. The large inductance aroused by the long probe is compensated by means of the electromagnetic coupling between the radiating patch and the U-shaped strip. This simple but effective technique to broaden the impedance bandwidth is a little bit similar to that of the design with the proximity-coupled strip feed [16]. The FR4 substrate is suspended above the horizontal ground plane ($G \times G$, 130 mm×130 mm) at a height of h_1 =13 mm. The radiating patch of 115.8 mm×115.8 mm and with a truncation of 22 mm at two diagonal corners is placed above the U-shaped strip at spacing of h_2 = 13 mm. The vertical ground of a height of H (28 mm) surrounds the radiating patch. The capacitance between the radiating patch and the ground is increased, as a result of size reduction. Air substrate is used in this configuration to achieve broader bandwidth, higher gain, and lower cost.

3 Simulation Result and Analysis

The proposed antenna is simulated by using Ansoft HFSS 3-D EM simulator, which based on finite element method (FEM). The simulation results show that the antenna has good impedance matching, excellent circular polarization, high antenna gain, and wide half power beam.

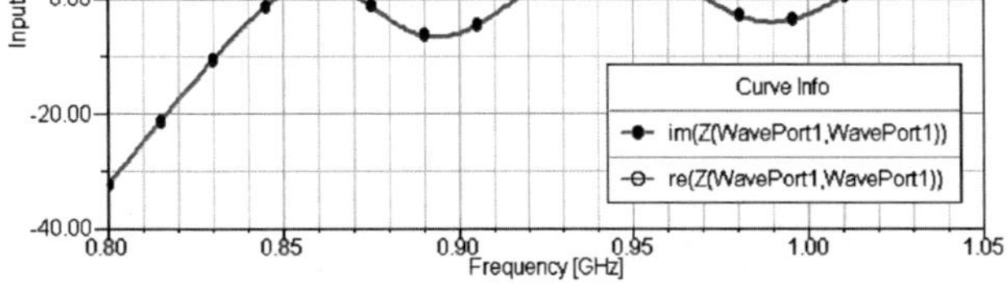

Fig. 2. Input impedance of the proposed antenna.

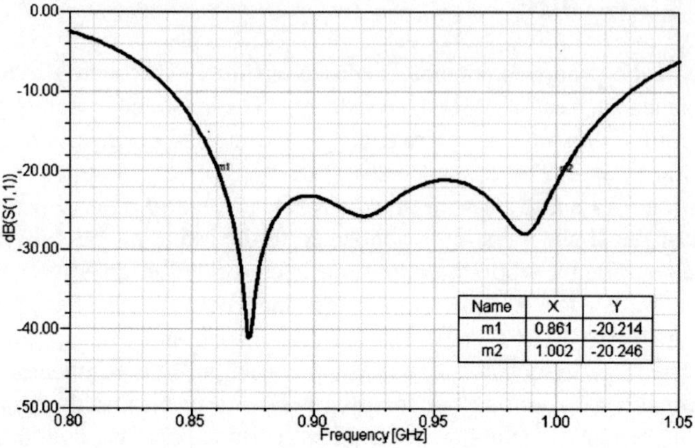

Fig. 3. S_{11} of the proposed antenna.

The input impedance against the frequency is given in Fig. 2. It is clearly observed that the input resistance varies around 50 Ω and the input reactance keeps small within a broad frequency range. The large reactance usually resulted from the long probe can effectively be cancelled out with the help of the strong electromagnetic coupling between the radiating patch and the U-shaped strip. The resultant impedance bandwidth for S_{11} < -20 dB reaches up to 15.1%, ranging from 861 to 1002 MHz as shown in Fig. 3.

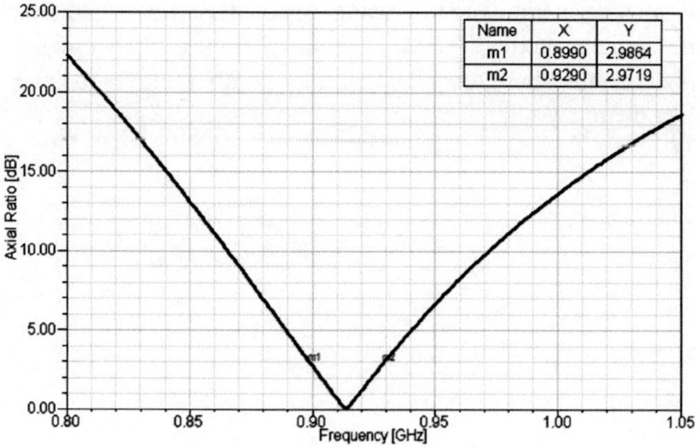

Fig. 4. AR of the proposed antenna.

As shown in Fig. 4, the 3-dB axial ratio bandwidth is about 30 MHz from 899 to 929 MHz, which can cover the American UHF RFID bandwidth.

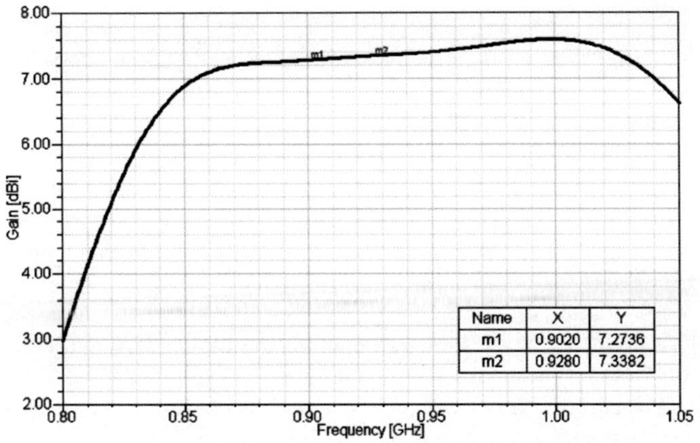

Fig. 5. Simulated gain of the proposed antenna.

Fig. 5 shows the antenna gain. It is clearly observed that the antenna gain is about 7.3 dBi in the operating bandwidth of 902-928MHz.

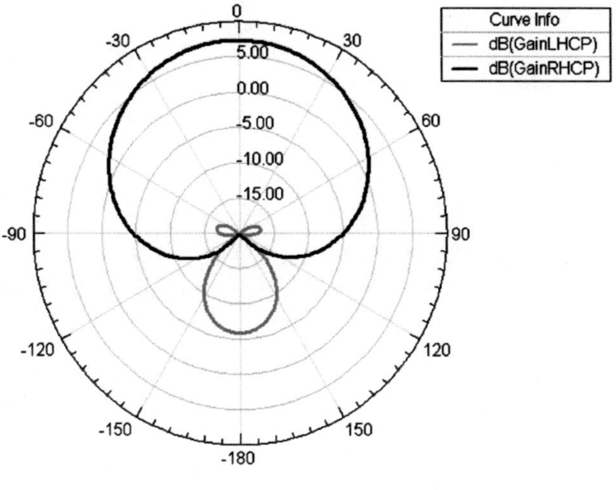

Fig. 6. Simulated radiation patterns of the proposed antenna.

Fig. 6 is the radiation pattern of the proposed CP antenna at 915 MHz. In the broadside direction (near 0°), the CP isolation is more than 25 dB, implying a good CP performance. The half power beamwidth is about 86°, which is also obtained from the Fig. 6. Compared to the other CP patch antenna, the half power beam is wider. It is also observed from the Fig. 6 that the front-to-back ratio of the proposed antenna is about 13 dB.

4 Conclusion

A miniaturized single-fed CP patch antenna is presented. The antenna is designed to operate at 902-928 MHz, which is compatible with American and Chinese UHF RFID reader. Broadband CP operation is achieved by using a thick air-layer substrate (about 0.08λ0) and selecting an optimal size of the truncated corners. By means of the U-shaped strip feed technique, the VSWR of the patch antenna is less than 1.2. In addition, the structure of the proposed antenna is simple and low-cost. Certainly, good impedance matching, excellent circular polarization, high antenna gain, wide beam and small size make the antenna suitable for portable UHF RFID reader applications.

Acknowledgments. This work was partly supported by the National Natural Science Foundation of China (Grant No. 61071044), the Traffic Applied Basic Research Project of the Ministry of Transport of China (Grant No. 2010-329-225-030) and the Fundamental Research Funds for the Central Universities.

References

1. Wang, Z.B., Fang, S.J., Fu, S.Q., Jia, S.L.: Single-Fed Broadband Circularly Polarized Stacked Patch Antenna With Horizontally Meandered Strip for Universal UHF RFID Applications. IEEE Transactions on Microwave Theory and Techniques 59, 1066–1073 (2011)
2. Chen, Z.N., Qing, X.M., Chung, H.L.: A Universal UHF RFID Reader Antenna. IEEE Transactions on Microwave Theory and Techniques 57, 1275–1282 (2009)
3. Nasimuddin, Chen, Z.N., Qing, X.M.: Asymmetric-Circular Shaped Slotted Microstrip Antennas for Circular Polarization and RFID Applications. IEEE Transactions on Antennas and Propagation 58, 3821–3828 (2010)
4. Barthel, H.: Regulatory status for using RFID in the UHF spectrum. EPCGlobal, `http://www.epcglobalinc.org/tech/freq_reg/RFID_at_UHF_Regulations_20090318.pdf`
5. Wang, Z.B., Fang, S.J., Fu, S.Q., Lü, S.W.: Dual-Band Probe-Fed Stacked Patch Antenna for GNSS Applications. IEEE Antennas and Wireless Propagation Letters 8, 100–103 (2009)
6. Lau, P.Y., Yung, K.K.O., Yung, E.K.N.: A Low-Cost Printed CP Patch Antenna for RFID Smart Bookshelf in Library. IEEE Transactions on Industrial Electronics 57, 1583–1589 (2010)
7. Nikitin, P.V., Rao, K.V.S.: Helical Antenna for Handheld UHF RFID Reader. In: IEEE International Conference on RFID, pp. 166–173. IEEE Press, New York (2010)
8. Yeh, S.A., Chen, H.M., Lin, Y.F., Yang, Z.Z., Chen, C.H.: Circularly Polarized Crossed Dipole Antennas for Handheld RFID Reader. In: International Conference on Applications of Electromagnetism and Student Innovation Competition Awards (AEM2C), pp. 134–138. IEEE Press, New York (2010)
9. Wang, Z.B., Fang, S.J., Fu, S.Q.: Broadband Stacked Patch Antenna with Low VSWR and Low Cross-Polarization. ETRI Journal 32, 618–621 (2010)
10. Sharma, P.C., Gupta, K.C.: Analysis and Optimized Design of Single Feed Circularly Polarized Microstrip Antennas. IEEE Transactions on Antennas and Propagation 31, 949–955 (1983)

11. Chen, W.S., Wu, C.K., Wong, K.L.: Novel Compact Circularly Polarized Square Microstrip Antenna. IEEE Transactions on Antennas and Propagation 49, 340–342 (2001)
12. Wang, Z.B., Fang, S.J., Fu, S.Q., Li, X.M.: Circularly Polarized Antenna with U-Shaped Strip for RFID Reader Operating at 902-928 MHz. In: International Symposium on Signals, Systems and Electronics, pp. 546–548. IEEE Press, New York (2010)
13. Lee, J.M., Kim, N.S., Pyo, C.S.: A Circular Polarized Metallic Patch Antenna for RFID Reader. In: Asia-Pacific Conference on Communications, pp. 116–118. IEEE Press, New York (2005)
14. Sun, J.S., Lee, Y.C.: Circularly Polarized Stacked Antenna With a High-Impedance Ground Plane for RFID Reader Applications. Microwave and Optical Technology Letters 50, 2248–2250 (2008)
15. Wang, Z.B., Fang, S.J., Fu, S.Q.: A Low Cost Miniaturized CP Antenna for UHF Radio Frequency Identification Reader Applications. Microwave and Optical Technology Letters 51, 2382–2384 (2009)
16. Meshram, M.K.: Analysis of L-strip Proximity Fed Rectangular Microstrip Antenna for Mobile Base Station. Microwave and Optical Technology Letters 49, 1817–1824 (2007)

A Good Anti-robust Algorithm with Map Watermarking

Wang Lele, Liu Wei, and Dong Yinghong

College of Computer Science and Technology, Changchun University
Changchun, Jilin Province, China
wanglele@126.com

Abstract. In this paper, the color feather of the watermark embedded in the map, then use the existence of relevant testing and inspect testing to verify the watermark, and finally through a variety of robustness tests and found that this algorithm has good shear resistance.

Keywords: Color Feature, robustness, watermark.

1 Introduction

With the rapid development of multi-media technology and digital web communication, the security of the digital information becomes more and more important. The technology of watermark is applied in every field of the security of information.

People studied all kinds of Watermark embedding, but few people study the Watermark embedding of the map watermark. Because the map is large, when we are using map, sometimes we take the whole map, sometimes we cut a part of the map. Thus we destroy the watermark which is embedded into the map. Used by changing the value of two adjacent edges on to achieve the purpose of embedding watermark to improve the robustness of the map, particularly the shear resistance. The author take the less the color of the map and the boundary between regions of different colors in this line of obvious feature into account.

2 The Existing Embedding Algorithm of Watermark on the Map

In recent years, many people have been interested in the study of Watermarking Algorithm on the map, but few of them is completed, such as the text-based geometric normalization algorithm[1] and Vector map watermarking algorithm[2].

3 A Map Watermark Embedding Algorithm Based on Color Feature

The research indicates that the map has a large amount of information; variety of color range is not so obvious with clear boundaries, etc. In order to embed watermark, we consider changing the distance between the edges to embed watermark. That is the

W. Zhang (Ed.): Advanced Technology in Teaching, AISC 163, pp. 789–794.
springerlink.com

location of the point moving sidelines. In order to bring the vision to reduce differences, the watermark sequence using the value of a random vector while we are in the embedded watermark, so that we can embed watermark algebra 0 in a certain region.

3.1 Color Extraction

When analyzing the characteristics of the color map, we extract color in a map with clear color and obvious color boundaries we extract original image to the single color, (see Fig. 1).

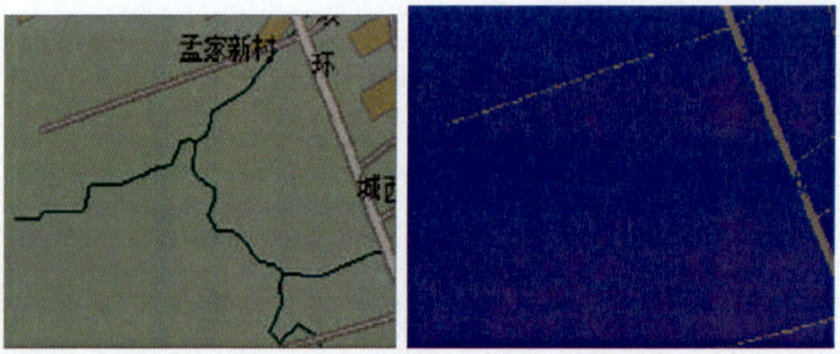

(a) (b)

Fig. 1. (a) the original map (b) the map to chart the effect of color extracted

3.2 Extraction of Edge Points

After making color extraction, we used the method of translation [4] extract the edge points, (see Fig. 2).

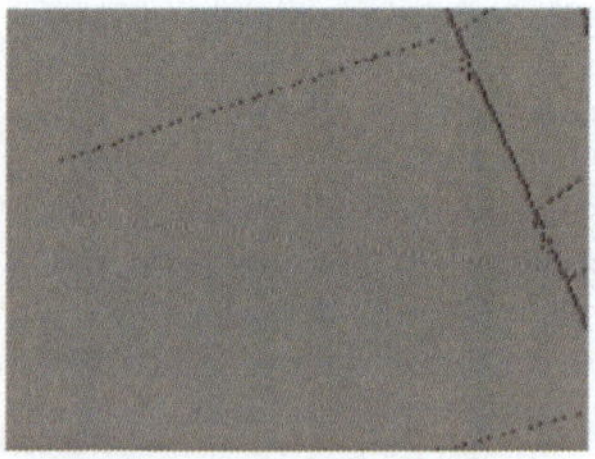

Fig. 2. Edge points on the map after extracting the effect of map

3.3 Watermark Embedding

We prepared the edge points of its embedding after the extraction of edge points. For each edge point, line coordinates of x have a displacement of w. Each displacement w is randomly chosen from the *(-1, 0, 1)*, this is to ensure that each edge has a minimum displacement. Because for each pixel is the smallest displacement of a pixel. In order to image the pixels do not introduce new after embedding the watermark image, Therefore, So the elements w in the value to 0 to make it satisfy the algebra. Among them, Where, m is a selected color in the region the number of edge pixels. To ensure its algebra and 0 each w is selected in the *(-1, 0, 1)*, We set for every three pixels values all the time may be one of *{(1,0,-1),(1,-1,0),(0,1,-1),(0,-1,1),(-1,1,0),(-1,0,1)}*. (See Fig.3)

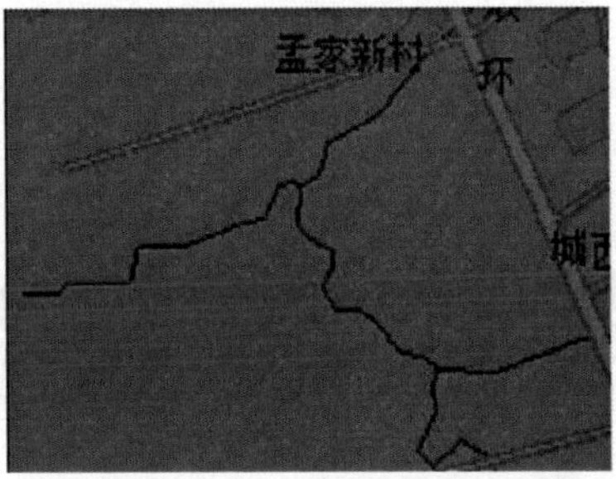

Fig. 3. Map watermarked renderings

In this paper, we change the position of the edge to achieve the purpose of watermarking. We may see the change of the watermarked image magnification on the edge points and we can see that the edge line of pixels with small changes, (see Fig.4).

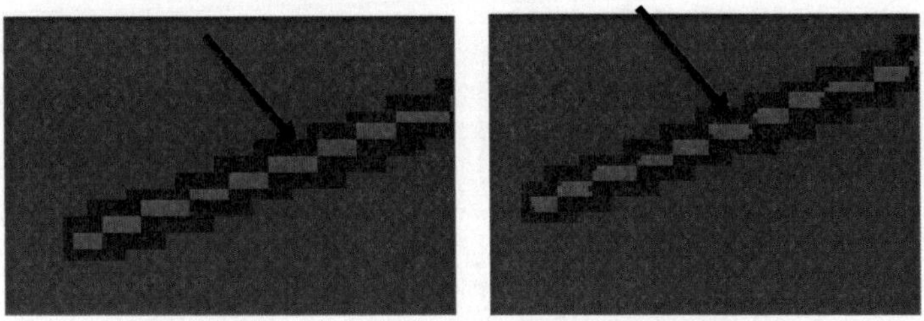

(a) (b)

Fig. 4. (a) Map Enlarge map, (b) Embedded after the Enlargement

4 Map Watermarking Based on Color Feature Detection Algorithm

In this paper, watermark detection is divided into two parts. First, we determine whether the host image and the Watermark image are of relevance .Second, we determine whether the watermark image watermark sequence (-1, 0, 1) and should come.

4.1 Correlation Detection

When detecting the watermark, we define a variable *sum*, Using *sum* to represent the image and the watermark image changes between the edge points.

$$sum = \begin{cases} sum+1, image1.GetPixel(x,y).R = image2.GetPixel(x'+1,y').R, \\ sum-1, image1.GetPixel(x,y).R = image2.GetPixel(x'-1,y').R \\ sum+0, image1.GetPixel(x,y).R = image2.GetPixel(x',y').R \end{cases} \quad (1)$$

Among, *image1.GetPixel.(x,y).R* on behalf of the host image the color of edge points, *image2.GetPixel.(x,y).R* on behalf of the watermark image the color of edge points, *(x,y)∈image1,(x',y') ∈image2.*

The host image and the watermark image, we compare two map relative points. If the host image and watermark of a point in the left image pixel same as pixel value, *sum* to minus 1; If the host image and watermark of a point in the right image pixel same as pixel value, *sum* to plus 1; If there is no change between two points, we assign 0 to *sum*. Statistical results of *sum* and *nodecount* is the ratio of the number of edge points for the *N1*;

$$N1=abs\ (sum/nodecount) \quad (2)$$

Where *nodecount* is the number of edge points, the value of the ratio in the ideal state should be 0.Since taking into account the existence of errors, after study, we found that when the ratio of no more than 0.01, we consider that the watermark image and host image has a certain relevance.

4.2 Detection of the Existence

Using (3), we define a variable *sum*, use *sum* to represent the host image and the watermark image changes between the edge points.

$$sum' = \begin{cases} sum' + 1, image\ 1.GetPixel\ (x,y).R = image\ 2.GetPixel\ (x'+1,y').R, \\ sum' + 1, image\ 1.GetPixel\ (x,y).R = image\ 2.GetPixel\ (x'-1,y').R \\ sum' + 0, image\ 1.GetPixel\ (x,y).R = image\ 2.GetPixel\ (x',y').R \end{cases} \quad (3)$$

Where, *image1.GetPixel.(x,y).R* representative of the host image in the color of the edge points, *image2.GetPixel.(x,y).R* representative of the watermark image in the color of the edge points,*(x,y)∈image1,(x',y') ∈image2.*

If the host image and watermark image of a point in the left pixel in the pixel values are the same, *sum* plus 1; if the host image and watermark image of a point in the right pixel in the same pixel value, *sum* plus 1 too; If there is no change between two points, we assign to x 0. The results of such statistics *sum* and *nodecount* is the ratio of the number of edge points for the *N2*.

$$N2=abs(sum/nodecount) \quad (4)$$

Where x is the number of edge points. Since the embedded watermark, in order to ensure that the process of embedding does not introduce new pixels, the text used in the watermark embedding each of the three adjacent points selected randomly a value between(-1,0,1). Then extract the watermark we use *sum* to store the time change in the number of points, if in accordance with the corresponding point is taken between the values of(-1,0,1), *sum* plus 1. Then calculated the ratio of x and the edge points, in an ideal state, the ratio should be 0.6666. Taking into account the existence of error and the watermarked image after the interference in the attack, after many experiments we believe that the ratio of more than 0.60 when we consider that the images are protected by the watermark image.

For the map is, we compare the host graph and the watermark graph, calculate the value of *N1* and *N2*, *N1* is 0.00388, *N2* is 0.66213.We say that it is protected by the watermark image.

5 Performance Testing Algorithm

Experiment 1: Cut the watermark detection after treatment

In order to cut attack in the verification, we chose the white road area to conduct embedding. But the road in this picture is not uniformly distributed in the experiment; it may cause excessive shear part of most of the embedded watermark cut off. In such circumstances we may have some errors when doing cropping experiment, but does not affect our results. (see Fig.5)

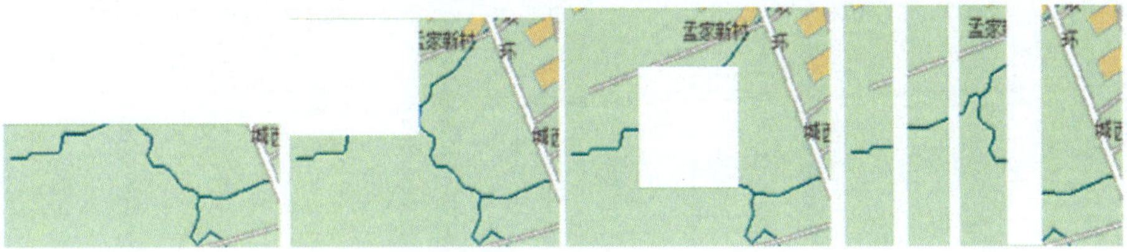

(a) (b) (c) (d)

Fig. 4. (a) (b) (c) (d) For the watermarked map with the four conditions of shear

Color feature based watermarking algorithm is very good resistance to shear capacity, Through the map image after embedding watermark four different cropping attack, we calculate the *N1* and *N2* values, obtained the following results shown in table 1.

Table 1. Watermark image after the test in different shear experimental results

Correlation value	N1	N2
Fig.4(a)	0.00582	0.7029
Fig.4(b)	0.00583	0.6759
Fig.4(c)	0.00388	0.6699
Fig.4(d)	0.00388	0.6339

Thus, this study of the robustness of the algorithm for a variety of attack effect is ideal, particularly for cropping. In order to increase their non-visible features, we can adjust the watermark embedding strength reduced to compensate.

6 Summary

After much experimentation, we found that the traditional watermarking algorithms are usually considered on the basis of the host image into a Watermark image. This approach tends to introduce new ones in the image pixels. Because the color map represents important geographic information, such traditional methods in the process of embedding the map become less relevant. Using this algorithm, it can improve the robustness and invisibility of balance, It guarantees the quality of the image, it also improve the ability to resist malicious attacks, especially in the performance of the shear.

References

1. Masry, M.A.: A Watermarking Algorithm for Mapand Chart Images. In: The Proceedings of the SPIE Conference on Security, Steganography and Watermarking of Multimedia Contents VII (January 2005)
2. Li, Y.-Y., Xu, L.-P.: Copyright protection of the vector map using the digital watermark. Journal of Xian University 31(5), 719–723 (2004)
3. Wang, X., Lin, H., Bao, H.: A robust watermarking algorithm for vector digital mapping. Computer Aided Design and Computer Graphics 16(10), 1377–1381 (2004)
4. Sun, S.-H., Lu, Z.-M., Niu, X.-M.: Digital watermarking technology and applications, pp. 137–308. Science Press, Beijing (2004)

An Information Retrieval Methods Based on the Sample of Feature Extraction

Wang Xiaohu[1], Wang Lele[2], and Li Nianfeng[2]

[1] College of Computer Science and Engineering, Changchun University of Technology
[2] College of Computer Science and Technology, Changchun University
Changchun, Jilin Province, China
{wangxiaohu,wanglele}@126.com

Abstract. In this paper, characteristics of the sample extraction method of retrieval mode, the first such as the size of the sample into subsets, each subset of the statistical model to extract features, and retain the support of its emergence as the weight of its characteristic pattern. Because this algorithm using only a few statistics, so the search speed greatly improved.

Keywords: search, sample, statistics.

1 Introduction

In recent years, Internet has developed rapidly, which brings together a wealth of information. However, because Web is unstructured, dynamic, organization is complex, to the user search data caused great difficulties. Information to solve the network information retrieval disorder and chaos of a basic method, but also makes the Internet under the search has become very crucial [1]. The current on-line retrieval major search engines and browsers to take a combination of these two information retrieval mechanisms are unable to meet the user to retrieve information for relevance and accuracy requirements. Order from a large number of semi-structured data found in the available models, Web mining came into being, and made some research results [2], [3], [4].

2 Web Mining

Web information retrieval is to come from the information retrieval technology and its most essential features of the system on the Web document collection and a collection of matching the needs of users and choose [5], [6], [7], [8].

Web mining can be divided into three categories: Web content mining, Web Structure Mining, Web usage mining [5], [9], [10], [11], [12]. Web content mining refers to the contents of the file from the Web, and descriptions of information for potentially valuable knowledge or model of the process; Web Structure Mining is the organizational structure and links to WWW relations derived knowledge. Mainly through the structure of the Web site analysis, deformation and induction will be classified Web pages in order to facilitate the search of information; Web usage

W. Zhang (Ed.): Advanced Technology in Teaching, AISC 163, pp. 795–800.
springerlink.com

mining is on the server when users access Web access to records left by mining, mining of the object is the server log information, also known as Web log mining.

3 Web Hyperlink Structure Mining

3.1 Page-Rank

Page-rank algorithm [8] is based on the assumption that this kind of thinking:

[1] A page is referenced several times that many pages have links pointing to it, and then this page is very important;
[2] A page even though not many references, but was a important page references, then this page may also be important;
[3] The importance of a page to be uniformly distributed and delivered to its reference page.

S. Brin and L. Page for computing algorithm authoritative page. Using (1), Calculated as follows:

$$R(i) = (1 - c) + c * \sum j \in B(i)\, R(j)/N(j) \tag{1}$$

They think that the best value of c is 0.85. Where B (i) on behalf of pages pointing to page i set. N (j) is the page j in the number of hyperlinks pointing to other pages. R (i) that the authority of degree i for the page. According to this principle, Page and Brin with keyword search and other text-based technology work together to improve the quality check.

Page-rank algorithm advantages: as long as the operator in advance of the Page-rank good value, when retrieval do not have to re-calculate, reducing the online time. Disadvantages: It retrieves comprehensive, to count the Page-rank a web page we should calculate the value of all the pages in the document set Page-rank value, compute too much; its Page-rank values and retrieval topic of irrelevant.

3.2 HITS

HITS [13] was first proposed in 1999 by Kleinberg. He believes that each page has two Ranking: authority Ranking (depends on the page pointing to it), the center Ranking (depends on the page it points to someone else). Its realization process is as follows:

[1] A text-based search engine is the result of a search set of R (Root Set);
[2] The page pointed to a collection of R and R of other pages point to come in form collection contains a collection of S;
[3] All the authority of the page-level, center Ranking, all set the initial value 1, then in accordance with the following algorithm:

Repeat until convergence

For all Page i in S, $a_i = \sum j \in B(i)\, h_j$; $h_i = \sum j \in F(i)\, a_j$

Normalize : $\sum i \ a_i^2 = 1$; $\sum i \ h_i^2 = 1$;

End

Where B (i) on behalf of all the pages pointing to page i set; F (i) on behalf of the page i points to page collection. Finally, according to the size of the authority Ranking to return to the user.

HITS algorithm of advantages: it is localized and involves only a small portion, usually a few thousand pages; its value and the search topic; this value is a page to a search topic relative to the weight. Disadvantages: the search topic have diffusion and generalization phenomenon, poor retrieval performance of these topic.

3.3 SALSA

SALSA algorithm [14] is to calculate the chain to reach a page out of roaming in the probability of chain web pages and into the chain into the chain of roaming in a Web page to reach the probability, then the probability of probability order a large number of output hub and authority web pages.

SALSA algorithm advantages: computation is extremely simple. Disadvantages: Enter the web page (the root set) is with the topic of many irrelevant; not filtered ad links, navigation links and other garbage links; to the weight of each link assigned inappropriate.

3.4 Hyperlink Similarity Function

Hyperlink similarity function [15] is the link between the use of Web pages of information to calculate the similarity method. To make use of it from three pages of information: the shortest path between two pages in length; the common ancestor of both the number of pages; two pages the number of common children. It hyperlinks to the similarity calculated as follows:

$$S_{links} = w_d * S_{des} + w_a * S_{anc} + w_s * S_{spl}$$

Where w_d 、 w_a 、 w_s is the weighted value, S_{des} is public offspring calculated similarity, S_{anc} is common ancestor of the calculated similarity, S_{spl} is the shortest path length calculated similarity.

Hyperlink similarity function advantages: the link is not complete, if the document set is small, the cross-hyperlinks to a lot of documentation set, which are cross-hyperlinked documentation set is ignored, the performance on the better than other methods.

4 An Information Retrieval Methods Based on the Sample of Feature Extraction

4.1 Design Idea

First, by using the existing search engines, such as based on HITS for Google, Alta Vista index-based part of speech as well as ontology-based Yahoo! Generated based on the result of the initial collection, then select the samples in this collection, the samples were imaged to 0 / 1 string form, such as the size of the sample into a subset extracted by statistical characteristics of each subset of the model and retain the support of its emergence as the weight of its mode characteristics. In order not to sample set of model features found in the lack of diversity, we extract multiple model features, calculating the weights of each feature vector to return is greater than the value of the characteristics of a given Wsup, Finally, the model features excavated, searched the initial set, returned to the user (see Fig. 1).

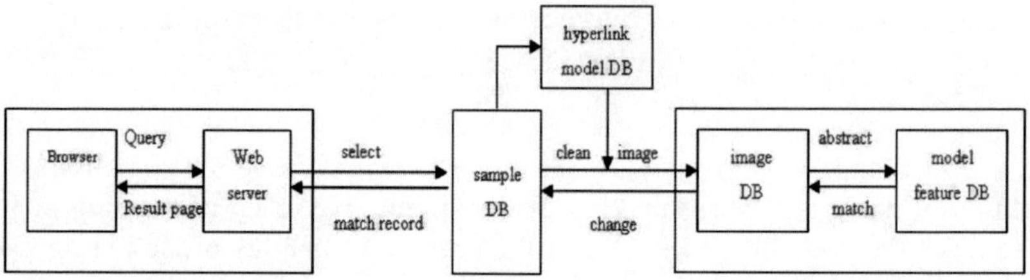

Fig. 1. The process of information retrieval

4.2 Algorithm

get Root set, select S

absorb from S to M, its structure is <Id, Add, Sup>

clear S, S-->T by M

T-->{ T_1, T_2, T_3, ……}, calculate w_{ij} :

For j = 1 to n do

$W_{ij} = \sum x \in T_i \; x_j^{(m)}$;

Normalize $T_i : \sum j \in T_i \; w_{ij}^2 = 1$;

If $w_{ij} > W_{sup}$ then add to W

Search by W with M, then return the result

Description:

[1] Get the initial set, options for the sample set S.

[2] Where M is the hyperlink model library, M extracted from the S, define its structure is such as <Id, Add, Sup>, integer value of Id to address identification, String type of Add to address content and the integer value Sup on behalf of this address by the frequency, that is support.

[3] Cleaning S, change to T by M training, T is the image set with 0 / 1 structure.

[4] Statistics for each subset of the weights of its components. W_{ij} are the focus of the first sub i j th component of the weight. To a sub-set of T_i, for example, contains m records, put the records were recorded as $x^{(1)}$, $x^{(2)}$, $x^{(3)}$, ……, $x^{(m)}$. For each record also contains n-component, these components are denoted by x1, x2, x3, ……, xn.

[5] Standardized the weights w_{ij} in T_i

[6]The weight is greater than the given support threshold model features added to the model feature set W.

[7] Return the result.

Advantage of this method is that it does not require the same algorithm as the previous simulation Web site topology, using clustering, computing Web page

Ranking and the correlation between Web pages, it just uses a number of statistical, performance improved; and this algorithm to retrieve information associated with the user, with better targeted and personalized. Drawback is that results directly affected the quality of the sample collection effects, and the number of sample classification, pattern extraction and some other parameters of the threshold can not be automatically generated, need to specify.

5 Problems and Future Research Directions

With the continuous development of WWW, and the number of series-level Web page growth, more and more difficult to retrieve information online. Hyperlink current Web page to find the authority faces the following problems:

[1] Not every hyperlink represents recognition, such as advertising;
[2]The commercial or competitive considerations, very few Web pages will point to their field of competition authority of the page, that is significant authority of the Web, usually can not directly mutual recognition;
[3] Authority of the few pages describing the category of special self-descriptions.

Priorities for future work is to improve the effectiveness and efficiency of existing systems, such as adding the concept of search technology, more Chinese encoding format, the major is to enhance the hyperlink of the (semantic) expression [16] and according to the use of hyperlinks law to find hyperlinks to (semantic) type [17] ,[18].

References

1. Brewer, E.A.: When Everything is Searchable. Communication of the ACM 44(3), 53–55 (2001)
2. Cooley, R., Mobasher, B., Srivastava, J.: Information and Pattern D is covery on WWW. In: Proceeding of International Conference on Tools with Artificial Intelligence, pp. 558–567. IEEE Press, New port Beach (1997)
3. Borges, J.A., Levene, M.: Data Mining of User Navigation Patterns. In: Masand, B., Spiliopoulou, M. (eds.) WebKDD 1999. LNCS (LNAI), vol. 1836, pp. 92–112. Springer, Heidelberg (2000)
4. Srivastava, J., Cooley, R., Deshpande, M., et al.: Web Usage Mining: Discovery and Applications of Usage Patterns from Web Data. SIGKDD Explorations 1(2), 12–23 (2000)
5. Wang, J.-C., Pan, J.-G., Zhang, F.-Y.: Research on Web Text Mining. Journal of Computer Research and Development 37(5), 513–520 (2000)
6. Zhao, D.-Q.: Data Mining:Principles, Methods and Application. New Technology of Library and Information Service (6), 41–44 (2000)
7. Baeza Yates, R., Ribeiro Neto, B.: Modern Information Retrieval. Addison Wesley, England (1999)
8. Brin, S., Page, L.: The Anatomy of a Large-Scale Hypertexual Web Search Engine, `http://www7.scu.edu.cn/programme/fullpapers/1921/com1921.htm`
9. Kosala, R., Blockeel, H.: Web Mining Research: A Survey. SIGKDD Exploration 2(1), 1–15 (2000)

10. Han, J.-W., Meng, X.-F., Wang, J., Li, S.-E.: Research on Web Mining:A Survey. Journal of Computer Research and Development 38(4), 405–414 (2001)
11. Deng, Y., Li, M.: Research on Web Mining and Tools. Computer Engineering and Applications (20), 64–65 (2001)
12. Wang, S., Gao, W., Li, J.-T.: Web Mining. Computer Science 27(4), 28–31 (2000)
13. Kleinberg, J.M.: Authoriative Sources in a Hyperlinked Environment. Journal of the ACM 46(5), 604–632 (1999)
14. Lempel, R., Moran, S.: The Stochastic Approach for Link-structure Analysis(SALSA) and the TKC Effect. In: 9th International World Wide Web Conference, Amsterdam, Netherlands (May 2000)
15. Weiss, R.: A Hierarchical Network Search Engine that Exploits Content-Link Hypertext Clustering. In: Proceedings of the Seventh ACM Conference on Hypertext, Washington, DC, http://www.psrg.lcs.mit.edu/ftpdir/papers/hypertext96.ps
16. Frei, H.P., Stieger, D.: The Use of Semantic Links in Hypertext Information Retrieva. Information Processing & Management 31(1), 1–13 (1995)
17. Spertus, E.: ParaSite: Mining structure information on the Web. Computer Network and ISDN System 29, 1205–1215 (1997)
18. Allan, J.: Automatic Hypertext Link Typing. In: Proc. for the Hypertext 1996 Conf., Washington, D.C., USA, vol. 3, pp. 42–52 (1996)
19. Chen, D.-Q.: Research on Web Structure Mining. Information Studies: Theory & Application 26(1), 59–61 (2003)

Use Case-Based Service-Oriented Analysis and Modeling

Wenjun Liu[1], Meitao Fu[1], Sanqiang Luo[2], and Dongcheng Zou[3]

[1] Nanchang Institute of Technology
Nanchang 330099, China
[2] Center of Network, Yichun University, 336000, China
[3] Dalian Dong'an Auto-Control System Co., 116600, China
myprivacy@sina.com, 18907955801@189.cn

Abstract. The drawbacks of Service-Oriented Analysis and Design (SOAD) methodologies are discussed and a new SOAD method based on Use Case (UC) approach is presented in this paper, therefore, a service model is designed to facilitate service identification from Use Case script. Moreover, a process to achieve high quality service identification is introduced as a guideline to an analyst work. Through the analysis of the method basic characteristics and its illustration, we can conclude that the method has several advantages compared to other methods which are based on business process modeling and enterprise architecture.

Keywords: Service-Oriented Architecture (SOA), Service-Oriented Analysis and Design (SOAD), Service identification, Use Case (UC).

1 Introduction

Service-Oriented Architecture (SOA) is a flexible set of design principles used during the phases of systems development and integration in computing. Most of the current research on SOA is focused on implementation issue, this is in no way sufficient to construct commercial-strength enterprise applications.

Based on survey of some SOAD methodologies, this paper introduces an UC-Based SOAD method. The method focuses on the analysis phase of a SOA project, namely, identification and description of service based on application requirements. The method has some advantages to identify services, such as, loose coupling and high cohesion.

The remainder of the paper is organized as follows: section 2 introduces related work about SOAD. Section 3 discuss Use Case technologies. In Section 4, SOAD method is introduced. In section 5, we illustrate the efficiency of the method to identify services in an application domain. In Section 6, we analyze how it accords with current principles in SOA. And in the last section, we conclude the characteristics of the method and future research.

2 Related Work

Service Oriented Modeling and Architecture (SOMA) is widely applied to service analysis process in SOA projects. Business processes which are in a variety of styles

W. Zhang (Ed.): Advanced Technology in Teaching, AISC 163, pp. 801–806.
springerlink.com

are primary input of their SOAD process. However, business process descriptions commonly bear some difficulties, such as: absence of formal semantics; too much message-oriented; and too little process oriented.

The others methodologies mostly provides some guidelines and criteria for evaluating current SOA applications.

In summary, current works on SOAD largely present what activities should be carried out, rather than specifying how they can be conducted. Or, they mostly depend on Business Process Modeling and Enterprise Architecture. In fact, in order to develop a small and medium size application system in a small and medium size enterprise, it is either unnecessary or a burden to design its Business Process Modeling and Enterprise Architectures as a start point.

Besides, Use Case (UC) as a requirement acquisition method is discussed and applied to SOA requirement. But, they do not give a guideline how to identify services from Use Case. They only explain how to use it as a supplement to business process model.

Therefore, as far as software application system development is concerned, we provide a SOAD method, which is based on Use Case requirement acquisition technology. It is a lightweight method compared to the above mentioned some methods.

3 Use Case Requirement Acquisition Technology

A use case in software engineering and system engineering is a description of a system's behavior as it responds to a request that originates from outside of that system. A use case should:

- Describe what the system shall do for the actor to achieve a particular goal.
- Include no implementation-specific language.
- Be at the appropriate level of detail.
- Not include detail regarding user interfaces and screens. This is done in user-interface design, which references the use case and its business rules.

If we can make full use of the information in UC model to identify and describe the services in the expected system, we will succeed to convert traditional usage of Use Case into services analysis and design in SOA. This is our effort in the paper.

4 An Soad Approach Based on Use Case

4.1 Use Case Script

A use case describes a set of sequences, not just a single sequence. It is appropriate to describe the primary flow path and alternate flow path separately. However, the flow path is never independent to each other. In fact, it is a control process which has a start point with initial condition and a few end points with a successful post condition, as shown in Figure 1.

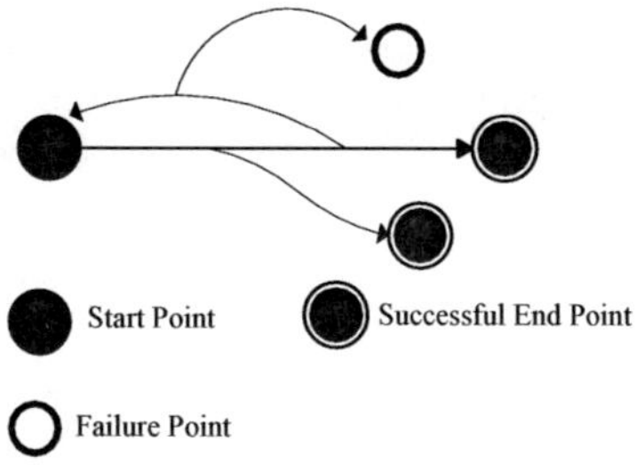

Fig. 1. Use Case behavioral sequence

4.2 A Service Model Design Based on Use Case

In order to identify and describe services from information which is provided by Use Case, we firstly design a service model, which is showed in Figure 2.

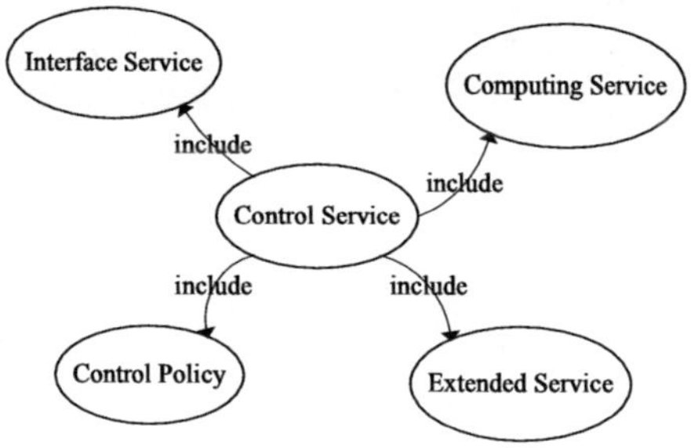

Fig. 2. A Service Model

- Interface Service

The function of this service is only to give participant an interface to gather their input and forward it to Control Service. It can be viewed as a virtual service.

- Computing Service

This Service represents a computing task taken up by computer which is an important part of system business logic. Compared to Interface Service, it is a real service which can be designed and implemented by some SOA technologies.

- Extended Service

In Use Case script, when a step ends and an extending point occurs, it will form an alternate flow path or failure flow path, this scenario can be described with Extended Service. Extended Service describes a sequence of sentence from its extending point to the path end or the flow path return to primary flow path.

- Control Service

It is a composite service which represents the Use Case because it takes the whole Use Case as service. It is made up of Interface Service, Computing Service and

Extended Service. In fact, it describes the Use Case primary flow path, while the extending flow path is described by Extended Service. Control Service adopts Control Policy to describe their internal structure.

- Control Policy

It is applied to describe the structure in Control Service or Extended Service. In fact, Control Policy describes which Extending Service to adopt.

Interface Service and Computing Service can be called atomic service.

4.3 Use Case Based SOAD Approach Analysis

According to the service model, now we introduce a process as guideline to analyst while using the SOAD method based on Use Case as following:

1) Elaborate the aimed Use Case script. Confirm Use Case script to description criterion.
2) Combine too fine-grained steps and reduce extending points as much as possible. The task will guarantee the expected service granularity.
3) Along the Use Case primary flow path, identify Interface Services and Computing Services.
4) Based on the flow path's extending point and control policy identified as above, identify and describe Extended services.
5) View the Use Case as a service and take the primary flow path as service process.
6) Iteratively and recursively apply step 3-5 to every extended service until all extended services internal structure are described.
7) Apply the services description as above to design and implement every service and provide the Use Case business objective.

5 An Illustration of the Approach

In the section, we give a "student search course's scores" illustration to test and verify our SOAD method. In every E-campus system, searching courses' scores is necessary business requirement. Generally, a student visits E-campus and enters his or her account and password registered in the system, the system check the user's account and password with database, and send back a page screen to the student. If the account and password is correct, then the student can continue his or her next operation, otherwise, repeat previous operation again. If the student can continue his or her operation, he or she should enter search criterion and make a request to the E-campus system, the system makes response after getting request information. In the process, some business policies will be checked, such as: user authority, the scores that he or she can look. Even the list of courses that he or she failed or passed.

In the requirement phase, if we apply Use Case to capture expected system feature, we will construct the diagram as figure 3 as follows:

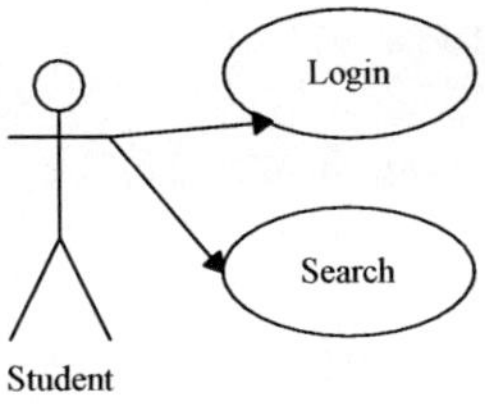

Fig. 3. Login & Search Use Case Diagram

The "Login & Search" Use Case script can be described as below.

1) Primary flow path:

a) Student submits search criterion to the system.
b) The system checks the user's session, to see whether he or she has passed login validation or not.
c) The system returns search results page screen to the user.

2) The failure flow path: 2a: the "searching score service" is not open at the time.

a) System displays that the function is not open yet by the time .
b) The student cancel system searching scores service.

3) The alternate flow path: 2b: the user has not passed login validation.

a) System displays that the user has not passed login validation, he or she can not search course's score.
b) The student enter correct account and password and submit for login validation.

In the script, "2a" and "2b" denote that they are two extending point derived from "step 2" in primary flow path.

Next, we adopt our SOAD method described in section 4 above to identify the service:

First, along the Use Case primary flow path, we identify Interface Services and Computing Services, such as:

- Name: getUserId, type: Interface Service, input: Null, output: UserId, Control Policy: Null
- Name: getCourseById, type: Interface Service, input: Null, output: Course, Control Policy: CheckUserAccount
- Second, along failure path: 2a, we identify Services as follows:
- Name: displayOpenDate, type: Computing Service, input: Null, output: search score service open time, Control Policy: Null
- Name: cancelSearch, type: Interface Service, input: Null, output: Null, Control Policy: Null

It is difficult to apply WSDL and BEPL to describe services in the phase because they are usually used in design and implementation phase. Of course, the description model should be the benefit of description with WSDL in design phase.

6 Conclusions and Future Work

In the paper, we firstly analyze the drawbacks of current SOAD methodologies. After that, we introduce a SOAD method based on Use Case. Therefore, a service model is designed to facilitate identifying services from Use Case script. Moreover, a process is provided as a guideline to aid analyst to accomplish high-quality service identification with the method.

As a formal or semi-formal service description model is necessary in the analysis phase, but right now it is difficult to describe the service with current language, such as WSDL. Designing an appropriate description language is necessary and challenge work in the future.

References

1. Schmidt, D., Stal, M., Rohnert, H., Buschmann, F.: Pattern-Oriented Software Architecture, Patterns for Concurrent and Networked Objects, vol. 2. John Wiley & Sons (September 2000); Specification (February 2006), `http://www.bpmn.org/`
2. Mattson, T., Sanders, B., Massingill, B.: Patterns for parallel programming. Addison-Wesley Professional (2004)
3. Schmidt, D.C.: Strategized locking, thread-saft interface, and scoped locking: Patterns and idioms for simplifying multi-threaded c++ components. C++ Report 11, 2–6 (1999)
4. Zalewski, J.: Real-time software architectures and design patterns: Fundamental concepts and their consequences, in Consequences. Annual Reviews in Control 2001, 133–146 (1999)
5. Ferrari, A.: JPVM: Network parallel computing in Java. Concurrency: Practice and Experience 10(11-13), 985–992 (1998)
6. Bergenti, F., Poggi, A.: Improving uml designs using automatic design pattern detection. In: Proc. 12th International Conf. Software Eng. and Knowledge Eng. (SEKE 2000), pp. 336–343 (2000)

Performance Study for Java Virtual Machine in Embedded Systems

Liu Wenjun

School of Information Engineering, Nanchang Institute of Technology (NIT), Nanchang 330099, China
myprivacy@sina.com

Abstract. Real time specification for Java (RTSJ) is intended for Real-time embedded applications. In this study, we present the architecture of a typical JVM, from the first generation, Interpreter architecture, to newest architecture, Just in time with Hotspot mechanism. Also, the structure of RTSJ can affect to the JVM for embedded devices have been analyzed.

Keywords: Real time, Virtual Machine, Java, RTSJ.

1 Introduction

Real-time specification for Java (RTSJ) defines how real-time behavior must occur within Java technology, APIs and semantic enhancements, this allows Java code developers to correctly reason about and control the temporal behavior of the applications. They are as follows:

- Time & Timers
- Memory Management Schemes
- Direct Access to Physical Memory
- Real-time Threads, Scheduling, and Synchronization
- Asynchronous Events Handling & Asynchronous Transfer of Control

However, from a JVM developer's perspective, it is not easy to apply those specifications to the real applications. The designs obstacles of a Real-time JVM are analyzed in this study through examine the architecture of a typical JVM and its components.

Section 2 discusses the Java Virtual Machine (JVM) architecture. Section 3 introduces challenges in JVM. And we get our conclusion in Section 4.

2 The JVM Architecture

As illustrated in Figure 1, the main components of a typical Java virtual machine are: Class loaders, Execution Engine, Garbage collector, Runtime data areas. They are described as follows:

- Class loaders are used to dynamically load application and library classes from a variety of sources such as the local file system and the network.

W. Zhang (Ed.): Advanced Technology in Teaching, AISC 163, pp. 807–811.
springerlink.com

- The execution engine is the heart of the virtual machine. It executes bytecode instructions loaded through class loaders. There are various existing types of execution engines. The details about those architectures will be analyzed in the section below.
- Garbage collector automatically reclaims the memory used by objects that are no longer referred to by the application that is running.

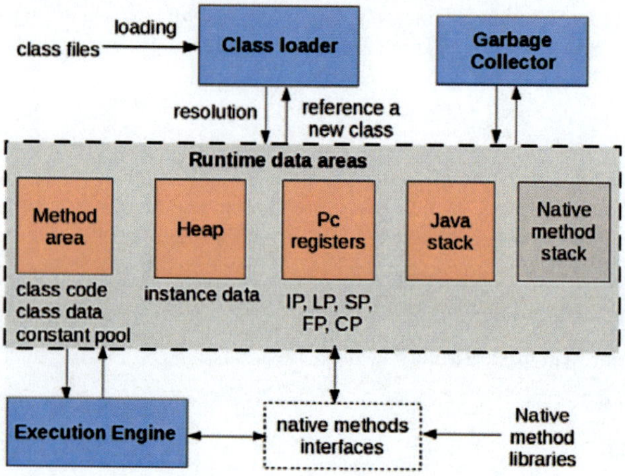

Fig. 1. The typical JVM Architecture

2.1 Execution Engine

1) Interpreter

The traditional execution engine of JVM is the Interpreter. In this architecture, Java bytecodes gradually feed into the execution engine, and are translated into native codes afterward. The main advantage of this method is its simplicity and ease to be implement. The disadvantage is its slow speed since every byte code is translated to many according native codes.

2) Just In Time (JIT)

A JIT compiler translates Java bytecodes to machine language instructions at the runtime. A JIT compiler compiles Java methods on demand. The JIT only compiles a code path when it knows that the code path is about to executed (as its name, just-in-time).

3) Dynamic Adaptive Compilation (DAC)

Some virtual machines use both a JIT compiler and an interpreter execution engine. The interpreter is used to execute the application initially. When the interpreter is executing, a profile process is also performed in order to determine which methods are used the most in the applications. Once such a method is identified, it is compiled at runtime. The strategy also helps gain the performance benefit associated with JIT compilation for those methods that are executed with most frequency, see Figure 2.

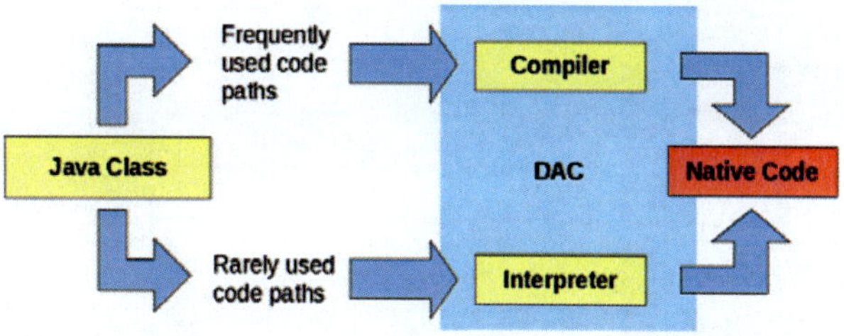

Fig. 2. The two way code paths in the DAC architecture

4) Ahead of Time Compilation (AOT)
In the AOT architecture, the whole java program is compiled to native code before execution process starts. That means all bytecodes must be compiled to native code, not just only the code path like JIT.

Table 1. Shows the general trend in start-up performance and steady-state performance between JIT and AOT

	Dynamic (JIT)	**Static (AOT)**
Start-up	Tunable, but not so good	Best
Steady-state	Best	Good
Interactive	Not so good	Good
Deterministic	Tunable, but not best	Best

2.2 Java Class Library (JCL)

The size and the structure of a Java Runtime Library contributes important role to performance of a Java virtual machine. This is because not only the time needed for loading JRL but also JRL structure can affect the code path of Java programs during the runtime.

Since Java class library contribute a big part in the total size of Java architecture. It is subjective of many researchers to reduce the size of a Java system. Table 2 shows the size and the number of the classes and of some well known JCLs.

Table 2. The size of some well-known Java class libraries

Name of JVM	**Size (MB)**	**Classes**
SUN JVM 1.6	41.708	16482
SUN JVM 1.4	26.311	9435
IBM JVM 1.4	22.152	15075
KVM/CLDC	0.179	116

2.3 Method Calls and Dynamic Bytecode Frequency

The numbers of method calls of a program is affected by the structure of the Java runtime library. Figure 3 shows the difference between method calls in Sun JVM 1.4.2 in the Linux version and the Windows version for programs.

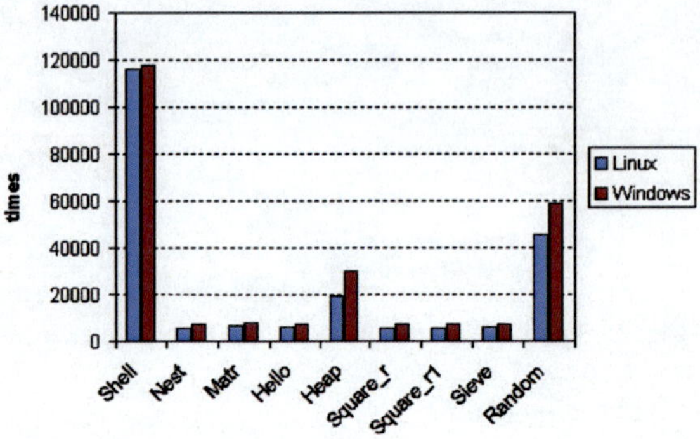

Fig. 3. The difference between method calls in Sun JVM 1.6.0 in the Linux version and the Windows version

The dynamic loading mechanism of a JVM, the way bytecodes are organized, can heavily affect to its performance. Table 3 shows the most dynamic bytecode frequency produced by KVM/CLDC – the popular JVM from Sun for embedded devices, in use for the "HelloWorld" program.

3 Challenges in Implementing a Real-Time JVM

Because of the constraints of embedded systems, the small size of a JVM is a compulsory requirement. Reducing the size of a JVM can be done either by reducing the sizes of its components or system class library or both of them.

Table 3. The highest dynamic bytecode execution frequency of the "HelloWorld" in KVM/CLDC

No.	Bytecode name	Numbers	Frequency (%)
1	aload_()	161	11.91
2	iload_1	93	6.88
3	invokespecial_fast	86	6.36
4	aload_1	70	5.18
5	return	61	4.51
6	iload_2	55	4.07
7	getfield_fast	45	3.33
8	invokestatic_fast	45	3.33

To reduce the system class library, we can apply some techniques: suppressing classes file, or introducing a small set of Java system class.

Another way to reduce the size of a Real-time JVM is to employ the ATO execution engine as it is applied in simpleRTJ[5] and Ovm[2]. However, it also eliminates the portability characteristic of the Java technology.

The RTSJ compliance features that mean the Real-time JVM must support real-time priorities, high-resolution timers, high-resolution clocks, and direct access to physical memory. The more Real-time facilities a JVM has, the more blurred the distinction between a JVM and an Oss duties is. The question of whether Java should be used in embedded real-time systems or not again arises.

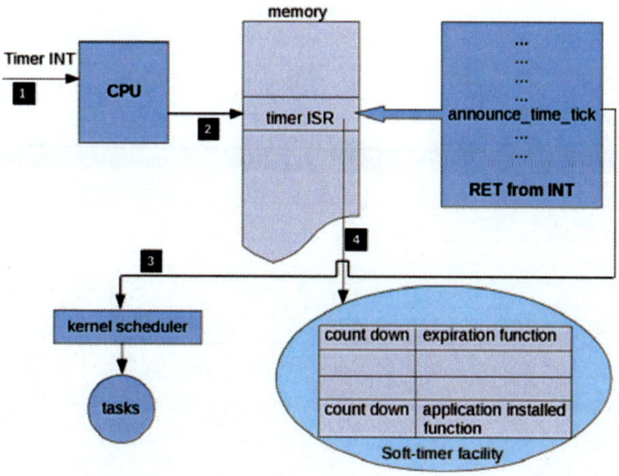

Fig. 4. Steps in servicing the timer interrupt in a typical Operating System

4 Conclusions and Future Work

The paper attempts to provide a better insight in designing a Real-time JVM by analyzing a typical JVM architecture and its components. Also, the technical challenges in designing a Real-time JVM for Real-time embedded systems are presented.

References

1. http://high-res-timers.sourceforge.net
2. Li, Q., Yao, C.: Real-time Concepts for Embedded Systems. CMP Books (2003)
3. La, V.Q.: Design virtual machine for java processing for a small embedded microprocessor core. Master's thesis, Offenburg University (2007)
4. Baker, J., Cunei, A., Flack, C., Pizlo, F., Prochazka, M., Vitek, J., Armbuster, A., Pla, E., Holmes, D.: Real-time java in avionics applications. In: Proceedings of the 12th IEEE Real-Time and Embedded Technology and Applications Symposium (2005)
5. http://www.rtjcom.com

A Study on Influence of Design Pattern in Shared Memory Multiprocessor Application

Liu Wenjun

School of Information Engineering, Nanchang Institute of Technology (NIT),
Nanchang 330099, China
myprivacy@sina.com

Abstract. With the emergence of multicore processors, parallel software is beginning to be used in the domain of application development. In this work, behavior of 2 software design patterns in a shared memory parallel environment is investigated regarding different aspects of parallelization including scalability, parallelizability and workload distribution.

Keywords: Design pattern, Parallelization, Multicore.

1 Introduction

As multicore processors become widespread parallelization is becoming an important concern for application software. It is more or less possible to use existing parallel programming models like PVM and MPI with multicore systems. However it is still necessary to merge methodologies currently used for developing application software and/or high performance software.

In our studies, experiments are performed on the behavior of observer, abstract factory and decorator in a shared memory multiprocessor system. The primary objective of our studies is to reason about the improvements that can be performed at design stage of software development by investigating the behavior of certain object oriented software structures in parallel environment.

Section 2 discusses the related work and quality concerns of object oriented software in multicore environments. Section 3 introduces design pattern implementations that are used and explains experiments that are performed over them to identify their parallel behavior. And we get our conclusions in Section 4.

2 Related Work

One of the most important concepts for today's object oriented software development is software design patterns by which the ossified solutions of frequently faced problems in object oriented design are abstracted by simple recipes of programming and design. Parallelization can be introduced without losing the focus on quality concerns about the software at hand. Singleton is an example to an improvement over software design patterns regarding parallelization/concurrency concerns. If the

W. Zhang (Ed.): Advanced Technology in Teaching, AISC 163, pp. 813–817.
springerlink.com

singleton pattern in its basic form [1] is used in a parallel system, programmer may end up in many singleton objects which contradict with the goal of the pattern. To overcome this, object creation portion of the code inside constructor can be treated as a critical section and that portion can be synchronized.

This is a typical problem of changing domains, which produces a flaw in software reliability even in a small, well-defined solution for a known problem in sequential domain. Looking from a broader perspective, many new problems may arise in many different quality aspects of former object oriented structures. This is our motivation behind investigating software design pattern behavior in parallel platforms.

3 Experiments on Patterns

Our experiments are performed in a 16 Intel 2.6GHz Xeon processored system with an operating system of Linux kernel 2.6 running on it. For the majority of the experiments, objects of the patterns are programmed as separate threads, which were assigned to processors explicitly under control of the programmer or automatically by the system scheduler depending on the type of the experiment.

3.1 Experiments on Observer

In this experiment the effect of multiple processors over a pool of identical workloaded observers is studied in order to see if distribution of observer objects upgrade the overall performance of the system.

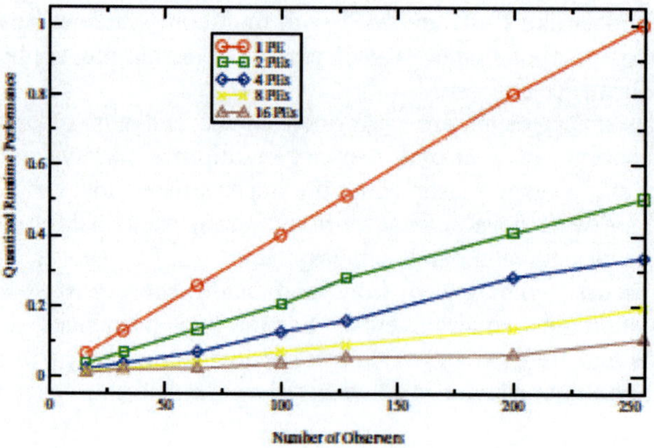

Fig. 1. Performance of the system for different number of Processing Elements

As seen in the Figure 1, pattern scales perfectly in the multiprocessor environment. In this example we can see that it makes sense to distribute each observer object during implementation for this pattern. Implementing observers as separate threads and distributing the objects to multiple processors improves pattern performance in a scalable way[6].

3.2 Experiments on Abstract Factory

The subject of our second set of experiments is "'Abstract Factory'" [1], we performed tests on more complicated structures than observer where different trade-offs can be included to see how factories behave under different kinds of configurations.

In order to prevent race conditions, we have implemented factories as critical sections since all of the creation requests go through them. Here, the bottleneck is the number of factories because it determines the number of possible parallel executions in practical. The upper bound for number of concrete factories is set to eight and number of processors are bound to four in factory experiments for simplicity.

1) Parallelization of Concrete Factories

In the first experiment, the effect of multiple processors over a pool of identical workloaded objects to be created are studied in order to see if distribution of concrete factories upgrade the overall performance of the system. Throughout the experiments, four different kinds of concrete factories are used to handle equal number of object creation requests from the clients which are distributed over different number of processors.

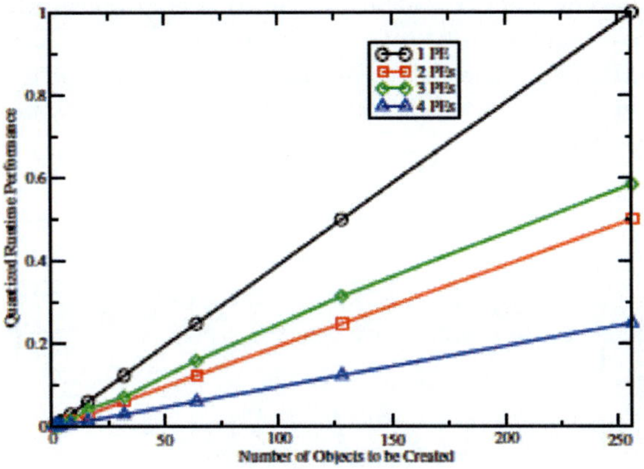

Fig. 2. Performance of the system for different number of Processing Elements

In Figure 2, pattern seems to scale well in a multiprocessor environment except in three processor case. For abstract factory pattern we again see that under certain circumstances it is not trivial to distribute the object creation process via concrete factories of the system.

2) Effects of Number of Concrete Factories in Parallelization

In the previous section we've ended up with the situation where a 2 processor system outperformed a 3 processor system even though the number of parallelized components were constant. We begin our experiments by fixing number of processors and experiment on effects of number of concrete factories(critical sections) in the system.

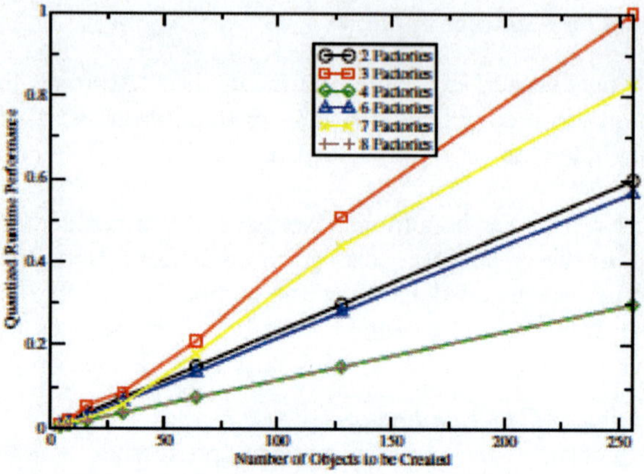

Fig. 3. Performance of the system for different number of Concrete Factories

In Figure 3 we can see that the performance of a four processor system with concrete factories creating idle workloaded objects, each assigned to a specific processing element. In the plot, systems with four and eight factories perform the best while systems with two and six concrete factories perform mediocre with respect to others and systems with three and seven concrete factories are clearly outperformed. In the second phase of our experiments we fix the number of objects to be created and analyze the relationship between the number of processors and number of concrete factories in the system.

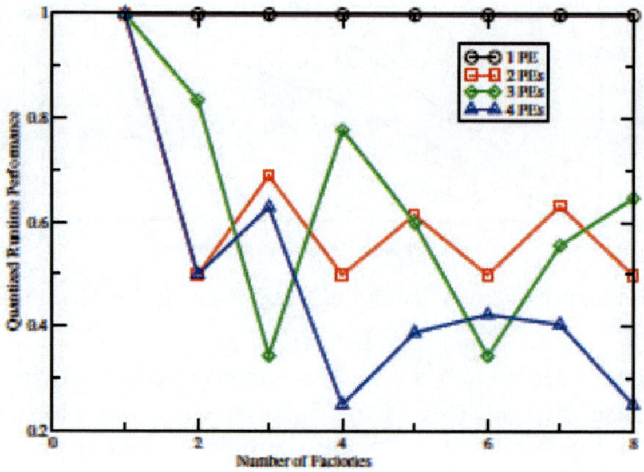

Fig. 4. Effects of Processing Element-Concrete Factory Relationship over the System Performance

In Figure 4 the peaks in the system performance always point to the number of factories which are multiples of number of processing elements in the system. For a two system with two processing elements, optimal number of concrete factories are multiples of two while for a three processor system this number becomes multiples of three and for a four processor system it becomes multiples of four. It is important here to remember object distribution is made with standard scheduling algorithm where each object that is requested to be created, its creator factory is assigned to the processor in a cyclic sequential way, more formally like in Equation (1).

$$affinity(Q_i) = i \mod k \tag{1}$$

where is function that sets the affinity for each object to be created in the system and k is the number of processing elements available. This leads us to the conclusion to bind and not to migrate much frequently accessed synchronized objects to specific processors.

4 Conclusions and Future Work

Main goal of our studies is to find the quality attributes of object oriented software on multicore systems. Our experiments explore important points of software design patterns on refactoring for multicore systems. Experimental results show that it is possible to implement a parallelization strategy during design stage of object oriented software for multicore systems.

We will use our results to investigate more general structures in object oriented software design that becomes important in parallelization.

References

1. Gamma, E., Helm, R., Johnson, R., Vlissides, J.: Design Patterns: Elements of Reusable Object-Oriented Software. Addison-Wesley Professional (1994)
2. Schmidt, D., Stal, M., Rohnert, H., Buschmann, F.: Pattern-Oriented Software Architecture, Patterns for Concurrent and Networked Objects, vol. 2. John Wiley & Sons (September 2000)
3. Mattson, T., Sanders, B., Massingill, B.: Patterns for parallel programming. Addison-Wesley Professional (2004)
4. Schmidt, D.C.: Strategized locking, thread-saft interface, and scoped locking: Patterns and idioms for simplifying multi-threaded c++ components. C++ Report 11, 2–6 (1999)
5. Zalewski, J.: Real-time software architectures and design patterns: Fundamental concepts and their consequences, in Consequences. Annual Reviews in Control 2001, 133–146 (1999)
6. Ferrari, A.: JPVM: Network parallel computing in Java. Concurrency: Practice and Experience 10(11-13), 985–992 (1998)
7. Bergenti, F., Poggi, A.: Improving uml designs using automatic design pattern detection. In: Proc. 12th International Conf. Software Eng. and Knowledge Eng. (SEKE 2000), pp. 336–343 (2000)

Land Quality Comprehensive Evaluation on Land Consolidation of Discrepancy Topography

Qu Chenxiao[1], Meng Qingxiang[2], Lin Yan[3], and Cai Congfa[1]

[1] College of Resources and Environment,
Huazhong Agricultural University Wuhan 430070, China
cfcai@public.wh.hb.cn
[2] College of Resources Environment,
Henan Agricultural University, Zhengzhou 450002, China
[3] College of Urban and Environment Science,
Central China Normal University, Wuhan 430079, China

Abstract. According to the characteristics of land consolidation project, this paper established the quality indicators of land consolidation from micro-topography, soil, water, landscape and other factors. The weights of these indexes were reckoned by AHP, and then the land quality based on the land consolidation project about Baiyang town and Guqiao township were evaluated, which both are hilly area in Henan province. The results showed that discrepancy topography had been changed significantly before and after land consolidation in the level of spatial distribution, size and overall quality. Land consolidation had a great impact on land quality, and project area in Baiyang town was up to 75.7% and Guqiao township to 37.3%.

Introduction

"Land Consolidation", originating in Germany Since the mid-16th century and through years of practice in the Netherlands, Finland and other European countries, has accumulated a wealth of experience, and gradually form a complete system [1]. Land consolidation in its true sense was formally proposed and operated till the late 90s of last century in china, marked by establishing Land Consolidation and Rehabilitation Center of National Land Agency in 1998 (renamed Ministry of Land and Resources in 1999) and launching the first land consolidation project in 2001. Land consolidation was being generally launched nationwide with billions of RMB invested[2-5]. The important factors including land use type, soil physical and chemical properties, road accessibility, drainage and irrigation conditions would been improved through measures of land leveling, farmland capital construction, land regularization and parcel merging[6-9], and the same as the effective area and quality of arable land after Land Consolidation project, Bindraban [10] pointed out that the interaction of LQI determined land quality, and also had scale effect. Ryder[11] and Leibig & Doran[12] found that there were a high degree of agreement between farmers view and experiment test results at farm and site scales when they studied in

*Corresponding author.

W. Zhang (Ed.): Advanced Technology in Teaching, AISC 163, pp. 819–828.
springerlink.com

the U.S.; Jinliang Ouyang[13] also reached a similar conclusion that they emphasized the importance of the behavior and views of peasant households in the evaluation of land quality.

Compared with its foreign counterparts, land quality research was started rather late in our country, and a comprehensive study system has not been formed. The studies mainly stayed in the construction of index system, which were not only weak in the mechanism and practical application but also lacked of long-term fixed-point monitoring data for assessment. From the standpoint of the department, Ministry of Land and Resources proposed agricultural land regulations [14].

1 Project Outline, Research Method and Date

1.1 Project Outline

Project I is located in Baiyang township Yiyang county in the southeast of Henan Province, involving 3 administrative villages of Tugou, Honggou and Longwo. Micro-topography is hill, and agrotype is red clay which is adhesive and with poor tilth and poor nutrient. Agriculture production is undeveloped and irrigation system leaves much to be desired. Layout of the field road, with large occupying area, is very unreasonable and most roads are poorly maintained, which hindered the development of agricultural production. The project is a key provincial investment for land consolidation with a construction period of 1 year and was completed in 2006. Construction scale was 153.00hm^{2,} including new land 32.84hm^2 which accounts for 21.46%.

Project II, is located in Guqiao Township belonging to Changge County in Henan Province, Involving 2 administrative villages of Caiwang and Xuwangzhao. The whole region is on the transition belt between the mountainous area in west of Henan and the plain regions in southeast of Henan. Northwest was the landscape High-Low southeast, showing a slow tilt changes. Groundwater is the main source of industrial and agricultural water use for the scarce surface water resource. Natural vegetation in the field characterized with secondary nature, is mainly covered with weed. Irrigation system is imperfect. Agro type is Fluvo-aquic soil with weak structure, permeability and poor capacity of water and nutrient. The project also is a key national investment for land consolidation with a construction period of 1 year and was completed in 2006. The construction scale was 214.59hm^2, including new land 24.6hm^2 which accounts for 11.24%.

1.2 Data Information Collection and Processing

Data information was obtained through the investigation of relevant units and local people, field observation, soil samples collection and laboratory analysis. Text information consisted of project feasibility research report, planning and design report and final acceptance report. Graphic information included 1:5000 land use maps before and after land consolidation and topographic maps of the project area.

After digitizing by MAPGIS, ARC/INFO, the village road maps before and after land consolidation was constructed on the basis of present land consolidation maps, the planning maps and the implemented land use maps; the irrigation guaranty ration maps and the parcel merging maps were drafted combining with the fieldwork and household surveys; the slope maps were created from the topographical maps; and the soil depth, the soil texture and soil organic matter dot distribution maps were drafted. Then the land use structure changes before and after consolidation was analyzed and land use degree were calculated, finally the changes in land quality were worked out.

2 Land Quality Evaluation on Land Consolidation Project

2.1 Establishment of Evaluation Index System

Before and after land consolidation project was due to differences in land quality caused by project construction. The differences of land quality before and after land consolidation were due to project construction. land consolidation works included land leveling, irrigation and water conservation, the field path and the shelter forest construction. The land leveling project generally included the cubic meter of earth and stone to excavate, the backfill, the transportation, the land leveling and so on. The topographical condition was meaningful to farmland quality evaluation for the macroscopic region. To the hilly area land arrangements project, the land leveling project would change the micro-relief landform inevitably in project area, and then influenced land quality.

According to the land quality evaluation indicator selection principle and above-mentioned analysis and combining with characteristic of land arrangements project, this study selected 4 factors, 11 indices, and established a three levels land quality evaluation indicator system (Table 1).

Table 1. The evaluation index system of land quality

Object	item	index
Land Quality Impact Assessment of Land Consolidation A	Micro-topography factors B1	Flatness of land C1 Production convenient degree C2 Entire degree of field block gauge C3
	Soil factors B2	Topsoil thickness C4 Soil fertility level C5 Land utilization ratio C6
	Water Factors B3	The degree of the Water source guarantee C7 The degree of irrigation guaranteesC8 The degree of draining water guarantees C9
	Landscape factors B4	Landscape productivity C10 Landscape aesthetic degree C11

The first layer was object, land quality impact assessment of land consolidation. The second layer was item, including micro-topography, soil, water, landscape and other factors, and the third layer was the index level (index), that was what each of the criterion level indexes.

Table 2. Index Classification

Evaluation Index	Index description method	Index characteristics and Classification standards	Rank
Flatness of land	May describe the land leveling situation, flatness of land is high, and high ease of mechanization and irrigation.	Slope <10,levelling degree is high[69]	1
		Slope 10—30,levelling degree is little high[69]	2
		Slope 30—50，leveling degree is mid	3
		Slope >50,levelling degree is bad	4
Production convenient degree	Reflect the production convenient situation before and after arrangement qualitative description	convenient	1
		inconvenient	2
Entire degree of field block gauge	Reflect the fragmented plots before and after arrangement breaking situation qualitative description	regular	1
		Irregular	2
Topsoil thickness	Soil indicators can be measured, its value is too small affecting crop growth quantitative description	25—40cm	1
		15—25cm	2
		Less than 15cm	3
Soil fertility level	Reflect the degree of soil fertility improvement after arrangement qualitative description	Fertile soil	1
		The soil fertility level is little high	2
		The fertility level is little low	3
		The soil is barren, the fertility level is low	4
Land utilization ratio	Reflect the degree of the land utilization before and after arrangement quantitative description	>90%, Land utilization ratio Is high	1
		85%—90%，little high	2
		80%—85%，middle	3
		〈80%，little low	4
The degree of the Water source guarantee	Reflect the guarantee situation of the irrigation water source before and after arrangement qualitative description	The irrigation water source is sufficient	1
		The irrigation water source is limit	2
		No irrigation water source	3
The degree of irrigation guarantee	The conditions of irrigation can be reflected by data, the more too close to 100%, the guaranteed degree is higher quantitative description with the percentage quota	The guarantee degree of irrigation is more than 90%	1
		75%—90%	2
		Less than 75%	3
The degree of draining water guarantees	Drainage conditions can be described by the combined effect of topography and drainage	Have perfecting facilities of draining away water, can discharge in time	1
		The drainage facility is ordinary, the	2

2.2 Evaluation Index Classification

In land quality evaluation of the land arrangements' target description, some evaluation indicators such as topsoil thickness and so on could be described quantificationally, and some evaluation indicators such as production convenience and so on needed to be given the qualitative description. The evaluation Index Classification for details was shown in Table 2.

2.3 Determination of Evaluation Index Weight

The determination of the weight of farmland quality evaluation factors, should been given according to the actual state of the farmland in project area and Combined with actual agricultural production affection extent of the evaluation factors, the weight of the evaluation factors were determined with the hierarchical analysis method. May also use regression analysis, principal component analysis, Delphi method and so on.

2.4 Determination of the Evaluation Factors Index Score

In the selected land arrangement of land quality evaluation index system, some factors could be partitioned quantitatively, some factors could only be partitioned qualitatively. In order to operating conveniently, four levels of evaluation factors were assigned with 100-10.

Table 3. Factor level-Score contraposition

	100	90	80	70	60	50	40	30	20	10
Flatness of land	1		2		3					
Production convenient degree	1		2							
Entire degree of field block gauge	1				2					
Topsoil thickness	1		2		3					
Soil fertility level	1		2		3		4			
Land utilization ratio	1		2		3		4			
The degree of the Water source guarantee	1		2		3					
The degree of irrigation guarantee	1		2		3					
The degree of draining water guarantees	1		2		3					
Landscape productivity	1		2		3		4			
Landscape aesthetic degrees	1		2		3					

2.5 The Standardization of Evaluation Factors Index

The study on land quality influence of land consolidation is aimed at revealing the implementation influence type and degree to project area. There were few studies about land arrangements' land quality influence, and there was not the unification evaluation system at present. Simultaneously, the different evaluation system's evaluating indicator's standardization used different method. This study based on status background value of the project area before the land arrangements, and then compared with index values after land consolidation. The land quality target value before land consolidation defaults 1.

Index standardized calculation method: $F = X_{after} / X_{before}$

After index value standardization, the F value was greater than 1, showing land quality index value enhanced; The F value was equal to 1, which explained that the land quality index value did not change; The F value was less than 1, which explained that the land quality index value reduces.

2.6 Determination of Evaluation Model

Comprehensive evaluation model: $E = \sum_{i=1}^{n} F_i * W_i$

In formula: E was the comprehensive evaluation value of land quality; Fi was standard valuesof the index i; Wi was the weight of the index i.

3 Results of Evaluation

According to the above methods, through field surveys and data analysis of the project areas, firstly assign and standardize the land quality indicators, and then assign different Index Weight, finally calculate the comprehensive evaluation results

3.1 Evaluation Results of Land Quality in Guqiao Township of Changge

3.1.1 Factors

Micro-topography factors: Generally, land consolidation region in Guqiao township of Changge was plain, but micro-topography undulating. The land consolidation projects were mainly about land leveling, and re-planning the layout of the plots. After land consolidation finished, flatness of land was improved, and the size and layout of the field plot was reasonable, in favor of the lighting for field crops, the facility for agricultural mechanization and irrigation drainage and increasing the effect of the wind defense.

Soil factors: the project of the land formation effectively improved the soil plow layer thickness, soil ability to retain water and nutrients, and the content of soil organic matter. Meanwhile, the land consolidation project would change brick and tile kilns, waste ponds and other terrain into arable land, which increased the effective area of cultivated land , improved the efficiency of the project area land, and finally enhanced the quality of the all project area of the land.

Water factors: The length of low-pressure pipeline buried in project area was 15.75km; the drainage channel was 10.547km, and flood control standard was up to against once-in-50 years flood. Through the construction of water conservancy

Table 4. The evaluation index system of land quality and Standard value (I)

object	item	indictor	Before consolidation	After consolidation	Standard
Results of the quality assessment of consolidated land A	Micro-topographic factors B1	Flatness C_1	80	100	1.25
		Facilitation C_2	80	100	1.25
		Regularity C_3	60	100	1.67
	Soil factors B2	Topsoil thickness C_4	80	100	1.25
		Fertility C_5	60	80	1.33
		Land utilization C_6	80	100	1.25
	Water factors B3	Water level of assurance C_7	60	100	1.67
		Irrigation level of assurance C_8	60	100	1.67
		Drainage level of assurance C_9	60	100	1.67
	Landscape factors B4	Landscape productivity C_{10}	80	100	1.25
		Landscape aesthetics C_{11}	50	90	1.8

projects, the margin of the water level of the project area was effectively improved, irrigation guarantee rate was improved to 100%, and the flood control capacity has also been enhanced.

Landscape factors: Through land consolidation, the project area was changed into high-quality farmland which had high flatness, high soil ability to retain water and fertilizer, high level of mechanization, appropriate sale, highly efficient irrigation and drainage. The food production capacity and production capacity in the project area were also improved. Scientific planning of terraces, rational distribution of agricultural and farmland protection forest road construction had improved the infrastructure of the project area. And it also made the terraces flat, the drainage interlinked; the roads connected into a network of forests, changed the former messy land utilization and promoted the agricultural landscape aesthetic degree.

3.1.2 Results of Comprehensive Assessment

Table 5. The result of Comprehensive evaluation (I)

	Category Index				The result of Comprehensive evaluation
	Micro-topographic factors	Soil factors	Water factors	Landscape factors	
After consolidation	1.318	1.282	1.667	1.433	1.373

As is shown in Table 5, after land consolidation, quality factors of the project land have shown increasing trend, especially to water factors which had increased as much as 66.7%. The quality of integrated land had significantly been improved as much as 37.3%. It indicated that the management of the project areas had an obvious positive effect on land quality.

3.2 Results of the Quality Assessment of Land Consolidation in Baiyang Town, Yiyang County

Table 6. The evaluation index system of land quality and Standard value

object	item	index	Before finishing	After finishing	Standard
Results of the quality assessment of land consolidation A	Micro-topographic factors B1	Flatness C1	40	100	2.5
		Facilitation C2	80	100	1.25
		Regularity C3	60	100	1.67
	Soil factors B2	Topsoil thickness C4	60	80	1.33
		Fertility C5	40	60	1.5
		Land utilization C6	80	100	1.25
	Water factors B3	Water level of assurance C7	60	100	1.67
		Irrigation level of assurance C8	60	100	1.67
		Drainage level of assurance C9	100	100	1
	Landscape factors B4	Landscape productivity C10	40	80	2
		Landscape aesthetics C11	50	90	1.8

3.2.1 Factors

Micro-topography factors: the land consolidation project, located in Yiyang county Poplar town, was hilly and micro-terrain showed fluctuations. Through land consolidation, the terraces were constructed , the former slope become smooth piece of paddy field and the layout of agricultural land and roads production were re-planned and adjusted. After land consolidation, flatness of land was improved and land size and layout became reasonable, which contribute field crops to lighting; and new roads were built to facilitate the farmers for mechanized production and to improve the ease of agricultural production.

Soil factors: by converting steep land into terraced fields, the risk of soil erosion reduced, the soil ability to retain water and nutrients was improved effectively, which enhanced the soil organic matter content. Meanwhile, the land such as mud flat or waste land were flatted into arable land , increasing the effective area of cultivated land, improving the project area's land use internsityand enhancing the overall quality of the project area of the land.

Water factors: two stations and two Pumping wells were newly built, which made the project area with sufficient water to meet irrigation requirements and to improve the water level of assurance. Meanwhile, the new agricultural drainage, which was 7.09 km long, made each plot irrigated better, increased drought tolerance, and improved irrigation guarantee rate. Natural conditions of drainage were very good in project area, so there were no constructions on the drainage.

Landscape factors: After land consolidation finished, the original steep slopes were transformed into well-structured trapezoidal field, and there was a new farmland shelterbelt. Project area of slope cropland was changed into land which had high formation and the modest size. These enhanced the capacity of the soil retain water and nutrients and drought, increased the level of mechanization of agricultural production, and greatly improved the food production capacity and production capacity in the project area. Construction of terraces, the re-layout of the road and the construction of farmland, made terraces regular, network distribution uniform, and windbreak structure good. These also changed the disordered land-utilization before consolidation. The project area of the landscape features was substantially improved.

3.2.2 Results of Comprehensive Assessment

As shown in Table 7, after land consolidation, quality factors of the project land had shown increasing trend, especially in micro-topography and landscape which were as high as 99.3% and 93.3%. The quality of integrated land had significantly improved, as much as 75.7%. It indicated that the management of the project areas had a obvious positive effect on land quality.

Table 7. The result of Comprehensive evaluation

	Category Index				The result of Comprehensive evaluation
	Micro-topographic factors	Soil factors	Water factors	Landscape factors	
After finished	1.993	1.379	1.589	1.933	1.757

4 Conclusion and Discussion

4.1 Conclusion

It's one of the main objectives of land consolidation to improve land productivity. Empirical research showed that: with the implementation of land consolidation and analysis of various factors, the project significantly improved the land quality, factors of which had shown increasing trend. In Guqiao township of Change, the plain area, water factors increased particularly by as much as 66.7%, and the land quality raised by as much as 37.3%; and In Baiyang township of Yiyang, the hilly region, especially factors of micro-topography and landscape increased by as much as 99.3% and 93.3% respectively, and the land quality improved as much as 75.7%.

4.2 Discussion

Studies on land quality of land consolidation belonged to a new work, with the characteristics of the relatively small range of subjects, the data of quantitative indicators being difficult to obtain, big variations in land quality before and after land consolidation and so on. The land quality evaluation method used by this paper had better reflected on the influence extent over land consolidation to land quality, but qualitative indexes held quite large proportion, making the final evaluation result strong subjective because of relatively small scope of the evaluation, the data of quantitative indexes difficult to obtain and land quality with large fluctuations of the objects. Therefore, land quality research of land consolidation should be paid much attention to avoid evaluation subjectively and establish a more scientific system of land quality indicators, to ensure an objective evaluation results.

Due to its nonlinearity, integrity and complexity, land quality assessment and investigation mathematical methods and modern technology to carry out[15]. ①land quality evaluation method in land consolidation: Based on principal component analysis, modified weighting method, comprehensive evaluation method and multi-angle single-factor evaluation method, summarize relative standards for land quality evaluation and then establish operational land quality assessment appropriate for land consolidation; ② land quality investigation indicators and methods in land consolidation: Based on the analysis of land quality property influenced by land consolidation, construct the land quality index system combined with "3S" technology and ground investigation and explore technological methods of land quality in land consolidation by use of "3S" technologies.

References

1. Wang, J., Yu, L., Luo, M., Zhai, G.: The overview of Land Consolidation. Areal Research and Development 2, 8–11 (2003)
2. Lu, X.: General theory of land consolidation strategies. Transactions of the Chinese Society of Agricultural Engineering 18(1), 1–5 (2002)
3. Gao, X.: Theory and Practice of Land Consolidation. Geilogical Publishing House, Beijing (2003)
4. Wu, Z., Liu, M., Davis, J.: Land consolidation and productivity in Chinese household crop production. China Economic Review 16, 28–49 (2005)
5. Planning Division in Ministry of Land and Resources, Land Consolidation and Rehabilitation Center. Planning examples for land development and rehabilitation. Geilogical Publishing House, Beijing (2001)
6. Crecente, R., Alvarez, C., Fra, U.: Economic, Social and Environmental Impact of Land Consolidation in Galicia. Land Use Policy (19), 135–147 (2002)
7. Coelho, J.C.: A systems approach for the estimation of the effects of land consolidation projects (LPCs): a model and its application. Agricultural Systems 68(3), 179–195 (2001)
8. Ishii, A.: The methods to consolidate scattered tenanted lots into large rice paddy lots by the land consolidation projects in Japan. Paddy Water Environ. (3), 225–233 (2005)
9. Planning department of country land administration ministry, et al. Land consolidation reference at home and broad, pp. 291–380. Chinese land publishing house, Beijing (1998)
10. Bindraban, P.S., Stoorvogel, J.J., Jansen, D.M.: Land quality indicators for sustainable land management: proposed method for yield gap and soil nutrient balance. Agriculture, Ecosystems and Environments 81, 102–103 (2000)
11. Reder: Land evaluation for steepland agriculture in the Dominican republic. The Geographical Journal 160(1), 74–86 (1994)
12. Leibig, M.A., Doran, J.W.: Evaluation of Point-scale assessments of soil quality. Journal of Soil and Water Conservation 54(2), 510–518 (1999)
13. Ouyang, J., Yu, Z., Zhang, F.: The soil quality changes and peasant behavior analysis based on eco-economic regions. Acta Ecologica Sinica 23(6), 1147–1155 (2003)
14. Ministry of Land and Resources(PRC), Regulations for gradation on agricultural land. Standards Press of China, Beijing (2003)
15. Wang, J.: Implications for Land Consolidation Research of China from Land Quality Development. Areal Research and Development 25(6), 108–111 (2006)

The Effect of a Ratings-Based Approach to Measuring Corporate Social Responsibility

Jun Ma

College of Business Administration
Nanchang Institute of Technology
Nanchang, China
majun5101227@163.com

Abstract. The ratings-based methodology employed by Socially Responsible Investment (SRI) indices in their selection processes excludes many corporations by creating limited-membership lists. This received ratings-based structure is yet to offer an incentive for most of the excluded corporations to invest in improving their levels of Corporate Social Responsibility (CSR). We, therefore, investigate under what circumstances a ratings-based method for assessing CSR could provide an incentive to firms excluded from SRI indices to invest in CSR. In this article, we attempt to offer a theoretical reply to this question. We argue that when all firms are publicly ranked according to SRI index parameters, such indices can indeed create a market incentive for increased investment by firms in improving their performance in the area of social responsibility. We further show that this incentive tapers off as the amount of investment required exceeds a certain point or if the amount of payback on that investment fails to reach a certain threshold.

Keywords: Corporate Social Responsibility, Social Responsible Investment, Corporate Social Performance, Corporate Governance Code.

1 Introduction

A recent survey of Corporate Social Responsibility (CSR) theory covers: the relationship between a business and its larger environment; the attitudes of corporate upper-echelons toward CSR; the effects of board members' demographic and non-demographic characteristics on their inclination to support CSR strategy and policy; and the link between a firm's commitment to CSR and its financial performance. The central focus of this article lies within the fourth dimension, theorizing on the possible link between the incentive firm has to obtain an SRI index rating and its willingness to invest in CSR. We note that the measurement systems used by socially responsible investment (SRI) indices vary (the unit of analysis is in the micro level), as do the rating methods that they employ (the unit of analysis is in the macro level), and that both offer inconclusive results. Most studies focus on the micro level, that is, on how a firm's level of CSR is measured, an example being an examination of the weighting system used by the Domini Social Index. By contrast, at the macro level, a few studies focus on which firms are included or excluded from the indices. We expand on the latter, to which we turn now.

W. Zhang (Ed.): Advanced Technology in Teaching, AISC 163, pp. 829–837.
springerlink.com

The question we tackle in this article is: How can a ratings-based method for assessing CSR provide an incentive to firms excluded from SRI indices to invest in CSR?

The major goal of this article is to propose that, having failed to achieve any published ranking, it makes no difference, whether excluded companies perform little in the way of CSR or a lot. In either case, their efforts remain unknown and, therefore, can have little impact on investor decision-making. Thus, businesses excluded from SRI lists are left with no incentive, whatsoever to improve on their CSR, and thus the marketplace is deprived of growth with respect to the number of firms exhibiting CSR in the marketplace. Our proposition is based on our theoretical model, which gauges the effect of SRI on firm behavior.

The structure of the article is as follows. Section "Literature Review and Hypotheses" mainly introduces literature review and hypotheses of the research. Section "Research Method" consists of a brief survey of the methodology used by some SRI indices to measure CSR and Corporate Governance Code (CGC). Section "Theoretical Model" presents our theoretical model, which describes the effect of a ratings-based approach to measuring firm social responsibility on firm behavior in the realms of CSR and CGC. Section "Conclusion and discussion" of the article discuses the economic merits of firms investing in improving their social responsibility rating as revealed by our model.

2 Literature Review and Hypotheses

2.1 Literature Review

Investment in the stocks of firms, which commit themselves to better environmental or social ends (i.e., CSR), as well as in the stocks of corporations, which adopt best practices (e.g., CGC), has been growing steadily (Camejo,2002).By way of illustration, several important indices, such as the Domini Social Index, choose to screen out firms connected with specific industries, such as the tobacco industry and production of alcohol, and to include firms which commit to environmental or social actions. Thus, an investment trend that takes CSR and firm best practices seriously is underway and, contrary to ''market wisdom'', such investments perform well in the stock market over time, for example, the Domini Social Index outperformed the S&P 500(Camejo,2002).

According to Schuler and Cording (2006), consumers consider it important to know about the social actions of firms. A suggestion made by their proposed model, which links the corporate social performance of a firm to its financial performance, is that information provided by external sources about the social performance of a firm will have a higher intensity impact on existing or potential consumers than will information provided by the firm itself. Their model also suggests that the greater the extent of information diffusion regarding a firm's social actions, the greater will be the intensity of its impact on these consumers.

Similarly, knowledge about the social performance of a firm is quite important to investors, individual and institutional alike. Institutional investment selection based on corporate social performance is quite significant in both the U.K. and the U.S. Institutional investors in the U.K. are subject to a set of regulatory, institutional and social pressures to encompass social performance in investment selection (Cox et al., 2004). In the U.S., SRI is no longer an option for institutional investors, but an imperative, argues Prakash Sethi (2005), since SRI decreases the long-term level of risk on the investment and concerns the long-term survival and growth of the firm. Given the sizes of the funds controlled by institutional investors, this fact impacts the structure of corporate ownership. The link between SRI and changing ownership structures is quite remarkable and, indeed, SRI has an impact on the conduct of businesses by the mere fact that the role it plays in the market leaves no room for real choice. Thus, SRI by institutional investors drives corporations to adopt a strategy of CSR in the market (Cox et al., 2004; Prakash Sethi, 2005; Solomon et al., 2002).

Furthermore, survey results indicate that the phenomenon of SRI positively affects investor attitudes toward socially responsible conduct by corporations (Fischhoff et al., 2001), as well as affecting corporate performance and the improvement of corporate environmental reporting. Environmental reporting is, indeed, used in the market to analyze the conduct of a corporation and, in turn, to provide external impetus to improve performance (O'Rourke, 2003).The merit of the reported data lies in them improving measurement credibility with respect to a firm's performance in the area of corporate responsibility, given that "what gets measured gets managed" (Dillenburg et al., 2003).

We note that the trend toward SRI is constantly being evaluated through assessment of the performance of SRI indices in comparison with the performance of other indices (O'Rourke, 2003). As a result of this public exposure, SRI indices are in a position to influence investor decision-making. Yet, it is apparent that a little to no research attention has been directed toward investigating the effect such indices have on the firms excluded from them, and thus on the market's ability to encourage CSR among all firms.

2.2 Hypotheses

Our hypothesis is that the CSR ranking methodology currently used by SRI indices fails to encourage firms excluded from these indices to invest in CSR. Our proposed theoretical model of the effect that a firm's social responsibility rating has on firm utility illustrates this problem situation and offers a clue to its resolution. We propose a model, which is within the scope of CSR and SRI. The SRI indices are used to rank the CSR of a limited selection of traded businesses. Thus, for this limited group of firms, financial gains are linked directly to SRI in a traded firm. The literature on SRI connects the financial performance of the firm to its CSR. We examine an additional measure: the commitment of a firm to CSR regardless of financial gains.

3 Research Method

Here, we undertake a brief survey of the methodology used by some SRI indices to measure CSR and CGC. In contradistinction with Friedman's view (i.e., that externalities should not be internalized), the triple bottom line measurement of business social responsibility includes environmental and social parameters in addition to financial considerations. Thus, the assessment of a profitable business may be contingent on the impact it attempts to make on all these parameters. Among the financial indices that have taken up this suggestion are the Domini 400 Social Index (limited to 400 companies), the FTSE4Good Index Series (covers the largest 50 or 100 companies appearing in the more-inclusive FTSE4Good benchmark index), and the Dow Jones Sustainability Index (includes 200 companies, and aims to represent the top 10% of leading companies committed to sustainable practices). Thus, all these indices select a chosen group of business organizations, which commit themselves to socially responsible management more than do other businesses.

In order to illustrate the ranking methodology used with respect to business corporations, consider the Domini 400 Social Index, as measured by KLD Research. KLD seeks to maintain the composition of the index at approximately 250 S&P companies, 100 non-S&P companies chosen for sector diversification and market capitalization, and 50 additional companies with exemplary social and environmental records. Its ranking method utilizes a number of exclusionary and qualitative social screens. The exclusionary screens eliminate companies involved in specific industries, namely: adult entertainment, alcohol, tobacco, firearms, gambling, nuclear power, and military weapons, from the index. Companies that do not meet KLD's financial screens (relating to market capitalization, earnings, liquidity, stock price, and debt to equity ratio) are also ineligible for inclusion in the index. The remaining companies are then evaluated (in the context of their industry and sector, as well as in relation to the broader market) with respect to a number of issues, each of which is assigned a different weight in determining the company's overall ranking. KLD assesses the following issues: Community Relations, Corporate Governance, Diversity, Employee Relations, Environment, Human rights and Product Quality and Safety. For each of the issues, KLD measures several positive parameters (strengths) and several negative parameters (concerns) .For example, with respect to the Environment, the strengths are in the areas of: clean energy, beneficial products and services, pollution prevention, recycling, and others. The concerns are: hazardous waste, regulatory problems, ozone depleting chemicals, substantial emissions, agricultural chemicals, climate change, and others .

Since, KLD maintains the list at 400 companies by adding a new company to the list only in order to replace a removal, and since it further maintains a specific compositional balance within the list, it is clear that many socially and environmentally responsible companies may not make it to the list. Indeed, this is explicitly acknowledged by KLD, who note (in their FAQs sheet) that 'The DSI is not meant to be the 400 best companies nor is it the only 400 companies that meet social criteria'. However, the import of this statement (and the similar state of affairs that exists with respect to other tradable SRI lists) is that companies that fail to be included, for whatever reason, become more-or-less invisible to investors in terms of their CSR efforts.

4 Theoretical Model

Assume $U(\Pi\ , S)$ is the firm utility, where Π is the firm's profit and S is a firm's social responsibility rating, i.e., the number of points a firm has on asocial responsibility index (higher points=higher rating). We assume that S designates the commitment of a firm to social responsibility beyond financial gains.

For simplicity, we define linear utility so that:

$$U(\Pi\ ,\ S) = a\Pi + (1-a)S\ ,\quad 0 \le a \le 1, \tag{1}$$

where a and $1-a$ are weighting factors for the firm's profit and for its social responsibility rating, respectively. If $a=1$, the firm derives no utility from its social responsibility rating.

Since, social responsibility has a financial cost, we assume that S negatively affects firm profit Π. On the other hand, S positively affects company image and reputation. An increasingly positive image and reputation are mediator variables that positively affect the financial performance of the firm.

For simplicity, we describe the profit function as follows:

$$\Pi = M + cS^2 - dS \tag{2}$$

where M is the firm's profit, excluding the effect of its social responsibility rating. Thus M relates to sales, production costs, research and development costs, etc.

The profit function takes into account both the negative effect of social responsibility on profit (because of the investment it requires from the firm; this is reflected in the scalar $-d$) and its positive effect on profit (due to its positive effects on image and reputation; these are reflected in the scalar $+c$).Fig 1 shows profit as a function of social responsibility rating.

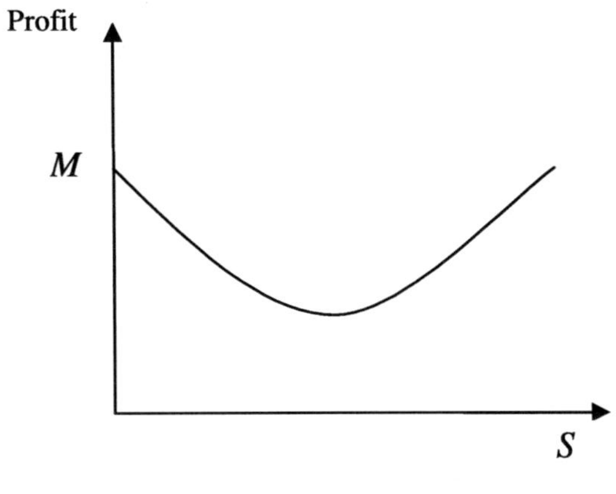

Fig. 1. Effect of a firm's social responsibility rating on its profit

From (1) and (2) we obtain:

$$U(\Pi\ ,\ S) = a(M + cS^2 - dS) + (1-a)S\ ,\ 0 \le a \le 1 \tag{3}$$

Using the equation for marginal utility, we can derive the derivative of the social responsibility term, which then leads to the minimum utility.

$$\frac{\partial U}{\partial S} = 2acS - ad + 1 - a = 0 \tag{4}$$

Rearranging yields an equation for S:

$$S = \frac{ad + a - 1}{2ac} = \frac{d}{2c} - \frac{1-a}{2ac} \tag{5}$$

Next, we find the social responsibility (S) value at which the utility is greater than its value at $S = 0$ (no investment in social responsibility), which we term the 'balance point'

$$U(\Pi\ ,\ S) > U(\Pi\ ,\ 0) \tag{6}$$

or

$$a(M + cS^2 - dS) + (1-a)S > aM \tag{7}$$

which, on rearrangement, yields:

$$S > \frac{a + ad - 1}{ac} = \frac{1}{c} + \frac{d}{c} - \frac{1}{ac} = \text{Balance Point} \tag{8}$$

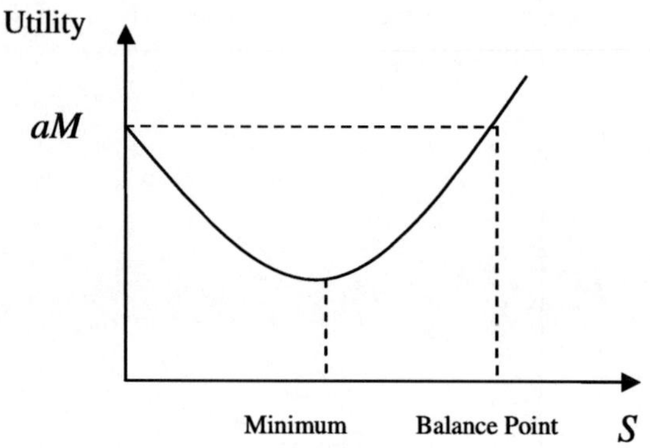

Fig. 2. The utility function

Note: Minimum= $\frac{d}{2c} - \frac{1-a}{2ac}$, Balance Point= $\frac{1}{c} + \frac{d}{c} - \frac{1}{ac}$

Fig 2 shows the company's utility as a function of S. The figure also indicates the S of minimum utility and the balance point, mentioned in (5) and (7), respectively.

An increase in c (the positive effect of social responsibility on image and reputation) means an increase in the positive effect of S (the firm's social responsibility rating) on its profit and a decrease in the balance point. This, in turn, means that more companies will be willing to invest in social responsibility in order to increase their utility, since less investment in social responsibility is needed to maintain the firm's utility at the point where $S = 0$ (see (7)). By contrast, if c decreases, the reverse will hold true, with the result that fewer companies will be willing to invest in social responsibility.

By contrast, an increase in d (the negative effect of social responsibility on profit) or in a (the weighting factor for the profit component of a firm's utility) increases the balance point. Such an increase raises firm costs and lowers the utility a firm receives from investing in improving its social responsibility rating (S). Consequently, profit decreases and more investment in social responsibility is needed to maintain the firm's utility at the point where $S = 0$.

Let's describe, at the macro level unit of analysis, two rating methods: a "partial rating" and a "full rating". The former method rates a select group of firms, an example being the Domini Social Index, while the latter method rates all firms in a given market. The positive effect of c on a firm in a partially rated market is lower that its effect on a firm in a fully rated market. This is because, in a fully rated market, the ranking of a firm affects its reputation and may increase its sales and so positively affect its profit Π. In a fully rated market, every firm would have an incentive to invest in social responsibility and to improve on its performance in that area. Yet, in the former case of a partially rated market, a firm excluded from the rated group has no incentive whatsoever to increase its investment in socially responsible behavior.

An indication of the magnitude of the barrier such investment funds place before unranked firms was described by O'Rourke, who asserts: "By promoting a particular fund as being socially, environmentally, and even financially responsible–this then begs the question as to why these criteria are not applied to all funds. By naming a fund as ethical and responsible–does it not imply that all the other funds are somehow unethical and irresponsible." For example, consider a firm that would have attained a ranking of 600 had it not been excluded from the Domini Social Index. Such a firm has no incentive to work towards a ranking of 450, because either ranking, be it 600 or 450, remains unknown to the public and so the improvement makes no impact on the firm's reputation. This prevents the firm from reaping benefits from its ranking and reduces its incentive to work towards improving its ranking. Yet a firm in a market, which is "fully rated" does have an incentive to change it's ranking, since an improvement from a ranking of 600 to 450 is known to the public and has an impact on the firm's reputation.

5 Conclusion and Discussion

5.1 Conclusions

Earlier work on the measurement of SRI indices focused on issues concerning the parameters used to formulate a rating. These parameters affect firms already included or close to inclusion within the indices. Our model elucidates the circumstances under which a ratings-based method for assessing CSR could offer an incentive to firms excluded from SRI indices to invest in CSR. We find that if all firms are publicly ranked according to SRI index parameters, then the investment a firm makes in improving its performance in the area of social responsibility generates a payoff in terms of improvements in the firm's public image and reputation, with consequent positive flow-on effects to profit. Thus, when all firms are ranked, market forces provide all firms with an incentive to invest in improving their SRI index ranking. We further find that this incentive tapers off as the amount of investment required for a firm to improve performance with respect to social responsibility exceeds a certain point or if the amount of utility a firm derives from its social responsibility rating fails to reach a certain threshold. However, none of these benefits accrue in a partially rated market, such as exists at present.

The model proposed here may be empirically tested in a market in which a corporate social index changes its rating methodology.

5.2 Limitations and Directions for Future Research

A possible shortcoming of the model is the existence of government regulations. By way of (extreme) example: if government were to fully subsidize CSR, then the balance point would be such as to render all ranking of socially responsible behavior irrelevant. However, so long as ranking methodologies continue to approximate those in current use, our model suggests that if all traded firms are ranked, then a firm's level of CSR, as measured by SRI index parameters, will impact its market performance.

Acknowledgment. It is a project supported by the National Natural Science Foundation of China (Grant No. 70962004).

References

1. Fischhoff, B., Nadai, A., Fischhoff, I.: Investing in Frankenfirms: Predicting Unacceptable Risks. The Journal of Psychology and Financial Markets 2(2), 100–111 (2001)
2. Prakash Sethi, S.: Investing in Socially Responsible Companies is a Must for Public Pension Funds–Because There is No Better Alternative. Journal of Business Ethics 56, 99–129 (2005)
3. Cox, P., Brammer, S., Millington, A.: An Empirical Examination of Institutional Investor Preferences for Corporate Social Performance. Journal of Business Ethics 52, 27–43 (2004)
4. Davis, K.: The Case for and Against Business Assumption of Social Responsibilities. The Academy of Management Review 16, 312–322 (1973)
5. Cowton, C.J.: Managing Financial Performance at an Ethical Investment Fund. Accounting, Auditing & Accountability 17(2), 249–275 (2004)
6. Hawley, J.T., Williams, A.T.: Can Universal Owners Be Socially Responsible Investors? In: Camejo, P. (ed.) The SRI Advantage: Why Socially Responsible Investing has Outperformed Financially, pp. 151–173. New Society Publishers, Canada (2002)
7. Cowton, C.J.: Accounting and Financial Ethics: From Margin to Mainstream. Business Ethics: A European Review 8(2), 99–107 (1999)
8. Ibrahim, N.A., Howard, D., Angelidis, J.P.: Board Members in the Service Industry: An Empirical Examination of the Relationship Between Corporate Social Responsibility Orientation and Directorial Type. Journal of Business Ethics 47, 393–401 (2003)
9. Bromley, D.B.: Psychological Aspects of Corporate Identity, Image and Reputation. Corporate Reputation Review 3(2), 240–252 (2000)
10. Orlitzky, M., Schmidt, F.L., Rynes, S.L.: Corporate Social Performance and Financial Performance. Organization Studies 24(3), 403–441 (2003)
11. O'Rourke, A.: The Message and Method of Ethical Investment. Journal of Cleaner Production 11, 683–693 (2003)
12. Camejo, P.: The SRI Advantage. New Society Publishers, Canada (2002)

13. Robins, F.: The Challenge of TBL:A Responsibility to Whom? Business and Society Review 111(1), 1–14 (2006)
14. Schuler, D.H., Cording, M.: A Corporate Social Performance—Corporate Financial Performance Behavioral Model for Consumers. The Academy of Management Review 31(1), 540–559 (2006)
15. Wood, D.J.: Corporate Social Performance Revisited. The Academy of Management Review 16(4), 691–718 (1991)
16. Solomon, J., Solomon, A., Norton, S.: Socially Responsible Investment in the UK: Drivers and Current Issues. Journal of General Management 27(3), 1–13 (2002)
17. Sparks, R., Cowton, C.J.: The Maturing of Socially Responsible Investment; A Review of the Developing Link with Corporate Social Responsibility. Journal of Business Ethics 52, 45–57 (2004)
18. Valor, C.: Corporate Social Responsibility and Corporate Citizenship: Towards Corporate Accountability. Business and Society Review 110(2), 191–212 (2005)

Equilibrium Price and Remark on Welfare in Electronic Commerce

Jun Ma

College of Business Administration
Nanchang Institute of Technology
Nanchang, China
majun5101227@163.com

Abstract. Although electronic commerce is regarded as a revolution that has transformed the structure of business and the mechanisms of economic systems, research on price competition and strategic behaviors in electronic commerce has been conducted through theoretical approaches, yielding conflicting results. Considering the development of technologies supporting electronic commerce, there have been few analytical analyses. In addition, previous empirical analyses have focused on where products are priced higher (or lower) rather than finding reasons for those differences. This paper analyzes pricing strategies between conventional offline firms and online firms through a theoretical approach, and draws some strategic implications with some remarks related to welfare aspects. Finally, this paper discusses the limitations of this study and highlights some directions for future research.

Keywords: Equilibrium price, Remark on welfare, Pricing strategies, Price level, Electronic commerce.

1 Introduction

The Internet has fundamentally changed the environment of business, offering business and consumers a powerful communication channel and making it possible for these two entities to come together in more efficient ways by creating new marketplaces. The Internet offers consumers greater benefits from increased information and lower transaction costs, including search costs, and a wider set of choices than those available in the traditional economic environment. It also improves the consumer's bargaining position with vendors, both online and via traditional channels. From the corporate point of view, the Internet provides opportunities to access global markets without having to incur large entry costs or having to keep sizable inventories, as well as improving targeted advertising and sales efforts.

While recent research in electronic commerce has been conducted in a variety of areas, there has been little theoretical research on price competition and strategic behaviors. The purpose of this study is to analyze pricing strategies between conventional offline firms and online firms through a theoretical approach and empirical analysis, and draw some important implications. In addition we make some remarks related to welfare aspects. Especially, the empirical study analyzes whether

W. Zhang (Ed.): Advanced Technology in Teaching, AISC 163, pp. 839–847.
springerlink.com

factors such as the size of market and product characteristics are related to price differences between offline and online channels through statistically testing the main results of the analytical analysis rather than just finding which products are higher in prices depending on channels.

The analysis of this paper presents several findings. The analysis of competition between conventional offline and online firms reveals why the equilibrium prices are different in electronic commerce despite products being physically homogeneous. The findings indicate that as more consumers have access to the Internet, both offline and online prices drop. If convenience associated with the online purchase becomes greater, the online price tends to exceed the offline price. In addition, the analysis finds that the profits of the online firm can increase as more consumers have Internet access if the online firm is relatively more efficient.

The paper is organized as follows: Section "Background and Literature Review" mainly introduces background and literature review of the research. Section "Research Methodology" presents a duopoly model between a conventional offline firm and a pure online firm, then explores equilibrium prices and draws some implications and provides remarks on the welfare of the producer and the consumer. Section "Conclusions and Limitations" provides conclusions and discussions of future research.

2 Background and Literature Review

2.1 Electronic Commerce

The concept of E-commerce first appeared in the 1970s. However, there is not a unified, authority, most people acceptable definition for E-commerce in academia. But some international organizations, research institutions, companies, specialists according to their own understanding defined E-commerce from different angles.

The International Chamber of Commerce convened the World Commerce objectives for soul pending issues in Paris on November 6th and 7th 1997. The experts in the fields of global business, information technology, law and government discussed the definition of E-commerce. They gave us E-commerce definitions from the point of view of commercial: E-commerce refers to the electronic trading activities. From the angle of coverage is defined by electronic trading parties face transactions rather than through trade or direct way to interview any form of business transactions. From the viewpoint of technology, E-commerce is defined as a collection of various technologies, including exchange data such as electronic data interchange and electronic mail, data acquisition such as database sharing, electronic bulletin board and automatically capture data such as bar code etc.

HP company defined E-commerce by electronic means to achieve the business trade activity of a kind of way, E-commerce transactions with electronic means to finish goods and services, business and exchange between advertiser and customer.

2.2 Differences between Offline and Online Retail

There are many substantial differences between offline and online retail environments. For example, the conventional offline retailer is based on a physical store where the vendor interacts with the customer. The merchandise is displayed in stores where customers can examine individual products, possibly sample them, and then, if desired, purchase and take them home immediately. From this viewpoint, customers incur some costs related to transportation and time consumption. Conventional physical retailers have several advantages over their pure online competitors, such as owning an established brand name, distribution infrastructure, and the shopping trip offering a physical shopping experience. However, there are also a number of drawbacks associated with the physical-retail model, including high investment cost and a limited number of hours and days of operation.

An online retailer comprises some or all of the following features. Depending on the product types, online retailing allows the customer to access online information about the product, place an order, pay for the product, and in the case of digital products(such as software, music, video, etc.),have them delivered instantaneously through the Internet. Because of the nature of online retailing, a number of advantages over traditional physical retailers can be identified, such as wide reach, exhaustive product selections, little infrastructure requirements, unlimited hours of operation, and a high degree of scalability. However, the online retailer also has drawbacks in terms of e-fulfillment issues arising from the physical delivery of real orders in the virtual world.

2.3 Empirical Result on Price Level

Kauffman and Walden and Wiseman review and effectively organize economic research issues related to Internet Commerce. Bakos and Harrington analyze the relation between search costs and product price in electronic marketplaces through a circular city model. Balasubramanian shows a strategic analysis of competition between a conventional offline store and a direct mail store focusing on the role of information and market coverage. Lal and Sarvary discuss possibilities of lower price competition in Internet commerce.

However, it should be noted that research in the analysis of competition between offline and online markets, thus far, has been conducted through empirical approaches, yielding conflicting results. Table 1 shows an overview of empirical results with respect to price level and classifies the studies in three categories: higher or lower prices in electronic commerce or inconclusive results.

Lee found that prices in online auction markets tend to be higher than prices in conventional auction markets for used cars sold from 1986 to 1995. Bailey compared the prices for books, CDs, and software sold in conventional offline and on the Internet channels in 1996 and 1997.As with Lee, Bailey found higher prices in online channels for each product category. He suggests that the higher prices he observed may be due to the possibility that many online stores add value in the form of convenience and other characteristics that consumers are willing to pay for(especially busy, affluent consumers). Erevelles et al. found that the prices of vitamins for Internet retailers were higher than for traditional offline retailers. Other studies also show results that parallel those shown in Table 1.

Table 1. Empirical Result on Price Level

Higher prices in online market	Lower prices in online market	Inconclusive
Lee	Brynjolfsson and Smith	Clay et al.
Bailey	Bakos et al.	Pan et al.
Arbeiterkammer Wien	Morton et al.	
Rolland and Srinivasan	Ancarani and Shankar	
Frank		

Brynjolfsson and Smith examined prices for books and CDs sold through conventional and the Internet channels in 1998 and 1999.Unlike Bailey, they found that the Internet had lower unit prices. They argue that this is because the online market has grown and has become efficient. Morton et al. studied dealer pricing of automobiles in California and they found that online prices are lower than offline prices although the difference was only 2%. Ancarani and Shankar found that the prices of books and CDs in Italy were lower at online retailers than at offline retailers by about 4–6%.

Clay et al., however, did not find any significant differences in the two channels for books, noting that the unit prices in online and physical bookstores during the week of April 19, 1999 were the same. Pan et al. found that some online prices were higher than offline prices, while other online prices were lower, depending on the products.

3 Research Methodology

3.1 Duopoly Model between Offline Firm and Online Firm

We analyzed pricing strategies between a conventional offline firm and a pure online firm using a theoretical approach by constructing a simple duopoly. Two standard models of spatial differentiation, ''on the line'' and ''on the circle'', are well known in terms of analytical methodology. We choose the commonly used spatial competition model of hotelling. We proposed a simplified version of the market and instead of assessing n offline firms, we assumed a single firm in the offline market. This approach is valid, since the offline firms in the circular city model (and in the absence of any web-based seller) end up selling at the same equilibrium price and being located at equal distances from each other. This equilibrium price depends on their production cost and the unit transportation cost of the buyers. One can safely assume that the linear-city model is equivalent to a small segment of the market covered by one of the many firms in the more general circular-city model.

We considered a linear city of length $\bar{s}$ where there is a conventional offline firm at the end of the city. There is also a pure online firm that sells the same good with no physical location. The unit production cost of both firms is equal to C^2.Consumers are distributed uniformly along the city. In other words, letting s denote the distance from the offline firm, which is located at 0, s is uniformly distributed on $[0, s]$. The distance here can represent different preferences such as the opportunity cost of time, the implicit cost of inconvenience, as well as the real cost of travel. Generally, these transportation costs are interpreted as degree of preferences. We assumed that the location of an offline firm is at 0.Thus a consumer located at 0 has maximum valuation of the good, V, and others preferences are decreasingly differentiated according to s.

Each consumer consumes one or zero units of the good. A fraction of consumers at each location point, m, have access to the Internet, while others, $1-m$, do not have access to the Internet. A consumer with access to the Internet may buy the good from the offline firm, in which case the consumer has to travel to the store and pay transportation cost ts, where t is the transportation cost per unit of length. If the consumer buys the good from the online firm, the consumer incurs cost, a, which may be search cost and other costs related to quality uncertainty, security risk, and delivery cost. These costs represent online specific inconvenience costs (monetary disutility) incurred when consumers buy from online retailers, and can affect the consumer's willingness to purchase from online retailers. We assume that consumers are homogeneous with respect to these costs, though they can be heterogeneous. Then, the utility of a consumer with Internet access located at s is:

$s = V - P_A - ts$ if he buys from the offline firm

$s = V - P_B - a$ if he buys from the online firm

$s = 0$ if he does not buy,

where P_A and P_B are the prices charged by the offline and online firms,respectively.Fig.1 depicts a typical equilibrium in the online market, where the net utilities of representative consumers are drawn. Note that in the equilibrium, both the offline and online firms compete with each other and coexist in the market.

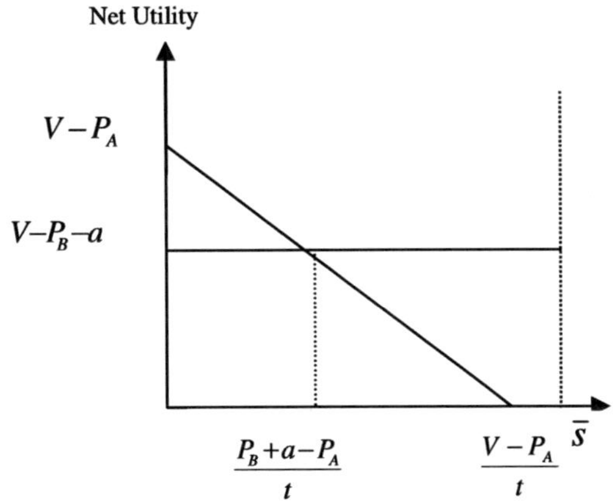

Fig. 1. The choice of consumers with access

The equilibrium is characterized by the following:

a) $P_A \leq P_B + a$

b) $V - P_B - a \geq 0$

$$\bar{s} \geq \frac{V - P_A}{t} \tag{1}$$

The first condition implies that neither the option of buying from the offline firm nor that of buying from the online firm completely dominates the other. In other words, letting $\hat{s} = \frac{P_B + a - P_A}{t}$, consumers with $s < \hat{s}$ prefer the offline firm while consumers with $s < \hat{s}$ prefer the online firm. The second condition means that consumers can earn nonnegative utility by buying from the online firm. Thus the online firm can attract consumers. The third condition implies that the city is so long that some consumers without access to the Internet will opt not to buy the good. In this type of equilibrium, the demand functions are given by:

$$D_A(P_A, P_B) = \frac{m(P_B + a - P_A)}{t} + \frac{(1-m)(V - P_A)}{t}$$

$$D_B(P_A, P_B) = m(\bar{s} - \frac{P_B + a - P_A}{t}) \tag{2}$$

3.2 Equilibrium Prices

We now examine equilibrium prices and then draw some implications. Given the demand functions in (2), the profit functions are as follows:

$$\Pi_A(P_A, P_B) = (P_A - C)(\frac{m(P_B + a - P_A)}{t} + \frac{(1-m)(V - P_A)}{t})$$

$$\Pi_B(P_B, P_A) = m(P_B - C)(\bar{s} - \frac{P_B + a - P_A}{t}) \tag{3}$$

The first-order conditions from (3) are

$$\frac{\partial \Pi_A}{\partial P_A} = \frac{m(P_B + a - P_A)}{t} + \frac{(1-m)(V - P_A)}{t} - \frac{P_A - C}{t} = 0$$

$$\frac{\partial \Pi_B}{\partial P_B} = m(\bar{s} - \frac{P_B + a - P_A}{t}) - \frac{m(P_B - C)}{t} = 0 \tag{4}$$

Define $X \equiv t\bar{s} + a$ to represent transaction costs of the offline and online firms. Also, define $k \equiv \frac{V - C}{X}$ and $r \equiv \frac{t\bar{s}}{X}$. In other words, k represents the ratio of valuation relative to transaction costs and r is the relative inefficiency of the offline firm in terms of the transaction costs. The equations in (4) can then be rearranged as follows:

$$m(P_B + a) - 2P_A + C + (1-m)(kX + C) = 0$$

$$2(P_B + a) - (P_A + C + X) = 0 \tag{5}$$

Letting P_A^* and P_B^* be the equilibrium prices, from the reaction functions in (5) we have

$$P_A^* = C + \frac{(m + 2k(1-m))X}{4-m}$$

$$P_B^* + a = C + \frac{(2 + k(1-m))X}{4-m} \tag{6}$$

Finally, these prices in (6) should satisfy the conditions given in (1), which gives the following conditions in terms of the underlying parameter:

$$\frac{2}{3} \le k \le \frac{m + (4-m)r}{2+m} \text{ and } a < \frac{X}{3} \tag{7}$$

From here forward, we will assume that these conditions in (7) hold.

Equilibrium prices as a function of m are depicted in Fig.2, where $P_A^* = P_B^*$ at m^* (if it exists).

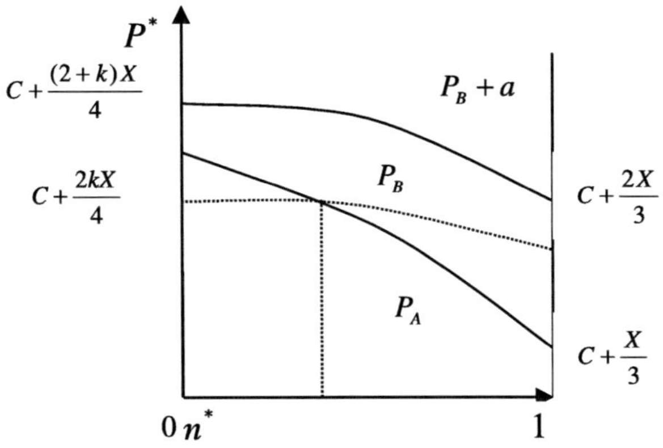

Fig. 2. Equilibrium prices in Electronic Commerce

3.3 Some Remarks on Welfare

Traditional economic views suggest that Internet growth reduces search costs for consumers and makes markets more competitive. This implies that the Internet may lead to consumer welfare gains, potentially at the expense of supplier profits. However, the general conclusion is that online price dispersion is no less than offline price dispersion and that market characteristics explain a significant portion of online price dispersion. Furthermore, numerous empirical studies, shown in Table 1, find that many online prices are not lower than offline prices. Lal and Saravary assert that the Internet can lower price competition and lead to reduced consumer search, even if it is more expensive than traditional distribution channels. This is in line with the trend that as the

Internet grows, online retailers have tried to make frictions (including search cost, differentiation strategies even they sell same products) in order to reduce price competition. This suggests that even with Internet growth, there are transaction costs still existing in the online market. In addition, the online market is beginning to occupy a greater market share of the retail market. In this section, we examine this issue, focusing on strategic interaction between offline and online firms.

4 Conclusions and Limitations

4.1 Conclusions

In this paper, we have analyzed pricing strategies between conventional offline firms and pure online firms through analytic and empirical analyses. We have drawn some important conclusions through analytic analyses. First, prices, both offline and online, drop as more consumers have access to the Internet. Second, as the online market matures or more consumers are connected to the Internet, the prices in online firms tend to be higher than in conventional offline firms. Third, if convenience associated with the online purchase increases, the online price tends to exceed the offline price. Finally, the profit of a typical offline firm decreases as more consumers gain access to the Internet, while the profit of a typical online firm can increase if the online firm is relatively efficient.

4.2 Limitations and Future Studies

However, our study has some limitations. For future analytical study, we anticipate conducting research on strategic interaction between offline and online firms when the offline firm enters the online market. Also, for the future empirical study, it is considered that the research on price differences depending on the types of retailers such as specialty stores, general malls, depart stores as well as hybrid retailer which products are sold through conventional and Internet distribution channels. And if it possible, it is needed to analyze more extensive product categories and determine how the prices vary between offline and online firms according to product categories. And we did not assess the relationship between prices and electronic market maturity, thus, in order to do so, one would ideally follow a time series of prices and Internet usage.

Acknowledgment. It is a project supported by Social Science Planning Project of Jiangxi Province (Grant No. 08YJ64).

References

1. Strader, T.J., Shaw, M.J.: Characteristics of electronic markets. Decision Support Systems 21, 185–198 (1997)
2. Pan, X., Ratchford, B.T., Shankar, V.: Can price dispersion in online markets be explained by differences in e-Tailer service quality? Journal of the Academy of Marketing Science 30(4), 83–103 (2002)

3. Morton, F.S., Zettelmeyer, F., Risso, J.S.: Internet car retailing. Journal of Industrial Economics 49(4), 501–519 (2001)
4. Lee, H.G.: Do electronic marketplaces lower the price of goods. Communications of the ACM 41(1), 73–80 (1998)
5. Lal, R., Sarvary, M.: When and how is the Internet likely to decrease price competition? Marketing Science 18(4), 485–503 (1999)
6. Kauffman, R., Walden, E.A.: Economics and electronic commerce: survey and directions for research. International Journal of Electronic Commerce 5(4), 5–116 (2001)
7. Enders, A., Jelassi, T.: The converging business models of Internet and bricks-and-mortar retailers. European Management Journal 18(5), 542–550 (2000)
8. Brynjolfsson, E., Smith, M.: Frictionless commerce? A comparison of Internet and conventional firms. Management Science 46(4), 563–585 (1999)
9. Balasubramanian, S.: Mail versus mall: a strategic analysis of competition between direct marketers and conventional firms. Marketing Science 17(3), 188–195 (1998)
10. Bakos, J.Y.: The emerging role of electronic marketplaces on the Internet. Communications of the ACM 41(8), 35–42 (1998)
11. Bakos, J.Y.: Reducing buyer search costs: implications for electronic marketplaces. Management Science 43(12), 1613–1630 (1997)
12. Smith, M., Bailey, J., Brynjolfsson, E.: Understanding Digital Markets: Review and Assessment, M.I.T., Working Paper (1999)
13. Brynjolfsson, E., Smith, M.D.: Frictionless commerce? A comparison of Internet and conventional retailers. Management Science 46(4), 563–585 (2000)
14. DeLone, W.H., McLean, E.R.: Information systems success: the quest for the dependent variable. Information Systems Research 3(1), 60–95 (1992)
15. Kaplan, R.S., Norton, D.P.: The balanced scorecard—measures that drive performance. Harvard Business Review, 71–79 (January–February 1992)
16. Nath, R., Akmanligil, M., Hjelm, K., Sakaguchi, T., Schultz, M.: Electronic commerce and internet: issues, problems and perspectives. International Journal of Information Management 18(2), 91–101 (1998)
17. Malone, T., Yates, J., Benjamin, R.: Electronic market and electronic hierarchies. Communications of the ACM 30(6), 484–497 (1987)

Equipments Requirement Forecasting Analytical Model of Air-Defence Brigade Considered RMS Factors

Shi Junbin[1], Zhang Hong[1], Shi Yanbin[2], and Yao Fushan[1]

[1] Dept. Control EngineeringChangchun Institute of Engineering Technology
Changchun, 130117, China
{Shijunbin_80,shiyanbin_80}@163.com
[2] Dept. Electronic Engineering Aviation University of Air force
Changchun, 130022, China

Abstract. During the mission of equipment operation, its qualities (Reliability, Maintainability, and Supportability) impact the number equipment. This paper establishes the requirement forecasting's integrated framework that considered RMS factors. And take the Air-defense brigade for example, the equipments requirement forecasting analytical model is established that Considered RMS factors.

Keywords: RMS, Equipments requirement, Analytical model.

1 Introduction

The expenditure trends to be much larger in the modern war and the economic strength become an important factor that influences the result of the war. How to give attention to both of the military and economic benefit becomes a serious problem that regarded by many countries.

Establishment the reasonable model for forecasting the requirement of equipment during the modern war is an important question for discussion, which use mission completion capability and minimum cost as restriction and target correspondingly. Besides the capability of the equipment, that the qualities (Reliability, Maintainability, and Supportability) of equipment also are primary factors that impact the number of equipment during operating mission.Under such background, combined with the air-defense brigade weapon system, the paper studied the models based on the theory of complex mission system and the framework for model establishment.

2 The Requirement Forecasting's Integrated Framework That Considered RMS Factors

This part mainly concluded when we established the requirement forecasting model that considered RMS factors, what works we must did. Firstly, analyze the operating and supporting of the equipment combat unit, and secondly establish the operating mission model of the equipment combat unit(ECU), finally build the mission completion capability model according to the logistic relationship during the mission. The figure 1 describes the framework of the requirement model.

W. Zhang (Ed.): Advanced Technology in Teaching, AISC 163, pp. 849–855.
springerlink.com

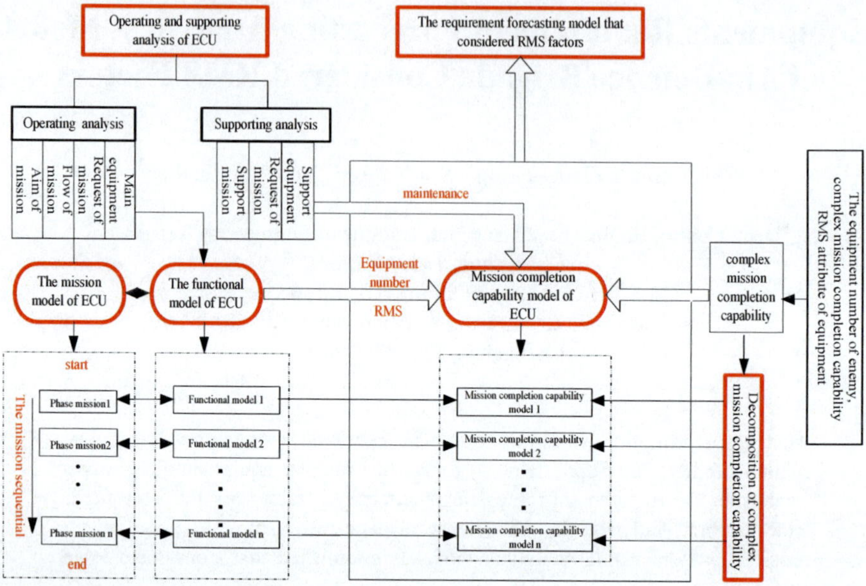

Fig. 1. The integrated framework of equipment combat unit equipment requirement model that considered RMS

2.1 The Operating Mission Model of EUC

Operating mission is defined as training or combat mission of EUC during certain time. The descriptions contain dependability time, mission intensity, maintenance request during time of war, environment condition, support condition, etc. The operating mission can be divided into some phased mission by schedule, whose relevant configuration changes during time periods (phase), but during the single phase, the configuration is fixed comparatively. Figure 2 describes the mission model of ECU.

2.2 The Functional Model of ECU

When establish the functional model of EUC, the weapon systems are decomposed by stepwise from top to bottom. We can consider the process of establishment functional model as the method which solve the problem from top to bottom, this method research the logistic relationship among functional system for some functions according to certain logistic combined system.

The functional model of EUC is built by three level structures, and the three level structures is weapon system, single equipment, the functional system of single equipment. Our purpose is to analyze the logistic relationship among various functional systems, so we can analyze the affection of RMS to the mission more correctly, and assure the relationship of the functional systems during phased mission. Figure 3 describe the establishment method of functional model, the up half part of figure 3 shows the process of decomposition of weapon system, and the down half part shows the logistic relationship among functional system.

2.3 The Mission Completion Capability Model of ECU

Follow three factors affect the mission completion, they are:

- Are the equipments available or not when phased mission start?
- Are the equipments is dependable or not during the mission?
- Are the equipments capable or not when equipments are in good condition?

Use A_0 to describe the equipments' availability, use parameter of dependability to describe the equipments' dependability, use parameter of precision, the rate of damage to describe the equipments' capability. Figure 4 shows the affection factors of mission completion capability.

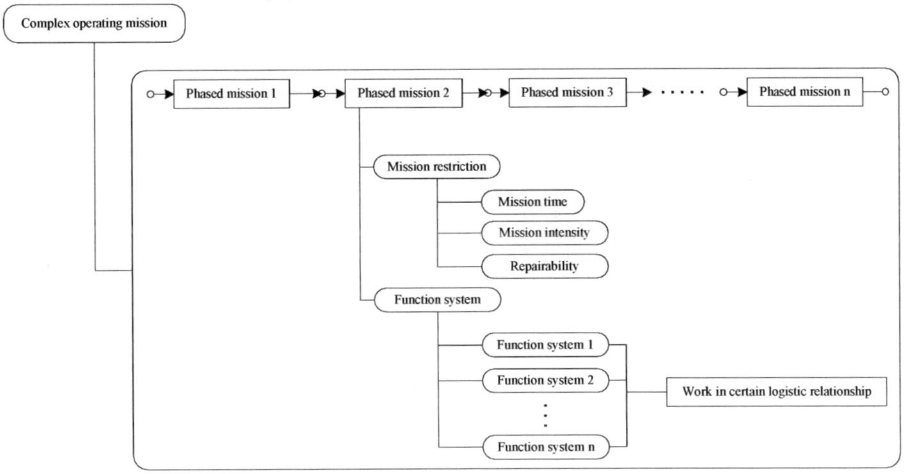

Fig. 2. Decomposition of complex mission

The mission completion capability of phased mission I is:

$$P_i = A_{oi} \cdot D_i \cdot C_i$$

Where: P_i——The mission completion capability of phase mission I;

A_{oi}——The A_0 of phase mission I;

D_i——The dependability of phase mission I;

C_i——The capability of equipments in phase mission I.

The follow formula describe the relationship between complex mission and phase mission :

$$P = \prod_{i=1}^{n} P_i = \prod_{i=1}^{n} A_{oi} \cdot D_i \cdot C_i$$

3 The Modeling Method of Equipment Requirement Forecasting Model

3.1 The Phase Model

According to the mission analysis, the mission model of ECU is established, and according to the logistic relationship of equipments, the mission time, the equipment requirement model is built for every phase mission. Finally, according to the inheritable relationship of number, the equipment number at the start of the complex mission is figured out.

The modeling method of the phase model is to establish the mission completion capability model of each phase mission, according to the request of the last phase mission completion capability, the number of equipment at start time of the last phase mission is figured out; use this number to be the equipment number at finish time of the n-1 phase mission, according to the request of the n-1 phase mission completion capability, the number of equipment at start time of the n-1 phase mission is figured out, analogized like this, the number of equipment at the start time is figured out.

3.2 The Integrated Model

Integrate each phase mission completion capability model and according to the mission completion capability request figure out the equipment number at the start time of complex mission.

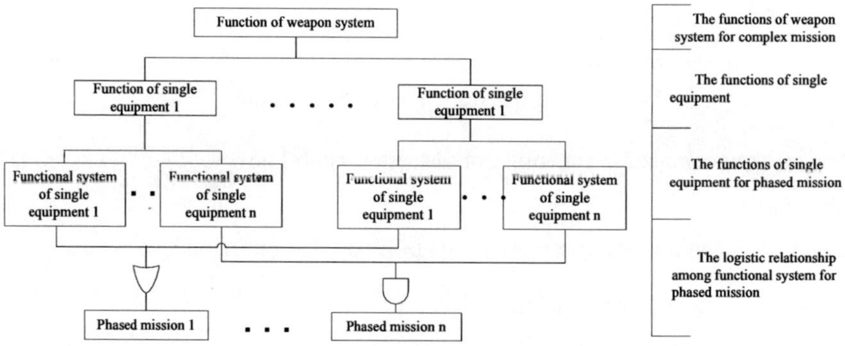

Fig. 3. Modeling method of functional model

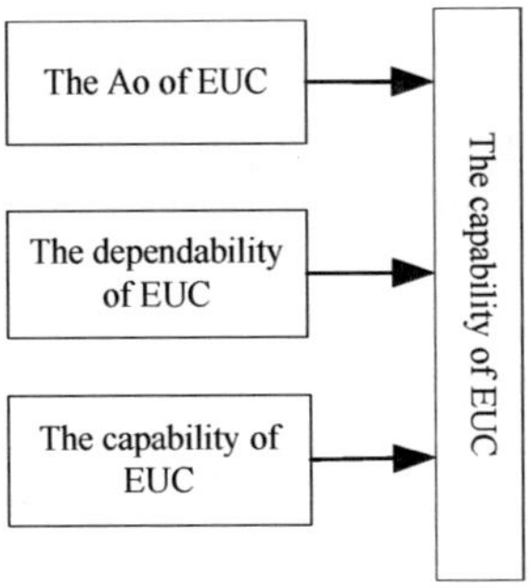

Fig. 4. The affection factors of mission completion capability

4 The Equipment Requirement Forecasting of Air-Defense Brigade

4.1 The Operating Mission Model of Air-Defense Brigade

According to the mission flow, the mission of the air-defense brigade is divided into follow phase missions:

- deport-preparation phase
- march phase
- long-distance transport phase
- march phase
- deployment and test phase
- operating phase

4.2 The Phase Model of Air-Defense Brigade

Take the march phase for example, according to the relationship among same equipment and different equipments, the equipment requirement forecasting model is established:

$$\begin{cases} P_4' \geq P_4^* \\ Min(C_S \cdot N_{S4} + C_F \cdot N_{F4} + C_Y \cdot N_{Y4} + C_{PM} \cdot N_{PM4} + C_{ZH} \cdot N_{ZH4} + C_{GP} \cdot N_{GP4}) \end{cases} \tag{1}$$

4.3 The Integrated Model of Air-Defense Brigade

According to the each phase mission completion capability model, the integrated model is established, like the follow formula:

$$\begin{cases} P' = \prod_{i=1}^{6} P_i \geq P^* \\ N_{D5} \geq N_{D6} \\ N_{GM5} \geq N_{GM5} \\ N_{S1} \geq N_{S2} \geq N_{S3} \geq N_{S4} \geq N_{S5} \geq N_{S6} \\ N_{F1} \geq N_{F2} \geq N_{F3} \geq N_{F4} \geq N_{F5} \geq N_{F6} \\ N_{Y1} \geq N_{Y2} \geq N_{Y3} \geq N_{Y4} \geq N_{Y5} \geq \lceil N_{D5}/4 \rceil \\ N_{PM1} \geq N_{PM2} \geq N_{PM3} \geq N_{PM4} \geq N_{PM5} \geq N_{PM6} \\ N_{ZH1} \geq N_{ZH2} \geq N_{ZH3} \geq N_{ZH4} \geq N_{ZH5} \geq N_{ZH6} \\ N_{GP1} \geq N_{GP2} \geq N_{GP3} \geq N_{GP4} \geq N_{GP5} \geq N_{GP6} \\ Min(C_D \cdot N_{D5} + C_{GM} \cdot N_{GM5} + \ldots\ldots + C_{GP} \cdot N_{GP6}) \end{cases} \quad (2)$$

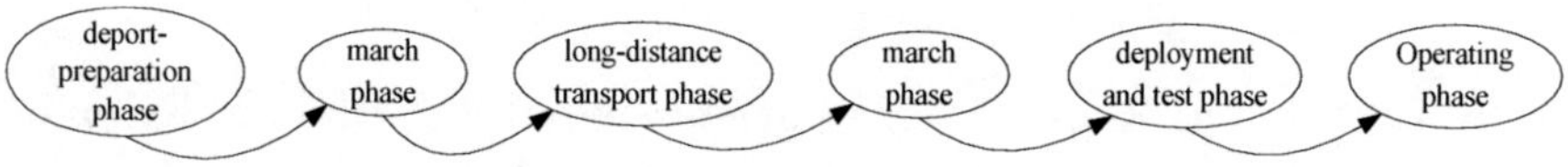

Fig. 5. The mission model of air-defense

5 Conclusion

Substitute data of RMS into phase model and integrated model, the result is following.

Table 1. Results of others phase based phase model

	S	F	D	Y	PM	ZH	GP	Mission cost
Deployment and test phase	2	4	5	2	4	4	34	
March phase	2	4	—	3	4	5	36	
Long-distance transport phase	2	4	—	3	4	5	36	22168
March phase	2	4	—	4	4	5	37	
Depot-preparation phase	2	4	—	5	5	6	39	

Table 2. Result of complex mission based integrated model

The mission capability	**0.9**						
Equipment	S	F	D	Y	PM	ZH	GP
Number	2	4	4	1	4	6	40
Mission cost	20480						

The result is calculated under the environment of Matlab 7.1.

References

1. Nie, C., Zhang, L.: Research to PMS Mission Sustainability Simulating Models. Journal of System Simulation (2008) (in Chinese)
2. Nie, C.-L., Zhang, L.: PMS Mission Sustainability Mathematic Evaluating Models. Journal of Ordnance Engineering College (2007) (in Chinese)
3. Chen, Y., Yu, Y.: Research on Ammo Supply Model for Complex Mission. Fire Control and Command Control (2006) (in Chinese)
4. Shi, J., Yu, Y.: Research on Mission Reliability of Combat Unit Complex Task Based on PMS. Ordnance Industry Automation (2007) (in Chinese)
5. Xing, L.: Analysis of generalized phased-mission system reliability, performance, and sensitivity. IEEE Transations on Reliability (2002)
6. Bondavalli, A., Mura, I., Nelli, M.: Analytical modeling and evaluation of phased-mission systems for space applications. IEEE (1997)

Design of Wireless Monitor System Based on CC2430 and GPRS

Xiuli Zeng, Zheng Yao, and Xuemei Song

College of Computer and Automatic Control
Hebei United University
Tangshan, China
xll@heut.edu.cn, Xuemei@126.com

Abstract. Wireless monitoring system consists of three parts: data monitoring nodes, gateway and remote monitoring center. The wireless gateway is communication center in the monitoring network between those data monitoring nodes (ZigBee network) and remote monitoring center (GPRS network). This paper makes a research on a kind of mobile nodes based on the MSP430 and CC2430 dual architectures, and researches on a wireless gateway for the wireless monitoring system and designs its software and hardware system. The experimental results show that the locations of nodes has lower computational complexity, higher positioning accuracy and stronger robustness.

Keywords: monitoring, wireless sensor network, S3C2430, GPRS.

1 Introduction

Environment monitoring is an important source of information for the protection and management. Traditional monitoring methods have the problem of long monitoring period, high strength, poor pertinence, slow data collection, limited monitoring range of waters and easiness to damage to the environment of monitoring region. It often runs in which people can't be close to, even in remotely dangerous environment, where the nodes survive with the limited battery energy. It is a burgeoning special wireless sensor networks behind Fieldbus and industrial Ethernet. Wireless network technology is a new communication technology for information exchange between devices, and it will play an important role for improving product quality, lowering production costs, improving energy efficiency in the industrial enterprise. Based on WSNs the environment monitoring system has the advantages of network self-organization, wide monitor range, low system cost and small environmental impact, thus it can be easy to realize to monitor environment.

2 DESIGN of MonitorSYSTEM

This paper researches on and designed a wireless gateway for WSNs oriented environment monitoring. The monitoring system includes data monitoring nodes, wireless gateway and remote monitoring center in three parts. The gateway uses the

W. Zhang (Ed.): Advanced Technology in Teaching, AISC 163, pp. 857–864.
springerlink.com

MSP430F1611 as host processor, CC2430 as a co-processor, with ZigBee technology and GPRS technology to realize two-way, highly efficient and stable communication between data monitoring node and gateway, the gateway and the remote monitoring center.

2.1 MCU Host Controller

The wireless gateway receives commands from the remote monitoring center and sends them to the appointed data monitoring node, at the same time it receives the command responses from the data monitoring node back to remote monitoring center. The wireless gateway also needs to respond to the orders of setting up its own state and inquiring itself from the remote monitoring center and sends the results back to the remote monitoring center.

The water environment monitoring system based on WSNs deals with and transmits a large amount of data, with high data transmission reliability and security. In the water environmental monitoring system based on wireless sensor networks, the wireless gateway should works outdoors, so the CPU power consumption and stability are important. The system uses the 16-bit rich-resource ultra-low power MSP430F1611 [1].

System hardware design uses MSP430F1611 as the main processor to realize controlling and command treatment in the gateway; CC2430 is used as a co-processor to realize data communication based on the ZigBee protocol between gateway and data monitoring sub-network; GPRS module is used to realize remote data communication between the remote monitoring center and gateway. System hardware block diagram is shown in Figure 1.

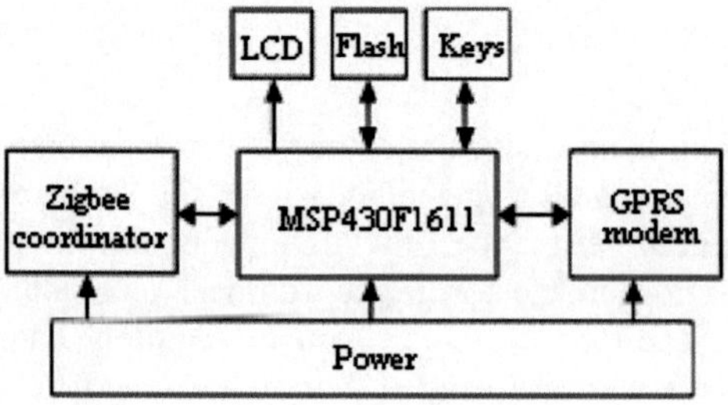

Fig. 1. System hardware block diagram

System software design uses μC/OS-II embedded operating system as the software platform of MSP430F1611 to improve the real-time performance of the system; ZigBee 2004 stack from Chengdu Wireless Dragon Information Technology company is used as a software platform of CC2430 module.

3 Design of Hardware Structure

3.1 ZigBee Module Design Based on CC2430

GPRS connects with MSP430F1611 through the serial port, MSP430F1611 sends AT commands to GPRS modem to control data transmission through GPRS Modem.

GPRS MODEM is applied to transmit the video data to the network. It can read the video data from the cache and makes it into the package, and then convert to the streaming file on the basis of RTP format. Finally, it transmits the data to the network through the built-in streaming media sever. The system can work independently without the assistance of computers to achieve the collecting and encoding of the digital video signal directly. The GPRS is a new type service for carrying data which is developed based on GSM system network, and it has many advantages, such as wide coverage, high speed data transmission, high-quality communications, always-on-line and billing by flow. It supports TCP/IP protocol, and it can be connected with the Internet directly without being transferred through PSTN net Client. Client can interview Video capture device and real-time view image data which is transmitted by GPRS wireless network through GPRS intelligent terminal and standard streaming media player procedure according to the IP address [2].

ZigBee module design directly uses CC2430 module from Chengdu Wireless Dragon Technology, in order to reduce the difficulty of RF circuit in the design of hardware module. CC2430 contains 8K SRAM and 64K Flash, so it no longer needs to increase memory[4]. The UART 1 interface circuit between module and MSP430F1611 is shown in figure 2.

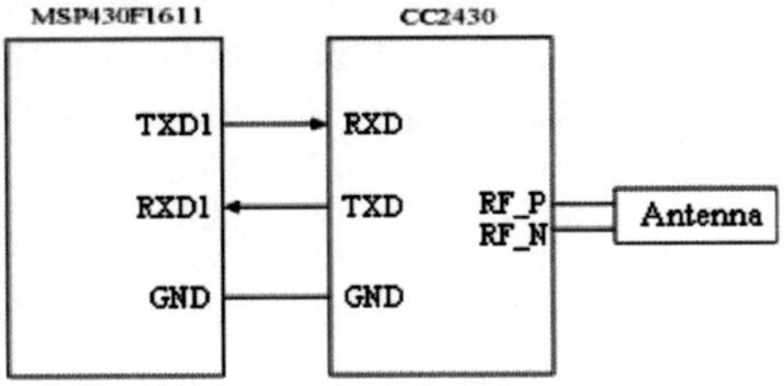

Fig. 2. Interface circuit of ZigBee module

3.2 Power Module Design

Because of the high peak current of GPRS modem and its strong interference on other modules when communicating, the power design uses high power isolated chips in order to reduce he interference by GPRS Modem on other parts.

Power module design is shown in figure 3. Power module uses LM2596 and TPS79533 dual power chips. As the gateway works outdoors, an external power supply can be 6 Ni-MH batteries (7.2V) or two lithium batteries (7.4V) [3].

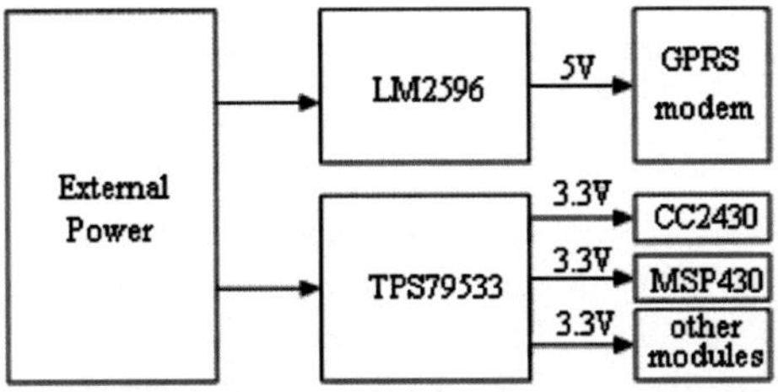

Fig. 3. Power module

3.3 Other Module

LCD module uses LCM19264.MSP430 connects with LCM19264 directly by general purpose IO port. The gateway uses 8Mbits SOIC8 packaged AT45DB081D as data memory through the SPI interface. A JTAG circuit is used according to the circuit design which TI recommends. Due to the small number of keys, the design uses the independent keyboard to simplify the hardware design of the system.

4 System Software Design

The system mainly uses the C language for programming. As a result of dual processor structure, the software design includes CC2430 modules software design and MSP430 module software design.

4.1 CC2430 Module Design Software

CC2430 module software consists of the ZigBee protocol stack, applications and board support package. Consumption of the energy in the system has two main aspects, the one is hardware, it is necessary to choose low-power equipments; on the other hand, the software, including the working modes and protocol stack. Wireless Networks for industrial automation and process Automation has three types, sleep mode, wake-up mode, and working model. The consumption of nodes in different modes has large difference. The combination of these modes could effectively implement energy-saving. Wireless Networks for industrial automation and process Automation defines data link layer, network layer and application layer only [4]. The protocol structure of zigbee is shown in figure 4.

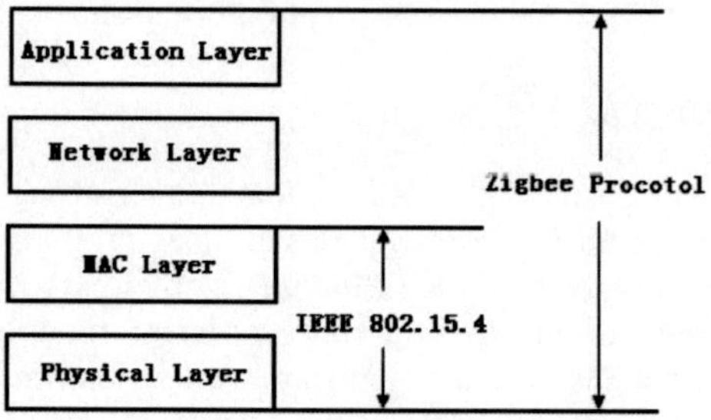

Fig. 4. The protocol structure of zigbee

The main purpose of using BSP in CC2430 is to complete package hardware, provide a wide range of function calls to the upper ZigBee protocol stack and applications.

The main mission of CC2430 module application is to build and maintain networks, at the same time undertake data transfer mission between MSP430 and ZigBee networks. CC2430 module does not deal with any specific data and is only responsible for transmitting data. Among them, the serial data transceivers are carried out through the serial port interrupt. ZigBee software level diagram is shown in figure 5.

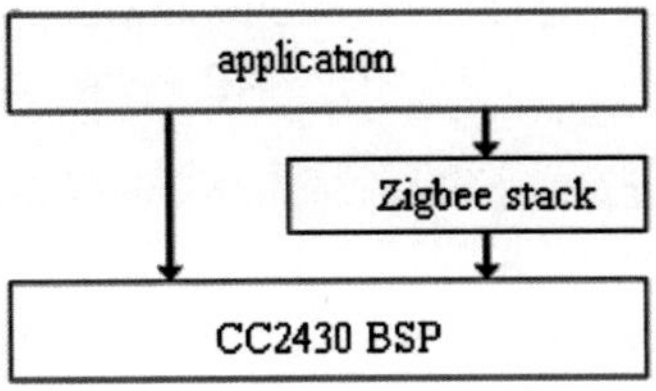

Fig. 5. ZigBee module software level diagram

4.2 MSP430 Module Software Design

In order to improve the reliability of system, enhance real-time performance and simplify systems programming, we transplant μC / OS-II embedded operating system to MSP430. Entire MSP430 module software level diagram is shown in figure 6.

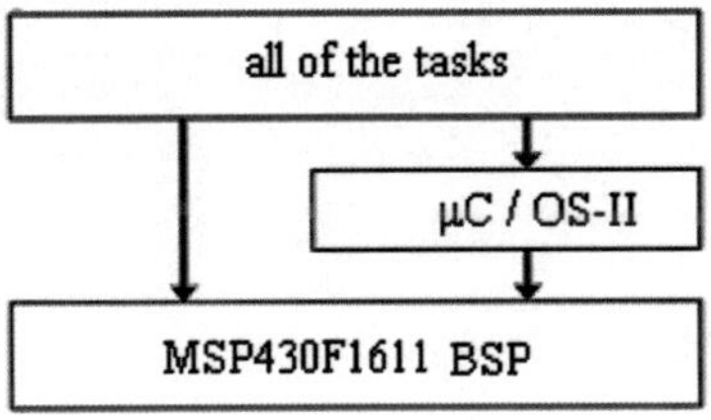

Fig. 6. MSP430 module software level diagram

μC / OS-II requires interrupt from time to time as its time tick interrupt. The time tick interrupt can be generated by the watchdog timer of MSP430F1611 and can be set the cycle of 32 ms.

MSP430 software module consists of 6 tasks and the corresponding serial transceiver interrupt service subroutine. The main tasks are: key scanning task, LCD display task, ZigBee module communication task, GPRS module communications task, data process task, flash task.

4.3 Other Module Design

Key scanning task, LCD display task and flash task are activated by data process task. Key scanning task is responsible for finding key state, if the button is depressed, the task will activate the data process task to deal with. LCD display task mainly indicates the status of gateway and provides user interface. Flash task reads and writes flash memory.

5 Communication task

The tasks and interrupt service subroutine deliver the message of data or keys to the data process task, and data process task notifies the other tasks to deal with message.

The other tasks and interrupt service subroutine, through a unified message mailbox communicate with data process task and data process task activates other tasks by Semaphore [5].

The communication diagram of each task and the corresponding serial interrupt service subroutine is shown in figure 7.

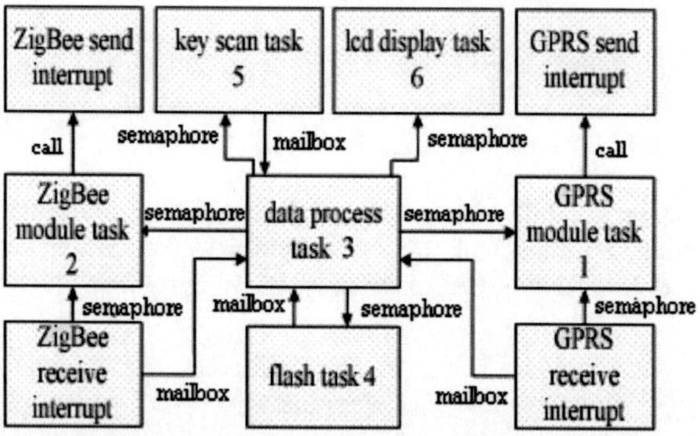

Fig. 7. The communication diagram of each task

5.1 GPRS Module Communication

GPRS module communication task is responsible for setting up GPRS network and communicating with the remote monitoring center through GPRS modem. The GPRS modem is only passive to orders from the remote monitoring center, so GPRS modem should be given an activated phone to activate GPRS modem. If the GPRS modem in one minute has no data to be transmitted, it will automatically enter the sleep state. You need to periodically (such as 50 seconds) send a heartbeat package to the remote monitoring center to maintain the GPRS connection. Task and the corresponding GPRS serial port transmitting and receiving interrupt service subroutine.

5.2 ZigBee Module Communication

ZigBee module communication task has the second highest priority. Through the task, gateway sends the commands to the data monitoring node. That is, if the data process task sends a signal to ZigBee task, ZigBee task will be activated to send the corresponding command.

Because of the weakness of ZigBee 2004 stack and MSP430 and CC2430 use 2 wire serial port for communication, when the CC2430 closes the global interruption, at that time if MSP430 sends a continuous data to CC2430, it will inevitably lead not to responding to the interruption and losing the data. In order to resolve this problem, we only have to use retransmission mechanisms.

5.3 Data Process Task

Data process task is mainly to analyze the command from the remote monitoring center and data from the data monitoring node to activate the corresponding tasks for processing.

Data process task calls corresponding module to handle in accordance with the command number in command frame: If it is needed to maintain the GPRS connection, data process task needs to be regularly through GPRS send a heartbeat packet to maintain the GPRS connection; if no need to maintain the connection, just to stop sending heartbeat packets; if data process task received a message from CC2430 monitoring node that node was found in (or out), it is needed to update node status table [6]. The flow chart of data process task is shown in figure 8.

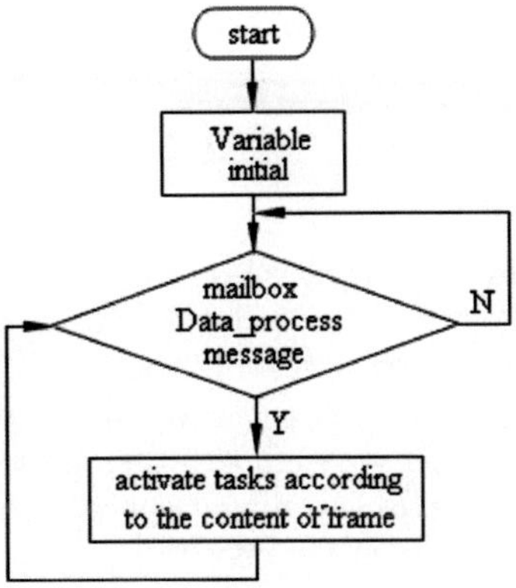

Fig. 8. The flow chart of data process task

5.4 Data Frame Format

The communication protocol in the water environmental monitoring system mainly deals with the design of communication frames, as the dual-CPU communication, in order to ensure reliable transmission of data, it should use the confirm frame. The communication frame includes the data frame and the command frame (GPRS transmission do not use the confirm mechanism, just directly sends the data frame and the command frame). The communications frame format are shown in Table 1 and the confirmable frame format is shown in Table 2.

Table 1. The communication frame format

Header	*2 Byte*
Kind of frame	*1 Byte*
Length of frame	*1 Byte*
Entity of frame	*1~65 Byte*
Destination	*1 Byte*
Source	*1 Byte*
Frame check	*1 Byte*

Table 2. The confirmable frame format

Header	***2 Byte***
Kind of frame	***1 Byte***
Length of frame	***1 Byte***
Entity of frame	***1 Byte***
Destination	***1 Byte***
Source	***1 Byte***
Frame check	***1 Byte***

6 Conclusion

The wireless monitoring system oriented WSNs has a wide monitoring range and low cost. Wireless gateway in the entire monitoring network acts as a bridge between the wireless sensor network and the GPRS network and is the communications center for the entire system. This paper researches and develops the wireless gateway based on dual processor structure applied ZigBee and GPRS wireless technology, elaborates its software and hardware design. In practicality, it operates stably and reliably and well meets the need of remote real time water environment monitoring system, it also has wide application prospect.

References

1. Jiang, P.: Survey on Key Technology of WSN-Based Wetland Water Quality Remote Real-Time Monitoring System. Chinese Journal of Sensors and Actuators 20(1), 183–186 (2007)
2. Dai, L., Zhang, W.-Y., Qian, J.: Design and implementation of an embedded wireless video supervising system. Journal of Hefei University of Technology (Natural Science) 31(1), 5–8 (2008)
3. Tanabe, I., Sasaki, H., Zheng, L.: The Implementation of Ultra Low Power IEEE802.15.4 Module and Its Application for Batteryless Wireless Sensors. Networked Sensing Systems (6), 93–96 (2008)
4. Du, J.-B., Wei, C.-J.: Research about wireless real-time monitor system based on GPRS. Journal of Shanxi Normal University (Natural Science Edition) 22(4), 40–42 (2008)
5. Tang, Y.-P., Zhao, X.-H., Zhou, Z.-S.: Elevator video surveillance system based in wireless communication. Computer Engineering 34(3), 263–265 (2008)
6. Jiang, P.: WSN-Based Wetland Water Quality data video Monitoring System. Chinese Journal of Sensors and Actuators 21(2), 244–248 (2009)

A Novel Method of Multi-view Virtual Image Synthesis for Auto-stereoscopic Display

Xi Ming[1,2], Xue Jiufei[1,2], Wang Lianghao[1,2], Li Dongxiao[1,2], and Zhang Ming[1,2]

[1]Institute of Information and Communication Engineering,
Zhejiang University, Hangzhou 310027, China
[2]Zhejiang Provincial Key Laboratory of Information Network Technology,
Hangzhou 310027, China
sxdfxm2004@126.com

Abstract. In this paper, an efficient depth image based view rendering method for high quality multi-view virtual images synthesis is proposed, which utilizes two color images with their associated depth maps. The proposed method consists of four main steps. The first step is virtual camera parameters design by using shift-sensor camera model to avoid the keystone distortion during virtual view synthesis. Secondly, the mixed color pixels around the boundary of image objects is detected and skipped in the following step, which can eliminate the "ghost" artifacts. Then the two input color images are rendered to multiple virtual view positions on the basis of depth-image-based rendering (DIBR) and merged into one image. Finally, a depth-aided inpainting algorithm is used to fill the disoccluded areas. Experiment results are shown to demonstrate the superior performance of the proposed method compared to other traditional DIBR methods.

Keywords: multi-view synthesis, parameters design, mixed pixel detection, depth-aided inpainting, shift-seosor model, depth-image-based rendering(DIBR).

1 Introduction

The development of auto-stereoscopic display technology demonstrates the demand of images of multiple views, generally more than 8 views. However, it is impossible to transmit multiple views simultaneously constrained to the bandwidth. High quality images of multiple virtual views need to be synthesized by algorithm on display side. European project "ATTEST" proposed a new "2D+Z" video format, which contains one color image and its corresponding per-pixel dense depth map. From "2D+Z" video format, depth-image-based rendering [1] (DIBR) can generate image of any virtual desired view theoretically, but it still has three problems [2], which are visibility, re-sampling and disocclusion. The disocclusion problem is considered to be the most significant and difficult one and lots of algorithms have been developed to solve this problem [3]-[11]. A simple method is to copy the value of edge pixel to the hole area directly or apply horizontal interpolation, but it will induce artifacts [3]. Preprocessing depth image by filters [4]-[5] may produce objects transformation. Inpainting algorithms [6]-[8] will provide visually acceptable reconstruction.

W. Zhang (Ed.): Advanced Technology in Teaching, AISC 163, pp. 865–873.
springerlink.com

But when the distance between reference view and virtual view increases during generation of multi-view images, the disocclusion problem will become so severe that the above mentioned methods cannot give comfortable restore result. Another approach [9] can obtain excellent rendering effect by utilizing Layered-depth-image (LDI) [10] as input, which consists of a number of color layers and its associated depth layers. However, the procedure of creating LDI is computationally complex and quite time-consuming and the transmission bandwidth of LDI also increases drastically. The mismatch between objects edge of the color image and that of the depth map will cause artifact, which is perceived as "ghost" effect. Preprocessing depth map by dilation operation [11] can reduce this artifact.

In our proposed method, we use two "2D+Z" videos of two widely separated views as input, which will be rendered to the same virtual view position. Most percentage of exposed hole pixels of one virtual image will be filled with pixels from another one and the increment of bandwidth is acceptable due to the correlation of two views. Additionally, we can obtain this input simply by using two capture cameras and stereo matching process. Before the rendering step, we propose a new method for virtual view parameters design and an operation of mixed color pixels detecting and skipping to eliminate the "ghost" artifact. In the end, a depth-aided inpainting algorithm [8] is applied to fill the remaining holes.

The remainder of this paper is organized as follows. Section 2 describes the related work. Section 3 illustrates the details of our method. Section 4 presents the experiment results. And our conclusions are given in section 5. Section 6 is the acknowledgement.

2 Related Work

2.1 DIBR Technique

In DIBR algorithm, an image pixel $\tilde{m}$ is projected to a 3D point $\tilde{M}$ in real scene, which is re-projected to a pixel $\tilde{m}'$ in the image plane of desired camera. The procedure is as the following formula [2].

$$Z'\tilde{m}' = ZA'RA^{-1}\tilde{m} + A't \tag{1}$$

Where Z and Z' respectively are the depth value of 3D point $\tilde{M}$ in the coordinate of reference and desired camera respectively. The rotation matrix R and translation vector t denote the transform relation from the desire camera to the world coordinate system.

2.2 Shift-Sensor Camera Model

There are two kinds of traditional camera setting models [12]: one is "toed-in" setting model which will cause keystone distortion leading eye-strain, the other is parallel setting model which only provides cross disparity. Shift-sensor camera model [2] is a development of parallel camera model, which shifts a small distance h of the parallel positioned camera's CCD sensor. Shift-sensor model can provide convergence effect but not induce keystone distortion. The small shift h can be expressed as [2]

$$h = -t_x \frac{\alpha_u}{Z_C} \cdot \quad (2)$$

Where Z_C is the selected convergence distance, α_u represents the normalized focus length u axis. t_x is defined as follows.

$$t_x = \begin{cases} -t_c: \text{ on the left of referrence camera} \\ +t_c: \text{ on the right of referrence camera} \end{cases} \quad (3)$$

Where t_c is the absolute value of baseline between reference and virtual shift-sensor camera.

3 Proposed Method

The procedure of our proposed method is illustrated in Fig. 1. Two "2D+Z" videos of two widely separated views are utilized as the input. We choose one view as the main view and the other view as the assistant view.

3.1 Virtual Camera Parameters Design

In order to generate multi-view virtual images which are suitable for auto-stereoscopic display, we need to design a group of camera parameters for every view. Virtual camera model should be chosen carefully, because 3D perception is determined by the model of virtual multi-camera array. Our proposed method utilizes the shift-sensor camera model, which can keep advantages of both traditional models. Shift-sensor model supplies positive and negative parallax, which will provide both inside and outside screen perceptions without inducing any artifacts or objects transformation.

The parameters of virtual view camera consist of intrinsic and extrinsic matrixes. We design the intrinsic matrix first.

A_m and A_a denote the intrinsic matrix of main view and assistant view respectively. $[R_m | t_m]$ and $[R_a | t_a]$ represent the extrinsic matrix of main and assistant view respectively. B is the length of baseline. All these items are supposed to be known in advance, which will be used during the parameter design process.

Suppose we need to generate N virtual views between the two input views. Choosing the main view as the reference camera view, A_v^i which is the intrinsic matrix of virtual view of index i, can be calculated by the following formula according to shift-sensor model.

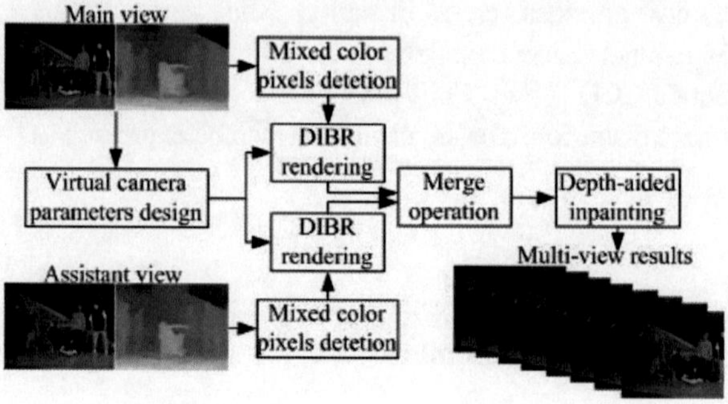

Fig. 1. Block diagram of the proposed multi-view synthesis method.

$$A_v^i = A_m + \begin{bmatrix} 0 & 0 & i \cdot h & 0 \\ 0 & 0 & 0 & 0 \\ 0 & 0 & 0 & 0 \end{bmatrix}, i = 1,2,\cdots,N-1 \tag{4}$$

h is defined by

$$h = -t_x \frac{\alpha_u}{Z_C} = \frac{B}{N} \cdot \frac{\alpha_u}{Z_C} \cdot sign(i) \cdot \tag{5}$$

α_u is the same item in (2). Z_C is the depth value of the convergence plane in shift-sensor setting model. $sign(\cdot)$ is a sign function. When the virtual view of index i is on the left of reference view, it equals 1, on the contrary, it values -1. This relation can be written down as

$$sign(\cdot) = \begin{cases} 1 : \text{on the left of reference view} \\ -1 : \text{on the right of reference view} \end{cases} . \tag{6}$$

Then the design of extrinsic matrix is illustrated. There is only horizontal shift and no rotation between the virtual view and the reference main view. So the rotate matrix of virtual view of index i which is denoted by R_v^i, is same as the main view. The translation vector of virtual view of index i is denoted by t_v^i. Then the extrinsic matrix of virtual view of index i is derived by the next equation.

$$\left[R_v^i \mid t_v^i\right] = \left[R_m \mid t_m + \begin{bmatrix} sign(i) \cdot i \cdot \dfrac{B}{N} \\ 0 \\ 0 \end{bmatrix} \right], \quad i = 1,2,\cdots,N-1 \tag{7}$$

Where t_m is the translation vector of main view. $sign(\cdot)$ is defined by (6).

3.2 Mixed Color Pixels Detection

The boundary pixels of objects will receive contribution from foreground and background color. It's very difficult to precisely classify these mixed pixels as foreground or background, which will cause inaccurate boundary matching between color image and depth image. The mismatch will induce contour artifact, which is visually perceived as "ghost" effect. Reference [13] uses boundary information layers and a matting algorithm to eliminate the "ghost" artifact, which is quite complicated and computationally time-consuming. Here, we propose a method to detect mixed color pixels, which will be skipped during the rendering process of virtual view.

There is no texture in the depth image, so we scan the depth image $D(u,v)$ and use a threshold T to detect the objects boundary in depth image. Then we extend the objects edge of one pixel width to W pixels width in the outside direction. This step can be expressed as

$$M(u+k,v)=\begin{cases}1, if \mid D(u,v)-D(u-1,v)\mid > T \\ 0, \qquad else\end{cases} \tag{8}$$

Where $M(u,v)$ represents the mask of mixed color pixels. If it equals 1, put the image pixel $I(u,v)$ into the mixed pixels' set Ω . k is defined by the following equation.

$$k=\begin{cases}0,-1,-2,\cdots,-(W-1), & if D(u,v)-D(u-1,v)>T \\ 0,1,2,\cdots,W-1 & , if D(u,v)-D(u-1,v)<-T \\ 0 & , \qquad else\end{cases} \tag{9}$$

3.3 Virtual View Rendering

Before the rendering process, color correction between main color image $I_m(u,v)$ and assistant $I_a(u,v)$ color image is needed. Because there may be some luminance difference between images captured from different view positions.

Then we use DIBR method expressed by (1) to render images of two input view to the virtual view. If a pixel belongs to the set of mixed color pixels Ω , we will skip this pixel. If it isn't in this set, we will do DIBR operation.

During the DIBR operation, we propose a register algorithm combined depth and distance information to handle the visibility and re-sampling problems. There are four registers in each pixel position of virtual view image, which are used to store the depth and distance of four nearest pixels projected from the reference image. The pixel value $I_v(u,v)$ of virtual view image is calculated as follows.

$$I_v(u,v)=\begin{cases}\sum_{j=1}^{N}\lambda_j I_j, & N\neq 0 \ and \ N\leq 4 \\ HOLE, & N=0\end{cases} \tag{10}$$

For every pixel position (u,v) of virtual image, its four registers only store warped pixel from reference image that located less than one pixel either in horizontal or vertical direction. N represents the numbers of warped pixels, which satisfy the

condition mentioned above. If N is larger than 4, we will delete the warped pixels with maximum depth from the registers. I_j is the value of reference pixel. λ_j is the normalized weight factor of the combination of distance and depth. *HOLE* flag means there is no warped pixel from reference image. In this situation we set the hole pixels with a white color(R=255 G=255 B=255).
λ_j is calculated by (11).

$$\lambda_j = \frac{\rho_j \omega_j}{\sum_{k=1}^{N} \rho_k \omega_k}; \quad \sum_{j=1}^{N} \lambda_j = 1, \quad \lambda_j > 0 \tag{11}$$

Where the weight factor of distance ω_j is expressed as (12). (x_j, y_j) is the projection position of warped pixel in virtual image plane.

$$\omega_j = \frac{1}{\sqrt{(x_j - u)^2 + (y_j - v)^2}} \tag{12}$$

The weight factor of depth ρ_j is expressed as

$$\rho_j = \begin{cases} 0, D_j > \mu_D \\ 1, D_j \le \mu_D \end{cases}. \tag{13}$$

Where D_j is the depth value of registered reference pixel. μ_D is the mean depth of all the registered reference pixels.

Until now, we get the virtual image $I_v^m(u,v)$ warped from main view and the virtual image $I_v^a(u,v)$ warped from assistant view. To diminish hole area, these two virtual images are merged together by utilizing a simple Z-buffer algorithm [14]. The merge operation is denoted as follows.

$$C_v(u,v) = \begin{cases} I_v^m(u,v), & \text{if } I_v^m(u,v) = NONHOLE \\ I_v^a(u,v), & \text{if } I_v^m(u,v) = HOLE \text{ and } I_v^a(u,v) = NONHOLE \\ I_v^m(u,v), & else \end{cases} \tag{14}$$

$C_v(u,v)$ denotes the image of virtual view after merge process. *NONHOLE* flag means there is at least one pixel warped to the pixel position (u,v) of virtual view.

3.4 Depth-Aided Inpainting

After the merge operation, there remain a few holes in merged virtual image $C_v(u,v)$ .Depth-aided inpainting algorithm [8], which takes the associated depth information of every pixel into account on the basis of exemplar-based inpainting [7], is used to cope with this problem in our proposed method.

4 Experiment Results

In this paper, we use the multi-view test sequences "breakdancers" provided by Microsoft Research Asia (MSR) [13]. This test sequences consist of 8 views spacing 20 centimeters in horizontal direction. #3 camera is chosen to be the main view in our method and #5 camera is the assistant view, so the length of baseline B equals 40 centimeters. Our "Super-D HDL46" auto-stereoscopic display equipment needs eight input views, that means $N=8$. Then the intrinsic and extrinsic matrixes of eight virtual views are obtained by utilizing our virtual parameters design method. We compare our proposed method with the conventional DIBR method, an asymmetric Gaussian filtering method [5], a depth-aided inpainting method [8].

Take the 4th virtual view position as an example, Fig. 2 shows the rendering results of the four compared methods.

Improvements have been made by the other three methods to solve the re-sampling and visibility problems of traditional DIBR method. Asymmetric Gaussian filtering method causes serious geometric distortion and transformation of objects as shown in Fig. 3(a). Although depth-aided inpainting method can fill the disoccuded areas with texture exemplar from background, there still exist apparent artifacts in disocclusion areas as shown in Fig. 3(b). As shown in Fig. 3(c), among the three methods our proposed method can provide a high image quality without inducing any other defects.

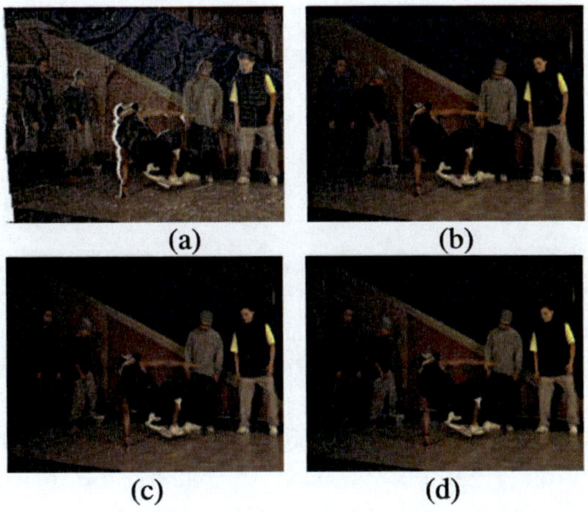

Fig. 2. Images of 4th virtual view position synthesized through four methods: (a) Conventional DIBR method, (b) Asymmetric Gaussian filtering method, (c) Depth-aided inpainting method, (d) Proposed method.

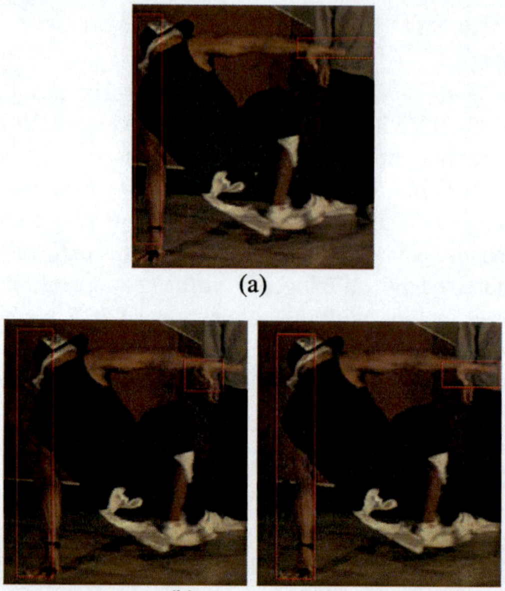

(a)

(b) (c)

Fig. 3. Enlarged segments (top-left point (370,324), bottom-right point (716,694)) from the 4th virtual views synthesized through the three methods: (a) Asymmetric Gaussian filtering method, (b) Depth-aided inpainting method, (c) Proposed method.

Furthermore, we input eight virtual images generated by the proposed method into our "Super-D HDL46" auto-stereoscopic display to evaluate the 3D effect of virtual images. Observers can feel strong 3D visual perception.

5 Conclusions

In this paper, we present a new virtual multi-view images synthesis method for auto-stereoscopic display, which designs virtual view parameters on the basis of shift-sensor model and generates virtual views from two widely separated views according to the designed parameters. The mixed color pixels of objects boundary can be skipped by our proposed detection. The combination of Z-buffer merge process and depth-aided inpainting algorithm is used to handle the disocclusion problem. Experiment results show that our proposed method can provide a sound image quality and strong 3D perception.

Acknowledgements. This work was supported by the National Natural Science Foundation of China (Grant No. 60802013, 60971061, 61072081), and the National Science and Technology Major Project of the Ministry of Science and Technology of China (Grant No. 2009ZX01033-001-007, 2009ZX03004-004-04).

References

1. Fehn, C.: A 3D-TV system based on video plus depth information. Signals, Systems and Computers 2, 1529–1533 (2003)
2. Fehn, C.: Depth-image-based rendering (DIBR), compression and transmission for a new approach on 3D-TV. In: Proceedings of SPIE Stereoscopic Displays and Virtual Reality Systems XI, vol. 5291, pp. 93–104 (2004)
3. Vázquez, C., Tam, W.J., Speranza, F.: Stereoscopic Imaging: Filling disoccluded areas in depth image-based rendering. In: Three-Dimensional TV, Video, and Display V, SPIE, vol. 6392, p. 63920D-12 (2006)
4. Tam, W.J., Alain, G., Zhang, L., Martin, T., Renaud, R.: Smoothing depth maps for improved stereoscopic image quality. In: Three Dimensional TV, Video and Display III, SPIE, vol. 5599, pp. 162–172 (2004)
5. Zhang, L., Tam, W.J.: Stereoscopic image generation based on depth images for 3D TV. IEEE Transactions on Broadcasting 51(2), 191–199 (2005)
6. Bertalmio, M., Sapiro, G., Caselles, V., Ballester, C.: Image inpainting. In: Proceedings of ACM Conference on Computer Graphics (SIGGRAPH), pp. 417–424 (2000)
7. Criminisi, A., Perez, P., Toyama, K.: Region filling and object removal by exemplar-based image inpainting. IEEE Transactions on Image Processing 13(9), 1200–1212 (2004)
8. Luo, K., Li, D.-X., Feng, Y.-M., Zhang, M.: Depth-aided inpainting for disocclusion restoration of multi-view images using depth-image-based rendering. Journal of Zhejiang University - Science A 10(12), 1738–1749 (2009)
9. Yoon, S.-U., Ho, Y.-S.: Multiple color and depth video coding using a hierarchical representation. IEEE Transactions on Circuits and Systems for Video Technology 17(11), 1450–1460 (2007)
10. Shade, J., Gortler, S., He, L.-W., Szeliski, R.: Layered Depth Images. In: Proc. of ACM SIGGRAPH 1998, pp. 231–242 (July 1998)
11. Oh, K.-J., Yea, S., Ho, Y.-S.: Hole filling method using depth based in-painting for view synthesis in free viewpoint television and 3-D video. In: Picture Coding Symposium, pp. 1–4 (May 2009)
12. Woods, A.J., Docherty, T.: Image distortions in stereoscopic video systems. In: Proc. of SPIE Stereoscopic Displays and Applications, vol. 1915, pp. 36–48 (February 1993)
13. Zitnick, C.L., Kang, S.B., Uyttendaele, M., Winder, S., Szeliski, R.: High-quality video view interpolation using a layered representation. In: ACM SIGGRAPH and ACM Transactions on Graphics, pp. 600–608 (August 2004)
14. Greene, N., Kass, M., Miller, G.: Hierarchical Z-buffer visibility. In: Proceedings of the SIGGRAPH, pp. 231–240 (September 1993)

Binding Mechanism of a MN's Multi-interface Sharing a Unique Network Prefix on PMIPv6

Wen Wei[1], Wang Lingjiao[2,*], and Guo Hua[2]

[1] Faculty of Information of Engineering
Jiangxi University of Science and Technology
Ganzhou, China
[2] School of Information Engineering
Xiangtan University
Xiangtan, China
ijlgbg@sina.com, xtu_wlj@126.com

Abstract. According to the basic Proxy Mobile IPv6 (PMIPv6) specification, more resource shall be consumed because of multiple copies of every packet being forwarded to a Mobile Node (MN), which may result in network congestion and some security issues for a MN's multi- interface sharing a unique network prefix. To tackle the issues and enhance reliability, an integrated binding mechanism is proposed. It can confirm the matching relation between an IF-ID and its IP address for all scenarios which are in lined with a MN's multi-interface sharing with a unique prefix. The proposed mechanisms are future proof in that they can be used in conjunction with any possible radio access technologies and provide well expansibility. Generally, the time delay which the LMA spends forwarding the first packet to the exact MAG is rather short. Furthermore, the mechanism can apply to more interfaces described in this paper.

Keywords: binding mechanism, LMA, MAG, PMIPv6, BCE.

1 Introduction

Mobility management, which solves the "roaming" and "mobile" issues, is one of the principal services in the next generation network (NGN) where the IP technology is the basis of the new public telecommunication network [1]. IETF presents Client mobile IPv6 (CMIPv6) and Proxy mobile IPv6 (PMIPv6) acting as the mobility management mechanism in Internet. CMIPv6 is a host-based mobility management protocol [2-3] and PMIPv6 is used as a network based local mobility management protocol and does not require the MN to be involved in the exchange of signaling messages [4-5]. These telecommunication operators prefer to the PMIPv6 specification because it can easily controlled by them.

In RFC 5213, the Per-MN-Prefix model is supported but the Shared-Prefix model is not supported. However, it presents that the Shared-Prefix model is a being scenario

* Corresponding author.

W. Zhang (Ed.): Advanced Technology in Teaching, AISC 163, pp. 875–882.
springerlink.com

and should be researched in the future. MN may own multi-interface and it is also known as multihoming MN [6]. This scenario where each interface of a multihoming MN is assigned a unique prefix has no any issue when the multiple interfaces attach to a PMIPv6 domain [7]. However, another scenario where a same unique prefix is assigned to all interfaces of a multihoming MN has man y issues which are presented in [8]. However, the solution is not put forward for some of these issues. If the management is dealt with the conventional PMIPv6 specification, more resource should be consumed because the MN's LMA will forward multiple copies of every packet destined for the MN. More seriously, it may result in network congestion or network failure. Simultaneously, it may also bring about some security issues because the forwarded packets may be transmitted to some incorrect MAGs. To tackle the issues, an enhancing integrated binding mechanism is proposed for improving reliability.

The remaining of this paper is organized as follows. Section 2 goes into details about the operation procedure of a multi-interface MN sharing a unique prefix in a PMIPv6 network, Section 3 presents enhanced integrated binding mechanism of a MN's multi-interface sharing a network prefix on PMIPv6. The MSC of An enhanced integrated binding mechanism is shown in Section 4. Section 5 presents the merits of the proposed mechanism and gives the conclusion remarks.

2 Network Model and Operation

2.1 Network Architecture

For clarifying the binding mechanisms, an assumptive network topology is illustrated in Fig. 1.

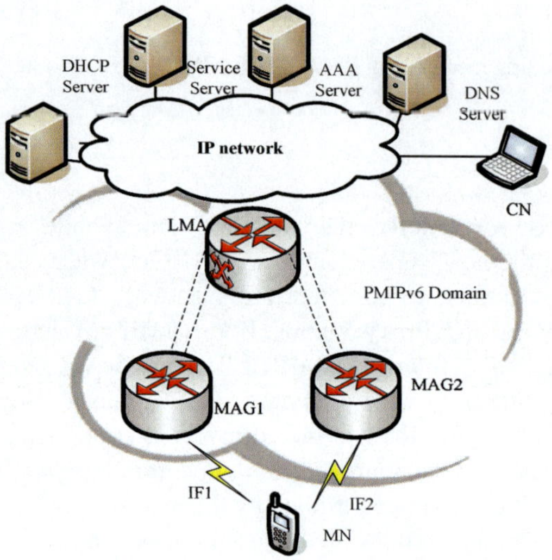

Fig. 1. Topology of a MN's muti-interface sharing a unique prefix

Table 1. BCEs of a unique prefix for all interfaces

MN-NAI	IF-ID	CoA	HNP
NAI	IF1-ID	MAG1.Address	MN.HNP1
NAI	IF2-ID	MAG2.Address	MN.HNP1

For clarifying the binding mechanisms, an assumptive network topology is illustrated in Fig. 1. For simplicity, it is assumed that the MN owns two interfaces. The correspondent node (CN) is the communication party of the MN. The DHCP Server and the DNS Server should be required for the address configuration of the MN. The AAA server must be contained for solving authorization, authentication and accounting. Some service may be provided to various users by service servers. The PMIPv6 domain is constituted by a LMA and two MAGs. One interface (IF1) of the MN connects to the MAG1, the other interface (IF2) attaches to MAG2. A MAG tracks the MN's movements to and from the access link and signals to the MN's LMA for registering the attachment information.

2.2 PMIPv6 Operation

The core function entity is composed by a LMA, which maintains the reachability status of the MN and undertakes a anchor point for Home Network Prefix (HNP) of a MN on a topology architecture, and a MAG, which is a mobility management entity which implements mobility management of a MN attaching to its access link, is responsible for detecting the mobility of a MN on its access link and sends a Proxy Binding Update (PBU) message to a LMA for registering the accessing interface.

The PMIPv6 base specification provides mobility management for all kinds of MNs. The scenarios supported by the PMIPv6 specification consist of the unique prefix for each interface of MN and the same unique prefix for all interfaces of MN. In the latter scenarios, the Shared- Prefix model is utilized to finish address configuration and data forward. In this paper, our research focuses on the multi-interface MN rather than multiple MNs.

AS is shown in Fig. 1, the event that the IF1 attaches to the MAG1 triggers the MAG1 sends a PBU message, which includes Network Access Identifier (NAI) and IF1 Identifier (IF1-ID), to the LMA, and then the LMA creates the first Binding Cache Entry for the MN if it accepts the PBU message. Similarly, the LMA creates the second BCE for the MN if it accepts the PBU message from the MAG2 under the condition that the MN's IF2 attaches to the MAG2. Because the IF-ID of the two PBU messages is different and the handover indication of the PBU message from the MAG2 indicates a new access, the LMA does not delete the first BCE. Therefore, the BCEs at LMA should be shown in Table 1. The LMA shall forwards the packet to the MN when it receives a packet destined for the MN according to the BCEs.

2.3 Problem

According to PMIPv6 base specification, LMA forwards a packet by matching the address prefix of the packet. Therefore, the multi-interface MN attaching to a PMIPv6 domain shall also abide by the rule. If the LMA receives a packe, which comes from a CN and its destination address is the IP address of the IF1, it shall utilizes the destination address to match the entries of BCEs. Because the match is that the prefix of IF1 address matches the HNP options of BCEs, the result of the matching is that two BCEs meet the matching requirement. Therefore, the LMA shall forward the packet to the MAG1 and the MAG2. Upon receiving the packet, the MAG2 decapsulates it and finds that none of entry in the Binding Update List Entry (BULE) matches the IP destination address of the IP packet and the MAG1 shall find the matching entry among its BULE and forward the packet to the IF 1. This results in the squander of expensive network resource. Security problem may happen because of the incorrect transmission.

3 Proposed an Enhanced Binding Mechanism

Two binding mechanisms have been proposed for solving the problems which emerge on the scenario of a same unique prefix for all interfaces of MN in a PMIPv6 domain in [9-10]. These mechanisms can avoid the network resource waste and the latent security problems because they can send a packet to an exact MAG. However, they can't solely implement the learning interface address. Therefore, an enhanced binding mechanism is proposed for solving the problem. It should be reminded that the mechanisms proposed are described according to the Fig. 1.

3.1 Binding Mechanism on Data Initiation

The reactive binding mechanism is that a LMA learns the relation between an IF-ID and its IP address after it has received a packet, and then forwards the packet to the exact MAG according to the BCE being updated by the relation.

When all or part interfaces of a MN attach to different MAG respectively in a PMIPv6 domain, every MAG shall be triggered to initiate the PMIPv6 registration procedure and updates its BULE. A LMA accepts the registration request and creates a BCE being labeled as "pending".

Upon receiving a packet from a CN, the LMA matches the IP address according to the matching principle. If a BCE matches the destination IP address and whose status is an "active" indication, it forwards the packet according to the BCE. If the statuses of the matching BCEs are "pending", the LMA caches the packet and triggers the reactive binding mechanism to learn the IP addresses of IF-IDs in the matching BCEs according to [9], and then find an exact tunnel which will be utilized to forward the cached packets and some imminent packets to an exact MAG.

3.2 Binding Mechanism on Registration Initiation

In the multi-interface sharing a unique prefix scenario, a LMA initiates a learning procedure to know the IP address of the registering interface by appending a new flag bit or an option in PBU/PBA message for rapidly forwarding packets to the exact MAG when a MAG initiates a PMIPv6 registration procedure. The learning procedure is named for the predictive binding mechanism on registration initiation [10].

In the specific scenario, the MAG sends a PBU message to a LMA for registering the new attachment when an interface accesses to a link of a MAG, and then the LMA creates a new BCE for the IF-ID and sets its status for "pending" after receiving the

PBU message and accepting the registration request. The LMA sends back a PBA message to the MAG. The PBA message includes a HNP which is assigned to the interface for the configuration of IP address about the interface, a new flag bit which is utilized to notify the MAG acquiring the IP address of the new registering IF-ID. The MAG advertises the HNP to the interface. The MN configures the IP address of the interface after it has received the HNP. The MAG initiates the learning procedure of the IP address of the interface by some appropriate mechanisms in []. And then, the MAG sends a new PBU including the IP address of the interface to the LMA once again. LMA receives the PBU and updates the BCE of the IF-ID by substituting the IP address for the HNP and its status is altered for "active". Subsequently, the LMA forwards packets to its destination according to the new BCEs. The match principle is identical to the one described in the reactive binding mechanism.

3.3 Enhanced Integrated Binding Mechanism

For the multi-interface sharing with a unique prefix, operator may prefer the predictive binding mechanism because its lower delay. However, the mechanism may be failure under some circumstances. Therefore an enhanced integrated mechanism is presented in this paper.

According to the predictive binding mechanism, a MAG should learn the IP address of the interface which accesses to its link. The MAG should initiate the second PMIPv6 registration procedure when it has known the IP address of the interface. Under some circumstances, the MAG can't learn the IP address of the accessed interface. Therefore it can't notify its LMA of the IP address information and the LMA keeps the status of the BCE on the interface in "pending". Subsequently, if the LMA receives a packet destined for the interface whose status is "pending", it should initiate the reactive binding mechanism to learn the IP address of the interface.

The new BCE is shown in table 2 after the enhanced binding mechanism. In fact, the interface addresses can be lay by one of two modes. One is the mode shown in table 2, the other is that a new item is appended to the BCE in table 1.

4 MSC of Enhanced Integrated Binding Mechanism

According to the fundamental principle of the integrated binding mechanism and the network architecture in Fig. 1, its MSC is illustrated in Fig.2 and the description of the MSC is simply depicted because the reactive mechanism and the predictive mechanism have been described in detail at anterior narrative.

Table 2. BCEs of a unique prefix for all interfaces

MN-NAI	IF-ID	CoA	HNP
NAI	IF1-ID	MAG1.Address	IF1-IP Add
NAI	IF2-ID	MAG2.Address	IF2-IP Add

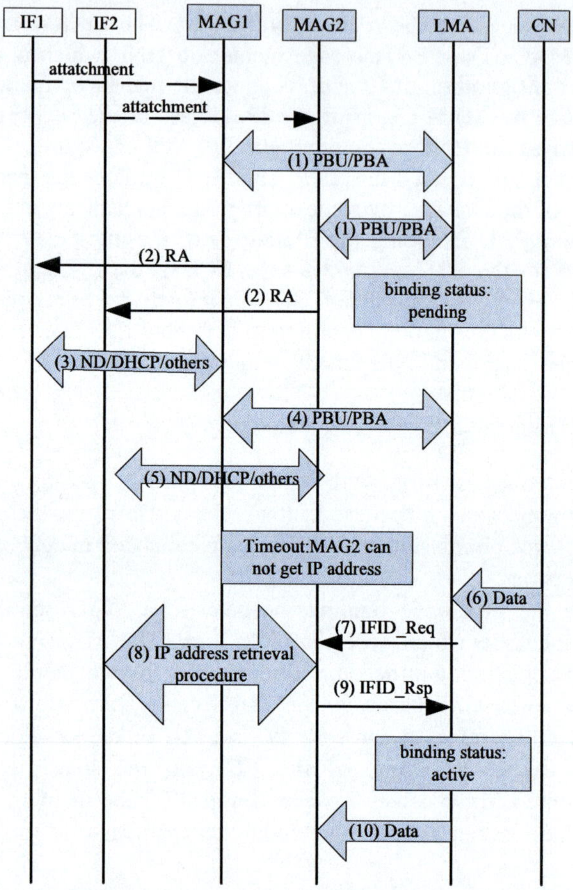

Fig. 2. MSC of enhanced binding mechanism

(1) When the IF1 and the IF2 access to the MAG1 and the MAG2 respectively, each of the two MAGs sends a PBU message to the LMA for the PMIPv6 registration procedure. Upon receiving the PBU message, the LMA accepts the two PMIPv6 registering request and assigns a unique HNP to the two interfaces. In addition, the LMA creates two BCEs for the two interfaces and sets the status of the two BCEs for "pending". It also response the PBU messages by sending two PBA messages including a extended flag bit "U" or a new extended option which imply the MAGs to get and notify the IP addresses of the two interfaces.

(2) Upon the receipt of the extend PBA messages, the two MAGs advertise the RA message including the assigned HNP respectively. And then the two interfaces configure their IP address according to the HNP in the received RA message.

(3-4) The MAG1 gets the IP address of the IF1 by the appropriate learning mechanisms, and sends a PBU message including theIP address to the LMA for registering the binding once again. The LMA accepts the request and updates the BCE of the IF1-ID. The status of the updated BCE is "active" now.

(5) Similarly, The MAG2 also learns the IP address of the IF2 by utilizing the recommended mechanisms. However, the MAG may not obtain the IP address thanks to some causes. Therefore it can't initiate a new registration procedure once again, and the status of the IF2-ID is still "pending" at the LMA.

(6) When the LMA receives a packet destined for the IF2, it looks for the BCEs at LMA and finds none of BCE matching the destination address from the packet header according to the improved matching principle. Therefore, it caches the packet and initiates the learning procedure of the reactive binding mechanism to get the IP address of IF2 and interfaces of other matching BCEs whose status is "pending".

(7) The LMA sends an IFID_Req message to the MAG2 for acquiring the IP address of the IF2.

(8) Upon receiving the IFID_Req message, the MAG2 initiates the IP address retrieval procedure. If the MAG2 obtains the IP address, it should update its BULE.

(9) The MAG2 builds up the IFID_Rsp message according to the learned IP address and sends it to the LMA. The LMA updates the BCE of the IF2 by utilizing the IP address and sets its status for "active".

(10) The LMA forwards packets to the MAG2 according to the updated BCEs.

The enhanced integrated binding mechanism, which assembles the reactive binding mechanism and the predictive binding mechanism, can have the LMA know the IP address of an interface which accesses its PMIPv6 domain. The mechanism possesses the merits as follow. It can confirm the matching relation between an IF-ID and its IP address for all scenarios which are in lined with a MN's multi-interface sharing with a unique prefix. Moreover it can be applied to the situation that the IP address of an interface brings about change. Generally, the time delay which the LMA spends forwarding the first packet to the exact MAG is rather short. Furthermore, the mechanism can apply to more than interfaces described in this paper.

5 Conclusion

For the scenario where a multi-interface MN shares with a unique prefix in a PMIP domain, the paper presents an integrated binding mechanism to solve the resource profligacy and the latent security danger from the scenario according the exiting protocols.

The proposed enhanced integrated binding mechanism is present for solving the issue when the predictive binding mechanism can't learn the IP address under some situations. First, the LMA also sets the status of the new registering BCE for "pending". But it adds a new flag bit "U" or extends a new option to request the corresponding MAG resolving the IP address of interface in the PBA message responding to the PBU message. The MAG initiates a new PMIPv6 registration procedure to notify the interface address. Upon obtain the IP address of the interface, the LMA updates the BCE of the interface. If the procedure is failure and the receiving a packet destined for an interface whose binding status is "pending", the LMA will send an IFID_Req message obtain the IP address of the interfaces whose HNPs match the destination IP address of the packet. Upon the IP address information from the IFID_Rsp message, the LMA updates the BCEs and sets their status for "active". And then the packet is forwarded to the exact MAG according to the updated BCEs.

In the proposed enhanced binding mechanisms, the transmission delay of the enhanced integrated binding mechanism is equal to that of the predictive binding mechanism under most circumstance. A flag bit or an extended option is added to the PBA message and reuses PBU message to include the IP address of an interface with the HNP option whose length option is set "128". Every BCE of a LMA are appended a new status option whose value is "pending" or "active". Simultaneously, its HNP may be replaced by the IP address of an interface or appends a new HOA option to a BCE. The BCE change of a LMA is an implementation event. The enhanced integrated binding mechanism adapts to all situations. Generally, it does not make use of the reactive binding mechanism until the predictive binding mechanism can't learn the IP address of an interface.

Acknowledgment. This work was supported in part by a grant from Xiangtan University project NO.09XZX21 and Jiangxi University of Science and Technology Research Foundation project NO. jxxj10046.

References

1. Nguyen, T.H., Sadiku, M.N.O.: Next generation networks. IEEE Potentials 21(2), 6–8 (2002)
2. van Nguyen, H., Roh, S.H., Ryu, J.K., Park, S.: A Modification for Fast Handover in Hierarchical Mobile IPv6. In: The 9th International Conference on Advanced Communication Technology, February 12-14, vol. 1, pp. 815–818 (2007)
3. Johnson, D., Perkins, C., Arkko, J.: Mobility Support in IPv6, draft-ietf-mext -rfc3775bis-03.txt, March 9 (2009)
4. Kim, H.G., Yu, M.J., Lee, J.M., Yu, Y.H., Choi, S.G.: Network Based Global Mobility Management Scheme in NGN. In: Fourth International Conference on Networked Computing and Advanced Information Management, NCM 2008, September 2-4, vol. 2, pp. 547–553 (2008)
5. Giaretta, G.: Interactions between PMIPv6 and MIPv6: scenarios and related issues, draft-ietf-netlmm-mip-interactions-00, October 24 (2008)
6. Gundavelli, S., Leung, K., Devarapalli, V., Chowdhury, K., Patil, B.: Proxy Mobile IPv6, draft-ietf-netlmm-proxymip6-18.txt (RFC 5213), May 30 (2008)
7. Li, Y., Kum, D.-W., Cho, Y.-Z.: Multihoming Support Scheme for Network Mobility Based on Proxy Mobile IPv6. In: ISECS International Colloquium on Computing, Communication, Control, and Management, CCCM 2008, August 3-4, vol. 2, pp. 635–639 (2008)
8. Jeyatharan, M., Ng, C., Hirano, J.: Multiple Interfaced Mobile Nodes in NetLMM, draft-jeyatharan-netlmm-multi-interface-ps-02, June 17 (2008)
9. Wang, L., Guo, H., Su, Y., Liu, C.: A Reactive Learning Mechanism for Multihoming MN on PMIPv6. In: CAR 2010 (March 2010)
10. Wang, L., Mi, Z., Peng, R., Guo, H., Lei, Z.: A Predictive Binding Mechanism on Multi-Interface MN in PMIPv6. In: WiCom 2010 (September 2010)

Author Index